AF251453

Systems & Control: Foundations & Applications

Ali Saberi
Anton A. Stoorvogel
Peddapullaiah Sannuti

Filtering Theory

*With Applications to Fault Detection,
Isolation, and Estimation*

Birkhäuser
Boston • Basel • Berlin

Ali Saberi
School of Electrical Engineering
  and Computer Science
Washington State University
Pullman, WA 99164
USA

Peddapullaiah Sannuti
Department of Electrical
  and Computer Engineering
Rutgers University
Piscataway, NJ 08854-8058
USA

Anton A. Stoorvogel
Department of Mathematics
  and Computer Science
Eindhoven University of Technology
5600 MB Eindhoven
The Netherlands

and

Department of Electrical Engineering,
  Mathematics and Computer Science
Delft University of Technology
2600 GA Delft
The Netherlands

Mathematics Subject Classification: 60G35, 93E10, 93E11

Library of Congress Control Number: 2006936424

ISBN-10: 0-8176-4301-X   e-ISBN-10: 0-8176-4564-0
ISBN-13: 978-0-8176-4301-0   e-ISBN-13: 0-978-0-8176-4564-9

Printed on acid-free paper.

www.birkhauser.com   (SB)

This book is dedicated to:

Ann, Ingmar, Ula, Dmitri, and Mirabella   (Ali Saberi)

My parents   (Anton A. Stoorvogel)

Jayanth   (Pedda Sannuti)

Contents

Preface

As soon as we (AS and PS) completed writing the book on H_2 *Optimal Control*, another task of equal magnitude was laid to our charge. This task was to work on *filtering* and related topics. This book releases us from this charge. In this endeavor, we are fortunate to have found a capable person in our friend and colleague (AAS) who helped us release our burden.

The subject of *filtering* is indeed vast and immense, much more so than the subject of H_2 *Optimal Control*. In this work, we have tried to present what we believe to be the fundamental issues of filtering. The book is not intended to give a chronological development of filtering from a historical point of view. A vast number of books already do so. Our intent here is to develop from our perspective the complete theory of filtering and various design methodologies associated with it along with their practical implementations. In this respect, we present here a state-of-the-art view of exact and almost input-decoupled filtering, H_2, and H_∞ filtering and inverse filtering issues, and include an application of filtering and inverse filtering to fault detection, isolation, and estimation. Most of the work reported here arose out of the research conducted by one or more of us and sometimes in collaboration with our students and colleagues.

Supposedly, young F. Scott Fitzgerald proclaimed in 1920 right after his first novel (*This Side of Paradise*) that, "an author ought to write for the youth of his own generation, the critics of the next, and the school masters of ever afterward". It is very presumptuous of us to say the same. Nevertheless, we strived to do so. Also, it is said that one must do his/her work and renounce the fruits of it to the ONE and MANY that pervade the universe. We have done our work. Let its fruits be fruits of many.

Our intended audience includes practicing engineers, graduate students, and researchers in filtering, signal processing, and control. An appropriate background for this book is a first graduate course in state-space methods as well as a first graduate course in filtering.

No work of this magnitude and nature can be undertaken without many sacrifices. Our families are the ones who sacrificed the time we could have spent with them otherwise. Needless to say that we owe a debt of gratitude to our families, and it is natural that we dedicate this book to them. We certainly also owe a debt of gratitude to our editor, Dr. Tamer Başar, and the editorial staff at Birkhäuser. Our special thanks go to the copy editor for a meticulous editing that improved the text.

Ali spent countless number of hours brooding over the manuscript of this book at Bucer's, the great coffee house of Moscow, Idaho. The melodious atmosphere at Bucer's nurses many bruised souls to their vitality. Ali acknowledges the contribution of all the good people of Bucer's.

The publication of this book marks over two decades of collaboration between AS and PS. Each would like to acknowledge this long partnership and friendship and express their hopes that this will continue undiminished for many more years.

Ali Saberi
Washington State University, Pullman, Washington, U.S.A.

Anton A. Stoorvogel
Eindhoven University of Technology, Eindhoven, The Netherlands
Delft University of Technology, Delft, The Netherlands

Peddapullaiah Sannuti
Rutgers University, Piscataway, New Jersey, U.S.A.

1
Introduction

1.1 Introduction

Estimation theory and, in the same breath, *filtering theory* is vast and rich in the literature and is central to a wide variety of disciplines, including control, communications, and signal processing. Also, it is relevant to such diverse areas as statistics, economics, bioengineering, and operations research. The terms *estimation* and *filtering* evoke many and varied responses among engineers and scientists. In its primary level, "estimation" is the process of arriving at a value for a desired and unknown variable from certain observations or measurements of other variables related to the desired one but contaminated with *noise*. Although one could trace the origins of estimation back to ancient times, Karl Friederich Gauss is generally acknowledged to be the forefather of what is now referred to as estimation theory. In his quest to predict the motions of planets and comets from telescopic measurements, Gauss at the age of 18 formulated the now well-known method of *least squares*. In modern times and in particular in the second half of the twentieth century, filtering theory has become synonymous with estimation theory mainly in the engineering literature. It might look odd that the term *"filter"* would apply to an *"estimator"*. In its common use, a "filter" is a physical device that can separate the wanted and unwanted fractions of a mixture. In electronics, a filter is seen as a circuit with a frequency-selective behavior, and thus can attenuate certain undesired components of the input signal and pass to the output certain desired components of the input signal. This notion of separating a signal into certain desired and undesired components can be stretched further in *signal processing*, where a signal contaminated with noise is enhanced or reconstructed by eliminating noise as much as possible. In the 1930s and 1940s, such a notion of filtering was stretched even further to separate the desired *signals* from *noise*, both of which were characterized by their covariance functions. The pioneering work of Kolmogorov and Wiener used a statistical characterization of the probability distributions of signals and noise in forming an optimal estimate of the signal, given the sum of the signal and noise. In the early 1960s, Kalman changed the then conventional formulation of the problem, in which the covariance of the signal process is given, to a new formulation in which a *model* for the signal process is given and views the signal as the output of a linear finite-dimensional dynamical system driven by a white noise process. Thus, Kalman championed a filter (*which is now affectionately called the Kalman filter*) to solve what is now known as the *linear quadratic Gaussian estimation problem*. Ever since the seminal work of Kalman, the concept of filtering has broadened in several ways, and the research in the field of filtering and estimation has been pursued on a variety of fronts. Indeed, with the advent of Kalman filtering, the word *filtering* assumed a meaning that is well beyond the original idea of separating the components of a mixture. Moreover, it has come to include the solution of an *inversion problem* in which one knows how to represent the measurable variables as functions of the variables of principle interest, then inverts this functional relationship, and in doing so estimates the independent variables as inverted functions of the measured variables. Such an *inverse filtering* problem occurs in many practical applications where a signal of interest passes through a system whose measured output is corrupted by noise. When the system under consideration is a linear system, the measured output signal, although corrupted by noise, is the *convolution* of the desired input signal with the impulse response of the system, which is either partially or fully known. In this context, *inverse filtering* is also known as *deconvolution*, and it plays an important role, especially in the areas of signal processing and communication.

Thus, the word *filter* is viewed now as having a variety of functions, some of which can be summarized as follows:

- A filter *estimates* instantaneously or with a specified delay an unknown signal, normally a function of state and input of a system using the measured outputs.

- A filter *separates* a signal into its component signals and, in particular, separates the desired signal from the noise.

- A filter *solves* an inversion problem in which one seeks to estimate the intended inputs of a system from the measured outputs of it.

- A filter is *used* as a learning tool by which a certain signal is tracked, estimated, and predicted.

Filters and, in particular, Kalman filters are now used ubiquitously in several diverse areas. The principal uses of Kalman filtering have been in *modern* control systems, in the tracking and navigation of all sorts of vehicles, and in the prediction of future behavior. The list of Kalman filter applications is endless and includes not only satellite navigation, trajectory estimation, guidance, video and laser tracking systems, and radars, but also oil drilling, water and air quality control, geodetic surveys, and many others. Indeed it is not an overstatement to assert that Kalman filter with its many extended formats represents one of the most widely and demonstrably useful results that emerged in the twentieth century. It has enabled humankind to do many things that could not have been done otherwise, and it has become as indispensable as silicon in the makeup of electronic systems.

1.2   Filtering problems

To give a precise meaning to the word *filtering*, consider the block diagram of a typical setup as depicted in Figure 1.1. In Figure 1.1, a given system is excited by two kinds of input signals: an intended excitation signal and noise. The noise signals arise from a variety of causes, such as unknown external input signals, plant noise, model uncertainties of the given system, as well as measurement or observation noise. The job of the *filter* is to estimate the desired output of the given system by using the measured output of the system. Obviously, to design a filter that does such an estimation, one has to have an associated *performance index* that is a function of the estimation error, which is the difference between the desired and the estimated output. The job of *filter design* is to render the chosen performance index as small as possible. The performance index or measure can be defined in several ways. One remarkably common performance index is the integral square of the estimation error in which small values of the error are weighed relatively less than large values of the error. Such a problem has been labeled as the least-squares estimation problem, least (minimum) mean square estimation, or filtering problem. A filter that can achieve the minimum of the chosen performance index is called an *optimal filter*. Instead of seeking an optimal filter, typically one could also seek to render the performance index less than a prescribed value.

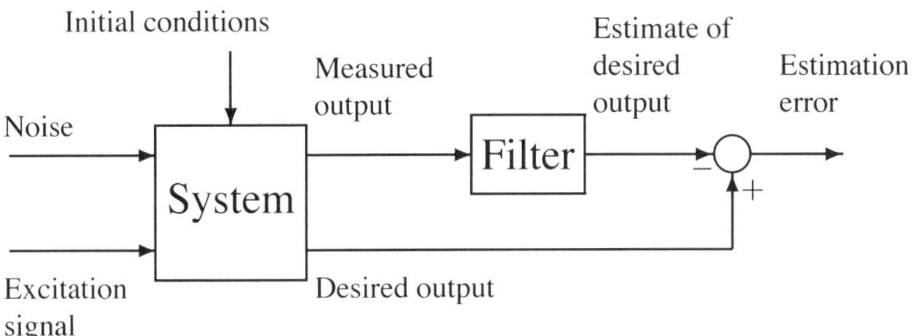

Figure 1.1: Block diagram of a typical setup

In estimation problems, one encounters three different divisions:

- Depending on whether we need to estimate the future, the current, or the past of the signals, we deal with a *prediction, filtering*, or *smoothing (delayed filtering)* problem. Nevertheless, independent of whether it is a prediction, filtering, or smoothing (delayed filtering) problem, one often calls it a filtering problem.

- In another type of classification, depending on whether observations are made over a finite or a semi-infinite interval of time, one deals with a filtering problem having a finite or infinite horizon of observations that are, respectively, referred to as finite or infinite horizon filtering problems.

- One could also divide the filtering problems encountered in the literature into four categories depending on the assumptions made on the noise characteristics.

  (*i*) In the first category, certain statistical assumptions (typically knowledge of covariance functions and power spectral densities) are made on the noise process.

  (*ii*) In the second category, no statistical assumptions are made, and the noise is essentially considered as unknown except that it has a finite root-mean-square (RMS) value.

  (*iii*) In the third category, noise signals can be divided into two subsets. As in the first category, certain statistical assumptions are made on the noise signals belonging to one set, while as in the second category, no assumptions are made on the noise signals belonging to the other set.

  (*iv*) For each of the above three categories, one can create another subcategory by adding other unknown input signals which contain a linear combination of sinusoidal signals, each of which has an unknown amplitude and phase but known frequency.

This book studies various classes of filtering problems and the corresponding filters that solve them. Various classes of filtering problems are defined based on different assumptions made on the noise characteristics and on different performance measures (indexes) that are adapted. Each class of filtering problem is studied in depth regarding (1) the existence of a filter that solves it, (2) uniqueness of such a filter, and (3) designing such a filter or filters while showing the flexibility in assigning poles of the filter or filters.

The first filtering problem arises by seeking that the RMS value of the estimation error be zero (i.e., by seeking that the estimation error tend asymptotically to zero as time progresses to infinity) irrespective of the nature of noise or input to the given system. Such a problem demands exact estimation, and it dictates that the transfer matrix from the inputs to the estimation error be identically zero. In other words, in such a problem, we seek a filter that estimates the desired output in such a way that the error in the estimation of the desired output is completely decoupled from the input(s). For this reason, we can call the problem we pose here the *exact input-decoupling filtering problem*, or, for short, the *EID filtering problem*, and the filters that solve such a problem the *exact input-decoupling filters* or EID filters. EID filtering is studied in depth in Chapter 7.

The EID filtering problem stated above demands a severe performance measure, namely that the transfer matrix from the input to the estimation error be identically zero. As such, the EID filtering problem is not always solvable. It is natural then to think of methods of relaxing the performance requirements so that the solvability conditions can possibly be weakened and thus allow us to deal with a larger class of systems. There are several ways by which the performance requirements can be weakened. We plan to relax the performance requirements progressively layer by layer to form a hierarchy of problems as outlined below.

We can introduce the first layer of relaxing the performance requirements by seeking that the RMS value of the estimation error be *"almost zero"*, or, equivalently, *"arbitrarily small"* or *"as small as desired"*, instead of being identically zero. To be more precise, we try to find a family of filters parameterized by some positive ε such that when applied to the system, the RMS value of the estimation error converges to zero as $\varepsilon \downarrow 0$. Such a problem can be termed as an almost-input-decoupled (AID) filtering problem or, for short, an AID filtering problem. The filters that solve the AID filtering problem can be termed not surprisingly as AID filters.

In AID filtering, we primarily focus on two cases depending on the assumptions made on the noise characteristics, whether we have known noise statistics or unknown noise statistics. Clearly, the noise can be modeled as a stochastic process. Whenever the noise is a stochastic process, all signals, i.e., the state, the measured output, and the desired output of the given plant or system, are naturally stochastic processes. In one framework, the noisy input to the given system can be modeled as a zero mean wide sense stationary white noise stochastic process of power spectral density (PSD) equal to an identity matrix, which can be called simply as white noise of unit intensity. In fact, one can assume without much loss of generality the input as any zero mean wide sense stationary stochastic process of known PSD, which is not necessarily a white noise. In this regard, one can recall easily the well-known fact that one can always generate a wide sense stationary stochastic process as the output of a linear time-invariant system driven by white noise of unit intensity so that the PSD of such a generated stochastic process approximates arbitrarily closely any known PSD of a wide sense stationary stochastic process. The needed linear time-invariant system to do so can always be appended to the given system. As such, there is not much loss of generality in assuming the input to be zero mean white noise of unit intensity.

Thus, in our study of AID filtering, as a first case we can model the input as a zero mean wide sense stationary white noise of unit PSD. Then, we seek a family of parameterized filters such that when applied to the system, the RMS norm of the estimation error signal converges to zero as the parameter tends to zero. By the definition of the H_2 norm of a transfer matrix, this is equivalent to demanding that the H_2 norm of the transfer function from the noise input to the estimation error be arbitrarily small. As such, such an AID filtering problem is referred to as the H_2 AID filtering problem, and the family of filters that solves such a problem as the family of H_2 AID filters. H_2 AID filtering is studied in depth in Chapter 8.

Unlike the first case discussed, we can assume no statistical information on the input except that it has a finite RMS norm. Under such an unknown input, we seek to render the ratio of the RMS value of the error to the RMS value of the input arbitrarily small. By the definition of the H_∞ norm of a transfer matrix, this is equivalent to demanding that the H_∞ norm of the transfer function from the noise input to the estimation error be arbitrarily small. As such, such an AID filtering problem is referred to as the H_∞ AID filtering problem, and the family of filters that solves such a problem as the family of H_∞ AID filters. H_∞ AID filtering is studied in depth in Chapter 9.

As discussed, AID filtering seeks to have the RMS norm of the estimation error signal as small as *desired*. We can relax such a performance requirement even further by seeking that the RMS norm of the error signal be as small as *possible* rather than as small as desired. This requirement leads us to optimally input-decoupling (OID) filtering. As in AID filtering, we can divide the OID filtering problems into two categories depending on the assumptions made on the noise characteristics. At first, we can follow the direction set by AID filtering under white noise input. That is, we can assume that the input to the given system is a white noise of unit intensity and seek to make the RMS norm of the error signal as small as *possible*. Once again, by the definition of the H_2 norm of a transfer matrix, this is equivalent to demanding that the H_2 norm of the transfer function from the noise input to the estimation error be minimized. As such, such an OID filtering problem under white noise input is referred to as an H_2 OID filtering problem, and the filters that solve such a problem as the H_2 OID filters. H_2 OID filtering is studied in depth in Chapter 10.

What we called an H_2 OID filtering problem is indeed the celebrated Kalman filtering problem. Most available literature on Kalman filtering deals only with what are known as *regular* filtering problems, which rely on a crucial assumption that all measured observations of the given system are fully corrupted by the white Gaussian noise process. On the other hand, many practical applications exist where a, what can be called *singular*, situation arises. Such cases arise, for instance, when observations are corrupted by *colored* noise or when some or all observations are being modeled as noise free. Also, the formalism of the Wiener filtering problem when transformed to a Kalman filtering problem leads to a singular situation. This book considers general singular filtering problems.

Another concept highly tied to H_2 OID filtering is the suboptimally input-decoupled (SOID) filtering problem. In the absence of a formal definition of suboptimality, any filter that is not optimal can be construed as a suboptimal filter. A good definition of suboptimality can be given through the notion of attaining an RMS norm of the error signal arbitrarily close to its infimum. In this regard, a sequence or a family of filters can be called suboptimal filters if one can select a filter from the family such that the resulting RMS norm of the error signal is within an arbitrarily specified value from its infimum. The filtering problem that arises in this regard can be termed as an H_2 SOID filtering problem, and the family of filters that solves such a problem as the family of H_2 SOID filters. H_2 SOID filtering is studied in depth in Chapter 10.

In the H_2 OID and H_2 SOID filtering considered above, we assume that the input to the given system is a white noise of unit intensity. We can follow another direction in which no statistical information about the input is available except that it has a finite RMS value. In this connection, for a given filter, we can express the performance by the smallest number γ for which the RMS norm of the error signal for any input is always less than or equal to γ times the RMS norm of the input. Then, one can pose a filtering problem as the problem of finding a filter that can achieve the smallest possible value for γ, which can be denoted by γ_{sp}^* or γ_p^* depending on whether we consider the class of strictly proper or proper filters. Clearly then, an optimal filter achieves the minimum possible performance, namely, γ_{sp}^* or γ_p^*. In the case of a suboptimal filter, for a specified number $\gamma > \gamma_{sp}^*$ or $\gamma > \gamma_p^*$, one seeks a filter that achieves the RMS norm of the error signal for any input less than or equal to γ times the RMS norm of the input. Such a suboptimal filter can be termed as a γ-level suboptimal filter. One can, of course, seek in general to design an optimal filter or a γ-level suboptimal filter. However, essentially most available literature bypasses seeking optimal filters and focuses only on γ-level suboptimal filters. A primary reason to do so is that for general systems no elegant analytic formula exists that enables one to compute the infimum performance measure γ_{sp}^* or γ_p^*. Only numerical approximations of computing γ_{sp}^* or γ_p^* exist, and this obviously implies that seeking γ-level suboptimal filters is very natural. Also, there is a secondary reason to seek such suboptimal filters, namely, that the existence conditions for an optimal filter are prohibitively complex. Moreover, for engineering applications, often suboptimal filters suffice.

By the definition of the H_∞ norm of the transfer function of a system, the γ-level suboptimal filtering problem we posed above is equivalent to demanding that the H_∞ norm of the transfer function from the noise input to the estimation error be less than γ. As such, such a problem is referred to as a γ-level H_∞ suboptimally input-decoupling filtering problem or, for short, as a γ-level H_∞ SOID filtering problem. The filter that solves such a problem can obviously be called the γ-level H_∞ SOID filter. These are studied in depth in Chapter 11.

All the filtering problems introduced above are reconsidered with additional unknown input signals containing a linear combination of sinusoidal signals, each of which has an unknown amplitude and phase but known frequency. We have used the qualifier "generalized" to refer to each of the above problems when such additional inputs are present. For instance, the H_2 OID, H_2 SOID, and γ-level H_∞ SOID filtering problems described above when additional sinusoidal inputs of unknown amplitude and phase but known frequency are present are, respectively, referred to as generalized H_2 OID, generalized H_2 SOID, and generalized γ-level H_∞ SOID filtering problems. The generalized H_2 OID and the generalized H_2 SOID filtering are studied in depth in Chapter 12, whereas the generalized γ-level H_∞ SOID filtering is studied in depth in Chapter 13.

In the OID filtering problems we discussed above, we assume that either the input to the system is a white noise of unit intensity or no statistical information about it is available except that it has a finite RMS value. Obviously, one can have a mixed case in which a part of the input is modeled as a white noise and the other part as a signal for which no statistical information is available except that it has a finite RMS value. The literature contains some such mixed filtering work, including the one authored by us [77]. However, research on mixed filtering problems is still progressing and is not complete. At this time, the research is not mature enough to be dealt with in this book.

All the filtering problems enumerated above are discussed in detail in this book for both continuous- and discrete-time systems. In particular, the solvability condition for each problem is developed. Also, the conditions for the uniqueness of a filter that solves a given problem are given. Whenever a problem is solvable, explicit design methods to construct a filter, or a family of filters, that solves a given problem, are discussed. Moreover, the available flexibility in placing the poles of a filter at desired locations is also pointed out. Regarding the architecture of filters, full-order strictly proper, full-order proper, as well as reduced-order proper filters are considered.

As is easily understandable, various types of faults arise in industrial processes because of malfunctions of internal components of a process, as well as those in measurement sensors and control actuators attached to the process. We apply the methods of filtering discussed above to fault detection, fault isolation, and fault estimation. Various fault detection, fault isolation, and fault estimation problems are formulated and studied in depth in Chapters 14 and 15.

2
Preliminaries

In this chapter we bring together the notations and acronyms used in this book and various definitions and facts related to matrices, linear spaces, linear operators, norms of deterministic as well as stochastic signals, and norms of linear time- or shift-invariant systems.

2.1   A list of symbols

Throughout this book we shall adopt the following conventions and notations:

$\mathbb{R}$	set of real numbers
$\mathbb{R}^+$	set of positive real numbers
Z^+	set of non-negative integers
$\mathbb{C}$	entire complex plane
$\mathbb{C}^-$	open left-half complex plane
$\mathbb{C}^0$	imaginary axis
$\mathbb{C}^+$	open right-half complex plane
$\mathbb{C}^{-0}$	closed left-half complex plane
$\mathbb{C}^\ominus$	set of complex numbers inside the unit circle
$\mathbb{C}^\circ$	unit circle
$\mathbb{C}^\oplus$	set of complex numbers outside the unit circle
$\mathbb{C}^\otimes$	set of complex numbers inside and on the unit circle
I	an identity matrix
I_k	identity matrix of dimension $k \times k$
A'	transpose of A
A^*	complex conjugate transpose of A
$\lambda(A)$	set of eigenvalues of A
$\sigma_{\max}(A)$	maximum singular value of A

$\sigma_{\min}(A)$ — minimum singular value of A

$\rho(A)$ — spectral radius of A that is equal to $\max_i |\lambda_i(A)|$

$\operatorname{trace} A$ — trace of A

$\ker A$ — the null space of A

$N(A)$ — the null space of A

$\operatorname{im} A$ — the range space of A

$R(A)$ — the range space of A

$\langle A \mid \operatorname{im} B \rangle$ — $\displaystyle\sum_{i=0}^{n-1} \operatorname{im} A^i B$, the controllability subspace of the pair (A, B)

$\langle \ker C \mid A \rangle$ — $\displaystyle\bigcap_{i=0}^{n-1} \ker CA^i$, the unobservable subspace of the pair (A, C)

$\mathcal{V}^\perp$ — orthogonal complement of a subspace $\mathcal{V}$ in $\mathbb{R}^n$

$\mathbb{R}[s]$ — ring of polynomials with real coefficients

$\mathbb{R}^{n \times m}[s]$ — set of all $n \times m$ matrices with coefficients in $\mathbb{R}[s]$

$\mathbb{R}(s)$ — field of rational functions with real coefficients

$\mathbb{R}^{n \times m}(s)$ — set of all $n \times m$ matrices with coefficients in $\mathbb{R}(s)$

2.2 A list of acronyms

ADD   Almost Disturbance Decoupling
ADDPS   Almost Disturbance Decoupling Problem via State feedback
AFDIE   Almost Fault Detection, Isolation, and Estimation
AID   Almost Input Decoupling
CARE   Continuous-time Algebraic Riccati Equation
CLMI   Continuous-time Linear Matrix Inequality
CQMI   Continuous-time Quadratic Matrix Inequality
CSS   Chen, Saberi, and Sannuti
DARE   Discrete-time Algebraic Riccati Equation
DDPS   Disturbance Decoupling Problem via State feedback
DLMI   Discrete-time Linear Matrix Inequality
DQMI   Discrete-time Quadratic Matrix Inequality
EDD   Exact Disturbance Decoupling
EDD_{li} Exact Disturbance Decoupling left invertible
EDD_{nli} Exact Disturbance Decoupling non-left invertible
EDDSPP   Exact Disturbance Decoupling with Simultaneous Pole Placement
EFDIE   Exact Fault Detection, Isolation, and Estimation
EID   Exact Input Decoupling

GDARE   Generalized Discrete-time Algebraic Riccati Equation
GDARI   Generalized Discrete-time Algebraic Riccati Inequality
OID   Optimally Input-Decoupling
PAFDIE   Partial Almost Fault Detection, Isolation, and Estimation
PEFDIE   Partial Exact Fault Detection, Isolation, and Estimation
psd   Positive Semi-Definite
PSD   Power Spectral Density
RMS   Root Mean Square
SCB   Special Coordinate Basis
SOID   Sub-Optimally Input-Decoupling

2.3   Matrices, linear spaces, and linear operators

In this section, we recall certain fundamental facts and properties of matrices, linear spaces, and linear operators that are relevant to this book. We have done so for the ease of readers and to establish the related notations used throughout the book.

We say a matrix A is said to be injective or surjective if A is of full column or row rank, respectively. By $\operatorname{rank}_{\mathcal{K}}$, we denote the rank of a matrix whose entries are in the field $\mathcal{K}$. We shall write *rank* only for the case when $\mathcal{K} = \mathbb{R}$ or $\mathcal{K} = \mathbb{C}$. Moreover, we use the term *normal rank* or *normrank* for $\operatorname{rank}_{\mathcal{K}}$ whenever $\mathcal{K} = \mathbb{R}(s)$. We note that if $A \in \mathbb{C}^{m \times n}$, we have that $\operatorname{im} A = \ker(A^*)^{\perp}$.

Let us next recall what we mean by the multiplicity structure of an eigenvalue of a matrix $A \in \mathbb{R}^{n \times n}$. The multiplicity structure of an eigenvalue λ of A comprises all sizes of the Jordan blocks corresponding to λ in the Jordan form of A; or equivalently, it comprises all degrees of the elementary divisors of the polynomial matrix $sI - A$ (elementary divisors of a polynomial matrix are defined on page 36). The multiplicity structure of λ is written as $\{m_1, \cdots, m_k\}$ where k is the number of Jordan blocks associated with λ, and $m_1, \cdots, m_k$ are the dimensions of such Jordan blocks. If all m_i, $i = 1, \cdots, k$, are unity, i.e. the multiplicity structure of λ is $\{1, \cdots, 1\}$, then we say λ is a simple eigenvalue of A.

Next, we recall from [102] certain results regarding the eigenvalues of the sum of two matrices and the product of two matrices.

For $A \in \mathbb{C}^{n \times n}$, let

$$F(A) := \{x^* A x \mid x \in \mathbb{C}^n \text{ and } x^* x = 1\}.$$

Then, $F(A)$ is a closed bounded convex set containing all eigenvalues of A. Now, consider also $B \in \mathbb{C}^{n \times n}$. Then the following hold*:

---

*If S_1 and S_2 are certain sets of complex numbers, we denote by $S_1 + S_2$, $S_1 S_2$, and S_1/S_2, respectively, the set of all numbers of the form $s_1 + s_2$, $s_1 s_2$, and s_1/s_2 for $s_i \in S_i$ (in the case of S_1/S_2, we assume that $0 \notin S_2$).

(i) If λ is an eigenvalue of $A + B$, then $\lambda \in F(A) + F(B)$.

(ii) Let $0 \neq F(B)$. If λ is an eigenvalue of $B^{-1}A$, then $\lambda \in F(A)/F(B)$.

(iii) Let A be arbitrary and B be Hermitian and positive semi-definite. If λ is an eigenvalue of AB, then $\lambda \in F(A)F(B)$.

(iv) Let A be Hermitian and B be Hermitian and positive definite. Then AB has as many positive, vanishing, and negative eigenvalues as A does.

If M is a subspace of $\mathbb{C}^n$, then we define the *orthogonal projector* P_M of $\mathbb{C}^n$ onto M by $P_M u = u$ if $u \in M$ and $P_M u = 0$ if $u \in M^{\perp}$. We note that $I - P_M = P_{M^{\perp}}$.

A matrix $U \in \mathbb{C}^{n \times n}$ is said to be a unitary matrix if $U^* = U^{-1}$. For a matrix $A \in \mathbb{C}^{m \times n}$, the generalized inverse of A (or Moore–Penrose inverse of A) is defined to be a unique matrix $A^{\dagger}$ in $\mathbb{C}^{n \times m}$ such that

(a) $AA^{\dagger}$ is an orthogonal projection onto im A, and

(b) $A^{\dagger}A$ is an orthogonal projection onto $(\ker A)^{\perp} = \text{im } A^*$.

Another equivalent definition for a generalized inverse of $A \in \mathbb{C}^{m \times n}$ is a unique matrix $A^{\dagger}$ in $\mathbb{C}^{n \times m}$ such that

(i) $AA^{\dagger}A = A$.

(ii) $A^{\dagger}AA^{\dagger} = A^{\dagger}$.

(iii) $AA^{\dagger}$ is a symmetric matrix.

(iv) $A^{\dagger}A$ is a symmetric matrix.

Some basic properties of the generalized inverse of $A \in \mathbb{C}^{m \times n}$ are listed as follows:

- $(A^{\dagger})^{\dagger} = A$.

- $(A^{\dagger})^* = (A^*)^{\dagger}$.

- If $\lambda \in \mathbb{C}$, $(\lambda A)^{\dagger} = \lambda^{\dagger}A^{\dagger}$, where $\lambda^{\dagger} = \frac{1}{\lambda}$ if $\lambda \neq 0$ and $\lambda^{\dagger} = 0$ if $\lambda = 0$.

- $A^* = A^*AA^{\dagger} = A^{\dagger}AA^*$.

- $(A^*A)^{\dagger} = A^{\dagger}(A^*)^{\dagger}$.

- $A^{\dagger} = (A^*A)^{\dagger}A^* = A^*(AA^*)^{\dagger}$.

- $(UAV)^{\dagger} = V^*A^{\dagger}U^*$, where U and V are unitary matrices.

- im A = im $AA^{\dagger}$ = im AA^*.

- im $A^{\dagger}$ = im A^* = im $A^{\dagger}A$ = im A^*A.

- $\operatorname{im}(I - AA^\dagger) = \ker AA^\dagger = \ker A^* = \ker A^\dagger = (\operatorname{im} A)^\perp$.

- $\operatorname{im}(I - A^\dagger A) = \ker A^\dagger A = \ker A = (\operatorname{im} A^*)^\perp$.

- If $B \in \mathbb{C}^{n \times p}$, then $(AB)^\dagger = (P_{R(A^*)} B)^\dagger (A P_{R(B)})^\dagger$.

- If $A^* A B B^* = B B^* A^* A$, then $(AB)^\dagger = B^\dagger A^\dagger$.

- If $A = BC$, where $B \in \mathbb{C}^{m \times r}$ and $C \in \mathbb{C}^{r \times n}$, while $r = \operatorname{rank} A = \operatorname{rank} B = \operatorname{rank} C$, then $A^\dagger = C^*(CC^*)^{-1}(B^*B)^{-1}B^*$.

The following necessary and sufficient conditions for a partitioned matrix to be *positive semi-definite* and *positive definite* are useful. Consider an arbitrarily partitioned Hermitian matrix Q:

$$Q = \begin{pmatrix} Q_{11} & Q_{12} \\ Q_{12}^* & Q_{22} \end{pmatrix}.$$

Then Q is *positive semi-definite* if and only if

$$\begin{cases} Q_{22} \geqslant 0 \\ Q_{12} = Q_{12} Q_{22}^\dagger Q_{22} \\ Q_{11} \geqslant Q_{12} Q_{22}^\dagger Q_{12}^* \end{cases}$$

or, equivalently, Q is *positive semi-definite* if and only if

$$\begin{cases} Q_{11} \geqslant 0 \\ Q_{12} = Q_{11} Q_{11}^\dagger Q_{12} \\ Q_{22} \geqslant Q_{12}^* Q_{11}^\dagger Q_{12}. \end{cases}$$

Similarly, Q is *positive definite* if and only if

$$\begin{cases} Q_{22} > 0 \\ Q_{11} > Q_{12} Q_{22}^{-1} Q_{12}^* \end{cases}$$

or, equivalently

$$\begin{cases} Q_{11} > 0 \\ Q_{22} - Q_{12}^* Q_{11}^{-1} Q_{12} > 0. \end{cases}$$

Let us next discuss the addition of subspaces and the associated notations. Suppose $\mathcal{X}$, $\mathcal{Y}$, and $\mathcal{Z}$ are some subspaces of $\mathbb{R}^n$ or $\mathbb{C}^n$. Then,

$$\mathcal{X} + \mathcal{Y} = \{ x + y \mid x \in \mathcal{X}, y \in \mathcal{Y} \}.$$

If $\mathcal{Z} = \mathcal{X} + \mathcal{Y}$ and $\mathcal{X} \cap \mathcal{Y} = \{0\}$, then $\mathcal{Z}$ is called the direct sum of $\mathcal{X}$ and $\mathcal{Y}$, and in this case, $\mathcal{Z}$ is written as $\mathcal{X} \oplus \mathcal{Y}$. Consider a subspace $\mathcal{X}$ in $\mathbb{R}^n$. Then, the orthogonal complement $\mathcal{X}^\perp$ of the subspace $\mathcal{X}$ is defined as

$$\mathcal{X}^\perp = \{ u \in \mathbb{R}^n \mid \langle u, v \rangle = 0 \text{ for every } v \in \mathcal{X} \}.$$

Let $\mathcal{X}$ and $\mathcal{Y}$ be two nontrivial subspaces of $\mathbb{R}^n$. If the inner product of x and y is zero for all $x \in \mathcal{X}$ and $y \in \mathcal{Y}$, then the two subspaces $\mathcal{X}$ and $\mathcal{Y}$ are said to be orthogonal, and this is denoted by $\mathcal{X} \perp \mathcal{Y}$. Also, if $\mathcal{Z} = \mathcal{X} + \mathcal{Y}$ and $\mathcal{X} \perp \mathcal{Y}$, then we denote $\mathcal{Z} = \mathcal{X} \oplus_\perp \mathcal{Y}$.

Next, for a matrix $M \in \mathbb{R}^{m \times n}$, the linear transformation $M\mathcal{X}$ is defined as

$$M\mathcal{X} := \{Mx \mid x \in \mathcal{X}\}.$$

Also, for a matrix $N \in \mathbb{R}^{n \times m}$,

$$N^{-1}\mathcal{X} := \{z \in \mathbb{R}^m \mid Nz \in \mathcal{X}\}.$$

The following relations will be useful in algebraic manipulations regarding subspaces:

$$
\begin{aligned}
\mathcal{X} \cap (\mathcal{Y} + \mathcal{Z}) &\supseteq (\mathcal{X} \cap \mathcal{Y}) + (\mathcal{X} \cap \mathcal{Z}) \\
\mathcal{X} + (\mathcal{Y} \cap \mathcal{Z}) &\subseteq (\mathcal{X} + \mathcal{Y}) \cap (\mathcal{X} + \mathcal{Z}) \\
(\mathcal{X}^\perp)^\perp &= \mathcal{X} \\
(\mathcal{X} + \mathcal{Y})^\perp &= \mathcal{X}^\perp \cap \mathcal{Y}^\perp \\
(\mathcal{X} \cap \mathcal{Y})^\perp &= \mathcal{X}^\perp + \mathcal{Y}^\perp \\
M(\mathcal{X} \cap \mathcal{Y}) &\subseteq M\mathcal{X} \cap M\mathcal{Y} \\
M(\mathcal{X} + \mathcal{Y}) &= M\mathcal{X} + M\mathcal{Y} \\
N^{-1}(\mathcal{X} \cap \mathcal{Y}) &= N^{-1}\mathcal{X} \cap N^{-1}\mathcal{Y} \\
N^{-1}(\mathcal{X} + \mathcal{Y}) &\supseteq N^{-1}\mathcal{X} + N^{-1}\mathcal{Y}.
\end{aligned}
\tag{2.1}
$$

Also, let $\mathcal{V}$ be a subspace of dimension m. Then we have

$$M\mathcal{X} \subseteq \mathcal{V} \iff M'\mathcal{V}^\perp \subseteq \mathcal{X}^\perp \tag{2.2}$$

$$(M^{-1}\mathcal{V})^\perp = M'\mathcal{V}^\perp. \tag{2.3}$$

Let $A = \mathbb{R}^{n \times n}$. Then $\mathcal{T}$, a subspace of $\mathbb{R}^n$, is an A-invariant subspace if

$$A\mathcal{T} \subseteq \mathcal{T}.$$

The following properties of an A-invariant subspace are useful:

(*i*) A subspace $\mathcal{T}$ with T a matrix such that $\mathcal{T} = \operatorname{im} T$ is A-invariant if and only if a matrix X exists such that

$$AT = TX.$$

(*ii*) Let $\mathcal{T}$ be an A-invariant subspace. Then a similarity transformation L exists such that

$$\tilde{A} := L^{-1}AL = \begin{pmatrix} \tilde{A}_{11} & \tilde{A}_{12} \\ 0 & \tilde{A}_{22} \end{pmatrix}, \quad \text{and} \quad \mathcal{T} = L^{-1}\begin{pmatrix} I \\ 0 \end{pmatrix}$$

with $\tilde{A}_{11} \in \mathbb{R}^{h \times h}$, where $h := \dim \mathcal{T}$.

The proofs of the above relations are simple and can be found in standard books on vector spaces.

Consider a matrix $A \in \mathbb{R}^{n \times n}$ and an A-invariant subspace $\mathcal{T} \subseteq \mathbb{R}^{n \times n}$. Then the *restriction* of A to $\mathcal{T}$ is the linear map $\rho : \mathcal{T} \to \mathcal{T}$ defined by

$$\rho(x) = Ax \quad \text{for all } x \in \mathcal{T}.$$

The restriction of A to $\mathcal{T}$ is usually denoted by $A \,|\, \mathcal{T}$.

Next, we would like to recall some elementary concepts regarding modal subspaces. We first develop some notations used in continuous-time systems. Consider a matrix $A \in \mathbb{R}^{n \times n}$. Let $\alpha(s)$ denote the characteristic polynomial of A and factor it as $\alpha(s) = \alpha_-(s) \cdot \alpha_+(s)$, where $\alpha_-(s)$ has all its roots in the open left-half complex plane $\mathbb{C}^-$ and $\alpha_+(s)$ has all its roots in the closed right-half complex plane $\mathbb{C}^{-0}$. Then the stable and unstable modal subspaces of $\mathbb{R}^n$ related to A are

$$X_-(A) = \ker \alpha_-(A),$$
$$X_+(A) = \ker \alpha_+(A).$$

It is easy to show that $X_-(A)$ is spanned by the real and the imaginary part of the generalized eigenvectors of A corresponding to the eigenvalues in $\mathbb{C}^-$. Similarly, $X_+(A)$ is spanned by the real and imaginary parts of the generalized eigenvectors of A corresponding to the eigenvalues in $\mathbb{C}^0 \cup \mathbb{C}^+$. These two modal subspaces are complementary; i.e., they are independent and their sum is $\mathbb{R}^n$; thus,

$$\mathbb{R}^n = X_-(A) \oplus X_+(A).$$

Standard numerical linear algebra can be used to compute the bases for modal subspaces. For example, one can transform A via orthogonal transformation T to a real Schur form

$$T'AT = \begin{pmatrix} A_- & \star \\ 0 & A_+ \end{pmatrix}, \tag{2.4}$$

where the eigenvalues of A_- and A_+ are, respectively, located in $\mathbb{C}^-$ and $\mathbb{C}^0 \cup \mathbb{C}^+$, and $\star$ denotes some matrix that is not necessarily zero. If we partition T in conformity with the partitioning on the right-hand side of (2.4),

$$T = \begin{pmatrix} T_1 & T_2 \end{pmatrix},$$

then it is obvious that the columns of T_1 form a basis for $X_-(A)$. That is,

$$X_-(A) = \operatorname{im} T_1.$$

Analogously, we develop some notations used in discrete-time systems. Consider a matrix $A \in \mathbb{R}^{n \times n}$. Let $\alpha(z)$ denote the characteristic polynomial of A and factor it as $\alpha(z) = \alpha_\ominus(z) \cdot \alpha_\odot(z)$, where $\alpha_\ominus(z)$ has all its roots within the unit circle $\mathbb{C}^\ominus$ in the complex plane and $\alpha_\odot(z)$ has all its roots on or outside the unit circle $\mathbb{C}^{\ominus} \cup \mathbb{C}^{\oplus}$. Then the stable and unstable modal subspaces of $\mathbb{R}^n$ related to A are

$$X_{\ominus}(A) = \ker \alpha_{\ominus}(A),$$
$$X_{\odot}(A) = \ker \alpha_{\odot}(A).$$

It is easy to show that $X_{\ominus}(A)$ is spanned by the generalized real eigenvectors of A corresponding to the eigenvalues in $\mathbb{C}^{\ominus}$. Similarly, $X_{\odot}(A)$ is spanned by the generalized real eigenvectors of A corresponding to the eigenvalues in $\mathbb{C}^{\circ} \cup \mathbb{C}^{\oplus}$. These two modal subspaces are complementary; i.e., they are independent and their sum is $\mathbb{R}^n$; thus,

$$\mathbb{R}^n = X_{\ominus}(A) \oplus X_{\odot}(A).$$

Again, as in the continuous-time case, standard numerical linear algebra can be used to compute the bases for modal subspaces.

2.4   Norms of deterministic signals

Many measures are used to describe the size of a signal. The measures of size are called norms. In this section, we recall some of the common norms for persistent or transient continuous-time (discrete-time) vector signals. We consider continuous-time vector signals $y : \mathbb{R}^{+} \rightarrow \mathbb{R}^n$ and discrete-time vector signals $y : \mathbb{Z}^{+} \rightarrow \mathbb{R}^n$.

Definition 2.1 (L_1, L_2, and L_∞ (ℓ_1, ℓ_2, and ℓ_∞) norms) *For a vector-valued continuous-time signal y, we define L_1, L_2, and L_∞ norms, respectively, as*

$$\|y\|_1 := \int_0^\infty \sum_{i=1}^n |y_i(t)| \, dt,$$

$$\|y\|_2 := \left(\int_0^\infty \sum_{i=1}^n y_i^2(t) \, dt \right)^{\frac{1}{2}},$$

$$\|y\|_\infty := \sup_{t \geq 0} \max_{1 \leq i \leq n} |y_i(t)|.$$

Analogously, for a vector-valued discrete-time signal y, we define ℓ_1, ℓ_2, and ℓ_∞ norms, respectively, as

$$\|y\|_1 \quad := \quad \sum_{k=0}^{\infty} \sum_{i=1}^{n} |y_i(k)|,$$

$$\|y\|_2 \quad := \quad \left(\sum_{k=0}^{\infty} \sum_{i=1}^{n} y_i^2(k) \right)^{\frac{1}{2}},$$

$$\|y\|_\infty \quad := \quad \sup_{k\geq 0} \ \max_{1\leq i \leq n} |y_i(k)|.$$

The square of the L_2 or ℓ_2 norm of a signal y is commonly termed as the total energy in the signal y. In many areas of engineering, the energy or square of the L_2 (ℓ_2) norm is used as a measure of the size of a transient signal y that decays to zero as time progresses towards infinity. By Parseval's theorem, $\|y\|_2$ can also be computed in the frequency domain as follows: for the continuous-time case,

$$\|y\|_2 = \left(\frac{1}{2\pi} \int_{-\infty}^{\infty} Y(j\omega)^* Y(j\omega) d\omega \right)^{1/2},$$

where Y is the Fourier transform of y; similarly, for the discrete-time case,

$$\|y\|_2 = \left(\frac{1}{2\pi} \int_{-\pi}^{\pi} Y(e^{j\omega})^* Y(e^{j\omega}) d\omega \right)^{1/2},$$

where Y is the z-transform of y.

Definition 2.2 *(**Deterministic RMS norm**) A continuous-time signal y for which the following limit is well-defined and finite:*

$$\lim_{T\to\infty} \frac{1}{T} \int_{0}^{T} y(t)' y(t) dt,$$

is called an RMS signal. The root-mean-square *(RMS)* value *of such a continuous-time signal y is defined as*

$$\|y\|_{\mathrm{RMS}} = \left(\lim_{T\to\infty} \frac{1}{T} \int_{0}^{T} y(t)' y(t) dt \right)^{1/2}. \tag{2.5}$$

Similarly, a discrete-time signal y for which the following limit is well-defined and finite:

$$\lim_{T\to\infty} \frac{1}{T} \sum_{k=0}^{T} y(k)' y(k),$$

is called an RMS signal. The RMS value *of such a discrete-time signal y is defined as*

$$\|y\|_{\mathrm{RMS}} = \left(\lim_{T \to \infty} \frac{1}{T} \sum_{k=0}^{T} y(k)'y(k) \right)^{1/2}. \tag{2.6}$$

The square of the RMS norm[†] of y is commonly termed as the average power of the signal y. Often in engineering, the RMS norm or average power is defined for a signal y that is persistent. We note that the RMS norm is a steady-state measure of a signal and is not affected by any transients.

Remark 2.3 *It is obvious that an L_2 (ℓ_2) signal has a zero RMS value. Also, an L_1 signal does not necessarily have a finite or well-defined RMS value, whereas, in contrast, an ℓ_1 signal always has a zero RMS value. Finally for an L_∞ (ℓ_∞) signal, the RMS value need not be well defined.*

2.5   Norms of stochastic signals

For a vector signal that is modeled as a wide-sense stationary or an asymptotically wide-sense stationary vector stochastic process (random sequence), the common measure of size is the RMS norm. We recall below the needed definition.

Definition 2.4 *(Stochastic RMS norm) For a wide-sense stationary vector stochastic process y with a bounded variance, we define*

$$\|y\|_{\mathrm{RMS}} = \left(E[y(t)'y(t)] \right)^{1/2}. \tag{2.7}$$

Analogously, for a wide-sense stationary vector random sequence y with a bounded variance, we define

$$\|y\|_{\mathrm{RMS}} = \left(E[y(k)'y(k)] \right)^{1/2}. \tag{2.8}$$

Here $E[.]$ denotes the expectation. For stochastic processes (random sequences) that are only asymptotically wide-sense stationary as time goes to infinity [i.e., for asymptotically wide-sense stationary processes (random sequences)], (2.7) and (2.8) need to be rewritten as

$$\|y\|_{\mathrm{RMS}} = \left(\lim_{t \to \infty} E[y(t)'y(t)] \right)^{1/2}, \tag{2.9}$$

[†]We would like to remark that the RMS norm is a pseudo-norm because the RMS norm of any energy or transient signal is zero.

and

$$\|y\|_{\text{RMS}} = \left(\lim_{k \to \infty} E[y(k)'y(k)] \right)^{1/2}, \tag{2.10}$$

respectively.

Note that in (2.7) and (2.8), the result is independent of t or k because the stochastic process (random sequence) is wide-sense stationary.

We note that if y is an ergodic stochastic process (random sequence), then the deterministic RMS norm of any realization of the stochastic process (random sequence) y is equal to the stochastic RMS norm of y with probability one.

Also, we note that the RMS value of a wide-sense stationary process y can be expressed in terms of its autocorrelation matrix $R_y(\tau)$,

$$R_y(\tau) := E[y(t)y'(t+\tau)],$$

or its power spectral density (PSD) $S_y(\omega)$,

$$S_y(\omega) := \int_{-\infty}^{\infty} R_y(\tau)e^{-j\omega\tau}d\tau.$$

That is,

$$\|y\|_{\text{RMS}} = \left(\text{trace}[R_y(0)] \right)^{1/2} = \left(\frac{1}{2\pi} \text{trace}\left[\int_{-\infty}^{\infty} S_y(\omega)d\omega \right] \right)^{1/2}.$$

Similarly, for a wide-sense stationary random sequence y, let the autocorrelation matrix be

$$R_y(n) := E[y(k)y'(k+n)],$$

and the power spectral density (PSD) be

$$S_y(\omega) := \sum_{n=-\infty}^{\infty} R_y(n)e^{-j\omega n}, \quad -\pi \leq \omega \leq \pi.$$

Then,

$$\|y\|_{\text{RMS}} = \left(\text{trace}[R_y(0)] \right)^{1/2} = \left(\frac{1}{2\pi} \text{trace}\left[\int_{-\pi}^{\pi} S_y(\omega)d\omega \right] \right)^{1/2}.$$

2.6   Norms of linear time- or shift-invariant systems

We recalled above the definitions of norms of signals. A notion related to the size of a signal is the gain of a transfer function of a linear time- or shift-invariant system. As in the case of a signal, once again, various norms are used to measure the size of a transfer function. In this section, we recall the definitions of certain such norms. Also, we recall methods of computing them.

Two well-known classic norms of linear time- or shift-invariant systems are the H_2 norm (which is the RMS value of the response of a system to white noise input of unit PSD) and the H_∞ norm (which is the RMS gain of the system). The definitions of these norms are recalled below.

Definition 2.5 (H_2 norm of a continuous-time system) *Consider a continuous-time system Σ having a $q \times \ell$ stable transfer function G. Then **the H_2 norm of the continuous-time system** Σ or, equivalently, of the transfer matrix G is defined as*

$$\|G\|_2 = \left(\frac{1}{2\pi} \operatorname{trace} \left[\int_{-\infty}^{\infty} G(j\omega)G^*(j\omega)d\omega \right] \right)^{1/2}. \tag{2.11}$$

We assign ∞ to the H_2 norm of an unstable continuous-time system.

We note that the H_2 norm is induced by an inner product; i.e., we have

$$\|G\|_2 = \langle G, G \rangle$$

with the inner product defined by

$$\langle G_1, G_2 \rangle = \frac{1}{2\pi} \operatorname{trace} \left[\int_{-\infty}^{\infty} G_1(j\omega)G_2^*(j\omega)d\omega \right].$$

By Parseval's theorem, the H_2 norm of the transfer matrix G can equivalently be defined as

$$\|G\|_2 = \left(\operatorname{trace} \left[\int_0^{\infty} g(t)g'(t)dt \right] \right)^{1/2}, \tag{2.12}$$

where $g(t)$ is the inverse Laplace transform of the transfer matrix or the unit impulse (Dirac distribution) response of the associated linear system. Thus, $\|G\|_2 = \|g\|_2$. It is also known that $\|G\|_2$ can be expressed in terms of the singular values of the matrix G at each frequency,

$$\|G\|_2 = \left(\frac{1}{2\pi} \int_{-\infty}^{\infty} \sum_{i=1}^{\min\{q,\ell\}} \sigma_i^2(G(j\omega))d\omega \right)^{1/2} \tag{2.13}$$

where $\sigma_i(G(j\omega))$ is the ith singular value of $G(j\omega)$.

Remark 2.6 (*Stochastic interpretation of the H_2 norm of a continuous-time system*) *Let us consider a continuous-time system with a stable transfer function G. Let the input w to the system be a wide-sense stationary stochastic process. Let z be the corresponding output. It is well known that*

$$S_z(\omega) = G(j\omega)S_w(\omega)G'(-j\omega), \qquad (2.14)$$

where S_w and S_z are the PSDs of $w(t)$ and $z(t)$, respectively. Then, the H_2 norm of $G(s)$ can be interpreted as the RMS value of the output z when the given system is driven by zero mean white noise with unit PSD. Note that formally white noise with unit PSD does not exist, but the above can be formalized using Brownian motion and stochastic differential equations.

Remark 2.7 *Note that the H_2 norm of a stable continuous-time system or transfer function $G(s)$ is finite if and only if it is strictly proper.*

Definition 2.8 (*H_2 norm of a discrete-time system*) *Consider a discrete-time system having a $q \times \ell$ stable transfer function G. Then the H_2 norm of G is defined as*

$$\|G\|_2 = \left(\frac{1}{2\pi}\operatorname{trace}\left[\int_{-\pi}^{\pi} G(e^{j\omega})G^*(e^{j\omega})d\omega\right]\right)^{1/2}. \qquad (2.15)$$

We assign ∞ to the H_2 norm of an unstable discrete-time system.

Again, we note that the H_2 norm is induced by an inner product; i.e., we have

$$\|G\|_2 = \langle G, G\rangle$$

with the inner product defined by

$$\langle G_1, G_2\rangle = \left(\frac{1}{2\pi}\operatorname{trace}\left[\int_{-\pi}^{\pi} G_1(e^{j\omega})G_2^*(e^{j\omega})d\omega\right]\right)^{1/2}.$$

Again, by Parseval's theorem, $\|G\|_2$ can equivalently be defined as

$$\|G\|_2 = \left(\operatorname{trace}\left[\sum_{k=0}^{\infty} g(k)g'(k)\right]\right)^{1/2} \qquad (2.16)$$

where g is the inverse z-transform of the transfer matrix or the unit impulse response of the associated linear system. Thus, $\|G\|_2 = \|g\|_2$.

Remark 2.9 *(Stochastic interpretation of the H_2 norm of a discrete-time system) Let us consider a discrete-time system with a stable transfer function $G(z)$. Let the input $w(k)$ to the system be a wide-sense stationary random sequence. Let $z(k)$ be the corresponding output. Then, it is well known that*

$$S_z(\omega) = G(e^{j\omega})S_w(\omega)G'(e^{-j\omega}), \quad -\pi \le \omega \le \pi \qquad (2.17)$$

where S_w and S_z are the PSDs of w and z, respectively. Then, once again the H_2 norm of G can be interpreted as the RMS value of the output z when the given system is driven by a zero mean white noise random sequence having unit variance.

State-space method for computing the H_2 norm: We present here briefly some results on the computation of the H_2 norm of a transfer function matrix when its realization is given in a state-space form (for details, see [8]). Consider the transfer function G of a continuous-time system described in a packed notation:

$$G(s) = \left[\begin{array}{c|c} A & B \\ \hline C & D \end{array}\right],$$

with A Hurwitz-stable. Let W_{obs}^c denote the observability grammian of the pair (A, C) and W_{con}^c the controllability grammian of (A, B). Note that D needs to be zero for a finite H_2 norm. We note that W_{obs}^c and W_{con}^c are the unique solutions of continuous-time Lyapunov equations:

$$A'W_{\text{obs}}^c + W_{\text{obs}}^c A + C'C = 0,$$
$$AW_{\text{con}}^c + W_{\text{con}}^c A' + BB' = 0.$$

The H_2 norm of $G(s)$ can now be computed by

$$\|G\|_2 = \text{trace}[B'W_{\text{obs}}^c B]$$
$$= \text{trace}[C W_{\text{con}}^c C'].$$

The computation of the H_2 norm of a transfer function G of a discrete-time system described in a packed notation:

$$G(z) = \left[\begin{array}{c|c} A & B \\ \hline C & D \end{array}\right],$$

with A Schur-stable, can be given along the same lines. That is, let W_{obs}^d and W_{con}^d be the unique solutions of discrete-time Lyapunov equations:

$$A'W_{\text{obs}}^d A - W_{\text{obs}}^d + C'C = 0,$$
$$AW_{\text{con}}^d A' - W_{\text{con}}^d + BB' = 0.$$

The H_2 norm of G can now be computed by

$$\|G(z)\|_2 = \text{trace}[B' W_{\text{obs}}^d B + D' D]$$
$$= \text{trace}[C W_{\text{con}}^d C' + DD'].$$

Definition 2.10 *(The H_∞ norm) Consider a continuous-time system having a $q \times \ell$ stable transfer function G. Then the H_∞ norm of G is defined as*

$$\|G\|_\infty := \sup_\omega \ \sigma_{\max}[G(j\omega)]. \tag{2.18}$$

Similarly, consider a discrete-time system having a $q \times \ell$ stable transfer function G. Then the H_∞ norm of G is defined as

$$\|G\|_\infty := \sup_{-\pi \leq \omega \leq \pi} \ \sigma_{\max}[G(e^{j\omega})]. \tag{2.19}$$

For a continuous-time system having a stable transfer function G, let w and z be energy signals that are, respectively, the input and the corresponding output of the given system. Similarly, for a discrete-time system having a stable transfer function G, let w and z be energy signals that are, respectively, the input and the corresponding output. Then it is easy to see that $\|G\|_\infty$ has the following interpretation for both continuous-time and discrete-time systems (where $\|\cdot\|_2$ denotes the L_2 and ℓ_2 norm, respectively):

$$\|G\|_\infty = \sup_{w \neq 0} \frac{\|z\|_2}{\|w\|_2}.$$

Also, when the input and the corresponding output (i.e., w and z) are power signals or wide-sense stationary stochastic processes (random sequences), the H_∞ norm of G turns out to coincide with its RMS gain, namely,

$$\|G\|_\infty = \|G\|_{\text{RMS gain}} = \sup_{\|w\|_{\text{RMS}} \neq 0} \frac{\|z\|_{\text{RMS}}}{\|w\|_{\text{RMS}}}.$$

We have the following remarks.

Remark 2.11 *An important property of the H_∞ norm, for both continuous-time and discrete-time systems, is that it is submultiplicative. That is, for transfer matrices G_1 and G_2, we have*

$$\|G_1 G_2\|_\infty \leq \|G_1\|_\infty \|G_2\|_\infty.$$

Remark 2.12 *It is interesting to contrast the H_2 and H_∞ norms. Consider a transfer matrix H. Then the fact that $\|H\|_\infty < \alpha$ for some $\alpha > 0$ implies that*

$$\|Hu\|_{\mathrm{RMS}} \leqslant \alpha \text{ for any input } u \text{ with} \|u\|_{\mathrm{RMS}} \leqslant 1.$$

In contrast, the H_2 norm bound specification $\|H\|_2 \leqslant \alpha$ implies that

$$\|Hu\|_{\mathrm{RMS}} \leqslant \alpha \text{ when input } u \text{ is a white noise with unit intensity.}$$

State-space method for computing the continuous-time H_∞ norm:

Regarding the H_∞ norm computation for continuous-time, there is a simple method to determine whether the inequality specification $\|G\|_\infty < \gamma$ is satisfied. To state this, given $\gamma > 0$, we define the matrix

$$M_\gamma = \begin{pmatrix} A + BR^{-1}D'C & \gamma^{-2}BR^{-1}B' \\ -C'(I + DR^{-1}D')C & -(A + BR^{-1}D'C)' \end{pmatrix}$$

where $R := \gamma^2 I - D'D > 0$. Then, we have

$$\|G\|_\infty < \gamma \iff M_\gamma \text{ has no imaginary eigenvalues, and } \sigma_{\max}(D) < \gamma. \quad (2.20)$$

The above discussion provides a simple bisection algorithm that enables us to compute the H_∞ norm numerically to any degree of numerical accuracy. In the following algorithm, the first three steps represent initialization, whereas the last step represents the bisection principle:

(*i*) Set $i = 0$, and set $\gamma_\ell = \|D\|$.

(*ii*) Choose any $\gamma_0 > \gamma_\ell$.

(*iii*) Use (2.20) to test whether $\|G\|_\infty < \gamma_i$. If so, $\gamma_u = \gamma_i$ and continue with step (*iv*). Otherwise, set $\gamma_{i+1} = 2\gamma_i$ and $i = i + 1$ and continue with step (*iii*).

(*iv*) Set $\gamma = (\gamma_u + \gamma_\ell)/2$. Use (2.20) to test whether $\|G\|_\infty < \gamma$. If so, set $\gamma_u = \gamma$ and otherwise set $\gamma_\ell = \gamma$, and then repeat step (*iv*).

We observe from step (*iv*) that $\|G\|_\infty$ is within the interval $[\gamma_\ell, \gamma_u]$. After each iteration, the size of the interval divides itself into half. Hence, one can stop the iterations when the desired level of accuracy is reached.

The matrix M_γ defined above is a Hamiltonian matrix associated with an H_∞^1 algebraic Riccati equation (see Chapter 4 and, in particular, Section 4.1.2). Therefore, the lack of imaginary axis eigenvalues can also be expressed in terms of a stabilizing solution of an algebraic Riccati equation or linear matrix inequality. This is also shown in Theorem 11.45. For more details concerning the computation of the H_∞ norm, we refer to [7, 10].

State-space method for computing the discrete-time H_∞ norm:

Similar to the continuous-time, for discrete-time, there is a simple method to determine whether the inequality specification $\|G\|_\infty < \gamma$ is satisfied. To state this, given $\gamma > 0$, we define the matrix pencil

$$M_\gamma(z) = \begin{pmatrix} zI - A & -B & 0 \\ C'C & C'D & I - zA' \\ D'C & D'D - \gamma^2 I & -zB' \end{pmatrix}$$

where $R := \gamma^2 I - D'D > 0$. Then, we have

$\|G\|_\infty < \gamma \iff M_\gamma(z)$ has no zeros on the unit circle, and

$$\sigma_{\max}(D + C(I - A)^{-1}B) < \gamma.$$

The above theorem provides a simple bisection algorithm that enables us to compute the H_∞ norm numerically to any degree of numerical accuracy, which is completely similar to the continuous-time case.

The matrix pencil M_γ defined above is associated with a discrete-time H_∞^1 algebraic Riccati equation (see Chapter 4 and, in particular, Section 4.2). Therefore, the lack of zeros on the unit circle can also be expressed in terms of a stabilizing solution of an algebraic Riccati equation or linear matrix inequality. This is shown in Theorem 11.63.

The computation of the H_∞ norm of transfer matrix of a discrete-time system through a bisection algorithm then follows similarly to the continuous-time case.

3

A special coordinate basis (SCB) of linear multivariable systems

3.1 Introduction

What is called the special coordinate basis (SCB) of a system plays a dominant role throughout this book. Hence, a clear understanding of the SCB is essential. The purpose of this chapter is to recall the SCB as well as its properties pertinent to this book. The SCB originated in [80, 82, 83] and was crystallized for strictly proper systems in [81] and for proper systems in [74]. Our presentation of SCB here omits all the proofs that can be found in the literature.

We would like to say boldly here that, given a linear time-invariant system, the corresponding SCB displays explicitly the architectural mapping of the internal workings of the system. By this we mean that the SCB identifies all pertinent structural elements of a system and their functions, and most significantly, it also displays the interconnections among all such structural elements. As such over time, since it was introduced about 20 years ago, the SCB of a system emerged as a powerful tool for almost all aspects of analysis and design of linear time-invariant systems. It played a major role in several research papers and books; for instance, our previous work [71, 75, 76] and this book demonstrates the power of the SCB. As will be clear to the readers, the influence of the SCB will be felt amply throughout this book and often from different angles.

3.2 SCB

We proceed now to develop the SCB of a linear time-invariant system Σ_* characterized by a quadruple (A, B, C, D). Let the dynamic equations of Σ_* be

$$\Sigma_* : \begin{cases} \sigma x = Ax + Bu \\ y = Cx + Du, \end{cases} \tag{3.1}$$

where σ is an operator indicating the time derivative $\frac{d}{dt}$ for continuous-time systems and a forward unit time shift for discrete-time systems. Also, $x \in \mathbb{R}^n$ is the state, $u \in \mathbb{R}^m$ is the control, and $y \in \mathbb{R}^p$ is the output. Without loss of generality, we assume that $(B' \quad D')'$ and $(C \quad D)$ are of full rank. Next, it is simple to verify that nonsingular transformations U and V exist such that

$$UDV = \begin{pmatrix} I_{m_0} & 0 \\ 0 & 0 \end{pmatrix},$$

(3.2)

where m_0 is the rank of matrix D. Hence, hereafter, without loss of generality, it is assumed that the matrix D has the form given on the right-hand side of (3.2). As such, without loss of generality, we can focus on a system Σ_* of the form

$$\Sigma_* : \begin{cases} \sigma x &= Ax + \begin{pmatrix} B_0 & \hat{B}_1 \end{pmatrix}\begin{pmatrix} u_0 \\ \hat{u}_1 \end{pmatrix} \\ \begin{pmatrix} y_0 \\ \hat{y}_1 \end{pmatrix} &= \begin{pmatrix} C_0 \\ \hat{C}_1 \end{pmatrix}x + \begin{pmatrix} I_{m_0} & 0 \\ 0 & 0 \end{pmatrix}\begin{pmatrix} u_0 \\ \hat{u}_1 \end{pmatrix}, \end{cases}$$

(3.3)

where the matrices B_0, $\hat{B}_1$, C_0, and $\hat{C}_1$ have appropriate dimensions.

We recall the following theorem and the properties that follow.

Theorem 3.1 *For any given system Σ_* characterized by the matrix quadruple (A, B, C, D), there exist*

(i) unique coordinate free non-negative integers $n_{a-}(\Sigma_*)$, $n_{ao}(\Sigma_*)$, $n_{a+}(\Sigma_*)$, $n_b(\Sigma_*)$, $n_c(\Sigma_*)$, $n_d(\Sigma_*)$, $m_d \leqslant m - m_0$, and q_i, $i = 1, \ldots, m_d$, and

(ii) nonsingular state, output, and input transformations Γ_s, Γ_o, and Γ_i that take the given Σ_* into the SCB that displays explicitly both the finite and the infinite zero structures of Σ_*.

The SCB is described by the following set of equations:

$$\sigma x_a^- = A_{aa}^- x_a^- + B_{a0}^- y_0 + L_{ad}^- y_d + L_{ab}^- y_b,$$

(3.4)

$$\sigma x_a^0 = A_{aa}^0 x_a^0 + B_{a0}^0 y_0 + L_{ad}^0 y_d + L_{ab}^0 y_b,$$

(3.5)

$$\sigma x_a^+ = A_{aa}^+ x_a^+ + B_{a0}^+ y_0 + L_{ad}^+ y_d + L_{ab}^+ y_b,$$

(3.6)

$$\sigma x_b = A_{bb} x_b + B_{b0} y_0 + L_{bd} y_d,$$

(3.7)

$$\sigma x_c = A_{cc} x_c + B_{c0} y_0 + L_{cd} y_d$$
$$\qquad + B_c(E_{ca}^- x_a^- + E_{ca}^0 x_a^0 + E_{ca}^+ x_a^+ + E_{cb} x_b) + B_c u_c,$$

(3.8)

$$y_0 = C_{0a}^- x_a^- + C_{0a}^0 x_a^0 + C_{0a}^+ x_a^+ + C_{0b} x_b + C_{0c} x_c + C_{0d} x_d + u_0,$$

(3.9)

$$y_b = C_b x_b,$$

(3.10)

and for each $i = 1, \ldots, m_d$,

$$\sigma x_i = A_{q_i} x_i + L_{i0} y_0 + L_{id} y_d$$
$$\qquad + B_{q_i}\left(u_i + E_{ia} x_a + E_{ib} x_b + E_{ic} x_c + \sum_{j=1}^{m_d} E_{ij} x_j \right),$$

(3.11)

$$y_i = C_{q_i} x_i,$$

(3.12)

where

$$x = \Gamma_s \tilde{x}, \qquad y = \Gamma_o \tilde{y}, \qquad u = \Gamma_i \tilde{u},$$

$$\tilde{x} = \begin{pmatrix} x_a \\ x_b \\ x_c \\ x_d \end{pmatrix}, \qquad x_a = \begin{pmatrix} x_a^- \\ x_a^0 \\ x_a^+ \end{pmatrix}, \qquad x_d = \begin{pmatrix} x_1 \\ x_2 \\ \vdots \\ x_{m_d} \end{pmatrix},$$

$$\tilde{y} = \begin{pmatrix} y_0 \\ y_d \\ y_b \end{pmatrix}, \qquad y_d = \begin{pmatrix} y_1 \\ y_2 \\ \vdots \\ y_{m_d} \end{pmatrix},$$

$$\tilde{u} = \begin{pmatrix} u_0 \\ u_d \\ u_c \end{pmatrix}, \qquad u_d = \begin{pmatrix} u_1 \\ u_2 \\ \vdots \\ u_{m_d} \end{pmatrix}.$$

The x_i $(i = 1, \ldots, m_d)$ together form x_d and, similarly, the y_i $(i = 1, \ldots, m_d)$ together form y_d, and

$$y_d = C_d x_d, \quad \text{where} \quad C_d = \begin{pmatrix} C_{q_1} & 0 & \cdots & 0 \\ 0 & \ddots & \ddots & \vdots \\ \vdots & \ddots & \ddots & 0 \\ 0 & \cdots & 0 & C_{q_{m_d}} \end{pmatrix}. \tag{3.13}$$

Here the states x_a^-, x_a^0, x_a^+, x_b, x_c, and x_d are of dimension $n_{a-}(\Sigma_*)$, $n_{ao}(\Sigma_*)$, $n_{a+}(\Sigma_*)$, $n_b(\Sigma_*)$, $n_c(\Sigma_*)$, and $n_d(\Sigma_*) = \sum_{i=1}^{m_d} q_i$, respectively, whereas x_i is of dimension q_i for each $i = 1, \ldots, m_d$. The control vectors u_0, u_d, and u_c are, respectively, of dimensions m_0, m_d, and $m_c = m - m_0 - m_d$, whereas the output vectors y_0, y_d, and y_b are, respectively, of dimensions $p_0 = m_0$, $p_d = m_d$, and $p_b = p - p_0 - p_d$. The matrices A_{q_i}, B_{q_i}, and C_{q_i} have the following form:

$$A_{q_i} = \begin{pmatrix} 0 & I_{q_i - 1} \\ 0 & 0 \end{pmatrix}, \quad B_{q_i} = \begin{pmatrix} 0 \\ 1 \end{pmatrix}, \quad C_{q_i} = \begin{pmatrix} 1 & 0 & \cdots & 0 \end{pmatrix}. \tag{3.14}$$

(Obviously for the case when $q_i = 1$, we have $A_{q_i} = 0$, $B_{q_i} = 1$, and $C_{q_i} = 1$.) Assuming that the x_i are arranged such that $q_i \leqslant q_{i+1}$, the matrix L_{id} has the particular form

$$L_{id} = \begin{pmatrix} L_{i1} & L_{i2} & \cdots & L_{i\,i-1} & 0 & 0 & \cdots & 0 \end{pmatrix}.$$

The last row of each L_{id} is identically zero. Furthermore, the pair (A_{cc}, B_c) is controllable, and the pair (C_b, A_{bb}) is observable. Moreover, for continuous-time systems, we have $\lambda(A_{aa}^-) \in \mathbb{C}^-$, $\lambda(A_{aa}^0) \in \mathbb{C}^0$, $\lambda(A_{aa}^+) \in \mathbb{C}^+$, whereas for discrete-time systems, we have $\lambda(A_{aa}^-) \in \mathbb{C}^\ominus$, $\lambda(A_{aa}^0) \in \mathbb{C}^\circ$, and $\lambda(A_{aa}^+) \in \mathbb{C}^\oplus$.

We have the following remarks.

Remark 3.2 *A software package to generate the SCB of any given system is given by Lin et al. [43], [44].*

A block diagram of the SCB of Theorem 3.1 is given in Figures 3.1–3.4. Figure 3.1 expresses the zero dynamics. Figure 3.2 presents dynamics that, as we will see later on, is present if and only if the system Σ_* is not left invertible. Figure 3.3 presents dynamics that, as we will see later on, is present if and only if the system Σ_* is not right invertible. Finally, the dynamics in Figure 3.4 is related to the infinite zero structure. In this last figure, a signal given by a double-edged arrow is some linear combination of outputs y_i, $i = 0$ to m_d , where as the signal given by the double-edged arrow with a solid dot in it is some linear combination of all states.

We note the following intuitive points regarding the SCB:

(*i*) For $i \neq 0$, the variable u_i controls the output y_i through a stack of q_i integrators, whereas x_i is the state associated with those integrators between u_i and y_i. Moreover, (A_{q_i}, B_{q_i}) and (C_{q_i}, A_{q_i}), respectively, form controllable and observable pairs. This implies that all states x_i are both controllable and observable.

(*ii*) The output y_b and the state x_b are not directly influenced by any inputs; however, they could be indirectly controlled through the output y_d. Moreover, (C_b, A_{bb}) forms an observable pair. This implies that the state x_b is observable.

(*iii*) The state x_c is directly controlled by the input u_c, but it does not directly affect any output. Moreover, (A_{cc}, B_c) forms a controllable pair. This implies that the state x_c is controllable.

(*iv*) The state x_a is neither directly controlled by any input nor does it directly affect any output.

We can rewrite the SCB given by Theorem 3.1 in a more compact form as a

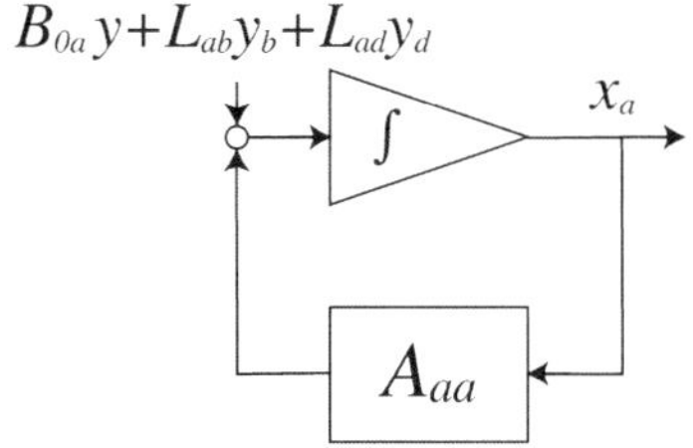

Figure 3.1: Zero dynamics

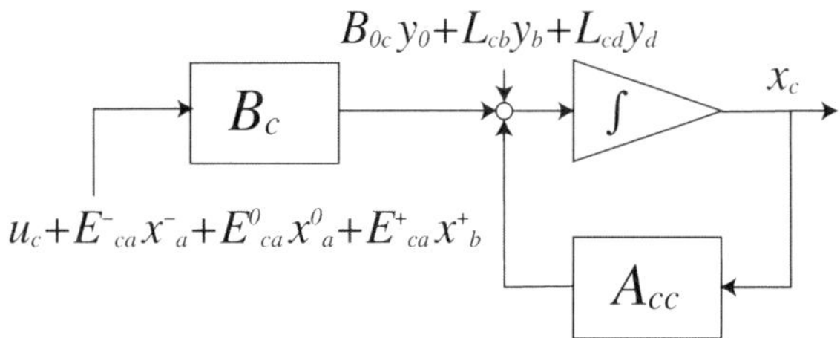

Figure 3.2: Non-left-invertible dynamics

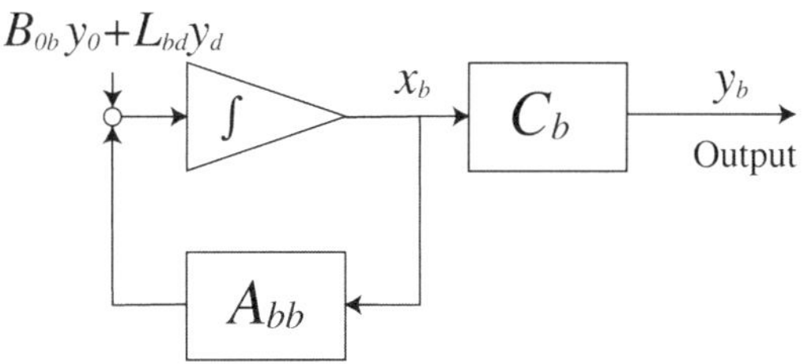

Figure 3.3: Non-right-invertible dynamics

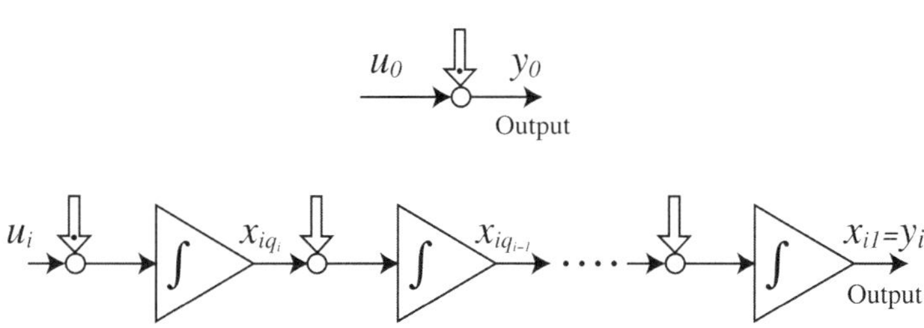

Figure 3.4: Infinite zero structure

system characterized by the quadruple $(\tilde{A}, \tilde{B}, \tilde{C}, \tilde{D})$ given by

$$\tilde{A} := \Gamma_s^{-1}(A - B_0 C_0)\Gamma_s =$$

$$\begin{pmatrix} A_{aa}^- & 0 & 0 & L_{ab}^- C_b & 0 & L_{ad}^- C_d \\ 0 & A_{aa}^0 & 0 & L_{ab}^0 C_b & 0 & L_{ad}^0 C_d \\ 0 & 0 & A_{aa}^+ & L_{ab}^+ C_b & 0 & L_{ad}^+ C_d \\ 0 & 0 & 0 & A_{bb} & 0 & L_{bd} C_d \\ B_c E_{ca}^- & B_c E_{ca}^0 & B_c E_{ca}^+ & B_c E_{cb} & A_{cc} & L_{cd} C_d \\ B_d E_{da}^- & B_d E_{da}^0 & B_d E_{da}^+ & B_d E_{db} & B_d E_{dc} & A_{dd} \end{pmatrix}, \quad (3.15)$$

where

$$A_{dd} = \begin{pmatrix} A_{q_1} & 0 & \cdots & 0 \\ 0 & \ddots & \ddots & \vdots \\ \vdots & \ddots & \ddots & 0 \\ 0 & \cdots & 0 & A_{q_{m_d}} \end{pmatrix} + L_{dd} C_d, \quad (3.16a)$$

$$B_d = \begin{pmatrix} B_{q_1} & 0 & \cdots & 0 \\ 0 & \ddots & \ddots & \vdots \\ \vdots & \ddots & \ddots & 0 \\ 0 & \cdots & 0 & B_{q_{m_d}} \end{pmatrix}, \quad (3.16b)$$

$$\tilde{B} := \Gamma_s^{-1}\begin{pmatrix} B_0 & \hat{B}_1 \end{pmatrix}\Gamma_i = \begin{pmatrix} B_{a0}^- & 0 & 0 \\ B_{a0}^0 & 0 & 0 \\ B_{a0}^+ & 0 & 0 \\ B_{b0} & 0 & 0 \\ B_{c0} & 0 & B_c \\ B_{d0} & B_d & 0 \end{pmatrix}, \quad (3.17)$$

$$\tilde{C} := \Gamma_o^{-1}\begin{pmatrix} C_0 \\ \hat{C}_1 \end{pmatrix}\Gamma_s = \begin{pmatrix} C_{0a}^- & C_{0a}^0 & C_{0a}^+ & C_{0b} & C_{0c} & C_{0d} \\ 0 & 0 & 0 & 0 & 0 & C_d \\ 0 & 0 & 0 & C_b & 0 & 0 \end{pmatrix}, \quad (3.18)$$

and

$$\tilde{D} := \Gamma_o^{-1} D \Gamma_i = \begin{pmatrix} I_{m_0} & 0 & 0 \\ 0 & 0 & 0 \\ 0 & 0 & 0 \end{pmatrix}. \quad (3.19)$$

In the above equations, if one needs expanded expressions for the matrices E_{da}^-, E_{da}^0, E_{da}^+, E_{db}, and E_{dc}, they can easily be obtained from (3.11). Also, one of the elements in $\tilde{A}$, namely, $B_c E_{cb}$ can equivalently be replaced by $L_{cb} C_b$ (and we do so whenever needed). Note that we always have that (A_{cc}, B_c) and (A_{dd}, B_d) are controllable while (C_b, A_{bb}) is observable.

Admittedly, the SCB of Theorem 3.1 looks complicated with all its innate decompositions of state, output, and input variables. However, as illustrated below and as will be evident throughout the book, the SCB of a linear system displays clearly the underlying structure of it. In fact, the proofs of several theorems, lemmas, and properties stated in later chapters will be hard to follow without having endured the intricacies of the SCB.

We next state some pertinent properties of the SCB of a linear system; each main property is stated in a subsection devoted to it. The properties discussed below are true for both continuous- and discrete-time systems. However, sometimes, for convenience of writing, we use the notations commonly used for continuous-time systems. The reader can easily decipher the corresponding notations for discrete-time systems. For clarity, whenever it is needed, we repeat our discussion for discrete-time systems.

3.2.1 Observability (detectability) and controllability (stabilizablity)

In this subsection, we examine the issues related to observability, detectability, controllability, and stabilizability of a system via its SCB. Note that we simply use detectability and stabilizability, which for continuous-time systems refers to $\mathbb{C}^-$-detectability and $\mathbb{C}^-$-stabilizability, whereas for discrete-time systems, this refers to $\mathbb{C}^\ominus$-detectability and $\mathbb{C}^\ominus$-stabilizablity.

We have the following property.

Property 3.3 *We note that (C_b, A_{bb}) and (C_{q_i}, A_{q_i}) form observable pairs. Unobservability can arise only in the variables x_a and x_c. In fact, the given system Σ_* is observable (detectable) if and only if $(C_{\mathrm{obs}}, A_{\mathrm{obs}})$ is observable (detectable), where*

$$A_{\mathrm{obs}} = \begin{pmatrix} A_{aa} & 0 \\ B_c E_{ca} & A_{cc} \end{pmatrix}, \qquad A_{aa} = \begin{pmatrix} A_{aa}^- & 0 & 0 \\ 0 & A_{aa}^0 & 0 \\ 0 & 0 & A_{aa}^+ \end{pmatrix},$$

$$C_{\mathrm{obs}} = \begin{pmatrix} C_{0a} & C_{0c} \\ B_d E_{da} & B_d E_{dc} \end{pmatrix}, \qquad C_{0a} = \begin{pmatrix} C_{0a}^- & C_{0a}^0 & C_{0a}^+ \end{pmatrix},$$

$$E_{da} = \begin{pmatrix} E_{da}^- & E_{da}^0 & E_{da}^+ \end{pmatrix}, \qquad E_{ca} = \begin{pmatrix} E_{ca}^- & E_{ca}^0 & E_{ca}^+ \end{pmatrix}.$$

Similarly, (A_{cc}, B_c) and (A_{q_i}, B_{q_i}) form controllable pairs. Basically, the variables x_a and x_b determine the controllability of the system. In fact, Σ_* is control-

lable (stabilizable) if and only if $(A_{\mathrm{con}}, B_{\mathrm{con}})$ is controllable (stabilizable), where

$$A_{\mathrm{con}} = \begin{pmatrix} A_{aa} & L_{ab}C_b \\ 0 & A_{bb} \end{pmatrix}, \quad B_{\mathrm{con}} = \begin{pmatrix} B_{a0} & L_{ad} \\ B_{b0} & L_{bd} \end{pmatrix},$$

$$B_{a0} = \begin{pmatrix} B_{a0}^- \\ B_{a0}^0 \\ B_{a0}^+ \end{pmatrix}, \quad L_{ab} = \begin{pmatrix} L_{ab}^- \\ L_{ab}^0 \\ L_{ab}^+ \end{pmatrix}, \quad L_{ad} = \begin{pmatrix} L_{ad}^- \\ L_{ad}^0 \\ L_{ad}^+ \end{pmatrix}.$$

3.2.2 Left- and right-invertibility

In this subsection, we examine the invertibility properties of Σ_* via its SCB. Let us first recall from [51] the definition of right and left invertibility.

Definition 3.4 *Consider a linear system Σ_*.*

- *Let u_1 and u_2 be any inputs to the system Σ_*, and let y_1 and y_2 be the corresponding outputs (for the same initial conditions). Then Σ_* is said to be left invertible, if $y_1(t) = y_2(t)$ for all $t \geq 0$ implies that $u_1(t) = u_2(t)$ for all $t \geq 0$.*

- *The system Σ_* is said to be right invertible if, for any $y_{ref}(t)$ defined on $[0, \infty)$, a $u(t)$ and a choice of $x(0)$ exist such that $y(t) = y_{ref}(t)$ for all $t \in [0, \infty)$.*

- *The system Σ_* is said to be invertible if the system is both left and right invertible.*

Remark 3.5 *One can easily deduce the following:*

- *(i) Σ_* is right invertible if and only if its transfer function matrix is a surjective rational matrix.*

- *(ii) Σ_* is right invertible if and only if the rank of $P_{\Sigma_*}(s) = n + p$ for all but finitely many $s \in \mathbb{C}$, where the polynomial matrix $P_{\Sigma_*}(s)$ is the Rosenbrock system matrix of Σ_* defined as*

$$P_{\Sigma_*}(s) := \begin{pmatrix} sI - A & -B \\ C & D \end{pmatrix}.$$

- *(iii) Σ_* is left invertible if and only if its transfer function matrix is an injective rational matrix.*

- *(iv) Σ_* is left invertible if and only if the rank of $P_{\Sigma_*}(s) = n + m$ for all but finitely many $s \in \mathbb{C}$.*

We have the following property connecting these properties to the special co-ordinate basis:

Property 3.6 *The given system Σ_* is right invertible if and only if x_b and hence y_b are nonexistent ($n_b = 0$, $p_b = 0$), left invertible if and only if x_c and hence u_c are nonexistent ($n_c = 0$, $m_c = 0$), and invertible if and only if both x_b and x_c are nonexistent. Moreover, Σ_* is degenerate if and only if it is neither left nor right invertible.*

3.2.3 Finite zero structure

In this subsection, we recall first the definition of invariant zeros of a system and their generalized associated right state and input zero direction chains, and then we discuss how SCB exhibits them in its structure.

The invariant zeros of a system Σ_* that is characterized by (A, B, C, D) are defined via the Smith canonical form of the Rosenbrock system matrix $P_{\Sigma_*}(s)$. Let us first briefly recall the Smith canonical form for any polynomial matrix $P(s) \in \mathbb{R}^{n \times m}[s]$. It is well known [25] that for any $P(s) \in \mathbb{R}^{n \times m}[s]$, unimodular* matrices $U(s) \in \mathbb{R}^{n \times n}[s]$ and $V(s) \in \mathbb{R}^{m \times m}[s]$ and a matrix $\Psi(s) \in \mathbb{R}^{n \times m}[s]$ of the form

$$\Psi(s) = \begin{pmatrix} \psi_1(s) & 0 & \cdots & \cdots & \cdots & 0 \\ 0 & \ddots & \ddots & & & \vdots \\ \vdots & \ddots & \psi_r(s) & \ddots & & \vdots \\ \vdots & & \ddots & 0 & \ddots & \vdots \\ \vdots & & & \ddots & \ddots & 0 \\ 0 & \cdots & \cdots & \cdots & 0 & 0 \end{pmatrix},$$

exist such that

$$P(s) = U(s)\Psi(s)V(s).$$

Here $\Psi(s)$ is called the Smith canonical form of $P(s)$ when the $\psi_i(s)$ are monic polynomials with the property that $\psi_i(s)$ divides $\psi_{i+1}(s)$ for $i = 1, \ldots, r-1$, and r is the normal rank of the matrix $P(s)$. The polynomials $\psi_i(s)$ are called the invariant factors of $P(s)$. Their product $\psi(s) = \psi_1(s)\psi_2(s)\cdots\psi_r(s)$ is called the zero polynomial of $P(s)$. Each invariant factor $\psi_i(s)$, $i = 1, 2, \ldots, r$, can be written as a product of linear factors

$$\psi_i(s) = (s - \lambda_{i1})^{\alpha_{i1}}(s - \lambda_{i2})^{\alpha_{i2}} \cdots (s - \lambda_{ik_i})^{\alpha_{ik_i}}, \quad i = 1, 2, \ldots, r,$$

---

*A polynomial matrix in $\mathbb{R}^{n \times m}[s]$ that is invertible with a polynomial inverse is called unimodular.

where $\lambda_{ik} \neq \lambda_{i\ell}$ ($k \neq \ell$) are complex numbers and α_{ik} ($k, \ell \in \{1, \ldots, k_i\}$) are positive integers. Then the complete set of factors, $(s - \lambda_{ik})^{\alpha_{ik}}$, $k = 1, 2, \ldots, k_i$, and $i = 1, 2, \ldots, r$, are called the *elementary divisors* of the polynomial matrix $P(s)$.

Now we are ready to recall the definition of the invariant zeros [47] of Σ_*.

Definition 3.7 *The roots of the zero polynomial $\psi(s)$ of the (Rosenbrock) system matrix $P_{\Sigma_*}(s)$ are called the invariant zeros of Σ_*.*

Remark 3.8 *It is obvious from the above definition that an alternative way of defining an invariant zero of Σ_* is as follows: $\lambda \in \mathbb{C}$ is called an invariant zero of Σ_* if the rank of $P_{\Sigma_*}(\lambda)$ is strictly smaller than the normal rank of $P_{\Sigma_*}(s)$. Note that the normal rank is defined as the rank of a polynomial or rational matrix in all but finitely many $s \in \mathbb{C}$.*

The SCB of Theorem 3.1 shows explicitly the invariant zeros of the system. To be more specific, we have the following property.

Property 3.9 *Invariant zeros of Σ_* are the eigenvalues of A_{aa}. Moreover, for continuous-time systems, the invariant zeros that are in $\mathbb{C}^-$, $\mathbb{C}^0$, and $\mathbb{C}^+$ are, respectively, the eigenvalues of A_{aa}^-, A_{aa}^0, and A_{aa}^+. Similarly, for discrete-time systems, the invariant zeros that are in $\mathbb{C}^{\ominus}$, $\mathbb{C}^0$, and $\mathbb{C}^{\oplus}$ are, respectively, the eigenvalues of A_{aa}^-, A_{aa}^0, and A_{aa}^+.*

For continuous-time systems, if all invariant zeros of a system Σ_* are in $\mathbb{C}^-$, then we say Σ_* is minimum phase; otherwise, Σ_* is said to be non-minimum phase. Those invariant zeros that are in $\mathbb{C}^-$ are called the stable invariant zeros. Also, those that are not in $\mathbb{C}^-$ are called the unstable invariant zeros. Analogously, for discrete-time systems, if all the invariant zeros of a system Σ_* are in $\mathbb{C}^{\ominus}$, then we say Σ_* is minimum phase; otherwise, Σ_* is said to be of non-minimum phase. Those invariant zeros that are in $\mathbb{C}^{\ominus}$ are called the stable invariant zeros. Also, those that are not in $\mathbb{C}^{\ominus}$ are called the unstable invariant zeros.

The following definition introduces the notions of algebraic and geometric multiplicities [37] of an invariant zero and its multiplicity structure.

Definition 3.10 *The algebraic multiplicity ρ_z of an invariant zero z is defined as the degree of the product of the elementary divisors of $P_{\Sigma_*}(s)$ corresponding to z. Likewise, the geometric multiplicity ν_z of an invariant zero z is defined as the number of the elementary divisors of $P_{\Sigma_*}(s)$ corresponding to z. Moreover, the* *invariant zero z is said to have a simple structure if its algebraic and geometric multiplicities are equal. Otherwise, it is referred to as an invariant zero with nonsimple structure.*

Given an invariant zero z, let $n_{z,i}$ be the degree of $(s - z)$ in the invariant factor $\Psi_i(s)$ of the Rosenbrock system matrix. Then the multiplicity structure of an invariant zero is defined as

$$S_z^* = \{n_{z,1}, n_{z,2}, \ldots, n_{z,v_z}\}. \tag{3.20}$$

If $n_{z,1} = n_{z,2} = \cdots = n_{z,v_z} = 1$, then we say z is a simple invariant zero of the given system Σ_*.

To discuss the multiplicity structure of an invariant zero as displayed by the SCB, we recall next the classic concept of the Jordan form of a general matrix A and the concept of the multiplicity structure of the eigenvalue of a matrix A. Given any matrix A of dimension $n \times n$, we can always find a nonsingular transformation matrix X (see [25]) such that

$$X^{-1}AX = J = \begin{pmatrix} J_1 & 0 & \cdots & 0 \\ 0 & J_2 & \ddots & \vdots \\ \vdots & \ddots & \ddots & 0 \\ 0 & \cdots & 0 & J_k \end{pmatrix}, \tag{3.21}$$

where J_i, $i = 1, \ldots, k$, are some $n_i \times n_i$ Jordan blocks,

$$J_i = \begin{pmatrix} \lambda_i & 1 & 0 & \cdots & 0 \\ 0 & \ddots & \ddots & \ddots & \vdots \\ \vdots & \ddots & \ddots & \ddots & 0 \\ \vdots & & & \ddots & \ddots & 1 \\ 0 & \cdots & \cdots & 0 & \lambda_i \end{pmatrix}. \tag{3.22}$$

We note that

$$\sum_{i=1}^{k} n_i = n.$$

Then, the geometric multiplicity v_λ of an eigenvalue $\lambda \in \lambda(A)$ is the number of Jordan blocks in (3.21) associated with λ and the algebraic multiplicity ρ_λ is the total number of repetitions of λ in $\lambda(A)$. Equivalently, the algebraic multiplicity is equal to the sum of the number of rows of all Jordan blocks associated with λ.

We recall also the following definition.

Definition 3.11 *For any given $\lambda \in \lambda(A)$, let there be v_λ Jordan blocks of A [see (3.21) and (3.22)] associated with λ. Let $n_{\lambda,1}, n_{\lambda,2}, \ldots, n_{\lambda,v_\lambda}$ be the dimensions of the corresponding Jordan blocks. Then we say λ is an eigenvalue of A with multiplicity structure S_λ^*,*

$$S_\lambda^* = \{n_{\lambda,1}, n_{\lambda,2}, \ldots, n_{\lambda,v_\lambda}\}. \tag{3.23}$$

If $n_{\lambda,1} = n_{\lambda,2} = \cdots = n_{\lambda,v_\lambda} = 1$, then we say λ is a simple eigenvalue of A.

The invariant factor $\Psi_i(s)$ of a matrix A is the monic polynomial of lowest degree such that for each eigenvalue λ with $v_\lambda \geq i$, $\Psi_i(s)$ has $n_{\lambda,i}$ zeros in λ.

Remark 3.12 *We note that algebraic multiplicity ρ_λ satisfies*

$$\rho_\lambda = n_{\lambda,1} + n_{\lambda,2} + \cdots + n_{\lambda,v_\lambda}.$$

We recall next the following classic concepts of generalized eigenvectors and the eigenvector chain associated with an eigenvalue of a matrix.

Definition 3.13 *A vector x is said to be a generalized eigenvector of grade k associated with an eigenvalue λ of a matrix A if and only if*

$$(A - \lambda I)^k x = 0 \quad and \quad (A - \lambda I)^{k-1} x \neq 0.$$

Remark 3.14 *The generalized eigenvector of grade one (i.e., $k = 1$) is the standard eigenvector associated with an eigenvalue of a matrix.*

Definition 3.15 *Let vector x be a generalized eigenvector of grade k associated with an eigenvalue λ of a matrix A. Let*

$$\begin{aligned}
x_k &= x \\
x_{k-1} &= (A - \lambda I)V &= (A - \lambda I)x_k \\
x_{k-2} &= (A - \lambda I)^2 V &= (A - \lambda I)x_{k-1} \\
&\;\;\vdots &\vdots \\
x_1 &= (A - \lambda I)^{k-1} V &= (A - \lambda I)x_2.
\end{aligned}$$

The set of vectors $\{x_1, x_2, \cdots, x_k\}$ is called a chain of generalized eigenvectors of length k associated with an eigenvalue λ.

For an eigenvalue λ with the multiplicity structure S_λ^* as given in (3.23), there are ν_λ chains of generalized eigenvectors with lengths $n_{\lambda,1}, n_{\lambda,2}, \cdots, n_{\lambda,\nu_\lambda}$. The total number of generalized eigenvectors in these chains equals the algebraic multiplicity ρ_λ. Moreover, these ρ_λ generalized eigenvectors are linearly independent.

We are now ready to discuss the invariant zeros together with their multiplicity structure of the system Σ_* as displayed by the SCB.

Property 3.16 *Consider the system Σ_* with its associated SCB. Then, z is an invariant zero of Σ_* with multiplicity structure S_z^* if and only if z is an eigenvalue of A_{aa} with multiplicity structure S_z^*.*

We need to recall next the notion of the right state and input zero directions and left state and input zero directions [37] associated with an invariant zero of a system. We focus first on the right state and input zero directions associated with an invariant zero for a left invertible system Σ_* (left invertibility is discussed in Definition 3.4).

Definition 3.17 *Consider an invariant zero z with a simple structure of a left-invertible system Σ_*. Then the associated right state and input zero directions, $x_z \neq 0$ and u_z, of Σ_* are defined as those that satisfy the condition*

$$P_{\Sigma_*}(z)\begin{pmatrix} x_z \\ u_z \end{pmatrix} = \begin{pmatrix} zI - A & -B \\ C & D \end{pmatrix}\begin{pmatrix} x_z \\ u_z \end{pmatrix} = 0.$$

Some papers in the literature extend the above definition to non-left-invertible systems. This is incorrect as argued in [70].

Whenever an invariant zero has a nonsimple multiplicity structure, a concept of generalized right state and input zero direction[†] chain associated with that invariant zero exist. A proper definition of this is given in [70]. At this time we would like to point out that although some researchers (e.g., see [47] and [89] among others) define the generalized right state and input zero direction chains as x_R^j and w_R^j, $j = 1, \cdots, \rho_z - \sigma_z$, satisfying

$$P_{\Sigma_*}(z)\begin{pmatrix} x_R^j \\ w_R^j \end{pmatrix} = -\begin{pmatrix} x_R^{j-1} \\ 0 \end{pmatrix}, \quad j = 1, \ldots, \rho_z - \sigma_z. \tag{3.24}$$

However, this is also incorrect for non-left-invertible systems as argued in [70].

[†]Generalized right state and input zero directions are also called pseudo-right state and input zero directions.

In what follows we identify the right state and input zero direction chain associated with an invariant zero z of a general system whether it is left invertible or not, and whether z has a simple multiplicity structure or not. However, as discussed, in the absence of a precise definition that is not based on any special coordinate basis, we caution that the Property 3.18 can be viewed either as a definition or as a property.

Let us start by defining the eigenvector chain associated with an eigenvalue of the matrix A_{aa}. Given an invariant zero z of the system Σ_* (i.e., the eigenvalue z of the matrix A_{aa}), for each $i = 1$ to ν_z, a set of vectors in $\mathbb{R}^{n_a}$ that satisfies the following condition (3.25) is the eigenvector chain of A_{aa} associated with the invariant zero z:

$$A_{aa}x^z_{i1} = zx^z_{i1}, \quad \text{and} \quad (A_{aa} - zI_{n_a})x^z_{ij+1} = x^z_{ij} \ , \quad j = 1,\ldots,n_{z,i} - 1. \tag{3.25}$$

We have the following property regarding the right state and input zero direction chain associated with an invariant zero of a system.

Property 3.18

    (*i*) *For each* $i = 1$ *to* ν_z, *a set of vectors in* $\mathbb{R}^n$ *given in* (3.26) *is the **generalized right state zero direction chain** of* Σ_* *associated with the invariant zero* z

$$x^z_{ij} = \Gamma_s \begin{pmatrix} x^z_{ij} \\ 0 \\ \vdots \\ 0 \end{pmatrix} \qquad j = 1 \text{ to } n_{z,i}. \tag{3.26}$$

    Also, x^z_{i1} *is the right state zero direction of* Σ_* *associated with* z.

    (*ii*) *For each* $i = 1$ *to* ν_z, *a set of vectors* w^z_{ij}, $j = 1$ *to* $n_{z,i}$, *in* $\mathbb{R}^m$ *as given in* (3.27) *is the **generalized right input zero direction chain** of* Σ_* *associated with the invariant zero* z:

$$w^z_{ij} = -\Gamma_o \begin{pmatrix} E_{da} \\ E_{dc} \end{pmatrix} x^z_{ij}, \tag{3.27}$$

    where E_{da} *is as defined in Property 3.3. Also,* w^z_{i1} *is said to be the right input zero direction of* Σ_* *associated with* z.

The following property gives a dynamical interpretation of finite zero structure of a system. It is formulated for continuous-time systems. An analogous formulation is valid for discrete-time systems as well.

Property 3.19 *(Dynamical interpretation of finite zero structure) For a system Σ_* that is not necessarily left-invertible, given that the initial condition, $x(0) = x_{i\alpha}^z$ for any $\alpha \le n_{z,i}$ and the input*

$$u = \sum_{j=1}^{\alpha} \frac{w_{ij}^z\, t^{\alpha-j} \exp(zt)}{(\alpha - j)!} \quad \text{for all } t \ge 0, \tag{3.28}$$

where z is any invariant zero of the system and $n_{z,i} \in S_z^*$, we have

$$y \equiv 0$$

and

$$x(t) = \sum_{j=1}^{\alpha} \frac{x_{ij}^z\, t^{\alpha-j} \exp(zt)}{(\alpha - j)!} \quad \text{for all } t \ge 0. \tag{3.29}$$

However, for left-invertible systems, the conditions $x(0) = x_{i\alpha}^z$ and (3.28) together imply and are implied by $y \equiv 0$ for all $t \ge 0$.

One can define the left state and input zero direction chain associated with an invariant zero of Σ_* as follows.

Definition 3.20 *The left state and input zero direction chain associated with each invariant zero of Σ_* are defined as the corresponding right state and input zero direction chain of the dual system Σ_{*d}.*

Remark 3.21 *We can connect the structural invariant indices lists of Morse [49] to SCB. In particular, the list $\mathfrak{I}_1$ of Morse is exactly equal to the invariant factors of A_{aa}.*

Next, we would like to recall the definition of the input decoupling zeros and the output decoupling zeros of a system.

Definition 3.22 *The zeros of the matrix pencil*

$$\begin{pmatrix} \lambda I - A & -B \end{pmatrix},$$

i.e., the values of λ for which the above pencil loses rank, are called the input decoupling zeros of Σ_*. They are also referred to as the input decoupling zeros of the pair (A, B).

The zeros of the matrix pencil

$$\begin{pmatrix} \lambda I - A \\ C \end{pmatrix},$$

i.e., the values of λ for which the above pencil loses rank, are called the output decoupling zeros of Σ_*. They are also referred to as the output decoupling zeros of the pair (C, A).

Remark 3.23 *In the literature, input decoupling zeros are also referred to as uncontrollable eigenvalues, whereas the output decoupling zeros are referred to as unobservable eigenvalues.*

Note that as we have done for invariant zeros, we can also associate a multiplicity structure with an input- or output-decoupling zero. The precise definition should be obvious from the above and hence is not included here.

The following property shows how input decoupling zeros and output decoupling zeros of Σ_* are displayed by the SCB.

Property 3.24 *Consider a system Σ_* with corresponding special coordinate basis. Define A_{con}, B_{con}, A_{obs}, and C_{obs} according to Property 3.3.*

   (i) *The input decoupling zeros of Σ_* are the input decoupling eigenvalues of the pair $(A_{\mathrm{con}}, B_{\mathrm{con}})$. Also, input decoupling zeros of (A_{bb}, L_{bd}) are contained in the set of input decoupling zeros of Σ_*. Some of the input decoupling zeros of Σ_* could be contained among its invariant zeros.*

   (ii) *The output decoupling zeros of Σ_* are the output decoupling eigenvalues of the pair $(C_{\mathrm{obs}}, A_{\mathrm{obs}})$. Also, output decoupling zeros of $(B_d E_{dc}, A_{cc})$ are contained in the set of output decoupling zeros of $(C_{\mathrm{obs}}, A_{\mathrm{obs}})$. Some of the output decoupling zeros of Σ_* could be contained among its invariant zeros.*

Remark 3.25 *As it is obvious from Property 3.24, it is crucial to realize that the input decoupling zeros and the output-decoupling zeros of a system need not be the invariant zeros.*

3.2.4   Infinite zero structure

In this subsection, we examine the infinite zero structure of a system and how it is displayed by the SCB. Let us first recall some pertinent information from the literature. Infinite zeros are defined either in association with root-locus theory or as Smith–McMillan zeros of the transfer function at infinity.

Let us first view the infinite zeros from the viewpoint of root-locus theory. For this, consider a strictly proper system Σ_* subject to a high-gain feedback $u = \rho y$

for a scalar gain ρ. It can then be shown (see, e.g., Hung and MacFarlane [31]) that the unbounded closed-loop poles of the feedback system can be listed as

$$s_{j\ell}(\rho) = \rho^{1/\nu_j}\,\eta_{j\ell} + \zeta_{j\ell}(\rho) \quad \text{for} \quad \ell = 1,\dots,\nu_j, \quad j = 1,\dots,m,$$

where

$$\lim_{\rho\to\infty} \rho^{-1/\nu_j}\,\zeta_{j\ell}(\rho) = 0.$$

Here $s_{j\ell}$ is termed an infinite zero of order ν_j. Actually, until recently, the infinite zeros defined this way were considered to be fictitious objects introduced for the convenience of visualization.

Let us next consider the infinite zeros from the viewpoint of Smith–McMillan theory. To define the zero structure of the system Σ_* at infinity, one can use the familiar Smith–McMillan description of the zero structure at finite frequencies of the corresponding transfer matrix H, which need neither to be square nor strictly proper. A rational matrix $H(s)$ possesses an infinite zero of order k when $H(1/z)$ has a finite zero of precisely that order at $z = 0$; see [19], [64], [68], and [99].

The number of zeros at infinity together with their orders indeed defines an infinite zero structure. It is important to note that for strictly proper transfer matrices, the above two definitions of the infinite zeros and their structure are consistent.

Owens [54] related the orders of the infinite zeros of the root-loci of a square system with a nonsingular transfer function matrix to the $\mathcal{C}^*$ structural invariant indices list $\mathfrak{I}_4$ of Morse [49]. This connection reveals that the *structure at infinity is in fact the topology of inherent integrations between the input and the output variables*. The SCB of Theorem 3.1 explicitly shows this topology of inherent integrations. The following property pinpoints this.

Property 3.26 *Let $\overline{q}_0 = m_0$. Let $\overline{q}_j$ be an integer such that exactly $\overline{q}_j$ elements of q_i, $i = 1,\dots,m_d$, are equal to j. Also, let σ be an integer such that $\overline{q}_j = 0$ for all $j > \sigma$. Then there are $\overline{q}_0$ infinite zeros of order 0, and $j\overline{q}_j$ number of infinite zeros of order j, for $j = 1,\dots,\sigma$. Moreover, the $\mathcal{C}^*$ structural invariant indices list $\mathfrak{I}_4$ of Morse is given by*

$$\mathfrak{I}_4 = \{\overbrace{0,0,\dots,0}^{\overline{q}_0},\ \overbrace{1,1,\dots,1}^{\overline{q}_1},\ \dots,\ \overbrace{\sigma,\sigma,\dots,\sigma}^{\overline{q}_\sigma}\}.$$

Remark 3.27 *The state vector x_d in the SCB of a system is nonexistent if and only if the given system does not have infinite zeros of order greater than or equal to one.*

3.2.5 Geometric subspaces

In this subsection, we connect some classic subspaces from the geometric theory of linear systems to SCB. That is, in what follows, we show certain interconnections between the decomposition of the state space as done by the SCB, and various invariant subspaces from the geometric theory. To do so, we recall the first two subspaces. The subspaces $\mathcal{V}_g(\Sigma_*)$ and $\mathcal{S}_g(\Sigma_*)$ are classic subspaces and are crucial elements of geometric theory of linear systems. Also, later on we recall two more subspaces, $\mathcal{V}_\lambda(\Sigma_*)$ and $\mathcal{S}_\lambda(\Sigma_*)$, which are recently introduced in the context of H_∞ theory.

Definition 3.28 *Consider a linear system Σ_* characterized by the matrix quadruple (A, B, C, D). Then,*

(i) *The $\mathbb{C}_g$-stabilizable weakly unobservable subspace $\mathcal{V}_g(\Sigma_*)$ is defined as the largest subspace of $\mathbb{R}^n$ for which a matrix F exists such that the subspace is $(A+BF)$-invariant, contained in $\ker(C+DF)$, whereas the eigenvalues of $(A + BF)|\mathcal{V}_g$ are contained in $\mathbb{C}_g \subseteq \mathbb{C}$.*

(ii) *The $\mathbb{C}_g$-detectable strongly controllable subspace $\mathcal{S}_g(\Sigma_*)$ is defined as the smallest subspace of $\mathbb{R}^n$ for which a matrix K exists such that the subspace is $(A + KC)$-invariant, contains $\mathrm{im}(B + KD)$, and is such that the eigenvalues of the map that is induced by $(A + KC)$ on the factor space $\mathbb{R}^n / \mathcal{S}_g$ are contained in $\mathbb{C}_g \subseteq \mathbb{C}$.*

For the case when $\mathbb{C}_g = \mathbb{C}$, $\mathcal{V}_g$ and $\mathcal{S}_g$ are, respectively, denoted by $\mathcal{V}^*$ and $\mathcal{S}^*$; also, for the case when $\mathbb{C}_g = \mathbb{C}^-$, $\mathcal{V}_g$ and $\mathcal{S}_g$ are, respectively, denoted by $\mathcal{V}^-$ and $\mathcal{S}^-$, whereas for the case $\mathbb{C}_g = \mathbb{C}^{-0}$, $\mathcal{V}_g$ and $\mathcal{S}_g$ are, respectively, denoted by $\mathcal{V}^{-0}$ and $\mathcal{S}^{-0}$. Analogously, for the case when $\mathbb{C}_g = \mathbb{C}^\ominus$, $\mathcal{V}_g$ and $\mathcal{S}_g$ are, respectively, denoted by $\mathcal{V}^\ominus$ and $\mathcal{S}^\ominus$, whereas for the case $\mathbb{C}_g = \mathbb{C}^\otimes$, $\mathcal{V}_g$ and $\mathcal{S}_g$ are, respectively, denoted by $\mathcal{V}^\otimes$ and $\mathcal{S}^\otimes$.

Moreover, let a $\mathbb{C}_g$ be chosen such that it has no common elements with the set of invariant zeros of Σ_*. Then the corresponding $\mathcal{V}_g(\Sigma_*)$, which is always independent of the particular choice of such a $\mathbb{C}_g$ is referred to as the strongly controllable subspace $\mathcal{R}^*(\Sigma_*)$.

Remark 3.29 *We note that $\mathcal{V}_g(\Sigma_*)$ and $\mathcal{S}_g(\Sigma_*)$ are dual in the sense that*

$$\mathcal{V}_g(\Sigma_{*d}) = \mathcal{S}_g(\Sigma_*)^\perp,$$

where Σ_{*d} is the dual system of Σ_*.

Moreover, it can be shown that $\mathcal{R}^*(\Sigma_*)$ equals $\mathcal{V}^*(\Sigma_*) \cap \mathcal{S}^*(\Sigma_*)$.

Remark 3.30 *It is easy to observe that $\mathcal{V}_g(\Sigma_*)$ and $\mathcal{S}_g(\Sigma_*)$ are invariant under state feedback and output injection.*

Remark 3.31 *We should note that if (A, B) is $\mathbb{C}_g$-stabilizable, then for $\mathcal{V}_g(\Sigma_*)$, a matrix F exists that satisfies the conditions stated in Definition 3.28 and, moreover, $A + BF$ is $\mathbb{C}_g$-stable. An analogous comment can be made for $\mathcal{S}_g(\Sigma_*)$.*

Remark 3.32 *It is easily shown that the subspaces $\mathcal{V}_g(\Sigma_*)$ and $\mathcal{S}_g(\Sigma_*)$ satisfy the following:*

$$\begin{pmatrix} A \\ C \end{pmatrix} \mathcal{V}_g(\Sigma_*) \subseteq \left(\mathcal{V}_g(\Sigma_*) \oplus \{0\} \right) + \mathrm{im} \begin{pmatrix} B \\ D \end{pmatrix} \tag{3.30}$$

and

$$\ker \begin{pmatrix} C & D \end{pmatrix} \cap \begin{pmatrix} A & B \end{pmatrix} \left(\mathcal{S}_g(\Sigma_*) \oplus \mathbb{R}^m \right) \subseteq \mathcal{S}_g(\Sigma_*). \tag{3.31}$$

By now it is clear that the SCB decomposes the state space into several distinct parts. In fact, the state space $\mathcal{X}$ is decomposed as

$$\mathcal{X} = \mathcal{X}_a^- \oplus \mathcal{X}_a^0 \oplus \mathcal{X}_a^+ \oplus \mathcal{X}_b \oplus \mathcal{X}_c \oplus \mathcal{X}_d .$$

Here $\mathcal{X}_a^-$ is related to the stable invariant zeros, i.e., to the eigenvalues of A_{aa}^-, which are the stable invariant zeros of Σ. Similarly, $\mathcal{X}_a^0 \oplus \mathcal{X}_a^+$ is related to the unstable invariant zeros of Σ. On the other hand, $\mathcal{X}_b$ is related to right invertibility, i.e. the system is right invertible if and only if $\mathcal{X}_b = \{0\}$, whereas $\mathcal{X}_c$ is related to left invertibility; i.e., the system is left invertible if and only if $\mathcal{X}_c = \{0\}$. The latter two equivalence are true provided that, as assumed before, $(B' \quad D')'$ and $(C \quad D)$ are of full rank. Finally, $\mathcal{X}_d$ is related to zeros of Σ at infinity.

We focus next on certain interrelationships between the SCB and some basic ingredients of the geometric control theory.

We can now point out certain interpretations of various components of the state-space as decomposed by the SCB. We have the following property.

Property 3.33 *Consider a system Σ_* that has already been transformed in the special coordinate basis.*

(i) $\mathcal{X}_a^- \oplus \mathcal{X}_a^0 \oplus \mathcal{X}_a^+ \oplus \mathcal{X}_c$ *is equal to $\mathcal{V}^*(\Sigma_*)$.*

(ii) $\mathcal{X}_a^- \oplus \mathcal{X}_c$ *is equal to $\mathcal{V}^-(\Sigma_*)$ or $\mathcal{V}^\ominus(\Sigma_*)$ for continuous- and discrete-time systems, respectively.*

(iii) $\mathcal{X}_a^- \oplus \mathcal{X}_a^0 \oplus \mathcal{X}_c$ *is equal to $\mathcal{V}^{-0}(\Sigma_*)$ or $\mathcal{V}^\otimes(\Sigma_*)$ for continuous- and discrete-time systems, respectively.*

(iv) $\mathcal{X}_c \oplus \mathcal{X}_d$ *is equal to $\mathcal{S}^*(\Sigma_*)$.*

(v) $\mathcal{X}_a^+ \oplus \mathcal{X}_c \oplus \mathcal{X}_d$ *is equal to $\mathcal{S}^{-0}(\Sigma_*)$ or $\mathcal{S}^\otimes(\Sigma_*)$ for continuous- and discrete-time systems, respectively.*

(vi) $\mathcal{X}_a^0 \oplus \mathcal{X}_a^+ \oplus \mathcal{X}_c \oplus \mathcal{X}_d$ is equal to $\mathcal{S}^-(\Sigma_*)$ or $\mathcal{S}^\ominus(\Sigma_*)$ for continuous- and discrete-time systems, respectively.

(vii) $\mathcal{X}_c$ is equal to $\mathcal{R}^*(\Sigma_*)$.

Remark 3.34 *In view of Property 3.6, it is obvious that Σ_* is left invertible if and only if*

$$\mathcal{R}^*(\Sigma_*) = 0 \text{ and } \begin{pmatrix} B \\ D \end{pmatrix} \text{ is injective}$$

or equivalently

$$\mathcal{V}^*(\Sigma_*) \cap B \ker D = 0 \text{ and } \begin{pmatrix} B \\ D \end{pmatrix} \text{ is injective.}$$

Similarly, Σ_* is right invertible if and only if

$$\mathcal{V}^*(\Sigma_*) + \mathcal{S}^*(\Sigma_*) = \mathbb{R}^n \text{ and } \begin{pmatrix} C & D \end{pmatrix} \text{ is surjective}$$

or equivalently

$$\mathcal{S}^*(\Sigma_*) + C^{-1} \operatorname{im} D = \mathbb{R}^n \text{ and } \begin{pmatrix} C & D \end{pmatrix} \text{ is surjective.}$$

We recall now two more geometric subspaces $\mathcal{V}_\lambda(\Sigma_*)$ and $\mathcal{S}_\lambda(\Sigma_*)$ that were introduced in [85].

Definition 3.35 *For any $\lambda \in \mathbb{C}$, we define*

$$\mathcal{V}_\lambda(\Sigma_*) = \left\{ \zeta \in \mathbb{C}^n \;\middle|\; \exists\, \omega \in \mathbb{C}^m \;:\; 0 = \begin{pmatrix} A - \lambda I & B \\ C & D \end{pmatrix} \begin{pmatrix} \zeta \\ \omega \end{pmatrix} \right\} \tag{3.32}$$

and

$$\mathcal{S}_\lambda(\Sigma_*) = \left\{ \zeta \in \mathbb{C}^n \;\middle|\; \exists\, \omega \in \mathbb{C}^{n+m} \;:\; \begin{pmatrix} \zeta \\ 0 \end{pmatrix} = \begin{pmatrix} A - \lambda I & B \\ C & D \end{pmatrix} \omega \right\}. \tag{3.33}$$

We note that the geometric subspaces $\mathcal{V}_\lambda(\Sigma_*)$ and $\mathcal{S}_\lambda(\Sigma_*)$ are associated with the right state zero directions of Σ_* if λ is an invariant zero of Σ_*. These subspaces can also be displayed by the SCB of Σ_* as given in the following property.

Property 3.36 *We have*

$$\mathcal{V}_\lambda(\Sigma_*) = \operatorname{im}\left\{ \Gamma_s \begin{pmatrix} X_{a\lambda} & 0 \\ 0 & 0 \\ 0 & X_{c\lambda} \\ 0 & 0 \end{pmatrix} \right\}, \tag{3.34}$$

where $X_{a\lambda}$ is a matrix whose columns form a basis for the subspace,

$$\{\zeta_a \in \mathbb{C}^{n_a} \mid (\lambda I - A_{aa})\zeta_a = 0\} \tag{3.35}$$

and

$$X_{c\lambda} = (A_{cc} + B_c F_c - \lambda I)^{-1} B_c, \tag{3.36}$$

with F_c being any appropriately dimensioned matrix subject to the constraint that $A_{cc} + B_c F_c$ has no eigenvalue at λ. We note that the existence of such an F_c is guaranteed as (A_{cc}, B_c) forms a controllable pair.

Also, we have

$$\mathcal{S}_\lambda(\Sigma_*) = \operatorname{im}\left\{ \Gamma_s \begin{pmatrix} \lambda I - A_{aa} & 0 & 0 & 0 \\ 0 & Y_{b\lambda} & 0 & 0 \\ 0 & 0 & I_{n_c} & 0 \\ 0 & 0 & 0 & I_{n_d} \end{pmatrix} \right\}, \tag{3.37}$$

where

$$\operatorname{im} Y_{b\lambda} = \ker C_b (A_{bb} + K_b C_b - \lambda I)^{-1}, \tag{3.38}$$

and where K_b is any appropriately dimensioned matrix subject to the constraint that $A_{bb} + K_b C_b$ has no eigenvalue at λ. We note that the existence of such a K_b is guaranteed as (C_b, A_{bb}) forms an observable pair.

Clearly, if λ is not an eigenvalue of A_{aa}, then we have

$$\mathcal{V}_\lambda(\Sigma_*) \subseteq \mathcal{R}^*(\Sigma_*) \tag{3.39}$$

and

$$\mathcal{S}_\lambda(\Sigma_*) \supseteq \mathcal{V}^*(\Sigma_*) + \mathcal{S}^*(\Sigma_*). \tag{3.40}$$

Next, we would like to note that $\mathcal{V}_\lambda(\Sigma_*)$ and $\mathcal{S}_\lambda(\Sigma_*)$ are dual in the sense that $\mathcal{V}_\lambda(\Sigma_* d) = \mathcal{S}_\lambda(\Sigma_*)^\perp$. Also, $\mathcal{S}_\lambda(\Sigma_*) = \mathcal{V}_\lambda(\Sigma_* d)^\perp$.

The subspaces $\mathcal{S}_\lambda(\Sigma_*)$ and $\mathcal{V}_\lambda(\Sigma_*)$ are the subspaces of $\mathbb{C}^n$ when we consider complex eigenvalues. For the almost disturbance decoupling problem as studied in Chapter 6, we need the following subspace:

$$\left(\mathcal{S}^*(\Sigma_*) + \mathcal{V}^{-0}(\Sigma_*) \right) \cap \{\cap_{\lambda \in \mathbb{C}^0} \mathcal{S}_\lambda(\Sigma_*)\}. \tag{3.41}$$

This space has the property that a vector x is in the subspace if and only if its complex conjugate x^* is in the subspace, and hence, this subspace of $\mathbb{C}^n$ has a basis of real-valued vectors; therefore, this space can be viewed as the complexification of subspaces of $\mathbb{R}^n$.

We have a precise characterization of this subspace. We factorize

$$X_a^0 = X_a^{01} \oplus X_a^{02}$$

such that X_a^{02} is defined by

$$X_a^{02} = \left\{ v \in \mathbb{R}^{n_{a\circ}} \,\middle|\, \exists \lambda \in \mathbb{C}^0 \text{ such that } v'A_{aa}^0 = \lambda v' \right\}.$$

Then we obtain

$$\left(\mathcal{S}^*(\Sigma_*) + \mathcal{V}^{-0}(\Sigma_*) \right) \cap \{\cap_{\lambda \in \mathbb{C}^0} \mathcal{S}_\lambda(\Sigma_*)\} = \mathcal{S}^*(\Sigma_*) + \mathcal{V}^-(\Sigma_*) + X_a^{01}$$
$$= X_a^- \oplus X_a^{01} \oplus X_c \oplus X_d.$$

Note that we will use later that the above structure results in a specific structure for A_{aa}^0,

$$A_{aa}^0 = \begin{pmatrix} A_{11} & A_{12} \\ 0 & A_{22} \end{pmatrix}, \quad B_{a0}^0 = \begin{pmatrix} B_1 \\ B_2 \end{pmatrix},$$

with

$$\left[\begin{pmatrix} A_{11} & A_{12} \\ 0 & A_{22} \end{pmatrix}, \begin{pmatrix} 0 \\ B_2 \end{pmatrix} \right]$$

controllable. The eigenvalues of A_{11} are contained in the set of eigenvalues of A_{22} with at least the same *geometric* multiplicity. Finally, A_{22} is diagonalizable.

3.2.6   *Miscellaneous properties of the SCB*

Several properties of linear multivariable time-invariant systems can trivially be visualized using the SCB of Theorem 3.1. We give below some of these properties.

Property 3.37 *The normal rank of Σ_* is equal to $m_d + m_0$. Moreover, it is easy to see the following:*

(*i*) normrank $P_{\Sigma_*}(s) = n + \text{normrank}[C(sI - A)^{-1}B + D]$.

(*ii*) *The normal rank of Σ_* is equal to p if and only if Σ_* is right invertible.*

(*iii*) *The normal rank of Σ_* is equal to m if and only if Σ_* is left invertible.*

Property 3.38

(*i*) Σ_* *is right invertible, and minimum phase* $\Rightarrow$ Σ_* *is stabilizable.*

(ii) Σ_* is left invertible, and minimum phase $\Rightarrow \Sigma_*$ is detectable.

(iii) Σ_* is invertible, and minimum phase $\Rightarrow \Sigma_*$ is stabilizable and detectable.

(iv) Σ_* is right invertible, and the invariant zeros are disjoint from the eigen-values (or unstable eigenvalues) of $A \Rightarrow \Sigma_*$ is controllable (stabilizable).

(v) Σ_* is left invertible, and its invariant zeros are disjoint from the eigenvalues (or unstable eigenvalues) of $A \Rightarrow \Sigma_*$ is observable (detectable).

(vi) Σ_* is invertible, and its invariant zeros are disjoint from the eigenvalues (or unstable eigenvalues) of $A \Rightarrow \Sigma_*$ is controllable and observable (stabilizable and detectable).

(vii) The feedthrough matrix D in Σ_* is injective $\Rightarrow \Sigma_*$ is left invertible and has no infinite zeros of order greater than or equal to one.

(viii) The feedthrough matrix D in Σ_* is surjective $\Rightarrow \Sigma_*$ is right invertible and has no infinite zeros of order greater than or equal to one.

We connected in Subsections 3.2.3 and 3.2.4 the lists $\mathfrak{I}_1$ and $\mathfrak{I}_4$ of Morse [49] to SCB. The following property connects the lists $\mathfrak{I}_2$ and $\mathfrak{I}_3$ of Morse to SCB.

Property 3.39

The list $\mathfrak{I}_2$ of Morse $=$ The controllability indices of the pair (A_{cc}, B_c).

The list $\mathfrak{I}_3$ of Morse $=$ The observability indices of the pair (C_b, A_{bb}).

We have also the following remark.

Remark 3.40 *The integers*

$$n_{a-}(\Sigma_*), n_{a\circ}(\Sigma_*), n_{a+}(\Sigma_*), n_b(\Sigma_*), n_c(\Sigma_*), n_d(\Sigma_*), m_d,$$
$$\text{and } q_i \, (i = 1, \ldots, m_d)$$

are structurally invariant with respect to state feedback and output injection. Moreover, the integers $n_a(\Sigma_*) = n_{a-}(\Sigma_*) + n_{a\circ}(\Sigma_*) + n_{a+}(\Sigma_*)$, $n_b(\Sigma_*)$, $n_c(\Sigma_*)$, and $n_d(\Sigma_*)$ are, respectively, equal to the number of elements in the lists $\mathfrak{I}_1, \mathfrak{I}_2, \mathfrak{I}_3,$ and $\mathfrak{I}_4$ of Morse. For further details, one can refer to [49] and [73].

3.2.7 Additional compact forms of the SCB

Finally, let us observe that, depending on some specific properties a given system satisfies, SCB can be written compactly in different formats. For convenience, we present below some such formats so that we can use them directly in later chapters as the need arises.

We will sometimes use the SCB in a more compact form where the special structure of x_d is not made explicit and where x_a^0 and x_a^+ are viewed together as x_a^{0+}. In this case, we get

$$\Gamma_s^{-1}(A - B_0 C_0)\Gamma_s = \begin{pmatrix} A_{aa}^- & 0 & L_{ab}^- C_b & 0 & L_{ad}^- C_d \\ 0 & A_{aa}^{0+} & L_{ab}^{0+} C_b & 0 & L_{ad}^{0+} C_d \\ 0 & 0 & A_{bb} & 0 & L_{bd} C_d \\ B_c E_{ca}^- & B_c E_{ca}^{0+} & B_c E_{cb} & A_{cc} & L_{cd} C_d \\ B_d E_{da}^- & B_d E_{da}^{0+} & B_d E_{db} & B_d E_{dc} & A_{dd} \end{pmatrix},$$

$$(3.42)$$

$$\Gamma_s^{-1}\begin{pmatrix} B_0 & \widehat{B}_1 \end{pmatrix}\Gamma_i = \begin{pmatrix} B_{a0}^- & 0 & 0 \\ B_{a0}^{0+} & 0 & 0 \\ B_{b0} & 0 & 0 \\ B_{c0} & 0 & B_c \\ B_{d0} & B_d & 0 \end{pmatrix},$$

$$(3.43)$$

$$\Gamma_o^{-1}\begin{pmatrix} C_0 \\ \widehat{C}_1 \end{pmatrix}\Gamma_s = \begin{pmatrix} C_{0a}^- & C_{0a}^{0+} & C_{0b} & C_{0c} & C_{0d} \\ 0 & 0 & 0 & 0 & C_d \\ 0 & 0 & C_b & 0 & 0 \end{pmatrix},$$

$$(3.44)$$

and

$$\Gamma_o^{-1}\begin{pmatrix} I_{m_0} & 0 \\ 0 & 0 \end{pmatrix}\Gamma_i = \begin{pmatrix} I_{m_0} & 0 & 0 \\ 0 & 0 & 0 \\ 0 & 0 & 0 \end{pmatrix}.$$

$$(3.45)$$

As discussed, the eigenvalues of A_{aa}^- and A_{aa}^{0+} are the invariant zeros of the given system Σ_*. Moreover, for continuous-time systems, the eigenvalues of A_{aa}^- are in the open left-half complex plane, whereas the eigenvalues of A_{aa}^{0+} are in the closed right-half complex plane. Similarly, for discrete-time systems, the eigenvalues of A_{aa}^- are within the unit circle, whereas the eigenvalues of A_{aa}^{0+} are on the unit circle or outside the unit circle.

If the given system Σ_* is left-invertible, then the decomposition in (3.42)–(3.45) simplifies because x_c is no longer present (see Property 3.6), and we obtain the following structure:

$$\Gamma_s^{-1}(A - B_0 C_0)\Gamma_s = \begin{pmatrix} A_{aa}^- & 0 & L_{ab}^- C_b & L_{ad}^- C_d \\ 0 & A_{aa}^{0+} & L_{ab}^{0+} C_b & L_{ad}^{0+} C_d \\ 0 & 0 & A_{bb} & L_{bd} C_d \\ B_d E_{da}^- & B_d E_{da}^{0+} & B_d E_{db} & A_{dd} \end{pmatrix},$$

$$(3.46)$$

$$\Gamma_s^{-1}\begin{pmatrix} B_0 & \hat{B}_1 \end{pmatrix}\Gamma_i = \begin{pmatrix} B_{a0}^- & 0 \\ B_{a0}^{0+} & 0 \\ B_{b0} & 0 \\ B_{d0} & B_d \end{pmatrix}, \tag{3.47}$$

$$\Gamma_o^{-1}\begin{pmatrix} C_0 \\ \hat{C}_1 \end{pmatrix}\Gamma_s = \begin{pmatrix} C_{0a}^- & C_{0a}^{0+} & C_{0b} & C_{0d} \\ 0 & 0 & 0 & C_d \\ 0 & 0 & C_b & 0 \end{pmatrix}, \tag{3.48}$$

and

$$\Gamma_o^{-1}\begin{pmatrix} I_{m_0} & 0 \\ 0 & 0 \end{pmatrix}\Gamma_i = \begin{pmatrix} I_{m_0} & 0 \\ 0 & 0 \\ 0 & 0 \end{pmatrix}. \tag{3.49}$$

Similarly, we will sometimes use the SCB in another more compact form where this time x_a^- and x_a^0 are viewed together as x_a^{-0}. In this case, we get

$$\Gamma_s^{-1}(A - B_0 C_0)\Gamma_s = \begin{pmatrix} A_{aa}^{-0} & 0 & L_{ab}^{-0}C_b & 0 & L_{ad}^{-0}C_d \\ 0 & A_{aa}^+ & L_{ab}^+C_b & 0 & L_{ad}^+C_d \\ 0 & 0 & A_{bb} & 0 & L_{bd}C_d \\ B_c E_{ca}^{-0} & B_c E_{ca}^+ & B_c E_{cb} & A_{cc} & L_{cd}C_d \\ B_d E_{da}^{-0} & B_d E_{da}^+ & B_d E_{db} & B_d E_{dc} & A_{dd} \end{pmatrix}, \tag{3.50}$$

$$\Gamma_s^{-1}\begin{pmatrix} B_0 & \hat{B}_1 \end{pmatrix}\Gamma_i = \begin{pmatrix} B_{a0}^{-0} & 0 & 0 \\ B_{a0}^+ & 0 & 0 \\ B_{b0} & 0 & 0 \\ B_{c0} & 0 & B_c \\ B_{d0} & B_d & 0 \end{pmatrix}, \tag{3.51}$$

$$\Gamma_o^{-1}\begin{pmatrix} C_0 \\ \hat{C}_1 \end{pmatrix}\Gamma_s = \begin{pmatrix} C_{0a}^{-0} & C_{0a}^+ & C_{0b} & C_{0c} & C_{0d} \\ 0 & 0 & 0 & 0 & C_d \\ 0 & 0 & C_b & 0 & 0 \end{pmatrix}, \tag{3.52}$$

and

$$\Gamma_o^{-1}\begin{pmatrix} I_{m_0} & 0 \\ 0 & 0 \end{pmatrix}\Gamma_i = \begin{pmatrix} I_{m_0} & 0 & 0 \\ 0 & 0 & 0 \\ 0 & 0 & 0 \end{pmatrix}. \tag{3.53}$$

Once again, as discussed, the eigenvalues of A_{aa}^{-0} and A_{aa}^0 are the invariant zeros of the given system Σ_*. Moreover, for continuous-time systems, the eigenvalues of A_{aa}^{-0} are in the closed left-half complex plane, whereas the eigenvalues of A_{aa}^+ are in the open right-half complex plane. Similarly, for discrete-time systems, the eigenvalues of A_{aa}^{-0} are on the unit circle or within the unit circle, whereas the eigenvalues of A_{aa}^+ are outside the unit circle.

If the given system Σ_* is left-invertible, then the decomposition in (3.50)–(3.53) simplifies because x_c is no longer present (see Property 3.6), and we obtain the following structure:

$$
\Gamma_s^{-1}(A - B_0 C_0)\Gamma_s =
\begin{pmatrix}
A_{aa}^{-0} & 0 & L_{ab}^{-0} C_b & L_{ad}^{-0} C_d \\
0 & A_{aa}^{+} & L_{ab}^{+} C_b & L_{ad}^{+} C_d \\
0 & 0 & A_{bb} & L_{bd} C_d \\
B_d E_{da}^{-0} & B_d E_{da}^{+} & B_d E_{db} & A_{dd}
\end{pmatrix},
\qquad (3.54)
$$

$$
\Gamma_s^{-1}\begin{pmatrix} B_0 & \hat{B}_1 \end{pmatrix}\Gamma_i =
\begin{pmatrix}
B_{a0}^{-0} & 0 \\
B_{a0}^{+} & 0 \\
B_{b0} & 0 \\
B_{d0} & B_d
\end{pmatrix},
\qquad (3.55)
$$

$$
\Gamma_o^{-1}\begin{pmatrix} C_0 \\ \hat{C}_1 \end{pmatrix}\Gamma_s =
\begin{pmatrix}
C_{0a}^{-0} & C_{0a}^{+} & C_{0b} & C_{0d} \\
0 & 0 & 0 & C_d \\
0 & 0 & C_b & 0
\end{pmatrix},
\qquad (3.56)
$$

and

$$
\Gamma_o^{-1}\begin{pmatrix} I_{m_0} & 0 \\ 0 & 0 \end{pmatrix}\Gamma_i =
\begin{pmatrix}
I_{m_0} & 0 \\
0 & 0 \\
0 & 0
\end{pmatrix}.
\qquad (3.57)
$$

4

Algebraic Riccati equations and matrix inequalities

Different types of algebraic equations or inequalities are encountered in many filtering and control problems. In this chapter, we study in detail some such equations or inequalities. In particular, we study what are known as algebraic Riccati equations, linear matrix inequalities, and quadratic matrix inequalities, all of which ensue in connection with both continuous- as well as discrete-time systems.

This chapter is organized as follows. We study in Section 4.1 algebraic Riccati equations that ubiquitously occur in connection with continuous-time filtering and control. A fundamental understanding of continuous-time algebraic Riccati equations (CAREs) is essential to look into different aspects of H_2 optimal filtering and control. This is true not only because of the significance of CAREs in their own right but also because, as will be seen in this chapter, the study of some other types of algebraic equations or inequalities can be converted to a study of appropriately defined associated CAREs. In this sense, the study of CAREs takes on an added significance. As a result, in Section 4.1, we carefully compile from the literature several important properties of CAREs. Three categories of CAREs are prominent, H_∞^1 CAREs, H_∞^2 CAREs, and H_2 CAREs. It is known that a CARE can have a variety of solutions that can be symmetric or nonsymmetric, sign definite or indefinite; also the number of solutions of a CARE can be either finite or infinite. The types of solutions that are of particular interest to H_2 optimal filtering and control are so-called real symmetric semi-stabilizing or stabilizing solutions, and real symmetric positive semi-definite or definite solutions. Hence, we focus our study on such solutions of a CARE. All in all, Section 4.1 concentrates on three principal aspects of a CARE: (1) properties of its solutions, (2) existence and uniqueness of (semi-)stabilizing or positive (semi-)definite solutions, and (3) algorithms to compute such solutions.

In Section 4.2, we concentrate on the study of algebraic Riccati equations that occur in connection with discrete-time optimal filtering and control; i.e., we concentrate on discrete-time algebraic Riccati equations (DAREs) and their generalized versions termed as GDAREs. As in continuous-time, three aspects of a DARE or a GDARE are of primary importance: (1) properties of its solutions, (2) existence and uniqueness of (semi-)stabilizing or positive (semi-)definite solutions, and (3) algorithms to compute such solutions. We focus on some of these aspects by a direct study of a given DARE or a GDARE, and we focus on some others indirectly via two very beneficial tools that are developed in this section. The first tool develops a one-to-one relationship between a solution of a DARE and a solution of an appropriately defined associated CARE, whereas similarly the second tool develops a one-to-one relationship between a solution of a GDARE and a solution of an appropriately defined associated DARE. Obviously, these tools enable us to use elegantly the results of the previous section on CAREs for DAREs as well as for GDAREs.

Similarly, in Sections 4.3 and 4.4, respectively, we concentrate on continuous-time linear matrix inequalities (CLMIs) and discrete-time linear matrix inequalities (DLMIs). After establishing some preliminary properties of a CLMI (respectively, a DLMI), the notions of positive semi-definite and positive definite rank minimizing solutions as well as semi-stabilizing and stabilizing solutions of a CLMI (respectively, a DLMI) are introduced. Furthermore, in the case of a DLMI, the notion of strongly rank minimizing solutions is introduced. Then, as in the previous sections, an important tool that shows a one-to-one relationship between a certain solution of a CLMI (respectively, a DLMI) and a solution of an associated CARE (respectively, an associated DARE) is developed. This tool helps us in a natural way to study the properties of various solutions of a CLMI (respectively, a DLMI). Also, it enables us to develop the existence conditions of such solutions and leads to the computational methods of obtaining them. Section 4.5, finally, covers continuous-time quadratic matrix inequalities (CQMIs). Our study of CQMIs parallels that of CLMIs. That is, after a preliminary study of it, notions of several types of solutions are introduced, followed by the development of an important tool that shows a one-to-one relationship between a certain solution of a CQMI and a solution of an associated CARE. Such a tool as before leads to the study of various properties of various solutions and development of existence conditions of such solutions and methods of computing them.

Thus, whether it is a DARE, a GDARE, a CLMI, a DLMI, or a CQMI, as in the case of a CARE, our study of it focuses on three aspects: (1) general properties of the equation or inequality, (2) existence and uniqueness of specific solutions relevant to filtering theory, and (3) algorithms to compute such solutions. To do so, a recurring notion followed in Sections 4.2 to 4.5 is to develop eventually a one-to-one relationship between a solution of a given DARE or a GDARE or a CLMI or a DLMI or a CQMI and a solution of an appropriately defined CARE. This clearly enables us to use elegantly the results on CAREs for DAREs, GDAREs, CLMIs, DLMIs, and CQMIs.

4.1   Continuous-time algebraic Riccati equations

What is called a continuous-time algebraic Riccati equation (CARE) is a certain quadratic matrix equation that is named after Count Jacopo Francesco Riccati, a famous Venetian mathematician who lived from 1676 until 1754. Since its introduction in control theory by R. E. Kalman [36] at the beginning of the 1960s, the

CARE has been found to have an impressive range of applications, such as H_2 and H_∞ optimal control theory, stability theory, stochastic filtering and stochastic control, and differential game theory. The purpose here is to give an expository survey of the so-called semi-stabilizing and stabilizing solutions as well as positive semi-definite and positive definite solutions of a CARE. Whereas this section considers CAREs, the next section considers DAREs.

The architecture of this section is as follows. After this basic introduction, we begin with the Subsection 4.1.1 that defines a general CARE and three specific classes of it that exhibit additional structure, namely, H_∞^1 CARE, H_∞^2 CARE, and H_2 CARE. The H_∞^1 CARE and the H_∞^2 CARE form two subclasses of the general CARE, and their intersection is exactly the class that is denoted as H_2-CARE. In Subsection 4.1.2, we introduce what is known as a Hamiltonian matrix and then study some properties exploring basically the relationship that exists between the Hamiltonian matrix and the CARE. A given CARE can have a variety of solutions, but our interest in this book is to study primarily the semi-stabilizing as well as stabilizing solutions. In this regard, in Subsection 4.1.3, we define as well as compile judiciously certain properties of real symmetric semi-stabilizing and stabilizing solutions. The properties studied here include the existence and uniqueness of such solutions. In Subsection 4.1.4, we will look at real symmetric positive semi-definite or positive definite solutions and their relationship with symmetric semi-stabilizing and stabilizing solutions. Once again, the properties obtained here include the existence and uniqueness of such solutions and are pertinent to our development in later chapters. Subsection 4.1.5 shows when the (semi-)stabilizing solution of an H_2 CARE is continuous with respect to perturbations in the parameters that define the H_2 CARE. Finally, Subsection 4.1.6 surveys some standard computational algorithms for obtaining the (semi-)stabilizing solution of a CARE. Thus, this section focuses on properties, existence and uniqueness, and algorithms to compute solutions of a CARE.

Basically, the material in this section is a judicious compilation from the literature. A good source that points out the recent and past literature on algebraic Riccati equations is a recent book by Bittanti et al. [5], and the references therein; see also [63], [66], [97], and [109].

4.1.1 Definition of a CARE and its subclasses

In this subsection, we present formally the definition of a CARE and its subclasses, the H_∞^1 CARE, the H_∞^2 CARE, and the H_2 CARE.

Definition 4.1 *The quadratic matrix equation for an unknown $n \times n$ matrix X of the form*

$$A'X + XA - XRX + Q = 0, \qquad (4.1)$$

*where A, R, and Q are $n \times n$ real matrices, with R and Q being symmetric, is called a **continuous-time algebraic Riccati equation (CARE)**.*

We will consider three special cases of CARE as follows:

- The H_2 **CARE** corresponds to $Q \geqslant 0$ and $R \geqslant 0$,

- The H_∞^1 **CARE** corresponds to $Q \geqslant 0$,

- The H_∞^2 **CARE** corresponds to $R \geqslant 0$.

In what follows, we first formally define the above three specific cases in the so-called general forms as they appear in filtering and control literature, and then we will show that these general forms can all be reduced to the cases mentioned above.

Definition 4.2 *The quadratic matrix equation in $X \in \mathbb{R}^{n \times n}$ of the form*

$$\bar{A}'X + X\bar{A} - (XB + S)\bar{R}^{-1}(B'X + S') + \bar{Q} = 0, \tag{4.2}$$

with

$$\begin{pmatrix} \bar{Q} & S \\ S' & \bar{R} \end{pmatrix} \geqslant 0, \tag{4.3}$$

is referred to as the H_2 CARE.

The condition that the matrix in (4.3) is positive semi-definite will play an important role, and hence, we will label this condition:

Definition 4.3 *The matrices $\bar{Q}$, $\bar{R}$, and S are said to satisfy* **Condition psd** *if*

$$\begin{pmatrix} \bar{Q} & S \\ S' & \bar{R} \end{pmatrix} \geqslant 0.$$

Under this condition, it follows that matrices $C \in \mathbb{R}^{p \times n}$ and $D \in \mathbb{R}^{p \times m}$ exist with $(C \quad D)$ of full rank such that

$$\begin{pmatrix} \bar{Q} & S \\ S' & \bar{R} \end{pmatrix} = \begin{pmatrix} C & D \end{pmatrix}' \begin{pmatrix} C & D \end{pmatrix}. \tag{4.4}$$

Note that by defining $A = \bar{A} - B\bar{R}^{-1}S'$ and $Q = \bar{Q} - S\bar{R}^{-1}S'$, we can eliminate the matrix S from (4.2) and obtain the equation

$$A'X + XA - XB\bar{R}^{-1}B'X + Q = 0. \tag{4.5}$$

Moreover, if we define $R = B\bar{R}^{-1}B'$, then we see that (4.2) is indeed a special case of the general CARE (4.1). The earlier definition of the H_2 CARE, namely,

(4.1) with $Q \geqslant 0$ and $R \geqslant 0$, is consistent with Definition 4.2 because (4.1) with $Q \geqslant 0$ and $R \geqslant 0$ can be rewritten as a CARE of the form (4.2) satisfying (4.3) and conversely.

Given (4.3), we can obtain the factorization (4.4) with D injective and obtain the H_2 CARE

$$\bar{A}'X + X\bar{A} - (XB + C'D)(D'D)^{-1}(B'X + D'C) + C'C = 0. \qquad (4.6)$$

We proceed next to define H_∞^1 CARE.

Definition 4.4 *The quadratic matrix equation in $X \in \mathbb{R}^{n \times n}$ of the form*

$$\bar{A}'X + X\bar{A} + XMX - (XB + S)\bar{R}^{-1}(B'X + S') + \bar{Q} = 0, \qquad (4.7)$$

with $M \geqslant 0$ and

$$\begin{pmatrix} \bar{Q} & S \\ S' & \bar{R} \end{pmatrix} \geqslant 0, \qquad (4.8)$$

is referred to as the H_∞^1 CARE.

Obviously, the H_2 CARE is a special case of the H_∞^1 CARE as it is obtained by setting $M = 0$. Moreover, if we define

$$A = \bar{A} - B\bar{R}^{-1}S', \quad Q = \bar{Q} - S\bar{R}^{-1}S', \text{ and } R = B\bar{R}^{-1}B' - M,$$

then we see that (4.7) is a special case of the general CARE (4.1). The earlier definition of the H_∞^1 CARE, namely, (4.1) with $Q \geqslant 0$, is consistent with Definition 4.4 because (4.1) with $Q \geqslant 0$ can be rewritten as a CARE of the form (4.7) satisfying (4.8) and conversely.

We can again use the factorization (4.4) with D injective together with the factorization $M = EE'$ and obtain the equation

$$\bar{A}'X + X\bar{A} + XEE'X - (XB + C'D)(D'D)^{-1}(B'X + D'C) + C'C = 0. \quad (4.9)$$

Finally, we introduce next the third special case of CARE, namely, H_∞^2 CARE.

Definition 4.5 *The quadratic matrix equation in $X \in \mathbb{R}^{n \times n}$ of the form*

$$\bar{A}'X + X\bar{A} - (XB + S)\bar{R}^{-1}(B'X + S') + \bar{Q} - N = 0, \qquad (4.10)$$

with $N \geqslant 0$ and

$$\begin{pmatrix} \bar{Q} & S \\ S' & \bar{R} \end{pmatrix} \geqslant 0, \qquad (4.11)$$

is referred to as the H_∞^2 CARE.

Obviously, the H_2 CARE is also a special case of the H_∞^2 CARE as it is obtained by setting $N = 0$. The earlier definition of the H_∞^2 CARE, namely, (4.1) with $R \geqslant 0$, is consistent with Definition 4.5 because (4.1) with $R \geqslant 0$ can be rewritten as a CARE of the form (4.10) satisfying (4.11) and conversely.

Moreover, if we define

$$A = \bar{A} - B\bar{R}^{-1}S', \quad Q = \bar{Q} - S\bar{R}^{-1}S' - N, \text{ and } R = B\bar{R}^{-1}B',$$

then we see that (4.10) is again a special case of the general CARE (4.1).

We can again use the factorization

$$\begin{pmatrix} \bar{Q} & S \\ S' & \bar{R} \end{pmatrix} = \begin{pmatrix} C_1' \\ D' \end{pmatrix} \begin{pmatrix} C_1 & D \end{pmatrix} \tag{4.12}$$

with D injective together with the factorization $N = C_2'C_2$ and obtain the equation

$$\bar{A}'X + X\bar{A} - (XB + C_1'D)(D'D)^{-1}(B'X + D'C_1) + C_1'C_1 - C_2'C_2 = 0. \tag{4.13}$$

As noted, a CARE can have different representations. Also, we noted earlier a CARE of the form (4.2) can be reduced to the form of general CARE (4.1). The following lemma relates some properties of the parameters of these two representations.

Lemma 4.6 *Consider a system* $(\bar{A}, B, C, D)$ *with D injective and define*

$$A = \bar{A} - B(D'D)^{-1}D'C, \quad R = B(D'D)^{-1}B', \quad Q = C'C - C'D(D'D)^{-1}D'C.$$

Then we have

   (i) $\langle \ker Q \mid A \rangle = \mathcal{V}^*(\bar{A}, B, C, D)$.

   (ii) The $\mathbb{C}_g$ unobservable subspace of (Q, A) is equal to $\mathcal{V}_g(\bar{A}, B, C, D)$.

   (iii) (A, R) is $\mathbb{C}_g$-stabilizable if and only if $(\bar{A}, B)$ is $\mathbb{C}_g$-stabilizable.

Proof : We note that $\mathcal{V}^*$ and $\mathcal{V}_g$ are invariant under a preliminary state feedback. Define $F = -(D'D)^{-1}D'C$, then,

$$\mathcal{V}^*(\bar{A}, B, C, D) = \mathcal{V}^*(A, B, C - DF, D),$$
$$\mathcal{V}_g(\bar{A}, B, C, D) = \mathcal{V}_g(A, B, C - DF, D).$$

Next, we note that for a subspace $\mathcal{V}$, we have that $(C - DF - DF_1)\mathcal{V} = \{0\}$ if and only if $(C - DF)\mathcal{V} = \{0\}$ and $DF_1\mathcal{V} = \{0\}$ because im D and im$(C - DF)$ are orthogonal. As D is injective, this implies that $F_1 \mid_\mathcal{V} = 0$. This result yields that $\mathcal{V}^*$ is the largest $(A + BF)$-invariant subspace contained in $\ker(C + DF)$. This result then immediately yields (i) because $Q = (C + DF)'(C + DF)$ and hence

$$\ker Q = \ker(C + DF).$$

Next we note that $\mathcal{V}_g$ is the largest $(A + BF)$-invariant subspace contained in $\ker(C + DF)$ such that the eigenvalues of $(A + BF) \mid_{\mathcal{V}_g}$ are contained in $\mathbb{C}_g$, which immediately yields (ii). (iii) immediately follows from the fact that a preliminary state feedback does not affect stabilizability. $\blacksquare$

We define next the dual continuous-time algebraic Riccati equation (the dual CARE) as

$$AY + YA' - YQY + R = 0. \tag{4.14}$$

We again can distinguish three special cases, as follows:

- The dual H_2 CARE is defined by

$$\bar{A}Y + Y\bar{A}' - (YC_1' + BD')(DD')^{-1}(C_1Y + DB') + BB' = 0 \tag{4.15}$$

  or equivalently by (4.14) with $Q \geqslant 0$ and $R \geqslant 0$.

- The dual H_∞^1 CARE is defined by

$$\bar{A}Y + Y\bar{A}' + YC_2'C_2Y - (YC_1' + BD')(DD')^{-1}(C_1Y + DB') + BB' = 0 \tag{4.16}$$

  or equivalently by (4.14) with $Q \geqslant 0$.

- The dual H_∞^2 CARE is defined by

$$\bar{A}Y + Y\bar{A}' - (YC_1' + BD')(DD')^{-1}(C_1Y + DB') + BB' - EE' = 0 \tag{4.17}$$

  or equivalently by (4.14) with $R \geqslant 0$.

The study of the above dual CAREs can obviously be reduced to that of the corresponding CAREs by dualization. Hence, no detailed discussion of these dual CAREs is given here.

4.1.2   The Hamiltonian matrix

All solutions of a CARE are closely associated with the invariant subspaces of a so-called Hamiltonian matrix H. In fact, it is known that it is possible to characterize any solution X of a CARE in terms of the invariant subspaces of an appropriately defined Hamiltonian matrix H. For this reason, we first introduce here the Hamiltonian matrix H and examine its properties in detail. More specifically, we will note that a one-to-one correspondence exists between the solution set of symmetric matrices satisfying a CARE and the set of n-dimensional H-invariant subspaces that are complementary to the n-dimensional subspace

$$\text{im} \begin{pmatrix} 0 \\ I \end{pmatrix}.$$

This correspondence in fact assigns the H-invariant subspace

$$\mathcal{S}_X := \text{im} \begin{pmatrix} I \\ X \end{pmatrix}$$

to the solution X of a CARE. This property reveals many relationships between H and the solutions of a CARE. This property can be used not only to develop the necessary and sufficient conditions for the existence of solutions of a CARE, but also to seek algorithms for constructing such solutions.

We introduce next the Hamiltonian matrix H.

Definition 4.7 *The $2n \times 2n$ matrix*

$$H = \begin{pmatrix} A & -R \\ -Q & -A' \end{pmatrix} \tag{4.18}$$

is called the Hamiltonian matrix associated with the CARE in (4.1).

It is prudent to recall now that two matrices V and W are similar if an invertible matrix Z exists such that $V = ZWZ^{-1}$. It is easily shown that the eigenvalues and the associated multiplicity structures are the same for two similar matrices.

Proposition 4.8 *H is similar to $-H'$. Hence, λ is an eigenvalue of H with a multiplicity structure $\{m_1, \cdots, m_k\}$ if and only if $-\lambda$ is an eigenvalue of H with the same multiplicity structure.*

Proof : Letting

$$J := \begin{pmatrix} 0 & -I \\ I & 0 \end{pmatrix}, \tag{4.19}$$

it is trivial to verify that

$$JHJ^{-1} = -H'. \tag{4.20}$$

This completes the proof because the eigenvalues and the associated multiplicity structures are the same for two similar matrices. ∎

Proposition 4.9 *The subspace*

$$\operatorname{im} \begin{pmatrix} I \\ X \end{pmatrix} \tag{4.21}$$

is an invariant subspace of H if and only if X satisfies the CARE (4.1). Moreover, in that case,

$$H \begin{pmatrix} I \\ X \end{pmatrix} = \begin{pmatrix} I \\ X \end{pmatrix} (A - RX). \tag{4.22}$$

Proof : By direct verification, we can see that X satisfies the CARE if and only if (4.22) is satisfied. It is clear that (4.22) immediately implies that (4.21) is an invariant subspace of the Hamiltonian. Conversely, if (4.21) is an invariant subspace of the CARE, then property (*i*) on page 14 implies that a matrix A_{cl} exists such that

$$H \begin{pmatrix} I \\ X \end{pmatrix} = \begin{pmatrix} I \\ X \end{pmatrix} A_{cl}.$$

But then using the structure of H, we note that $A_{cl} = A - RX$, and hence, (4.22) is satisfied. ∎

The above proposition establishes the direct relationship of solutions to the CARE and a specific class of invariant subspaces of the Hamiltonian matrix.

It is easy to verify the following proposition:

Proposition 4.10 *For any solution $X \in \mathbb{R}^{n \times n}$ of the CARE (4.1), we have*

$$H \begin{pmatrix} I & 0 \\ X & I \end{pmatrix} = \begin{pmatrix} I & 0 \\ X & I \end{pmatrix} \begin{pmatrix} A - RX & -R \\ 0 & -(A' - XR) \end{pmatrix}. \tag{4.23}$$

Remark 4.11 *Equation (4.23) establishes that if X is a solution of the CARE (4.1), then*

$$\lambda(H) = \lambda(A - RX) \cup \lambda(-A + RX).$$

4.1.3   *Stabilizing and semi-stabilizing solutions of a CARE*

A CARE can have a variety of solutions. However, we shall be concerned here primarily with semi-stabilizing and stabilizing solutions, which we formally define as follows:

Definition 4.12 *A solution of a CARE as in* (4.1) *is said to be a semi-stabilizing solution if* $A - RX$ *has all its eigenvalues in the closed left-half complex plane. Similarly, a solution of a CARE* (4.1) *is said to be a stabilizing solution if* $A - RX$ *has all its eigenvalues entirely in the open left-half complex plane, i.e., if* $A - RX$ *is Hurwitz-stable.*

For the H_2 CARE as defined in (4.6), the solution is said to be semi-stabilizing if the matrix

$$\bar{A} - B(D'D)^{-1}(B'X + D'C) \tag{4.24}$$

has its eigenvalues in the closed left-half complex plane. Similarly, it is called stabilizing if the matrix (4.24) has its eigenvalues in the open left-half plane.

For the H_∞^1 CARE as defined in (4.9), the solution is said to be semi-stabilizing if the matrix

$$\bar{A} - B(D'D)^{-1}(B'X + D'C) - EE'X \tag{4.25}$$

has its eigenvalues in the closed left-half complex plane. Similarly, it is called stabilizing if the matrix (4.25) has its eigenvalues in the open left-half plane.

Finally, for the H_∞^2 CARE as defined in (4.13), the solution is said to be semi-stabilizing if the matrix (4.24) has its eigenvalues in the closed left-half complex plane. Similarly, it is called stabilizing if the matrix (4.24) has its eigenvalues in the open left-half plane.

In view of the properties of the Hamiltonian H as explored in the previous subsection, the following proposition states the necessary and sufficient conditions for the existence of a stabilizing solution to a CARE.

Proposition 4.13 *Consider a CARE as in* (4.1). *Then it has a stabilizing solution* X *if and only if the following conditions hold:*

   (i) The associated Hamiltonian matrix (4.7) *has no eigenvalues on the imaginary axis* $\mathbb{C}^0$.

   (ii) The subspaces $X_-(H)$ *and* $\mathrm{im} \begin{pmatrix} 0 \\ I \end{pmatrix}$ *are complementary.*

Proof : It is a consequence of Propositions 4.9 and 4.10. ∎

The above proposition has two conditions for the existence of a stabilizing solution to a CARE. We would like to examine both conditions one at a time for the three specific CAREs, H_2 CARE, H_∞^1 CARE, and H_∞^2 CARE, which have some special structures. We focus first on condition (i).

Let us consider the H_2 CARE given in (4.6). We note that it can be represented also in the form of (4.1) by letting

$$A = \bar{A} - B(D'D)^{-1}D'C, \quad Q = C'C - C'D(D'D)^{-1}D'C,$$
$$\text{and } R = B(D'D)^{-1}B',$$

with $Q \geq 0$ and $R \geq 0$. We note also that the Hamiltonian H associated with an H_2 CARE in (4.6) is given by (4.18) with A, Q, and R as given above. Then, the following proposition characterizes when the Hamiltonian associated with an H_2 CARE has eigenvalues on $\mathbb{C}^0$.

Proposition 4.14 *Consider an H_2 CARE as in (4.6). Then a number λ on the imaginary axis $\mathbb{C}^0$ is an eigenvalue of the Hamiltonian H if and only if it is an uncontrollable eigenvalue of $(\bar{A}, B)$ or an invariant zero of the system $(\bar{A}, B, C, D)$.*

Proof : Let $\lambda = j\omega$ be an eigenvalue of H, and

$$0 \neq \begin{pmatrix} x \\ y \end{pmatrix} \in \mathbb{C}^{2n}$$

be the associated eigenvector. That is, let

$$\begin{pmatrix} A & -R \\ -Q & -A' \end{pmatrix} \begin{pmatrix} x \\ y \end{pmatrix} = j\omega \begin{pmatrix} x \\ y \end{pmatrix}. \tag{4.26}$$

Expansion of the above equation leads to

$$Ax - Ry = j\omega x, \tag{4.27}$$
$$-Qx - A'y = j\omega y. \tag{4.28}$$

Let v^* denote the complex conjugate transpose of v. Premultiplying (4.27) with y^* and (4.28) with x^*, we obtain

$$y^*Ax - y^*Ry = j\omega y^*x, \tag{4.29}$$
$$-x^*Qx - x^*A'y = j\omega x^*y. \tag{4.30}$$

The sum of (4.29) and (4.30) is given by

$$-x^*Qx - y^*Ry - [x^*A'y - (x^*A'y)^*] = j\omega[x^*y + (x^*y)^*]. \tag{4.31}$$

Observe that $x^*y + (x^*y)^*$ is a real number and that $x^*A'y - (x^*A'y)^*$ is a purely imaginary number. So, equation (4.31) implies that

$$-x^*Qx - y^*Ry = 0. \tag{4.32}$$

As $Q \geqslant 0$ and $R \geqslant 0$, this yields $R y = 0$ or, equivalently, $B' y = 0$ and

$$x^* Q x = (Cx + Dz)'(Cx + Dz) = 0,$$

where $z = -(D'D)^{-1} D'Cx$ and hence $Cx + Dz = 0$. Moreover, $Ax = \bar{A}x + Bz$. Using $Ry = 0$ and $Qx = 0$ in (4.27) and (4.28), we obtain

$$y^* A = j\omega y^* \quad \text{and} \quad y^* B = 0, \tag{4.33}$$

$$\bar{A}x + Bz = j\omega x \quad \text{and} \quad Cx + Dz = 0. \tag{4.34}$$

If $y \neq 0$, then (4.33) implies that $\lambda = j\omega$ is an uncontrollable eigenvalue of (A, B) or equivalently that of $(\bar{A}, B)$. If $x \neq 0$, then (4.34) implies that $\lambda = j\omega$ is invariant zero of $(\bar{A}, B, C, D)$. As we know that x and y cannot be both zero, either one of these cases must be satisfied.

The converse follows from the above procedure by reversing the arguments. This concludes our proof. ∎

The above proposition together with Proposition 4.8 establishes that the Hamiltonian matrix of an H_2 CARE (4.6) has a stable invariant subspace of dimension n if and only if $(\bar{A}, B)$ has no uncontrollable eigenvalues on the imaginary axis and the system $(\bar{A}, B, C, D)$ has no invariant zeros on the imaginary axis. In this case, we have

$$X_-(H) = \operatorname{im} T = \operatorname{im} \begin{pmatrix} T_1 \\ T_2 \end{pmatrix} \tag{4.35}$$

with $T_1 \in \mathbb{R}^{n \times n}$ and $T_2 \in \mathbb{R}^{n \times n}$, whereas T has full column rank.

Remark 4.15 *Proposition 4.14 is stated for an H_2 CARE as in (4.6). We can give a similar result for the H_2 CARE corresponding to (4.1), i.e., for (4.1) with $Q \geqslant 0$ and $R \geqslant 0$. Let $R = \bar{B}\bar{B}'$ and $Q = \bar{C}'\bar{C}$. Then a number λ on the imaginary axis $\mathbb{C}^0$ is an eigenvalue of the Hamiltonian H if and only if it is an uncontrollable eigenvalue of $(A, \bar{B})$ and/or an unobservable eigenvalue of $(\bar{C}, A)$.*

We consider now the H_∞^1 CARE of (4.9) and the H_∞^2 CARE of (4.13). We observe that the above geometric characterization whether the Hamiltonian has eigenvalues on the imaginary axis or not is not possible for an H_∞^1 CARE. As an example, consider

$$\bar{A} = 0, \quad B = 1, \quad E = \gamma, \quad C = \begin{pmatrix} 1 \\ 0 \end{pmatrix}, \quad D = \begin{pmatrix} 0 \\ 1 \end{pmatrix},$$

which yields the Hamiltonian

$$H = \begin{pmatrix} 0 & \gamma^2 - 1 \\ -1 & 0 \end{pmatrix}.$$

This H has eigenvalues on the imaginary axis for $\gamma \geq 1$ and no eigenvalues on the imaginary axis for $\gamma < 1$. A similar behavior can be observed for an H_∞^2 CARE of (4.13). That is, the geometric characterization whether the Hamiltonian has eigenvalues on the imaginary axis or not is not possible for an H_∞^2 CARE either. Again, as an example, consider

$$\bar{A} = 0, \quad B = 1, \quad C_1 = \begin{pmatrix} 1 \\ 0 \end{pmatrix}, \quad C_2 = \gamma, \quad D = \begin{pmatrix} 0 \\ 1 \end{pmatrix}$$

which yields a Hamiltonian with eigenvalues on the imaginary axis for $\gamma \geq 1$ and without eigenvalues on the imaginary axis for $\gamma < 1$.

We proceed now to examine the second condition of Proposition 4.13, namely, $X_-(H)$ should be complementary to the subspace

$$\mathrm{im} \begin{pmatrix} 0 \\ I \end{pmatrix}$$

or, equivalently, T_1 must be invertible [using the notation from (4.35)]. This would then imply that $X = T_2 T_1^{-1}$ is a stabilizing solution of the CARE.

We have the following result regarding H_2 CARE and H_∞^2 CARE.

Proposition 4.16 *Consider an H_2 CARE as in (4.2) or an H_∞^2 CARE as in (4.10). Assume that $(\bar{A}, B)$ is $\mathbb{C}^-$-stabilizable and the associated Hamiltonian H has no eigenvalues on the imaginary axis $\mathbb{C}^0$. Then*

$$X_-(H) \qquad and \qquad \mathrm{im} \begin{pmatrix} 0 \\ I \end{pmatrix} \; are\ complementary. \qquad (4.36)$$

The above proposition considers the H_2 CARE and H_∞^2 CARE. For the H_∞^1 CARE, a simple characterization of the complementarity condition as presented above cannot be obtained. To see this, consider

$$\bar{A} = 1, \quad B = 1, \quad E = 1, \quad C = \begin{pmatrix} 1 \\ 0 \end{pmatrix}, \quad D = \begin{pmatrix} 0 \\ 1 \end{pmatrix}.$$

The corresponding Hamiltonian is

$$H = \begin{pmatrix} 1 & 0 \\ -1 & -1 \end{pmatrix},$$

which has no eigenvalues on the imaginary axis. But we have

$$X_-(H) = \mathrm{im} \begin{pmatrix} 0 \\ 1 \end{pmatrix},$$

and hence, the complementarity condition is not satisfied.

Before we prove Proposition 4.16, we need certain notations and a preliminary lemma. Assume that H has no eigenvalues on the imaginary axis $\mathbb{C}^0$. Note that T as defined in (4.35) has the property that im T is H-invariant and H restricted to im T is asymptotically stable and, hence, an asymptotically stable matrix H_- exists such that

$$HT = TH_-. \tag{4.37}$$

We will also need the following lemma:

Lemma 4.17 *Consider a general CARE as in (4.1), and assume that the associated Hamiltonian H has no eigenvalues on the imaginary axis $\mathbb{C}^0$. Consider the matrices T_1 and T_2 as defined in (4.35). Then we have*

(i) $T_1'T_2$ *is symmetric:*

$$T_1'T_2 = T_2'T_1. \tag{4.38}$$

(ii) *Consider an H_2 CARE as in (4.2) or an H_∞^2 CARE as in (4.10). In that case,* ker T_1 *is invariant under H_-. That is, for all $x \in$ ker T_1, we have* $H_- x \in$ ker T_1.

Proof : Let

$$J := \begin{pmatrix} 0 & -I \\ I & 0 \end{pmatrix}.$$

By premultiplying (4.37) with $T'J$, we obtain

$$T'JHT = T'JTH_-. \tag{4.39}$$

Now as JH is symmetric, it follows that the right-hand side of (4.39) is symmetric as well. That is,

$$(T'JT)H_- = H_-'(T'J'T) = -H_-'(T'JT). \tag{4.40}$$

The above equation implies that

$$(T'JT)H_- + H_-'(T'JT) = 0. \tag{4.41}$$

We note that (4.41) is a Lyapunov equation. As H_- is Hurwitz stable, it follows from Lemma 4.162 that the unique solution of (4.41) is

$$T'JT = 0. \tag{4.42}$$

This equation is equivalent to (4.38), and this proves the first statement.

The proof of the second statement follows next. Combining (4.18) and (4.37), we obtain that

$$AT_1 - RT_2 = T_1 H_-. \tag{4.43}$$

Consider now an $x \in \ker T_1$. Premultiplying and postmultiplying (4.43), respectively, with $x'T_2'$ and x, we get

$$-x'T_2'RT_2x = x'T_2'T_1 H_-x. \tag{4.44}$$

Also from the first statement, we get

$$x'T_2'T_1 H_-x = x'T_1'T_2 H_-x = 0. \tag{4.45}$$

Combining (4.44) and (4.45), we obtain $x'T_2'RT_2x = 0$. As we are working with either the H_2 or the H_∞^2 CARE, we have $R \geqslant 0$ and, hence, we find that $RT_2x = 0$. Using this property after postmultiplying (4.43) with x, we get

$$AT_1x = T_1 H_-x.$$

As $x \in \ker T_1$, the above implies that $T_1 H_-x = 0$, and hence, $H_-x \in \ker T_1$. $\blacksquare$

We are now ready to prove Proposition 4.16.

Proof of Proposition 4.16 : We exploit again the structure established in (4.35). It is obvious that

$$X_-(H) \text{ and } \operatorname{im} \begin{pmatrix} 0 \\ I \end{pmatrix}$$

are complementary if and only if T_1 is invertible. We will prove by contradiction that T_1 is invertible. Assume that $\ker T_1$ is nonempty; i.e., T_1 is singular. As $\ker T_1$ is H_- invariant by Lemma 4.17, we find that H_- restricted to $\ker T_1$ has an eigenvalue λ and a corresponding eigenvector x; i.e., $H_-x = \lambda x$. As H_- is asymptotically stable, we have $\operatorname{Re} \lambda < 0$. Premultiplying (4.37) with $(0 \quad I)$, we get

$$\begin{pmatrix} 0 & I \end{pmatrix} \begin{pmatrix} A & -R \\ -Q & -A' \end{pmatrix} \begin{pmatrix} T_1 \\ T_2 \end{pmatrix} = \begin{pmatrix} 0 & I \end{pmatrix} \begin{pmatrix} T_1 \\ T_2 \end{pmatrix} H_-.$$

This result implies that

$$-QT_1 - A'T_2 = T_2 H_-. \tag{4.46}$$

Postmultiplying the above equation with x, we get

$$(A' + \lambda I)T_2x = 0. \tag{4.47}$$

Now as $x \in \ker T_1$, from the proof of Lemma 4.17, it is clear that we have $RT_2x = 0$, and as $R = B\bar{R}^{-1}B'$, we obtain that

$$B'T_2x = 0. \tag{4.48}$$

Keeping (4.47) and (4.48) together, we have

$$\begin{pmatrix} A' + \lambda I \\ B' \end{pmatrix} T_2 x = 0. \tag{4.49}$$

But the $\mathbb{C}^-$-stabilizability of (A, B) implies that

$$\mathrm{rank} \begin{pmatrix} A' + \lambda I \\ B' \end{pmatrix} = n \tag{4.50}$$

for any λ with $\mathrm{Re}(\lambda) \leq 0$. Hence, we have

$$T_2 x = 0. \tag{4.51}$$

We note that equations $T_1 x = 0$ and $T_2 x = 0$ together imply that $T x = 0$. Because T has full column rank, we obtain $x = 0$. But this is a contradiction. Hence, the result follows. ∎

Finally, the next proposition, among others, shows that a stabilizing solution of a CARE is symmetric and unique.

Proposition 4.18 *Consider a general CARE as in (4.1). Then we have the following:*

(i) *There is a one-to-one correspondence between the solution set of the CARE and the set of n-dimensional H-invariant subspaces that are complementary to the n-dimensional subspace*

$$\mathrm{im} \begin{pmatrix} 0 \\ I \end{pmatrix}. \tag{4.52}$$

In particular, for any H-invariant subspace $\mathcal{S}$ that is complementary to (4.52) a unique solution X of the CARE (4.1) exists such that

$$\mathcal{S} := \mathrm{im} \begin{pmatrix} I \\ X \end{pmatrix}. \tag{4.53}$$

Conversely, any solution X of the CARE (4.1) yields a subspace $\mathcal{S}$ according to (4.53) that is H-invariant and complementary to (4.52).

The restriction of H to $\mathcal{S}$ is given by $A - RX$; that is, the following relationship holds:

$$H \begin{pmatrix} I \\ X \end{pmatrix} = \begin{pmatrix} I \\ X \end{pmatrix} (A - RX). \tag{4.54}$$

(ii) *A stabilizing solution of the CARE (4.1) exists if and only if $X_-(H)$ is an n-dimensional subspace that is complementary to (4.52). In this case, a stabilizing solution X is symmetric and unique. Moreover,*

$$X = T_2 T_1^{-1}$$

with T_1 and T_2 as defined in (4.35).

Proof : We prove the first statement first.

(Necessity) If $\mathcal{S}$ is an n-dimensional subspace complementary to (4.52), then an $n \times n$ matrix X exists such that

$$\mathcal{S} = \operatorname{im}\begin{pmatrix} I \\ X \end{pmatrix}. \tag{4.55}$$

If, in addition, $\mathcal{S}$ is H-invariant, then an $n \times n$ matrix L exists such that

$$H\begin{pmatrix} I \\ X \end{pmatrix} = \begin{pmatrix} I \\ X \end{pmatrix} L$$

or

$$\begin{pmatrix} A & -R \\ -Q & -A' \end{pmatrix}\begin{pmatrix} I \\ X \end{pmatrix} = \begin{pmatrix} I \\ X \end{pmatrix} L.$$

This result implies that

$$A - RX = L, \tag{4.56}$$
$$-Q - A'X = XL. \tag{4.57}$$

It is simple to see from the above equations that X is a solution of the CARE (4.1).

(Sufficiency) Conversely, if X satisfies the CARE (4.1), then

$$\begin{pmatrix} A & -R \\ -Q & -A' \end{pmatrix}\begin{pmatrix} I \\ X \end{pmatrix} = \begin{pmatrix} I \\ X \end{pmatrix}(A - RX).$$

This equation shows that (4.55) is an H-invariant subspace and $A - RX$ is the matrix of $H|\mathcal{S}$ with respect to the basis given by the columns of

$$\begin{pmatrix} I \\ X \end{pmatrix}.$$

We now proceed to prove the second part.

The correspondence between $X_-(H)$ and the stabilizing solution of the CARE (4.1) follows from the first part of Proposition 4.18. To show that $X = T_2 T_1^{-1}$, we observe that

$$X_-(H) = \operatorname{im} T = \operatorname{im} \begin{pmatrix} T_1 \\ T_2 \end{pmatrix}.$$

It follows that T_1 is nonsingular because $X_-(H)$ is complementary to (4.52). Hence,

$$X_-(H) = \operatorname{im} \begin{pmatrix} T_1 \\ T_2 \end{pmatrix} = \operatorname{im} \begin{pmatrix} I \\ X \end{pmatrix} T_1,$$

where $X := T_2 T_1^{-1}$. This also implies that

$$X_-(H) = \operatorname{im} \begin{pmatrix} I \\ X \end{pmatrix}.$$

For uniqueness, we observe that

$$\operatorname{im} \begin{pmatrix} I \\ X_1 \end{pmatrix} = \operatorname{im} \begin{pmatrix} I \\ X_2 \end{pmatrix}$$

if and only if $X_1 = X_2$. To prove that X is symmetric, let us start with

$$X T_1 = T_2. \tag{4.58}$$

Premultiplying the above equation with T_1', we have

$$T_1' X T_1 = T_1' T_2. \tag{4.59}$$

Now take the transpose of (4.58) and postmultiply it with T_1, to obtain

$$T_1' X' T_1 = T_2' T_1. \tag{4.60}$$

Using Lemma 4.17, we obtain

$$T_1' T_2 = T_2' T_1.$$

This yields that (4.59) and (4.60) together imply that X is real and symmetric. ∎

Proposition 4.18 shows that the CARE (4.1) can have many solutions, but there is at most one that can be connected to $X_-(H)$ and that is hence stabilizing. The existence of this particular solution is guaranteed if and only if $X_-(H)$ is n-dimensional and complementary to (4.52).

The following proposition is a consequence of Propositions 4.8 and 4.18.

Proposition 4.19 *Consider a general CARE as given in (4.1). Then, as in Proposition 4.13, it has a stabilizing solution if and only if* (*i*) *The associated Hamiltonian matrix H has no eigenvalues on the imaginary axis $\mathbb{C}^0$.*

(*ii*) $X_-(H)$ *and* $\mathrm{im} \begin{pmatrix} 0 \\ I \end{pmatrix}$ *are complementary.*

Moreover, such a stabilizing solution is symmetric and unique. It is given by

$$X = T_2 T_1^{-1},$$

where

$$X_-(H) = \mathrm{im} \begin{pmatrix} T_1 \\ T_2 \end{pmatrix},$$

with $T_1 \in \mathbb{R}^{n \times n}$ and $T_2 \in \mathbb{R}^{n \times n}$.

Proof : It is obvious in view of Propositions 4.8 and 4.18. ∎

The following are some important corollaries. First we note that for the H_∞^2 CARE, the complementarity condition is automatically satisfied when $(\bar{A}, B)$ is $\mathbb{C}^-$-stabilizable.

Corollary 4.20 *Consider an H_2 CARE as in (4.2). Then it has a stabilizing solution if and only if*

(*i*) *the system $(\bar{A}, B, C, D)$ has no invariant zeros on the imaginary axis $\mathbb{C}^0$, and*

(*ii*) *$(\bar{A}, B)$ is $\mathbb{C}^-$-stabilizable.*

Moreover, such a stabilizing solution is symmetric and unique.

Proof : In view of Proposition 4.14, conditions (*i*) and (*ii*) guarantee a Hamiltonian without eigenvalues on the imaginary axis. Moreover, $X_-(H)$ and $\mathrm{im} \begin{pmatrix} 0 \\ I \end{pmatrix}$ are complementary according to (4.16). Then Proposition 4.19 guarantees the existence, uniqueness, and symmetry of the stabilizing solution. The converse implication follows along the same lines. ∎

Corollary 4.21 *Consider an H_∞^2 CARE as in (4.10). Then it has a stabilizing solution if and only if* (*i*) *The associated Hamiltonian matrix H has no eigenvalues on the imaginary axis $\mathbb{C}^0$.*

(*ii*) *$(\bar{A}, B)$ is $\mathbb{C}^-$-stabilizable.*

Moreover, such a stabilizing solution is symmetric and unique.

Proof : It follows from Propositions 4.19 and 4.16.   ∎

Remark 4.22 *We should note that for the H_∞^1 CARE, the complementarity condition is not automatically satisfied even when $(\bar{A}, B)$ is $\mathbb{C}^-$-stabilizable.*

Our discussion so far pertains to stabilizing solutions of a CARE. We turn our attention now to semi-stabilizing solutions of a CARE.

We discuss below an important property of a real symmetric semi-stabilizing solution of an H_∞^2 CARE. The characterization of the stabilizing solution via the Hamiltonian makes it obvious that in all cases, the stabilizing solution is unique. But we learn that under a mild condition of $(\bar{A}, B)$ being $\mathbb{C}^-$-stabilizable, a real symmetric semi-stabilizing solution of an H_∞^2 CARE, if it exists, is the largest of all possible real symmetric solutions of the H_∞^2 CARE. As the H_2 CARE is a special case of the H_∞^2 CARE, the same properties are true for it as well.

Proposition 4.23 *Consider an H_∞^2 CARE as in (4.10) or an H_2 CARE as in (4.2), and assume that $(\bar{A}, B)$ is $\mathbb{C}^-$-stabilizable. Then the following hold:*

(*i*) *A real symmetric semi-stabilizing solution, say X_{ss}, of the H_∞^2 CARE or H_2 CARE, if it exists, is larger than or equal to any other real symmetric solution X; i.e., $X_{ss} \geq X$.*

(*ii*) *A real symmetric semi-stabilizing solution of the H_∞^2 CARE or H_2 CARE, if it exists, is unique.*

The results corresponding to the above proposition are not true for the H_∞^1 CARE (4.9). To show that Part (*i*) of the above result does not hold for the H_∞^1 CARE, consider the example,

$$\bar{A} = -3, \quad B = -\sqrt{2}, \quad E = 2, \quad C = \begin{pmatrix} 0 \\ 2 \end{pmatrix}, \quad D = \begin{pmatrix} 1 \\ 0 \end{pmatrix}.$$

The corresponding CARE has two solutions $X = 2$ and $X = 1$. It is not the largest solution $X = 2$ that is stabilizing but the smallest $X = 1$ that is stabilizing.

To show that Part (*ii*) of the above result does not hold either for the H_∞^1 CARE (4.9), consider the example,

$$\bar{A} = \begin{pmatrix} 0 & 0 \\ 0 & 1 \end{pmatrix}, \quad B = \begin{pmatrix} 1 & 0 \\ 0 & 1 \end{pmatrix}, \quad E = \begin{pmatrix} 0 \\ \sqrt{2} \end{pmatrix}, \quad C = \begin{pmatrix} 0 & 0 \\ 0 & 0 \\ 0 & 1 \end{pmatrix}, \quad D = \begin{pmatrix} 1 & 0 \\ 0 & 1 \\ 0 & 0 \end{pmatrix}.$$

In this case, we have an infinite number of semi-stabilizing solutions of the given CARE. Let

$$X_1 = \begin{pmatrix} \lambda & \lambda \\ \lambda & \lambda - 1 \end{pmatrix}, \qquad X_2 = \begin{pmatrix} \lambda & -\lambda \\ -\lambda & \lambda - 1 \end{pmatrix}.$$

For any λ, both X_1 and X_2 yield a semi-stabilizing solution. Note that there is, however, only one solution whose kernel has the same dimension as the number of invariant zeros in the closed left-half plane.

Before we give the proof of Proposition 4.23, we consider the following lemma.

Lemma 4.24 *Consider three matrices F, R, and Q in $\mathbb{R}^{n \times n}$, where R and Q are symmetric positive semi-definite matrices whereas* $\mathrm{Re}\,\lambda(F) \leq 0$, *and the pair (F, R) is $\mathbb{C}^-$-stabilizable. Then a symmetric matrix $P \in \mathbb{R}^{n \times n}$ that satisfies*

$$F'P + PF = -PRP - Q \tag{4.61}$$

is positive semi-definite.

Proof : Suppose $z \in \mathbb{R}^n$ is such that $Pz = 0$. Then (4.61) and the fact $Q \geq 0$ imply that $Qz = 0$ and therefore $PFz = 0$. This implies that $\ker P \subseteq \ker Q$, and $\ker P$ is invariant under F. Therefore, an orthogonal matrix $S \in \mathbb{R}^{n \times n}$ exists such that

$$S'PS = \begin{pmatrix} 0 & 0 \\ 0 & P_2 \end{pmatrix},$$

with P_2 invertible and

$$S'QS = \begin{pmatrix} 0 & 0 \\ 0 & Q_2 \end{pmatrix}, \quad S'FS = \begin{pmatrix} F_1 & F_{12} \\ 0 & F_2 \end{pmatrix}, \quad \text{and } S'RS = \begin{pmatrix} R_1 & R_{12} \\ R'_{12} & R_2 \end{pmatrix}.$$

Pre- and postmultiplying (4.61) with S' and S, it reduces to

$$F'_2 P_2 + P_2 F_2 = -P_2 R_2 P_2 - Q_2$$

or, equivalently,

$$P_2^{-1} F'_2 + F_2 P_2^{-1} = -R_2 - P_2^{-1} Q_2 P_2^{-1}. \tag{4.62}$$

Note that we also have $R_2 \geq 0$, $Q_2 \geq 0$, $\operatorname{Re}\lambda(F_2) \leq 0$, and (F_2, R_2) $\mathbb{C}^-$-stabilizable. We show next that F_2 is actually Hurwitz-stable. This is shown by contradiction. Assume that F_2 has an eigenvalue on the imaginary axis, say $j\alpha$, with $\alpha \in \mathbb{R}$. Then

$$u^* F_2 = j\alpha u^*$$

for some $0 \neq u \in \mathbb{C}^n$. Then (4.62) yields

$$u^*(-R_2 - P_2^{-1} Q_2 P_2^{-1})u = 0. \tag{4.63}$$

Because R_2 and Q_2 are positive semi-definite, (4.63) implies that $u^* R_2 = 0$, and thus,

$$u^*\left(F_2 - j\alpha I \quad R_2 \right) = 0.$$

But this is a contradiction to the fact that (F_2, R_2) is $\mathbb{C}^-$-stabilizable. This concludes that F_2 is actually Hurwitz-stable. Now going back to (4.62), we can conclude from Lemma 4.162 that P_2^{-1} is positive definite. This then implies that P is positive semi-definite.   ∎

Proof of Proposition 4.23 :   Let us first consider item (i). Let X be any real symmetric solution of the given H_∞^2 CARE. This can be reduced to the general CARE by setting $A = \bar{A} - B\bar{R}^{-1}S'$, $Q = \bar{Q} - S\bar{R}^{-1}S'$, and $R = B\bar{R}^{-1}B'$. Define $P = X_{ss} - X$. Then we have

$$(A - RX_{ss})'P + P(A - RX_{ss}) = -PRP.$$

Observe that $\operatorname{Re}\lambda(A - RX_{ss}) \leq 0$ as X_{ss} is a semi-stabilizing solution. Moreover, the pair $(A - RX_{ss}, B)$ is $\mathbb{C}^-$-stabilizable as (A, B) is $\mathbb{C}^-$-stabilizable and

$$A - RX_{ss} = A - B\bar{R}^{-1}(B'X_{ss} + S').$$

Then Lemma 4.24 implies that $P \geq 0$, and hence, $X_{ss} \geq X$.

Next, the proof of item (ii) of Proposition 4.23 is obvious. After all, assume that there are two real symmetric semi-stabilizing solutions, X_{ss}^1 and X_{ss}^2, to the given H_∞^2 CARE. Then using item (i), one concludes that $X_{ss}^1 \geq X_{ss}^2$ and $X_{ss}^2 \geq X_{ss}^1$, which implies that $X_{ss}^1 = X_{ss}^2$.   ∎

Under the mild assumption of $(\bar{A}, B)$ being $\mathbb{C}^-$-stabilizable, we develop next the necessary and sufficient conditions for the existence of a semi-stabilizing solution to an H_2 CARE and to an H_∞^2 CARE. For an H_∞^1 CARE, the existence conditions for a semi-stabilizing solution are not known. However, we can define a new notion of a strongly semi-stabilizing solution of an H_∞^1 CARE and then give necessary and sufficient conditions for its existence.

We have the following proposition regarding an H_2 CARE.

Proposition 4.25 *Consider an H_2 CARE as in (4.6). Assume that the pair $(\bar{A}, B)$ is $\mathbb{C}^-$-stabilizable. Then it has a semi-stabilizing solution. Moreover, such a solution is unique and is the largest among all symmetric solutions. Furthermore, such a solution is a stabilizing solution if the system characterized by the quadruple $(\bar{A}, B, C, D)$ has no invariant zeros on the imaginary axis $\mathbb{C}^0$.*

Proof : We can reduce an H_2 CARE of the form (4.6) to the form (4.1) with $Q \geq 0$ and $R \geq 0$. As noted in Lemma 4.6, we have that (A, R) is $\mathbb{C}^-$-stabilizable. In a suitable basis, we have the following decomposition:

$$A = \begin{pmatrix} A_1 & A_2 \\ 0 & A_4 \end{pmatrix}, \quad R = \begin{pmatrix} R_{11} & R'_{21} \\ R_{21} & R_{22} \end{pmatrix}, \quad \text{and} \quad Q = \begin{pmatrix} Q_{11} & 0 \\ 0 & 0 \end{pmatrix},$$

where the pair (A_1, R_{11}) is $\mathbb{C}^-$-stabilizable while (Q_{11}, A_1) has no unobservable eigenvalues on the imaginary axis and the eigenvalues of A_4 are all on the imaginary axis. According to Remark 4.15, a stabilizing solution X_1 of the H_2 CARE

$$A'_1 X_1 + X_1 A_1 - X_1 R_{11} X_1 + Q_{11} = 0$$

exists. We therefore see that $A_1 - R_{11} X_1$ is Hurwitz-stable. Then note that

$$X = \begin{pmatrix} X_1 & 0 \\ 0 & 0 \end{pmatrix}$$

is a semi-stabilizing solution of the original CARE.

The fact that the semi-stabilizing solution is unique and the largest symmetric solution has been established in Proposition 4.23.   ■

Remark 4.26 *We will show in Subsection 4.1.4 that the semi-stabilizing solution of an H_2 CARE discussed in Proposition 4.25 is indeed positive semi-definite.*

We consider next an H_∞^2 CARE and characterize the existence of its semi-stabilizing solution.

Proposition 4.27 *Consider an H_∞^2 CARE as in (4.13). Assume that the pair $(\bar{A}, B)$ is $\mathbb{C}^-$-stabilizable. A symmetric semi-stabilizing solution of the H_∞^2 CARE exists if and only if a matrix $\bar{X}$ exists satisfying the linear matrix inequality,*

$$\begin{pmatrix} \bar{A}'\bar{X} + \bar{X}\bar{A} + C'_1 C_1 - C'_2 C_2 & \bar{X}B + C'_1 D \\ B'\bar{X}B + D'C_1 & D'D \end{pmatrix} \geq 0. \tag{4.64}$$

Moreover, such a semi-stabilizing solution is unique and is the largest among all symmetric solutions of the H_∞^2 CARE. Furthermore, this solution is a stabilizing solution if and only if the system characterized by the quadruple $(\bar{A}, B, \bar{C}, \bar{D})$ has no invariant zeros on the imaginary axis $\mathbb{C}^0$, where $\bar{C}$ and $\bar{D}$ are such that

$$\begin{pmatrix} \bar{C}' \\ \bar{D}' \end{pmatrix}\begin{pmatrix} \bar{C} & \bar{D} \end{pmatrix} = \begin{pmatrix} \bar{A}'\bar{X} + \bar{X}\bar{A} + C_1'C_1 - C_2'C_2 & \bar{X}B + C_1'D \\ B'\bar{X} + D'C_1 & D'D \end{pmatrix}.$$

Proof : We first note that any solution $\bar{X}$ of the CARE must also satisfy (4.64). The necessity of the existence of solution of (4.64) is then trivially established. Next, given a solution $\bar{X}$ of (4.64), we note that X is a solution of the CARE (4.13) if and only if $\hat{X} = X - \bar{X}$ satisfies

$$\bar{A}'\hat{X} + \hat{X}\bar{A} - (\hat{X}B + \bar{C}'\bar{D})(\bar{D}'\bar{D})^{-1}(B'\hat{X} + \bar{D}'\bar{C}) + \bar{C}'\bar{C} = 0 \qquad (4.65)$$

and

$$\bar{A} - B'(D'D)^{-1}(B'X + D'C_1) = \bar{A} - B'(\bar{D}'\bar{D})^{-1}(B'\hat{X} + \bar{D}'\bar{C}).$$

Hence, the existence of a semi-stabilizing solution of the original CARE (4.13) is guaranteed if and only if a semi-stabilizing solution of the H_2 CARE (4.65) exists. We can directly apply Proposition 4.25 to guarantee that a semi-stabilizing solution to (4.13) exists. The fact that that the semi-stabilizing solution is the largest symmetric solution of the CARE has already been established in Proposition 4.23.
∎

We need to study next the H_∞^1 CARE. As we noted by an example, an infinite number of semi-stabilizing solutions can exist. As a matter of fact, one can establish that the existence of more than one semi-stabilizing solution immediately guarantees that an infinite number of semi-stabilizing solutions exist. Also the specific ordering mentioned for the H_2 and H_∞^2 CAREs in the above two theorems are not valid for the H_∞^1 CARE. There is no known straightforward characterization of the existence of a semi-stabilizing solution of the H_∞^1 CARE. However, we can discuss here a particular type of solution called the strongly semi-stabilizing solution whose definition is as follows:

Definition 4.28 *Consider an H_∞^1 CARE of the form (4.9). Then a semi-stabilizing solution X of this CARE is called a **strongly semi-stabilizing solution** of it if it satisfies the following additional properties:*

- *The dimension of the kernel of X is equal to the number of invariant zeros of $(\bar{A}, B, C, D)$ in the closed left-half plane.*

- *The number of invariant zeros of $(\bar{A}, B, C, D)$ on the imaginary axis is equal to the number of eigenvalues of $A - RX$ on the imaginary axis.*

The H_2 CARE is a special case of the H_∞^1 CARE for which the semi-stabilizing solution is always unique and strongly semi-stabilizing. It is not so for a general CARE. That is, it might be that a semi-stabilizing solution exists but not a strongly semi-stabilizing solution. To see this, let

$$A = -1, B = 0, E = 1, C = \begin{pmatrix} 1 \\ 0 \end{pmatrix}, D = \begin{pmatrix} 0 \\ 1 \end{pmatrix}.$$

For these matrices, $X = 1$ is a semi-stabilizing solution that does not satisfy the second additional property of a strongly semi-stabilizing solution. On the other hand, for the H_∞^1 CARE associated with

$$A = 0, B = 1, E = 1, C = 0, D = 1,$$

any X is a semi-stabilizing solution of it but only $X = 0$ is a strongly semi-stabilizing solution.

Our next task is to characterize the existence of a strongly semi-stabilizing solution of an H_∞^1 CARE and to compute it. In fact, we obtain below a strongly semi-stabilizing solution of a given H_∞^1 CARE in terms of the stabilizing solution of another suitably defined H_∞^1 CARE. To do so, we use the fact (Corollary 4.36 that is to be developed in Subsection 4.1.4 combined with the first additional property of a strongly semi-stabilizing solution) that

$$\ker X = \mathcal{V}^{-0}(\overline{A}, B, C, D).$$

Let $\mathcal{W}$ be a subspace such that we obtain a decomposition of the state space $\mathbb{R}^n = \mathcal{V}^{-0} \oplus \mathcal{W}$. Then for the matrix $F = -(D'D)^{-1}D'C$, we have

$$\overline{A} + BF = \begin{pmatrix} A_{11} & A_{12} \\ 0 & A_{22} \end{pmatrix}, \quad B = \begin{pmatrix} B_1 \\ B_2 \end{pmatrix}, \quad E = \begin{pmatrix} E_1 \\ E_2 \end{pmatrix}, \quad C + DF = \begin{pmatrix} 0 & C_2 \end{pmatrix}.$$

But with respect to this basis, we have

$$X = \begin{pmatrix} 0 & 0 \\ 0 & X_1 \end{pmatrix}, \tag{4.66}$$

where X_1 satisfies the following H_∞^1 CARE:

$$A_{22}'X_1 + X_1 A_{22} + X_1 E_2 E_2' X_1 - (X_1 B_2 + C_2'D)(D'D)^{-1}(B_2'X_1 + D'C_2)$$
$$+ C_2'C_2 = 0. \tag{4.67}$$

It is clear from the second additional property of a strongly semi-stabilizing solution that X_1 must be a stabilizing solution of the H_∞^1 CARE (4.67) that can be obtained using the tools of Subsection 4.1.6. Then (4.66) yields the unique strongly semi-stabilizing solution of the original H_∞^1 CARE (4.9) with respect to the specific basis chosen in the state space.

We summarize the results of the above development by the following proposition.

Proposition 4.29 *Consider an H_∞^1 CARE of the form (4.9). Also, consider the H_∞^1 CARE (4.67) derived from the data of H_∞^1 CARE (4.9). Then, a strongly semi-stabilizing solution of the H_∞^1 CARE (4.9) exists if and only if a stabilizing solution of the H_∞^1 CARE (4.67) exists.*

Finally, the following proposition shows a property that we will exploit later in Chapter 11.

Proposition 4.30 *Consider an H_∞^1 CARE of the form (4.9). If a matrix $Y \geqslant 0$ exists such that*

$$G(Y) = \bar{A}'Y + Y\bar{A} - (YB + C'D)(D'D)^{-1}(B'Y + D'C)$$
$$+ YEE'Y + C'C \leqslant 0 \quad (4.68)$$

with $\ker G(Y) = \ker Y$ *and*

$$\bar{A} - B(D'D)^{-1}(B'Y + D'C) + EE'Y$$

has all its eigenvalues in the closed left-half plane, then a positive semi-definite, strongly semi-stabilizing solution of the H_∞^1 CARE (4.9) exists.

Proof : Without loss of generality, we can rewrite (4.68) for Y as

$$G(Y) = A'Y + YA - YRY + Q \leqslant 0,$$

where $A - RY$ has all its eigenvalues in the closed left-half plane. Moreover, $Q \geqslant 0$. It is easily verified that $\ker Y$ is an A-invariant subspace contained in $\ker Q$, and using this, we can obtain the following decomposition in a suitable basis:

$$A = \begin{pmatrix} \bar{A}_{11} & \bar{A}_{12} \\ 0 & \bar{A}_{22} \end{pmatrix}, \qquad Q = \begin{pmatrix} 0 & 0 \\ 0 & \bar{Q} \end{pmatrix}$$

$$R = \begin{pmatrix} \bar{R}_{11} & \bar{R}_{12} \\ \bar{R}_{21} & \bar{R}_{22} \end{pmatrix}, \qquad Y = \begin{pmatrix} 0 & 0 \\ 0 & \bar{Y} \end{pmatrix}$$

with $\bar{Y}$ invertible. We find

$$\bar{A}_{22}'\bar{Y} + \bar{Y}A_{22} - \bar{Y}R_{22}\bar{Y} + \bar{Q} < 0.$$

If we can find $\bar{X}$ such that

$$\bar{A}_{22}'\bar{X} + \bar{X}A_{22} - \bar{X}R_{22}\bar{X} + \bar{Q} = 0 \tag{4.69}$$

while $\bar{A}_{22} - R_{22}\bar{X}$ has all its eigenvalues in the closed left-half plane, then it is easily verified that

$$X = \begin{pmatrix} 0 & 0 \\ 0 & \bar{X} \end{pmatrix}$$

is a strongly semi-stabilizing solution of the H_∞^1 CARE (4.9).

We decompose the state space $\mathcal{X} = \mathcal{X}_1 \oplus \mathcal{X}_2$ such that $\mathcal{X}_1$ equals the $\mathbb{C}^{-0}$-unobservable subspace of $(\bar{C}, \bar{A}_{22})$, where $\bar{C}$ is such that $\bar{C}'\bar{C} = \bar{Q}$. In a basis associated with this decomposition, we then find

$$\bar{A}_{22} = \begin{pmatrix} A_{11} & A_{12} \\ 0 & A_{22} \end{pmatrix}, \quad \bar{R}_{22} = \begin{pmatrix} R_{11} & R_{12} \\ R_{21} & R_{22} \end{pmatrix}, \quad \bar{Q} = \begin{pmatrix} 0 & 0 \\ 0 & Q_{22} \end{pmatrix}.$$

But then if we decompose Y compatibly, we find

$$\bar{Y} = \begin{pmatrix} Y_{11} & Y_{12} \\ Y_{21} & Y_{22} \end{pmatrix}$$

and we can define

$$Z = \begin{pmatrix} Z_{11} & Z_{12} \\ Z_{21} & Z_{22} \end{pmatrix} = \bar{Y}^{-1}.$$

But then we find

$$Z_{22}A'_{22} + A_{22}Z_{22} - R_{22} + Z_{22}Q_{22}Z_{22} < 0$$

or, equivalently,

$$Z_{22}(-A_{22})' + (-A_{22})Z_{22} + R_{22} - Z_{22}Q_{22}Z_{22} > 0$$

while $(C_2, -A_{22})$ is $\mathbb{C}^-$-detectable where C_2 is any matrix such that $Q_{22} = C_2'C_2$. The latter follows because the subspace $\mathcal{X}_2$ is associated with the complement of the $\mathbb{C}^{-0}$-unobservable subspace of $(\bar{C}, \bar{A}_{22})$, which is equal to the undetectable subspace of $(\bar{C}, -\bar{A}_{22})$. By applying Proposition 4.27, we then find that a matrix V exists such that

$$V(-A_{22})' + (-A_{22})V + R_{22} - VQ_{22}V = 0$$

and such that $-A_{22} - VQ_{22}$ has all its eigenvalues in the open left-half plane. Moreover, Z_{22} is invertible and $V \geqslant Z_{22}$ (which follows directly from the proof of Proposition 4.27), and hence, V is invertible. But then

$$A'_{22}W + WA_{22} - WR_{22}W + Q_{22} = 0$$

for $W = V^{-1}$ and

$$A_{22} - R_{22}W = V(-A_{22} - Q_{22}V)V^{-1}$$

clearly has all its eigenvalues in the open left-half plane. Then it easily verified that

$$X = \begin{pmatrix} 0 & 0 \\ 0 & W \end{pmatrix}$$

is a strongly semi-stabilizing solution of the H_∞^1 CARE (4.69).   ∎

4.1.4 Positive semi-definite and positive definite solutions

In the previous subsection, we studied two important properties, namely, semi-stabilizing and stabilizing properties of a solution of a CARE. We introduce in this subsection two new additional properties, namely, positive semi-definiteness and positive definiteness of a solution of a CARE. It turns out that these two new properties are closely related to the semi-stabilizing and stabilizing properties introduced earlier. Our goal in this subsection is to explore such a relationship. This allows us to establish the necessary and sufficient conditions for the existence of a positive definite solution of an H_2 CARE as well as for a positive semi-definite solution of an H_2 CARE. In the process of doing so, we also examine the kernel of a positive semi-definite solution of an H_2 CARE and an H_∞^1 CARE.

The following property says that a real symmetric semi-stabilizing solution of an H_2 CARE, if it exists, is also positive semi-definite.

Proposition 4.31 *Consider an H_2 CARE as in (4.2), and assume that $(\overline{A}, B)$ is $\mathbb{C}^-$-stabilizable. Then the real symmetric semi-stabilizing solution of the H_2 CARE, if it exists, is positive semi-definite.*

Proof : Suppose the unique real symmetric semi-stabilizing solution X_{ss} of the H_2 CARE (4.2) exists. One can rewrite (4.2) as (4.5), and we obtain

$$A'_{cl} X_{ss} + X_{ss} A_{cl} = -X_{ss} B \overline{R}^{-1} B' X_{ss} - Q$$

with $A_{cl} = A - RX = A - B\overline{R}^{-1}(B'X_{ss} + S')$. As X_{ss} was a semi-stabilizing solution, we note that $\mathrm{Re}\,\lambda(H) \leq 0$. Moreover, (A, B) is $\mathbb{C}^-$-stabilizable and, hence, the pair $(A_{cl}, B\overline{R}^{-1}B')$ is $\mathbb{C}^-$-stabilizable. Then from Lemma 4.24, it follows that X_{ss} is positive semi-definite.   ∎

The following proposition is somewhat a reverse of Proposition 4.31.

Proposition 4.32 *Consider an H_2 CARE as in (4.2). A positive definite solution X, if it exists, is semi-stabilizing; i.e., all eigenvalues of $A - RX$ are located in the closed left-half plane. Moreover, all eigenvalues of $A - RX$ that are on the imaginary axis are simple; i.e., the multiplicity structure of an eigenvalue on the imaginary axis is $\{1, \ldots, 1\}$.*

Proof : The H_2 CARE (4.2) can be rewritten as a Lyapunov equation

$$(A - RX)'X + X(A - RX) = -Q - XRX.$$

Now that X is positive definite, and after applying Lemma 4.162, we find that $A - RX$ must be conditionally stable; i.e., all its eigenvalues are located in the closed left-half plane, and those on the imaginary axis are simple. ∎

The above two propositions connect solutions of an H_2 CARE which are positive (semi-)definite with solutions that are (semi-)stabilizing. As the following examples show, these connections are not true for an H_∞^1 CARE and for an H_∞^2 CARE. For the H_∞^1 CARE (4.9), we note that $X = 1$ is a positive definite solution that is not stabilizing when

$$\bar{A} = -2, \quad B = 1, \quad E = 2, \quad C = \begin{pmatrix} 1 \\ 0 \end{pmatrix}, \quad D = \begin{pmatrix} 0 \\ 1 \end{pmatrix}.$$

Similarly, for the H_∞^2 CARE (4.13), we have $X = 1$ as a positive definite solution that is not stabilizing when

$$\bar{A} = 2, \quad B = 1, \quad C_1 = \begin{pmatrix} 1 \\ 0 \end{pmatrix}, \quad C_2 = 2, \quad D = \begin{pmatrix} 0 \\ 1 \end{pmatrix}.$$

Having studied the positive semi-definite solution of an H_2 CARE, we study next the kernel of a positive semi-definite solution.

Proposition 4.33 *Consider an H_2 CARE as in (4.1) with $Q \geqslant 0$ and $R \geqslant 0$. Assume that (A, R) is $\mathbb{C}^-$-stabilizable. Also, let $\mathcal{U}$ denote the $\mathbb{C}^{-0}$ unobservable subspace of (Q, A). Then, a positive semi-definite solution X of the H_2 CARE, if it exists, has $\mathcal{U}$ as a subset of its null space. That is, $X\mathcal{U} = 0$.*

Proof : Let λ be an unobservable mode of (Q, A) with $\mathrm{Re}\,\lambda \leqslant 0$, and let v be the corresponding eigenvector of A; i.e., $v \in \mathcal{U}$ and

$$Av = \lambda v \qquad \text{and} \qquad Qv = 0.$$

We have

$$v^*(XA + A'X + Q - XRX)v = 2(\mathrm{Re}\,\lambda)v^*Xv - v^*XRXv = 0.$$

As X is positive semi-definite, the above equation implies that $Xv = 0$ if $\mathrm{Re}\,\lambda < 0$. If $\mathrm{Re}\,\lambda = 0$, then we have $v^*XRXv = 0$ or $(v^*X)R = 0$. Also, observe that

$$v^*(XA + A'X + Q - XRX) = (v^*X)A - \lambda v^*X = (v^*X)(A - \lambda I) = 0.$$

Thus, v^*X belongs to the uncontrollable subspace of (A, R). As (A, R) is $\mathbb{C}^-$-stabilizable, we have $v^*X = 0$ or $Xv = 0$.

If the space $\mathcal{U}$ is spanned by the eigenvectors of A, then the above implies $\mathcal{U} \subset \ker X$. In general, we also need to consider generalized eigenvectors. Assume that

$$Ax = \lambda x + y$$

with $y \in \mathcal{U} \cap \ker Q$. Then the above argument can also be used to yield that $x \in \ker X$. But then for a chain of generalized eigenvectors as defined in Definition 3.15 with $x_i \in \mathcal{U}$, we can recursively establish that $x_i \in \ker X$. As the generalized eigenvectors span $\mathcal{U}$, we find that $\mathcal{U} \subset \ker X$. ∎

Remark 4.34 *Proposition 4.33 pertains to an H_2 CARE as in (4.1). For an H_2 CARE as in (4.6), Proposition 4.33 can be stated as follows: Assume that $(\overline{A}, B)$ is $\mathbb{C}^-$-stabilizable. Then, a positive semi-definite solution X of the H_2 CARE (4.6), if it exists, has $\mathcal{V}^{-0}(\overline{A}, B, C, D)$ as a subset of its null space. That is, $X\mathcal{V}^{-0}(\overline{A}, B, C, D) = 0$. Obviously, this implies that a necessary condition for the existence of a positive definite solution of an H_2 CARE (4.6) is that the invariant zeros of the system $(\overline{A}, B, C, D)$ must be in the open right-half plane.*

Proposition 4.35 *Consider an H_∞^1 CARE as in (4.1) with $Q \geqslant 0$. Then the null space of any real symmetric solution X is an A-invariant subspace contained in the null space of Q; i.e., $\ker X \subset \langle \ker Q \mid A \rangle$.*

Proof : Let $x \in \ker X$. By multiplying the H_∞^1 CARE in (4.1) by x' and x on the left and right, respectively, we find that $x'Qx = 0$ and as $Q \geqslant 0$, we find that $Qx = 0$, and hence, $\ker X \subset \ker Q$. Now multiplying the H_∞^1 CARE by x on the right leads to $XAx = 0$; i.e., $Ax \in \ker X$. The latter establishes that $\ker X$ is A-invariant. This implies that $QA^i \ker X = 0$ for any integer i and, hence, the result. ∎

If the H_∞^1 CARE is of the form (4.9), then, using Lemma 4.6, we obtain the following corollary of the above proposition:

Corollary 4.36 *Consider an H_∞^1 CARE as in (4.9). Then the null space of any real symmetric solution X is contained in the weakly unobservable subspace $\mathcal{V}^*$ associated with the quadruple $(\overline{A}, B, C, D)$; i.e., $\ker X \subset \mathcal{V}^*(\overline{A}, B, C, D)$.*

Corollary 4.36 allows us to characterize exactly the kernel of a positive semi-definite solution of an H_2 CARE.

Corollary 4.37 *Consider an H_2 CARE as in (4.6). Assume that $(\overline{A}, B)$ is $\mathbb{C}^-$-stabilizable. Then, in accordance with Proposition 4.31, a real symmetric semi-stabilizing solution X of it, if it exists, is positive semi-definite. For such a semi-stabilizing solution X, we have $\mathcal{V}^{-0}(\overline{A}, B, C, D) = \ker X$, where $\mathcal{V}^{-0}$ denotes the $\mathbb{C}^{-0}$-stabilizable weakly unobservable subspace.*

Proof : We observe that for semi-stabilizing solutions, $\ker X$ is an A invariant subspace contained in $\mathcal{V}^*$ for which the eigenvalues of A restricted to the kernel of X must be in the closed left-half plane. This implies that $\ker X \subset \mathcal{V}^{-0}(\overline{A}, B, C, D)$. The reverse implication follows from Proposition 4.33. $\blacksquare$

The following proposition pertains to the existence of a solution of an H_2 CARE having both the features of being stabilizing and being positive definite.

Proposition 4.38 *Consider an H_2 CARE as in (4.1) with $Q \geqslant 0$ and $R \geqslant 0$. Then, a positive definite stabilizing solution of the H_2 CARE exists if and only if the pair (A, R) is $\mathbb{C}^-$-stabilizable and all unobservable eigenvalues of (Q, A) are in the open right-half plane.*

Proof :
(Necessity) Suppose that a positive definite stabilizing solution X exists. Then as $A - RX$ is Hurwitz-stable, it follows that the pair (A, R) must be $\mathbb{C}^-$-stabilizable. Also, from Proposition 4.33, it follows that the $\mathbb{C}^{-0}$ unobservable subspace of (Q, A) must be equal to $\{0\}$, and hence, unobservable eigenvalues of (Q, A) are all in the open right-half plane.

(Sufficiency) Let (A, R) be $\mathbb{C}^-$-stabilizable, and assume that all unobservable eigenvalues of (Q, A) are in the open right-half plane. Then, from Proposition 4.14 and Remark 4.15, it follows that the Hamiltonian matrix H has no eigenvalues on the imaginary axis. Hence, it follows that a stabilizing solution X of the given H_2 CARE exists. Assume that $\ker X$ is nonempty. As $\ker X$ is A-invariant, this implies that an eigenvector of A exists with eigenvalue λ contained in $\ker X$. Then $(A - RX)x = \lambda x$, and because $A - RX$ is Hurwitz-stable, we find that $\operatorname{Re} \lambda < 0$. Hence, A restricted to $\ker X$ is Hurwitz-stable, but combined with $\ker X \subset \ker Q$, this implies that λ is a stable unobservable eigenvalue. This result yields a contradiction, and hence, X is invertible. As we already established in Proposition 4.31 that a stabilizing solution is positive semi-definite, we find that X is positive definite. $\blacksquare$

If the H_2 CARE has the form (4.6), then we can also express the above conditions in terms of $(\overline{A}, B, C, D)$. We first note that a system has an anti-stable right-inverse (all poles in the open right-half plane) if and only if $\mathcal{V}^{-0} = \{0\}$ and $(B' \quad D')$ is right invertible. The latter is automatically satisfied because D is injective.

Corollary 4.39 *Consider an H_2 CARE as in (4.6). A stabilizing positive definite solution of this H_2 CARE exists if and only if (A, B) is $\mathbb{C}^-$-stabilizable and the system (A, B, C, D) has an anti-stable left inverse. Moreover, in this case, the positive definite solution is always unique.*

Proof : This is a combination of Proposition 4.38 with Lemma 4.6.   ∎

Having inter-related the positive semi-definiteness and positive definite features to the semi-stabilizing and stabilizing features of a solution of a CARE, we present next a theorem that portrays the necessary and sufficient conditions for the existence of a positive definite solution to an H_2 CARE.

Theorem 4.40 *Consider an H_2 CARE as in (4.1) with $Q \geqslant 0$ and $R \geqslant 0$. Then it has a positive definite solution if and only if the matrices A, Q, and R can be transformed to the following form using an appropriate basis:*

$$A = \begin{pmatrix} A_1 & 0 \\ 0 & A_2 \end{pmatrix}, \quad R = \begin{pmatrix} R_1 & 0 \\ 0 & 0 \end{pmatrix}, \quad and \quad Q = \begin{pmatrix} Q_1 & 0 \\ 0 & 0 \end{pmatrix},$$

where the pair (A_1, R_1) is $\mathbb{C}^-$-stabilizable, the pair $(Q_1, -A_1)$ is $\mathbb{C}^-$-detectable, and A_2 is diagonalizable and has all its eigenvalues on the imaginary axis. Moreover, the solution X takes the form

$$X = \begin{pmatrix} X_1 & 0 \\ 0 & X_2 \end{pmatrix},$$

where X_1 is unique and $A_1 - R_1 X_1$ is Hurwitz-stable. Furthermore, X_2 is one of uncountably many positive definite solutions of the equation

$$A_2' X_2 + X_2 A_2 = 0.$$

In the proof of the above theorem, the following results will be useful.

Proposition 4.41 *Assume that*

$$S_2' Z_3 + Z_3 S_2 \leqslant 0 \tag{4.70}$$

with S_2 diagonalizable and with all its eigenvalues on the imaginary axis. Then, we have

$$S_2' Z_3 + Z_3 S_2 = 0. \tag{4.71}$$

Proof : Suppose $S_2 \in \mathbb{R}^{k \times k}$. As S_2 is diagonalizable, a basis of $\mathbb{C}^k$ consisting of eigenvectors $\{v_1, \ldots, v_k\}$ exists. Let v_i be an eigenvector associated with an eigenvalue $j\omega_i$ of S_2,

$$v_i^*(S_2'Z_3 + Z_3 S_2)v_i = 0. \tag{4.72}$$

But combining (4.70) and (4.72), it follows that

$$(S_2'Z_3 + Z_3 S_2)v_i = 0,$$

for $i = 1, \ldots, k$, and because $\{v_1, \ldots, v_k\}$ span the whole space, we find that

$$S_2'Z_3 + Z_3 S_2 = 0. \qquad \blacksquare$$

Proposition 4.42 *For any A_2 with* $\mathrm{Re}\,\lambda(A_2) = 0$, *and A_2 diagonalizable, uncountably many symmetric positive definite X_2 exist that satisfy*

$$A_2'X_2 + X_2 A_2 = 0. \tag{4.73}$$

We would like to note that A_2 diagonalizable is actually a necessary condition for the solvability of (4.73) given that $\mathrm{Re}\,\lambda(A_2) = 0$.

Proof : A nonsingular transformation matrix T_2 exists such that $\tilde{A}_2 = T_2^{-1}A_2 T_2$ is in real Jordan canonical form. As A_2 is diagonalizable and has all its eigenvalues on the imaginary axis, we find that $\tilde{A}_2$ is skew-symmetric.
Defining $\tilde{X}_2 = T_2'X_2 T_2$, (4.73) can be rewritten as

$$\tilde{A}\tilde{X}_2 + \tilde{X}_2\tilde{A}' = 0.$$

From the fact that $\tilde{A} = -\tilde{A}'$, it follows then that $\tilde{X}_2 = \alpha I$ is a positive definite solution for all $\alpha > 0$. But then

$$X_2 = (T_2')^{-1}(\alpha I)T_2^{-1}$$

is a solution of (4.73) for every $\alpha > 0$. $\qquad \blacksquare$

Proof of Theorem 4.40 : (Necessity) Suppose that a positive definite solution X exists. Then, from Proposition 4.32, it follows that X is indeed a semi-stabilizing solution. This means that $\mathrm{Re}\,\lambda(A - RX) \leq 0$. Also, $A - RX$ can be rewritten through a change of basis as

$$A - RX = TST^{-1} = T\begin{pmatrix} S_1 & 0 \\ 0 & S_2 \end{pmatrix}T^{-1},$$

with T invertible, $\operatorname{Re} \lambda(S_1) < 0$, $\operatorname{Re} \lambda(S_2) = 0$, while S_2 is diagonalizable. Now the given H_2 CARE can be rewritten as

$$(A - RX)'X + X(A - RX) = -Q - XRX$$

or, equivalently, as

$$S'(T'XT) + (T'XT)S = -T'(Q + XRX)T. \qquad (4.74)$$

Let

$$T'XT = \begin{pmatrix} X_1 & X_{21} \\ X_{21}' & X_2 \end{pmatrix}. \qquad (4.75)$$

Then (4.74) can be expanded as

$$\begin{pmatrix} S_1'X_1 + X_1 S_1 & S_1'X_{21} + X_{21}S_2 \\ S_2'X_{21}' + X_{21}'S_1 & S_2'X_2 + X_2 S_2 \end{pmatrix} = -T'(Q + XRX)T \leqslant 0. \qquad (4.76)$$

It follows that

$$S_2'X_2 + X_2 S_2 \leq 0.$$

Next, as S_2 is diagonalizable and $\operatorname{Re} \lambda(S_2) = 0$, we obtain from Proposition 4.41 that

$$S_2'X_2 + X_2 S_2 = 0. \qquad (4.77)$$

In view of (4.76) and (4.71), we can conclude that

$$S_1'X_{21} + X_{21}S_2 = 0. \qquad (4.78)$$

Now that S_1' and $-S_2$ have disjoint sets of eigenvalues, it follows from Lemma 4.236 that $X_{21} = 0$. This leads to

$$S_1'X_1 + X_1 S_1 \leqslant 0. \qquad (4.79)$$

Now (4.74), (4.77), and (4.78) imply that

$$T'QT = \begin{pmatrix} Q_1 & 0 \\ 0 & 0 \end{pmatrix} \qquad (4.80)$$

for some $Q_1 \geq 0$. Next, (4.76) combined with (4.77) implies that

$$T'XRXT = \begin{pmatrix} G & 0 \\ 0 & 0 \end{pmatrix}$$

with $G \geqslant 0$. Using (4.75) with $X_{21} = 0$, we have

$$\begin{pmatrix} X_1 & 0 \\ 0 & X_2 \end{pmatrix} T^{-1} R (T^{-1})' \begin{pmatrix} X_1 & 0 \\ 0 & X_2 \end{pmatrix} = \begin{pmatrix} G & 0 \\ 0 & 0 \end{pmatrix},$$

and hence,

$$T^{-1}R(T^{-1})' = \begin{pmatrix} R_1 & 0 \\ 0 & 0 \end{pmatrix}$$

for a suitably chosen R_1. Now we have

$$S = T^{-1}(A - RX)T = T^{-1}AT - T^{-1}R(T^{-1})'T'XT.$$

This implies that

$$S = T^{-1}AT - \begin{pmatrix} X_1^{-1}G & 0 \\ 0 & 0 \end{pmatrix}.$$

Thus,

$$T^{-1}AT = \begin{pmatrix} S_1 + R_1X_1 & 0 \\ 0 & S_2 \end{pmatrix} := \begin{pmatrix} A_1 & 0 \\ 0 & A_2 \end{pmatrix}.$$

Hence, we can find a matrix T such that

$$T^{-1}AT = \begin{pmatrix} A_1 & 0 \\ 0 & A_2 \end{pmatrix}, \quad T^{-1}R(T^{-1})' = \begin{pmatrix} R_1 & 0 \\ 0 & 0 \end{pmatrix}, \quad \text{and } T'QT = \begin{pmatrix} Q_1 & 0 \\ 0 & 0 \end{pmatrix}.$$

Also, $\mathrm{Re}\,\lambda(A_2) = \mathrm{Re}\,\lambda(S_2) = 0$, $A_1 - B_1B_1'X = S_1$, and thus $A_1 - B_1B_1'X$ is Hurwitz-stable while the pair (A_1, B_1) is $\mathbb{C}^-$-stabilizable. Moreover, from Proposition 4.33, the pair $(-A_1, C_1)$ is $\mathbb{C}^-$-detectable. This concludes the necessity part of the proof of Theorem 4.40.
(Sufficiency) In view of Proposition 4.42, the sufficiency part of the proof of Theorem 4.40 follows trivially from Proposition 4.38. $\blacksquare$

The above results portray the necessary and sufficient conditions for the existence of a positive definite solution to an H_2 CARE. We focus next on the existence of a positive semi-definite solution to the same equation. We first have the following theorem.

Theorem 4.43 *Consider an H_2 CARE as in (4.1) with $Q \geqslant 0$ and $R \geqslant 0$. Then it has a positive semi-definite solution if and only if the matrices A, R, and Q can be transformed to the following form using an appropriate basis:*

$$A = \begin{pmatrix} A_1 & 0 \\ A_3 & A_2 \end{pmatrix}, \quad R = \begin{pmatrix} R_{11} & R_{21}' \\ R_{21} & R_{22} \end{pmatrix}, \quad \text{and } Q = \begin{pmatrix} Q_1 & 0 \\ 0 & 0 \end{pmatrix},$$

where the pair (A_1, R_{11}) is $\mathbb{C}^-$-stabilizable, and the pair (Q_1, A_1) is observable.

Proof : (Sufficiency) As (A_1, R_{11}) is $\mathbb{C}^-$-stabilizable and (Q_1, A_1) is observable, in accordance with Theorem 4.40, a positive definite solution X_1 to the H_2 CARE

$$A_1'X_1 + X_1A_1 - X_1R_{11}'X_1 + Q_1 = 0$$

exists. Next, let $X \in \mathbb{R}^{n \times n}$ be a positive semi-definite matrix defined as

$$X := \begin{pmatrix} X_1 & 0 \\ 0 & 0 \end{pmatrix}. \tag{4.81}$$

Then, it is straightforward to verify by substitution that in fact X is a solution of the H_2 CARE (4.1).

(Necessity) Let X be a positive semi-definite solution of the H_2 CARE (4.1). In a suitable basis, we have (4.81) with X_1 positive definite. As noted, $\ker X$ is A-invariant and contained in $\ker Q$. Therefore, in our new basis we have

$$A = \begin{pmatrix} \tilde{A}_1 & 0 \\ \tilde{A}_2 & \tilde{A}_4 \end{pmatrix}, \quad Q = \begin{pmatrix} \tilde{Q}_1 & 0 \\ 0 & 0 \end{pmatrix}, \quad R = \begin{pmatrix} \tilde{R}_{11} & \tilde{R}_{12} \\ \tilde{R}'_{12} & \tilde{R}_{22} \end{pmatrix}.$$

We can see that X_1 is a positive definite solution of the H_2 CARE:

$$\tilde{A}'_1 X_1 + X_1 \tilde{A}_1 - X_1 \tilde{R}'_{11} X_1 + \tilde{Q}_1 = 0.$$

As $\tilde{Q}_1$ is invertible, by using Theorem 4.40, we find that $(\tilde{A}_1, \tilde{R}_{11})$ is $\mathbb{C}^-$-stabilizable and all unobservable eigenvalues of $(\tilde{Q}_1, A_1)$ are in the closed right-half plane. Clearly, in a suitable basis we have

$$\tilde{Q}_1 = \begin{pmatrix} Q_1 & 0 \\ 0 & 0 \end{pmatrix}, \qquad \tilde{A}_1 = \begin{pmatrix} A_1 & 0 \\ \tilde{A}_{21} & \tilde{A}_{22} \end{pmatrix}$$

with (Q_1, A_1) observable. If we decompose $\tilde{R}_{11}, \tilde{R}_{21}$, and $\tilde{A}_2$ compatibly

$$\tilde{R}_{11} = \begin{pmatrix} R_{11} & \bar{R}'_{21} \\ \bar{R}_{21} & \bar{R}_{22} \end{pmatrix}, \qquad \tilde{R}_{21} = \begin{pmatrix} \bar{R}_{31} & \bar{R}_{32} \end{pmatrix}, \qquad \tilde{A}_2 = \begin{pmatrix} \tilde{A}_{31} & \tilde{A}_{32} \end{pmatrix},$$

then we can define

$$A_2 = \begin{pmatrix} \tilde{A}_{21} \\ \tilde{A}_{31} \end{pmatrix}, \qquad A_4 = \begin{pmatrix} \tilde{A}_{22} & 0 \\ \tilde{A}_{32} & \tilde{A}_4 \end{pmatrix},$$

and

$$R_{21} = \begin{pmatrix} \bar{R}_{21} \\ \bar{R}_{31} \end{pmatrix}, \qquad R_{22} = \begin{pmatrix} \bar{R}_{22} & \bar{R}'_{32} \\ \bar{R}_{32} & \tilde{R}_{22} \end{pmatrix}.$$

It is then easy to check that we have the structure as presented in the theorem. $\blacksquare$

An H_2 CARE is often written in the form of (4.2). For this H_2 CARE, the necessary and sufficient conditions for the existence of a positive semi-definite solution are formulated in the following theorem.

Theorem 4.44 *Consider an H_2 CARE as in (4.6). Then it has a positive semi-definite solution if and only if the pair (A_{bb}, B_{b0}) is $\mathbb{C}^-$-stabilizable. Here the matrices A_{bb} and B_{b0} are from the representation of the system Σ characterized by the quadruple $(\overline{A}, B, C, D)$ in the special coordinate basis.*

Proof : One can see the result easily when the system Σ characterized by the quadruple $(\overline{A}, B, C, D)$ is written in the special coordinate basis.   ∎

Remark 4.45 *The above theorem can be stated in geometric language. Namely, the H_2 CARE (4.6) has a positive semi-definite solution if and only if*

$$X_-(A) + \langle A \mid \text{im } B \rangle + \mathcal{V}^*(\Sigma) = \mathbb{R}^n,$$

where $X_-(A)$ is the stable modal subspace of $\mathbb{R}^n$ related to A, $\langle A \mid \text{im } B \rangle$ is the controllable subspace, and $\mathcal{V}^*(\Sigma)$ represents the weakly unobservable subspace of the system Σ as defined in Section 3.2.

4.1.5   Continuity properties

We studied in the previous subsection the existence of a (semi)-stabilizing solution of a CARE and its relationship with the positive (semi)-definite solution. We focus here on the continuity of the (semi)-stabilizing solution of a CARE with respect to change of parameters.

We consider first an H_2 CARE. From the results of the previous subsection, the existence of a stabilizing solution of an H_2 CARE is guaranteed when $(\overline{A}, B)$ is $\mathbb{C}^-$-stabilizable and $(\overline{A}, B, C, D)$ has no invariant zeros on the imaginary axis. Assume that the parameters are perturbed. In other words, we have a family of systems $(\overline{A}^\varepsilon, B^\varepsilon, C^\varepsilon, D^\varepsilon)$ such that, for all $\varepsilon \in [0, \delta)$, $(\overline{A}^\varepsilon, B^\varepsilon)$ is $\mathbb{C}^-$-stabilizable and $(\overline{A}^\varepsilon, B^\varepsilon, C^\varepsilon, D^\varepsilon)$ has no invariant zeros on the imaginary axis. Assume that

$$\lim_{\varepsilon \to 0} \overline{A}^\varepsilon = \overline{A}^0, \quad \lim_{\varepsilon \to 0} B^\varepsilon = B^0, \quad \lim_{\varepsilon \to 0} C^\varepsilon = C^0, \quad \lim_{\varepsilon \to 0} D^\varepsilon = D^0. \quad (4.82)$$

Let X^ε be the stabilizing solution associated with the parameters

$$(\overline{A}^\varepsilon, B^\varepsilon, C^\varepsilon, D^\varepsilon).$$

We have the following result, which can also be found in [21].

Theorem 4.46 *Consider a family of systems $(\overline{A}^\varepsilon, B^\varepsilon, C^\varepsilon, D^\varepsilon)$ such that, for all $\varepsilon \in [0, \delta)$, $(\overline{A}^\varepsilon, B^\varepsilon)$ is $\mathbb{C}^-$-stabilizable and $(\overline{A}^\varepsilon, B^\varepsilon, C^\varepsilon, D^\varepsilon)$ has no invariant zeros on the imaginary axis. Assume that (4.82) is satisfied. Let X^ε be the stabilizing solution of the H_2 CARE (4.6) associated with the parameters $(\overline{A}^\varepsilon, B^\varepsilon, C^\varepsilon, D^\varepsilon)$. Then we have*

$$\lim_{\varepsilon \to 0} X^\varepsilon = X^0. \quad (4.83)$$

Proof : Consider the Hamiltonian H^ε associated with the H_2 CARE for the parameters $(\bar{A}^\varepsilon, B^\varepsilon, C^\varepsilon, D^\varepsilon)$. The Hamiltonian depends continuously on ε. Moreover, by assumption, the subspace $X_-(H^\varepsilon)$ also has a fixed dimension n in the limit. Then, from linear algebra, it is well known that $X_-(H^\varepsilon)$ depends continuously on ε. But then as

$$X_-(H^\varepsilon) = \begin{pmatrix} I \\ X^\varepsilon \end{pmatrix},$$

we find that X^ε depends also continuously on ε, and therefore, (4.83) is satisfied. $\blacksquare$

It is obvious that the above result is also true for the H_∞^1 and H_∞^2 CAREs provided that we are guaranteed that the stabilizing solutions of these equations exist for all ε (including the limiting case $\varepsilon = 0$).

An immediate question arises of whether a similar result is valid for the semi-stabilizing solution.

Theorem 4.47 *Consider a family of systems* $(\bar{A}^\varepsilon, B^\varepsilon, C^\varepsilon, D^\varepsilon)$ *such that, for all* $\varepsilon \in [0, \delta)$, $(\bar{A}^\varepsilon, B^\varepsilon)$ *is* $\mathbb{C}^-$-*stabilizable and that* (4.82) *is satisfied. Let* X^ε *be the semi-stabilizing solution of the* H_2 *CARE* (4.6) *associated with the parameters* $(\bar{A}^\varepsilon, B^\varepsilon, C^\varepsilon, D^\varepsilon)$. *Then we have*

$$\lim_{\varepsilon \to 0} X^\varepsilon = X^0. \tag{4.84}$$

Proof : We first observe that the solution X_ε of the CARE is bounded. This can be seen by noting that a stabilizing feedback F exists such that $A_0 + B_0 F$ is asymptotically stable. But then, for all ε small enough, $A_\varepsilon + B_\varepsilon F$ is asymptotically stable. But then interpreting X_ε as the optimal cost of a linear quadratic control problem, it is easily seen that this cost is bounded from above by the cost associated with the suboptimal feedback $u = Fx$. On the other hand, the optimal cost is bounded from below by 0.

It is trivial to see next that any convergent subsequence converges to a semi-stabilizing solution of the CARE associated with the quadruple $(\bar{A}^0, B^0, C^0, D^0)$. As the semi-stabilizing solution is unique, it must converge to X_0. A bounded sequence where all convergent subsequences have the same limit is convergent itself. Hence, we find that (4.84) is satisfied. $\blacksquare$

Note that the above can be extended in a straightforward manner to the H_∞^2 CARE provided we are guaranteed that the semi-stabilizing solution exists for all ε, including in the limit for $\varepsilon = 0$. The extension for the H_∞^1 CARE is problematic because in this case, we are no longer guaranteed that the semi-stabilizing solution is unique.

4.1.6 Algorithms for the computation of stabilizing solutions

In this subsection, we look at another important aspect, namely, the computational aspect of obtaining the stabilizing solution of a CARE. We briefly review three classes of algorithms. These are transform methods, Newton's method, and methods based on a matrix sign function. Both the transform and the sign function methods use the solution concept as was given in Proposition 4.19. That is, they first find a basis for the stable modal subspace of the associated Hamiltonian matrix and then obtain the stabilizing solution. These two methods differ in the way they compute a basis for the stable modal subspace of the Hamiltonian.

Transform methods (eigenvector solution methods, Schur-based methods)

A textbook approach for obtaining the stabilizing solution of a CARE is that of Potter [63] as given in Proposition 4.19. This approach involves transformation of the Hamiltonian matrix to its Jordan form. More specifically, one needs to compute a matrix of eigenvectors T to perform the transformation

$$T^{-1}\begin{pmatrix} A & -R \\ -Q & -A' \end{pmatrix} T = \begin{pmatrix} -\Lambda & 0 \\ 0 & \Lambda \end{pmatrix},$$

where $T \in \mathbb{R}^{2n \times 2n}$ and $-\Lambda$ is composed of Jordan blocks corresponding to eigenvalues in the open left-half plane only. Standard necessary assumptions for the existence of the stabilizing solution guarantee that the Hamiltonian matrix has no eigenvalues on the imaginary axis $\mathbb{C}^0$, and then obviously the Hamiltonian matrix has n left-half plane eigenvalues and n right-half plane eigenvalues. Now we partition T as

$$T = \begin{pmatrix} T_1 & T_2 \end{pmatrix} = \begin{pmatrix} T_{11} & T_{21} \\ T_{12} & T_{22} \end{pmatrix},$$

where $T_{ij} \in \mathbb{R}^{n \times n}$, for all i, j, and $T_i \in \mathbb{R}^{2n \times n}$, for all i. Then it follows from Proposition 4.19 that the stabilizing solution X of a CARE can be obtained by solving a system of linear equations:

$$X = T_{12} T_{11}^{-1}.$$

Several numerical difficulties are associated with this approach when the Hamiltonian matrix has multiple or near-multiple eigenvalues. For a cogent discussion of numerical difficulties associated with the numerical determination of the Jordan form, we refer interested readers to the classic paper of Golub and Wilkinson [28].

To alleviate these numerical difficulties, Schur methods were proposed by Laub [42]. The proposed procedure makes use of the fact that the stabilizing solution of a CARE requires only the computation of a matrix U_1 such that

$$\operatorname{im} U_1 = \operatorname{im} T_1.$$

In view of this observation, the approach in [42] uses the ordered Schur decomposition of Stewart [91] to compute an orthogonal basis for the stable modal subspace of the Hamiltonian matrix. More specifically, an orthogonal matrix U of "Schur vectors" is computed so that

$$\begin{pmatrix} U_{11} & U_{21} \\ U_{12} & U_{22} \end{pmatrix}' \begin{pmatrix} A & -R \\ -Q & -A' \end{pmatrix} \begin{pmatrix} U_{11} & U_{21} \\ U_{12} & U_{22} \end{pmatrix} = \begin{pmatrix} S_{11} & S_{12} \\ 0 & S_{22} \end{pmatrix},$$

with $S_{ij} \in \mathbb{R}^{n \times n}$ and $U_{ij} \in \mathbb{R}^{n \times n}$, where S_{11} is a quasi-upper triangular matrix with eigenvalues in $\mathbb{C}^-$ and S_{22} is a quasi-upper triangular matrix with eigenvalues in $\mathbb{C}^+$. The key observation is that

$$\mathrm{im} \begin{pmatrix} U_{11} \\ U_{12} \end{pmatrix} = \mathrm{im} \begin{pmatrix} T_{11} \\ T_{12} \end{pmatrix} = X_-(H).$$

Next, the stabilizing solution of a CARE can be computed by solving a system of linear equations

$$X = U_{12} U_{11}^{-1}.$$

The principal difficulty with this method is the ordering of eigenvalues in matrix S. This problem can be overcome in a numerically stable way (see [91] and [22]).

The Schur-based methods for the computation of a basis for a relevant invariant subspace of the Hamiltonian matrix are currently among the most reliable and efficient algorithms to obtain the stabilizing solution of a CARE.

The problem with transform methods in general is that they are difficult to parallelize.

Newton's Method

Newton's method is an iterative method that involves the solution of a linearized form of a CARE, namely, a Lyapunov equation. It requires an initial guess X_0 for which $A - R X_0$ is a Hurwitz-stable matrix. Newton's method is typically used for iterative refinement of the solutions computed by orthogonal transformation methods, but for a limited class of problems, it may be appropriate as the primary solution method. In the following, we shall briefly discuss the theory underlying Newton's method.

Suppose we have an approximation X_0 to the solution X of a CARE. Write X as

$$X = X_0 + (X - X_0).$$

Then the CARE (4.1) can be rewritten as

$$(A - R X_0)' X + X(A - R X_0) = -X_0 R X_0 - Q + (X - X_0) R (X - X_0).$$

Assuming that $X - X_0$ is "small" enough to ignore the second-order term $(X - X_0) R (X - X_0)$, we obtain the linearized equation (Lyapunov equation):

$$(A - R X_0)' X + X(A - R X_0) = -X_0 R X_0 - Q.$$

The idea is to set up conditions under which X_1, the solution of the above Lyapunov equation, would be a better approximation to X than the original X_0. That is,

$$\|X - X_1\| < \|X - X_0\|.$$

Newton's method is simply an extension of this idea; specifically, we perform a repeated application of this idea. Starting with an initial guess X_0, we solve the Lyapunov equation for $k \geq 0$,

$$(A - RX_k)'X_{k+1} + X_{k+1}(A - RX_k) = -X_k RX_k - Q.$$

It was shown in [40] and [79] that if $A - RX_0$ is Hurwitz-stable, then for $k \geq 1$,

(i) $A - RX_k$ is Hurwitz-stable,

(ii) $0 \leq X \leq X_{k+1} \leq X_k \leq \cdots \leq X_1$,

(iii) $\lim_{k \to \infty} X_k = X$.

The usual approach to solve the Lyapunov equation appearing at each step of Newton's method is to use the standard QZ algorithm.

For further details and more recent refinements of the theory of Newton's method, we refer interested readers to [39].

Matrix Sign Function Methods

The main difficulty with the transform methods is that they are very difficult to parallelize. These difficulties have led to an increased interest in iterative algorithms based on matrix sign functions. The matrix sign function algorithms are designed for solving CAREs with dense and unstructured coefficient matrices. They separate the invariant subspaces corresponding to the negative and positive real part eigenvalues without actually computing the eigenvalues. In the following, we shall provide some very brief introductory material on the basic theory of sign function methods. We restrict our attentions to an H^2_∞ CARE. The advantage of an H^2_∞ CARE over a general CARE is that we can a priori characterize when a stabilizing solution exists, that is, when (A, R) stabilizing and the Hamiltonian does not have eigenvalues on the imaginary axis. For a general CARE, we also need to check the complementarity condition (4.36), which in principle requires the computation of $X_-(H)$.

We first start with the definition of a matrix sign function, which is a generalization of the sign of scalars.

Definition 4.48 *Let $H \in \mathbb{R}^{m \times m}$ with no eigenvalues on the imaginary axis $\mathbb{C}^0$. Moreover, let the Jordan decomposition of H be*

$$H = T(D + N)T^{-1},$$

*where $D = \operatorname{diag}(\lambda_i)$ contains the eigenvalues of H, and N is a nilpotent matrix. Then the **matrix sign function** of H is defined as the matrix*

$$Z = \operatorname{sgn} H = T \operatorname{diag}\{\operatorname{sgn}(\lambda_i)\}T^{-1}.$$

Here

$$\operatorname{sgn}(\lambda_i) = \begin{cases} +1 & \text{if } \operatorname{Re}\lambda_i > 0, \\ -1 & \text{if } \operatorname{Re}\lambda_i < 0. \end{cases}$$

From this definition, it is obvious that Z is diagonalizable with eigenvalues $+1$ or -1 and solves the equation

$$Z^2 - I = 0.$$

The sign function of a matrix may be computed by iterative methods. One of the simplest is Newton's method applied to the equation $Z^2 - I = 0$, namely,

(*i*) Set $Z_0 = H$.

(*ii*) Set $Z_{k+1} = (Z_k + Z_k^{-1})/2$ for $k > 0$.

Then we have

$$\lim_{k \to \infty} Z_k = \operatorname{sgn} H.$$

The convergence of this simple scheme can be enhanced by various scaling strategies. For further results on this, interested readers are referred to [2] and [11].

Having explored the matrix sign function and iterative methods of computing it, we now proceed to show how the sign function can be used in the computation of the stabilizing solution of a CARE. The following theorem is due to Roberts [67].

Theorem 4.49 *Consider an H_∞^2 CARE as in (4.1) with $R \geq 0$. Assume that the associated Hamiltonian matrix H has no eigenvalues on the imaginary axis $\mathbb{C}^0$, and that (A, B) is $\mathbb{C}^-$-stabilizable. Let*

$$Z = \begin{pmatrix} Z_{11} & Z_{12} \\ Z_{21} & Z_{22} \end{pmatrix} = \operatorname{sgn} H,$$

where $Z_{ij} \in \mathbb{R}^{n \times n}$. Then X, the unique symmetric stabilizing solution of the H_∞^2 CARE, is the solution of a linear equation

$$\begin{pmatrix} Z_{12} \\ Z_{22} + I \end{pmatrix} X = -\begin{pmatrix} Z_{11} + I \\ Z_{21} \end{pmatrix}. \tag{4.85}$$

Proof : Let

$$T = \begin{pmatrix} I & Y \\ 0 & I \end{pmatrix} \begin{pmatrix} I & 0 \\ -X & I \end{pmatrix} = \begin{pmatrix} I - YX & Y \\ -X & I \end{pmatrix},$$

where X is the stabilizing solution of

$$A'X + XA - XRX + Q = 0$$

and Y satisfies the Lyapunov equation

$$(A - RX)Y + Y(A - RX)' = -R.$$

We note that T^{-1} can be computed explicitly as

$$T^{-1} = \begin{pmatrix} I & 0 \\ X & I \end{pmatrix} \begin{pmatrix} I & -Y \\ 0 & I \end{pmatrix} = \begin{pmatrix} I & -Y \\ X & I - XY \end{pmatrix}.$$

We observe next that

$$THT^{-1} = T \begin{pmatrix} A & -R \\ -Q & -A' \end{pmatrix} T^{-1} = \begin{pmatrix} A - RX & 0 \\ 0 & -(A - RX)' \end{pmatrix}.$$

Thus,

$$\operatorname{sgn} H = T^{-1} \operatorname{sgn} \begin{pmatrix} A - RX & 0 \\ 0 & -(A - RX)' \end{pmatrix} T$$

$$= T^{-1} \begin{pmatrix} -I & 0 \\ 0 & I \end{pmatrix} T$$

$$= \begin{pmatrix} -I + 2YX & -2Y \\ -2X + 2XYX & I - 2XY \end{pmatrix}.$$

Hence,

$$I + \operatorname{sgn} H = \begin{pmatrix} 2YX & -2Y \\ -2X + 2XYX & 2I - 2XY \end{pmatrix}$$

or

$$\begin{pmatrix} Z_{11} + I & Z_{12} \\ Z_{21} & Z_{22} + I \end{pmatrix} = \left(\begin{pmatrix} 2Y \\ 2(XY - I) \end{pmatrix} X \quad \begin{pmatrix} -2Y \\ 2(I - XY) \end{pmatrix} \right),$$

which leads to

$$\begin{pmatrix} Z_{12} \\ Z_{22} + I \end{pmatrix} X = - \begin{pmatrix} Z_{11} + I \\ Z_{21} \end{pmatrix}.$$

Note that because $\operatorname{rank}(I + \operatorname{sgn} H) = n$, it is easily seen that the solution is unique. The proof is now complete. ∎

Theorem 4.49 can be applied to a general CARE, and not merely to an H_∞^2 or H_2 CARE, as long as the existence of a unique symmetric stabilizing solution to it is guaranteed. In general we need to check the complementarity condition (4.36). The following theorem from [38] shows that the existence of a unique stabilizing solution (in particular, the complementarity condition) is actually equivalent to the existence of a unique solution to the linear (matrix sign) equation (4.85).

Theorem 4.50 *Consider a CARE as in (4.1). Then it has a unique symmetric stabilizing solution if and only if the Hamiltonian matrix H has no eigenvalues on the imaginary axis $\mathbb{C}^0$ and*

$$\text{rank} \begin{pmatrix} Z_{12} \\ Z_{22} + I \end{pmatrix} = n.$$

Proof: (Necessity) Suppose that a symmetric stabilizing solution X_s of the CARE in (4.1) exists. Then it follows from Remark 4.11 that the Hamiltonian matrix H has no eigenvalues on the imaginary axis. Now let $v \in \mathbb{C}^n$ be in

$$\ker \begin{pmatrix} Z_{12} \\ Z_{22} + I \end{pmatrix}.$$

Then, using the same notations as in the proof of Theorem 4.49, we have

$$\begin{pmatrix} Z_{11} + I & Z_{12} \\ Z_{21} & Z_{22} + I \end{pmatrix} \begin{pmatrix} 0 \\ v \end{pmatrix} = \begin{pmatrix} 0 \\ 0 \end{pmatrix},$$

which yields

$$\begin{pmatrix} 2Y_s X_s & -2Y_s \\ 2(X_s Y_s - I)X_s & -2(X_s Y_s - I) \end{pmatrix} \begin{pmatrix} 0 \\ v \end{pmatrix} = \begin{pmatrix} 0 \\ 0 \end{pmatrix},$$

where Y_s is the solution of the Lyapunov equation

$$(A - RX_s)Y_s + Y_s(A - RX_s)' = -R.$$

This result means that

$$Y_s v = 0 \quad \text{and} \quad X_s Y_s v - v = 0,$$

which in turn implies that $v = 0$, and hence,

$$\ker \begin{pmatrix} Z_{12} \\ Z_{22} + I \end{pmatrix} = 0.$$

(Sufficiency) Assume that

$$\operatorname{rank}\begin{pmatrix} Z_{12} \\ Z_{22} + I \end{pmatrix} = n.$$

Then (4.85) in Theorem 4.49 has a unique solution, say X_s. Next, it can be shown that

$$\begin{pmatrix} I \\ X_s \end{pmatrix}$$

forms a basis for $X_-(H)$. Hence, it follows that X_s is the symmetric stabilizing solution to the CARE (4.1). $\blacksquare$

4.1.7 Algorithms for the computation of semi-stabilizing solutions

Consider an H_2 CARE of the form (4.6). Assume that (A, B) is $\mathbb{C}^-$-stabilizable. Then by Proposition 4.25, we know a semi-stabilizing solution exists. To compute this unique semi-stabilizing solution, we use the fact (from Corollary 4.37) that

$$\ker X = \mathcal{V}^{-0}(\overline{A}, B, C, D).$$

Let $\mathcal{W}$ be a subspace such that we obtain a decomposition of the state space $\mathbb{R}^n = \mathcal{V}^{-0} \oplus \mathcal{W}$. Then for the matrix $F = -D(D'D)^{-1}$, we have

$$\overline{A} + BF = \begin{pmatrix} A_{11} & A_{12} \\ 0 & A_{22} \end{pmatrix}, \quad B = \begin{pmatrix} B_1 \\ B_2 \end{pmatrix}, \quad C + DF = \begin{pmatrix} 0 & C_2 \end{pmatrix}.$$

But with respect to this basis, we have

$$X = \begin{pmatrix} 0 & 0 \\ 0 & X_1 \end{pmatrix} \tag{4.86}$$

and

$$A_{22}' X_1 + X_1 A_{22} - (X_1 B_2 + C_2' D)(D'D)^{-1}(B_2' X_1 + D'C_2) + C_2' C_2 = 0. \tag{4.87}$$

We know, by construction, that (A_{22}, B_2, C_2, D) has no eigenvalues on the imaginary axis. Therefore, X_1 must be a stabilizing solution of the above equation that can be obtained using the tools of the previous section. Then (4.86) yields the unique semi-stabilizing solution of the original CARE (4.6) with respect to the specific basis chosen in the state space.

The next step is to consider the H_∞^2 CARE (4.10). Assume that (A, B) is $\mathbb{C}^-$-stabilizable. It is clear that any solution of the H_∞^2 CARE must satisfy

$$\begin{pmatrix} \overline{A}'X + X\overline{A} + \overline{Q} - N & XB + SXB' + S' & \overline{R} \end{pmatrix} \geqslant 0.$$

The set of matrices satisfying the above linear matrix inequality is a convex set, and finding any one solution of this equation is a convex feasibility problem that can be solved efficiently (see, for instance, [9]). Let X_0 be one particular solution of this linear matrix inequality. We factorize

$$\left(\bar{A}'X_0 + X_0\bar{A} + \bar{Q} - N \quad X_0 B + S X_0 B' + S' \quad \bar{R} \right) = \begin{pmatrix} C' \\ D' \end{pmatrix} \begin{pmatrix} C & D \end{pmatrix}.$$

But then $\bar{X} = X - X_0$ satisfies an H_2 CARE:

$$\bar{A}'\bar{X} + \bar{X}\bar{A} - (\bar{X}B + C'D)(D'D)^{-1}(B'\bar{X} + D'C) + C'C = 0.$$

It is easily checked that X is a semi-stabilizing solution of our H_∞^2 CARE if and only if $\bar{X}$ is a semi-stabilizing solution of the above H_2 CARE. By Proposition 4.25, we know a semi-stabilizing solution exists. Moreover, we can use the previously developed tools for H_2 CARE to find the semi-stabilizing solution.

Methods to compute the semi-stabilizing solution of an H_∞^1 CARE still need to be discussed. Note that this is, in general, still an open question; that is, for instance, complicated by the fact that the semi-stabilizing solution might not be unique. However, as discussed in Proposition 4.29, computationally the strongly semi-stabilizing solution of an H_∞^1 CARE as in (4.9) can be obtained in terms of the stabilizing solution of a suitably defined H_∞^2 CARE as given in (4.67). As such, the computation of the strongly semi-stabilizing solution of an H_∞^1 CARE reduces to the computation of the stabilizing solution of an H_∞^2 CARE.

4.2 Standard and generalized discrete-time algebraic Riccati equations

Like a CARE in the continuous-time case, the discrete-time algebraic Riccati equation (DARE) and its generalized version abbreviated as a GDARE are important in connection with discrete-time filtering and control problems. The intent of this section is to study DAREs and GDAREs. As in the previous section, our study focuses on three aspects: (1) properties of certain types of solutions of DAREs and GDAREs, such as semi-stabilizing, stabilizing, positive semi-definite, and positive definite solutions that are relevant to H_2 optimal control and filtering; (2) existence and uniqueness of such solutions; and (3) algorithms to compute such solutions. Such a study is facilitated by elegantly developing two primary tools: the first tool develops a one-to-one relationship between a solution of a DARE and a solution of an appropriately defined associated CARE, whereas similarly the second tool develops a one-to-one relationship between a solution of a GDARE and a solution of an appropriately defined associated DARE. Obviously, the first tool enables us to convert the tasks of computing, learning, or understanding the properties of a solution of a DARE to similar tasks on the associated CARE, whereas in a similar way, the second tool enables us to convert the tasks of computing, learning, or understanding the properties of a solution of a GDARE to similar tasks on the associated DARE. Thus, the results of the previous section can elegantly be used for DAREs as well as for GDAREs. Some of the results in this section are based on the results reported in Chen et al. [16] and Stoorvogel and Saberi [94], whereas others have never been reported in the open literature.

The architecture of this section is as follows. After this basic introduction, in Subsection 4.2.1, we define a DARE and its subclasses H_2 DARE, H_∞^1 DARE, and H_∞^2 DARE. Also, we define a generalized version of a DARE called GDARE and its subclasses H_2 GDARE, H_∞^1 GDARE, and H_∞^2 GDARE. Moreover, we define here formally the terminology of semi-stabilizing and stabilizing solutions of a DARE or a GDARE. We examine in Subsection 4.2.2 the basic structure of a GDARE. In Subsection 4.2.3, we connect the solutions of a DARE to deflating subspaces. This section is somewhat analogous to Subsection 4.1.2, which develops the inter-relationship that exists between an Hamiltonian matrix and a CARE. In Subsection 4.2.4, given a DARE, we define an associated CARE and then develop an important tool that connects the solutions of the given DARE to those of its associated CARE. Having done so, in Subsection 4.2.5, either by using the above-mentioned tool or directly, we compile judiciously certain properties of real symmetric semi-stabilizing solutions and stabilizing solutions of a DARE, and then interrelate them with the properties of symmetric positive semi-definite or positive definite solutions of the same DARE. The properties studied here include the existence and uniqueness of semi-stabilizing, stabilizing, positive semi-definite, and positive definite solutions. Subsection 4.2.6 studies the continuity property of the (semi-)stabilizing solution of an H_2 DARE. Analogous to Subsection 4.2.4, in Subsection 4.2.7, we define an associated DARE of a given GDARE and then develop another important tool that connects the solutions of the given GDARE to those of its associated DARE. Using such a tool, in Subsection 4.2.8, we then compile various properties of real symmetric semi-stabilizing solutions and stabilizing solutions of a GDARE and then interrelate them with the properties of symmetric positive semi-definite or positive definite solutions of the same GDARE. Finally, Subsection 4.2.9 shows when a (semi-)stabilizing solution of an H_2 GDARE is continuous with respect to perturbations in the parameters that define the H_2 GDARE.

4.2.1 Definitions

Definition 4.51 Let $A \in \mathbb{R}^{n \times n}$, $B \in \mathbb{R}^{n \times m}$, $Q \in \mathbb{R}^{n \times n}$, $R \in \mathbb{R}^{m \times m}$, and $S \in \mathbb{R}^{n \times m}$ with Q and R being symmetric. Then the constrained quadratic matrix equation for an unknown $n \times n$ matrix X given by

$$\det(R + B'XB) \neq 0 \tag{4.88a}$$

and

$$X = A'XA - (A'XB + S)(R + B'XB)^{-1}(B'XA + S') + Q \tag{4.88b}$$

is called a **discrete-time algebraic Riccati equation (DARE)**.

Just like in continuous time, as given below, we can distinguish three special cases of the above DARE.

Definition 4.52 *Let* $A \in \mathbb{R}^{n \times n}$, $B \in \mathbb{R}^{n \times m}$, $Q \in \mathbb{R}^{n \times n}$, $R \in \mathbb{R}^{m \times m}$, *and* $S \in \mathbb{R}^{n \times m}$ *with* Q *and* R *being symmetric. Then the constrained matrix equation for an unknown* $n \times n$ *matrix* X *given by*

$$X = A'XA - (A'XB + S)(R + B'XB)^{-1}(B'XA + S') + Q, \tag{4.89a}$$

$$R + B'XB > 0, \tag{4.89b}$$

and

$$\begin{pmatrix} Q & S \\ S' & R \end{pmatrix} \geq 0 \tag{4.89c}$$

is called an H_2 *DARE.*

If only (4.89a) *is satisfied with* $Q \geq 0$ *and* $R + B'XB$ *invertible, we will call it an* H_∞^1 *DARE.*

If only (4.89a) *and* (4.89b) *are satisfied, we will call it an* H_∞^2 *DARE.*

Definition 4.53 *A solution of the DARE in* (4.88) *(or a solution to one of the special cases from Definition 4.52) is said to be a semi-stabilizing solution if* $A - B(R + B'XB)^{-1}(B'XA + S')$ *has all of its eigenvalues either on the unit circle or within the unit circle of the complex plane, i.e., in* $\mathbb{C}^\otimes$. *Similarly, a solution of the DARE is said to be a stabilizing solution if* $A - B(R + B'XB)^{-1}(B'XA + S')$ *has all its eigenvalues inside the unit circle of the complex plane, i.e., in* $\mathbb{C}^\ominus$.

A more general DARE that has been used in the optimal control literature is defined next.

Definition 4.54 *Let* $A \in \mathbb{R}^{n \times n}$, $B \in \mathbb{R}^{n \times m}$, $Q \in \mathbb{R}^{n \times n}$, $R \in \mathbb{R}^{m \times m}$, *and* $S \in \mathbb{R}^{n \times m}$ *with* Q *and* R *being symmetric. Then the constrained matrix equation for an unknown* $n \times n$ *matrix* X *given by*

$$X = A'XA - (A'XB + S)(R + B'XB)^{\dagger}(B'XA + S') + Q \tag{4.90a}$$

and

$$\ker[R + B'XB] \subseteq \ker[A'XB + S] \tag{4.90b}$$

*is called a **general DARE** or **GDARE**.*

In the above definition, $S^\dagger$ denotes the Moore–Penrose generalized inverse of S. A generalized inverse (a matrix $S^\dagger$ satisfying $SS^\dagger S = S$ and $S^\dagger SS^\dagger = S^\dagger$) is in general not unique. The Moore–Penrose generalized inverse is the unique generalized inverse such that $SS^\dagger$ is symmetric. However, in the above definition, we have $TS^\dagger V$ with $\ker S \subset \ker T$ and $\mathrm{im}\, S \subseteq \mathrm{im}\, V$. In this case, it is easy to show that $TS^\dagger V$ is uniquely determined independent of our choice for the generalized inverse whenever $\mathrm{im}\, V \subseteq \mathrm{im}\, S$ and $\ker T \subseteq \ker S$. From (4.90b), we can then show that the equality in (4.90a) is independent of our choice for the generalized inverse.

Remark 4.55 *The subspace inclusion in (4.90b) is obviously automatically satisfied whenever $R + B'XB$ is invertible. In this case, the GDARE in (4.90) reduces to the DARE in (4.88).*

We introduce next three special cases of the GDARE that satisfies $R + B'XB \geq 0$. These three cases are the generalizations of the three cases defined in Definition 4.52.

Definition 4.56 *Let $A \in \mathbb{R}^{n \times n}$, $B \in \mathbb{R}^{n \times m}$, $Q \in \mathbb{R}^{n \times n}$, $R \in \mathbb{R}^{m \times m}$, and $S \in \mathbb{R}^{n \times m}$ with Q and R being symmetric. Then the constrained matrix equation for an unknown $n \times n$ matrix X given by*

$$X = A'XA - (A'XB + S)(R + B'XB)^\dagger(B'XA + S') + Q, \tag{4.91a}$$

$$\ker[R + B'XB] \subseteq \ker[A'XB + S], \tag{4.91b}$$

$$R + B'XB \geq 0, \tag{4.91c}$$

and

$$\begin{pmatrix} Q & S \\ S' & R \end{pmatrix} \geq 0 \tag{4.91d}$$

is called an H_2 GDARE.

If only (4.91a) and (4.91b) are satisfied with $Q \geq 0$, we will call it an H_∞^1 GDARE.

If only (4.91a), (4.91b), and (4.91c) are satisfied, we will call it an H_∞^2 GDARE.

Definition 4.57 *Consider a matrix*

$$N(z, X) := \begin{pmatrix} M(z) \\ L(X) \end{pmatrix}, \tag{4.92}$$

where

$$M(z) := \begin{pmatrix} zI - A & -B \end{pmatrix}$$

and

$$L(X) := \begin{pmatrix} Q + A'XA - X & A'XB + S \\ B'XA + S' & B'XB + R \end{pmatrix}. \tag{4.93}$$

Then a solution X of the GDARE in (4.90) (or a solution to one of the special cases from definition 4.56) is said to be a semi-stabilizing solution if the rank of $N(z, X)$ is equal to its normal rank for all z outside the unit circle of the complex plane, i.e., for all $z \in \mathbb{C}^{\oplus}$. Similarly, a solution X of the GDARE is said to be a stabilizing solution if the rank of $N(z, X)$ is equal to its normal rank for all z on or outside the unit circle, i.e., for all $z \in \mathbb{C}^{\oplus} \cup \mathbb{C}^{\circ}$.

Remark 4.58 *Obviously, Definition 4.57 reduces to Definition 4.53 whenever the matrix $R + B'XB$ is nonsingular.*

Remark 4.59 *Definition 4.57 can be rewritten as follows. A solution X of a GDARE is said to be a stabilizing (respectively, semi-stabilizing) solution if all eigenvalues of the matrix*

$$A - B(B'XB + R)^{\dagger}(B'XA + S') - B\left[I - (B'XB + R)^{\dagger}(B'XB + R)\right]F$$

are inside the unit circle (respectively, inside or on the unit circle) for some suitably chosen matrix F.

The distinction between an H_2 and an H_{∞}^2 DARE is whether the matrices Q, R, and S in (4.88) satisfy a certain structural condition. To avoid stating that condition repetitively, we describe it below and label it as Condition psd, where "psd" stands for positive semi-definite.

Definition 4.60 (Condition psd) *The matrices Q, R, and S are said to satisfy Condition psd if*

$$\begin{pmatrix} Q & S \\ S' & R \end{pmatrix} \geq 0.$$

Under this condition, it follows that matrices $C \in \mathbb{R}^{p \times n}$ and $D \in \mathbb{R}^{p \times m}$ exist with $(C \quad D)$ of full rank such that

$$\begin{pmatrix} Q & S \\ S' & R \end{pmatrix} = \begin{pmatrix} C' \\ D' \end{pmatrix}\begin{pmatrix} C & D \end{pmatrix}. \tag{4.94}$$

4.2.2   Basic structure of a GDARE

To start our study of DAREs, let us observe that the only difference between a DARE and a GDARE as formulated in our Definitions 4.51 and 4.54 is whether $R + B'XB$ is nonsingular. On the other hand, as will be established in this subsection, a revealing fundamental property of a GDARE is that either *all* or *none* of the solutions have the property that the corresponding $R + B'XB$ is nonsingular. That is, a given GDARE cannot have one solution, say X_1, for which $R + B'X_1B$ is nonsingular and another solution, say X_2, for which $R + B'X_2B$ is singular. Thus, in one case, the study of a GDARE is exactly the same as that of a DARE, whereas in another case, it is not. The distinction between the two cases depends on the normal rank of a rational matrix $\hat{H}$, to be introduced soon. Similarly, unlike the H^1_∞ GDARE, we note that both H_2 GDARE and an H^2_∞ GDARE require the condition $R + B'XB \geq 0$. Again, as will be seen soon, a revealing fact is that either *all* or *none* of the solutions of a GDARE satisfy the condition $R + B'XB \geq 0$ depending on the properties of the rational matrix $\hat{H}$.

To proceed with, let us first define the rational matrix $\hat{H}$:

$$\hat{H}(z) := \begin{pmatrix} zB'(I - zA')^{-1} & I \end{pmatrix} \begin{pmatrix} Q & S \\ S' & R \end{pmatrix} \begin{pmatrix} (zI - A)^{-1}B \\ I \end{pmatrix}. \qquad (4.95)$$

Remark 4.61 *Consider the case when the matrices Q, R, and S satisfy the Condition psd. Then, we note that $\hat{H}(z) = G'(z^{-1})G(z)$, where $G(z)$ is the transfer function of the system characterized by the matrix quadruple (A, B, C, D).*

The following identity will be useful.

Lemma 4.62 *For any symmetric matrix X, we have*

$$\hat{H}(z) = \hat{H}(z, X) := \begin{pmatrix} zB'(I - zA')^{-1} & I \end{pmatrix} L(X) \begin{pmatrix} (zI - A)^{-1}B \\ I \end{pmatrix}$$

with $L(X)$ defined by (4.93).

Proof : The identity can be verified by writing a realization in descriptor form for $\hat{H}$ and $\hat{H}(\cdot, X)$. ∎

We have the following lemma from [94].

Lemma 4.63 *Consider a GDARE as in (4.90). Let X be any symmetric solution of it. Then the following hold:*

(*i*) $\widehat{H}(z)$ *has full normal rank if and only if* $B'XB + R$ *is invertible.*

(*ii*) *The inertia* of* $B'XB + R$ *is equal to the inertia of* $\widehat{H}(z)$ *for all but finitely many z on the unit circle.*

(*iii*) $B'XB + R \geqslant 0$ *if and only if* $\widehat{H}(z) \geqslant 0$ *for all points z on the unit circle.*

Proof : We will use Lemma 4.62. Assume that a symmetric matrix X that satisfies the GDARE in (4.90) is given. Then it can be checked straightforwardly that

$$V'(z^{-1})L(X)V(z) = \begin{pmatrix} \widehat{H}(z) & 0 \\ 0 & 0 \end{pmatrix}, \tag{4.96}$$

where $V(z)$ is given by

$$V(z) := \begin{pmatrix} (zI - A)^{-1}B & I \\ I & -(B'XB + R)^{\dagger}(B'XA + S') \end{pmatrix} \tag{4.97}$$

and, as before, $S^{\dagger}$ denotes the Moore–Penrose generalized inverse of a matrix S. Note that $V(z)$ is square and invertible for almost all z. Hence, (4.96) implies that the rank of $L(X)$ equals the normal rank of $\widehat{H}(z)$. As X satisfies the GDARE, the rank of $L(X)$ is also equal to the rank of $B'XR + R$. This guarantees in particular that $\widehat{H}(z)$ has full normal rank if and only if $B'XB + R$ is invertible. Moreover, for all but finitely many points on the unit circle, the inertia of $L(X)$ is equal to the inertia of $\widehat{H}(z)$. As the Schur complement of $B'XB + R$ in $L(X)$ is zero, we find that the inertia of $B'XB + R$ equals the inertia of $\widehat{H}(z)$ for all but finitely many points on the unit circle. In particular, we have that $\widehat{H}(z) \geqslant 0$ on the unit circle if and only if $B'XB + R \geqslant 0$. ∎

The last point in the above lemma is basically a special case of the second point but listed separately because it plays an important role in our development here. Note that the above lemma implies that a necessary condition for the existence of a solution to a GDARE is that the inertia of $\widehat{H}(z)$ is independent of z except for some possible singularities. A necessary condition for the existence of a solution with $B'XB + R \geqslant 0$ (i.e., a solution to an H^2_{∞} GDARE) is that $\widehat{H}(z) \geqslant 0$ for all z on the unit circle. Finally, note that $\widehat{H}(z)$, being of full normal rank, guarantees that the generalized inverse in (4.90a) is a normal inverse and that (4.90b) is automatically satisfied, and thus, a GDARE reverts back to a DARE.

**The inertia of a matrix is defined as the triple of the number of eigenvalues outside the unit circle, the number of eigenvalues on the unit circle, and the number of eigenvalues inside the unit circle.*

4.2.3   Solutions of a DARE and deflating subspaces

We first consider the case that $\hat{H}$, defined by (4.95), has full rank. From Theorem 4.63, we know that in this case, if a solution exists to a given GDARE, then it must be that $B'XB + R$ is invertible. Hence, we can study (4.88) instead of (4.90) (i.e., a GDARE in this case reduces to a DARE). Moreover, an obvious necessary condition for the existence of solutions of the DARE is that we can find a symmetric matrix Z such that $R + B'ZB$ is invertible.

Using such a matrix Z, it is easy to check that a matrix X is a solution of the DARE (4.88b) if and only if

$$\begin{pmatrix} \tilde{A} & 0 \\ -\tilde{Q} & I \end{pmatrix} \begin{pmatrix} I \\ X \end{pmatrix} = \begin{pmatrix} I - LZ & L \\ -\tilde{A}'Z & \tilde{A}' \end{pmatrix} \begin{pmatrix} I \\ X \end{pmatrix} A_{cl}, \tag{4.98}$$

where

$$L = B(R + B'ZB)^{-1}B',$$
$$\tilde{A} = A - B(R + B'ZB)^{-1}(B'ZA + S'),$$
$$\tilde{Q} = Q + A'ZA - (A'ZB + S)(R + B'ZB)^{-1}(B'ZA + S'),$$
$$A_{cl} = A - B(R + B'XB)^{-1}(B'XA + S').$$

Given a matrix pencil $H_1 - zH_2$, we note that a subspace $\mathcal{V}$ is called a deflating subspace whenever

$$\dim\{H_1\mathcal{V} + H_2\mathcal{V}\} \leqslant \dim \mathcal{V}.$$

If H_2 is invertible, then $\mathcal{V}$ is a deflating subspace if and only if $\mathcal{V}$ is an invariant subspace of $H_2^{-1}H_1$. A similar comment can be made if H_1 is invertible. Note that if $\mathcal{V}$ is one-dimensional, then for a vector v in $\mathcal{V}$, we have

$$H_1 v = \lambda H_2 v \quad \text{or} \quad H_2 v = 0, \tag{4.99}$$

and if H_2 is invertible, we see that v is an eigenvector of $H_2^{-1}H_1$. For this reason, a λ satisfying (4.99) is often called a generalized eigenvalue provided either $H_1 v$ or $H_2 v$ is unequal to zero. In the case $H_2 v = 0$, the associated generalized eigenvalue is said to be ∞. Generalized eigenvalues are exactly the zeros of a pencil; i.e., the values of z for which $H_1 - zH_2$ loses rank. $z = \infty$ is called a zero of the pencil if $sH_1 - H_2$ loses rank for $s = 0$. Finally, a $2n$-dimensional matrix pencil $H_1 - zH_2$, which satisfies the property that

$$H_1 \begin{pmatrix} 0 & I_n \\ -I_n & 0 \end{pmatrix} H_1' = H_2 \begin{pmatrix} 0 & I_n \\ -I_n & 0 \end{pmatrix} H_2',$$

is called a symplectic pencil. It is easy to see that this implies that λ is a zero if and only if λ^{-1} is a zero. Note that this is in some sense the discrete-time analog of the Hamiltonian matrix.

Using some properties of matrix pencils defined in Section 4.C, we see that (4.98) states that

$$\mathcal{V} = \operatorname{im} \begin{pmatrix} I \\ X \end{pmatrix} \tag{4.100}$$

is a deflating subspace of the matrix pencil

$$\begin{pmatrix} \tilde{A} & 0 \\ -\tilde{Q} & I \end{pmatrix} - z \begin{pmatrix} I - LZ & L \\ -\tilde{A}'Z & \tilde{A}' \end{pmatrix} \tag{4.101}$$

such that the zeros of the pencil restricted to $\mathcal{V}$ are the eigenvalues of the matrix A_{cl}, which are therefore finite. Note that this is a symplectic pencil. Moreover, it is a regular pencil by noting that the pencil can be constructed as a Schur-complement of $R + B'ZB$ of the matrix pencil:

$$\begin{pmatrix} A - zI & 0 & B \\ Z - Q - A'ZA & I - zA' & -A'ZB - S \\ B'ZA + S' & zB' & R + B'ZB \end{pmatrix}. \tag{4.102}$$

After reordering of rows and columns in (4.102), we obtain the matrix

$$\begin{pmatrix} A - zI & B & 0 \\ -Z + Q + A'ZA & A'ZB + S & zA' - I \\ B'ZA + S' & R + B'ZB & zB' \end{pmatrix}.$$

Using Lemma 4.62, it is then easy to check that this matrix has full rank because we assumed $\hat{H}$ has full rank.

If the matrix $\tilde{A}$ is invertible, we find that the matrix

$$\begin{pmatrix} \tilde{A} & 0 \\ -\tilde{Q} & I \end{pmatrix} \begin{pmatrix} I - LZ & L \\ -\tilde{A}'Z & \tilde{A}' \end{pmatrix}^{-1}$$

is a symplectic matrix in the sense that λ is an eigenvalue of this matrix if and only if λ^{-1} is an eigenvalue of this matrix. Moreover, λ and λ^{-1} have the same multiplicity structure. In this case, (4.100) is just an invariant subspace and we see immediately the connection to the Hamiltonian matrix and its role for CAREs. Unfortunately, if $\tilde{A}$ is not invertible, we have to use symplectic pencils and deflating subspaces instead of matrices and invariant subspaces.

In the above, we have obtained the following lemma.

Lemma 4.64 *Assume that $\hat{H}$, given by (4.95), has full normal rank. A subspace $\mathcal{V}$ of the form (4.100) is a deflating subspace of the regular matrix pencil (4.101) such that the matrix pencil restricted to $\mathcal{V}$ has only finite zeros if and only if X is a solution of (4.88).*

Proof : If X is a solution of (4.88), then it is easy to show that $\mathcal{V}$ is a deflating subspace of (4.101) such that the matrix pencil restricted to $\mathcal{V}$ has only finite zeros.

For the converse, we note that if X satisfies (4.98), then the first row implies that

$$\operatorname{im}(I + B(R + B'ZB)^{-1}B'(X - Z))$$
$$\supset \operatorname{im}(A - B(R + B'ZB)^{-1}(B'ZA + S')).$$

Some straightforward manipulations then yield (4.90b). This makes it easy to see that X satisfies (4.90a) as well. As $\widehat{H}$ has full normal rank, Theorem 4.63 then tells us that $B'XB + R$ is invertible and, hence, X satisfies (4.88). $\blacksquare$

The matrix Z plays an important role in the above formulation but can be chosen rather arbitrarily. We will transform our problem to get rid of the matrix Z. Note that doing so has the disadvantage that we lose the structure of a symplectic pencil (and its analogy with the continuous-time Hamiltonian pencil) and in an increase in dimension. Therefore, the above characterization with the presence of Z has some clear advantages, but for a deeper understanding of the structure of the problem, it is attractive to remove this matrix Z.

We define the matrix pencil:

$$\begin{pmatrix} A & 0 & B \\ -Q & I & -S \\ S' & 0 & R \end{pmatrix} - z\begin{pmatrix} I & 0 & \mu B \\ 0 & A' & -\mu S \\ 0 & -B' & \mu R \end{pmatrix}. \tag{4.103}$$

Note that this pencil is no longer symplectic. The following lemma establishes that $\widehat{H}$ has full normal rank if and only if this pencil is regular.

Lemma 4.65 *A μ exists such that the matrix*

$$\begin{pmatrix} I - \mu A & -\mu B & 0 \\ Q & S & -\mu I + A' \\ S & R & B' \end{pmatrix} \tag{4.104}$$

has full rank if and only if $\widehat{H}$ has full normal rank. In this case, (4.104) has full rank for all but finitely many μ. As a consequence, $\widehat{H}$ has full normal rank if and only if the pencil (4.103) is regular.

Proof : First choose μ such that A has no eigenvalue at μ or μ^{-1}. We define $U = B'(\mu I - A')^{-1}Q + S'$ and $V = (\mu I - A')^{-1}[Q(\mu^{-1} - A)^{-1}B + S]$, and we get

$$
\begin{pmatrix} I - \mu A & -\mu B & 0 \\ Q & S & -\mu I + A' \\ S' & R & B' \end{pmatrix}
\begin{pmatrix} I & 0 & (\mu^{-1}I - A)^{-1}B \\ 0 & 0 & I \\ 0 & (A' - \mu I)^{-1} & V \end{pmatrix}
$$

$$
= \begin{pmatrix} (I - \mu A) & 0 & 0 \\ 0 & I & 0 \\ 0 & -B'(\mu I - A')^{-1} & I \end{pmatrix}
\begin{pmatrix} I & 0 & 0 \\ Q & I & 0 \\ U & 0 & \hat{H}(\mu^{-1}) \end{pmatrix}.
$$

Hence, we see that (4.104) has full rank if and only if μ is such that $\hat{H}(\mu^{-1})$ has full rank, which immediately yields the first part of the lemma.

We choose $\lambda \in \mathbb{C}$ such that $\lambda \mu \neq 1$. Then we obtain

$$
\begin{pmatrix} A & 0 & B \\ -Q & I & -S \\ S' & 0 & R \end{pmatrix}
- \lambda \begin{pmatrix} I & 0 & \mu B \\ 0 & A' & -\mu S \\ 0 & -B' & \mu R \end{pmatrix} =
$$

$$
\begin{pmatrix} \lambda I & 0 & 0 \\ 0 & -I & 0 \\ 0 & 0 & I \end{pmatrix}
\begin{pmatrix} I - \lambda^{-1}A & -\lambda^{-1}B & 0 \\ Q & S & -\lambda^{-1}I + A' \\ S' & R & B' \end{pmatrix}
\begin{pmatrix} I & 0 & 0 \\ 0 & 0 & (1 - \lambda\mu)I \\ 0 & \lambda I & 0 \end{pmatrix},
$$

which implies that the pencil (4.103) is regular if the matrix (4.104) has full rank for $\mu = \lambda^{-1}$. Therefore, it is immediate that the pencil is regular if and only if $\hat{H}$ has full normal rank.   ∎

For the pencil (4.103), we will study deflating subspaces of the form

$$
\mathcal{V} = \mathrm{im} \begin{pmatrix} I \\ X \\ P \end{pmatrix}. \tag{4.105}
$$

The paper [33] connected solutions of a (4.91a) to deflating subspaces of (4.103) for $\mu = 0$. In [41, Theorem 15.2.2], a similar result was also obtained for $\mu = 0$ and an extra invertibility requirement on the invariant subspace. We generalize their results and have the following result.

Theorem 4.66 *Assume that the rational matrix H has full normal rank. Choose μ such that the matrix in (4.104) has full rank. Let the pencil (4.103) be given and define L by*

$$
L(X) := \begin{pmatrix} Q + A'XA - X & A'XB + S \\ B'XA + S' & B'XB + R \end{pmatrix}. \tag{4.106}
$$

(*i*) *If a symmetric matrix X is such that the rank of $\mathbf{L}(X)$ is equal to m, then a matrix P exists such that $\mathcal{V}$ defined by (4.105) is a deflating subspace of (4.103). Conversely, if (4.105) is a deflating subspace of (4.103), then X is such that the rank of $\mathbf{L}(X)$ is equal to m.*

(*ii*) *If a matrix X satisfies (4.88), then a matrix P exists such that $\mathcal{V}$ defined by (4.105) is a deflating subspace of (4.103) and the zeros of the pencil restricted to $\mathcal{V}$ are finite. Conversely, if (4.105) is a deflating subspace of (4.103) and the eigenvalues of the pencil restricted to $\mathcal{V}$ are finite, then X is a solution of (4.88).*

Proof : Assume $\mathbf{L}(X)$ has rank m. Let L_1 be square but not necessarily invertible and L_2 be such that

$$\mathbf{L}(X)\begin{pmatrix} L_1 \\ L_2 \end{pmatrix} = 0, \qquad \begin{pmatrix} L_1 \\ L_2 \end{pmatrix} \text{ injective.} \qquad (4.107)$$

We know that (4.104) has full rank, and therefore,

$$\begin{pmatrix} I - \mu A & -\mu B & 0 \\ Q & S & -\mu I + A' \\ S' & R & B' \end{pmatrix} \begin{pmatrix} L_1 \\ L_2 \\ 0 \end{pmatrix} = \begin{pmatrix} L_1 - \mu A L_1 - \mu B L_2 \\ 0 \\ 0 \end{pmatrix}$$

must be injective; hence $L_1 - \mu A L_1 - \mu B L_2$, which is a square matrix, must be invertible. We choose

$$V_1 = A L_1 + B L_2, \quad V_2 = L_1 \text{ and } P = L_2(L_1 - \mu A L_1 - \mu B L_2)^{-1}.$$

Then it is easily checked that

$$\begin{pmatrix} A & 0 & B \\ -Q & I & -S \\ S' & 0 & R \end{pmatrix} \begin{pmatrix} I \\ X \\ P \end{pmatrix} V_2 = \begin{pmatrix} I & 0 & \mu B \\ 0 & A' & -\mu S \\ 0 & -B' & \mu R \end{pmatrix} \begin{pmatrix} I \\ X \\ P \end{pmatrix} V_1. \qquad (4.108)$$

We know that $V_2 - \mu V_1 = (L_1 - \mu A L_1 - \mu B L_2)$ is invertible, and hence, $V_1 - z V_2$ is a regular pencil. By Lemma 4.169, $\mathcal{V}$ defined by (4.105) is then a deflating subspace for (4.103).

To prove the converse in part (*i*), we assume that $\mathcal{V}$ is a deflating subspace of the pencil (4.103). But in this case, we know that matrices V_1 and V_2 exist with $(V_1' \ V_2')$ surjective such that

$$\begin{pmatrix} A & 0 & B \\ -Q & I & -S \\ S' & 0 & R \end{pmatrix} \begin{pmatrix} I \\ X \\ P \end{pmatrix} V_2 = \begin{pmatrix} I & 0 & \mu B \\ 0 & A' & -\mu S' \\ 0 & -B' & \mu R \end{pmatrix} \begin{pmatrix} I \\ X \\ P \end{pmatrix} V_1. \qquad (4.109)$$

But after premultiplication with the matrix

$$W = \begin{pmatrix} I & 0 & 0 \\ -A'X & I & 0 \\ B'X & 0 & I \end{pmatrix},$$

we obtain that

$$L(X)\begin{pmatrix} V_2 \\ P(V_2 - \mu V_1) \end{pmatrix} = 0.$$

Moreover,

$$\begin{pmatrix} V_2 \\ P(V_2 - \mu V_1) \end{pmatrix}$$

is an injective matrix. After all, if $V_2 x = 0$ and $\mu P V_1 x = 0$ for some $x \neq 0$, then we obtain from (4.109) that $V_1 x = 0$, which is in contradiction with the fact that $V_1 - z V_2$ is a regular pencil. Hence, the rank of $L(X)$ is less than or equal to m. However, Lemma 4.62 together with the assumption that $\hat{H}$ is of full rank guarantees that the rank of $L(X)$ is at least m.

For part (ii), we note that X satisfies (4.88) if and only if the rank of $L(X)$ is equal to the rank of $B'XB + R$, which is then invertible. Moreover, this is equivalent to the requirement that in (4.107) we can choose $L_1 = I$. On the other hand, $\mathcal{V}$ is a deflating subspace of the pencil (4.103) such that the zeros of the pencil restricted to $\mathcal{V}$ are finite if and only if (4.108) is satisfied with $V_2 = I$. The same steps as in the proof of part (i) but with $V_2 = L_1 = I$ then yield a proof of part (ii).   ■

Remark 4.67 *Suppose we have the matrix pencil (4.103). We can ask ourselves whether we can derive a result equivalent to Theorem 4.66 for the case when the rational matrix $\hat{H}$ is no longer of full rank. This can indeed be done, and we refer interested readers to [94].*

The above development leads to the following results regarding (semi-)stabilizing solutions of DAREs.

Theorem 4.68 *Assume that $\hat{H}$ has full normal rank. A stabilizing solution of the DARE as in Definition 4.51, if it exists, is unique. Moreover, if a semi-stabilizing solution exists, it is actually a stabilizing solution if and only if*

$$\begin{pmatrix} zI - A & -B & 0 \\ Q & S & I - zA' \\ S' & R & -zB' \end{pmatrix} \tag{4.110}$$

has full rank for all z on the unit circle.

Proof : A stabilizing solution of the DARE is clearly unique because solutions of it have a one-to-one relation with deflating subspaces of the symplectic pencil (4.101). As the symplectic pencil has at most n stable eigenvalues, a stable n-dimensional subspace of the pencil is unique, and hence, the associated solution to the DARE is unique.

The matrix pencil (4.103) is regular. Therefore, a semi-stabilizing solution of the DARE is necessarily stabilizing if the matrix pencil (4.103) has no zeros on the unit circle. It is easy to see that if (4.103) has a zero λ (with $\lambda\mu \neq 1$), then (4.110) has a nonempty kernel for $z = \lambda$. Hence, if (4.110) has full rank for all z on the unit circle, then the matrix pencil (4.103) has no zeros on the unit circle. ■

Remark 4.69 *If Q, R, and S satisfy Condition psd and we make the factorization in (4.94), then it can be checked that the zeros of the pencil (4.110) are exactly the invariant zeros and their inverses of the system (A, B, C, D).*

4.2.4   Connections between a DARE and its associated CARE

In this subsection, at first, our main goal is to establish a one-to-one relationship between a solution of a DARE and that of an appropriately defined associated CARE. This relationship has a long history. As far as we know, the first result for the H_2 DARE was presented in [96]. For the H_∞ DARE, this result was obtained in [16]. This relationship, as we pointed out, has an immediate consequence: Namely, it enables us to convert various tasks related to a DARE, such as computing a solution, verifying the existence of a positive definite or a semi-definite solution, or learning or understanding some other properties of a solution, to similar tasks on the associated CARE. As we have extensively studied and developed several important properties of certain solutions of a CARE, the advantage in transforming any given task related to a DARE to that of an associated CARE is then transparent. Using such an advantage, we next enumerate without proof a number of relevant properties and existence conditions of certain solutions of a DARE. In a similar manner, several other properties of DAREs that are not enumerated here can be developed.

In this subsection, for simplicity, we assume that matrix A has no eigenvalues at -1. Note that if this is not satisfied, then we choose F such that $\widetilde{A} = A + BF$ has no eigenvalues in -1 and we note that X is a solution of the DARE (4.88) if and only if X is a solution of the following DARE:

$$X = \widetilde{A}'X\widetilde{A} - (\widetilde{A}'XB + \widetilde{S})(R + B'XB)^{-1}(B'XA + \widetilde{S}') + \widetilde{Q}, \qquad (4.111)$$

where

$$\widetilde{S} = S - F'R, \qquad \widetilde{Q} = Q - SF - F'S' + F'RF.$$

By working with (4.111), we can then establish the same results.

As the matrix A has no eigenvalues at -1, we can define

$$\bar{A} := (A + I)^{-1}(A - I),$$
$$\bar{B} := 2(A + I)^{-2}B,$$
$$\bar{S} := -Q(A + I)^{-1}B + S,$$
$$\bar{R} := R + B'(A' + I)^{-1}Q(A + I)^{-1}B - S'(A + I)^{-1}B - B'(A' + I)^{-1}S. \tag{4.112}$$

Given a DARE, we define next an associated CARE.

Definition 4.70 *Given a DARE as in Definition 4.51, we define an* **associated CARE** *as*

$$\tilde{X}\bar{A} + \bar{A}'\tilde{X} - (\tilde{X}\bar{B} + \bar{S})\bar{R}^{-1}(\bar{B}'\tilde{X} + \bar{S}') + Q = 0. \tag{4.113}$$

Theorem 4.71 *Consider a DARE as in* (4.88) *and its associated CARE as in* (4.113). *Assume that A has no eigenvalues at -1. Also, consider two $n \times n$ matrices X and $\tilde{X}$ related by $X = 2(A' + I)^{-1}\tilde{X}(A + I)^{-1}$. Then the following two statements are equivalent:*

(*i*) *X is a symmetric solution to the DARE as in* (4.88) *and the matrix pencil*

$$\begin{pmatrix} Q & S & I - zA' \\ S & R & -zB' \\ zI - A & -B & 0 \end{pmatrix} \tag{4.114}$$

   has no zero at -1.

(*ii*) *$\tilde{X}$ is a symmetric solution to the associated CARE in* (4.113), *and the matrix $\bar{A} - \bar{B}\bar{R}^{-1}(\bar{B}'\tilde{X} + \bar{S}')$ has no eigenvalue at 1.*

Remark 4.72 *Note that the matrix $\bar{R}$ defined in* (4.112) *is nonsingular if and only if the matrix pencil in* (*i*) *has no zero at -1. Note that this condition is quite natural because the above connection is related to a bilinear transform, and therefore, if the symplectic pencil has an eigenvalue at -1, then this would imply that the associated continuous-time Hamiltonian would need to have an eigenvalue at ∞ that clearly is not possible. Conversely, when starting with the continuous-time Hamiltonian, it is not a problem that eigenvalues are mapped to infinity because a symplectic pencil can of course have eigenvalues at infinity. But as noted before, only deflating subspaces connected to finite eigenvalues are coupled to the solutions of the CARE. Therefore, in this case, we have to make sure that the invariant subspace of the Hamiltonian is not connected to the point 1, which is expressed in condition* (*ii*). *Note that $\bar{A} - \bar{B}\bar{R}^{-1}(\bar{B}'\tilde{X} + \bar{S}')$ having no eigenvalue at 1 can be shown to be equivalent to $B'XB + R$ being invertible provided $\tilde{X}$ satisfies the above CARE.*

Proof : Now, let us consider the following reductions:

$$A'XA - X + Q$$
$$= 2A'(A' + I)^{-1}\tilde{X}(A + I)^{-1}A - 2(A' + I)^{-1}\tilde{X}(A + I)^{-1} + Q$$
$$= (A' + I)^{-1}(2A'\tilde{X}A - 2\tilde{X})(A + I)^{-1} + Q$$
$$= (A' + I)^{-1}[(A' + I)\tilde{X}(A - I) + (A' - I)\tilde{X}(A + I)](A + I)^{-1} + Q$$
$$= \tilde{X}(A - I)(A + I)^{-1} + (A' + I)^{-1}(A' - I)\tilde{X} + Q$$
$$= \tilde{X}\bar{A} + \bar{A}'\tilde{X} + Q. \tag{4.115}$$

Moreover, it follows from (4.112) that

$$A = (I + \bar{A})(I - \bar{A})^{-1}$$
$$B = 2(I - \bar{A})^{-2}\bar{B}$$
$$\bar{S} = -Q(I - \bar{A})^{-1}\bar{B} + S$$
$$X = (I - \bar{A}')\tilde{X}(I - \bar{A})/2$$
$$\bar{R} = R + \bar{B}'(I - \bar{A}')^{-1}Q(I - \bar{A})^{-1}\bar{B} - S'(I - \bar{A})^{-1}\bar{B} - \bar{B}'(I - \bar{A}')^{-1}S.$$

Finally,

$$R + B'XB = \bar{R} + \bar{B}'(I - \bar{A}')^{-1}(\bar{A}'\tilde{X} + \tilde{X}\bar{A} + Q)(I - \bar{A})^{-1}\bar{B}$$
$$+ (\bar{S}' + \bar{B}'\bar{X})(I - \bar{A})^{-1}\bar{B} + \bar{B}'(I - \bar{A}')^{-1}(\bar{S} + \bar{X}\bar{B}). \tag{4.116}$$

The above equation implies, after some algebraic manipulations, that $R + B'XB$ is invertible if and only if

$$\begin{pmatrix} \bar{A}'\tilde{X} + \tilde{X}\bar{A} + Q & \bar{S} + \bar{X}\bar{B} & I - \bar{A}' \\ \bar{S}' + \bar{B}'\bar{X} & \bar{R} & \bar{B}' \\ I - \bar{A} & B & 0 \end{pmatrix} \tag{4.117}$$

is invertible.

$(i) \Rightarrow (ii)$ Let us start with the following trivial equality:

$$(A' + I)X(A + I) - (A' + I)XA - A'X(A + I) + A'XA - X = 0.$$

The above equality implies that

$$X - XA(A + I)^{-1} - (A' + I)^{-1}A'X$$
$$+ (A' + I)^{-1}A'XA(A + I)^{-1} - (A' + I)^{-1}X(A + I)^{-1} = 0.$$

Then we have

$$
\begin{aligned}
\bar{R} &= R + B'(A' + I)^{-1}Q(A + I)^{-1}B - S'(A + I)^{-1}B - B'(A' + I)^{-1}S \\
&= R + B'(A' + I)^{-1}Q(A + I)^{-1}B - S'(A + I)^{-1}B - B'(A' + I)^{-1}S \\
&\quad + B'XB - B'XA(A + I)^{-1}B - B'(A' + I)^{-1}A'XB \\
&\quad + B'(A' + I)^{-1}A'XA(A + I)^{-1}B - B'(A' + I)^{-1}X(A + I)^{-1}B \\
&= R + B'XB - (B'XA + S')(A + I)^{-1}B - B'(A' + I)^{-1}(A'XB + S) \\
&\quad + B'(A' + I)^{-1}(A'XA + Q - X)(A + I)^{-1}B \\
&= R + B'XB - (B'XA + S')(A + I)^{-1}B - B'(A' + I)^{-1}(A'XB + S) \\
&\quad + B'(A' + I)^{-1}(A'XB + S)(R + B'XB)^{-1}(B'XA + S')(A + I)^{-1}B \\
&= [I - B'(A' + I)^{-1}(A'XB + S)(R + B'XB)^{-1}] \\
&\quad \times (R + B'XB)[I - (R + B'XB)^{-1}(B'XA + S')(A + I)^{-1}B].
\end{aligned}
$$

Note that we have used (4.88) for the fourth equality.

The condition that the matrix pencil in (i) has no zero at -1 implies that $\bar{R}$ is invertible. We obtain

$$
\begin{aligned}
R + B'XB &= [I - B'(A' + I)^{-1}(A'XB + S)(R + B'XB)^{-1}]^{-1}\bar{R} \\
&\quad \times [I - (R + B'XB)^{-1}(B'XA + S')(A + I)^{-1}B]^{-1}. \quad (4.118)
\end{aligned}
$$

Hence,

$$
\begin{aligned}
(S + A'XB)&(R + B'XB)^{-1}(B'XA + S') \\
&= (A'XB + S)[I - (R + B'XB)^{-1}(B'XA + S')(A + I)^{-1}B]\bar{R}^{-1} \\
&\quad \times [I - (R + B'XB)^{-1}(B'XA + S')(A + I)^{-1}B]'(B'XA + S') \\
&= [A'XB + (X - A'XA - Q)(A + I)^{-1}B + S]\bar{R}^{-1} \\
&\quad \times [A'XB + (X - A'XA - Q)(A + I)^{-1}B + S]' \\
&= [(A'X + X - Q)(A + I)^{-1}B + S]\bar{R}^{-1} \\
&\quad \times [(A'X + X - Q)(A + I)^{-1}B + S]' \\
&= [(A' + I)X(A + I)(A + I)^{-2}B - Q(A + I)^{-1}B + S]\bar{R}^{-1} \\
&\quad \times [(A' + I)X(A + I)(A + I)^{-2}B - Q(A + I)^{-1}B + S]' \\
&= (\tilde{X}\bar{B} + \bar{S})\bar{R}^{-1}(\bar{B}'\tilde{X} + \bar{S}').
\end{aligned}
$$

Again, we have used (4.88) for the second equality. Finally, the above equation together with (4.88) and (4.115) imply that

$$
\tilde{X}\bar{A} + \bar{A}'\tilde{X} - (\tilde{X}\bar{B} + \bar{S})\bar{R}^{-1}(\bar{B}'\tilde{X} + \bar{S}') + Q = 0.
$$

Finally, note that the condition that $B'XB + R$ is invertible implies that (4.117) is invertible. It is not hard to check that this, combined with the above CARE, implies that $\bar{A} - \bar{B}\bar{R}^{-1}(\bar{B}'\tilde{X} + \bar{S}')$ has no eigenvalue at 1.

$(ii) \Rightarrow (i)$

Using (4.116) and the associated CARE (4.113), we have

$$
\begin{aligned}
R + B'XB &= \bar{R} + \bar{B}'(I - \bar{A}')^{-1}(\tilde{X}\bar{B} + \bar{S}) + (\tilde{X}\bar{B} + \bar{S})'(I - \bar{A})^{-1}\bar{B} \\
&\quad + \bar{B}'(I - \bar{A}')^{-1}(\tilde{X}\bar{B} + \bar{S})\bar{R}^{-1}(\tilde{X}\bar{B} + \bar{S})'(I - \bar{A})^{-1}\bar{B} \\
&= [I + \bar{R}^{-1}(\tilde{X}\bar{B} + \bar{S})'(I - \bar{A})^{-1}\bar{B}]'\bar{R} \\
&\quad \times [I + \bar{R}^{-1}(\tilde{X}\bar{B} + \bar{S})'(I - \bar{A})^{-1}\bar{B}].
\end{aligned}
$$

The associated CARE together with $\bar{A} - \bar{B}\bar{R}^{-1}(\bar{B}'\tilde{X} + \bar{S}')$ having no eigenvalue at 1 implies that (4.117) is invertible. But then this implies that $B'XB + R$ is invertible. Thus, we can rewrite the above equation as

$$
\begin{aligned}
\bar{R} = [I + \bar{B}'(I - \bar{A}')^{-1}(\tilde{X}\bar{B} + \bar{S})\bar{R}^{-1}]^{-1}(R + B'XB) \\
\times [I + \bar{R}^{-1}(\tilde{X}\bar{B} + \bar{S})'(I - \bar{A})^{-1}\bar{B}]^{-1}.
\end{aligned}
$$

We have the following reductions:

$$
\begin{aligned}
(\tilde{X}\bar{B} &+ \bar{S})\bar{R}^{-1}(\tilde{X}\bar{B} + \bar{S})' \\
&= (\tilde{X}\bar{B} + \bar{S})[I + \bar{R}^{-1}(\tilde{X}\bar{B} + \bar{S})'(I - \bar{A})^{-1}\bar{B}] \\
&\quad \times (R + B'XB)^{-1}[I + \bar{R}^{-1}(\tilde{X}\bar{B} + \bar{S})'(I - \bar{A})^{-1}\bar{B}]'(\tilde{X}\bar{B} + \bar{S})' \\
&= [\tilde{X}\bar{B} + \bar{S} + (\tilde{X}\bar{B} + \bar{S})\bar{R}^{-1}(\tilde{X}\bar{B} + \bar{S})'(I - \bar{A})^{-1}\bar{B}] \\
&\quad \times (R + B'XB)^{-1} \\
&\quad \times [\tilde{X}\bar{B} + \bar{S} + (\tilde{X}\bar{B} + \bar{S})\bar{R}^{-1}(\tilde{X}\bar{B} + \bar{S})'(I - \bar{A})^{-1}\bar{B}]' \\
&= [\tilde{X}\bar{B} - Q(I - \bar{A})^{-1}\bar{B} + (\tilde{X}\bar{A} + \bar{A}'\tilde{X} + Q)(I - \bar{A})^{-1}\bar{B} + S] \\
&\quad \times (R + B'XB)^{-1} \\
&\quad \times [\tilde{X}\bar{B} - Q(I - \bar{A})^{-1}\bar{B} + (\tilde{X}\bar{A} + \bar{A}'\tilde{X} + Q)(I - \bar{A})^{-1}\bar{B} + S]' \\
&= [(I + \bar{A}')\tilde{X}(I - \bar{A})^{-1}\bar{B} + S](R + B'XB)^{-1} \\
&\quad \times [\bar{B}'(I - \bar{A}')^{-1}\tilde{X}(I + \bar{A}) + S'] \\
&= (A'XB + S)(R + B'XB)^{-1}(B'XA + S').
\end{aligned}
$$

Again, we have used (4.113) to get the third equality. Finally, it follows from (4.113), (4.115), and the above equation that

$$
A'XA - (A'XB + S)(R + B'XB)^{-1}(B'XA + S') + Q - X = 0.
$$

The proof of Theorem 4.71 is now complete.   ■

Theorem 4.71 characterizes in general the relationship between the solutions of a DARE in (4.88) and those of its associated CARE (4.113). In the following Theorem 4.74, we will establish the relationship between a stabilizing solution of a

DARE and that of its associated CARE. We basically have three types of DAREs: H_2, H_∞^1, and H_∞^2 DAREs. We made the same distinctions in continuous-time. It is therefore interesting to see whether the above connection between the CAREs and DAREs preserves this subdivision. In other words, we enquire whether an H_2 DARE can be mapped to an H_2 CARE and conversely. Similar questions can be formulated for H_∞^1 and H_∞^2 DAREs and CAREs. The answers for the cases are positive and are a direct consequence of the following lemma.

Lemma 4.73 *Using the notation of Theorem (4.71), we have*

$$\begin{pmatrix} Q & S \\ S' & R \end{pmatrix} \geqslant 0$$

if and only if

$$\begin{pmatrix} Q & \bar{S} \\ \bar{S}' & \bar{R} \end{pmatrix} \geqslant 0.$$

Moreover, $B'XB + R \geqslant 0$ if and only if $\bar{R} \geqslant 0$.

Proof : We have

$$\begin{pmatrix} I & -(A+I)^{-1}B \\ 0 & I \end{pmatrix} \begin{pmatrix} Q & S \\ S' & R \end{pmatrix} \begin{pmatrix} I & 0 \\ -B'(A'+I)^{-1} & I \end{pmatrix} = \begin{pmatrix} Q & \bar{S} \\ \bar{S}' & \bar{R} \end{pmatrix}.$$

This equation immediately yields the first result. The second result follows from (4.118). ∎

Theorem 4.74 *Consider a DARE as in (4.88) and its associated CARE (4.113). Assume that A has no eigenvalues at -1. Also, consider two $n \times n$ matrices X and $\widetilde{X}$ related by $X = 2(A' + I)^{-1}\widetilde{X}(A + I)^{-1}$. Then the following two statements are equivalent:*

(i) *X is a stabilizing solution of the DARE in (4.88) implying that the matrix*

$$A_{cl} := A - B(R + B'XB)^{-1}(B'XA + S') \tag{4.119}$$

has all its eigenvalues in $\mathbb{C}^\ominus$.

(ii) *$\widetilde{X}$ is a stabilizing solution of the associated CARE (4.113) implying that the matrix*

$$\widetilde{A}_{cl} := \widetilde{A} - \widetilde{R}\widetilde{X} \tag{4.120}$$

has all its eigenvalues in $\mathbb{C}^-$.

Proof : $(i) \Rightarrow (ii)$

By Theorem 4.68, we immediately find that the pencil (4.114) has no zero at -1. Therefore, according to Theorem 4.71, we know that $\widetilde{X}$ satisfies the CARE. Let

$$Z := I - (R + B'XB)^{-1}(B'XA + S')(A + I)^{-1}B.$$

Noting that

$$\det Z = \det[I - (R + B'XB)^{-1}(B'XA + S')(A + I)^{-1}B]$$
$$= \det[I - B(R + B'XB)^{-1}(B'XA + S')(A + I)^{-1}]$$
$$= \det[I + A_{cl}] \det[(A + I)^{-1}],$$

it follows that Z is nonsingular provided that the eigenvalues of A_{cl} are all inside the unit circle $\mathbb{C}^{\ominus}$. Following the arguments from the proof of Theorem 4.71, we have

$$\begin{aligned}
\widetilde{A}_{cl} &= \widetilde{A} - \widetilde{R}\widetilde{X} \\
&= \overline{A} - \overline{B}\overline{R}^{-1}(\widetilde{X}\overline{B} + \overline{S})' \\
&= \overline{A} - \overline{B}Z^{-1}(R + B'XB)^{-1}(B'XA + S') \\
&= (A + I)^{-1}(A - I) - 2(A + I)^{-2}B \\
&\quad \times [I - (R + B'XB)^{-1}(B'XA + S')(A + I)^{-1}B]^{-1} \\
&\quad \times (R + B'XB)^{-1}(B'XA + S') \\
&= (A + I)^{-1}\{A - I - 2[I + A - B(R + B'XB)^{-1}(B'XA + S')]^{-1} \\
&\quad \times B(R + B'XB)^{-1}(B'XA + S')\} \\
&= (A + I)^{-1}(A_{cl} + I)^{-1}\{[I + A - B(R + B'XB)^{-1}(B'XA + S')] \\
&\quad \times (A - I) - 2B(R + B'XB)^{-1}(B'XA + S')\} \\
&= (A + I)^{-1}(A_{cl} + I)^{-1}(A_{cl} - I)(A + I).
\end{aligned}$$

This result implies that the eigenvalues of $\widetilde{A}_{cl}$ are all in $\mathbb{C}^-$ provided that the eigenvalues of A_{cl} are all inside $\mathbb{C}^{\ominus}$. This statement follows by noting that if λ is an eigenvalue of A_{cl}, then $(\lambda + 1)^{-1}(\lambda - 1)$ is an eigenvalue of $\widetilde{A}_{cl}$.

$(ii) \Rightarrow (i)$

Note that because $\overline{A} - \overline{B}\overline{R}^{-1}(\overline{B}'\widetilde{X} + \overline{S}')$ is $\mathbb{C}^-$-stable, it obviously has no eigenvalue at 1. Therefore, Theorem (4.71) yields that X satisfies the DARE. The rest follows from reversing the above arguments for the implication "$(i) \Rightarrow (ii)$".

■

4.2.5 *Properties, existence, and computation of various types of solutions of a DARE*

As we noted, Theorems 4.71 and 4.74 lay an important road map to examine the properties of various types of solutions of a DARE, to study the existence conditions of them, and to compute them in terms of similar tasks on an appropriately defined associated CARE. There is, however, one technical problem. We can only connect solutions of a DARE to the solutions of the said CARE when the pencil (4.114) has no zero at -1. In other words, only under the specific requirement that the pencil (4.114) has no zero at -1 do we have a one-one connection between the set of all solutions of the DARE and the set of all solutions of the CARE. It is this difficulty that prevents us from *immediately* obtaining the discrete-time results from the corresponding properties of the associated continuous-time results. As in the case of CAREs, in this subsection, (a) we study the semi-stabilizing, stabilizing, positive semi-definite and positive definite properties of a solution of a DARE, (b) we develop the existence conditions for such solutions, and (c) we also develop methods of computing them. We obtain some of the properties of a DARE directly rather than via the associated CARE, whereas others are derived by using the associated CARE especially when the above said technical problem does not play any major role.

We have the following proposition regarding the semi-stabilizing property of a solution of a DARE, which is derived directly rather than via the associated CARE.

Proposition 4.75 *Consider an H_∞^2 DARE or an H_2 DARE as in Definition 4.52. Assume that the pair (A, B) is $\mathbb{C}^\ominus$-stabilizable and $\hat{H}$ has full normal rank. Then the following hold:*

  (i) *A real symmetric semi-stabilizing solution X_{ss} of the H_∞^2 DARE or H_2 DARE, if it exists, is larger than or equal to any real symmetric solution X; i.e., $X_{ss} \geq X$.*

  (ii) *A real symmetric semi-stabilizing solution X_{ss} of the H_∞^2 DARE or H_2 DARE, if it exists, is unique.*

  (iii) *A real symmetric semi-stabilizing solution X_{ss} of the H_2 DARE, if it exists, is positive semi-definite; i.e., $X_{ss} \geq 0$.*

Proof : Let a real symmetric semi-stabilizing solution X_{ss} and a real symmetric solution X of the H_∞^2 DARE exist satisfying (4.89a) and (4.89b). Note that any real symmetric solution X also satisfies the linear matrix inequality (LMI):

$$\begin{pmatrix} -X + A'XA + Q & A'XB + S \\ S' + B'XA & R + B'XB \end{pmatrix} \geq 0. \tag{4.121}$$

We define

$$\begin{pmatrix} \bar{Q} & \bar{S} \\ \bar{S}' & \bar{R} \end{pmatrix} = \begin{pmatrix} -X + A'XA + Q & A'XB + S \\ S' + B'XA & R + B'XB \end{pmatrix} \geq 0.$$

It follows from (4.89b) that $\bar{R} > 0$. Let us next define

$$\bar{X} = X_{ss} - X.$$

This $\bar{X}$ satisfies the DARE:

$$\bar{X} = A'\bar{X}A - (A'\bar{X}B + \bar{S})(\bar{R} + B'\bar{X}B)^{-1}(B'\bar{X}A + \bar{S}') + \bar{Q}, \qquad (4.122)$$

or equivalently, it satisfies the DARE:

$$\begin{cases} \bar{X} = A'_{cl}\bar{X}A_{cl} + \begin{pmatrix} I & F' \end{pmatrix}\begin{pmatrix} \bar{Q} & \bar{S} \\ \bar{S}' & \bar{R} \end{pmatrix}\begin{pmatrix} I \\ F \end{pmatrix}, \\ F = (\bar{R} + B'\bar{X}B)^{-1}(B'\bar{X}A + \bar{S}'), \\ A_{cl} = A - BF. \end{cases} \qquad (4.123)$$

It is straightforward to observe that

$$A_{cl} = A - B(R + B'X_{ss}B)^{-1}(B'X_{ss}A + S').$$

As such, A_{cl} has all its eigenvalues inside or on the unit circle because X_{ss} is a semi-stabilizing solution. We show next that all (generalized) eigenvectors associated with the eigenvalues on the unit circle of A_{cl} are contained in ker $\bar{X}$. We prove this only for the eigenvectors, and similar analysis can be developed for the generalized eigenvectors (if there are any). Note that we can rewrite A_{cl} as

$$A_{cl} = (I + B\bar{R}^{-1}B'\bar{X})^{-1}\tilde{A} \qquad (4.124)$$

with $\tilde{A} = A - B\bar{R}^{-1}\bar{S}$. Assume that $A_{cl}x = \lambda x$ with $|\lambda| = 1$. From the DARE (4.123), we obtain

$$x^*\begin{pmatrix} I & F' \end{pmatrix}\begin{pmatrix} \bar{Q} & \bar{S} \\ \bar{S}' & \bar{R} \end{pmatrix}\begin{pmatrix} I \\ F \end{pmatrix}x = 0,$$

where * denotes complex conjugate transpose, which implies that $(\bar{S}' + \bar{R}F)x = 0$. The latter yields $\tilde{A}x = A_{cl}x$, and hence,

$$B\bar{R}^{-1}B'\bar{X}x = 0.$$

This equation in turn yields $B'\bar{X}x = 0$. Then the DARE (4.123) yields also

$$\bar{X}x = \lambda A'_{cl}\bar{X}x = \lambda A'\bar{X}x.$$

If $\bar{X}x \neq 0$, the above equation implies that $(\lambda^*)^{-1}$ is an eigenvalue of A with the left eigenvector as $(\bar{X}x)^*$. In view of the fact that $B'\bar{X}x = 0$, we observe then that $(\lambda^*)^{-1}$ is an uncontrollable eigenvalue of the pair (A, B). This yields a contradiction. Hence, $\bar{X}x = 0$. Then in a suitable basis, we obtain the following:

$$\tilde{A} = \begin{pmatrix} A_{11} & 0 \\ A_{21} & A_{22} \end{pmatrix}, \qquad \bar{X} = \begin{pmatrix} X_{11} & 0 \\ 0 & 0 \end{pmatrix}, \qquad B = \begin{pmatrix} B_1 \\ B_2 \end{pmatrix},$$

with A_{11} having no eigenvalues on the unit circle and X_{11} satisfying the DARE:

$$X_{11} = A'_{11} X_{11} A_{11} - A'_{11} X_{11} B_1 (\bar{R} + B_1 X_{11} B'_1)^{-1} B_1 X_{11} A_{11},$$

whereas the eigenvalues of

$$A_{11,cl} = A_{11} - B_1 (\bar{R} + B_1 X_{11} B'_1)^{-1} B_1 X_{11} A_{11}$$

are all in the open unit disk. To obtain the above DARE, we have used the fact that $\bar{Q} - \bar{S} \bar{R}^{-1} \bar{S}' = 0$. We obtain

$$X_{11} = A'_{11,cl} X_{11} A_{11,cl} + F' \bar{R} F.$$

From Lemma 4.234 we see that $X_{11} \geqslant 0$, and hence, $X_{ss} \geqslant X$. This proves part (i) of Proposition 4.75.

The proof of part (ii) of Proposition 4.75 follows from the fact that the semi-stabilizing solution must be the largest solution, and hence, it must obviously be unique.

Finally, for an H_2 DARE, we observe that $X = 0$ satisfies (4.121), and hence, the above development implies that $X_{ss} \geqslant 0$. ∎

The following proposition follows simply from the proof of Proposition 4.75.

Proposition 4.76 *Consider an H_∞^2 DARE as in (4.89). Assume that the pair $(\bar{A}, B)$ is $\mathbb{C}^{\ominus}$-stabilizable and $\hat{H}$ has full normal rank.*

A semi-stabilizing solution of the H_∞^2 DARE exists if and only if a matrix $\bar{X}$ exists satisfying the linear matrix inequality:

$$\begin{pmatrix} -\bar{X} + A'\bar{X}A + Q & A'\bar{X}B + S \\ B\bar{X}A + S' & R \end{pmatrix} \geqslant 0. \tag{4.125}$$

Moreover, such a semi-stabilizing solution is unique and is the largest among all symmetric solutions. Finally, X is a semi-stabilizing solution of the H_∞^2 DARE if and only if $\hat{X} = X - \bar{X}$ is a semi-stabilizing solution of the H_2 DARE:

$$\hat{X} = A'\hat{X}\bar{A} - (A'XB + \bar{S})(B'\hat{X}B + \bar{R})^{-1}(B'\hat{X}A + \bar{S}') + \bar{Q} = 0, \tag{4.126}$$

where

$$\begin{pmatrix} \bar{Q} & \bar{S} \\ \bar{S}' & \bar{R} \end{pmatrix} = \begin{pmatrix} -\bar{X} + A'\bar{X}A + Q & A'\bar{X}B + S \\ B\bar{X}A + S' & R \end{pmatrix} \geqslant 0.$$

Proof : It follows from the proof of Proposition 4.75 where we use Theorem 4.84 for the existence of a semi-stabilizing solution of the H_2 DARE (4.126). ∎

Remark 4.77 *Clearly, in view of Proposition 4.76, computing a semi-stabilizing solution of an H_∞^2 DARE can be reduced to finding a semi-stabilizing solution of an associated H_2 DARE.*

The following proposition is derived directly and not via the associated CARE. It pertains to the positive semi-definite property of a solution of an H_2 DARE.

Proposition 4.78 *Consider an H_2 DARE as in (4.89) with the matrices Q, R, and S satisfying the Condition psd, and the pair (A, B) being $\mathbb{C}^\ominus$-stabilizable. Also, let $\mathcal{U}$ denote the $\mathbb{C}^\otimes$ unobservable subspace of*

$$(Q - SR^{-1}S', A - BR^{-1}S).$$

Then, a positive semi-definite solution X of the H_2 DARE, if it exists, has $\mathcal{U}$ as a subset of its null space. That is, $X\mathcal{U} = 0$.

Proof : The proof follows along the same lines as that of Proposition 4.33 but is given here for completeness.

Let X be a positive semi-definite solution of the H_2 DARE (4.89). We have

$$X = \tilde{A}'X\tilde{A} + \tilde{Q} - \tilde{A}'XB(B'XB + R)^{-1}B'X\tilde{A} \qquad (4.127)$$

with

$$\tilde{A} = A - BR^{-1}S \quad \text{and} \quad \tilde{Q} = Q - S'R^{-1}S.$$

Also, clearly $(\tilde{A}, B)$ is $\mathbb{C}^\ominus$-stabilizable. Assume that $\tilde{A}x = \lambda x$ and $\tilde{Q}x = 0$. We find that

$$(1 - |\lambda|^2)x^*Xx = -|\lambda|^2 x^*XB(B'XB + R)^{-1}B'Xx.$$

When $|\lambda| < 1$, we find from $X \geqslant 0$ and $B'XB + R \geqslant 0$ that $Xx = 0$. On the other hand, when $|\lambda| = 1$, we find that $B'Xx = 0$. But applying x to the right-hand side of (4.127) then yields

$$Xx = \lambda \tilde{A}Xx.$$

But then combined with $B'Xx = 0$, this would yield that $(\tilde{A}, B)$ is not $\mathbb{C}^\ominus$-stabilizable, which yields a contradiction unless $Xx = 0$.

As in the proof of Proposition 4.33, we can also establish that generalized eigenvectors of A inside $\mathcal{U}$ are also contained in $\ker X$. Hence, the result.   ∎

Proposition 4.78 helps us to construct the semi-stabilizing solution of an H_2 DARE. In this regard, we first note that we can use the relationship with the associated CARE provided the pencil (4.114) has no zero at -1. Let us rewrite the H_2 DARE (4.89) as

$$X = \tilde{A}'X\tilde{A} + \tilde{Q} - \tilde{A}XB(B'XB + R)^{-1}B'X\tilde{A}, \qquad (4.128)$$

with

$$\tilde{A} = A - BR^{-1}S \quad \text{and} \quad \tilde{Q} = Q - S'R^{-1}S.$$

But then according to Proposition 4.78 in a suitable basis, we obtain the following:

$$\tilde{A} = \begin{pmatrix} A_{11} & 0 \\ A_{21} & A_{22} \end{pmatrix}, \quad \bar{X} = \begin{pmatrix} X_{11} & 0 \\ 0 & 0 \end{pmatrix}, \quad B = \begin{pmatrix} B_1 \\ B_2 \end{pmatrix}, \quad \tilde{Q} = \begin{pmatrix} Q_{11} & 0 \\ 0 & 0 \end{pmatrix},$$

where the pair (A_{11}, Q_{11}) has no unobservable eigenvalues on the unit circle, and moreover, X_{11} satisfies the H_2 DARE:

$$X_{11} = A_{11}'X_{11}A_{11} - A_{11}'X_{11}B_1(R + B_1X_{11}B_1')^{-1}B_1X_{11}A_{11} + Q_{11}, \quad (4.129)$$

while the eigenvalues of

$$A_{11,cl} = A_{11} - B_1(\bar{R} + B_1X_{11}B_1')^{-1}B_1X_{11}A_{11}$$

are all in the closed unit disk. To see the nature of the solution X_{11} of (4.129), let us assume that a λ and an x exist such that $A_{11,cl}x = \lambda x$ with $|\lambda| = 1$. We have

$$X_{11} = A_{cl}'X_{11}A_{cl} + A_{11}'X_{11}B_1(R + B_1X_{11}B_1')^{-1}B_1X_{11}A_{11} + Q_{11},$$

and applying x and x' to the left and right, respectively, we find that $Q_{11}x = 0$ and $B_1X_{11}A_{11}x = 0$. But then $Ax = \lambda x$ and $Q_{11}x = 0$, which is in contradiction with (Q_{11}, A_{11}) having no unobservable eigenvalues on the unit circle. Therefore, $A_{11,cl}$ has no eigenvalues on the unit circle, and hence, X_{11} is a stabilizing solution of the H_2 DARE.

We can easily compute the stabilizing solution of a DARE. For stabilizing solutions, the connection between DAREs and CAREs is as established in Theorem 4.74. Therefore, computing X_{11} can be established through this connection as well as by computing a stable deflating subspace of a symplectic pencil as established in Subsection 4.2.3.

We use next the road map that exists between a DARE and its associated CARE. We do so to study the existence conditions for some specific types of solutions, say stabilizing solutions or positive semi-definite stabilizing solutions, of a DARE. This can be done by first knowing the existence conditions for the corresponding types of solutions of the associated CARE and then translating such conditions appropriately for the given DARE. Clearly, it has a limitation in that only solutions of the DARE for which $\bar{R}$ is invertible can be connected to the associated CARE. But in deriving Theorem 4.74, we have seen that stabilizing solutions of the DARE always yield a matrix $\bar{R}$, which is invertible, and therefore, we have a complete one-to-one relationship between the stabilizing solutions of a DARE and the stabilizing solutions of the associated CARE. Using this property, the following theorems can easily be obtained without much work.

Theorem 4.79 *Consider an H_2 DARE as in (4.89) with the pair (A, B) being $\mathbb{C}^{\ominus}$-stabilizable. Define C and D according to (4.94), and let the system represented by (A, B, C, D) have no invariant zeros on the unit circle and be left invertible. Then the H_2 DARE of (4.89) has a unique stabilizing solution. Moreover, this solution is positive semi-definite and is the largest real symmetric solution of it.*

Proof : According to Remark 4.61, we know that the rational matrix $\hat{H}$ has full rank because (A, B, C, D) is left-invertible. We also note that according to Remark 4.69, the matrix pencil (4.110) has no invariant zeros on the unit circle. It requires some work, but the fact that (4.110) has no invariant zeros on the unit circle is equivalent to the fact that the Hamiltonian of the associated CARE has no eigenvalues on the imaginary axis. The rest of the proof is left to the reader. ∎

Theorem 4.80 *Assume that $\hat{H}$ has full normal rank. Consider an H_2 DARE as in (4.89) with $S = 0$. Then, it has a positive definite solution if and only if, with respect to an appropriate basis, the matrices A, B, and Q are of the form*

$$A = \begin{pmatrix} A_1 & 0 \\ 0 & A_2 \end{pmatrix}, \quad B = \begin{pmatrix} B_1 \\ 0 \end{pmatrix}, \quad \text{and } Q = \begin{pmatrix} Q_{11} & 0 \\ 0 & 0 \end{pmatrix},$$

where the pairs (A_1, B_1) and $(Q_{11}, -A_1)$ are $\mathbb{C}^{\ominus}$-stabilizable and detectable, respectively, while A_2 is diagonalizable and has all its eigenvalues on the unit circle $\mathbb{C}^{\circ}$. Moreover, the solution X then takes the form

$$X = \begin{pmatrix} X_1 & 0 \\ 0 & X_2 \end{pmatrix},$$

where X_1 is unique, $A_1 - B_1(R + B_1' X_1 B_1)^{-1} B_1' X_1 A_1$ is Schur-stable, and X_2 is any one of the solutions of

$$X_2 - A_2' X_2 A_2 = 0.$$

Proof : This can be established using the same arguments as in the continuous-time case; for details, see the proof of Theorem 4.40. ∎

Theorem 4.81 *Assume that $\hat{H}$ has full normal rank. Consider an H_2 DARE as in (4.89) with $S = 0$. Then, it has a positive semi-definite solution if and only if,* *using an appropriate basis, the matrices A, B, and Q can be transformed to the form*

$$A = \begin{pmatrix} A_1 & 0 \\ A_3 & A_2 \end{pmatrix}, \quad B = \begin{pmatrix} B_1 \\ B_2 \end{pmatrix}, \quad \text{and} \quad Q = \begin{pmatrix} Q_{11} & 0 \\ 0 & 0 \end{pmatrix},$$

where the pair (A_1, B_1) is $\mathbb{C}^{\ominus}$-stabilizable and the pair (Q_{11}, A_1) is observable.

Proof : We first note that if $X \geqslant 0$ is a solution of the H_2 DARE, then we have

$$X = (A - BF)'X(A - BF) + Q + F'(R + B'XB)F + F'B'XBF,$$

where $F = (R + B'XB)^{-1}(B'XA + S')$. As $R + B'XB$ is invertible, the above equation immediately yields that whenever $Xx = 0$, then also $Fx = 0$. This in turn implies that $XAx = 0$. In other words, we have that $\ker X$ is A-invariant. The rest of the proof, with obvious modifications, is the same as the proof in the continuous-time case; see the proof of Theorem 4.43. ∎

Remark 4.82 *Theorem 4.81 can also be stated in geometric language. Namely, the H_2 DARE in (4.89) with $S = 0$ has a positive semi-definite solution if and only if*

$$X_{\ominus}(A) + \langle A \mid \text{im } B \rangle + \langle \ker Q \mid A \rangle = \mathbb{R}^n.$$

Here, $X_{\ominus}(A)$ is the stable modal subspace of $\mathbb{R}^n$ related to A, $\langle A \mid \text{im } B \rangle$ is the controllable subspace of the pair (A, B), and $\langle \ker Q \mid A \rangle$ is the unobservable subspace of (Q, A).

The above condition can easily be extended to the general case where the condition $S = 0$ is not necessarily satisfied. We obtain that a positive semi-definite solution exists if and only if

$$X_{\ominus}(A) + \langle A \mid \text{im } B \rangle + \mathcal{V}^*(\Sigma) = \mathbb{R}^n,$$

where Σ is the system associated with the quadruple (A, B, C, D), where C and D are as in (4.94).

Remark 4.83 *The condition $S = 0$ in Theorems 4.80 and 4.81 is not essential and can easily be removed. To see this, define the following:*

$$R(X) := X - A'XA + (A'XB + S)(R + B'XB)^{-1}(B'XA + S') - Q,$$

$$\widetilde{R}(X) := X - \widetilde{A}'X\widetilde{A} + \widetilde{A}'XB(R + B'XB)^{-1}B'X\widetilde{A} - \widetilde{Q},$$

$$A_x := A - B(R + B'XB)^{-1}(B'XA + S'),$$

$$\widetilde{A}_x := \widetilde{A} - B(R + B'XB)^{-1}B'X\widetilde{A},$$

where $\tilde{A} := A - BF$ and $\tilde{Q} = Q + SF$ with F such that $S + F'R = 0$. It is straightforward to check that $\tilde{Q}$ is symmetric and $\tilde{Q} \geqslant 0$ using (4.89c).

Then, it is not difficult to verify that $\boldsymbol{R}(X) = \tilde{\boldsymbol{R}}(X)$ and $A_x = \tilde{A}_x$. Thus, one can assume without any loss of generality that $S = 0$.

We develop next sufficient conditions for the existence of a semi-stabilizing solution to an H_2 DARE and then give a method of constructing it. We have the following theorem.

Theorem 4.84 *Consider an H_2 DARE as in (4.89). Assume that $\hat{H}$ has full normal rank and the pair (A, B) is $\mathbb{C}^{\ominus}$-stabilizable. Then it has a semi-stabilizing solution. Moreover, such a solution is unique, positive semi-definite, and is the largest among all symmetric solutions.*

Proof : It follows from Proposition 4.78, Theorem 4.80, and Remark 4.83.   ∎

The H_∞^1 DARE remains to be studied. As in the continuous-time case, a semi-stabilizing solution of an H_∞^1 DARE need not be unique. Therefore, like in the continuous-time case, we can introduce the notion of a strongly semi-stabilizing solution and study it. There is, however, a specific difficulty that needs to be mentioned with the discrete-time. In continuous time, we factorized the quadratic term in (4.1) as

$$R = M - B\bar{R}^{-1}B',$$

with $M \geqslant 0$ and $\bar{R} > 0$ and we obtain the structure in (4.4) with $S = 0$. However, in discrete time, the structure of the H_∞^1 DARE is determined by the intertia of $B'XB + R$. Assume for simplicity that we have (4.89a) with $S = 0$. We know the inertia is independent of the specific solution X of the DARE. Then it is clear that we can factorize $B'XB + R$ as

$$B'XB + R = \begin{pmatrix} B_1' \\ B_2' \end{pmatrix} X \begin{pmatrix} B_1' \\ B_2' \end{pmatrix} + \begin{pmatrix} R_1 & 0 \\ 0 & R_2 \end{pmatrix},$$

with

$$B_1'XB_1 + R_1 > 0,$$

$$B_2'XB_2 + R_2 - B_2'XB_1(B_1'XB_1 + R_1)^{-1}B_1'XB_2 < 0.$$

But it is not clear whether this factorization can be done independent of the solution X of the DARE. Obviously, if the factorization depends on X, then it is impossible to use this step in deriving a computational tool to compute X. However, fortunately, the kind of H_∞^1 DARE that is ubiquitously used in H_∞ optimal control and filtering indeed has such a factorization as given by the following H_∞^1 DARE:

$$X = A'XA + C'C - \begin{pmatrix} B_1'XA + D_1'C \\ B_2'XA + D_2'C \end{pmatrix}' \times$$

$$\begin{pmatrix} D_1'D_1 + B_1'XB_1 & D_1'D_2 + B_1'XB_2 \\ D_2'D_1 + B_2'XB_1 & D_2'D_2 - D_3'D_3 + B_2'XB_2 \end{pmatrix}^{-1} \begin{pmatrix} B_1'XA + D_1'C \\ B_2'XA + D_2'C \end{pmatrix},$$

$$(4.130)$$

subject to

$$B_1'XB_1 + D_1'D_1 > 0 \tag{4.131a}$$

and

$$B_2'XB_2 + D_2'D_2 - D_3'D_3$$
$$- (B_2'XB_1 + D_2'D_1)(B_1'XB_1 + R_1)^{-1}(B_1'XB_2 + D_1'D_2) < 0. \tag{4.131b}$$

For the above H_∞^1 DARE, obviously, X is a semi-stabilizing solution if the following matrix:

$$A - \begin{pmatrix} B_1 \\ B_2 \end{pmatrix}' \times$$

$$\begin{pmatrix} D_1'D_1 + B_1'XB_1 & D_1'D_2 + B_1'XB_2 \\ D_2'D_1 + B_2'XB_1 & D_2'D_2 - D_3'D_3 + B_2'XB_2 \end{pmatrix}^{-1} \begin{pmatrix} B_1'XA + D_1'C \\ B_2'XA + D_2'C \end{pmatrix}$$

has all its eigenvalues in the closed unit disk. Also, X is called a stabilizing solution if the above matrix has all its eigenvalues in the open unit disk.

The following definition introduces the notion of a strongly semi-stabilizing solution of an H_∞^1 DARE.

Definition 4.85 *Consider an H_∞^1 DARE of the form (4.130) subject to the conditions (4.131). A semi-stabilizing solution X of the H_∞^1 DARE is called a **strongly semi-stabilizing solution** of it if the following two additional properties are satisfied:*

- *The dimension of the kernel of X is equal to the number of invariant zeros of (A, B_1, C, D_1) in the closed unit disk.*

- *The number of invariant zeros of (A, B_1, C, D_1) on the unit circle is equal to the number of eigenvalues of A_{cl} on the unit circle.*

Our next task is to characterize the existence of a strongly semi-stabilizing solution of an H_∞^1 DARE and to compute it. In fact, we obtain below a strongly semi-stabilizing solution of a given H_∞^1 DARE in terms of the stabilizing solution of another suitably defined H_∞^1 DARE. We first have the following proposition.

Proposition 4.86 *Consider an H_∞^1 DARE of the form (4.130) subject to the conditions (4.131). Assume that the pair (A, B) is $\mathbb{C}^\ominus$-stabilizable. Then, a positive semi-definite solution X of it, if it exists, has $\mathcal{V}^\otimes(A, B_1, C, D_1)$ as a subset of its null space. That is, $X\mathcal{V}^\otimes = 0$.*

Proof : The proof follows along the same lines as the proof of Proposition 4.78 as soon as we realize that whenever X satisfies the H_∞^1 DARE, it implies that

$$\begin{pmatrix} -X + A'XA + C'C & AXB_1 + C'D_1 \\ B_1XA + D_1'C & D_1'D_1 + B_1'XB_1 \end{pmatrix} \geq 0. \qquad \blacksquare$$

Using the above proposition, we can establish that a positive semi-definite strongly semi-stabilizing solution must have the following form. We use the fact (from the above proposition combined with the first additional property of a strongly semi-stabilizing solution) that

$$\ker X = \mathcal{V}^\otimes(A, B_1, C, D_1).$$

Again the use of the special coordinate basis as described in Chapter 3 provides us with the tools to establish that, in a suitable basis, we obtain a decomposition of the state space $\mathbb{R}^n = \mathcal{V}^\otimes \oplus \mathcal{W}$, where we used that $\mathcal{V}^\otimes = \mathcal{X}_a^- \oplus \mathcal{X}_a^0 \oplus \mathcal{X}_c$ and we defined $\mathcal{W} = \mathcal{X}_a^+ \oplus \mathcal{X}_b \oplus \mathcal{X}_d$. Then for a matrix F such that $D_1'(C + D_1F) = 0$, we have

$$\bar{A}+BF = \begin{pmatrix} A_{11} & A_{12} \\ 0 & A_{22} \end{pmatrix}, \; B_1 = \begin{pmatrix} B_{11} \\ B_{21} \end{pmatrix}, \; B_2 = \begin{pmatrix} B_{12} \\ B_{22} \end{pmatrix}, \; C+D_1F = \begin{pmatrix} 0 & C_2 \end{pmatrix}.$$

But with respect to this basis, we have

$$X = \begin{pmatrix} 0 & 0 \\ 0 & X_1 \end{pmatrix}. \tag{4.132}$$

Moreover, X_1 satisfies the following H_∞^1 DARE:

$$X_1 = A_{22}'X_1A_{22} + C'C - \begin{pmatrix} B_{21}'X_1A_{22} \\ B_{22}'X_1A_{22} + D_2'C \end{pmatrix}' \times$$

$$\begin{pmatrix} D_1'D_1 + B_{21}'X_1B_{21} & D_1'D_2 + B_{21}'X_1B_{22} \\ D_2'D_1 + B_{22}'X_1B_{21} & D_2'D_2 - D_3'D_3 + B_{22}'X_1B_{22} \end{pmatrix}^{-1} \begin{pmatrix} B_{21}'X_1A_{22} \\ B_{22}'X_1A_{22} + D_2'C \end{pmatrix}, \tag{4.133}$$

subject to

$$B_{21}' X_1 B_{21} + D_1' D_1 > 0 \tag{4.134a}$$

and

$$B_{22}' X_1 B_{22} + D_2' D_2 - D_3' D_3$$
$$- (B_{22}' X_1 B_{21} + D_2' D_1)(B_{21}' X_1 B_{21} + R_1)^{-1}(B_{21}' X_1 B_{22} + D_1' D_2) < 0. \tag{4.134b}$$

It is clear from the second additional property of a strongly semi-stabilizing solution that X_1 must be a stabilizing solution of the H_∞^1 DARE (4.133) subject to the conditions (4.134). Then (4.132) yields the unique semi-stabilizing solution of the original H_∞^1 DARE (4.130) subject to the conditions (4.131).

We summarize the results of the above development by the following proposition.

Proposition 4.87 *Consider an H_∞^1 DARE of the form (4.130) subject to the conditions (4.131). Also, consider the H_∞^1 DARE (4.133) is derived from the data of H_∞^1 DARE (4.130) and is subject to the conditions (4.134). Then, a strongly semi-stabilizing solution of the H_∞^1 DARE (4.130) exists if and only if a stabilizing solution of the H_∞^1 DARE (4.133) exists.*

Finally, the following proposition shows a property that we will exploit later in Chapter 11.

Proposition 4.88 *Consider an H_∞^1 DARE of the form (4.130). If a matrix Y exists such that*

$$Y > A'YA + C'C - \begin{pmatrix} B_1'YA + D_1'C \\ B_2'YA + D_2'C \end{pmatrix}'$$
$$\begin{pmatrix} D_1'D_1 + B_1'YB_1 & D_1'D_2 + B_1'YB_2 \\ D_2'D_1 + B_2'YB_1 & D_2'D_2 - D_3'D_3 + B_2'YB_2 \end{pmatrix}^{-1} \begin{pmatrix} B_1'YA + D_1'C \\ B_2'YA + D_2'C \end{pmatrix},$$

subject to

$$B_1'YB_1 + D_1'D_1 > 0 \tag{4.135}$$

and

$$B_2'YB_2 + D_2'D_2 - D_3'D_3$$
$$- (B_2'YB_1 + D_2'D_1)(B_1'YB_1 + R_1)^{-1}(B_1'YB_2 + D_1'D_2) < 0, \tag{4.136}$$

then a strongly semi-stabilizing solution of the H_∞^1 DARE (4.130) exists.

Proof : Note that one way of proving this result is to use the connection between CAREs and DAREs and rely on the continuous-time result in Proposition 4.30. We will present here a direct proof for the discrete-time result, but we only sketch the main ideas and leave some details to the reader.

We first note that X is a strongly semi-stabilizing solution of the H_∞^1 DARE (4.130) if and only if $\widetilde{X} = X - Y$ satisfies

$$\widetilde{X} = A'\widetilde{X}A + L_1 - (A'\widetilde{X}\bar{B} + L_2)(\bar{B}'\widetilde{X}\bar{B} + L_3)^{-1}(\bar{B}'\widetilde{X}A + L_2'), \quad (4.137)$$

where

$$\bar{B} = \begin{pmatrix} B_1 \\ B_2 \end{pmatrix}, \qquad \bar{D} = \begin{pmatrix} D_1 \\ D_2 \end{pmatrix}, \qquad \bar{D}_3 = \begin{pmatrix} 0 \\ D_3 \end{pmatrix},$$

and

$$\begin{pmatrix} L_1 & L_2 \\ L_2' & L_3 \end{pmatrix} = \begin{pmatrix} -Y + A'YA + C'C & A'Y\bar{B}' + C'\bar{D} \\ \bar{B}'YA + \bar{D}'C & \bar{B}Y\bar{B} + \bar{D}'\bar{D} \end{pmatrix}.$$

Note that (4.135) and (4.136) guarantee that we can decompose

$$L_3 = \begin{pmatrix} L_{31} & L_{32} \\ L_{32}' & L_{33} \end{pmatrix}$$

with $L_{31} > 0$ and $L_{33} - L_{32}'L_{31}^{-1}L_{32} < 0$ while

$$L_4 = L_1 - L_2L_3^{-1}L_2' < 0.$$

We can then rewrite (4.137) as

$$\widetilde{X} = A_{cl}'\widetilde{X}A_{cl} + L_4 - A_{cl}'\widetilde{X}\bar{B}(\bar{B}'\widetilde{X}\bar{B} + L_3)^{-1}\bar{B}'\widetilde{X}A_{cl}. \quad (4.138)$$

Obviously, either A_{cl} is invertible or not. If A_{cl} is not invertible, a reduction in a suitable basis exists. That is,

$$A_{cl} = \begin{pmatrix} 0 & A_2 \\ 0 & A_3 \end{pmatrix}, \quad L_4 = \begin{pmatrix} L_{41} & L_{42} \\ L_{42}' & L_{43} \end{pmatrix},$$

then it is easy to verify that

$$\widetilde{X} = L_4 + \begin{pmatrix} 0 & 0 \\ 0 & \widetilde{X}_{22} \end{pmatrix}$$

and we can then derive a DARE for $\widetilde{X}_{22}$ having the same structure as (4.138) with the corresponding A_{cl} matrix invertible.

Therefore, we can always use a reduction technique to obtain a DARE of the form (4.138) with the corresponding A_{cl} invertible. But in this case, using the classic inverse lemma, we can rewrite (4.138) as

$$\widetilde{X} = A_{cl}\widetilde{X}(\bar{B}L_3^{-1}\bar{B}'\widetilde{X} + I)^{-1}A_{cl} + L_4,$$

and with $\widetilde{X}$ invertible, we can even get

$$\widetilde{X} = A_{cl}(\bar{B}L_3^{-1}\bar{B}' + \widetilde{X}^{-1})^{-1}A_{cl} + L_4.$$

It is then easily verified that we can use similar tools as in the proof of Proposition 4.30 to obtain a complete proof. ∎

4.2.6 Continuity properties of the H_2 DARE

Theorem 4.84 gives the sufficient conditions for the existence of a unique semi-stabilizing solution of an H_2 DARE. In this subsection, we examine the continuity of such a solution with respect to the parameters Q, R, and S. To start with, we parameterize the matrices Q, R, and S with a scalar parameter ε and rewrite the H_2 DARE of (4.89) as

$$R^\varepsilon + B'X^\varepsilon B > 0 \tag{4.139a}$$

and

$$X^\varepsilon = A'X^\varepsilon A - (A'X^\varepsilon B + S^\varepsilon)(R^\varepsilon + B'X^\varepsilon B)^{-1}(B'X^\varepsilon A + (S^\varepsilon)') + Q^\varepsilon, \tag{4.139b}$$

where the matrices Q^ε, R^ε, and S^ε satisfy the Condition psd and hence permit the factorization

$$\begin{pmatrix} Q^\varepsilon & S^\varepsilon \\ (S^\varepsilon)' & R^\varepsilon \end{pmatrix} = \begin{pmatrix} (C^\varepsilon)' \\ (D^\varepsilon)' \end{pmatrix} \begin{pmatrix} C^\varepsilon & D^\varepsilon \end{pmatrix}. \tag{4.140}$$

Assuming that the pair $(A^\varepsilon, B^\varepsilon)$ is $\mathbb{C}^\ominus$-stabilizable and the system with realization $(A^\varepsilon, B^\varepsilon, C^\varepsilon, D^\varepsilon)$ is left-invertible, it follows from Theorem 4.84 that for each ε a unique semi-stabilizing solution of the H_2 DARE (4.139) exists. Assume that

$$\lim_{\varepsilon \to 0} A^\varepsilon = A^0, \quad \lim_{\varepsilon \to 0} B^\varepsilon = B^0, \quad \lim_{\varepsilon \to 0} C^\varepsilon = C^0, \quad \lim_{\varepsilon \to 0} D^\varepsilon = D^0. \tag{4.141}$$

We have the following result.

Theorem 4.89 *Consider the H_2 DARE given in (4.139). Let $\varepsilon \in [0, \delta)$. Assume that $(A^\varepsilon, B^\varepsilon)$ is $\mathbb{C}^\ominus$-stabilizable. Let C^ε and D^ε be defined as in (4.140). Assume that (4.141) is satisfied. Let X^ε be the semi-stabilizing solution of the H_2 DARE (4.139). Then we have*

$$\lim_{\varepsilon \to 0} X_{ss}^\varepsilon = X_{ss}^0. \tag{4.142}$$

Proof : It follows the same arguments as in the proof of Theorem 4.46. The only issue is whether

$$\lim_{\varepsilon \to 0} (B^\varepsilon)'X_{ss}^\varepsilon B^\varepsilon + R^\varepsilon$$

is invertible. But this follows from Lemma 4.63. After all we can argue that the limit $\overline{X}_{ss}$ of a convergent subseries of X_{ss}^{ε} is a solution of the general DARE. But as we already know that X_{ss}^0 is a solution of the DARE with $(B^0)'X_{ss}^0 B^0 + R^0$ invertible, we know from Lemma 4.63 that $\overline{X}_{ss}$ must also be a solution of the DARE with $(B^0)'\overline{X}_{ss}B^0 + R^0$ invertible because the rank of this matrix is the same for all solutions of the DARE.   ■

4.2.7 Connections between a GDARE and its associated DARE

The previous subsection explored the properties, existence, and computation of various types of solutions of a DARE. One tool that helped in this endeavor is the fact that there is a one-to-one connection between various types of solutions of a DARE and similar solutions of a suitably defined CARE. To follow such a theme for GDAREs, in this subsection, we develop a one-to-one connection between various types of solutions of a GDARE and similar solutions of a suitably defined DARE. As in the previous subsection, such a connection between GDAREs and DAREs has an immediate consequence: Namely, it enables us to convert various tasks related to a GDARE, such as computing a solution, verifying the existence of a positive definite or a semi-definite solution, or learning or understanding some other properties of a solution, to similar tasks on the associated DARE. As we have extensively studied and developed several important properties of certain solutions of a DARE, the advantage in transforming any given task related to a GDARE to that of an associated DARE is then transparent.

Our development in this subsection is easier to comprehend if two different cases of GDAREs are considered depending on the rank of the rational matrix $\widehat{H}$ defined in (4.95).

Case 1 [$\widehat{H}$ has full normal rank]

We note that $\widehat{H}$ having full normal rank is equivalent to the normal rank of $\widehat{H}$ being m. Using the relationship mentioned in Remark 4.61, we also note that for the classic special case of the H_2 GDARE, the assumption that $\widehat{H}$ has full normal rank is equivalent to the assumption that the system characterized by the quadruple (A, B, C, D) is left invertible, where C and D are defined according to (4.94). As was shown in Lemma 4.63, under the assumption that $\widehat{H}$ has full normal rank, any solution of a GDARE must be such that $B'XB + R$ is invertible. Then, as mentioned, whenever $B'XB + R$ is invertible, the subspace inclusion in (4.90b) is automatically satisfied, and thus, the GDARE in (4.90) reduces to the DARE in (4.88).

Case 2 [$\widehat{H}$ does not have full normal rank]

In this subsection, we restrict our attention to the case of an H_2 GDARE when $\widehat{H}$ does not have full normal rank. Note that this case deals with the situation when the system characterized by the quadruple (A, B, C, D) is not necessarily left invertible. Our goal here is to obtain an H_2 DARE associated with the given H_2 GDARE such that the solutions of the H_2 DARE have a one-to-one relationship to the solutions of the H_2 GDARE. As such, the tasks of computing, learning, or understanding the properties of a solution of an H_2 GDARE reduce to similar tasks for the associated H_2 DARE. As discussed, the tasks one needs to perform on a DARE can be converted to similar tasks on an associated CARE. In this manner, one can study an H_2 GDARE in terms of an associated H_2 DARE, which can be studied in terms of another associated CARE.

Using the special coordinate basis, it is obvious that a suitable basis for the state, input and output spaces and a suitable preliminary feedback exist such that in these bases the system matrices exhibit a special structure. In particular, we choose a basis in the state space $\mathcal{X}_1 \oplus \mathcal{X}_2$ such that $\mathcal{X}_1 = \mathcal{R}^*(\Sigma)$ and a basis in the input space $\mathcal{U}_1 \oplus \mathcal{U}_2$ such that $\mathcal{U}_1 = B^{-1}\mathcal{R}^*(\Sigma) \cap \ker D$. With respect to this basis, we obtain

$$A_F = A + BF = \begin{pmatrix} A_{11} & A_{12} \\ 0 & A_{22} \end{pmatrix}, \quad B = \begin{pmatrix} B_{11} & B_{12} \\ 0 & B_{22} \end{pmatrix},$$

$$C_F = C + DF = \begin{pmatrix} 0 & C_2 \end{pmatrix}, \quad D = \begin{pmatrix} 0 & D_2 \end{pmatrix}, \quad (4.143)$$

with the property that (A_{11}, B_{11}) is controllable, $C_2'D_2 = 0$, and $(A_{22}, B_{22}, C_2, D_2)$ is left-invertible. As will be exploited extensively later on, a matrix X is a solution of the H_2 GDARE if and only if

$$\begin{pmatrix} -X + A'XA + C'C & A'XB + C'D \\ B'XA + D'C & D'D + B'XB \end{pmatrix} \geq 0$$

and

$$\mathrm{rank} \begin{pmatrix} -X + A'XA + C'C & A'XB + C'D \\ B'XA + D'C & D'D + B'XB \end{pmatrix} = \mathrm{rank}\, D'D + B'XB.$$

Using this characterization together with

$$\begin{pmatrix} I & F' \\ 0 & I \end{pmatrix} \begin{pmatrix} -X + A'XA + C'C & A'XB + C'D \\ B'XA + D'C & D'D + B'XB \end{pmatrix} \begin{pmatrix} I & 0 \\ F & I \end{pmatrix}$$

$$= \begin{pmatrix} -X + A_F'XA_F + C_F'C_F & A_F'XB \\ B'XA_F & D'D + B'XB \end{pmatrix} \geq 0, \quad (4.144)$$

we obtain that X is also a solution of the H_2 GDARE:

$$X = A_F'XA_F - A_F'XB(R + B'XB)^\dagger B'XA_F + C_F'C_F, \quad (4.145a)$$

$$\ker[R + B'XB] \subseteq \ker A_F'XB, \quad (4.145b)$$

$$R + B'XB \geq 0. \tag{4.145c}$$

Using the basis as described above together with the decomposition

$$X = \begin{pmatrix} X_{11} & X_{12} \\ X_{21} & X_{22} \end{pmatrix},$$

we obtain from (4.144) that

$$\begin{pmatrix} A'_{11}X_{11}A_{11} - X_{11} & A'_{11}X_{12}A_{22} - X_{12} & \star & A'_{11}X_{11}B_{11} \\ A'_{22}X_{21}A_{11} - X_{21} & \star & \star & \star \\ \star & B'_{11}X_{11}A_{12} + B'_{11}X_{12}A_{22} & \star & \star \\ B'_{11}X_{11}A_{11} & \star & \star & B'_{11}X_{11}B_{11} \end{pmatrix} \geq 0, \tag{4.146}$$

where $\star$ indicates some matrix of not much importance for us at this time. Also, by extracting a principal submatrix of (4.146), we get

$$\begin{pmatrix} A'_{11}X_{11}A_{11} - X_{11} & A'_{11}X_{11}B_{11} \\ B'_{11}X_{11}A_{11} & B'_{11}X_{11}B_{11} \end{pmatrix} \geq 0. \tag{4.147}$$

As (A_{11}, B_{11}) is controllable, a matrix F_s exists such that $A_{11} + B_{11}F_s$ is Schur stable. By postmultiplying (4.147) with

$$\begin{pmatrix} I & 0 \\ F_s & 0 \end{pmatrix}$$

and premultiplying the resultant with the transpose of the above matrix, we get

$$(A_{11} + B_{11}F_s)'X_{11}(A_{11} + B_{11}F_s) - X_{11} \geq 0.$$

In view of Lemma 4.163, the above implies that $X_{11} \geq 0$. Similarly, as the pair (A_{11}, B_{11}) is controllable, a matrix F_a exists such that $A_{11} + B_{11}F_a$ is Schur anti-stable (all eigenvalues outside the unit circle). By postmultiplying (4.147) with

$$\begin{pmatrix} I & 0 \\ F_a & 0 \end{pmatrix}$$

and premultiplying the resultant with the transpose of the above matrix, we get

$$(A_{11} + B_{11}F_a)'X_{11}(A_{11} + B_{11}F_a) - X_{11} \geq 0.$$

Again, in view of Lemma 4.163, the above implies that $X_{11} \leq 0$. As both $X_{11} \geq 0$ and $X_{11} \leq 0$, we must have $X_{11} = 0$. Then, after evaluating the inequality (4.146) with $X_{11} = 0$, we immediately conclude that

$$A'_{11}X_{12}A_{22} - X_{12} = 0 \quad \text{and} \quad B'_{11}X_{12}A_{22} = 0.$$

This implies that

$$(A_{11} + B_{11}F)' X_{12} A_{22} - X_{12} = 0$$

for any F. As (A_{11}, B_{11}) is controllable, a matrix F_r exists such that $A_{11} + B_{11}F_r$ and A_{22} have no eigenvalues in common. Then, in view of Lemma 4.237, we can conclude that $X_{12} = 0$. It is then easily shown that X_{22} is a solution of the H_2 GDARE:

$$
\begin{aligned}
X_{22} = {}& A_{22}' X_{22} A_{22} + C_2' C_2 \\
& - A_{22}' X_{22} B_{22} (D_2' D_2 + B_{22}' X_{22} B_{22})^\dagger B_{22}' X_{22} A_{22}, \quad \text{(4.148a)}
\end{aligned}
$$

$$\ker[D_2' D_2 + B_{22}' X_{22} B_{22}] \subseteq \ker A_{22}' X_{22} B_{22}, \quad \text{(4.148b)}$$

$$D_2' D_2 + B_{22}' X_{22} B_{22} \geq 0. \quad \text{(4.148c)}$$

As the system $(A_{22}, B_{22}, C_2, D_2)$ is left-invertible, it follows from a direct application of Lemma 4.63 that $D_2' D_2 + B_{22}' X_{22} B_{22}$ is invertible, and therefore, we have an H_2 DARE.

Now we are ready to define an H_2 DARE associated with the H_2 GDARE of Definition 4.56.

Definition 4.90 *Consider the matrices A_{22}, B_{22}, C_2, and D_2 defined in (4.143). Then the quadratic matrix equation for an unknown matrix X_{22} given by*

$$
\begin{aligned}
A_{22}' X_{22} A_{22} - {}& X_{22} + C_2' C_2 \\
& - A_{22}' X_{22} B_{22} (B_{22}' X_{22} B_{22} + D_2' D_2)^{-1} B_{22}' X_{22} A_{22} = 0, \quad \text{(4.149a)}
\end{aligned}
$$

$$B_{22}' X_{22} B_{22} + D_2' D_2 > 0, \quad \text{(4.149b)}$$

*is called the **associated H_2 DARE** for the H_2 GDARE (4.91).*

The above arguments yield the following theorems, which establish a one-to-one relationship between the solutions of the H_2 GDARE and its associated H_2 DARE.

Theorem 4.91 *Consider an H_2 GDARE as in (4.91) and its associated H_2 DARE as in (4.149). Also, consider two real symmetric matrices X and X_{22} that, in the bases used in (4.143), are related by*

$$X = \begin{pmatrix} 0 & 0 \\ 0 & X_{22} \end{pmatrix}. \quad \text{(4.150)}$$

Then the following two statements are equivalent:

(i) X is a real symmetric solution of the H_2 GDARE in (4.91).

(ii) X_{22} is a real symmetric solution of the associated H_2 DARE in (4.149).

Theorem 4.92 *Consider an H_2 GDARE as in (4.91) and its associated H_2 DARE as in (4.149). Also, consider two symmetric matrices X and X_{22} that, in the bases used in (4.143), are related by (4.150). Then the following two statements are equivalent:*

(i) X is a stabilizing solution of the H_2 GDARE in (4.91).

(ii) X_{22} is a stabilizing solution of the associated H_2 DARE in (4.149).

Moreover, the following two statements are equivalent:

(i) X is a semi-stabilizing solution of the H_2 GDARE in (4.91).

(ii) X_{22} is a semi-stabilizing solution of the associated H_2 DARE in (4.149).

We now consider H_∞^2 GDAREs. The connection between an H_∞^2 GDARE and an H_∞^2 DARE can be established by again using the fact that the H_∞^2 DARE is in fact nothing else than a shifted H_2 DARE.

Let $\bar{X}$ satisfy

$$\begin{pmatrix} \bar{Q} & \bar{S} \\ \bar{S}' & \bar{R} \end{pmatrix} := \begin{pmatrix} -\bar{X} + A'\bar{X}A + Q & A'\bar{X}B + S \\ B\bar{X}A + S' & R \end{pmatrix} \geqslant 0.$$

As any solution of the H_∞^2 DARE must satisfy this inequality, the existence of $\bar{X}$ is without loss of generality. But then X satisfies the H_∞^2 GDARE if and only if $\hat{X} = X - \bar{X}$ is a solution of the H_2 GDARE:

$$\hat{X} = A'\hat{X}\bar{A} - (A'XB + \bar{S})(B'\hat{X}B + \bar{R})^\dagger (B'\hat{X}A + \bar{S}') + \bar{Q} = 0.$$

Moreover, semi-stabilizing (stabilizing) solutions are mapped to semi-stabilizing (respectively, stabilizing) solutions. The connection of an H_2 GDARE to an H_2 DARE then also yields a connection between an H_∞^2 GDARE to an H_2 DARE.

For the H_∞^1 GDARE, the situation is again more complicated. We consider a special structure as also studied in Definition 4.85:

$$X = A'XA + C'C - \begin{pmatrix} B_1'XA + D_1'C \\ B_2'XA + D_2'C \end{pmatrix}'$$

$$\begin{pmatrix} D_1'D_1 + B_1'XB_1 & D_1'D_2 + B_1'XB_2 \\ D_2'D_1 + B_2'XB_1 & D_2'D_2 - D_3'D_3 + B_2'XB_2 \end{pmatrix}^\dagger \begin{pmatrix} B_1'XA + D_1'C \\ B_2'XA + D_2'C \end{pmatrix}, \quad (4.151)$$

subject to

$$B_1'XB_1 + D_1'D_1 \geqslant 0 \qquad\qquad (4.152a)$$

and

$$B_2' X B_2 + D_2' D_2 - D_3' D_3$$
$$- (B_2' X B_1 + D_2' D_1)(B_1' X B_1 + R_1)^\dagger (B_1' X B_2 + D_1' D_2) < 0. \quad (4.152b)$$

Our next task is to characterize the existence of a strongly semi-stabilizing solution of an H_∞^1 GDARE and to compute it. In fact, we obtain below a strongly semi-stabilizing solution of a given H_∞^1 GDARE in terms of the stabilizing solution of another suitably defined H_∞^1 DARE. We first have the following proposition.

Proposition 4.93 *Consider an H_∞^1 GDARE of the form* (4.151) *subject to the conditions* (4.152). *Assume that the pair (A, B) is $\mathbb{C}^\ominus$-stabilizable. Then, a positive semi-definite solution X of it, if it exists, has $\mathcal{V}^\otimes(A, B_1, C, D_1)$ as a subset of its null space. That is, $X \mathcal{V}^\otimes = 0$.*

Proof : This uses the same arguments as the proof of Proposition 4.86. ∎

Again the use of the special coordinate basis as described in Chapter 3 provides us with the tools to establish that in a suitable basis, we obtain a decomposition of the state space $\mathbb{R}^n = \mathcal{V}^\otimes \oplus \mathcal{W}$ where we used that $\mathcal{V}^\otimes = \mathcal{X}_a^- \oplus \mathcal{X}_a^0 \oplus \mathcal{X}_c$ and we defined $\mathcal{W} = \mathcal{X}_a^+ \oplus \mathcal{X}_b \oplus \mathcal{X}_d$. Then for a matrix F such that $D_1'(C + D_1 F) = 0$, we have

$$A + B_1 F = \begin{pmatrix} A_{11} & A_{12} \\ 0 & A_{22} \end{pmatrix}, B_1 = \begin{pmatrix} B_{11} & B_{12} \\ 0 & B_{22} \end{pmatrix}, B_2 = \begin{pmatrix} B_{31} \\ B_{32} \end{pmatrix},$$

$$C + D_1 F = \begin{pmatrix} 0 & C_2 \end{pmatrix}, D = \begin{pmatrix} 0 & D_2 \end{pmatrix}, \quad (4.153)$$

with the property that (A_{11}, B_{11}) is controllable, $C_2' D_2 = 0$, and $(A_{22}, B_{22}, C_2, D_2)$ is left-invertible. Using Proposition 4.93, one obtains that a positive semi-definite matrix X is a solution of the H_∞^1 GDARE if and only if

$$X = \begin{pmatrix} 0 & 0 \\ 0 & X_{22} \end{pmatrix},$$

and X_{22} satisfies an associated H_∞^1 DARE:

$$X_{22} = A_{22}' X_{22} A_{22} + C'C - \begin{pmatrix} B_{22}' X_{22} A_{22} \\ B_{32}' X_{22} A_{22} + D_2'C \end{pmatrix}' \times$$

$$\begin{pmatrix} D_1' D_1 + B_{22}' X_{22} B_{22} & D_1' D_2 + B_{22}' X_{22} B_{32} \\ D_2' D_1 + B_{32}' X_{22} B_{22} & D_2' D_2 - D_3' D_3 + B_{32}' X_{22} B_{32} \end{pmatrix}^{-1} \times$$

$$\begin{pmatrix} B_{22}' X_{22} A_{22} \\ B_{32}' X_{22} A_{22} + D_2'C \end{pmatrix}.$$

subject to

$$B_{22}' X_{22} B_{22} + D_1' D_1 > 0$$

and

$$B_{32}' X_{22} B_{32} + D_2' D_2 - D_3' D_3$$
$$- (B_{32}' X_{22} B_{22} + D_2' D_1)(B_{22}' X_{22} B_{22} + R_1)^{-1}(B_{22}' X_{22} B_{32} + D_1' D_2) < 0.$$

4.2.8   *Properties, existence, and computation of various types of solutions of a GDARE*

Theorems 4.91 and 4.92 provide the tools to study certain types of solutions of a given H_2 GDARE in terms of a study of similar solutions of its associated H_2 DARE (4.149). By using these tools, in this subsection, (1) we study the semi-stabilizing, stabilizing, positive semi-definite, and positive definite properties of a solution of a GDARE; (2) we develop the existence conditions for such solutions, and (3) we develop methods of computing them. The results of this subsection are direct consequences of Theorems 4.91 and 4.92 and do not need much effort to obtain them.

We start with the following proposition, which makes use of Proposition 4.75.

Proposition 4.94 *Consider an H_2 GDARE satisfying (4.91) or an H_∞^2 GDARE satisfying only the first three conditions (4.91a), (4.91b), and (4.91c), where the pair (A, B) is $\mathbb{C}^\ominus$-stabilizable. Then the following hold:*

(i) *A real symmetric semi-stabilizing solution X_{ss}, if it exists, is unique.*

(ii) *A real symmetric semi-stabilizing solution X_{ss}, if it exists, is larger than or equal to any real symmetric solution; i.e., $X_{ss} \geq X$.*

(iii) *A real symmetric semi-stabilizing solution X_{ss} of the H_2 GDARE, if it exists, is positive semi-definite; i.e., $X_{ss} \geq 0$.*

Proof : It is left to the reader. ∎

Theorem 4.68 gives a condition when semi-stabilizing solutions are actually stabilizing solutions. The matrix pencil in that theorem is regular if and only if $\hat{H}$ has full rank. Hence, if we are dealing with GDAREs, this matrix pencil might be a singular pencil. Nevertheless, exploiting the relationship between the H_2 GDAREs and H_2 GDAREs and the associated decomposition of the system matrices, we can obtain the following theorem in a straightforward fashion.

Theorem 4.95 *A semi-stabilizing solution of an H_2 GDARE (4.91), if it exists, is a stabilizing solution if and only if the rank of the matrix pencil*

$$\begin{pmatrix} Q & S & I - zA' \\ S' & R & -zB' \\ zI - A & -B & 0 \end{pmatrix}$$

is equal to its normal rank for all z on the unit circle $\mathbb{C}^{\circ}$.

For the H_∞^2 GDARE, the relationship with an H_∞^2 DARE yields the same result. However, for the H_∞^1 GDARE, although correct, it is much more complicated to establish the result because we have no method in general to reduce it to an H_∞^1 DARE unless we impose extra conditions such as (4.152) in addition to (4.151).

Again, Theorems 4.91 and 4.92 provide the tools to study the existence conditions for a specific type of solution of a given H_2 GDARE in terms of the existence conditions for a similar solution of its associated H_2 DARE (4.149). In this regard, without much effort, we can arrive at the following theorems.

Theorem 4.96 *Consider an H_2 GDARE as in (4.91). Let C and D be defined according to (4.94). A positive semi-definite solution of (4.91) exists if and only if*

$$X_\ominus(A) + \langle A \mid \operatorname{im} B \rangle + \mathcal{V}^*(\Sigma) = \mathbb{R}^n.$$

Here Σ is a system with realization (A, B, C, D) and state-space dimension n, $X_\ominus(A)$ is the stable modal subspace of $\mathbb{R}^n$ related to A, $\langle A \mid \operatorname{im} B \rangle$ is the controllable subspace of the pair (A, B), and $\mathcal{V}^*(\Sigma)$ as defined in Section 3.2 represents the weakly unobservable subspace.

Proof : First of all, we note that the H_2 GDARE as in (4.91) has a positive semi-definite solution if and only if its associated H_2 DARE (4.149) has a positive semi-definite solution. The rest is a straightforward combination of the relationship between the H_2 GDARE and its associated H_2 DARE and Remark 4.82. ∎

By exploiting the relationship between the H_2 GDARE as in (4.91) and its associated H_2 DARE (4.149), we can also establish the following two theorems.

Theorem 4.97 *Consider an H_2 GDARE as in (4.91) with the pair (A, B) being $\mathbb{C}^{\ominus}$-stabilizable. Define C and D by (4.94), and let the system represented by (A, B, C, D) have no invariant zeros on the unit circle. Then the given H_2 GDARE has a unique stabilizing solution. Moreover, this solution is positive semi-definite and is the largest real symmetric solution of it.*

Theorem 4.98 *Consider an H_2 GDARE as in (4.91) with the pair (A, B) being $\mathbb{C}^{\ominus}$-stabilizable. Then, it has a unique semi-stabilizing solution. Moreover, this solution is positive semi-definite and larger than any other real symmetric solution of it.*

4.2.9 *Continuity properties of the H_2 GDARE*

Theorem 4.98 guarantees that, when the pair (A, B) is $\mathbb{C}^{\ominus}$-stabilizable, a unique semi-stabilizing solution of the H_2 GDARE exists. In this subsection, we examine the continuity of such a solution with respect to its parameters Q, R, and S. To start with, we parameterize the matrices Q, R, and S with a scalar parameter ε, and we rewrite the H_2 GDARE of (4.91) as

$$X^{\varepsilon} = A'X^{\varepsilon}A - (A'X^{\varepsilon}B + S^{\varepsilon})(R^{\varepsilon} + B'X^{\varepsilon}B)^{\dagger}(B'X^{\varepsilon}A + (S^{\varepsilon})') + Q^{\varepsilon}, \quad (4.154a)$$

$$\ker[R^{\varepsilon} + B'X^{\varepsilon}B] \subseteq \ker[A'X^{\varepsilon}B + S^{\varepsilon}], \quad (4.154b)$$

$$R^{\varepsilon} + B'X^{\varepsilon}B \geq 0, \quad (4.154c)$$

and

$$\begin{pmatrix} Q^{\varepsilon} & S^{\varepsilon} \\ (S^{\varepsilon})' & R^{\varepsilon} \end{pmatrix} \geq 0. \quad (4.154d)$$

Because of (4.154d), the matrices Q^{ε}, R^{ε}, and S^{ε} permit the factorization:

$$\begin{pmatrix} Q^{\varepsilon} & S^{\varepsilon} \\ (S^{\varepsilon})' & R^{\varepsilon} \end{pmatrix} = \begin{pmatrix} (C^{\varepsilon})' \\ (D_{\varepsilon})' \end{pmatrix} \begin{pmatrix} C^{\varepsilon} & D^{\varepsilon} \end{pmatrix}. \quad (4.155)$$

Assuming that the pair $(A^{\varepsilon}, B^{\varepsilon})$ is $\mathbb{C}^{\ominus}$-stabilizable, it follows from Theorems 4.84 and 4.92 that for each ε, a unique semi-stabilizing solution of the above H_2 GDARE exists. Assume that

$$\lim_{\varepsilon \to 0} A^{\varepsilon} = A^{0}, \quad \lim_{\varepsilon \to 0} B^{\varepsilon} = B^{0}, \quad \lim_{\varepsilon \to 0} C^{\varepsilon} = C^{0}, \quad \lim_{\varepsilon \to 0} D^{\varepsilon} = D^{0}. \quad (4.156)$$

Let X_{ss}^{ε} be the semi-stabilizing solution associated with the parameters $(A^{\varepsilon}, B^{\varepsilon}, C^{\varepsilon}, D^{\varepsilon})$. These conditions are basically equivalent to the conditions leading up the continuity result in Theorem 4.89 except that we have an H_2 GDARE now. The following example shows that the above conditions are **not** sufficient to guarantee the continuity of the semi-stabilizing solution of it.

Example 4.99 Consider a family of systems $(A^{\varepsilon}, B^{\varepsilon}, C^{\varepsilon}, D^{\varepsilon})$ with $A^{\varepsilon} = -1/2$, $B^{\varepsilon} = \varepsilon$, $C^{\varepsilon} = 1$, and $D^{\varepsilon} = \varepsilon^{2}$. Let X^{ε} be the semi-stabilizing solution of the H_2 GDARE (4.154) with Q^{ε}, S^{ε} and R^{ε} defined according to (4.155). Then, we have

$$\lim_{\varepsilon \to 0} X^{\varepsilon} = 1 \neq \tfrac{4}{3} = X^{0}.$$

The following theorem identifies two special cases where the continuity of the semi-stabilizing solutions of H_2 GDAREs can be guaranteed.

Theorem 4.100 *Consider the H_2 GDARE given in (4.154). Let $\varepsilon \in [0, \delta)$. Assume that $(A^\varepsilon, B^\varepsilon)$ is $\mathbb{C}^\ominus$-stabilizable. Let C^ε and D^ε be defined as in (4.155). Assume that (4.156) is satisfied. Let X^ε be the semi-stabilizing solution of the H_2 GDARE (4.154). Then we have*

$$\lim_{\varepsilon \to 0} X^\varepsilon = X^0$$

if either one of the following conditions is satisfied:

(i) A and B do not depend on ε, i.e., $A^0 = A^\varepsilon$ and $B^0 = B^\varepsilon$ for all ε and the matrix (4.155) is increasing in ε for $\varepsilon \in [0, \delta]$.

(ii) A $\delta > 0$ exists such that the normal rank of

$$\hat{H}(z) := \left(z(B^\varepsilon)'(I - z(A^\varepsilon)')^{-1} \quad I \right) \begin{pmatrix} Q^\varepsilon & S^\varepsilon \\ (S^\varepsilon)' & R^\varepsilon \end{pmatrix} \begin{pmatrix} (zI - A^\varepsilon)^{-1} B^\varepsilon \\ I \end{pmatrix}$$

is independent of ε for all $\varepsilon \in [0, \delta]$.

Proof : The condition that (4.155) is increasing in ε implies that the semi-stabilizing solution X^ε of the H_2 GDARE is increasing in ε, which can be seen by interpreting X^ε as the cost function of an associated linear quadratic control problem (see [75]) or through results such as Proposition 4.76. Hence, $\bar{X} = \lim_{\varepsilon \to 0} X^\varepsilon$ exists. Because X^ε is increasing in ε, we have $\bar{X} \geq X^0$. Second, using Proposition 4.143 and the fact that X^0 is a semi-stabilizing solution of the H_2 GDARE for $\varepsilon = 0$, we find that X_0 is the largest solution of the associated linear matrix inequality. It is easy to establish that $\bar{X}$ is a solution of the linear matrix inequality and hence $\bar{X} \leq X_0$. But as $\bar{X} \leq X_0$ and $\bar{X} \geq X_0$, we have $\bar{X} = X_0$ as required.

The continuity of X^ε in the case when condition *(ii)* is satisfied follows along the same lines as in the proofs of Theorems 4.46 and 4.89. ∎

4.3   Continuous-time linear matrix inequalities

As we shall see later on, CAREs play prominent roles in solving *regular H_2* and H_∞ optimal control and filtering problems. However, in singular H_2 optimal control or filtering, the roles of CAREs are taken over by so-called continuous-time linear matrix inequalities (CLMIs). The introduction of a CLMI in singular H_2

optimal control and filtering is closely tied to the notion of *dissipation inequality*, originating from what are called *dissipative systems* [62,103]. In fact, Schumacher in his work [88] refers to the CLMI as a dissipative matrix inequality. However, here we regard the CLMI as separate from the dissipation inequality despite the fact that both notions are conceptually the same.

Here, after first defining a CLMI, we define what are termed as rank minimizing solutions of a CLMI that are pertinent to our study of H_2 optimal control and filtering. Our goals are (1) to learn certain relevant properties of rank minimizing solutions, (2) to develop the conditions under which certain kinds of rank minimizing solutions exist, and (3) to find methods of computing appropriate rank minimizing solutions. Following the philosophy of the previous section, we achieve these goals by establishing a connection between the rank minimizing solutions of a given CLMI and the solutions of an appropriately defined associated H_2 CARE, and then studying such an H_2 CARE.

We introduce the CLMI in the following definition.

Definition 4.101 *Let $\bar{A} \in \mathbb{R}^{n \times n}$, $B \in \mathbb{R}^{n \times m}$, $\bar{Q} \in \mathbb{R}^{n \times n}$, $\bar{R} \in \mathbb{R}^{m \times m}$, and $S \in \mathbb{R}^{n \times m}$ with $\bar{Q}$ and $\bar{R}$ being symmetric. The matrix inequality, for an unknown $n \times n$ matrix X of the form*

$$L(X) \geq 0, \tag{4.157}$$

where

$$L(X) := \begin{pmatrix} \bar{Q} + \bar{A}'X + X\bar{A} & XB + S \\ B'X + S' & \bar{R} \end{pmatrix},$$

*is called a **continuous-time linear matrix inequality (CLMI)**. Moreover, when X satisfies (4.157), it is referred to as a solution of the CLMI. Note that $L(X) \geq 0$ implies that $\bar{R} \geq 0$; i.e., $\bar{R}$ cannot be indefinite.*

We denote the set of all real symmetric solutions of the CLMI in (4.157) as Γ; i.e.,

$$\Gamma := \left\{ X \in \mathbb{R}^{n \times n} \mid X = X' \text{ and } L(X) \geq 0 \right\}. \tag{4.158}$$

As in Section 4.2, we regularly require the matrices $\bar{Q}, \bar{R}$, and S in (4.157) to satisfy a certain structural condition known as Condition psd. The precise definition for Condition psd is given in Definition 4.60. For clarity, this is repeated below.

Definition 4.102 *The matrices $\bar{Q}, \bar{R}$, and S are said to satisfy **Condition psd** if*

$$\begin{pmatrix} \bar{Q} & S \\ S' & \bar{R} \end{pmatrix} \geq 0.$$

Under this condition, it follows that matrices C and D of appropriate dimensions exist with $(C \quad D)$ of full rank such that

$$\begin{pmatrix} \bar{Q} & S \\ S' & \bar{R} \end{pmatrix} = \begin{pmatrix} C' \\ D' \end{pmatrix} \begin{pmatrix} C & D \end{pmatrix}. \tag{4.159}$$

A relevant set of solutions of a CLMI in the context of H_2 theory is a set of what are called rank minimizing solutions. To develop a definition for such solutions, we first need to state some properties of CLMIs.

To start with we observe that for every $X \in \Gamma$, real matrices C_x and D_x exist such that

$$L(X) = \begin{pmatrix} C'_x \\ D'_x \end{pmatrix} \begin{pmatrix} C_x & D_x \end{pmatrix}, \tag{4.160}$$

and such that $(C_x \quad D_x)$ is of full rank. Then we can define a system Σ_x characterized by the matrix quadruple $(\bar{A}, B, C_x, D_x)$. The transfer function of Σ_x is then given by

$$H_x(s) := C_x(sI - \bar{A})^{-1} B + D_x. \tag{4.161}$$

We have the following lemma, which is closely related to a discrete-time result we presented in Lemma 4.62:

Lemma 4.103 *Let $X \in \Gamma$ and $\rho(X)$ be the normal rank of $H_x(s)$. Then we have*

$$\rho(X) = \rho := \text{normrank } \hat{H}(s), \tag{4.162}$$

where $\hat{H}(s)$ is defined as

$$\hat{H}(s) := \begin{pmatrix} B'(-sI - \bar{A}')^{-1} & I \end{pmatrix} \begin{pmatrix} \bar{Q} & S \\ S' & \bar{R} \end{pmatrix} \begin{pmatrix} (sI - \bar{A})^{-1} B \\ I \end{pmatrix}.$$

Proof : We note that $H'_x(-s) H_x(-s) = \hat{H}(s)$. Consequently, $\rho(X) = \rho$. ∎

Remark 4.104 *It follows from Lemma 4.103 that the normal rank of $H_x(s)$ is independent of X for any $X \in \Gamma$.*

The following lemma provides a lower bound on the rank of $L(X)$ for any $X \in \Gamma$.

Lemma 4.105 *For any $X \in \Gamma$, we have*

$$\text{rank}\{L(X)\} \geq \rho, \tag{4.163}$$

and, moreover, the equality holds if and only if Σ_x is right-invertible.

Proof : Observe that

$$\text{rank } L(X) = \text{rank} \begin{pmatrix} C_x & D_x \end{pmatrix} \geq \text{normrank } H_x(s) = \rho.$$

Hence, $\text{rank}\{L(X)\} \geq \rho$. For equality, we recall from Property 3.37 that

$$\text{rank} \begin{pmatrix} C_x & D_x \end{pmatrix} = \text{normrank } H_x(s)$$

if and only if Σ_x is right-invertible. Hence, $\text{rank } L(X) = \rho$ if and only if Σ_x is right-invertible. ■

Now we are ready to define the set of rank minimizing solutions of a CLMI.

Definition 4.106 *A solution $X \in \Gamma$ is said to be rank minimizing if*

$$\text{rank } L(X) = \rho. \tag{4.164}$$

Moreover, we denote the set of all rank minimizing solutions of the CLMI in (4.157) as $\Gamma_{\min}$; i.e.,

$$\Gamma_{\min} := \{ \, X \in \Gamma \mid \text{rank } L(X) = \rho \, \}. \tag{4.165}$$

Also, the set of all positive semi-definite rank minimizing solutions of the CLMI in (4.157) is defined as $\Gamma_{\min}^{psd}$:

$$\Gamma_{\min}^{psd} := \{ \, X \in \Gamma_{\min} \mid X \geq 0 \, \}. \tag{4.166}$$

Similarly, the set of all positive definite rank minimizing solutions of the CLMI in (4.157) is defined as $\Gamma_{\min}^{pd}$:

$$\Gamma_{\min}^{pd} := \{ \, X \in \Gamma_{\min} \mid X > 0 \, \}. \tag{4.167}$$

The following lemma shows that under a very weak condition in which the pair $(\bar{A}, B)$ is $\mathbb{C}^-$-stabilizable, the set $\Gamma_{\min}$ is nonempty.

Lemma 4.107 *Consider a CLMI as in (4.157) with the matrices $\bar{Q}$, $\bar{R}$, and S satisfying the Condition psd, and the pair $(\bar{A}, B)$ being $\mathbb{C}^-$-stabilizable. Then, at least one rank minimizing solution to the CLMI exists; i.e., $\Gamma_{\min}$ is nonempty.*

Proof : The proof follows from Proposition 4.116 presented later on.   ∎

Remark 4.108 *The condition of Lemma 4.107 can be weakened further. In fact, in [21], it is shown that at least one rank minimizing solution to the CLMI in (4.157) exists, provided:*

- *$(\overline{A}, B)$ is sign-controllable; i.e., the uncontrollable eigenvalues of $(\overline{A}, B)$, say $\lambda_1, \lambda_2, \cdots, \lambda_k$, are such that $\lambda_i + \lambda_j \neq 0$ for any i and j.*

- *The CLMI in (4.157) has at least one solution.*

Obviously, in the case when Condition psd is satisfied, $X = 0$ is a solution of the CLMI and the second condition is always satisfied. The first condition is obviously weaker than the requirement of $(\overline{A}, B)$ being $\mathbb{C}^-$-stabilizable.

We next have the following additional definition regarding semi-stabilizing and stabilizing solutions.

Definition 4.109 *Consider a matrix pencil*

$$N(s, X) := \begin{pmatrix} M(s) \\ L(X) \end{pmatrix},$$

where $M(s) := (sI - \overline{A} \quad -B)$. Then a solution X of the CLMI is said to be a semi-stabilizing solution if $X \in \Gamma_{\min}$ and if all finite zeros of the matrix pencil $N(s, X)$ are in the closed left-half plane; i.e., in $\mathbb{C}^{-0}$. Similarly, a solution X of the CLMI is said to be a stabilizing solution if $X \in \Gamma_{\min}$ and if all finite zeros of $N(s, X)$ are in the open left-half plane; i.e., in $\mathbb{C}^-$. The set of all semi-stabilizing solutions of a CLMI is denoted by $\Gamma_{\min}^{ss}$. Similarly, the set of all stabilizing solutions of a CLMI is denoted by $\Gamma_{\min}^{s}$.

Remark 4.110 *For the case when the matrices $\overline{Q}, \overline{R}$, and S satisfy the Condition psd, and when the matrix D is injective, Definition 4.109 for a semi-stabilizing or a stabilizing solution can be simplified. That is, $X \in \Gamma$ is a semi-stabilizing solution if the matrix $\overline{A} - B(D'D)^{-1}(B'X + D'C)$ has all its eigenvalues in the closed left-half complex plane $\mathbb{C}^{-0}$. Similarly, $X \in \Gamma$ is a stabilizing solution if the matrix $\overline{A} - B(D'D)^{-1}(B'X + D'C)$ has all its eigenvalues in the open left-half complex plane $\mathbb{C}^-$.*

4.3.1 Connections between a CLMI and its associated CARE

In this subsection, we examine the rank minimizing solutions of a CLMI. To simplify our presentation, we assume that the matrices $\bar{Q}$, $\bar{R}$, and S in the given CLMI satisfy the Condition psd. If the matrices $\bar{Q}$, $\bar{R}$, and S of the given CLMI do not satisfy the Condition psd and if at least one solution of the CLMI exists, then one can perform a shifting transformation on the given CLMI in order to obtain another CLMI in which the corresponding matrices $\bar{Q}$, $\bar{R}$, and S satisfy the Condition psd. To see this, let $\hat{X}$ be a solution of $L(X) \geq 0$. Factorize $L(\hat{X})$ as

$$L(\hat{X}) = \begin{pmatrix} C'_{\hat{X}} \\ D'_{\hat{X}} \end{pmatrix} \begin{pmatrix} C_{\hat{X}} & D_{\hat{X}} \end{pmatrix},$$

where $\begin{pmatrix} C_{\hat{X}} & D_{\hat{X}} \end{pmatrix}$ is of full rank with $C_{\hat{X}}$ and $D_{\hat{X}}$ having appropriate dimensions. Next, define a new $L(X)$, namely $L_s(X)$,

$$L_s(X) := \begin{pmatrix} \bar{A}'X + X\bar{A} + C'_{\hat{X}}C_{\hat{X}} & XB + C'_{\hat{X}}D_{\hat{X}} \\ B'X + D'_{\hat{X}}C_{\hat{X}} & D'_{\hat{X}}D_{\hat{X}} \end{pmatrix}. \tag{4.168}$$

Observe that $L(X) = L_s(X - \hat{X})$, and thus, $L(X) \geq 0$ if and only if $L_s(X - \hat{X}) \geq 0$. Thus, we note that X is a solution, or a rank minimizing solution, of the CLMI $L(X) \geq 0$ if and only if $X - \hat{X}$ is, respectively, a solution, or a rank minimizing solution, of the new CLMI $L_s(X) \geq 0$. Hence, the set of all solutions as well as the set of all real symmetric rank minimizing solutions of the CLMI in (4.157) can be obtained from those of $L_s(X) \geq 0$. Also, comparing $L_s(X)$ in (4.168) with $L(X)$ in (4.157), one notes that $L_s(X)$ is a special case of $L(X)$ for which the matrices $\bar{Q}$, $\bar{R}$, and S satisfy the Condition psd. It is trivial to verify that this relationship between the solutions of the original and the shifted CLMIs also extends to the case when we consider (semi-)stabilizing solutions. As such, we can assume basically without loss of generality that the matrices $\bar{Q}$, $\bar{R}$, and S satisfy the Condition psd.

Our goals, as mentioned, are indeed to study the properties of rank-minimizing solutions of a CLMI, develop the conditions under which they exist, and find the methods of constructing such solutions. These goals are facilitated by developing a connection between the rank minimizing solutions of a CLMI and certain solutions of an appropriately defined associated H_2 CARE. We develop such a connection here. We consider two cases. The first one deals with a CLMI when $\bar{R} > 0$, the so-called regular case, whereas the second case treats a CLMI when $\bar{R} \geq 0$, the so-called singular case.

Case 1 [Regular case]: $\bar{R} > 0$

We first would like to recall Proposition 4.27, which shows that the existence of a semi-stabilizing solution of an H^2_∞ CARE is directly related to the existence of a solution of an associated CLMI. The solution of this CLMI can then be used to transform the CLMI into a new one that satisfies Condition psd using the shifting argument presented in the beginning of this subsection. A similar shifting argument can be used to transform an H^2_∞ CARE into a H_2 CARE as is clear from the proof of Proposition 4.27.

We have the following theorem.

Theorem 4.111 *Consider a CLMI as in (4.157) with the matrices $\bar{Q}$, $\bar{R}$, and S in it satisfying the Condition psd. Also, let $\bar{R} > 0$. Define the following associated H_2 CARE:*

$$\bar{A}'X + X\bar{A} - (XB + S)\bar{R}^{-1}(B'X + S') + \bar{Q} = 0. \tag{4.169}$$

Then the following hold:

(i) The set of all real symmetric solutions of the H_2 CARE in (4.169) is equal to the set of all rank minimizing solutions $\Gamma_{\min}$ of the CLMI in (4.157).

(ii) The set of all real symmetric stabilizing solutions of the H_2 CARE in (4.169) is equal to the set of all stabilizing solutions $\Gamma^s_{\min}$ of the CLMI in (4.157).

(iii) The set of all real symmetric semi-stabilizing solutions of the H_2 CARE in (4.169) is equal to the set of all semi-stabilizing solutions $\Gamma^{ss}_{\min}$ of the CLMI in (4.157).

Proof : Observe that

$$\begin{pmatrix} I & -(XB + S)\bar{R}^{-1} \\ 0 & I \end{pmatrix} L(X) \begin{pmatrix} I & 0 \\ -\bar{R}^{-1}(B'X + S') & I \end{pmatrix} = \begin{pmatrix} R(X) & 0 \\ 0 & \bar{R} \end{pmatrix}.$$

As $\rho = m$, we have

$$\operatorname{rank} L(X) = m + \operatorname{rank} R(X) = \rho + \operatorname{rank} R(X).$$

This equation immediately yields the connection between the rank minimizing solutions of the CLMIs and the solutions of the H_2 CAREs. To understand the relation between the (semi-)stabilizing solutions of the CLMIs and those of H_2 CAREs, we note the following:

$$\begin{pmatrix} I & 0 & B\bar{R}^{-1} \\ 0 & I & -(XB + S)\bar{R}^{-1} \\ 0 & 0 & I \end{pmatrix} N(s, X) \begin{pmatrix} I & 0 \\ -\bar{R}^{-1}(B'X + S') & I \end{pmatrix}$$

$$= \begin{pmatrix} sI - \bar{A} + B\bar{R}^{-1}(S' + B'X) & 0 \\ R(X) & 0 \\ 0 & \bar{R} \end{pmatrix}.$$

The results of Theorem 4.111 then follow almost immediately.   ∎

Case 2 [Singular case]: $\bar{R} \geq 0$

For this singular case, the method of developing a connection between the rank minimizing solutions of a CLMI and certain solutions of an appropriately defined associated H_2 CARE is complex. A very critical tool we use in this regard is to look at the quadruple $(\bar{A}, B, C, D)$ where the matrices C and D satisfy (4.159) as a system Σ_* and use a particular structure of this system Σ_*, namely its SCB form. If we assume that the system Σ_* that is characterized by $(\bar{A}, B, C, D)$ is already in the state-space coordinates of SCB, then from SCB theory, it follows that a matrix F exists such that $(\bar{A}, B, C, D)$ has the following structure, which we refer to as compact SCB form:

$$\bar{A} + BF = \begin{pmatrix} A_{11} & A_{12} \\ KC_1 & A_{22} \end{pmatrix}, \quad B = \begin{pmatrix} B_{11} & B_{12} \\ 0 & B_{22} \end{pmatrix},$$

$$C + DF = \begin{pmatrix} C_1 & C_2 \end{pmatrix}, \quad D = \begin{pmatrix} 0 & D_2 \end{pmatrix}, \quad (4.170)$$

where F is such that $(C + DF)'D = 0$ while (A_{11}, B_{11}) is controllable, $C_2'D_2 = 0$, $C_1'D_2 = 0$, and D_2 is left-invertible. Let us emphasize that, to obtain (4.170), we have first assumed that the state space $\mathcal{X}$ of the system Σ_* characterized by the quadruple $(\bar{A}, B, C, D)$ is already in SCB form and then decomposed $\mathcal{X}$ as $\mathcal{X}_1 \oplus \mathcal{X}_2$ with $\mathcal{X}_1 = \mathcal{S}^*(\Sigma_*)$.

The following lemma exploits the above compact form of the system Σ_* and enables us to move very close to our goal of connecting the given CLMI to an associated H_2 CARE. In fact, it tells us that any symmetric solution of the CLMI has a special structure. By exploiting such a structure, we can connect the given CLMI to a reduced CLMI, which corresponds to the regular case; i.e., its $\bar{R}$ matrix is positive definite. This obviously leads us to connect the given CLMI to an associated H_2 CARE.

Lemma 4.112 *Assume that a symmetric X is a solution of the CLMI (4.157) in which the matrices $\bar{Q}$, $\bar{R}$, and S satisfy the Condition psd and $\bar{R} \geq 0$. Let the matrices C and D be such that (4.159) is satisfied, and let Σ_* be a system characterized by $(\bar{A}, B, C, D)$. Also, let $\mathcal{S}^*(\Sigma_*)$ be as in Definition 3.28. Then we have $X \mathcal{S}^*(\Sigma_*) = 0$; i.e., whenever the system Σ_* is in the compact SCB form as given in (4.170), X can be written as*

$$X = \begin{pmatrix} 0 & 0 \\ 0 & X_{22} \end{pmatrix}. \quad (4.171)$$

Proof : Let F be as used in the decomposition (4.170). Next, we define

$$W(X) := \begin{pmatrix} I & F' \\ 0 & I \end{pmatrix} L(X) \begin{pmatrix} I & 0 \\ F & I \end{pmatrix}. \quad (4.172)$$

If $L(X) \geq 0$, then so is $W(X)$; i.e.,

$$W(X) = \begin{pmatrix} X\bar{A}_F + \bar{A}'_F X + C'_F C_F & XB \\ B'X & D'D \end{pmatrix} \geq 0, \qquad (4.173)$$

with $\bar{A}_F = \bar{A} + BF$ and $C_F = C + DF$. We claim that $B \ker D \subseteq \ker X$. Let $u \in \mathbb{R}^m$ be such that $Du = 0$. Then we have

$$\begin{pmatrix} 0' & u' \end{pmatrix} W(X) \begin{pmatrix} 0 \\ u \end{pmatrix} = 0.$$

Hence, as $W(X) \geq 0$, we find

$$W(X) \begin{pmatrix} 0 \\ u \end{pmatrix} = 0.$$

This implies that $XBu = 0$ and hence $B \ker D \subseteq \ker X$. Next we show that $\ker X$ is $(C_F, \bar{A}_F)$-invariant. Assume that $x \in \ker X \cap \ker C_F$. Then

$$x'(X\bar{A}_F + \bar{A}'_F X + C'_F C_F)x = 0.$$

We note now the fact that $W(X) \geq 0$ implies that $X\bar{A}_F + \bar{A}'_F X + C'_F C_F \geq 0$. Thus, the above implies that

$$(X\bar{A}_F + \bar{A}'_F X + C'_F C_F)x = X\bar{A}_F x = 0,$$

and therefore, $\bar{A}_F x \in \ker X$. This implies that $\bar{A}_F(\ker X \cap \ker C_F) \in \ker X$, and hence, $\ker X$ is $(C_F, \bar{A}_F)$-invariant. Let Σ_F be a system characterized by the quadruple $(\bar{A}_F, B, C_F, D)$. It is well known that $\mathcal{S}^*$ is not affected by a preliminary state feedback, and as, $C'_F D = 0$, it is easy to show on the basis of the definition that $\mathcal{S}^*(\Sigma_F)$ is the smallest space having the two properties, $B \ker D \subseteq \ker X$ and $\bar{A}_F(\ker X \cap \ker C_F) \in \ker X$. The second property is equivalent to $\ker X$ being $(C_F, \bar{A}_F)$-invariant (see [98, 110]). This leads to $\mathcal{S}^*(\Sigma_F) \in \ker X$ and, therefore, $X\mathcal{S}^*(\Sigma_*) = 0$ as required.   ∎

Now we are ready to relate the solutions of a given CLMI to the solutions of another associated CLMI that corresponds to the regular case.

Theorem 4.113 *Consider a CLMI as in (4.157) with the matrices $\bar{Q}$, $\bar{R}$, and S in it satisfying the Condition psd and with $\bar{R} \geq 0$. Let the matrices C and D be such that (4.159) is satisfied. Also, let the system Σ_* characterized by $(\bar{A}, B, C, D)$ be in the compact SCB form as given in (4.170). Define an associated CLMI to the CLMI given in (4.157):*

$$L_r(\bar{X}) := \begin{pmatrix} \bar{A}'_r \bar{X} + \bar{X}\bar{A}_r + C'_r C_r & \bar{X}B_r + C'_r D_r \\ B'_r \bar{X} + D'_r C_r & D'_r D_r \end{pmatrix} \qquad (4.174)$$

by using the quadruple $(\bar{A}_r, B_r, C_r, D_r)$ given by

$$\bar{A}_r = A_{22}, B_r = \begin{pmatrix} KC_1 L & B_{22} \end{pmatrix}, C_r = C_2, D_r = \begin{pmatrix} C_1 L & D_2 \end{pmatrix}, \qquad (4.175)$$

where L is a matrix such that rank $C_1 L = $ rank C_1 while $C_1 L$ is injective. Then the following hold:

(i) If X is given such that $L(X) \geqslant 0$, then we have

$$X := \begin{pmatrix} 0 & 0 \\ 0 & \bar{X} \end{pmatrix}, \qquad (4.176)$$

with $\bar{X}$ such that $L_r(\bar{X}) \geqslant 0$ and

$$\text{rank } L(X) = \text{rank } L_r(\bar{X}). \qquad (4.177)$$

Moreover,

$$\text{rank} \begin{pmatrix} M(s) \\ L(X) \end{pmatrix} = n_1 + \text{rank} \begin{pmatrix} M_r(s) \\ L_r(\bar{X}) \end{pmatrix} \qquad (4.178)$$

for all s, where $M_r(s) = (sI - \bar{A}_r \quad B_r)$ and n_1 is the number of columns of A_{11}.

(ii) Conversely, if $\bar{X}$ is given such that $L_r(\bar{X}) \geqslant 0$, then if we define X by (4.176), we have that $L(X) \geqslant 0$ and (4.177) and (4.178) are satisfied.

Proof : From the definition of $W(X)$ in (4.172), it follows that $L(X) \geq 0$ if and only if $W(X) \geq 0$. Thus, it is sufficient to investigate the properties of $W(X)$. Using (4.170), the special structure of X characterized in Lemma 4.112, we obtain

$$W(X) = \begin{pmatrix} C_1'C_1 & C_1'K'X_{22} + C_1'C_2 & 0 \\ X_{22}KC_1 + C_2'C_1 & A_{22}'X_{22} + X_{22}A_{22} + C_2'C_2 & X_{22}B_{22} \\ 0 & B_{22}'X_{22} & D_2D_2 \end{pmatrix}.$$

Next define

$$Z(X) = \begin{pmatrix} A_{22}'X_{22} + X_{22}A_{22} + C_2'C_2 & C_1'K'X_{22} + C_1'C_2 & X_{22}B_{22} \\ X_{22}KC_1 + C_2'C_1 & C_1'C_1 & 0 \\ B_{22}'X_{22} & 0 & D_2'D_2 \end{pmatrix}.$$

We can easily check that $Z(X) \geqslant 0$ if and only if $W(X) \geqslant 0$ and rank $W(X) = $ rank $Z(X)$. Note that as

$$G = \begin{pmatrix} I & 0 & 0 \\ 0 & L & 0 \\ 0 & 0 & I \end{pmatrix}$$

is such that rank $Z(X)G = \text{rank } Z(X)$, we find that $L(X) \geq 0$ if and only if $L_r(\bar{X}) \geq 0$ and rank $L(X) = \text{rank } L_r(\bar{X})$.

To proceed, we observe that the rank of matrix

$$\begin{pmatrix} M(s) \\ L(X) \end{pmatrix}$$

is equal to the rank of

$$U(s, X) := \begin{pmatrix} I & 0 & 0 \\ 0 & I & F' \\ 0 & 0 & I \end{pmatrix} \begin{pmatrix} M(s) \\ L(X) \end{pmatrix} \begin{pmatrix} I & 0 \\ F & I \end{pmatrix}.$$

Using the same decompositions as before and interchanging some columns and rows, we find that $U(s, X)$ is similar to

$$\begin{pmatrix} -A_{12} & sI - A_{11} & -B_{12} & -B_{11} \\ A'_{22}X_{22} + X_{22}A_{22} + C'_2C_2 & C'_1K'X_{22} + C'_1C_2 & X_{22}B_{22} & 0 \\ X_{22}KC_1 + C'_2C_1 & C'_1C_1 & 0 & 0 \\ B'_{22}X_{22} & 0 & D'_2D_2 & 0 \\ 0 & 0 & 0 & 0 \end{pmatrix}. \tag{4.179}$$

Let M be such that $C_1V = 0$ and $(L \quad V)$ is invertible. We know that $(A_{11}, B_{11}, C_1, 0)$ is strongly controllable, and hence,

$$\begin{pmatrix} sI - A_{11} & B_{11} \\ C_1 & 0 \end{pmatrix} = n_1 + \text{rank } C_1$$

for all s, where n_1 is equal to the number of columns of A_{11}. Given our definition of V, this yields that $((sI - A_{11})V \quad -B_{11})$ is surjective for every s, and hence, by postmultiplying (4.179) by

$$\begin{pmatrix} I & 0 & 0 & 0 \\ 0 & L & M & 0 \\ 0 & 0 & 0 & I \end{pmatrix},$$

we obtain (4.178) and the rest of the theorem follows. ∎

Theorem 4.113 connects the given CLMI (4.157), which is singular, to an associated CLMI (4.174), which is regular. In view of this result, it is straightforward to connect the given CLMI (4.157) to an associated H_2 CARE as discussed in the following theorem.

Theorem 4.114 *Consider a CLMI as in* (4.157) *with the matrices $\bar{Q}$, $\bar{R}$, and S in it satisfying the Condition psd and with $\bar{R} \geq 0$. Let the matrices C and D be such that* (4.159) *is satisfied. Also, let the system Σ_* characterized by $(\bar{A}, B, C, D)$ be in the compact SCB form as given in* (4.170). *Define an associated H_2 CARE to the CLMI* (4.157):*

$$R(\bar{X}) := \bar{X}\bar{A}_r + \bar{A}_r'\bar{X} + C_r'C_r - (\bar{X}B_r + C_r'D_r)(D_r'D_r)^{-1}(B_r'\bar{X} + D_r'C_r) = 0,$$
$$(4.180)$$

where the quadruple $(\bar{A}_r, B_r, C_r, D_r)$ is given by (4.175). *Let $\bar{X}$ be a solution of the H_2 CARE in* (4.180). *Next, define an $n \times n$ matrix X given by* (4.176). *Then we have the following:*

(i) *The set of all rank minimizing solutions $\boldsymbol{\Gamma}_{\min}$ of the CLMI* (4.157) *is given by*

$$\boldsymbol{\Gamma}_{\min} = \left\{ X \in \mathbb{R}^{n \times n} \mid X \text{ is given by } (4.176) \text{ with } \bar{X} \text{ being a real} \right.$$
$$\left. \text{symmetric solution of the } H_2 \text{ CARE in } (4.180) \right\}.$$

(ii) *The set of all positive semi-definite rank minimizing solutions $\boldsymbol{\Gamma}_{\min}^{psd}$ of the CLMI* (4.157) *is given by*

$$\boldsymbol{\Gamma}_{\min}^{psd} = \left\{ X \in \mathbb{R}^{n \times n} \mid X \in \boldsymbol{\Gamma}_{\min} \text{ with } \bar{X} \geq 0 \right\}.$$

(iii) *The set of all positive definite rank minimizing solutions $\boldsymbol{\Gamma}_{\min}^{pd}$ of the CLMI* (4.157) *is an empty set.*

(iv) *The set of all stabilizing solutions $\boldsymbol{\Gamma}_{\min}^{s}$ of the CLMI* (4.157) *is given by*

$$\boldsymbol{\Gamma}_{\min}^{s} = \left\{ X \in \mathbb{R}^{n \times n} \mid X \text{ is given by } (4.176) \text{ with } \bar{X} \text{ being a} \right.$$
$$\left. \text{stabilizing solution of the } H_2 \text{ CARE in } (4.180) \right\}.$$

(v) *The set of all semi-stabilizing solutions $\boldsymbol{\Gamma}_{\min}^{ss}$ of the CLMI* (4.157) *is given by*

$$\boldsymbol{\Gamma}_{\min}^{ss} = \left\{ X \in \mathbb{R}^{n \times n} \mid X \text{ is given by } (4.176) \text{ with } \bar{X} \text{ being a} \right.$$
$$\left. \text{semi-stabilizing solution of the } H_2 \text{ CARE in } (4.180) \right\}.$$

Proof : By Theorem 4.113, we have a $1 - 1$ connection between the solutions of the CLMI (4.157) and the solutions of the associated CLMI (4.174). We have D_r injective, and hence, this associated CLMI corresponds to the regular case. The results then follow from Theorem 4.111. ■

Remark 4.115 *Theorems 4.113 and 4.114 establish a mapping given by (4.176) between a variety of solutions of a CLMI and the corresponding solutions of an associated H_2 CARE. This mapping (4.176) was obtained by assuming that the system Σ_* characterized by the quadruple $(\bar{A}, B, C, D)$ is already in the state-space coordinates of compact SCB form (4.170). If it is not so, then from the SCB theory, by using the given quadruple $(\bar{A}, B, C, D)$, one can easily find a nonsingular matrix Γ_s such that the one-to-one mapping (4.176) is replaced by the following equation:*

$$X = (\Gamma_s^{-1})' \begin{pmatrix} 0 & 0 \\ 0 & \bar{X} \end{pmatrix} \Gamma_s^{-1}. \tag{4.181}$$

As such, the results of Theorems 4.113 and 4.114 are true if one views the one-to-one mapping as given by (4.181).

4.3.2 Properties, existence, and computation of various types of solutions of a CLMI

The previous section provides us the tools to study certain types of solutions of a given CLMI in terms of a study of similar solutions of its associated H_2 CARE. By using these tools, in this subsection, (1) we study the semi-stabilizing, stabilizing, positive semi-definite, and positive definite properties of a solution of a CLMI; and (2) we develop the existence conditions for such solutions. This also leads to methods of computing them.

We start with the following proposition.

Proposition 4.116 *Consider a CLMI as in (4.157) with the matrices $\bar{Q}$, $\bar{R}$, and S satisfying the Condition psd, and the pair $(\bar{A}, B)$ being $\mathbb{C}^-$-stabilizable. Then the following hold:*

(i) A real symmetric semi-stabilizing solution, say X_{ss}, exists and is larger than any solution X of the given CLMI; i.e., $X_{ss} \geq X$ for all $X \in \Gamma$.

(ii) The real symmetric semi-stabilizing solution X_{ss} is unique; i.e., $\Gamma_{\min}^{ss}$ contains exactly one element.

(iii) The real symmetric semi-stabilizing solution is positive semi-definite.

(iv) The real symmetric semi-stabilizing solution is stabilizing if and only if $(\bar{A}, B, C, D)$ has no invariant zeros on the imaginary axis.

Proof : First note that using Theorem 4.114, we can associate an H_2 CARE to this CLMI. By Theorem 4.25, this H_2 CARE has a unique semi-stabilizing solution, and hence, it follows from Theorem 4.114 that the CLMI also has a unique semi-stabilizing solution. This establishes part (*ii*).

To establish part (*i*), let X be an arbitrary solution of the CLMI and X_{ss} be a semi-stabilizing solution of the CLMI. Define C_x and D_x such that

$$L(X) = \begin{pmatrix} C_x' \\ D_x' \end{pmatrix} \begin{pmatrix} C_x & D_x \end{pmatrix}.$$

Then it is easy to verify that $\widetilde{X} = X_{ss} - X$ is a semi-stabilizing solution of the CLMI:

$$\begin{pmatrix} \bar{A}'\widetilde{X} + \widetilde{X}\bar{A} + C_x'C_x & \widetilde{X}B + C_x'D_x \\ B'\widetilde{X} + D_x'C_x & D_x'D_x \end{pmatrix} \geq 0.$$

But then by Theorem 4.114, if we assume a compact SCB form (4.170) basis for the system characterized by the quadruple $(\bar{A}, B, C_x, D_x)$, we have

$$X_{ss} - X = \widetilde{X} = \begin{pmatrix} 0 & 0 \\ 0 & \bar{X} \end{pmatrix}$$

with $\bar{X}$ a semi-stabilizing solution of an H_2 CARE. But then by Theorem 4.23, we have $\bar{X} \geq 0$ and, hence, $X_{ss} \geq X$. In general, whenever the system characterized by $(\bar{A}, B, C_x, D_x)$ is not in SCB compact form (4.170), in view of Remark 4.115, it follows again that $X_{ss} \geq X$. Equation (*iii*) follows from part (*i*) after we note that $X = 0$ is a solution of the CLMI.

We know P is a semi-stabilizing solution of the CLMI if and only if P is a semi-stabilizing solution of the associated H_2 CARE given by (4.180). For the H_2 CARE of this form, we know from Theorem 4.25 that a semi-stabilizing solution is a stabilizing solution if and only if $(\bar{A}_r, B_r, C_r, D_r)$ has no invariant zeros on the imaginary axis. From the construction it is easy to check that the invariant zeros of $(\bar{A}_r, B_r, C_r, D_r)$ are equal to the invariant zeros of $(\bar{A}, B, C, D)$, which completes the proof. ∎

Remark 4.117 *We can actually precisely characterize the kernel of this semi-stabilizing solution X_{ss}:*

$$\ker X_{ss} = \mathcal{S}^*(\Sigma_*) + \mathcal{V}^{-0}(\Sigma_*),$$

where Σ_* is a system characterized by the quadruple $(\bar{A}, B, C, D)$.

The following theorems that deal with the existence conditions for some specific types of solutions of a CLMI can easily be obtained without much work.

Theorem 4.118 *Consider a CLMI as in (4.157) with the matrices $\bar{Q}$, $\bar{R}$, and S satisfying the Condition psd. Then a positive semi-definite rank minimizing solution exists; i.e., $\boldsymbol{\Gamma}_{\min}^{psd}$ is nonempty, if and only if in an appropriate basis,*

$$\bar{A} = \begin{pmatrix} A_1 & 0 \\ A_2 & A_3 \end{pmatrix}, \quad B = \begin{pmatrix} B_1 \\ B_2 \end{pmatrix}, \quad C = \begin{pmatrix} C_1 & 0 \end{pmatrix},$$

with (A_1, B_1) being a $\mathbb{C}^-$-stabilizable pair.

Proof : It follows from Theorems 4.43 and 4.114.   ∎

Remark 4.119 *The condition given in Theorem 4.118 can be stated in terms of geometric language. That is, $\boldsymbol{\Gamma}_{\min}^{psd}$ is nonempty, if and only if*

$$X_-(\bar{A}) + \langle \bar{A} \mid \operatorname{im} B \rangle + \mathcal{V}^*(\Sigma) = \mathbb{R}^n.$$

Here, $X_-(\bar{A})$ is the stable modal subspace of $\mathbb{R}^n$ related to $\bar{A}$, $\langle \bar{A} \mid \operatorname{im} B \rangle$ is the controllable subspace of the pair $(\bar{A}, B)$, and $\mathcal{V}^*(\Sigma)$ as defined in Section 3.2 represents the weakly unobservable subspace of the system Σ characterized by the quadruple $(\bar{A}, B, C, D)$.

Theorem 4.120 *Consider a CLMI as in (4.157) with the matrices $\bar{Q}$, $\bar{R}$, and S satisfying the Condition psd and the pair $(\bar{A}, B)$ being $\mathbb{C}^-$-stabilizable. Then $X = 0$ is a unique positive semi-definite semi-stabilizing solution of it if and only if the system represented by $(\bar{A}, B, C, D)$ is right-invertible and has no invariant zeros in the open right-half plane $\mathbb{C}^+$.*

Proof : This can be checked by direct verification.   ∎

4.3.3   Continuity properties of CLMIs

Proposition 4.116 guarantees that a unique positive semi-definite and semi-stabilizing solution of a CLMI exists whenever the matrices $\bar{Q}$, $\bar{R}$, and S satisfy the Condition psd and when the pair $(\bar{A}, B)$ is $\mathbb{C}^-$-stabilizable. In this subsection, we examine the continuity of such a solution with respect to the parameters $\bar{Q}$, $\bar{R}$, and S. To start with, we parameterize the matrices $\bar{Q}$, $\bar{R}$, and S with a scalar parameter ε and rewrite the CLMI (4.157) as

$$L^\varepsilon(X^\varepsilon) := \begin{pmatrix} \bar{Q}^\varepsilon + (\bar{A}^\varepsilon)' X^\varepsilon + X^\varepsilon \bar{A}^\varepsilon & X^\varepsilon B^\varepsilon + S^\varepsilon \\ (B^\varepsilon)' X^\varepsilon + (S^\varepsilon)' & \bar{R}^\varepsilon \end{pmatrix} \geq 0, \qquad (4.182)$$

where for all ε, the matrices $\bar{Q}^\varepsilon$, $\bar{R}^\varepsilon$, and S^ε satisfy the Condition psd and the pair $(\bar{A}, B)$ is $\mathbb{C}^-$-stabilizable. The matrices $\bar{Q}^\varepsilon$, $\bar{R}^\varepsilon$, and S^ε permit the factorization:

$$\begin{pmatrix} \bar{Q}^\varepsilon & S^\varepsilon \\ (S^\varepsilon)' & \bar{R}^\varepsilon \end{pmatrix} = \begin{pmatrix} (C^\varepsilon)' \\ (D^\varepsilon)' \end{pmatrix} \begin{pmatrix} C^\varepsilon & D^\varepsilon \end{pmatrix}. \tag{4.183}$$

It follows from Proposition 4.116 that for each ε, a unique positive semi-definite semi-stabilizing solution of the CLMI (4.182) exists. Let this solution be denoted by X_{ss}^ε. Assume that

$$\lim_{\varepsilon \to 0} \bar{A}^\varepsilon = \bar{A}^0, \quad \lim_{\varepsilon \to 0} B^\varepsilon = B^0, \quad \lim_{\varepsilon \to 0} C^\varepsilon = C^0, \quad \lim_{\varepsilon \to 0} D^\varepsilon = D^0. \tag{4.184}$$

These conditions are basically equivalent to the conditions leading to the continuity result in Theorem 4.46 except that this time we have a CLMI.

The following example shows that the above conditions are **not** sufficient to guarantee the continuity of the semi-stabilizing solution of the CLMI.

Example 4.121 Consider a family of systems $(\bar{A}^\varepsilon, B^\varepsilon, C^\varepsilon, D^\varepsilon)$ with $\bar{A}^\varepsilon = -1$, $B^\varepsilon = \varepsilon$, $C^\varepsilon = 1$, and $D^\varepsilon = \varepsilon^2$. Let X^ε be the semi-stabilizing solution of the CLMI (4.182) with $\bar{Q}^\varepsilon$, S^ε, and $\bar{R}^\varepsilon$ defined according to (4.183). Then, we have

$$\lim_{\varepsilon \to 0} X^\varepsilon = 0 \neq \tfrac{1}{2} = X^0.$$

The following theorem identifies two special cases where the continuity of the semi-stabilizing solution of the CLMI can be guaranteed.

Theorem 4.122 *Consider the CLMI given in (4.182). Let $\varepsilon \in [0, \delta)$. Assume that $(A^\varepsilon, B^\varepsilon)$ is $\mathbb{C}^-$-stabilizable. Let C^ε and D^ε be defined as in (4.183). Assume that (4.184) is satisfied. Let X^ε be the semi-stabilizing solution of the CLMI (4.182). Then we have*

$$\lim_{\varepsilon \to 0} X^\varepsilon = X^0$$

if either one of the following conditions is satisfied:

(i) $\bar{A}$ and B do not depend on ε; i.e., $\bar{A}^0 = \bar{A}^\varepsilon$ and $B^0 = B^\varepsilon$ for all ε, and the matrix (4.183) is increasing in ε for $\varepsilon \in [0, \delta]$.

(ii) A $\delta > 0$ exists such that the normal rank of

$$G(s) := C^\varepsilon (sI - \bar{A}^\varepsilon)^{-1} B^\varepsilon + D^\varepsilon$$

is independent of ε for all $\varepsilon \in [0, \delta]$.

Proof : The proof basically follows along the same lines as in the proof of Theorem 4.100. ∎

4.4   Discrete-time linear matrix inequalities

In this section, we focus on discrete-time linear matrix inequalities (DLMIs). As in the case of CLMIs, we are primarily interested in what are called rank minimizing solutions of DLMIs. However, a subset of all rank minimizing solutions that we shall refer to as a set of strongly rank minimizing solutions is the set of solutions that are most pertinent to H_2 optimal control and filtering. As such, our focus here is on strongly rank minimizing solutions. Our goals in this section are, as in the previous sections, (1) to learn certain relevant properties of strongly rank minimizing solutions, (2) to develop the conditions under which certain kinds of strongly rank minimizing solutions exist, and (3) to find methods of computing appropriate strongly rank minimizing solutions. Following the philosophy of previous sections, we achieve these goals by establishing a connection between strongly rank minimizing solutions of a given DLMI and solutions of an appropriately defined associated DARE, and then, studying such a DARE.

To preserve the conceptual thought process, in this section, the notations used for several objects are the same as those in the section for CLMIs. However, whenever we refer to a particular result, we shall distinguish between CLMIs and DLMIs by quoting appropriate definitions or equations.

We introduce a DLMI in the following definition.

Definition 4.123 *Let $A \in \mathbb{R}^{n \times n}$, $B \in \mathbb{R}^{n \times m}$, $Q \in \mathbb{R}^{n \times n}$, $R \in \mathbb{R}^{m \times m}$, and $S \in \mathbb{R}^{n \times m}$ with Q and R being symmetric. The matrix inequality for an unknown $n \times n$ matrix X of the form*

$$L(X) \geq 0, \tag{4.185}$$

where

$$L(X) := \begin{pmatrix} Q + A'XA - X & A'XB + S \\ B'XA + S' & B'XB + R \end{pmatrix},$$

*is called a **discrete-time linear matrix inequality (DLMI)**. Moreover, when X satisfies (4.185), it is referred to as a solution of the DLMI. We denote the set of real symmetric solutions of the DLMI in (4.185) as Γ; i.e.,*

$$\Gamma := \left\{ X \in \mathbb{R}^{n \times n} \mid X = X' \text{ and } L(X) \geq 0 \right\}. \tag{4.186}$$

As in the previous sections, now and then, we require the matrices Q, R, and S in (4.185) to satisfy a certain structural condition. To avoid stating that condition repetitively, we describe it below and label it as Condition psd, where "psd" stands for positive semi-definite.

Definition 4.124 *(**Condition psd**) The matrices Q, R, and S are said to satisfy Condition psd if*

$$\begin{pmatrix} Q & S \\ S' & R \end{pmatrix} \geq 0.$$

Under this condition, it follows that matrices $C \in \mathbb{R}^{p \times n}$ and $D \in \mathbb{R}^{p \times m}$ with $(C \quad D)$ of full rank exist such that

$$\begin{pmatrix} Q & S \\ S' & R \end{pmatrix} = \begin{pmatrix} C & D \end{pmatrix}' \begin{pmatrix} C & D \end{pmatrix}. \tag{4.187}$$

Note that in a DARE and in particular in the relationship between a GDARE and a DARE, a certain transfer matrix played a crucial role. As it also plays a crucial role for DLMIs, we recapitulate the definition of this transfer matrix:

$$\hat{H}(z) := \left(zB'(I - zA')^{-1}, \, I \right) \begin{pmatrix} Q & S \\ S' & R \end{pmatrix} \begin{pmatrix} (zI - A)^{-1}B \\ I \end{pmatrix}. \tag{4.188}$$

It turns out that the set of solutions of a DLMI coincides with that of an appropriate general discrete-time algebraic Riccati inequality. This is explored in the following lemma. Note the strong connection to Proposition 4.76.

Lemma 4.125 *Consider a DLMI as in (4.185) and its solution set Γ as defined in (4.186). Then the solution set Γ coincides with the set of all real symmetric solutions of the general discrete-time algebraic Riccati inequality (H_2-GDARI) defined by*

$$B'XB + R \geq 0, \tag{4.189a}$$

$$\ker[B'XB + R] \subseteq \ker[A'XB + S], \tag{4.189b}$$

and

$$A'XA - X - (A'XB + S)(R + B'XB)^{\dagger}(B'XA + S') + Q \geq 0. \tag{4.189c}$$

Proof : Assume that $X \in \Gamma$. Inequality (4.189a) follows in an obvious way. To show equation (4.189b), let us choose $z = (z_1' \quad z_2')'$, where $z_2 \in \ker[B'XB + R]$ and $z_1 = -\alpha(A'XB + S)z_2$ for some positive parameter α. Obviously, for all $X \in \Gamma$, we have

$$z'L(X)z \geq 0 \text{ for all } \alpha > 0.$$

This implies that

$$z'L(X)z = \alpha^2 z_2'(B'XA + S')(Q + A'XA - X)(A'XB + S)z_2$$
$$- 2\alpha z_2'(B'XA + S')(A'XB + S)z_2 \geq 0.$$

If $(A'XB + S)z_2 \neq 0$, then the left-hand side of the above equation is negative for a sufficiently small α, which would yield a contradiction. Thus, we must have

$$(A'XB + S)z_2 = 0 \text{ for all } z_2 \in \ker[B'XB + R],$$

and this implies (4.189b). Finally (4.189c) is the Schur complement of $(R + B'XB)$ in $L(X)$.

To show that a solution X of (4.189) is such that $X \in \Gamma$ is straightforward. ∎

As in CLMIs, it turns out that a relevant set of solutions of a DLMI in the context of H_2 theory is a set of what are called rank minimizing solutions. To develop an appropriate definition for such solutions, as in continuous time, we first need to state some properties of DLMIs.

To start with, we observe that for every $X \in \Gamma$, real matrices C_x and D_x exist such that

$$L(X) = \begin{pmatrix} C_x & D_x \end{pmatrix}' \begin{pmatrix} C_x & D_x \end{pmatrix}, \tag{4.190}$$

and such that $(C_x \quad D_x)$ is of full rank. Then we can define a linear system Σ_x characterized by the matrix quadruple (A, B, C_x, D_x). The transfer function of Σ_x is then given by

$$H_x(z) := C_x(zI - A)^{-1}B + D_x. \tag{4.191}$$

We have the following lemma.

Lemma 4.126 *Let $X \in \Gamma$ and $\rho(X)$ be the normal rank of $H_x(z)$. Then we have*

$$\rho(X) = \rho := \text{normrank } \hat{H}(z), \tag{4.192}$$

where $\hat{H}(z)$ is as defined in (4.188).

Proof : This is a direct consequence of Lemma 4.62. ∎

The following lemma provides a lower bound on the rank of $L(X)$ for any $X \in \Gamma$.

Lemma 4.127 *For any $X \in \Gamma$, we have*

$$\text{rank } L(X) \geq \rho, \tag{4.193}$$

and moreover, the equality holds if and only if Σ_x is right invertible.

Proof : It is similar to the proof of Lemma 4.105. ∎

Now we are ready to define the set of rank minimizing solutions of a DLMI.

Definition 4.128 *A solution $X \in \Gamma$ is said to be rank minimizing if*

$$\operatorname{rank} L(X) = \rho. \tag{4.194}$$

Moreover, we denote the set of all rank minimizing solutions of the DLMI in (4.185) as $\Gamma_{\min}$; i.e.,

$$\Gamma_{\min} := \{\, X \in \Gamma \mid \operatorname{rank} L(X) = \rho \,\}. \tag{4.195}$$

We introduce next the notions of semi-stabilizing and stabilizing solutions of DLMIs. We would like to stress the close relationship to the definition of (semi-) stabilizing solutions of the GDARE in Definition 4.57.

Definition 4.129 *Consider a matrix*

$$N(z, X) := \begin{pmatrix} M(z) \\ L(X) \end{pmatrix},$$

where $M(z) := (zI - A \quad -B)$. Then a solution $X \in \Gamma$ is said to be a semi-stabilizing solution if it is a rank-minimizing solution and all finite zeros of the matrix pencil $N(z, X)$ are inside or on the unit circle; i.e., in $\mathbb{C}^{\otimes}$, and if the number of zeros at infinity of $N(z, X)$ is equal to the rank of $L(X)$. Similarly, a solution $X \in \Gamma_{\min}$ is said to be a stabilizing solution if all finite zeros of $N(z, X)$ are inside the unit circle; i.e., in $\mathbb{C}^{\ominus}$, and if the number of zeros at infinity of $N(z, X)$ is equal to the rank of $L(X)$. The set of all semi-stabilizing solutions of a DLMI is denoted by $\Gamma_{\min}^{ss}$. Similarly, the set of all stabilizing solutions of a DLMI is denoted by $\Gamma_{\min}^{s}$.

Remark 4.130 *Definition 4.129 can be rewritten as follows. A solution X of a DLMI is said to be a stabilizing (respectively, semi-stabilizing) solution if all eigenvalues of the matrix*

$$A - B(B'XB + R)^{\dagger}(B'XA + S') - B(I - (B'XB + R)^{\dagger}(B'XB + R))F$$

are inside the unit circle (respectively, inside and/or on the unit circle) for some suitably chosen matrix F.

As in the case of CLMIs, to facilitate our study of DLMIs, we would like to establish a connection between the rank minimizing solutions of a given DLMI and the solutions of an appropriately defined associated DARE. However, it turns out that unlike CLMIs, one cannot establish such a connection for all rank minimizing solutions of a given DLMI. Nevertheless, a connection can be established for a subset of all rank minimizing solutions, which we shall refer to as strongly rank minimizing solutions. Fortunately, these strongly rank minimizing solutions are actually the ones that are pertinent to H_2 optimal control theory and H_2 optimal filtering theory. We next have the following definition for strongly rank minimizing solutions.

Definition 4.131 *A solution* $X \in \Gamma$ *is said to be a strongly rank minimizing solution if*

$$\mathrm{rank}\, L(X) = \mathrm{rank}(B'XB + R). \tag{4.196}$$

Moreover, we denote the set of all strongly rank minimizing solutions of the DLMI (4.185) *as*

$$\mathscr{L}_{\min} := \{ X \in \Gamma \mid \mathrm{rank}\, L(X) = \mathrm{rank}(B'XB + R) \}.$$

Also, the set of all positive semi-definite strongly rank minimizing solutions of the DLMI (4.185) *is defined as* $\mathscr{L}_{\min}^{psd}$:

$$\mathscr{L}_{\min}^{psd} := \{ X \in \mathscr{L}_{\min} \mid X \geq 0 \}.$$

Similarly, the set of all positive definite strongly rank minimizing solutions of the DLMI (4.185) *is defined as* $\mathscr{L}_{\min}^{pd}$:

$$\mathscr{L}_{\min}^{pd} := \{ X \in \mathscr{L}_{\min} \mid X > 0 \}.$$

The following theorem discusses the conditions under which the set $\mathscr{L}_{\min}$ is nonempty.

Theorem 4.132 *Consider a DLMI as in* (4.185) *with the matrices* Q, R, *and* S *satisfying the Condition psd and the pair* (A, B) *being* $\mathbb{C}^{\ominus}$-*stabilizable. Then at least one strongly rank minimizing solution exists; i.e.,* $\mathscr{L}_{\min}$ *is nonempty.*

Proof : It is a consequence of Theorem 4.147, which will be presented later on.
∎

Now we are ready to show that $\mathscr{L}_{\min} \subseteq \Gamma_{\min}$. This is formulated as the following lemma.

Lemma 4.133

$$\mathscr{L}_{\min} \subseteq \Gamma_{\min}.$$

Proof : Let $X \in \mathcal{L}_{\min}$. Then from Theorem 4.137, which will be presented later on, we have X satisfying the general H_2 DARE (4.198). Then from Lemma 4.63, it is evident that the rank of $B'XB + R$ is equal to the normal rank of $\hat{H}$. Hence,

$$\operatorname{rank}(B'XB + R) = \rho \ \text{ for all } \ X \in \mathcal{L}_{\min}.$$

This implies that $X \in \Gamma_{\min}$. ∎

A natural question that arises next is under what conditions $\mathcal{L}_{\min}$ coincides with $\Gamma_{\min}$. This is answered in the next lemma.

Lemma 4.134

$$\mathcal{L}_{\min} = \Gamma_{\min}$$

if the matrix pencil

$$\begin{pmatrix} Q & S & I - zA' \\ S' & R & -zB' \\ zI - A & -B & 0 \end{pmatrix} \tag{4.197}$$

has exactly ρ zeros at infinity where ρ is as defined in Lemma 4.126.

Proof : We have

$$L_1(z) \begin{pmatrix} Q & S & I - zA' \\ S' & R & -zB' \\ zI - A & -B & 0 \end{pmatrix} R_1(z)$$

$$= \begin{pmatrix} Q + A'XA - X & S + A'XB & I - zA' \\ S' + B'XA & R + B'XB & -zB' \\ zI - A & -B & 0 \end{pmatrix},$$

where

$$L_1(z) := \begin{pmatrix} I & 0 & -(z^{-1}I + A')X/2 \\ 0 & I & -B'X/2 \\ 0 & 0 & I \end{pmatrix},$$

$$R_1(z) = \begin{pmatrix} I & 0 & 0 \\ 0 & I & 0 \\ -X(I + z^{-1}A)/2 & -z^{-1}XB/2 & I \end{pmatrix}.$$

We note that the above transformation does not affect the infinite zeros. As X satisfies the DLMI, matrices C_x and D_x exist such that

$$\begin{pmatrix} C'_x \\ D'_x \end{pmatrix} \begin{pmatrix} C_x & D_x \end{pmatrix} = \begin{pmatrix} Q + A'XA - X & A'XB + S' \\ S + B'XA & R + B'XB \end{pmatrix}.$$

Next, we continue with suitable pre- and postmultiplication without affecting the infinite zeros:

$$
L_2(z)\begin{pmatrix} Q & S & I - zA' \\ S' & R & -zB' \\ zI - A & -B & 0 \end{pmatrix} R_2(z) = \begin{pmatrix} 0 & 0 & I - zA' \\ 0 & \hat{H}(z) & 0 \\ zI - A & 0 & 0 \end{pmatrix},
$$

where

$$
L_2(z) := \begin{pmatrix} I & 0 & -C_x'C_x(zI - A)^{-1} \\ -zB'(I - A'z)^{-1} & I & -D_x'C_x(zI - A)^{-1} \\ 0 & 0 & I \end{pmatrix},
$$

$$
R_2(z) := \begin{pmatrix} I & (zI - A)^{-1}B & 0 \\ 0 & I & 0 \\ 0 & 0 & I \end{pmatrix}.
$$

Note that the latter transformation does not affect the infinite zeros provided that A is invertible. The latter matrix has exactly ρ infinite zeros provided $\hat{H}(z)$ does not lose rank at infinity. In other words, we have

$$
\text{rank } H(\infty) = \rho.
$$

But as $B'XB + R = D_x'D_x$ and

$$
H(\infty) = [D_z' - B'(A')^{-1}C_z']D_z,
$$

we obtain that rank $B'XB + R \geqslant \rho$. Because X is a rank minimizing solution, we already know that

$$
\rho = \text{rank } L(X) \geqslant \text{rank}(B'XB + R),
$$

and this immediately yields that X is a strongly rank minimizing solution.

Assume that A is not invertible. Because (A, B) is stabilizable, a matrix F exists such that $A_F = A + BF$ is invertible. But then X is also a rank minimizing solution of

$$
\begin{pmatrix} I & F' \\ 0 & 0 \end{pmatrix}\begin{pmatrix} Q + A'XA - X & S + A'XB \\ S' + B'XA & B'XB + R \end{pmatrix}\begin{pmatrix} I & 0 \\ F & 0 \end{pmatrix}
$$
$$
= \begin{pmatrix} Q_F + A_F'XA_F - X & S_F + A_F'XB \\ S_F' + B'XA_F & B'XB + R \end{pmatrix},
$$

where

$$
Q_F = Q' + F'S' + SF + F'RF, \qquad S_F = S + F'R.
$$

As A_F is invertible, we can then use the previous arguments to establish that X is actually a strongly rank minimizing solution of the DLMI.   ∎

Next, we would like to show that the stabilizing and semi-stabilizing solutions of a DLMI are in fact its strongly rank minimizing solutions. The following lemma shows this.

Lemma 4.135 *A semi-stabilizing or stabilizing solution of a DLMI is automatically a strongly rank minimizing solution. That is,*

$$\boldsymbol{\Gamma}_{\min}^{ss} \subseteq \boldsymbol{\mathscr{L}}_{\min}.$$

Proof : Assume that $X \in \boldsymbol{\Gamma}_{\min}^{ss}$; i.e., X is a semi-stabilizing or stabilizing solution of a DLMI. Then, we know by definition that

$$\begin{pmatrix} zI - A & -B \\ Q + A'XA - X & A'XB + S' \\ S + B'XA & R + B'XB \end{pmatrix}$$

has exactly ρ infinite zeros where $\rho = \text{normrank } \hat{H}(z)$ with $\hat{H}$ as defined by (4.95). As X satisfies the DLMI, matrices C_x and D_x exist such that

$$\begin{pmatrix} C_x' \\ D_x' \end{pmatrix} \begin{pmatrix} C_x & D_x \end{pmatrix} = \begin{pmatrix} Q + A'XA - X & A'XB + S' \\ S + B'XA & R + B'XB \end{pmatrix}.$$

We note that

$$\begin{pmatrix} I & 0 & 0 \\ -C_x'C_x(zI - A)^{-1} & I & 0 \\ -D_x'C_x(zI - A)^{-1} & 0 & I \end{pmatrix} \begin{pmatrix} zI - A & -B \\ C_x'C_x & C_x'D_x \\ D_x'C_x & D_x'D_x \end{pmatrix} \begin{pmatrix} I & (zI - A)^{-1}B \\ 0 & I \end{pmatrix}$$
$$= \begin{pmatrix} zI - A & -B \\ 0 & C_x'[C_x(zI - A)^{-1}B + D_x] \\ 0 & D_x'[C_x(zI - A)^{-1}B + D_x] \end{pmatrix}.$$

Hence, we see that $C_x(zI - A)^{-1}B + D_x$ can have no infinite zeros. In other words, rank $D_x = \rho$, but then

$$\rho = \text{rank } D_x'D_x = \text{rank}(B'XB + R),$$

which implies that X is a strongly rank minimizing solution. ∎

4.4.1   Connections between a DLMI and its associated DARE

In this section, we develop interconnections between a DLMI and its associated DARE. To start with, we show that the set of strongly rank minimizing solutions of a DLMI coincides with the set of real symmetric solutions of an appropriately defined associated H_2 GDARE. In view of Subsection 4.2.7, there is a relationship between an H_2 GDARE and its associated H_2 DARE. Exploiting such a relationship, we immediately obtain a one-to-one connection between strongly rank minimizing solutions of a given DLMI and real symmetric solutions of an appropriately defined associated H_2 DARE. In fact, this subsection shows that the reduction technique exploited in Subsection 4.2.7 can be directly exploited to get a better understanding of the solutions of a DLMI whether they are rank-minimizing or not. Such an understanding enables us (1) to learn certain relevant properties of rank minimizing solutions, (2) to develop the conditions under which certain kinds of rank minimizing solutions exist, and (3) to find methods of computing appropriate rank minimizing solutions.

We first connect the given DLMI to an associated H_2-GDARE as defined below.

Definition 4.136 *Consider a DLMI as in (4.185). Then its associated general H_2 algebraic Riccati equation (H_2-GDARE) is defined as*

$$X = A'XA - (A'XB + S)(R + B'XB)^{\dagger}(B'XA + S') + Q, \qquad (4.198\text{a})$$

$$\ker[B'XB + R] \subseteq \ker[A'XB + S], \qquad (4.198\text{b})$$

and

$$B'XB + R \geq 0. \qquad (4.198\text{c})$$

We have the following theorem.

Theorem 4.137 *Consider a DLMI as in (4.185) and its associated general H_2 GDARE as in (4.198). Then the set of all strongly rank minimizing solutions of the DLMI (4.185), namely $\mathcal{L}_{\min}$, coincides with the set of all real symmetric solutions of the associated general H_2 GDARE (4.198).*

Proof : According to Lemma 4.125, conditions (4.198b) and (4.198c) are satisfied for any $X \in \Gamma$. Next we note that

$$\begin{pmatrix} I & -(A'XB + S)W^{\dagger} \\ 0 & I \end{pmatrix} L(X) \begin{pmatrix} I & 0 \\ -W^{\dagger}(B'XA + S') & I \end{pmatrix}$$
$$\begin{pmatrix} A'XA - X - (A'XB + S)W^{\dagger}(B'XA + S') + Q & 0 \\ 0 & W \end{pmatrix}$$

with $W = B'XB + R$, where we have used that $V(I - W^\dagger W) = 0$ if $\ker W \subseteq \ker V$. Hence, the rank of $L(X)$ equals the sum of the rank of $B'XB + R$ and the rank of its Schur complement. Therefore, the Schur complement must be 0 implying (4.198a). $\blacksquare$

The above theorem obviously gives us a tool for computing solutions as well as learning some properties of solutions of a DLMI. In this regard, we would like to recall the relationship between an H_2 GDARE and an H_2 DARE that we established in Subsection 4.2.7. This immediately yields a connection between strongly rank minimizing solutions of a given DLMI and solutions of an appropriately defined H_2 DARE. This facilitates a better understanding of a given DLMI. We divide our development below into two parts based on the normal rank of $\hat{H}$ as defined in (4.188).

Case 1 [$\hat{H}$ has full normal rank]

We first recall that $\hat{H}(z)$, being full normal rank, means that the normal rank of $\hat{H}(z)$ is equal to m. Next, we note that for the classic special case when the matrices Q, R, and S satisfy the Condition psd, the assumption that $\hat{H}(z)$ has full normal rank is equivalent to the assumption that the system characterized by the quadruple (A, B, C, D) is left-invertible.

Our main result for this case is given in the following theorem.

Theorem 4.138 *Consider a DLMI as in* (4.185) *with $\hat{H}$ having full normal rank. Define an H_2 DARE by*

$$R + B'XB > 0 \qquad\qquad (4.199a)$$

and

$$X = A'XA - (A'XB + S)(R + B'XB)^{-1}(B'XA + S') + Q. \qquad (4.199b)$$

Then the following hold:

(i) *The set of all real symmetric solutions of the H_2 DARE in* (4.199) *coincides with the set of strongly rank minimizing solutions $\mathcal{L}_{\min}$ of the DLMI in* (4.185).

(ii) *The set of all positive semi-definite solutions of the H_2 DARE in* (4.199) *coincides with the set of positive semi-definite strongly rank minimizing solutions $\mathcal{L}_{\min}^{psd}$ of the DLMI in* (4.185).

(iii) *The set of all positive definite solutions of the H_2 DARE in* (4.199) *coincides with the set of all positive definite strongly rank minimizing solutions $\mathcal{L}_{\min}^{pd}$ of the DLMI in* (4.185).

(iv) *The set of all real symmetric stabilizing solutions of the H_2 DARE in* (4.199) *coincides with the set of all stabilizing as well as strongly rank minimizing solutions $\Gamma_{\min}^{s}$ of the DLMI in* (4.185).

(v) The set of all real symmetric semi-stabilizing solutions of the H_2 DARE in (4.199) coincides with the set of all semi-stabilizing as well as strongly rank minimizing solutions $\Gamma_{\min}^{ss}$ of the DLMI in (4.185).

Proof : It follows from Theorem 4.137 and Lemma 4.63. ∎

Case 2 [$\hat{H}(z)$ does not have full normal rank]

In this case, to simplify our presentation, we assume that the matrices Q, R, and S in the given DLMI satisfy the Condition psd. In the beginning of Subsection 4.3.1, we used a shifting argument to reduce the general case to a CLMI where Condition psd is satisfied. This can also be done in the discrete-time case. That is, when the the matrices Q, R, and S in the given DLMI do not satisfy the Condition psd, one can perform a shifting transformation on the given DLMI, whenever at least one solution exists, in order to obtain a new DLMI in which the corresponding matrices Q, R, and S satisfy the Condition psd. To see this, let $\hat{X}$ be a solution of $L(X) \geq 0$. Factorize $L(\hat{X})$ as

$$L(\hat{X}) = \begin{pmatrix} C'_{\hat{X}} \\ D'_{\hat{X}} \end{pmatrix} \begin{pmatrix} C_{\hat{X}} & D_{\hat{X}} \end{pmatrix},$$

where $(C_{\hat{X}} \quad D_{\hat{X}})$ has full row rank. Next, define a new DLMI

$$L_s(X) := \begin{pmatrix} C'_{\hat{X}} C_{\hat{X}} + A'XA - X & A'XB + C'_{\hat{X}} D_{\hat{X}} \\ B'XA + D'_{\hat{X}} C_{\hat{X}} & B'XB + D'_{\hat{X}} D_{\hat{X}} \end{pmatrix}. \tag{4.200}$$

Observe that $L(X) = L_s(X - \hat{X})$, and thus, $L(X) \geq 0$ if and only if $L_s(X - \hat{X}) \geq 0$. Thus, we note that X is a solution, a rank minimizing solution, or a strongly rank minimizing solution of the DLMI $L(X) \geq 0$ if and only if $X - \hat{X}$ is, respectively, a solution, a rank minimizing solution, or a strongly rank minimizing solution of the new DLMI $L_s(X) \geq 0$.

Next, comparing $L_s(X)$ in (4.200) with $L(X)$ in (4.185), one notes that $L_s(X)$ is a special case of $L(X)$ for which the matrices Q, R, and S satisfy the Condition psd. As such, without loss of generality, to start with one can assume that the matrices Q, R, and S satisfy the Condition psd.

As in CLMIs, for the singular case, the method of developing a connection between the rank minimizing solutions of a DLMI and certain solutions of an appropriately defined associated H_2 DARE is complex. Once again we look at the quadruple (A, B, C, D) where the matrices C and D satisfy (4.187) as a system Σ_* and use a particular structure of this system Σ_*, namely its SCB form. If we assume that the system Σ_* that is characterized by (A, B, C, D) is already in the state-space coordinates of SCB, then as in (4.170), we obtain the following compact SCB form:

$$A + BF = \begin{pmatrix} A_{11} & A_{12} \\ 0 & A_{22} \end{pmatrix}, B = \begin{pmatrix} B_{11} & B_{12} \\ 0 & B_{22} \end{pmatrix}, C + DF = \begin{pmatrix} 0 & C_2 \end{pmatrix},$$

$$D = \begin{pmatrix} 0 & D_2 \end{pmatrix}, \quad (4.201)$$

where F is such that $(C + DF)'D = 0$ while (A_{11}, B_{11}) is controllable and $(A_{22}, B_{22}, C_2, D_2)$ is left-invertible. Let us emphasize that, to obtain (4.201), we have first assumed that the state space $\mathcal{X}$ of the system Σ_* characterized by the quadruple (A, B, C, D) is already in SCB form, and then decomposed $\mathcal{X}$ as $\mathcal{X}_1 \oplus \mathcal{X}_2$ with $\mathcal{X}_1 = \mathcal{R}^*(\Sigma_*)$. We note next that

$$\begin{pmatrix} I & F' \\ 0 & I \end{pmatrix} L(X) \begin{pmatrix} I & 0 \\ F & I \end{pmatrix}$$

$$= \begin{pmatrix} -X + A_F'XA_F + C_F'C_F & A_F'XB \\ B'XA_F & D'D + B'XB \end{pmatrix} \geqslant 0 \quad (4.202)$$

with $A_F = A + BF$ and $C_F = C + DF$.

The following lemma, which is analogous to Lemmas 4.112, exploits the above compact form of the system Σ_* characterized by the quadruple (A, B, C, D) and enables us to move very close to our goal of connecting the given DLMI to an associated H_2 DARE. In fact, it tells us that any symmetric solution of the DLMI has a special structure. By exploiting such a structure, we can connect the given DLMI to a reduced CLMI that corresponds to the regular case; i.e., the corresponding $\hat{H}(z)$ has full normal rank. This obviously leads us to connect the given DLMI to an associated H_2 DARE.

Lemma 4.139 *Assume that a symmetric X is a solution of the DLMI (4.185) in which the matrices Q, R, and S satisfy the Condition psd. Let the matrices C and D be such that (4.187) is satisfied, and let a system Σ_* be characterized by (A, B, C, D). Also, let $\mathcal{R}^*(\Sigma_*)$ be as in Definition 3.28. Then we have $X\mathcal{R}^*(\Sigma_*) = 0$; i.e., whenever the system Σ_* is in the compact SCB form as given in (4.201), X can be written as*

$$X = \begin{pmatrix} 0 & 0 \\ 0 & X_{22} \end{pmatrix}. \quad (4.203)$$

Proof : This follows immediately from the arguments used in Subsection 4.2.7. ∎

Now we are ready to relate the solutions of a given DLMI to the solutions of another associated DLMI that corresponds to the regular case.

Theorem 4.140 *Consider a DLMI as in (4.185) for which the corresponding $\hat{H}(z)$ does not have full normal rank. Let the matrices Q, R, and S in it satisfy the Condition psd. Also, let the matrices C and D be such that (4.187) is satisfied, and let the system Σ_* characterized by (A, B, C, D) be in the compact SCB form as given in (4.201). Define an associated DLMI to the DLMI given in (4.185):*

$$L_r(\bar{X}) := \begin{pmatrix} A_{22}'\bar{X}A_{22} - \bar{X} + C_2'C_2 & A_{22}'\bar{X}B_{22} \\ B_{22}'\bar{X}A_{22} & B_{22}'\bar{X}B_{22} + D_2'D_2 \end{pmatrix} \geq 0, \quad (4.204)$$

by using the quadruple $(A_{22}, B_{22}, C_2, D_2)$ as given in (4.201). Then the following hold:

(i) If X is given such that $L(X) \geq 0$, then in the basis related to the decomposition (4.201), we have

$$X := \begin{pmatrix} 0 & 0 \\ 0 & \bar{X} \end{pmatrix} \quad (4.205)$$

with $\bar{X}$ such that $L_r(\bar{X}) \geq 0$ and

$$\operatorname{rank} L(X) = \operatorname{rank} L_r(\bar{X}). \quad (4.206)$$

Moreover,

$$\operatorname{rank}\begin{pmatrix} M(s) \\ L(X) \end{pmatrix} = n_1 + \operatorname{rank}\begin{pmatrix} M_r(s) \\ L_r(\bar{X}) \end{pmatrix} \quad (4.207)$$

for all s, where $M_r(s) = (sI - A_{22} \quad B_{22})$, while n_1 is the number of columns of A_{11}.

(ii) Conversely, if $\bar{X}$ is given such that $L_r(\bar{X}) \geq 0$ then if we define X by (4.205) we have that $L(X) \geq 0$ and (4.206) and (4.207) are satisfied.

Proof : The proof follows more or less directly from Lemma 4.139 and some manipulations similar to the continuous-time arguments as given in the proof of Theorem 4.113. ∎

Theorem 4.140 connects the given DLMI (4.185), which is singular, to an associated DLMI (4.204), which is regular. In view of this result, it is straightforward to connect the given DLMI (4.185) to an associated H_2 DARE as discussed in the following theorem.

Theorem 4.141 *Consider a DLMI as in* (4.185) *for which the corresponding* $\hat{H}(z)$ *does not have full normal rank. Let the matrices* Q, R, *and* S *in it satisfy the Condition psd. Also, let the matrices* C *and* D *be such that* (4.187) *is satisfied, and let the system* Σ_* *characterized by* (A, B, C, D) *be in the compact SCB form as given in* (4.201). *Define an associated* H_2 *DARE to the DLMI* (4.185):

$$A'_{22}\bar{X}A_{22} - \bar{X} + C'_2 C_2 - (A'_{22}\bar{X}B_{22} + C'_2 D_2)$$
$$\times (B'_{22}\bar{X}B_{22} + D'_2 D_2)^{-1}(B'_{22}\bar{X}A_{22} + D'_2 C_2) = 0, \quad (4.208a)$$

$$B'_{22}\bar{X}B_{22} + D'_2 D_2 > 0, \tag{4.208b}$$

where the matrix quadruple $(A_{22}, B_{22}, C_2, D_2)$ *is as in* (4.201). *Let* (4.201) *be a solution of the* H_2 *DARE in* (4.208). *Next, define an* $n \times n$ *matrix* X *given by* (4.203). *Then we have the following:*

(i) *The set of all strongly rank minimizing solutions* $\mathscr{L}_{\min}$ *of the DLMI in* (4.185) *is given by*

$$\mathscr{L}_{\min} = \{X \in \mathbb{R}^{n \times n} \mid X \text{ is given by } (4.203) \text{ with } \bar{X} \text{ a real}$$
$$\text{symmetric solution of the } H_2 \text{ DARE in } (4.208)\}.$$

(ii) *The set of all positive semi-definite strongly rank minimizing solutions* $\mathscr{L}_{\min}^{psd}$ *of the DLMI in* (4.185) *is given by*

$$\mathscr{L}_{\min}^{psd} = \{X \in \mathbb{R}^{n \times n} \mid X \in \mathscr{L}_{\min} \text{ with } \bar{X} \geq 0\}.$$

(iii) *The set of all positive definite strongly rank minimizing solutions* $\mathscr{L}_{\min}^{pd}$ *of the DLMI in* (4.185) *is an empty set.*

(iv) *The set of all stabilizing solutions* $\mathit{\Gamma}_{\min}^s$ *of the DLMI in* (4.185) *is given by*

$$\mathit{\Gamma}_{\min}^s = \{X \in \mathbb{R}^{n \times n} \mid X \in \mathscr{L}_{\min} \text{ with } \bar{X} \text{ a stabilizing}$$
$$\text{solution of the } H_2 \text{ DARE in } (4.208)\}.$$

(v) *The set of all semi-stabilizing solutions* $\mathit{\Gamma}_{\min}^{ss}$ *of the DLMI in* (4.185) *is given by*

$$\mathit{\Gamma}_{\min}^{ss} = \{X \in \mathbb{R}^{n \times n} \mid X \in \mathscr{L}_{\min} \text{ with } \bar{X} \text{ a semi-stabilizing}$$
$$\text{solution of the } H_2 \text{ DARE in } (4.208)\}.$$

Remark 4.142 *Theorems 4.140 and 4.141 establish a mapping given by* (4.205) *between a variety of solutions of a DLMI and the corresponding solutions of an associated* H_2 *DARE. This mapping* (4.205) *was obtained by assuming that the* *system Σ_* characterized by the quadruple (A, B, C, D) is already in the state-space coordinates of compact SCB form (4.201). If it is not so, then from the SCB theory, by using the given quadruple (A, B, C, D), one can easily find a nonsingular matrix Γ_s such that the one-to-one mapping (4.205) is replaced by the following equation:*

$$X = (\Gamma_s^{-1})' \begin{pmatrix} 0 & 0 \\ 0 & \bar{X} \end{pmatrix} \Gamma_s^{-1}. \tag{4.209}$$

As such the results of Theorems 4.140 and 4.141 are true if one views the one-to-one mapping as given by (4.209).

4.4.2 Properties, existence, and computation of various types of solutions of a DLMI

Previous section provides us the tools to study certain types of solutions of a given DLMI in terms of a study of similar solutions of its associated H_2 DARE. By using these tools, in this subsection, (1) we study the semi-stabilizing, stabilizing, positive semi-definite, and positive definite properties of a solution of a DLMI; and (2) we develop the existence conditions for such solutions. This also leads to methods of computing them.

We start with the following proposition.

Proposition 4.143 *Consider a DLMI as in (4.185) with the matrices Q, R, and S satisfying the Condition psd and the pair (A, B) being $\mathbb{C}^{\ominus}$-stabilizable. Then the following hold:*

(i) *A real symmetric semi-stabilizing solution, say X_{ss}, if it exists, is larger than any solution X of the given DLMI; i.e., $X_{ss} \geq X$ for all $X \in \Gamma$.*

(ii) *A real symmetric semi-stabilizing solution X_{ss}, if it exists, is unique; i.e., $\Gamma_{\min}^{ss}$ is at most a singleton set.*

(iii) *A real symmetric semi-stabilizing solution of it, if it exists, is also positive semi-definite.*

Proof : This follows along the same lines as the proof of the continuous-time result presented in Proposition 4.116.   ∎

Remark 4.144 *Theorem 4.95 yields that a semi-stabilizing solution of the DLMI* (4.185), *if it exists, is a stabilizing solution if and only if the matrix pencil*

$$\begin{pmatrix} Q & S & I - zA' \\ S' & R & -zB' \\ zI - A & -B & 0 \end{pmatrix}$$

has full rank for all z on the unit circle $\mathbb{C}^{\circ}$.

The following theorems that deal with the existence conditions for some specific types of solutions of a DLMI can easily be obtained without much work.

Theorem 4.145 *Consider a DLMI as in* (4.185) *with the matrices Q, R, and S satisfying the Condition psd. Define C and D according to* (4.187). *Then a positive semi-definite strongly rank minimizing solution exists; i.e., $\mathcal{L}^{psd}_{\min}$ is nonempty, if and only if the H_2 GDARE* (4.91) *has a positive semi-definite solution or, equivalently, if and only if*

$$X_{\ominus}(A) + \langle A \mid \operatorname{im} B \rangle + \mathcal{V}^*(\Sigma) = \mathbb{R}^n.$$

Here n is the dimension of the square matrix A, and $X_{\ominus}(A)$ is the stable modal subspace of $\mathbb{R}^n$ related to A, $\langle A \mid \operatorname{im} B \rangle$ is the controllable subspace of the pair (A, B), and $\mathcal{V}^*(\Sigma)$ as defined in Section 3.2 represents the weakly unobservable subspace of the system Σ characterized by the quadruple (A, B, C, D).

Proof : The proof follows from Theorems 4.137 and 4.96. ■

Theorem 4.146 *Consider a DLMI as in* (4.185) *with the matrices Q, R, and S satisfying the Condition psd and the pair (A, B) being $\mathbb{C}^{\ominus}$-stabilizable. Also, assume that the system represented by (A, B, C, D) has no invariant zeros on the unit circle $\mathbb{C}^{\circ}$. Then the given DLMI has a unique stabilizing solution. Moreover, this solution is positive semi-definite and is the largest among all real symmetric solutions of it.*

Proof : The proof follows from Theorems 4.137 and 4.97. ■

Theorem 4.147 *Consider a DLMI as in* (4.185) *with the matrices Q, R, and S satisfying the Condition psd and the pair (A, B) being $\mathbb{C}^{\ominus}$-stabilizable. Then a unique semi-stabilizing solution exists. Moreover, this solution is positive semi-definite, strongly rank minimizing, and the largest solution of the DLMI.*

Proof : Theorem 4.98 establishes that a semi-stabilizing, strongly rank-minimizing solution of the DLMI exists due to the $1 - 1$ correspondence between the solutions of the H_2 GDARE and the strongly rank minimizing solutions of the DLMI. That this is the largest solution is a consequence of Proposition 4.143. ∎

Theorem 4.148 *Consider a DLMI as in (4.185) with the matrices Q, R, and S satisfying the Condition psd and the pair (A, B) being $\mathbb{C}^{\ominus}$-stabilizable. Then $X = 0$ is the unique positive semi-definite semi-stabilizing solution if and only if the system represented by (A, B, C, D) is right-invertible, has no invariant zeros in $\mathbb{C}^{\otimes}$, and has no infinite zeros of order greater than or equal to one.*

Proof : This can be checked via direct verification and is the discrete-time equivalent of Theorem 4.120. ∎

4.4.3   Continuity properties of the DLMI

Theorem 4.147 guarantees that a unique positive semi-definite semi-stabilizing solution of a DLMI exists whenever the matrices Q, R, and S satisfy the Condition psd and when the pair (A, B) is $\mathbb{C}^{\ominus}$-stabilizable. In this subsection, we examine the continuity of such a solution with respect to the parameters Q, R, and S. To start with, we parameterize the matrices Q, R, and S with a scalar parameter ε and rewrite the linear matrix inequality (4.185) as

$$L^{\varepsilon}(X^{\varepsilon}) := \begin{pmatrix} Q^{\varepsilon} + A'X^{\varepsilon}A - X^{\varepsilon} & A'X^{\varepsilon}B + S^{\varepsilon} \\ B'X^{\varepsilon}A + (S^{\varepsilon})' & B'X^{\varepsilon}B + R^{\varepsilon} \end{pmatrix} \geq 0, \qquad (4.210)$$

where for all ε, the matrices Q^{ε}, R^{ε}, and S^{ε} satisfy the Condition psd, and the pair (A, B) is $\mathbb{C}^{-}$-stabilizable. The matrices Q^{ε}, R^{ε}, and S^{ε} permit the factorization:

$$\begin{pmatrix} Q^{\varepsilon} & S^{\varepsilon} \\ (S^{\varepsilon})' & R^{\varepsilon} \end{pmatrix} = \begin{pmatrix} (C^{\varepsilon})' \\ (D^{\varepsilon})' \end{pmatrix} \begin{pmatrix} C^{\varepsilon} & D^{\varepsilon} \end{pmatrix}. \qquad (4.211)$$

It follows from Theorem 4.147 that for each ε a unique positive semi-definite semi-stabilizing solution of the DLMI (4.210) exists. Let this solution be denoted by X_{ss}^{ε}. Assume that

$$\lim_{\varepsilon \to 0} A^{\varepsilon} = A^0, \quad \lim_{\varepsilon \to 0} B^{\varepsilon} = B^0, \quad \lim_{\varepsilon \to 0} C^{\varepsilon} = C^0, \quad \lim_{\varepsilon \to 0} D^{\varepsilon} = D^0. \quad (4.212)$$

These conditions are basically the equivalent to the conditions leading up to the continuity result in Theorem 4.100, except that this time we have a DLMI.

The above conditions are **not** sufficient to guarantee continuity of the semi-stabilizing solution of the DLMI. Example 4.99 gives an explicit counterexample.

The following theorem identifies two special cases where the continuity of the semi-stabilizing solution of the DLMI can be guaranteed.

Theorem 4.149 *Consider a family of systems $(A^\varepsilon, B^\varepsilon, C^\varepsilon, D^\varepsilon)$ such that, for all $\varepsilon \in [0, \delta)$, $(A^\varepsilon, B^\varepsilon)$ is $\mathbb{C}^\ominus$-stabilizable. Assume that (4.212) is satisfied. Let X^ε be the semi-stabilizing solution of the DLMI in (4.185) with Q^ε, S^ε, and R^ε defined according to (4.211). Then we have*

$$\lim_{\varepsilon \to 0} X^\varepsilon = X^0$$

if either one of the following conditions is satisfied:

(i) A and B do not depend on ε; i.e., $A^0 = A^\varepsilon$ and $B^0 = B^\varepsilon$ for all ε, and the matrix (4.211) is increasing in ε for $\varepsilon \in [0, \delta]$.

(ii) A $\delta > 0$ exists such that the normal rank of

$$G(z) := C^\varepsilon(zI - A^\varepsilon)^{-1} B^\varepsilon + D^\varepsilon$$

is independent of ε for all $\varepsilon \in [0, \delta]$.

Proof : The proof basically follows along the same lines as in the proof of Theorem 4.100. ∎

4.5   Continuous-time quadratic matrix inequalities

As mentioned in earlier sections, algebraic Riccati equations are primary tools used in solving the *regular* H_2 and H_∞ control and filtering problems. However, as algebraic Riccati equations do in the case of regular problems, quadratic matrix inequalities (QMIs) introduced by Stoorvogel [92] play a central role in solving the singular H_∞ control and filtering problems. It turns out that quadratic matrix inequalities have properties that are somewhat similar to those of linear matrix inequalities discussed earlier. We introduce now a quadratic matrix inequality in the following definition.

Definition 4.150 *Let $A \in \mathbb{R}^{n \times n}$, $B \in \mathbb{R}^{n \times m}$, $Q \in \mathbb{R}^{n \times n}$, $R \in \mathbb{R}^{m \times m}$, $S \in \mathbb{R}^{n \times m}$, and $E \in \mathbb{R}^{n \times l}$ with Q and R being symmetric. The matrix inequality for an unknown $n \times n$ matrix X of the form*

$$\mathcal{Q}(X) \geq 0, \tag{4.213}$$

where

$$\boldsymbol{Q}(X) := \begin{pmatrix} Q + A'X + XA + XEE'X & XB + S \\ B'X + S' & R \end{pmatrix},$$

*is called a **continuous-time quadratic matrix inequality (CQMI)**. Moreover, if X satisfies (4.213), then it is referred to as a solution of the CQMI.*

Remark 4.151 *Note that the CQMI in (4.213) when the quadratic term $XEE'X$ is dropped is precisely the same as the CLMI in (4.157).*

We denote the set of real symmetric solutions of the CQMI in (4.213) as $\boldsymbol{\Gamma}$; i.e.,

$$\boldsymbol{\Gamma} := \left\{ X \in \mathbb{R}^{n \times n} \mid X = X' \text{ and } \boldsymbol{Q}(X) \geq 0 \right\}. \tag{4.214}$$

Here we restrict ourselves to the case when the matrices Q, R, and S satisfy the Condition psd. That is, we assume that matrices $C \in \mathbb{R}^{p \times n}$ and $D \in \mathbb{R}^{p \times m}$ exist with $(C \quad D)$ of full rank such that

$$\begin{pmatrix} Q & S \\ S' & R \end{pmatrix} = \begin{pmatrix} C' \\ D' \end{pmatrix} \begin{pmatrix} C & D \end{pmatrix}.$$

This restriction is not necessary for our development; however, we have chosen it here merely for the sake of simplicity. Moreover, this special class of quadratic matrix inequalities is relevant to H_∞ control and filtering theory and, hence, to the context of this book. With the above restriction, the CQMI in (4.213) is rewritten as

$$\boldsymbol{Q}(X) = \begin{pmatrix} C'C + A'X + XA + XEE'X & XB + C'D \\ B'X + D'C & D'D \end{pmatrix} \geq 0. \tag{4.215}$$

To start with, we would like to make a comment that the proofs of various properties of CQMIs to be presented shortly are analogous to the proofs of similar properties of CLMIs. For this reason, we simply state the results and leave the proofs to the reader.

As in the case of CLMIs, a relevant set of solutions of a CQMI in the context of H_∞ theory is the set of rank minimizing solutions. As in earlier sections, before we develop the definition of rank minimizing solutions, we need to examine the following property.

Lemma 4.152 *Consider the CQMI in (4.215). Let the normal rank of $D + C(sI - A)^{-1}B$ be denoted by ρ. Then, for all $X \in \boldsymbol{\Gamma}$, the rank of $\boldsymbol{Q}(X)$ is always greater than or equal to ρ; i.e.,*

$$\operatorname{rank} \boldsymbol{Q}(X) \geq \rho \qquad \forall X \in \boldsymbol{\Gamma}.$$

Proof : It is left to the reader.   ■

Now, we are ready to define the set of all rank minimizing solutions of a CQMI.

Definition 4.153 *A solution $X \in \Gamma$ is said to be rank minimizing if*

$$\operatorname{rank} Q(X) = \rho. \tag{4.216}$$

Moreover, we denote the set of all rank minimizing solutions of the CQMI in (4.215) as $\Gamma_{\min}$; i.e.,

$$\Gamma_{\min} := \{ X \in \Gamma \mid \operatorname{rank} Q(X) = \rho \}. \tag{4.217}$$

Also, the set of all positive semi-definite rank-minimizing solutions of the CQMI in (4.215) is denoted by $\Gamma_{\min}^{psd}$:

$$\Gamma_{\min}^{psd} := \{ X \in \Gamma_{\min} \mid X \geq 0 \}. \tag{4.218}$$

Similarly, the set of all positive definite rank minimizing solutions of the CQMI in (4.215) is denoted by $\Gamma_{\min}^{pd}$:

$$\Gamma_{\min}^{pd} := \{ X \in \Gamma_{\min} \mid X \geq 0 \}. \tag{4.219}$$

We next have the following additional definition regarding semi-stabilizing and stabilizing solutions.

Definition 4.154 *Consider a matrix pencil*

$$N(s, X) := \begin{pmatrix} M(s, X) \\ Q(X) \end{pmatrix}, \tag{4.220}$$

where $M(s, X) := \begin{pmatrix} sI - A - EE'X & -B \end{pmatrix}$. Then a solution $X \in \Gamma_{\min}$ is said to be a semi-stabilizing solution if all finite zeros of the matrix pencil $N(s, X)$ are in the closed left-half plane, i.e., in $\mathbb{C}^{-0}$. Similarly, a solution $X \in \Gamma_{\min}$ is said to be a stabilizing solution if all finite zeros of $N(s, X)$ are in the open left-half plane, i.e., in $\mathbb{C}^{-}$. The set of all semi-stabilizing solutions of a CQMI is denoted by $\Gamma_{\min}^{ss}$. Similarly, the set of all stabilizing solutions of a CQMI is denoted by $\Gamma_{\min}^{s}$.

As indicated for the case of H_{∞}^{1} CAREs, the semi-stabilizing solution of the CQMI need not be unique. For later purposes, we need a stronger concept, which we will call strongly semi-stabilizing solutions of the CQMI in line with a similar definition for CAREs.

Definition 4.155 *Consider the CQMI* (4.215). *A solution* $X \in \boldsymbol{\Gamma}_{\min}$ *is said to be a strongly semi-stabilizing solution if*

- *X is semi-stabilizing.*

- *The kernel of X is equal to $\mathcal{V}^{-0}(A, B, C, D)$.*

- *The number of invariant zeros of (A, B, C, D) on the imaginary axis is equal to the number of zeros on the imaginary axis of the matrix pencil $N(s, X)$ given by* (4.220).

4.5.1   Connection between a CQMI and its associated CARE

In this subsection, we examine the rank minimizing solutions of a CQMI by developing a connection between a CQMI and its associated CARE. We consider two cases. The first one deals with a CQMI where the matrix D is injective, the so-called regular case; whereas the second one treats a CQMI, where D is not injective, the so-called singular case.

Case 1 [Regular case]

In this case, the matrix D is injective. We have the following theorem.

Theorem 4.156 *Consider a CQMI as in* (4.215) *when D is injective. Define an associated H_{∞}^{1} CARE:*

$$X\bar{A} + \bar{A}'X + C'C - (XB + C'D)(D'D)^{-1}(B'X + D'C) + XEE'X = 0.$$
(4.221)

Then the following hold:

(i) *The set of rank minimizing solutions $\boldsymbol{\Gamma}_{\min}$ of the CQMI in* (4.215) *is equal to the set of real symmetric solutions of the H_{∞}^{1} CARE in* (4.221).

(ii) *The set of positive semi-definite, rank minimizing solutions $\boldsymbol{\Gamma}_{\min}^{psd}$ of the CQMI in* (4.215) *is equal to the set of positive semi-definite solutions of the H_{∞}^{1} CARE in* (4.221).

(iii) *The set of positive definite, rank minimizing solutions $\boldsymbol{\Gamma}_{\min}^{pd}$ of the CQMI in* (4.215) *is equal to the set of positive definite solutions of the H_{∞}^{1} CARE in* (4.221).

(iv) *The set of stabilizing solutions $\boldsymbol{\Gamma}_{\min}^{s}$ of the CQMI in* (4.215) *is equal to the set of real symmetric stabilizing solutions of the H_{∞}^{1} CARE in* (4.221).

(v) *The set of semi-stabilizing solutions $\boldsymbol{\Gamma}_{\min}^{ss}$ of the CQMI in* (4.215) *is equal to the set of real symmetric semi-stabilizing solutions of the H_{∞}^{1} CARE in* (4.221).

(vi) *The set of strongly semi-stabilizing solutions $\boldsymbol{\Gamma}^{ss}_{\min}$ of the CQMI in (4.215) is equal to the set of strongly semi-stabilizing solutions of the H^1_∞ CARE in (4.221).*

Proof : It is similar to the proof of Theorem 4.111. ∎

Case 2 [Singular case]

For this singular case, as in CLMIs, developing a connection between a CQMI and its associated H^1_∞ CARE is somewhat complex. As before, we need to look at the quadruple (A, B, C, D) as a system Σ_* and use a particular structure of this system Σ_*, namely its SCB form. If we assume that the system Σ_* that is characterized by $(\bar{A}, B, C, D)$ is already in the state-space coordinates of SCB, then from SCB theory, it follows that a matrix F exists such that $(\bar{A}, B, C, D)$ has the following structure, which we again refer to as a compact SCB form:

$$\bar{A} + BF = \begin{pmatrix} A_{11} & A_{12} \\ KC_1 & A_{22} \end{pmatrix}, \quad B = \begin{pmatrix} B_{11} & B_{12} \\ 0 & B_{22} \end{pmatrix},$$

$$C + DF = \begin{pmatrix} C_1 & C_2 \end{pmatrix}, \quad D = \begin{pmatrix} 0 & D_2 \end{pmatrix}, \quad E = \begin{pmatrix} E_1 \\ E_2 \end{pmatrix}, \quad (4.222)$$

where F is such that $(C + DF)'D = 0$ while (A_{11}, B_{11}) is controllable, $C_2'D_2 = 0$, $C_1'D_2 = 0$, and D_2 is left-invertible. Let us emphasize again that, to obtain (4.222), we have first assumed that the state space $\mathcal{X}$ of the system Σ_* characterized by the quadruple $(\bar{A}, B, C, D)$ is already in SCB form and then decomposed $\mathcal{X}$ as $\mathcal{X}_1 \oplus \mathcal{X}_2$ with $\mathcal{X}_1 = \mathcal{S}^*(\Sigma_*)$.

The following lemma exploits the above compact form of the system Σ_* and enables us to move very close to our goal of connecting the given CQMI to an associated H^1_∞ CARE.

Lemma 4.157 *Consider the CQMI in (4.215), and let $X \in \boldsymbol{\Gamma}$. Also, let Σ_* be a system characterized by $(\bar{A}, B, C, D)$, and let $\mathcal{S}^*(\Sigma_*)$ be as in Definition 3.28. Then, $X\mathcal{S}^*(\Sigma_*) = 0$; i.e., whenever the system Σ_* is in the compact SCB form as given in (4.222), X can be written as*

$$X = \begin{pmatrix} 0 & 0 \\ 0 & X_{22} \end{pmatrix}. \quad (4.223)$$

Proof : It is left to the reader. ∎

Now we are ready to relate the solutions of a given CQMI to the solutions of another associated CQMI, which corresponds to the regular case.

Theorem 4.158 *Consider the CQMI* (4.215). *Let the system* Σ_* *characterized by* $(\bar{A}, B, C, D)$ *be in the compact SCB form as given in* (4.222). *Define an another associated CQMI:*

$$Q_r(\bar{X}) := \begin{pmatrix} \bar{A}'_r \bar{X} + \bar{X} \bar{A}_r + C'_r C_r + \bar{X} E_r E'_r \bar{X} & \bar{X} B_r + C'_r D_r \\ B'_r \bar{X} + D'_r C_r & D'_r D_r \end{pmatrix}, \quad (4.224)$$

by using the quintuple $(\bar{A}_r, B_r, C_r, D_r, E_r)$ *given by*

$$\bar{A}_r = A_{22}, B_r = \begin{pmatrix} KC_1 L & B_{22} \end{pmatrix}, C_r = C_2, D_r = \begin{pmatrix} C_1 L & D_2 \end{pmatrix}, E_r = E_2,$$
$$(4.225)$$

where L is a matrix such that $\operatorname{rank} C_1 L = \operatorname{rank} C_1$ *while $C_1 L$ is injective. Then, we have the following:*

(i) *If X is given such that* $Q(X) \geq 0$, *then in the above basis, we have*

$$X := \begin{pmatrix} 0 & 0 \\ 0 & \bar{X} \end{pmatrix} \quad (4.226)$$

with $\bar{X}$ such that $Q_r(\bar{X}) \geq 0$ *and*

$$\operatorname{rank} Q(X) = \operatorname{rank} Q_r(\bar{X}). \quad (4.227)$$

Moreover,

$$\operatorname{rank} \begin{pmatrix} M(s) \\ Q(X) \end{pmatrix} = n_1 + \operatorname{rank} \begin{pmatrix} M_r(s) \\ Q_r(\bar{X}) \end{pmatrix} \quad (4.228)$$

for all s, where $M_r(s) = (sI - \bar{A}_r \quad B_r)$ *and n_1 is the number of columns of A_{11}.*

(ii) *Conversely, if $\bar{X}$ is given such that* $Q_r(\bar{X}) \geq 0$, *then if we define X by* (4.226), *we have that* $Q(X) \geq 0$ *and* (4.227) *and* (4.228) *are satisfied.*

Theorem 4.158 connects the given CQMI (4.215), which is singular, to an associated CQMI (4.224), which is regular. In view of this result, it is straightforward to connect the given CQMI to an associated H^1_∞ CARE as discussed in the following theorem.

Theorem 4.159 *Consider the CQMI* (4.215) *when D is not injective. Define an associated H^1_∞ CARE by using the quadruple* $(\bar{A}_r, B_r, C_r, D_r, E_r)$ *as*

$$\bar{X} \bar{A}_r + \bar{A}'_r \bar{X} + C'_r C_r$$
$$- (\bar{X} B_r + C'_r D_r)(D'_r D_r)^{-1}(B'_r \bar{X} + D'_r C_r) + \bar{X} E_r E'_r \bar{X} = 0. \quad (4.229)$$

Let $\bar{X}$ be a solution of the H^1_∞ CARE in (4.229). Next, define a matrix X by

$$X := \begin{pmatrix} 0 & 0 \\ 0 & \bar{X} \end{pmatrix}. \tag{4.230}$$

Then we have the following:

(*i*) *The set of rank minimizing solutions $\boldsymbol{\Gamma}_{\min}$ of the CQMI in (4.215) is given by*

$$\boldsymbol{\Gamma}_{\min} = \left\{ X \in \mathbb{R}^{n \times n} \mid X \text{ is given by (4.230) with } \bar{X} \text{ being a real} \right.$$
$$\left. \text{symmetric solution of the } H^1_\infty \text{ CARE in (4.229)} \right\}.$$

(*ii*) *The set of positive semi-definite, rank minimizing solutions $\boldsymbol{\Gamma}^{psd}_{\min}$ of the CQMI in (4.215) is given by*

$$\boldsymbol{\Gamma}^{psd}_{\min} = \left\{ X \in \mathbb{R}^{n \times n} \mid X \in \boldsymbol{\Gamma}_{\min} \text{ with } \bar{X} \geq 0 \right\}.$$

(*iii*) *The set of positive definite, rank minimizing solutions $\boldsymbol{\Gamma}^{pd}_{\min}$ of the CQMI in (4.215) is an empty set.*

(*iv*) *The set of stabilizing solutions $\boldsymbol{\Gamma}^{s}_{\min}$ of the CQMI in (4.215) is given by*

$$\boldsymbol{\Gamma}^{s}_{\min} = \left\{ X \in \mathbb{R}^{n \times n} \mid X \text{ is given by (4.230) with } \bar{X} \text{ a stabilizing} \right.$$
$$\left. \text{solution of the } H^1_\infty \text{ CARE in (4.229)} \right\}.$$

(*v*) *The set of semi-stabilizing solutions $\boldsymbol{\Gamma}^{ss}_{\min}$ of the CQMI in (4.215) is given by*

$$\boldsymbol{\Gamma}^{ss}_{\min} = \left\{ X \in \mathbb{R}^{n \times n} \mid X \text{ is given by (4.230) with } \bar{X} \text{ a semi-stabilizing} \right.$$
$$\left. \text{solution of the } H^1_\infty \text{ CARE in (4.229)} \right\}.$$

(*vi*) *The set of strongly semi-stabilizing solutions $\boldsymbol{\Gamma}^{ss}_{\min}$ of the CQMI in (4.215) is given by*

$$\boldsymbol{\Gamma}^{ss}_{\min} = \left\{ X \in \mathbb{R}^{n \times n} \mid X \text{ is given by (4.230) with } \bar{X} \text{ a strongly semi-} \right.$$
$$\left. \text{-stabilizing solution of the } H^1_\infty \text{ CARE in (4.229)} \right\}.$$

Proof : It is similar to the proof of Theorem 4.114.   ∎

Remark 4.160 *The above theorems also cover Theorems 4.111 and 4.114 as a special case by setting $E = 0$.*

In view of Theorem 4.159, the problem of obtaining the existence conditions for semi-stabilizing, stabilizing, or positive semi-definite solutions of the CQMI reduces to obtaining the existence conditions for similar solutions of the H_∞^1 CARE in (4.229). Also, in the H_∞ literature, a pertinent solution of a CQMI that has been used often is a stabilizing solution. We note that from Theorems 4.19 and 4.159, it immediately follows that a stabilizing solution of a quadratic matrix inequality, if it exists, is also unique.

Remark 4.161 *Theorems 4.158 and 4.159 establish a mapping given by (4.230) between a variety of solutions of a CQMI and the corresponding solutions of an associated H_∞^1 CARE. This mapping (4.230) was obtained by assuming that the system Σ_* characterized by the quadruple $(\bar{A}, B, C, D)$ is already in the state-space coordinates of compact SCB form (4.222). If it is not so, then from the SCB theory by using the given quadruple $(\bar{A}, B, C, D)$, one can easily find a nonsingular matrix Γ_s, such that the one-to-one mapping (4.230) is replaced by the following equation:*

$$X = (\Gamma_s^{-1})' \begin{pmatrix} 0 & 0 \\ 0 & \bar{X} \end{pmatrix} \Gamma_s^{-1}. \tag{4.231}$$

As such the results of Theorems 4.158 and 4.159 are true if one views the one-to-one mapping as given by (4.231).

Finally, we would like to make a comment on discrete-time quadratic matrix inequalities (DQMIs). Although DQMIs can be discussed following the concepts introduced in this section for CQMIs, we note that nothing has yet been done in this regard in the existing literature. This may be because of the lack of a clear understanding as to the potential of such DQMIs. For this reason, we do not dwell on this topic here.

4.A   Linear matrix equations

The Lyapunov equation plays an important role in our analysis, and we recall here some results for this equation. We first present the continuous-time Lyapunov equation.

Lemma 4.162 *Consider the linear matrix equation with unknown X,*

$$A'X + XA + Q = 0, \tag{4.232}$$

*which is known as the **Lyapunov equation**.*

(*i*) *If A is Hurwitz-stable, then the solution X is unique. Moreover, whenever Q is positive semi-definite, we have X is positive semi-definite.*

(*ii*) *If X is positive definite and Q is positive semi-definite, then A is conditionally stable; i.e., all its eigenvalues are located in the closed left-half plane, and those on the imaginary axis are simple (in other words, the multiplicity structure of an eigenvalue on the imaginary axis is $\{1, \ldots, 1\}$).*

Proof : To establish part (*i*), we assume X_1 and X_2 both satisfy (4.232). Then $X_1 - X_2$ satisfies

$$A'(X_1 - X_2) + (X_1 - X_2)A = 0.$$

A simple recursion then implies that

$$(-A')^k (X_1 - X_2) = (X_1 - X_2)A^k$$

for all k, and using the series expansion of e^{At}, we obtain

$$e^{-A't}(X_1 - X_2) = (X_1 - X_2)e^{At}$$

for all t, and hence,

$$X_1 - X_2 = e^{A't}(X_1 - X_2)e^{At}$$

for all t. But then

$$X_1 - X_2 = \lim_{t \to \infty} e^{A't}(X_1 - X_2)e^{At} = 0,$$

where we exploited the stability of A. This implies the uniqueness of the solution. It can be easily verified that the unique solution of (4.232) is given by

$$X = \int_0^\infty e^{A't} Q e^{At}\, dt,$$

which is clearly positive semi-definite whenever Q is positive semi-definite. It remains to establish part (*ii*). Consider the system

$$\dot{x} = Ax. \tag{4.233}$$

Given a positive definite solution of the Lyapunov equation, we find

$$\frac{d}{dt}x^* X x = -x^* Q x,$$

and because Q is positive semi-definite, we find that $x^* X x$ is nonincreasing; therefore, for any initial condition (even when complex valued), the system has a bounded solution. This clearly implies that the eigenvalues must be in the closed left-half plane. Assume an eigenvalue on the imaginary axis is not simple. Then vectors $x_1 \neq 0$ and $x_2 \neq 0$ exist such that

$$Ax_2 = \lambda x_2 + x_1, \qquad Ax_1 = \lambda x_1,$$

but then it is easily verified that the initial condition $x(0) = x_2$ yields a solution $x(t) = e^{\lambda t}(x_2 + tx_1)$, which is not bounded, and hence, we have a contradiction. The proof of part (ii) is now complete. $\blacksquare$

Lemma 4.163 *Consider the linear matrix equation with unknown X,*

$$X = A'XA + Q, \tag{4.234}$$

*which is known as the discrete **Lyapunov equation**.*

(i) *If A is Schur-stable, then the solution X is unique. Moreover, whenever Q is positive semi-definite, we have that X is positive semi-definite.*

(ii) *If A is Schur-anti-stable (all eigenvalues outside the unit circle), then the solution X is unique. Moreover, whenever Q is positive semi-definite, we have that X is negative semi-definite.*

(iii) *If X is positive definite and Q is positive semi-definite, then A is conditionally stable; i.e., all its eigenvalues are located in the closed unit disk, and those on the unit circle are simple.*

Proof : To establish part (i), we assume that X_1 and X_2 both satisfy (4.232). Then $X_1 - X_2$ satisfies

$$X_1 - X_2 = A'(X_1 - X_2)A.$$

A simple recursion then implies that

$$X_1 - X_2 = (A')^k(X_1 - X_2)A^k$$

for all k. But then

$$X_1 - X_2 = \lim_{k \to \infty} (A')^k(X_1 - X_2)A^k = 0,$$

where we exploited the stability of A. This implies the uniqueness of the solution. It can be easily verified that the unique solution of (4.232) is given by

$$X = \sum_{k=0}^{\infty} (A')^k Q A^k,$$

which is clearly positive semi-definite whenever Q is positive semi-definite.

Part (*ii*) follows directly from part *i*. After all, a Schur-anti-stable matrix is by definition invertible with its inverse being Schur-stable. It is trivial to check that we have

$$-X = (A^{-1})'(-X)A^{-1} + Q,$$

and part *i* then immediately yields the required results.

It remains to establish part (*iii*). Consider the system

$$x(k+1) = Ax(k). \tag{4.235}$$

Given a positive definite solution of the Lyapunov equation, we find

$$x^*(k+1)Xx(k+1) = x^*(k)Xx(k) - x^*(k)Qx(k),$$

and as Q is positive semi-definite, we find that x^*Xx is nonincreasing; therefore, for any initial condition (even when complex valued), the system (4.233) has a bounded solution. This clearly implies that the eigenvalues must be in the closed unit disk. Assume an eigenvalue on the unit circle is not simple. Then vectors $x_1 \neq 0$ and $x_2 \neq 0$ exist such that

$$Ax_2 = \lambda x_2 + x_1, \qquad Ax_1 = \lambda x_1,$$

but then it is easily verified the initial condition $x(0) = x_2$ yields a solution $x(k) = \lambda^k x_2 + k\lambda^{k-1}x_1$, which is not bounded, and hence, we have a contradiction. The proof of part (*iii*) in now complete. ■

Lemma 4.164 *Consider the linear matrix equation with unknown X,*

$$A_1 X - XA_2 + Q = 0, \tag{4.236}$$

*which is known as the **Sylvester equation**.*

(i) *If A_1 and A_2 have no eigenvalues in common, then a solution X exists and is unique.*

(ii) *If A_1 and A_2 have at least one eigenvalue in common then matrices Q exist for which (4.236) has no solution. Moreover, there are an infinite number of solutions in case $Q = 0$.*

Proof : Consider the linear function $f : \mathbb{R}^{n \times n} \to \mathbb{R}^{n \times n}$ defined by

$$f(X) = A_1 X - XA_2.$$

To guarantee the solvability of (4.236), it suffices to establish that the mapping f is surjective. Uniqueness is a consequence of injectivity of this mapping. But clearly in this case, because f is a mapping from $\mathbb{R}^n$ into itself, f is injective if and only if f is surjective, and therefore, the existence of a unique solution to (4.236) is established as soon as we prove that f is injective. Assume that A_1 and A_2 have no eigenvalues in common. Consider the characteristic polynomial of A_1:

$$p(z) = \det(zI - A_1) = \prod_{i=1}^{n}(z - \lambda_i)$$

with λ_i as eigenvalues of A_1. Then the theorem of Cayley–Hamilton guarantees that $p(A_1) = 0$. On the other hand,

$$p(A_2) = \prod_{i=1}^{n}(A_2 - \lambda_i I)$$

is invertible because none of the λ_i is an eigenvalue of A_2. Assume that $f(X) = 0$. We have

$$A_1 X = X A_2,$$

and hence,

$$A_1^k X = X A_2^k$$

for all k. This implies that

$$p(A_1)X = Xp(A_2).$$

As $p(A_1) = 0$ and $p(A_2)$ is invertible, this implies that $X = 0$, which establishes the required injectivity of f.

It remains to consider the case when A_1 and A_2 have at least one eigenvalue in common. In that case, an eigenvalue λ exists with an associated left eigenvector x of A_2 and an associated right eigenvector y of A_1; i.e., we have $xA_2 = \lambda x$ and $A_1 y = \lambda y$. Note that x and y can be complex. But then

$$f(xy) = 0,$$

and hence, f is not injective. But then f is not surjective either, and this implies that (4.236) does not have a solution for all Q. Moreover, for $Q = 0$, we have that $\alpha(xy + \overline{xy})$ is a real solution of (4.236) for $Q = 0$ and for all α, and therefore, there are infinitely many solutions.   ∎

Lemma 4.165 *Consider the linear matrix equation with unknown X,*

$$X = A_1 X A_2 + Q, \tag{4.237}$$

*which is known as the discrete **Sylvester equation**.*

(i) *If no eigenvalue of A_1 is the reciprocal of an eigenvalue of A_2 and conversely, then a solution X exists and is unique.*

(ii) *If an eigenvalue of A_1 is the reciprocal of an eigenvalue of A_2, or conversely, then matrices Q exist for which (4.237) has no solution. Moreover, there are an infinite number of solutions in the case when $Q = 0$.*

Proof : We choose $\lambda \neq 0$ such that $I + \lambda A_2$ and $\lambda I + A_1$ are both invertible. Then the Sylvester equation can be rewritten as

$$(\lambda I + A_1)^{-1}(\lambda I - A_1)X - X(\lambda A_2 - I)(\lambda A_2 + I)^{-1}$$
$$= 2\lambda(\lambda I + A_1)^{-1}Q(\lambda A_2 + I)^{-1}.$$

This is a continuous-time Sylvester equation, and the result can then be obtained directly from Lemma 4.164. $\blacksquare$

4.B   Reduction to the case that H has full normal rank

The reduction technique outlined in Subsection 4.2.7 is based on a solution of the Riccati equation. However, when studying the Riccati equation in linear quadratic control and in particular the linear matrix inequality associated to it, then it is desirable to have a reduction technique available that is based on a solution of the linear matrix inequality instead of a reduction based on a solution of the Riccati equation. In this appendix, we will present a technique to reduce problems where H, as defined in (4.95), does not have full rank to the case where H has full rank based on a solution of the linear matrix inequality.

Consider the linear matrix inequality:

$$L(X) := \begin{pmatrix} Q + A'XA - X & A'XB + S \\ B'XA + S' & B'XB + R \end{pmatrix} \tag{4.238}$$

with the associated matrix pencil

$$\begin{pmatrix} zI - A & -B \\ Q + A'XA - X & A'XB + S \\ B'XA + S' & B'XB + R \end{pmatrix}, \tag{4.239}$$

which is used to define semi-stabilizing and stabilizing solutions. The matrices $Q, R,$ and S are said to satisfy **Condition psd** if

$$\begin{pmatrix} Q & S \\ S' & R \end{pmatrix} \geq 0. \tag{4.240}$$

This is clearly equivalent to the case when 0 is a solution of the linear matrix inequality. We first introduce a transformation that transforms the original DLMI into an DLMI where the associated matrices satisfy Condition psd.

We start with the basic assumption that at least one solution $\bar{X}$ of the linear matrix inequality exists; i.e., $\bar{X} \in \Gamma$. We can then factorize $L(\bar{X})$ as

$$L(\bar{X}) = \begin{pmatrix} C' \\ D' \end{pmatrix} \begin{pmatrix} C & D \end{pmatrix}. \tag{4.241}$$

We define a new discrete linear matrix inequality:

$$\bar{L}(X) := \begin{pmatrix} C'C + A'XA - X & A'XB + C'D \\ B'XA + D'C & B'XB + D'D \end{pmatrix} \geq 0. \tag{4.242}$$

Note that this new linear matrix inequality always has 0 as a solution, and hence, the associated matrices satisfy Condition psd. We have that

$$L(X) \geq 0 \text{ if and only if } \bar{L}(X - \bar{X}) \geq 0.$$

We define $\mathcal{R}^*(\Sigma)$ according to Definition 3.28. Hence, a matrix F exists such that $\mathcal{R}^*(\Sigma)$ is $(A + BF)$-invariant and contained in $\ker(C + DF)$. After incorporating the transformation associated with this preliminary feedback F, we define a new DLMI, which we will refer to as the shifted DLMI.

Definition 4.166 *The shifted DLMI associated with the DLMI (4.238) is defined as*

$$L^s(X) \geq 0, \tag{4.243}$$

where

$$L^s(X) := \begin{pmatrix} I & F' \\ 0 & I \end{pmatrix} \bar{L}(X) \begin{pmatrix} I & 0 \\ F & I \end{pmatrix} = \begin{pmatrix} \tilde{C}'\tilde{C} + \tilde{A}'X\tilde{A} - X & \tilde{A}'XB + \tilde{C}'D \\ B'X\tilde{A} + D'\tilde{C} & B'XB + D'D \end{pmatrix} \tag{4.244}$$

and $\tilde{A} = A + BF$ and $\tilde{C} = C + DF$.

Proposition 4.167 *Let $\bar{X}$ be a solution of the DLMI (4.238), and let F be such that $\mathcal{R}^*(\Sigma)$ is $(A + BF)$-invariant and contained in $\ker(C + DF)$, where Σ is the system with realization (A, B, C, D) with (C, D) defined by (4.241). Then we have, as follows:*

(i) X is a solution of the DLMI (4.238) if and only if $X - \bar{X}$ is a solution of the associated shifted DLMI (4.243).

(ii) *Let X be a solution of the DLMI. Then the rank of $L(X)$ equals the rank of $L^s(X - \bar{X})$. In particular, X is a rank minimizing solution of the linear matrix inequality (4.238) if and only if $X - \bar{X}$ is a rank minimizing solution of the associated shifted discrete linear matrix inequality (4.243).*

(iii) *Let X be a solution of the linear matrix inequality. The zeros of the matrix pencil (4.239) are equal to the zeros of the following matrix pencil:*

$$\begin{pmatrix} zI - \tilde{A} & -B \\ \tilde{C}'\tilde{C} + \tilde{A}'\tilde{X}\tilde{A} - \tilde{X} & \tilde{A}'\tilde{X}B + \tilde{C}'D \\ B'\tilde{X}\tilde{A} + D'\tilde{C} & B'\tilde{X}B + D'D \end{pmatrix},$$

where $\tilde{X} = X - \bar{X}$. In particular, X is a semi-stabilizing (stabilizing) solution of the DLMI (4.238) if and only if $X - \bar{X}$ is a semi-stabilizing (stabilizing) solution of the shifted DLMI (4.243).

The above proposition shows that without loss of generality, we can focus on a DLMI for which Condition psd is satisfied, and hence, from here on we assume $\bar{X} = 0$ and

$$\begin{pmatrix} C' \\ D' \end{pmatrix} \begin{pmatrix} C & D \end{pmatrix} = \begin{pmatrix} Q & S \\ S' & R \end{pmatrix}.$$

Next, we show how to reduce the DLMI for which $\hat{H}$ is not full rank to the case where $\hat{H}$ has full rank.

We start with the same basis as used in (4.143) on page 132, i.e., a basis in the state space $\mathcal{X}_1 \oplus \mathcal{X}_2$ such that $\mathcal{X}_1 = \mathcal{R}^*(\Sigma)$ and a basis in the input space $\mathcal{U}_1 \oplus \mathcal{U}_2$ such that $\mathcal{U}_1 = B^{-1}\mathcal{R}^*(\Sigma) \cap \ker D$. With respect to this basis, a matrix F exists such that A, B, C, and D have a special form:

$$A + BF = \begin{pmatrix} A_{11} & A_{12} \\ 0 & A_{22} \end{pmatrix}, B = \begin{pmatrix} B_{11} & B_{12} \\ 0 & B_{22} \end{pmatrix},$$

$$C + DF = \begin{pmatrix} 0 & C_2 \end{pmatrix}, D = \begin{pmatrix} 0 & D_2 \end{pmatrix}, \tag{4.245}$$

such that (A_{11}, B_{11}) is controllable, $C_2'D_2 = 0$, and $(A_{22}, B_{22}, C_2, D_2)$ left-invertible. With respect to this decomposition, we have

$$X = \begin{pmatrix} X_{11} & X_{12} \\ X_{21} & X_{22} \end{pmatrix}.$$

On page 133 after (4.146), it was argued that $X_{11} = 0$, $X_{12} = 0$, and $X_{21} = 0$. In other words, we only have to compute X_{22}. It is easy to see that the linear matrix inequality reduces to

$$L^r(X_{22}) = \begin{pmatrix} A_{22}'X_{22}A_{22} - X_{22} + C_2'C_2 & A_{22}'X_{22}B_{22} + C_2'D_2 \\ B_{22}'X_{22}A_{22} + D_2'C_2 & B_{22}'X_{22}B_{22} + D_2'D_2 \end{pmatrix} \geq 0.$$

$$\tag{4.246}$$

However, as $(A_{22}, B_{22}, C_2, D_2)$ is left-invertible, this is a linear matrix inequality such that the associated rational matrix H^r has full rank where

$$
H^r(z) = \left(B'_{22}(z^{-1}I - A'_{22})^{-1} \quad I \right) \begin{pmatrix} C'_2 C_2 & C'_2 D_2 \\ D'_2 C_2 & D'_2 D_2 \end{pmatrix} \begin{pmatrix} (zI - A_{22})^{-1} B_{22} \\ I \end{pmatrix}
$$
$$
= G'(-z)G(z), \tag{4.247}
$$

where G is the transfer matrix of $(A_{22}, B_{22}, C_2, D_2)$. This enables us to first derive results for the case that H has full rank and then use the above reduction step to derive results for the general case. The results of the above reduction scheme are put together in the following theorem.

Theorem 4.168 *Consider the linear matrix inequality* (4.238) *such that Condition psd, given in* (4.240), *is satisfied. Moreover, we assume that we have chosen the appropriate bases as described above. Then, the following hold:*

(i) *X is a solution of the DLMI* (4.238) *if and only if*

$$
X = \begin{pmatrix} 0 & 0 \\ 0 & X_{22} \end{pmatrix}
$$

and X_{22} is a solution of reduced DLMI (4.246).

(ii) *Let X be a solution of the DLMI. Then the rank of $L(X)$ equals the rank of $L^r(X_{22})$. In particular, X is a rank minimizing solution of the DLMI* (4.238) *if and only if X_{22} is a rank minimizing solution of the associated reduced DLMI* (4.246).

(iii) *Let X be a solution of the DLMI. The zeros of the matrix pencil* (4.239) *are equal to the zeros of the matrix pencil:*

$$
\begin{pmatrix} zI - A_{22} & -B_{22} \\ A'_{22}X_{22}A_{22} - X_{22} + C'_2 C_2 & A'_{22}X_{22}B_{22} + C'_2 D_2 \\ B'_{22}X_{22}A_{22} + D'_2 C_2 & B'_{22}X_{22}B_{22} + D'_2 D_2 \end{pmatrix}.
$$

In particular, X is a semi-stabilizing (stabilizing) solution of the DLMI (4.238) *if and only if X_{22} is a semi-stabilizing (stabilizing) solution of the reduced DLMI* (4.246).

4.C Matrix pencils and generalized eigenvalue problems

Matrix pencils and its properties presented in this appendix can be found in more detail in [25, 90].

Consider a matrix pencil $H_1 - zH_2$. We call the matrix pencil regular if the pencil is square and invertible for almost all λ. We call λ a zero of the matrix pencil if $H_1 - zH_2$ loses rank for $z = \lambda$. We call ∞ a zero if $H_1 - zH_2$ loses rank at infinity or equivalently $\mathrm{rank}\, H_2 < \mathrm{normrank}\, H_1 - zH_2$. Zeros of the matrix pencil can be equivalently defined as the zeros of the invariant factors of the (associated) Kronecker normal form. The (algebraic) multiplicity of a zero λ is defined as the sum of the multiplicities of the zero λ of each invariant factor.

For a regular pencil $H_1 - zH_2$, the multiplicity of a finite zero can be determined via the multiplicity of this zero of the polynomial $\det(H_1 - zH_2)$. The multiplicity of a zero at infinity is equal to the multiplicity of the zero at the origin of the polynomial $\det(sH_1 - H_2)$. For a singular pencil, the multiplicity of a finite zero λ can be determined by finding rational matrices $V(z)$ and $W(z)$ that are invertible in a neighborhood of $z = \lambda$ and a square rational matrix $T(z)$ with full normal rank such that

$$V(z)(H_1 - zH_2)W(z) = \begin{pmatrix} T(z) & 0 \\ 0 & 0 \end{pmatrix}.$$

Then the multiplicity of the zero at λ is equal to the multiplicity of the zero λ of the rational function $\det T(z)$. The multiplicity of a zero at ∞ of the matrix pencil $H_1 - zH_2$ is equal to the multiplicity of a zero at 0 of the matrix pencil $sH_1 - H_2$.

A regular $2n$-dimensional matrix pencil $H_1 - zH_2$ that satisfies the property that

$$H_1 \begin{pmatrix} 0 & I_n \\ -I_n & 0 \end{pmatrix} H_1' = H_2 \begin{pmatrix} 0 & I_n \\ -I_n & 0 \end{pmatrix} H_2',$$

where I_n denotes the $n \times n$-identity matrix, is called a symplectic pencil. It is easy to see that this implies that λ is a zero if and only if λ^{-1} is a zero.

A subspace $\mathcal{V}$ is called deflating with respect to the matrix pencil $H_1 - zH_2$ if

$$\dim \{H_1 \mathcal{V} + H_2 \mathcal{V}\} \leq \dim \mathcal{V}. \tag{4.248}$$

For regular pencils, we have an equality in (4.248). We have the following alternative characterization.

Lemma 4.169 $\mathcal{V}$ *is a deflating subspace if and only if a regular matrix pencil* $L_1 - zL_2$ *exists such that*

$$H_1 V L_2 = H_2 V L_1,$$

where V is an injective matrix such that $\mathrm{im}\, V = \mathcal{V}$.

If $H_1 - zH_2$ is a regular pencil, then the pencil $L_1 - zL_2$ is unique up to pre- and postmultiplication by invertible matrices.

The matrix pencil $H_1 - zH_2$ restricted to $\mathcal{V}$ is defined by $H_1|_{\mathcal{V}} - zH_2|_{\mathcal{V}}$. For regular pencils, the zeros of the pencil $L_1 - zL_2$ are the zeros of the pencil

$H_1 - zH_2$ restricted to $\mathcal{V}$. It is important to realize that for singular pencils, the zeros of the symplectic pencil $H_1 - zH_2$ restricted to $\mathcal{V}$ are not necessarily zeros of $H_1 - zH_2$. However, this latter property holds if for all but finitely many zeros the dimension of the kernel of $H_1 - zH_2$ is equal to the dimension of the kernel of $H_1|_{\mathcal{V}} - zH_2|_{\mathcal{V}}$.

5

Exact disturbance decoupling via state and full information feedback

5.1 Introduction

The exact disturbance decoupling (EDD) problem is to find a controller such that
the closed-loop transfer function from an exogenous disturbance signal to a controlled output is equal to zero. In classic as well as modern control theory, the
problems of EDD as well as almost disturbance decoupling (ADD) occupy central positions. Several important problems can be recast as either EDD or ADD
problems, for instance, H_2 optimal control, robust control, decentralized control,
noninteracting control, model reference, and tracking control. Several different
versions of the EDD and ADD problems have been investigated extensively for
the last two decades. In fact, it can be said that the development of the geometric
approach to system theory started with the first version of a disturbance decoupling problem. Among others, the prominent works in disturbance decoupling
include [1,3,32,50,87,95,104,111]. The material of this chapter is based mainly
on the work of Saberi et al. [75].

5.2 Problem formulation

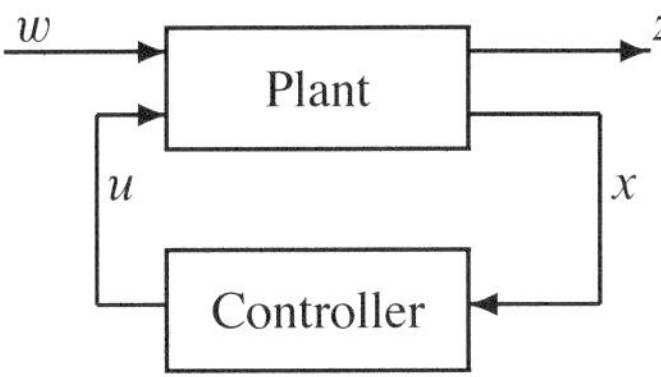

Figure 5.1: Closed-loop system $\Sigma \times \Sigma_c$

We are concerned here with the EDD problem via state feedback with internal
$\mathbb{C}_g$-stability (normally coined in the literature as EDDPS) and with the EDD problem via full information feedback with internal $\mathbb{C}_g$-stability. Consider Figure 5.1,

which depicts a closed-loop system comprising a plant and a state feedback controller. The plant is described by the system Σ:

$$\Sigma : \begin{cases} \sigma x = Ax + Bu + Ew \\ z \ = Cx + Du, \end{cases} \tag{5.1}$$

where σ is an operator indicating the time derivative $\frac{d}{dt}$ for continuous-time systems and a forward unit time shift for discrete-time systems. Here as usual $x \in \mathbb{R}^n$ is a state, $u \in \mathbb{R}^m$ is a control input, $w \in \mathbb{R}^\ell$ is an exogenous disturbance input, and $z \in \mathbb{R}^q$ is a controlled output. The given plant or system Σ is characterized by the matrix quintuple (A, B, C, D, E). In what follows, a subsystem of Σ characterized by the matrix quadruple (A, B, C, D) will play a significant role and we will denote it by Σ_{sub}.

The controller Σ_C can be a static or dynamic state feedback controller. A general proper dynamic state feedback controller is of the form:

$$\Sigma_C : \begin{cases} \sigma v = \ Jv + \ Lx \\ u \ = Mv + Nx. \end{cases} \tag{5.2}$$

The closed-loop system $\Sigma \times \Sigma_C$, i.e., the interconnection of the given system Σ and the controller Σ_C, can then be described by

$$\Sigma \times \Sigma_C : \begin{cases} \sigma x_{cl} = A_{cl}x_{cl} + E_{cl}w \\ z \ = C_{cl}x_{cl}, \end{cases} \tag{5.3}$$

where

$$x_{cl} = \begin{pmatrix} x \\ v \end{pmatrix}, \quad A_{cl} = \begin{pmatrix} A + BN & BM \\ L & J \end{pmatrix}, \quad E_{cl} = \begin{pmatrix} E \\ 0 \end{pmatrix},$$

and

$$C_{cl} = \begin{pmatrix} C + DN & DM \end{pmatrix}.$$

Thus, the closed-loop transfer function from the disturbance w to the controlled output z is given by

$$T_{zw}(\Sigma \times \Sigma_C) = C_{cl}(sI - A_{cl})^{-1}E_{cl}. \tag{5.4}$$

A special case of the controllers given in (5.2) is given by a static state feedback controller of the form:

$$\Sigma_C : u = -Fx. \tag{5.5}$$

Whenever a static feedback controller is used, $T_{zw}(\Sigma \times \Sigma_C)$ is given by

$$T_{zw}(\Sigma \times \Sigma_C) = (C - DF)(sI - A + BF)^{-1}E. \tag{5.6}$$

We will also consider another type of controller, namely, *full information feedback controllers*. In static full information feedback controllers, in addition to static state feedback, one also uses static feedback from the disturbance w whenever it is available. That is, static full information feedback controllers are of the form:

$$\Sigma_C : u = -Fx + Gw. \tag{5.7}$$

The closed-loop transfer function when the static full information feedback controller is used is given by

$$T_{zw}(\Sigma \times \Sigma_C) = (C - DF)(\sigma I - A + BF)^{-1}(BG + E) + DG. \tag{5.8}$$

In the case of dynamic feedback, we look at feedback laws of the following specific form:

$$\Sigma_C : \begin{cases} \sigma v = \ Jv + Lx \\ u \ \ = Mv + Nx + Gw. \end{cases} \tag{5.9}$$

The closed-loop transfer function when the dynamic full information feedback controller is used is given by

$$T_{zw}(\Sigma \times \Sigma_C) = C_{cl}(sI - A_{cl})^{-1}\begin{pmatrix} E + BG \\ 0 \end{pmatrix} + DG. \tag{5.10}$$

Note that the feedback law (5.9) can also be converted to pure disturbance feedback laws because knowing w, we can easily reconstruct x via an appropriate filter. However, this specific structure turns out to be convenient for our purposes.

We introduced above four different architectures for controllers, dynamic state feedback, static state feedback, static full information feedback, and dynamic full information feedback architectures. For any such architecture of controllers, we are now ready to formulate formally the EDD problem as given below.

Problem 5.1 Consider a continuous- or discrete-time system Σ as in (5.1). Also, consider any one of the four different architectures for controllers, dynamic state feedback, static state feedback, dynamic full information feedback, and static full information feedback architectures. For a given architecture of controllers, the **exact disturbance decoupling problem with internal $\mathbb{C}_g$-stability** for Σ, denoted by EDD problem, is the problem of finding, if it exists, a controller Σ_C such that the closed-loop system $\Sigma \times \Sigma_C$ as depicted in Figure 5.1 is $\mathbb{C}_g$-stable, while the resulting closed-loop transfer function $T_{zw}(\Sigma \times \Sigma_C)$ is constrained to be zero. We note that the EDD problem is characterized by the matrix quintuple (A, B, C, D, E).

Remark 5.2 *Note that we normally consider $\mathbb{C}_g = \mathbb{C}^-$ or at least $\mathbb{C}_g$ is a subset of $\mathbb{C}^-$ for continuous-time systems. Similarly, for discrete-time systems, we consider $\mathbb{C}_g = \mathbb{C}^\ominus$ or at least $\mathbb{C}_g$ is a subset of $\mathbb{C}^\ominus$. For this problem, this* *is the only difference between continuous- and discrete-time systems. Therefore, we formulate everything in terms of* $\mathbb{C}_g$ *with the assumption that* $\mathbb{C}_g \subseteq \mathbb{C}^-$ *for continuous-time systems and for discrete-time systems* $\mathbb{C}_g \subseteq \mathbb{C}^\ominus$.

In what follows, a controller that solves the EDD problem is called an EDD controller. For a given system, such an EDD controller is in general not unique. On the other hand, in practice, obtaining a controller that achieves EDD is not the only design goal. Practical control problems dictate several other considerations. If so, the nonuniqueness of an EDD controller can be a blessing to a designer as it lets one take into account some other design considerations. One of the foremost considerations in linear control system design is the flexibility or freedom available in assigning the closed-loop spectrum as desired. This prompts us to formulate a problem that will be coined as the EDD problem with simultaneous pole placement via state feedback. In trying to solve such a problem, one can ask a fundamental question: For a given system, what is the available freedom in the closed-loop pole assignment while still achieving EDD? In this regard, it turns out that one does not always have complete freedom in closed-loop pole assignment. In fact, for each specified EDD problem, a set of *fixed modes* exists termed as the set of EDD fixed modes. Such a set contains as its elements certain complex numbers that every EDD controller must assign among the closed-loop poles. In other words, to achieve EDD, the designer of a controller is forced to place some closed-loop poles at certain fixed locations, whereas the rest can be placed at will. This is one of the constraints of requiring EDD. Also, another constraint exists. It turns out that there are often some pole/zero cancellations in EDD controller design. In fact, the phenomenon of pole/zero cancellations is not unique to such a design. It also occurs in other control design schemes such as H_2 or H_∞ optimization. It is well known that pole/zero cancellations are not always desired; for example, for continuous-time systems, those near the imaginary axis in the left-half complex plane are always undesirable, and in fact some of them are often prohibited by practical considerations. We will establish that, in fact, it might not be necessary to cancel all left-half complex plane zeros to achieve EDD. Some cancellations can be avoided, and some others cannot. Similar comments can be made for discrete-time systems with respect to the unit disk. Hence, it is imperative that we ask another fundamental question: For a given system, what are the unavoidable pole/zero cancellations one must impose to achieve EDD? In this regard, for a given EDD problem, one needs to construct what can be termed as a set of EDD fixed decoupling zeros, which obviously *in general* is a subset of the set of EDD fixed modes, and which shows the minimum number and required locations of pole/zero cancellations in the given EDD problem. This is another constraint when requiring an EDD controller design.

It is transparent that the existing freedom and constraints are two adversary aspects of any design. A designer, while exploiting available design freedom, seeks a design that satisfies as closely as possible the specified practical considerations while yet honoring the imposed constraints dictated by the requirement of EDD.

As such, it is of immense help to a designer to construct or parameterize the set of all EDD controllers, while simultaneously characterizing both the sets of EDD fixed modes and of EDD fixed decoupling zeros. This leads us to first formulate the following tasks:

(*i*) to parameterize and construct the set of all possible EDD controllers,

(*ii*) to determine the set of EDD fixed modes, and

(*iii*) to determine the set of EDD fixed decoupling zeros.

Once we carry out the above three tasks, we can enquire as to how one can use the above characterizations to arrive at a practical design of a control system. Indeed the above characterizations can effectively be used by a designer in a number of directions so as to meet certain practical design specifications in addition to EDD. As discussed, one immediate application of the above characterizations is to solve the EDD problem with simultaneous pole placement via state feedback. Assume we have a prescribed region of the complex plane that is symmetric with respect to the real axis and contains at least one real number. One can establish that one can produce a state feedback design having the simultaneous property that the resulting closed-loop poles are in this prescribed region of the complex plane if and only if this prescribed region contains the set of EDD fixed modes. Thus, our next task is as follows:

(*iv*) to develop an algorithm to arrive at an EDD state feedback controller, which simultaneously places the closed-loop poles at the prescribed locations in the complex plane whenever it can be done.

Before we proceed with our development, we formally present the following notations and definitions.

Definition 5.3 *Consider the EDD problem that is characterized by a quintuple (A, B, C, D, E). Then, the sets of all EDD static state, dynamic state, static full information, and dynamic full information feedback controllers are, respectively, denoted by $F_s(A, B, C, D, E)$, $F_d(A, B, C, D, E)$, $F_{s,f}(A, B, C, D, E)$, and $F_{d,f}(A, B, C, D, E)$. We note that*

$$F_s(A, B, C, D, E) \subset F_d(A, B, C, D, E).$$

Similarly,

$$F_{s,f}(A, B, C, D, E) \subset F_{d,f}(A, B, C, D, E).$$

A comment about our notation is in order. Often, the dependence of the sets F_s, F_d, etc., on the quintuple (A, B, C, D, E) is not shown explicitly. This is done for brevity as we are concerned only with a state feedback control problem, which is simply characterized by the quintuple (A, B, C, D, E).

Definition 5.4 *Consider the EDD problem characterized by the quintuple (A, B, C, D, E). Then, a scalar $\lambda \in \mathbb{C}_g$ is said to be an EDD fixed mode if λ is a pole of the closed-loop system $\Sigma \times \Sigma_C$ for every controller of a particular type that one uses. The sets of all EDD fixed modes (including multiplicities) corresponding to the static state, dynamic state, static full information, and dynamic full information feedback controllers are, respectively, denoted by $\boldsymbol{\Omega}_s(A, B, C, D, E)$, $\boldsymbol{\Omega}_d(A, B, C, D, E)$, $\boldsymbol{\Omega}_{s,f}(A, B, C, D, E)$, and $\boldsymbol{\Omega}_{d,f}(A, B, C, D, E)$.*

Remark 5.5 *Obviously by enlarging the class of feedback controllers, one can never enlarge the set of fixed modes. Thus, we note that*

$$\boldsymbol{\Omega}_d(A, B, C, D, E) \subseteq \boldsymbol{\Omega}_s(A, B, C, D, E),$$
$$\boldsymbol{\Omega}_{s,f}(A, B, C, D, E) \subseteq \boldsymbol{\Omega}_s(A, B, C, D, E),$$
$$\boldsymbol{\Omega}_{d,f}(A, B, C, D, E) \subseteq \boldsymbol{\Omega}_{s,f}(A, B, C, D, E).$$

Definition 5.6 *Consider the EDD problem characterized by a quintuple (A, B, C, D, E). Then, a scalar $\lambda \in \mathbb{C}_g$ is said to be an EDD fixed decoupling zero if λ is either an input or an output decoupling zero (or both) of the closed-loop system $\Sigma \times \Sigma_C$ for every EDD controller of a particular type that one uses. The sets of all EDD fixed decoupling zeros corresponding to the static state, dynamic state, static full information, and dynamic full information feedback controllers are, respectively, denoted by $\boldsymbol{\Lambda}_d(A, B, C, D, E)$, $\boldsymbol{\Lambda}_s(A, B, C, D, E)$, $\boldsymbol{\Lambda}_{d,f}(A, B, C, D, E)$, and $\boldsymbol{\Lambda}_{s,f}(A, B, C, D, E)$.*

Remark 5.7 *Note that a controller that achieves EDD must obviously yield a closed-loop transfer matrix identical to zero. From this it can be easily concluded that the set of EDD fixed decoupling zeros corresponding to a particular controller coincides with the EDD fixed modes corresponding to the same controller. That is, $\boldsymbol{\Lambda}_d = \boldsymbol{\Omega}_d$, $\boldsymbol{\Lambda}_s = \boldsymbol{\Omega}_s$, $\boldsymbol{\Lambda}_{d,f} = \boldsymbol{\Omega}_{d,f}$, and $\boldsymbol{\Lambda}_{s,f} = \boldsymbol{\Omega}_{s,f}$. This begs the question, why in the first place did we introduce two separate sets for each controller, one for the set of EDD fixed modes and the other for the set of EDD fixed decoupling zeros. The answer is simple. We separately introduced the set of fixed modes and the set of fixed input decoupling zeros because we will define these sets in many different contexts, and it is only in this particular case of EDD that these two sets happen to be the same.*

Remark 5.8 *In what follows, we focus a lot of attention on the set of all EDD static state feedback controllers $\boldsymbol{F}_s(A, B, C, D, E)$. This is because it plays a dominant role in defining the other controllers as well. We also mention that we often view the set $\boldsymbol{F}_s(A, B, C, D, E)$ either as a collection of all EDD static state* *feedback controllers or depending on the context and by abuse of notation as a collection of all EDD static state feedback gains such that the control law $u = -Fx$ with $F \in F_s(A, B, C, D, E)$ is an EDD static state feedback controller. This should not cause any confusion. Whenever any concern of ambiguity exists, we state it clearly.*

5.3   Solvability conditions for EDD

In this section, we recall the solvability conditions for EDD. The conditions under which the EDD problem can be solved are normally stated in terms of a subspace of geometric theory of linear systems, namely the $\mathbb{C}_g$-stabilizable weakly unobservable subspace $\mathcal{V}_g$ as defined in Definition 3.28.

We have the following theorem pertaining to state feedback controllers.

Theorem 5.9 *Consider the continuous- or discrete-time system Σ as in (5.1). Also, consider a subsystem Σ_{sub} characterized by the quadruple (A, B, C, D). Then, the following hold:*

(i) *The EDD problem is solvable by a dynamic state feedback controller if and only if it is solvable by a static state feedback controller.*

(ii) *The EDD problem is solvable by a static state feedback controller if and only if (A, B) is $\mathbb{C}_g$-stabilizable and im $E \subseteq \mathcal{V}_g(\Sigma_{\mathrm{sub}})$.*

Proof : See the literature, for instance, [75, 95]. ∎

The necessary and sufficient conditions for the solvability of EDD problem are the same whether static or dynamic state feedback controllers are used. In Sections 5.4 and 5.5, we consider, respectively, the static and dynamic state feedback controllers and then study various aspects of EDD problem.

We have the following theorem pertaining to full information feedback controllers.

Theorem 5.10 *Consider the continuous- or discrete-time system Σ as in (5.1). Also, consider a subsystem Σ_{sub} characterized by the quadruple (A, B, C, D). Then, the following hold:*

(i) *The EDD problem is solvable by a dynamic full information feedback controller of the form (5.9) if and only if it is solvable by a static full information feedback controller of the form (5.7).*

(*ii*) *The EDD problem is solvable by a static full information feedback controller of the form (5.7) if and only if* (A, B) *is* $\mathbb{C}_g$*-stabilizable and*

$$\operatorname{im} E \subseteq \mathcal{V}_g(\Sigma_{\mathrm{sub}}) + B \ker D. \tag{5.11}$$

Proof : We can view the static full information feedback controllers of the form (5.7) as preliminary feedback laws of the type $u = Gw + v$ followed by static state feedback laws. Then any preliminary feedback law must have the property that, after applying this feedback law, the EDD problem for the resulting system must be solvable via static state feedback laws. It is clear that this is the case only if $DG = 0$, and according to Theorem 5.9, we need

$$\operatorname{im}(E + BG) \subseteq \mathcal{V}_g(\Sigma_{\mathrm{sub}}).$$

It is easy to see that this immediately yields (5.11).

If we have a dynamic full information feedback represented in the frequency domain by

$$U(s) = H_x(s)X(s) + H_w(s)W(s),$$

with H_x and H_w the transfer matrices from x and w, respectively, to u, then we can write

$$H_w(s) = H_{sp}(s) + G,$$

where G is a static matrix and H_{sp} is strictly proper. But then effect of the feedback can be equivalently described by

$$U(s) = H_x(s)X(s) + H_{sp}(s)[sX(s) - AX(s) - BU(s)] + GW(s),$$

which results in

$$U(s) = \tilde{H}(s)X(s) + GW(s).$$

We see that we equivalently use a static preliminary feedback $u = Gw + v$ followed by dynamic state feedback laws. It is clear that this is the case only if $DG = 0$, and according to Theorem 5.9, we need

$$\operatorname{im}(E + BG) \subseteq \mathcal{V}_g(\Sigma_{\mathrm{sub}}). \qquad \blacksquare$$

5.4   Static state feedback laws and associated fixed modes and fixed decoupling zeros

In this section, we are concerned with static state feedback laws of the form (5.5). Our interest is to develop an algorithm that generates the sets F_s, Ω_s, and Λ_s. We call such an algorithm the *EDD algorithm*. The *EDD algorithm* takes as its input data the set of five matrices that characterize the given EDD problem and then

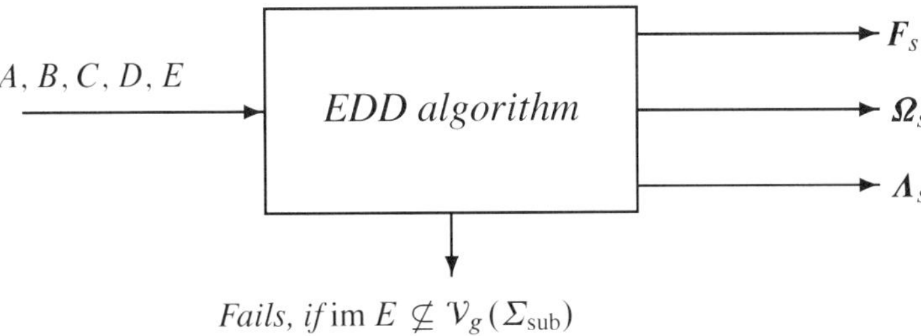

Figure 5.2: A block diagram interpretation of *EDD algorithm*

parameterizes and characterizes (or constructs) the set of all EDD state feedback gains F_s, the set of EDD fixed modes Ω_s, and the set of EDD fixed decoupling zeros Λ_s. This is illustrated in Figure 5.2. Thus, the *EDD algorithm* is a vehicle or a tool by which a designer can come up with a practically acceptable design of choice. Although the algorithm can be easily adapted for more general cases, we describe the *EDD algorithm* when the stability domain is equal to $\mathbb{C}_g$, where $\mathbb{C}_g$ is a subset of $\mathbb{C}^-$ or $\mathbb{C}^\ominus$ for continuous- or discrete-time systems, respectively.

To start with, it is prudent to recall the necessary and sufficient conditions under which an EDD problem with stability is solvable. As discussed in Theorem 5.9, a static feedback controller that solves the EDD problem exists if and only if the pair (A, B) is $\mathbb{C}_g$-stabilizable and im $E \subseteq \mathcal{V}_g(\Sigma_{\mathrm{sub}})$. To gain familiarity with the condition im $E \subseteq \mathcal{V}_g(\Sigma_{\mathrm{sub}})$, let us first examine one extreme case of it, namely, $E = 0$. Note that in the case when $E = 0$, no exogenous disturbance affects the system. Then, obviously, every controller that guarantees the internal $\mathbb{C}_g$-stability of the resulting closed-loop system is an EDD state feedback controller. The other extreme case occurs when im $E = \mathcal{V}_g(\Sigma_{\mathrm{sub}})$, and this case can be referred to as the *worst case* because it corresponds to the situation when a disturbance signal can affect the dynamics of the given system in the worst way while still satisfying the subspace inclusion condition im $E \subseteq \mathcal{V}_g(\Sigma_{\mathrm{sub}})$. As will be seen soon, the set of all EDD state feedback controllers for the worst case can be included in the set of all EDD state feedback controllers for the case when im E is strictly included in $\mathcal{V}_g(\Sigma_{\mathrm{sub}})$.

In the following subsections, we first consider the case when the subsystem Σ_{sub} is left-invertible, and then the case when Σ_{sub} is non-left-invertible. It turns out, however, that an EDD problem for Σ_{sub} that is not left-invertible can be converted by an appropriate prefeedback to an equivalent EDD problem with stability wherein Σ_{sub} is in fact left-invertible. The heart of the *EDD algorithm* is the decomposition of the subsystem Σ_{sub} so that both the finite and the infinite zero structures become transparent. This allows us to visualize clearly the condition im $E \subseteq \mathcal{V}_g(\Sigma_{\mathrm{sub}})$. The needed decomposition of Σ_{sub} can be done by representing it in SCB.

5.4.1 EDD algorithm—left-invertible case

In this subsection, we develop the *EDD algorithm* under the assumption that Σ_{sub} is left-invertible. For clarity, we refer to the *EDD algorithm* here as the EDD_{li} *algorithm* to signify that Σ_{sub} is left invertible. The EDD_{li} *algorithm* is divided into two basic steps. Step 1 consists of the computation of a pair of matrices (A_z, B_z) from the input data (A, B, C, D, E). Next, by using the pair (A_z, B_z), Step 2 characterizes and parameterizes F_s and determines Ω_s and Λ_s.

The EDD$_{li}$ Algorithm

Step 1: Computation of the pair (A_z, B_z).

In this step, we compute the pair (A_z, B_z). Our computations are divided into three substeps.

Step 1a: Representation of Σ_{sub} in SCB.

In this step, we first construct the SCB of Σ_{sub}, which is characterized by the quadruple (A, B, C, D). We note that the variable x_c in the SCB of Σ_{sub} is nonexistent as Σ_{sub} is assumed to be left-invertible. The needed development to construct the SCB is presented in Chapter 3, and the precise SCB that we need can be found in (3.46) to (3.49).

From Property 3.24 of SCB, it is simple to see that im $E \subseteq \mathcal{V}_g(\Sigma_{\text{sub}})$ implies that

$$\Gamma_s^{-1} E = \left((E_a^-)' \quad 0 \quad 0 \quad 0 \right)'. \tag{5.12}$$

Here we used that $\mathbb{C}_g \subseteq \mathbb{C}^-$ or $\mathbb{C}_g \subseteq \mathbb{C}^\ominus$ for continuous- or discrete-time systems, respectively.

If $\Gamma_s^{-1} E$ is not of the form (5.12), the EDD problem is not solvable via static state feedback controllers and the procedure of EDD_{li} *algorithm* stops at this point. Otherwise, it continues to the next step.

Step 1b: Decomposition of A_{aa}^-.

In this step, we decompose A_{aa}^- into two parts, one part is controllable via the disturbance w and the other part is unaffected by the disturbance. Consider the pair (A_{aa}^-, E_a^-). This pair need not be controllable, i.e., the disturbance w need not affect all modes of A_{aa}^-. Thus, we can compute a nonsingular transformation T_a such that

$$T_a^{-1} A_{aa}^- T_a = \begin{pmatrix} A_{aa}^{11} & A_{aa}^{12} \\ 0 & A_{aa}^{22} \end{pmatrix} \quad \text{and} \quad T_a^{-1} E_a^- = \begin{pmatrix} E_a^1 \\ 0 \end{pmatrix}, \tag{5.13}$$

where the pair (A_{aa}^{11}, E_a^1) is controllable. Also, let us partition

$$T_a^{-1} L_{ab}^- = \begin{pmatrix} L_{ab}^1 \\ L_{ab}^2 \end{pmatrix}, \quad T_a^{-1} L_{ad}^- = \begin{pmatrix} L_{ad}^1 \\ L_{ad}^2 \end{pmatrix}, \quad T_a^{-1} B_{a0}^- = \begin{pmatrix} B_{a0}^1 \\ B_{a0}^2 \end{pmatrix},$$

and

$$
E_{\bar{d}a}^- T_a = \begin{pmatrix} E_{da}^1 & E_{da}^2 \end{pmatrix}, \quad C_{0a}^- T_a = \begin{pmatrix} C_{0a}^1 & C_{0a}^2 \end{pmatrix}.
$$

Step 1c: Computation of A_z and B_z.

Based on the above development, we can finally form the matrices A_z and B_z as follows:

$$
A_z := \begin{pmatrix} A_{aa}^{22} & 0 & L_{ab}^2 C_b & L_{ad}^2 C_d \\ 0 & A_{aa}^{0+} & L_{ab}^{0+} C_b & L_{ad}^{0+} C_d \\ 0 & 0 & A_{bb} & L_{bd} C_d \\ B_d E_{da}^2 & B_d E_{da}^{0+} & B_d E_{db} & A_{dd} \end{pmatrix}, \quad B_z := \begin{pmatrix} B_{a0}^2 & 0 \\ B_{a0}^{0+} & 0 \\ B_{b0} & 0 \\ B_{d0} & B_d \end{pmatrix}.
$$

$$(5.14)$$

From Property 3.3 of SCB, it is simple to verify that the pair (A_z, B_z) is $\mathbb{C}_g$ stabilizable if and only if the pair (A, B) is $\mathbb{C}_g$ stabilizable. Thus, whenever (A, B) is $\mathbb{C}_g$-stabilizable, a gain F_z exists such that $\lambda(A_z - B_z F_z) \subset \mathbb{C}_g$.

Step 2: Characterization and parameterization of $\boldsymbol{F}_s$, $\boldsymbol{\Omega}_s$, and $\boldsymbol{\Lambda}_s$.

In this step, we first parameterize $\boldsymbol{F}_s$ in terms of a parameter F_z, which renders the matrix $A_z - B_z F_z$ $\mathbb{C}_g$-stable. To do so, we first define the set

$$
\boldsymbol{F}_z := \{ F_z \mid \lambda(A_z - B_z F_z) \subset \mathbb{C}_g \} \tag{5.15}
$$

and then partition $F_z \in \boldsymbol{F}_z$ to be compatible with the partitioning of A_z and B_z:

$$
F_z = \begin{pmatrix} F_{a0}^2 & F_{a0}^{0+} & F_{b0} & F_{d0} \\ F_{ad}^2 & F_{ad}^{0+} & F_{bd} & F_{dd} \end{pmatrix}. \tag{5.16}
$$

We define next a gain F as

$$
F = \Gamma_i \begin{pmatrix} C_{0a}^1 & C_{0a}^2 + F_{a0}^2 & C_{0a}^{0+} + F_{a0}^{0+} & C_{0b} + F_{b0} & C_{0d} + F_{d0} \\ E_{da}^1 & F_{ad}^2 & F_{ad}^{0+} & F_{bd} & F_{dd} \end{pmatrix} T_s^{-1}, \tag{5.17}
$$

where

$$
T_s = \Gamma_s \begin{pmatrix} T_a & 0 & 0 & 0 \\ 0 & I & 0 & 0 \\ 0 & 0 & I & 0 \\ 0 & 0 & 0 & I \end{pmatrix}.
$$

This leads us to parameterize $\boldsymbol{F}_s$ in terms of F_z as given by

$$
\boldsymbol{F}_s = \{ F \in \mathbb{R}^{m \times n} \mid F \text{ is given by (5.17) with } F_z \in \boldsymbol{F}_z \}. \tag{5.18}
$$

Furthermore, we can determine $\boldsymbol{\Omega}_s$ and $\boldsymbol{\Lambda}_s$ as

$$
\boldsymbol{\Omega}_s = \boldsymbol{\Lambda}_s = \lambda(A_{aa}^{11}) \cup \{\text{input decoupling zeros of } (A_z, B_z)\}. \tag{5.19}
$$

The fact that $\mathrm{im}\, E \subset \mathcal{V}_g(\Sigma_{\mathrm{sub}})$ implies that $\lambda(A_{aa}^{11}) \subset \mathbb{C}_g$.

This concludes the description of the *EDD$_{li}$ algorithm*. We have the following theorem.

Theorem 5.11 *Consider the EDD problem as defined by Problem 5.1 for a system Σ as in (5.1). Assume that the pair (A, B) is $\mathbb{C}_g$ stabilizable and that* $\operatorname{im} E \subseteq \mathcal{V}_g(\Sigma_{\mathrm{sub}})$. *Moreover, let Σ_{sub} be left-invertible. Then we have:*

Set of EDD state feedback controllers: *As claimed in (5.18), the set of all static state feedback gains F_s as defined in Definition 5.3 is given by*

$$F_s = \left\{ F \in \mathbb{R}^{m \times n} \mid F \text{ is given by (5.17) with } F_z \in F_z \right\}.$$

That is, the state feedback law $u = -Fx$ is in the set $F \in F_s$ as defined in (5.18) if and only if when applied to Σ, it is $\mathbb{C}_g$-admissible and the closed-loop transfer function $T_{zw}(\Sigma \times \Sigma_c) = (C - DF)(sI - A + BF)^{-1}E = 0$.

EDD fixed modes: *As claimed in (5.19), the set of all EDD fixed modes Ω_s as defined in Definition 5.4 is given by*

$$\Omega_s = \lambda(A_{aa}^{11}) \cup \left\{ \text{input decoupling zeros of } (A_z, B_z) \right\}.$$

That is, any EDD state feedback controller must assign the elements of Ω_s among the closed-loop poles. The rest of the closed-loop poles can be assigned at desired locations in $\mathbb{C}_g$ as long as the desired locations are self-conjugate, i.e., symmetric with respect to the real axis. It should be emphasized that that a state feedback gain $F \in F_s$ exists that results in a desired set of self-conjugate closed-loop poles if and only if the desired set of poles contains Ω_s.

EDD fixed decoupling zeros: *As claimed in (5.19), set of all EDD fixed decoupling zeros Λ_s as defined in Definition 5.6 is given by*

$$\Lambda_s = \lambda(A_{aa}^{11}) \cup \left\{ \text{input decoupling zeros of } (A_z, B_z) \right\}.$$

That is, regardless of the choice of F from F_s, the absolutely minimum number and locations of pole/zero cancellations in a resulting closed-loop transfer function are given by the set Λ_s.

Proof : See Section 5.A.   ∎

Remark 5.12 *As will become clear from our proof in the non-left-invertible case, the input-decoupling zeros of (A_z, B_z) are a subset of the input-decoupling zeros of (A, B). The reason for being only a subset comes from the fact that A_{aa}^{11} might also contain input-decoupling zeros of the original system. Actually, the input-decoupling zeros of (A_z, B_z) are precisely the input-decoupling zeros of*

$$\Big(A, (B \quad E)\Big). \tag{5.20}$$

At this time, some comments on the sets F_s, Ω_s, and Λ_s are in order. Given the quintuple (A, B, C, D, E), it is clear that the EDD_{li} *algorithm* explicitly yields all EDD state feedback controllers as expressed by the set F_s, the number and locations of all EDD fixed modes, as well as the EDD fixed decoupling zeros as expressed by the set $\Omega_s = \Lambda_s$. Let us note that in the given quintuple (A, B, C, D, E), the quadruple (A, B, C, D) prescribes the dynamic model Σ_{sub} of the given plant, whereas the matrix E prescribes how the disturbance w is coupled to the plant. Obviously, for any fixed dynamic model of the plant, the sets F_s and Ω_s and thus Λ_s, have a definite relationship with the matrix E. To examine this, let us recall again a condition for the existence of an EDD state feedback controller, namely, im $E \subseteq \mathcal{V}_g(\Sigma_{\mathrm{sub}})$. It is interesting to note that the size of the set F_s decreases while the size of the set Ω_s grows as im E varies from $\{0\}$ to $\mathcal{V}_g(\Sigma_{\mathrm{sub}})$. Hence, the sizes of F_s and Ω_s obtained for im $E = \{0\}$ are, respectively, the largest and the smallest possible ones, whereas the sizes of the same obtained for im $E = \mathcal{V}_g(\Sigma_{\mathrm{sub}})$ are, respectively, the smallest and the largest possible ones. Moreover, both the sets F_s and Ω_s have a *nested* property as stated and formalized in the following proposition.

Proposition 5.13 *Consider two different values for* E, *say* E_1 *and* E_2, *and let* im $E_1 \subseteq$ im $E_2 \subseteq \mathcal{V}_g(\Sigma_{\mathrm{sub}})$. *Let* F_{s1} *and* Ω_{s1} *be the sets corresponding to* F_s *and* Ω_s *for the case when* $E = E_1$. *Similarly, let* F_{s2} *and* Ω_{s2} *be the sets corresponding to* F_s *and* Ω_s *for the case when* $E = E_2$. *Then, we have*

$$F_{s2} \subseteq F_{s1} \text{ and } \Omega_{s1} \subseteq \Omega_{s2}. \tag{5.21}$$

Proof : Let us first consider the proof of the property that $F_{s2} \subseteq F_{s1}$. Given that im $E_1 \subseteq$ im E_2, we note that a matrix X exists such that $E_1 = E_2 X$. Then for any $F \in F_{s2}$, i.e.,

$$(C - DF)(sI - A + BF)^{-1} E_2 = 0,$$

we have

$$(C - DF)(sI - A + BF)^{-1} E_1 = (C - DF)(sI - A + BF)^{-1} E_2 X = 0,$$

which implies that $F \in F_{s1}$. Hence, $F_{s2} \subseteq F_{s1}$.

Let us next consider the proof of property $\Omega_{s1} \subseteq \Omega_{s2}$. As $F_{s2} \subseteq F_{s1}$, the set of fixed closed-loop poles when varying over the smaller set F_{s2} is obviously larger than or equal to the set of fixed closed-loop poles when varying over the larger set F_{s1}. Hence the result follows.   ∎

Next, following the same lines of Propositions 5.13, it is easy to show that when im $E = \{0\}$, then Ω_s is equal to the set of input decoupling zeros of Σ_{sub}, whereas in the case that im $E = \mathcal{V}_g(\Sigma_{\mathrm{sub}})$, the set Ω_s is equal to the set of stable invariant zeros of Σ_{sub} together with the input-decoupling zeros of (5.20). In general, when $\{0\} \neq \operatorname{im} E \subsetneq \mathcal{V}_g(\Sigma_{\text{sub}})$, the set Ω_s is equal to a subset of the stable invariant zeros and input-decoupling zeros of Σ_{sub}.

The importance of Theorem 5.11 and Proposition 5.13 cannot be overemphasized. The knowledge of the entire set of feedback controllers F_s makes it easier to take into account design criteria other than EDD. Also, the set of fixed modes Ω_s clearly points out what every EDD closed-loop system must include among its poles. We emphasize that an arbitrarily chosen $F \in F_s$ may induce pole/zero cancellations beyond that given by the set Λ_s (or equivalently by Ω_s). Thus, with the knowledge of Ω_s, a designer can easily ascertain whether any unwanted pole/zero cancellations or, as a matter of fact, any unwanted closed-loop poles are to be necessarily involved or not in a final design.

The following example illustrates our results.

Example 5.14 We consider an adapted version of Example 7.2.1 of [75]. Let the continuous-time EDD problem with $\mathbb{C}_g = \mathbb{C}^-$ be described by

$$
A = \begin{pmatrix} -1 & 2 & 0 & 0 & 1 \\ 0 & -1 & 0 & 0 & -3 \\ 0 & 0 & -2 & 5 & 0 \\ 0 & 0 & 0 & -2 & 0 \\ 1 & 4 & 0 & 0 & 2 \end{pmatrix}, \quad B = \begin{pmatrix} 0 & 0 \\ 0 & 0 \\ 1 & 0 \\ 1 & 0 \\ 0 & 1 \end{pmatrix}, \quad E = \begin{pmatrix} 1 & 0 \\ 0 & -1 \\ 0 & 0 \\ 0 & 0 \\ 0 & 0 \end{pmatrix},
$$

$$
C = \begin{pmatrix} 0 & 0 & 0 & 3 & 0 \\ 0 & 0 & 0 & 0 & 1 \end{pmatrix}, \quad D = \begin{pmatrix} 1 & 0 \\ 0 & 0 \end{pmatrix}.
$$

This system is already in the SCB and Σ_{sub} is left-invertible. In particular, we find that

$$
A_{aa}^- = \begin{pmatrix} -1 & 2 & 0 & 0 \\ 0 & -1 & 0 & 0 \\ 0 & 0 & -2 & 5 \\ 0 & 0 & 0 & -2 \end{pmatrix}, \quad E_a^- = \begin{pmatrix} 1 & 0 \\ 0 & -1 \\ 0 & 0 \\ 0 & 0 \end{pmatrix}.
$$

Using the decomposition of step 1b of the EDD_{li} algorithm, we obtain

$$
A_{aa}^{11} = \begin{pmatrix} -1 & 2 \\ 0 & -1 \end{pmatrix}, \quad E_a^1 = \begin{pmatrix} 1 & 0 \\ 0 & -1 \end{pmatrix}, A_{aa}^{22} = \begin{pmatrix} -2 & 5 \\ 0 & -2 \end{pmatrix}.
$$

This leads to A_z and B_z as

$$
A_z = \begin{pmatrix} -2 & 5 & 0 \\ 0 & -2 & 0 \\ 0 & 0 & 2 \end{pmatrix}, \quad B_z = \begin{pmatrix} 1 & 0 \\ 1 & 0 \\ 0 & 1 \end{pmatrix}.
$$

Then we get the set of EDD state feedback gains F_s as

$$F_s = \left\{ F \,\middle|\, F = \begin{pmatrix} 0 & 0 & F_{z11} & F_{z12} & F_{z13} \\ 1 & 4 & F_{z21} & F_{z22} & F_{z23} \end{pmatrix} \right\},$$

and the EDD fixed modes are given as

$$\Omega_s = \Lambda_s = \{-1, -1\}.$$

Clearly, this example is oversimplified and streamlined. It was not our intention here to provide an intrinsic EDD control problem appearing in practice. We merely sought an easy manner to illustrate Theorem 5.11.

As formalized in Theorem 5.11, among other things, the EDD_{li} algorithm constructs the set of all static state feedback controllers F_s. An important question that arises next is under what conditions F_s is a singleton. The following lemma deals with this. Also, it characterizes the resulting $\Lambda_s = \Omega_s$.

Lemma 5.15 *Consider the given system Σ as in (5.1) with $(C \quad D)$ surjective. Let Σ_{sub} be left-invertible. Also, assume that the pair (A, B) is $\mathbb{C}_g$-stabilizable and that im $E \subseteq \mathcal{V}_g(\Sigma_{\mathrm{sub}})$. Then, the EDD state feedback law is unique if and only if the following conditions hold:*

(i) The subsystem Σ_{sub} of Σ has all its invariant zeros in the set $\mathbb{C}_g$.

(ii) The subsystem Σ_{sub} is right-invertible.

(iii) The matrix D is invertible.

(iv) The pair $(A - BD^{-1}C, E)$ is controllable.

Moreover, under the above conditions,

- *$F_s = \{-D^{-1}C\}$, which is a singleton,*

- *$\Lambda_s = \Omega_s = \{\text{ Invariant zeros of } \Sigma_{\mathrm{sub}}\} = \lambda(A - BD^{-1}C).$*

Proof : Under the conditions given in the lemma, it is simple to see that the matrices A_z and B_z as in (5.14) of the EDD_{li} algorithm are nonexistent. Thus, the result is obvious from the construction procedure of the EDD_{li} algorithm. ∎

5.4.2 EDD algorithm—non-left-invertible case

Here we consider the case when Σ_{sub} is not left-invertible. As in the previous subsection, our goal here is to determine the set of all EDD state feedback gains F_s and its associated sets, namely the set of all EDD fixed modes Ω_s and the set of all EDD fixed decoupling zeros Λ_s. Our methodology here is as follows. We first use a prefeedback gain to stabilize the dynamics associated with the subspace $\mathcal{R}^*(\Sigma_{\text{sub}})$. This transforms the given EDD problem for a given system to an associated EDD problem for a new system whose data ensure that its subsystem Σ_{sub} is left-invertible. This methodology lets us determine the set of EDD fixed modes Ω_s and the set of EDD fixed decoupling zeros Λ_s. However, this methodology lets us parameterize and characterize **only** a subset (denoted here by F_s^{sub}) of **all** EDD state feedback gains F_s. This is a drawback of our method; it is attributable to the fact that one need not first stabilize a subsystem of a composite system in order to stabilize the composite system. One can trivially construct some examples to demonstrate this. On the other hand, to characterize the complete set F_s, one needs a new methodology different from that of ours. At this time this remains an open research problem. However, we would like to stress that we completely characterize the sets of all EDD fixed modes Ω_s and the set of all EDD fixed decoupling zeros Λ_s.

Our design algorithm here is a modification of the EDD_{li} algorithm. We refer to this here as EDD_{nli} algorithm to signify that Σ_{sub} is not left-invertible. The EDD_{nli} algorithm, as in the previous case, is divided into two basic steps. Step 1 consists of the computation of a pair of matrices $(\bar{A}_z, \bar{B}_z)$ from the input data (A, B, C, D, E). Next, by using the pair $(\bar{A}_z, \bar{B}_z)$, Step 2 characterizes and parameterizes F_s^{sub}.

Finally, we present a pair of matrices $(\check{A}_z, \check{B}_z)$ that helps us to parameterize the sets Ω_s and Λ_s. We divide Step 1 into four substeps. The main difference between the EDD_{li} and EDD_{nli} algorithms is in Step 1b where a certain prefeedback gain is introduced.

The EDD$_{nli}$ Algorithm

Step 1: Computation of the pair $(\bar{A}_z, \bar{B}_z)$.

In this step, we compute the pair $(\bar{A}_z, \bar{B}_z)$. As in the previous subsection, our computations are divided into four substeps.

Step 1a: Representation of Σ_{sub} in SCB.

As in the previous subsection, we first construct in this step the SCB of Σ_{sub}. As Σ_{sub} is not left-invertible, the variable x_c does exist in the SCB in this case. In the SCB, we have the structure as presented in (3.42)–(3.45).

From Property 3.24 of the SCB, it is simple to see that im $E \subseteq \mathcal{V}_g(\Sigma_{\text{sub}})$ implies that

$$\Gamma_s^{-1} E = \left((E_a^-)' \quad 0 \quad 0 \quad (E_c)' \quad 0 \right)'. \tag{5.22}$$

If $\Gamma_s^{-1} E$ is not of the form (5.22), the EDD problem is not solvable via static state feedback controllers and the procedure of EDD_{nli} *algorithm* stops at this point. Otherwise, it continues to the next step.

Step 1b: Determination of $\boldsymbol{F}_{\text{pre}}$.

Unlike in the previous subsection, here we first construct a set of static prefeedback gains $\boldsymbol{F}_{\text{pre}}$:

$$\boldsymbol{F}_{\text{pre}} := \left\{ \begin{pmatrix} 0 & 0 & 0 & 0 & 0 \\ 0 & 0 & 0 & 0 & 0 \\ F_{ca}^- & F_{ca}^{0+} & F_{cb} & F_c & F_{cd} \end{pmatrix} \middle| \ \lambda(A_{cc} - B_c F_c) \subset \mathbb{C}_g \right\}, \tag{5.23}$$

where F_{ca}^-, F_{ca}^{0+}, F_{cb}, and F_{cd} are arbitrary matrices with appropriate dimensions.

Step 1c: Another decomposition of state space.

We first decompose A_{aa}^- into two parts, one part is controllable via the disturbance w, and the other part is unaffected by the disturbance. Consider the pair (A_{aa}^-, E_a^-). This pair need not be controllable, i.e., the disturbance w need not affect all the modes of A_{aa}^-. Thus, we can compute a nonsingular transformation T_a such that

$$T_a^{-1} A_{aa}^- T_a = \begin{pmatrix} A_{aa}^{11} & A_{aa}^{12} \\ 0 & A_{aa}^{22} \end{pmatrix} \quad \text{and} \quad T_a^{-1} E_a^- = \begin{pmatrix} E_a^1 \\ 0 \end{pmatrix}, \tag{5.24}$$

where the pair (A_{aa}^{11}, E_a^1) is controllable. Also, we note that the pair

$$\begin{pmatrix} A_{aa}^- & 0 \\ B_c(E_{ca}^- - F_{ca}^-) & A_{cc}^c \end{pmatrix}, \quad \begin{pmatrix} E_a^- \\ E_c \end{pmatrix} \tag{5.25}$$

need not be controllable where $A_{cc}^c := A_{cc} - B_c F_c$ for a chosen $F_{\text{pre}} \in \boldsymbol{F}_{\text{pre}}$. Thus, for each $F_{\text{pre}} \in \boldsymbol{F}_{\text{pre}}$, we compute a nonsingular transformation T_{ac} such that

$$T_{ac}^{-1} \begin{pmatrix} A_{aa}^- & 0 \\ B_c(E_{ca}^- - F_{ca}^-) & A_{cc}^c \end{pmatrix} T_{ac} = \begin{pmatrix} A_{ac}^{11} & A_{ac}^{12} \\ 0 & A_{ac}^{22} \end{pmatrix} \tag{5.26}$$

and

$$T_{ac}^{-1} \begin{pmatrix} E_a^- \\ E_c \end{pmatrix} = \begin{pmatrix} E_{ac}^1 \\ 0 \end{pmatrix}, \tag{5.27}$$

where the pair (A_{ac}^{11}, E_{ac}^1) is controllable. Also, we partition

$$T_{ac}^{-1} \begin{pmatrix} L_{ab}^- C_b \\ B_c E_{cb} - B_c F_{cb} \end{pmatrix} = \begin{pmatrix} X_{acb}^1 \\ X_{acb}^2 \end{pmatrix}, \quad T_{ac}^{-1} \begin{pmatrix} L_{ad}^- C_d \\ L_{cd} C_d - B_c F_{cd} \end{pmatrix} = \begin{pmatrix} X_{acd}^1 \\ X_{acd}^2 \end{pmatrix},$$

$$T_{ac}^{-1}\begin{pmatrix} 0 \\ B_c \end{pmatrix} = \begin{pmatrix} B_{ca}^1 \\ B_{ca}^2 \end{pmatrix}, \quad T_{ac}^{-1}\begin{pmatrix} B_{a0}^- \\ B_{c0} \end{pmatrix} = \begin{pmatrix} B_{ac0}^1 \\ B_{ac0}^2 \end{pmatrix},$$

and

$$\begin{pmatrix} E_{da}^- & E_{dc} \end{pmatrix} T_{ac} = \begin{pmatrix} E_{dac}^1 & E_{dac}^2 \end{pmatrix}, \quad \begin{pmatrix} C_{0a}^- & C_{0c} \end{pmatrix} T_{ac} = \begin{pmatrix} C_{0ac}^1 & C_{0ac}^2 \end{pmatrix}.$$

Step 1d: Computation of $\bar{A}_z$ and $\bar{B}_z$.

Based on the above development, we can finally form the matrices $\bar{A}_z$ and $\bar{B}_z$ as follows:

$$\bar{A}_z := \begin{pmatrix} A_{aa}^{0+} & L_{ab}^{0+} C_b & 0 & L_{ad}^{0+} C_d \\ 0 & A_{bb} & 0 & L_{bd} C_d \\ B_{ca}^2 E_{ca}^{0+} & X_{acb}^2 & A_{ac}^{22} & X_{acd}^2 \\ B_d E_{da}^{0+} & B_d E_{db} & B_d E_{dac}^2 & A_{dd} \end{pmatrix}, \quad \bar{B}_z := \begin{pmatrix} B_{a0}^{0+} & 0 \\ B_{b0} & 0 \\ B_{aco}^2 & 0 \\ B_{d0} & B_d \end{pmatrix}.$$

$$(5.28)$$

From Property 3.3 of the SCB, it is simple to verify that the pair $(\bar{A}_z, \bar{B}_z)$ is $\mathbb{C}_g$-stabilizable if and only if the original pair (A, B) is $\mathbb{C}_g$-stabilizable. Thus, whenever (A, B) is $\mathbb{C}_g$ stabilizable, a gain F_z exists such that $\lambda(\bar{A}_z - \bar{B}_z F_z) \subset \mathbb{C}_g$.

Step 2: Characterization and parameterization of F_s^{sub}.

In this step, we first parameterize F_s^{sub} in terms of $F_{\mathrm{pre}} \in \boldsymbol{F}_{\mathrm{pre}}$ and F_z, which renders $\bar{A}_z - \bar{B}_z F_z$ $\mathbb{C}_g$-stable. To do so, we first define the set

$$\boldsymbol{F}_z := \{\, F_z \mid \lambda(\bar{A}_z - \bar{B}_z F_z) \subset \mathbb{C}_g \,\}. \tag{5.29}$$

We need to stress that the set $\boldsymbol{F}_z$ depends on the choice of $F_{\mathrm{pre}} \in \boldsymbol{F}_{\mathrm{pre}}$. We also partition $F_z \in \boldsymbol{F}_z$ to be compatible with the partitioning of $\bar{A}_z$ and $\bar{B}_z$,

$$F_z = \begin{pmatrix} F_{a0}^{0+} & F_{b0} & F_{ac0}^2 & F_{d0} \\ F_{ad}^{0+} & F_{bd} & F_{acd}^2 & F_{dd} \end{pmatrix}. \tag{5.30}$$

Also, we let

$$F = \Gamma_i \left(F_{\mathrm{pre}} + F_{\mathrm{scb}} T_s^{-1} \right) \Gamma_s^{-1}, \tag{5.31}$$

where

$$F_{\mathrm{scb}} = \begin{pmatrix} C_{0a}^{0+} + F_{a0}^{0+} & C_{0b} + F_{b0} & C_{0ac}^1 & C_{0ac}^2 + F_{ac0}^2 & C_{0d} + F_{d0} \\ F_{ad}^{0+} & F_{bd} & E_{dac}^1 & F_{acd}^2 & F_{dd} \\ 0 & 0 & 0 & 0 & 0 \end{pmatrix}$$

and

$$T_s = \begin{pmatrix} I & 0 & 0 & 0 & 0 \\ 0 & 0 & 0 & I & 0 \\ 0 & 0 & I & 0 & 0 \\ 0 & I & 0 & 0 & 0 \\ 0 & 0 & 0 & 0 & I \end{pmatrix} \begin{pmatrix} T_{ac} & 0 \\ 0 & I \end{pmatrix} \begin{pmatrix} I & 0 & 0 & 0 & 0 \\ 0 & 0 & 0 & I & 0 \\ 0 & 0 & I & 0 & 0 \\ 0 & I & 0 & 0 & 0 \\ 0 & 0 & 0 & 0 & I \end{pmatrix}.$$

This leads us to parameterize F_s^{sub} as

$$F_s^{\text{sub}} := \left\{ F \in \mathbb{R}^{m \times n} \mid F \text{ is given by (5.31) with } F_{\text{pre}} \in F_{\text{pre}} \text{ and } F_z \in F_z \right\}.$$
$$(5.32)$$

Step 3: Characterization and parameterization of Ω_s and Λ_s.

To obtain the fixed modes and the fixed decoupling zeros independent of the specific choice of the prefeedback, we define the following matrices:

$$\breve{A}_z := \begin{pmatrix} A_{aa}^{22} & 0 & L_{ab}^2 C_b & L_{ad}^2 C_d \\ 0 & A_{aa}^{0+} & L_{ab}^{0+} C_b & L_{ad}^{0+} C_d \\ 0 & 0 & A_{bb} & L_{bd} C_d \\ B_d E_{da}^2 & B_d E_{da}^{0+} & B_d E_{db} & A_{dd} \end{pmatrix}, \quad \breve{B}_z := \begin{pmatrix} B_{a0}^2 & 0 \\ B_{a0}^{0+} & 0 \\ B_{b0} & 0 \\ B_{d0} & B_d \end{pmatrix}.$$
$$(5.33)$$

Note the similarity with the same matrices defined for the case of left-invertible systems in (5.14). However, we stress that the formulas might be the same, but the expression is for a different type of system because the SCB for non-left-invertible systems is obviously different.

We define

$$\Omega_s^1 := \Lambda_s^1 := \lambda(A_{aa}^{11}),$$

where A_{aa}^{11} as defined in (5.24), and

$$\Omega_s^2 := \Lambda_s^2 = \left\{ \text{ input decoupling zeros of } (\breve{A}_z, \breve{B}_z) \right\},$$

where $\breve{A}_z$ and $\breve{B}_z$ are defined in (5.33). Finally,

$$\Omega_s = \Omega_s^1 \cup \Omega_s^2 = \Lambda_s^1 \cup \Lambda_s^2 = \Lambda_s.$$
$$(5.34)$$

This concludes the description of the EDD_{nli} algorithm. We should note that it only parameterizes a *subset* of all static state feedback controllers that achieve EDD. However, we do not lose flexibility in pole placement. As a matter of fact, the following theorem will establish that the set of fixed modes and fixed decoupling zeros when restricting attention to the state feedback laws in the set F_s^{sub} is equal to the set of EDD fixed modes and EDD fixed decoupling zeros.

Theorem 5.16 *Consider the EDD problem as defined by Problem 5.1 for a system Σ as in (5.1). Assume that the pair (A, B) is $\mathbb{C}_g$-stabilizable, and that $\text{im } E \subseteq \mathcal{V}_g(\Sigma_{\text{sub}})$. Moreover, assume that Σ_{sub} is not left-invertible. Then, we have:*

- *Any feedback gain in the set F_s^{sub} with F_s^{sub} as in (5.32) is a state feedback gain achieving EDD. That is, the state feedback law $u = -Fx$ with $F \in F_s^{\text{sub}}$, when applied to Σ, is $\mathbb{C}_g$-admissible and the closed-loop transfer function*

$$T_{zw}(\Sigma \times \Sigma_c) = (C - DF)(sI - A + BF)^{-1} E = 0.$$

- *The set of **EDD fixed modes** Ω_s as defined in Definition 5.4 is given by (5.34). That is, any EDD state feedback controller must assign the elements of Ω_s among the closed-loop poles. The rest of the closed-loop poles can be assigned at desired locations in $\mathbb{C}_g$ as long as the desired locations are self-conjugate, i.e., symmetric with respect to the real axis.*

  Moreover, a state feedback gain $F \in \boldsymbol{F}_s^{\mathrm{sub}}$ exists that results in a desired set of self-conjugate closed-loop poles if and only if the desired set of poles contain Ω_s.

- *The set of **EDD fixed decoupling zeros** Λ_s as defined in Definition 5.6 is given by (5.34). That is, any EDD state feedback controller must assign the elements of Λ_s among the decoupling zeros. The rest of the decoupling zeros can be assigned at desired locations in $\mathbb{C}_g$ as long as the desired locations are self-conjugate, i.e., symmetric with respect to the real axis.*

  Moreover, a state feedback gain $F \in \boldsymbol{F}_s^{\mathrm{sub}}$ exists that results in a desired set of self-conjugate decoupling zeros if and only if the desired set of decoupling zeros contain Λ_s.

Proof : The first part of Theorem 5.16 follows with some obvious modifications from the proof of Theorem 5.11. We first establish that the set of fixed modes and fixed decoupling zeros, when restricting attention to the state feedback laws in the set $\boldsymbol{F}_s^{\mathrm{sub}}$, are equal to Ω_s and Λ_s, respectively.

We note that the eigenvalues of A_{ac}^{11} that we cannot arbitrarily assign are the controllable eigenvalues of the pair (A_{aa}^-, E_a^-) (namely, the eigenvalues of A_{aa}^{11}), which form precisely the set $\Omega_s^1 = \Lambda_s^1$. Next, using T_{ac} as in (5.26), one can choose a suitable SCB transformation such that we have

$$BT_i = \begin{pmatrix} \bar{B}_0 & \bar{B}_d & \bar{B}_c \end{pmatrix},$$

$$T(A - BF_{\mathrm{pre}})T^{-1} = \begin{pmatrix} \begin{pmatrix} A_{ac}^{11} \\ 0 \\ 0 \\ 0 \\ B_d E_{dac}^1 \end{pmatrix} & \begin{pmatrix} 0 & 0 & A_{ac}^{12} & 0 \end{pmatrix} \\ & \bar{A}_z \end{pmatrix}, \qquad (5.35)$$

and

$$T\begin{pmatrix} B_0 & B_d \end{pmatrix} = \begin{pmatrix} \begin{pmatrix} B_{ac0}^1 & 0 \end{pmatrix} \\ \bar{B}_z \end{pmatrix}, \qquad TE = \begin{pmatrix} E_{ac}^1 \\ 0 \end{pmatrix}. \qquad (5.36)$$

We can then, via an obvious preliminary feedback, make the term containing $B_d E_{dac}^1$ equal to zero. Moreover (A_{ac}^{11}, E_{ac}^1), by construction, is controllable.

Then it is obvious that the input-decoupling zeros of $(\bar{A}_z, \bar{B}_z)$ are equal to the input decoupling zeros of

$$\left(A - \bar{B}_c F_1, (\bar{B}_0 \quad \bar{B}_d \quad E) \right),$$

where F_1 is such that $\bar{B}_c F_1 = B F_{\text{pre}}$. As we can vary F_1 arbitrarily, it is obvious that the only eigenvalues that we cannot influence are the input decoupling zeros of

$$\left(A, (B \quad E) \right).$$

As noted in Remark 5.12 for the left-invertible case, these input-decoupling zeros are equal to the input-decoupling zeros of (A_z, B_z) and are hence given by the set $\mathit{\Omega}_{s1}$.

Hence it is obvious that when varying over the set $\boldsymbol{F}_s^{\text{pre}}$, the fixed modes are equal to $\mathit{\Omega}_s$. Note that fixing $\boldsymbol{F}^{\text{pre}}$ fixes some additional modes beyond $\mathit{\Omega}_s$, but these are part of the eigenvalues of A_{cc}^c, which we can assign arbitrarily.

Next, we need to show that if we have a state feedback $u = -Fx$ that achieves EDD but is not in the set $\boldsymbol{F}_s^{\text{pre}}$, then we cannot further reduce the class of fixed modes and the class of input decoupling zeros. We can decompose this feedback as

$$\Gamma_i^{-1} F x = \begin{pmatrix} F_0 \\ F_d \\ F_c \end{pmatrix} \tilde{x}.$$

Then, if we fix $u_c = -F_c \tilde{x}$, we obtain a new system with the remaining inputs u_0 and u_d. This is obviously similar to what we did before. However, the prefeedback that we uses in the set $\boldsymbol{F}_s^{\text{pre}}$ is constrained in that $A_{cc} - B_c F_1$ is stable (in the notation as used before). At this moment, we obtain a system for which we are not sure whether the x_c dynamics are stable. But for the rest the arguments as used before apply. It is again obvious that $\mathit{\Omega}_s^1$ is among the fixed modes, whereas the input-decoupling zeros of (A_z, B_z) must be stable because we know that after the prefeedback $u_c = -F_c \tilde{x}$, the feedback $u_0 = -F_0 \tilde{x}$, $u_d = -F_d \tilde{x}$ achieves stability. But it is obvious from the structure in (5.35) and (5.36) (with F_{pre} representing our prefeedback $u_c = -F_c \tilde{x}$) that the input-decoupling zeros must always contain as a subset the input-decoupling zeros from the set $\mathit{\Omega}_s^2$.   ∎

Remark 5.17 *We would like to emphasize that the characterizations of the fixed modes and fixed decoupling zeros as presented above includes the associated multiplicity structure of the eigenvalues as defined in Definition 3.11.*

Remark 5.18 *The characterizations of the fixed modes and fixed decoupling zeros can also be defined in terms of the original system parameters without use of* *the SCB. This applies equally to the left-invertible case and the non-left-invertible case. As we have seen in the above proof, we have*

$$\Omega_s^2 := \Lambda_s^2 := \left\{ \text{ input decoupling zeros of } \left(A, (B \quad E) \right) \right\}.$$

On the other hand, we have

$$\Omega_s^1 := \Lambda_s^1 := \{ \text{ invariant zeros of } (A, B, C, D) \}$$
$$- \left\{ \text{ invariant zeros of } \left(A, (B \quad E), C, (D \quad 0) \right) \right\}.$$

The first set on the right is always larger than the second set, so the difference is taken by leaving out the elements of the second set from the elements of the first set using multiplicities. This characterization is similar to the results obtained in [48]. Note that this characterization does not enable us to characterize the full multiplicity structure of an eigenvalue but only its algebraic multiplicity, whereas the characterization given in our algorithm also enables us to characterize the full multiplicity structure of the fixed modes.

As in Lemma 5.15, we would like to examine next the conditions for the uniqueness of a static state feedback controller that achieves EDD when Σ_{sub} is not left-invertible. In this regard, a close examination of the EDD_{nli} algorithm reveals that a static state feedback controller that achieves EDD is never unique whenever Σ_{sub} is not left-invertible. The following lemma asserts this.

Lemma 5.19 *Consider the given system Σ as in (5.1) with $(C \quad D)$ surjective. Assume that the pair (A, B) is $\mathbb{C}_g$-stabilizable, and that $\operatorname{im} E \subseteq \mathcal{V}_g(\Sigma_{\text{sub}})$. Then, the EDD state feedback law is unique if and only if the following conditions hold:*

(i) The subsystem Σ_{sub} is invertible.

(ii) The subsystem Σ_{sub} of Σ has all its invariant zeros in the set $\mathbb{C}_g$.

(iii) The matrix D is invertible.

(iv) The pair $(A - BD^{-1}C, E)$ is controllable.

Moreover, under the above conditions,

- $F_s = \{ -D^{-1}C \}$*, which is a singleton,*

- $\Lambda_s = \Omega_s = \{ \text{ Invariant zeros of } \Sigma_{\text{sub}} \} = \lambda(A - BD^{-1}C).$

Proof : From the EDD_{nli} algorithm, it is obvious that whenever Σ_{sub} is not left-invertible, an EDD state feedback law is always nonunique. The rest follows from Lemma 5.15. ∎

5.4.3   An algorithm for EDD with pole placement

We characterized already in the last two subsections the set of static state feedback controllers that achieve EDD. Also, we characterized and realized a method of constructing the associated sets of EDD fixed modes and the EDD fixed decoupling zeros. One can envision several ways of using such characterizations for the purpose of designing a control system that meets simultaneously a number of secondary specifications while preserving EDD property. In this subsection, we would like to illustrate the power of characterizing the sets F_s and Ω_s by considering the EDD problem with simultaneous pole placement via static state feedback.

We have the following definition.

Problem 5.20 Consider the EDD problem as defined by Problem 5.1 for a system Σ as in (5.1). Also, consider a set Λ of n self-conjugate elements in the set $\mathbb{C}_g$. Then, the exact disturbance decoupling problem via static state feedback with internal $\mathbb{C}_g$-stability and with simultaneous pole placement via static state feedback (abbreviated as EDDSPP) is a problem of finding for the given system Σ an EDD static state feedback controller with a simultaneous property of placing the resulting closed-loop poles at the desired locations as given by the set Λ.

We have the following theorem regarding the solvability of the above-defined problem.

Theorem 5.21 *Consider the system Σ as in (5.1). Also, consider a set Λ of n desired self-conjugate closed-loop poles all in the set $\mathbb{C}_g$. Then, for the given system Σ, the EDDSPP with Λ as the desired set of closed-loop poles, is solvable if and only if*

    (i) (A, B) is $\mathbb{C}_g$-stabilizable,

    (ii) im $E \subseteq \mathcal{V}_g(\Sigma_{\text{sub}})$, and

    (iii) $\Omega_s(A, B, C, D, E) \subseteq \Lambda$.

Proof : It is obvious.   ∎

Assuming that the conditions given in the above theorem are satisfied, our next task is to develop an algorithm of arriving at a controller that solves the EDDSPP for a given system Σ and for a given set Λ of n desired closed-loop poles. The algorithm we develop is called the *EDDSPP algorithm*. As depicted in Figure 5.3, the *EDDSPP algorithm* has two sets of inputs, one is the matrix quintuple (A, B, C, D, E) which characterizes the given EDD problem, and another is the set Λ of n desired self-conjugate closed-loop poles. The output of the *EDDSPP algorithm* is a static state feedback control gain F that solves the given EDDSPP. The *EDDSPP algorithm* fails if im $E \not\subseteq \mathcal{V}_g(\Sigma_{\text{sub}})$ and/or if $\boldsymbol{\Omega}_s(A, B, C, D, E) \not\subseteq \Lambda$ because in this case, there is no solution to the given EDDSPP.

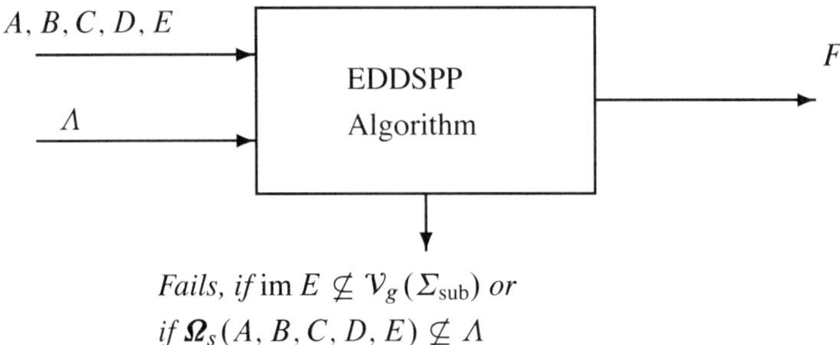

Figure 5.3: A block diagram interpretation of the *EDDSPP algorithm*

The *EDDSPP algorithm* is a special version of the EDD_{nli} *algorithm*; i.e., the *EDDSPP algorithm* is created from the EDD_{nli} *algorithm* by fixing certain parameters so that the closed-loop poles can be placed at certain desired locations. Let us expand on this. It is clear from the construction of EDD_{nli} *algorithm* that one can assign the closed-loop poles at any desired locations in $\mathbb{C}_g$ as long as the desired locations are symmetric with respect to the real axis (i.e., self-conjugate) and include the EDD fixed modes as given by $\boldsymbol{\Omega}_s(A, B, C, D, E)$. To do so, one properly assigns $\lambda(A_{cc} - B_c F_c) \cup \lambda(A_z - B_z F_z)$ by appropriately choosing F_c and F_z [see (5.23) and (5.29)]. This aspect of EDD_{nli} *algorithm* is the key in developing the *EDDSPP algorithm*, which is presented next.

The *EDDSPP algorithm*

Step 1:
Apply the EDD_{nli} *algorithm* to the quintuple (A, B, C, D, E), and compute the pair (A_z, B_z) as in (5.28). Then construct the set of EDD fixed modes, i.e., $\boldsymbol{\Omega}_s(A, B, C, D, E)$ as given by (5.34).
Step 2:
Check to verify that $\boldsymbol{\Omega}_s(A, B, C, D, E) \subseteq \Lambda$. If not, not all closed-loop poles can be assigned at the desired locations given by the set Λ. If

$$\boldsymbol{\Omega}_s(A, B, C, D, E) \subseteq \Lambda,$$

define Δ_s such that
$$\Delta_s \cup \boldsymbol{\Omega}_s(A, B, C, D, E) = \Lambda$$

and proceed to the next step.

Step 3:

Determine a gain F_c such that $\lambda(A_{cc} - B_c F_c)$ is contained in Δ_s. We note that this is always possible, but the choice of F_c is in most cases not unique.

Step 4:

With F_c as chosen in Step 3, use Step 1b of the EDD_{nli} *algorithm* to fix a prefeedback gain F_{pre} of the set $\boldsymbol{F}_{\mathrm{pre}}$ as given in (5.23).

Step 5:

Choose F_z such that

$$\lambda(A_{cc} - B_c F_c) \cup \lambda(\bar{A}_z - \bar{B}_z F_z) = \Delta_s$$

and together with F_{pre} as in Step 4, determine the static state feedback gain F using (5.30) and (5.31).

It is easy to verify that the static state feedback gain F determined by the above algorithm is in $\boldsymbol{F}_s(A, B, C, D, E)$, whereas the closed-loop poles are at the locations specified by Λ; i.e., $\lambda(A - BF) = \Lambda$.

Remark 5.22 *The above algorithm places closed-loop poles at certain desirable locations by exploiting the freedom available in selecting an appropriate state feedback gain via EDD_{nli} algorithm. By examining the EDD_{nli} algorithm, one also realizes that certain freedom exists in prescribing the closed-loop eigenvectors. Here we do not pursue making use of such a freedom.*

5.5 Dynamic state feedback laws and associated fixed modes and fixed decoupling zeros

Our goal in this section is to characterize and parameterize all EDD proper dynamic state feedback controllers $\boldsymbol{F}_d$ and their associated sets of EDD fixed modes $\boldsymbol{\Omega}_d$ as well as EDD fixed decoupling zeros $\boldsymbol{\Lambda}_d$. We first recall that an EDD proper dynamic state feedback controller exists if and only if the pair (A, B) is $\mathbb{C}_g$-stabilizable, and im $E \subseteq \mathcal{V}_g(\Sigma_{\mathrm{sub}})$. We note that these existence conditions are the same as those in the case when the $\mathbb{C}_g$-admissible controllers are of static state feedback type. Next, as in the previous section, we present our results separately for two different cases: (1) Σ_{sub} is left-invertible and (2) Σ_{sub} is not left-invertible.

5.5.1 Σ_{sub} *is left-invertible*

We assume throughout this subsection that Σ_{sub} is left-invertible. We start by constructing a set of parameterized controllers, which as we shall see is the same as the set of all EDD proper dynamic state feedback controllers $\boldsymbol{F}_d$. To do so, choose any fixed $F \in \boldsymbol{F}_s$, and define a set $\boldsymbol{Q}$ as*

$$\boldsymbol{Q} := \left\{ Q(s) \in \mathcal{RH}_\infty \mid Q(s) = W(s)(I - EE^\dagger)(sI - A + BF), \right.$$
$$\left. W(s) \in \mathcal{RH}_2 \right\}, \quad (5.37)$$

where, as usual, $E^\dagger$ is the Moore–Penrose generalized inverse of E. Next, define a set of parameterized controllers as

$$\Sigma_C : \quad \begin{cases} \dot{\xi} = (A - BF)\xi + By_1, \\ u = -Fx + y_1, \\ y_1 = Q(s)(x - \xi), \end{cases} \quad (5.38)$$

for $Q \in \boldsymbol{Q}$. We have the following theorem.

Theorem 5.23 *Consider the EDD problem with stability as defined by Problem 5.1 for a system Σ as in (5.1). Assume that the pair (A, B) is $\mathbb{C}_g$-stabilizable, and that* im $E \subseteq \mathcal{V}_g(\Sigma_{\text{sub}})$. *Moreover, let Σ_{sub} be left-invertible. Then, the set of parameterized controllers defined in (5.37) and (5.38) is equal to the set $\boldsymbol{F}_d$ as defined in Definition 5.3. That is, for any $W \in \mathcal{RH}_2$, the corresponding Σ_C is an EDD controller; moreover, any proper dynamic EDD controller can be written in the form of (5.38) for some $W \in \mathcal{RH}_2$.*

Proof : Assume that we have any proper dynamic state feedback controller that achieves EDD. Obviously we can parameterize all stabilizing controllers via the Youla parameterization. For our purposes, it is more convenient to use a variation of the standard Youla parameterization developed in [69]. In other words, because our controller is stabilizing, we are guaranteed that it can be written in the form (5.38) for some proper stable transfer matrix Q. We need to establish that the fact that the controller achieves EDD implies that $Q \in \boldsymbol{Q}$; i.e., Q is of the form given in (5.37) for some $W \in \mathcal{RH}_2$.

We first recall that F, being an element of the set $\boldsymbol{F}_s$, ensures that $A - BF$ is stable, and that

$$(C - DF)(sI - A + BF)^{-1}E \equiv 0. \quad (5.39)$$

Next, let a matrix quadruple (A_q, B_q, C_q, D_q) correspond to a state-space realization of Q. Then, after some simple algebraic manipulations, it follows that the controller (5.38) when applied to Σ_{sub} yields the closed-loop transfer function from w to z as

$$T_{zw}(F) = C_e(sI - A_e)^{-1}B_e, \quad (5.40)$$

---

*We recall that $\mathcal{RH}_\infty$ denotes the set of proper and stable rational matrices, whereas $\mathcal{RH}_2$ denotes the set of strictly proper and stable rational matrices.

where

$$A_e = \begin{pmatrix} A - BF & BC_q & BD_q \\ 0 & A_q & B_q \\ 0 & 0 & A - BF \end{pmatrix}, \quad B_e = \begin{pmatrix} E \\ 0 \\ E \end{pmatrix}, \tag{5.41}$$

and

$$C_e = \begin{pmatrix} C - DF & DC_q & DD_q \end{pmatrix}. \tag{5.42}$$

Then it is simple to verify that

$$\begin{aligned} T_{zw}(F) &= (C - DF)(sI - A + BF)^{-1}E \\ &\quad + [(C - DF)(sI - A + BF)^{-1}B + D]Q(s)(sI - A + BF)^{-1}E \\ &= \left[(C - DF)(sI - A + BF)^{-1}B + D\right]Q(s)(sI - A + BF)^{-1}E. \end{aligned} \tag{5.43}$$

By the assumption that Σ_{sub} is left-invertible, it follows that $T_{zw}(F) = 0$ if and only if

$$Q(s)(sI - A + BF)^{-1}E = 0. \tag{5.44}$$

Therefore, we know that in our case that (5.44) is satisfied. We define

$$W(s) = Q(s)(sI - A + BF)^{-1}$$

and we clearly have $W \in \mathcal{RH}_2$. Also, (5.44) implies that $W(s)E = 0$, and hence, we find that $W(s)(I - EE^\dagger) = Q(s)(sI - A + BF)^{-1}$, which implies that $Q(s) = W(s)(I - EE^\dagger)(sI - A + BF)$, and hence, $Q \in \mathcal{Q}$.

To prove the converse, we assume that we have a controller of the form (5.38) with $Q(s) \in \mathcal{Q}$. From the modified Youla parameterization, it is immediate that the controller is stabilizing. We note that if $Q \in \mathcal{Q}$, it is trivial to see that $Q(s)(sI - A + BF)^{-1}E = W(s)(I - EE^\dagger)E = 0$. Hence (5.44) holds. This in turn yields from (5.43) that the closed-loop transfer matrix is equal to zero, and hence, we achieve EDD. The proof is now complete. ∎

Remark 5.24 *It is worth noting that if* $(A_w, B_w, C_w, 0)$ *is a state-space realization of* W, *then the state-space realization of* Q *as in (5.37) is given by*

$$\left(A_w, [A_w B_w(I - EE^\dagger) - B_w(I - EE^\dagger)(A - BF)], C_w, C_w B_w(I - EE^\dagger)\right).$$

The following theorem characterizes the set of EDD fixed modes $\mathcal{Q}_d$ and the set of EDD fixed decoupling zeros Λ_d. In fact, it turns out that even by expanding the set of control laws from $\mathbb{C}_g$-admissible static state feedback laws to $\mathbb{C}_g$-admissible proper dynamic state feedback laws, the set of EDD fixed modes and hence the set of EDD fixed decoupling zeros remain the same.

Theorem 5.25 *Consider the EDD problem as defined by Problem 5.1 for a system Σ as in (5.1). Assume that the pair (A, B) is $\mathbb{C}_g$-stabilizable and that $\mathrm{im}\, E \subseteq \mathcal{V}_g(\Sigma_{\mathrm{sub}})$. Moreover, let Σ_{sub} be left-invertible. Then, we have*

$$\Omega_d = \Omega_s = \Lambda_d = \Lambda_s.$$

Proof : See Section 5.A.   ■

We have the following lemma regarding the uniqueness of a dynamic state feedback solution and the resulting Λ_s and Ω_s.

Lemma 5.26 *Consider the EDD problem as defined by Problem 5.1 for a system Σ as in (5.1). Assume that the pair (A, B) is $\mathbb{C}_g$-stabilizable and that $\mathrm{im}\, E \subseteq \mathcal{V}_g(\Sigma_{\mathrm{sub}})$. Moreover, let Σ_{sub} be left-invertible. Then, the EDD dynamic state feedback law is unique if and only if the following conditions hold:*

(i) The subsystem Σ_{sub} of Σ has all its invariant zeros in the set $\mathbb{C}_g$.

(ii) The subsystem Σ_{sub} is right-invertible.

(iii) The matrix D is invertible.

(iv) The pair $(A - BD^{-1}C, E)$ is controllable.

(v) $\mathrm{im}\, E = \mathbb{R}^n$.

Moreover, under the above conditions,

- *$F_d = \{-D^{-1}C\}$, which is a singleton,*

- *$\Lambda_s = \Omega_s = \{\,\text{Invariant zeros of } \Sigma_{\mathrm{sub}}\}$.*

Proof : The fact that F_d is a singleton implies that $Q(s) \equiv 0$. Then, we have $\mathrm{im}\, E = \mathbb{R}^n$. It is then simple to show that this and the condition

$$\mathrm{im}\, E \subseteq \mathcal{V}_g(\Sigma_{\mathrm{sub}})$$

imply the rest of the conditions (i) to (iv) of Lemma 5.26. Next, the converse part is obvious. Also, the remaining results follow directly from Lemma 5.15.   ■

Remark 5.27 *It is interesting to note that because of the extra condition $\mathrm{im}\, E = \mathbb{R}^n$, the condition under which an EDD proper dynamic state feedback controller is unique, is stronger than the condition for which an EDD static state feedback controller is unique (see Lemma 5.15).*

We would like to stress that a controller in a state-space representation is never unique in a strict sense. We can add stable unobservable or stable uncontrollable modes without affecting the fact that the system achieves stability and EDD. Finally, we can obviously apply basis transformations in the state space. In the above lemma, we discard these issues and consider two controllers identical when they have the same transfer matrix.

5.5.2 Σ_{sub} is not left-invertible

We assume throughout this subsection that Σ_{sub} is not left-invertible. Again, we start by constructing a set of parameterized controllers that, in this case, happens to be a subset of all EDD proper dynamic state feedback controllers $\boldsymbol{F}_d$. To do so, choose any fixed $F \in \boldsymbol{F}_s^{\mathrm{sub}}$, and define a set $\boldsymbol{\mathcal{Q}}$ as

$$\boldsymbol{\mathcal{Q}} := \big\{ Q(s) \in \mathcal{RH}_\infty \mid Q(s) = W(s)(I - EE^\dagger)(sI - A + BF),$$
$$W(s) \in \mathcal{RH}_2 \big\}. \quad (5.45)$$

Next, consider

$$\Sigma_C \quad : \quad \begin{cases} \dot{\xi} = (A - BF)\xi + By_1, \\ u = -Fx + y_1, \\ y_1 = Q(s)(x - \xi), \end{cases} \quad (5.46)$$

for some proper and stable $Q(s) \in \boldsymbol{\mathcal{Q}}$. We have the following theorem.

Theorem 5.28 *Consider the EDD problem as defined by Problem 5.1 for a system Σ as in (5.1). Assume that the pair (A, B) is $\mathbb{C}_g$-stabilizable and that im $E \subseteq \mathcal{V}_g(\Sigma_{\mathrm{sub}})$. Moreover, let Σ_{sub} be non-left-invertible. Then, any controller Σ_C in the set of parameterized controllers defined by (5.45) and (5.46) is an EDD controller.*

Proof : It is sufficient to show that the chosen controller Σ_C when applied to Σ solves the EDD problem. Again, let a matrix quadruple (A_q, B_q, C_q, D_q) correspond to a state-space realization of $Q(s)$. As in the proof of Theorem 5.23, when Σ_C is applied to Σ, after some simple algebraic manipulations, we obtain the closed-loop transfer function from w to z as given in (5.43). Next, for any $Q(s) \in \boldsymbol{\mathcal{Q}}$, it is trivial to see that

$$Q(s)(sI - A + BF)^{-1}E = W(s)(I - EE^\dagger)E = 0.$$

The proof is now complete. ∎

The following theorem characterizes the set of EDD fixed modes $\boldsymbol{\Omega}_d$ and the set of EDD fixed decoupling zeros $\boldsymbol{\Lambda}_d$. Again, as in the previous subsection, it turns out that $\boldsymbol{\Omega}_d = \boldsymbol{\Omega}_s = \boldsymbol{\Lambda}_d = \boldsymbol{\Lambda}_s$.

Theorem 5.29 *Consider the EDD problem with stability as defined by Problem 5.1 for a system Σ as in (5.1). Assume that the pair (A, B) is $\mathbb{C}_g$-stabilizable and that* im $E \subseteq \mathcal{V}_g(\Sigma_{\text{sub}})$. *Moreover, let Σ_{sub} be not left-invertible. Then, we have*

$$\Omega_d = \Omega_s = \Lambda_d = \Lambda_s.$$

Proof : It is left to the reader. ∎

As in Lemma 5.26, we would like to examine next the conditions for the uniqueness of a dynamic state feedback controller that achieves EDD when Σ_{sub} is not left-invertible. The following lemma addresses this issue.

Lemma 5.30 *Consider the EDD problem as defined by Problem 5.1 for a system Σ as in (5.1). Assume that the pair (A, B) is $\mathbb{C}_g$-stabilizable, and that* im $E \subseteq \mathcal{V}_g(\Sigma_{\text{sub}})$. *Then, the EDD dynamic state feedback law is unique if and only if the following conditions hold:*

(i) *The subsystem Σ_{sub} is invertible.*

(ii) *The subsystem Σ_{sub} of Σ has all its invariant zeros in the set $\mathbb{C}_g$.*

(iii) *The matrix D is invertible.*

(iv) *The pair $(A - BD^{-1}C, E)$ is controllable.*

(v) im $E = \mathbb{R}^n$.

Moreover, under the above conditions,

- $F_d = \{-D^{-1}C\}$, *which is a singleton,*

- $\Lambda_s = \Omega_s = \{$ *Invariant zeros of $\Sigma_{\text{sub}}\}$.*

Proof : From the EDD_{nli} algorithm, it is obvious that whenever Σ_{sub} is not left-invertible, an EDD state feedback law is always nonunique. The rest follows from Lemma 5.26. ∎

5.6 Static and dynamic full information feedback laws and associated fixed modes and fixed decoupling zeros

Theorem 5.10 discusses the solvability conditions of the EDD problem under static or dynamic full information feedback controllers. In this section, we determine the sets of all EDD static and dynamic full information feedback controllers $\boldsymbol{F}_{s,f}(A,B,C,D,E)$ and $\boldsymbol{F}_{d,f}(A,B,C,D,E)$ and the associated sets of EDD fixed modes $\boldsymbol{\Omega}_{s,f}(A,B,C,D,E)$ and $\boldsymbol{\Omega}_{d,f}(A,B,C,D,E)$. Note that for each type of controllers, the set of EDD fixed decoupling zeros coincides with the set of EDD fixed modes.

Following the line of thinking from the proof of Theorem 5.10, we can ascertain in what sense a preliminary feedback $u = Gw + v$ affects the sets of controllers as well as the sets of fixed modes. Let us first consider the case when Σ_{sub} is left-invertible. In this case, the matrix G is uniquely determined by the conditions that $DG = 0$ and

$$\mathrm{im}(E + BG) \subseteq \mathcal{V}_g(\Sigma_{\mathrm{sub}}). \tag{5.47}$$

Hence it follows immediately that

$$\boldsymbol{\Omega}_{s,f}(A,B,C,D,E) = \boldsymbol{\Omega}_s(A,B,C,D,E+BG), \tag{5.48a}$$

$$\boldsymbol{\Omega}_{d,f}(A,B,C,D,E) = \boldsymbol{\Omega}_d(A,B,C,D,E+BG), \tag{5.48b}$$

$$\boldsymbol{F}_{s,f}(A,B,C,D,E) = \boldsymbol{F}_s(A,B,C,D,E+BG), \tag{5.48c}$$

$$\boldsymbol{F}_{d,f}(A,B,C,D,E) = \boldsymbol{F}_d(A,B,C,D,E+BG). \tag{5.48d}$$

This leads to the following theorem.

Theorem 5.31 *Consider the continuous- or discrete-time system Σ as in (5.1). Assume that the subsystem Σ_{sub} characterized by the quadruple (A,B,C,D) is left-invertible. Also, assume that the EDD problem via static (or, equivalently, dynamic) full information feedback with internal $\mathbb{C}_g$-stability is solvable by a feedback of the form (5.7) [or of the form (5.9)]. Let G be a matrix that satisfies the conditions that $DG = 0$ and (5.47). Then, for any controller of the form (5.7) solving the EDD problem, (5.48) is true.*

Proof : As we said before stating the theorem, the theorem follows if a matrix G is uniquely determined by the conditions that $DG = 0$ and (5.47). The fact that the matrix G is unique follows from the fact that $B \ker D \subset \mathcal{S}(\Sigma_{\mathrm{sub}})$. If we have two matrices G_1 and G_2 such that $DG_1 = 0$ and $DG_2 = 0$ while

$$\mathrm{im}(E + BG_1) \subseteq \mathcal{V}_g(\Sigma_{\mathrm{sub}}) \quad \text{and} \quad \mathrm{im}(E + BG_2) \subseteq \mathcal{V}_g(\Sigma_{\mathrm{sub}}),$$

then

$$\operatorname{im} B(G_1 - G_2) \subseteq \mathcal{V}_g(\Sigma_{\mathrm{sub}}) \text{ and } D(G_1 - G_2) = 0.$$

But for a left-invertible system, we have that

$$B \ker D \cap \mathcal{V}_g(\Sigma_{\mathrm{sub}}) = \{0\}$$

and hence we must have $B(G_1 - G_2) = 0$. As for a left-invertible system $(B' \ D')$ must be surjective, we find that $G_1 = G_2$.   ∎

Let us consider next the case when Σ_{sub} is not left-invertible. Also, let

$$\mathcal{G} = \{G \mid DG = 0 \text{ and } (5.47) \text{ is true}\}.$$

Note that in this case, we can have more than one G that is an element of $\mathcal{G}$. That is, we can have more than one preliminary feedback of the type $u = Gw + v$ that results in a system for which the EDD problem with state feedback and internal stability is solvable. In particular, let two distinct matrices G_1 and G_2 be in $\mathcal{G}$. The question then is whether the resulting EDD problem has the same sets of fixed modes. In other words, we enquire whether the following equations are true:

$$\boldsymbol{\Omega}_s(A, B, C, D, E + BG_1) = \boldsymbol{\Omega}_s(A, B, C, D, E + BG_2),$$
$$\boldsymbol{\Omega}_d(A, B, C, D, E + BG_1) = \boldsymbol{\Omega}_d(A, B, C, D, E + BG_2).$$

Our enquiry shows that the above equations are true. The above leads to the following theorem.

Theorem 5.32 *Consider the continuous- or discrete-time system Σ as in (5.1). Assume that the subsystem Σ_{sub} characterized by the quadruple (A, B, C, D) is not left-invertible. Also, assume that the EDD problem via static (or, equivalently, dynamic) full information feedback with internal $\mathbb{C}_g$-stability is solvable by a feedback of the form (5.7) [or of the form (5.9)]. Then, the following hold:*

(i) Let a matrix G be any element of $\mathcal{G}$. We then have

$$\boldsymbol{\Omega}_{s,f}(A, B, C, D, E) = \boldsymbol{\Omega}_s(A, B, C, D, E + BG),$$
$$\boldsymbol{\Omega}_{d,f}(A, B, C, D, E) = \boldsymbol{\Omega}_d(A, B, C, D, E + BG).$$

(ii) We have

$$\boldsymbol{F}_{s,f}(A, B, C, D, E) = \bigcup_{G \in \mathcal{G}} \boldsymbol{F}_s(A, B, C, D, E + BG),$$
$$\boldsymbol{F}_{d,f}(A, B, C, D, E) = \bigcup_{G \in \mathcal{G}} \boldsymbol{F}_d(A, B, C, D, E + BG).$$

Proof : When Σ_{sub} is not left-invertible, the fact that

$$\text{im}(E + BG) \subseteq \mathcal{V}_g(\Sigma_{\text{sub}}),$$

implies that, in the SCB, we have

$$\Gamma_s^{-1}(E + BG) = \left((E_a^-)' \quad 0 \quad 0 \quad (E_c)' \quad 0 \right)'.$$

It is easily checked that given the conditions on G, different choices for G can only affect E_c whereas E_a^- is uniquely determined. It is then simply verified from the derivations of the fixed modes, fixed decoupling zeros and the EDD_{nli} algorithm that Ω_s and Ω_d are independent of the matrix E_c. The results of the theorem follow then immediately.

The characterization for $F_{s,f}(A, B, C, D, E)$ in part (ii) is trivial. Regarding the characterization of $F_{d,f}(A, B, C, D, E)$, we use the argument used earlier in the proof of Theorem 5.10 that an arbitrary dynamic full information feedback can be represented as a dynamic state feedback combined with a static disturbance feedback.   ∎

Remark 5.33 *Alternatively, the above equalities relating $\Omega_{s,f}$ and Ω_s immediately follow from Remark 5.18.*

5.A   Proofs of Theorems 5.11 and 5.25

In this section, we prove certain facts mentioned in the EDD_{li} *algorithm* and Theorems 5.11 and 5.25. The concept behind the EDD_{li} *algorithm* is simple and straightforward. By constructing the SCB of Σ_{sub}, one recognizes that a certain subsystem Σ_z, namely the one characterized by the pair (A_z, B_z), is not affected by the disturbance at all and as such can be separated from the rest of Σ_{sub}. Thus, denoting the gain which stabilizes Σ_z by F_z, and taking into account the interconnections between Σ_z and the rest of Σ_{sub}, one can easily parameterize the set of all static state feedback controllers F_s that solves the EDD problem for the given system Σ.

5.A.1   Proof of Theorem 5.11

Proof of Theorem 5.11 :
Part 1: It is straightforward to verify by some simple calculations that for any F of the form (5.17), we have $(C - DF)(sI - A + BF)^{-1}E \equiv 0$. It follows then that the control law $u = -Fx$ with F as in (5.17) achieves EDD.

Conversely, if a state feedback $u = -Fx$ achieves EDD, then obviously $(C - DF)(sI - A + BF)^{-1}E \equiv 0$. Without loss of generality but for simplicity of presentation, we assume that Σ_{sub} is in the form of SCB with A_{aa}^- partitioned as in Step 1b of the EDD_{li} *algorithm*. Let us define $\mathcal{V}_s := \langle A - BF \mid \mathrm{im}\, E \rangle$; i.e., $\mathcal{V}_s$ is the smallest $(A - BF)$-invariant subspace containing $\mathrm{im}\, E$. Thus, $\mathcal{V}_s \subseteq \ker(C - DF)$, and by definition $\mathcal{V}_s \subseteq \mathcal{V}_g(\Sigma_{\mathrm{sub}})$, which is given by

$$\mathcal{V}_g(\Sigma_{\mathrm{sub}}) = \mathrm{span}\left\{ \begin{pmatrix} I & 0 \\ 0 & I \\ 0 & 0 \\ 0 & 0 \\ 0 & 0 \end{pmatrix} \right\}.$$

Hence, a similarity transformation T exists such that

$$T^{-1}(A - BF)T = \begin{pmatrix} A_{cc} & A_{c\bar{c}} \\ 0 & A_{\bar{c}\bar{c}} \end{pmatrix}, \quad T^{-1}E = \begin{pmatrix} E_c \\ 0 \end{pmatrix}, \tag{5.49}$$

and

$$(C - DF)T = \begin{pmatrix} 0 & C_{\bar{c}} \end{pmatrix}, \quad V_s = \mathrm{im}\left\{ T\begin{pmatrix} I \\ 0 \end{pmatrix} \right\}, \tag{5.50}$$

where (A_{cc}, E_c) is controllable. It is now straightforward to verify that T can be chosen as the following form:

$$T = \begin{pmatrix} T_* & 0 & 0 & 0 \\ 0 & I & 0 & 0 \\ 0 & 0 & I & 0 \\ 0 & 0 & 0 & I \end{pmatrix}, \tag{5.51}$$

where T_* is of dimension $\dim \mathcal{V}_g(\Sigma_{\mathrm{sub}}) \times \dim \mathcal{V}_g(\Sigma_{\mathrm{sub}})$. Let

$$F = \begin{pmatrix} C_0 \\ 0 \end{pmatrix} + \begin{pmatrix} F_{a0}^1 & F_{a0}^2 & F_{a0}^{0+} & F_{b0} & F_{d0} \\ F_{ad}^1 & F_{ad}^2 & F_{ad}^{0+} & F_{bd} & F_{dd} \end{pmatrix}. \tag{5.52}$$

We note that (5.49)–(5.52) imply that

$$T_*^{-1}\begin{pmatrix} A_{aa}^{11} - B_{a0}^1 F_{a0}^1 & A_{aa}^{12} - B_{a0}^1 F_{a0}^2 \\ -B_{a0}^2 F_{a0}^1 & A_{aa}^{22} - B_{a0}^2 F_{a0}^2 \end{pmatrix}T_* = \begin{pmatrix} A_{cc} & A_{c\bar{c}}^* \\ 0 & A_{\bar{c}\bar{c}}^* \end{pmatrix} \tag{5.53}$$

and

$$T_*^{-1}\begin{pmatrix} E_a^1 \\ 0 \end{pmatrix} = \begin{pmatrix} E_c \\ 0 \end{pmatrix}. \tag{5.54}$$

We know that the set $\mathcal{V}_s$ is invariant, and hence, states in $\mathcal{V}_s$ do not affect the state x_d. Moreover we have $\mathcal{V}_s \subset \ker(C - DF)$. Hence we need

$$\begin{pmatrix} -F_{a0}^1 & -F_{a0}^2 \\ B_d(E_{da}^1 - F_{ad}^1) - B_{d0}F_{a0}^1 & B_d(E_{da}^2 - F_{ad}^2) - B_{d0}F_{a0}^2 \end{pmatrix} T_* = \begin{pmatrix} 0 & \star \end{pmatrix},$$

(5.55)

where again $\star$ denotes a matrix of not much interest. Here we note that (5.53))–(5.55) imply that the system characterized by the matrix quadruple,

$$\left(\begin{pmatrix} A_{aa}^{11} & A_{aa}^{12} \\ 0 & A_{aa}^{22} \end{pmatrix}, \begin{pmatrix} E_a^1 \\ 0 \end{pmatrix}, \begin{pmatrix} -F_{a0}^1 & -F_{a0}^2 \\ B_d(E_{da}^1 - F_{ad}^1) & B_d(E_{da}^2 - F_{ad}^2) \end{pmatrix}, 0 \right), \quad (5.56)$$

has a transfer matrix equal to 0. Then the controllability of the pair (A_{aa}^{11}, E_a^1) implies that $F_{a0}^1 \equiv 0$. Noting that B_d is injective, it follows then that $F_{ad}^1 = E_{da}^1$ and thus F must be of the form given by (5.17).

Part 2: It follows simply from the construction of the EDD_{li} algorithm.

Part 3: As noted, the fact that a controller achieves disturbance decoupling implies that the closed-loop transfer matrix is equal to zero. Hence, we immediately find that the EDD zeros must be equal to the set of EDD fixed modes. Hence, the result follows. $\blacksquare$

5.A.2   Proof of Theorem 5.25

By the definition, it is trivial to see that $\mathit{\Omega}_d \subseteq \mathit{\Omega}_s$ and $\Lambda_d \subseteq \Lambda_s$.

   Conversely, let us consider any given EDD dynamic controller with the state-space realization:

$$\Sigma_C \; : \; \begin{cases} \dot{v} = A_{\mathrm{cmp}}\, v + B_{\mathrm{cmp}}\, x \\ u = C_{\mathrm{cmp}}\, v + D_{\mathrm{cmp}}\, x. \end{cases} \qquad (5.57)$$

As Σ_C is an EDD controller, we have $T_{zw}(\Sigma \times \Sigma_C) = 0$. Then it is simple to verify that

$$u = -F_{au}\, x_{au} := -\begin{pmatrix} C_{\mathrm{cmp}} & D_{\mathrm{cmp}} \end{pmatrix} x_{au} \qquad (5.58)$$

is an EDD static state feedback law for the following auxiliary system:

$$\Sigma_{au} \; : \; \begin{cases} \dot{x}_{au} = A_{au}\, x_{au} + B_{au}\, u + E_{au}\, w \\ z_{au} = C_{au}\, x_{au} + D_{au}\, u, \end{cases} \qquad (5.59)$$

where

$$x_{au} = \begin{pmatrix} v \\ x \end{pmatrix}, \quad A_{au} = \begin{pmatrix} A_{\mathrm{cmp}} & B_{\mathrm{cmp}} \\ 0 & A \end{pmatrix}, \quad B_{au} = \begin{pmatrix} 0 \\ B \end{pmatrix}, \quad E_{au} = \begin{pmatrix} 0 \\ E \end{pmatrix},$$

and

$$C_{au} = \begin{pmatrix} 0 & C \end{pmatrix}, \quad D_{au} = D.$$

We first observe that the input decoupling zeros of the pair (A, B) are also among the input decoupling zeros of (A_{au}, B_{au}). Next, without loss of generality, we again assume that the matrix quadruple (A, B, C, D) is in the form of SCB with A_{aa}^- partitioned as in Step 1b of the EDD_{li} algorithm; i.e., we have

$$
A_{au} - \begin{pmatrix} 0 & 0 \\ 0 & B_0 C_0 \end{pmatrix}
$$

$$
= \begin{pmatrix}
A_{cmp} & B_{cmp}^1 & B_{cmp}^2 & B_{cmp}^{0+} & B_{cmp}^b & B_{cmp}^d \\
0 & A_{aa}^{11} & A_{aa}^{12} & 0 & L_{ab}^1 C_b & L_{ad}^1 C_d \\
0 & 0 & A_{aa}^{22} & 0 & L_{ab}^2 C_b & L_{ad}^2 C_d \\
0 & 0 & 0 & A_{aa}^{0+} & L_{ab}^{0+} C_b & L_{ad}^{0+} C_d \\
0 & 0 & 0 & 0 & A_{bb} & L_{bd} C_d \\
0 & B_d E_{da}^1 & B_d E_{da}^2 & B_d E_{da}^{0+} & B_d E_{db} & A_{dd}
\end{pmatrix},
$$

$$
B_{au} = \begin{pmatrix}
0 & 0 \\
B_{a0}^1 & 0 \\
B_{a0}^2 & 0 \\
B_{a0}^{0+} & 0 \\
B_{b0} & 0 \\
B_{d0} & B_d
\end{pmatrix}, \quad
E_{au} = \begin{pmatrix}
0 \\
E_a^1 \\
0 \\
0 \\
0 \\
0
\end{pmatrix},
$$

$$
C_{au} = \begin{pmatrix}
0 & C_{0a}^1 & C_{0a}^2 & C_{0a}^{0+} & C_{0b} & C_{0d} \\
0 & 0 & 0 & 0 & 0 & C_d \\
0 & 0 & 0 & 0 & C_b & 0
\end{pmatrix}, \quad
D_{au} = \begin{pmatrix}
I_{m_0} & 0 \\
0 & 0 \\
0 & 0
\end{pmatrix}.
$$

Then following the proof of the SCB Theorem (see Sannuti and Saberi [81], Appendix A.2) and some simple algebra, one can compute a nonsingular state transformation Γ_{au} such that

$$
\Gamma_{au}^{-1}\left(A_{au} - \begin{pmatrix} 0 \\ B_0 \end{pmatrix} \begin{pmatrix} 0 & C_0 \end{pmatrix} \right) \Gamma_{au}
$$

$$
= \begin{pmatrix}
A_{cmp}^{a+} & 0 & 0 & 0 & \star & L_{ab}^{a+} C_b & L_{ad}^{a+} C_d \\
0 & A_{cmp}^{a-} & \star & \star & 0 & L_{ab}^{a-} C_b & L_{ad}^{a-} C_d \\
0 & 0 & A_{aa}^{11} & A_{aa}^{12} & 0 & L_{ab}^1 C_b & L_{ad}^1 C_d \\
0 & 0 & 0 & A_{aa}^{22} & 0 & L_{ab}^2 C_b & L_{ad}^2 C_d \\
0 & 0 & 0 & 0 & A_{aa}^{0+} & L_{ab}^{0+} C_b & L_{ad}^{0+} C_d \\
0 & 0 & 0 & 0 & 0 & A_{bb} & L_{bd} C_d \\
0 & 0 & B_d E_{da}^1 & B_d E_{da}^2 & B_d E_{da}^{0+} & B_d E_{db} & A_{dd}
\end{pmatrix},
$$

$$
\Gamma_{au}^{-1} B_{au} =
\begin{pmatrix}
B_{a0}^{a+} & 0 \\
B_{a0}^{a-} & 0 \\
B_{a0}^{1} & 0 \\
B_{a0}^{2} & 0 \\
B_{a0}^{0+} & 0 \\
B_{b0} & 0 \\
B_{d0} & B_{d}
\end{pmatrix}, \qquad
\Gamma_{au}^{-1} E_{au} =
\begin{pmatrix}
0 \\
0 \\
E_{a}^{1} \\
0 \\
0 \\
0 \\
0
\end{pmatrix},
$$

and

$$
C_{au} \Gamma_{au} =
\begin{pmatrix}
0 & 0 & C_{0a}^{1} & C_{0a}^{2} & C_{0a}^{0+} & C_{0b} & C_{0d} \\
0 & 0 & 0 & 0 & 0 & 0 & C_{d} \\
0 & 0 & 0 & 0 & 0 & C_{b} & 0
\end{pmatrix},
$$

where $\lambda(A_{\mathrm{cmp}}^{a+}) \subset \mathbb{C}^{+} \cup \mathbb{C}^{0}$ and $\lambda(A_{\mathrm{cmp}}^{a-}) \subset \mathbb{C}_{g}$. Moreover, the transformed system is in the form of SCB. Then following the same line of reasoning as in Part 1 of the proof of Theorem 5.11 in Subsection 5.A.1, it can be shown that $T_{zw}(\Sigma \times \Sigma_{C}) = 0$ implies that

$$
F_{au} \Gamma_{au} := \begin{pmatrix} C_{\mathrm{cmp}} & D_{\mathrm{cmp}} \end{pmatrix} \Gamma_{au} :=
\begin{pmatrix}
\star & \star & C_{0a}^{1} & \star & \star & \star & \star \\
\star & \star & E_{da}^{1} & \star & \star & \star & \star
\end{pmatrix},
$$

and hence $\boldsymbol{\Omega}_{s} \subseteq \lambda(A_{au} - B_{au} F_{au})$. Similarly, following the same arguments as in Part 3 of the proof of Theorem 5.11 of Subsection 5.A.1, one can show that the elements of $\boldsymbol{\Lambda}_{s}$ are among the decoupling zeros of the closed-loop system. As Σ_{C} can be any member of $\boldsymbol{F}_{d}$, it follows that $\boldsymbol{\Omega}_{s} \supseteq \boldsymbol{\Omega}_{d}$ and $\boldsymbol{\Lambda}_{s} \supseteq \boldsymbol{\Lambda}_{d}$. The proof of Theorem 5.25 is now complete.

6

Almost disturbance decoupling via state and full information feedback

6.1 Introduction

The objective of the exact disturbance decoupling (EDD) problem studied in Chapter 5 was to find a controller such that the controlled output was completely decoupled from an exogenous disturbance signal. In many cases, the EDD problem is not solvable and the next case to investigate is the almost disturbance decoupling (ADD) problem whose objective is to find a sequence of controllers such that the controlled output can be arbitrarily well decoupled from the disturbance by choosing an appropriate member of this sequence. To our knowledge, this problem was first introduced by J. L. Willems in [105]. A solution for this problem by state feedback was later obtained by J. C. Willems in [106, 107]. Many researchers have contributed to the understanding of ADD. In [72], the solvability conditions for different versions of the ADD problem can be found for the most general case.

Our first concern here is to formulate a precise criterion for ADD. EDD is unambiguous, but in the case of ADD, we need to show in what sense the coupling between the disturbance and the controlled output is small. Two criteria have been studied in the literature:

- In H_2 *almost disturbance decoupling* (H_2 ADD), we try to make the variance of the controlled output arbitrarily small given an exogenous disturbance that is a white noise stochastic process. This is equivalent to trying to make the H_2 norm of the transfer matrix from the disturbance to the controlled output arbitrarily small.

- In H_∞ *almost disturbance decoupling* (H_∞ ADD), we try to make the power of the controlled output arbitrarily small for all exogenous disturbances whose power is bounded by 1. This is equivalent to trying to make the H_∞ norm of the transfer matrix from the disturbance to the controlled output arbitrarily small.

Also, in connection with ADD, a large difference exists between continuous- and discrete-time systems. In EDD, the difference is marginal and we only have to use the appropriate stability domain, either $\mathbb{C}^-$ or $\mathbb{C}^\ominus$. In ADD, we will see that there are intrinsic differences between the continuous and discrete time because of the lack of a high gain design in discrete time.

Saberi and Sannuti and their coworkers introduced in [45,46] direct methods of design for H_2 ADDPS, rather than Riccati-based design, in order to better use the flexibility available in design. For H_∞ ADDPS while excluding imaginary axis invariant zeros, these direct methods were initially derived in [55,56]. The most general results were presented in [13,14].

In this chapter we first recall from literature the solvability conditions for ADD, and then we develop design procedures to determine the sequences of controllers that achieve ADD with an added feature of assigning certain finite closed-loop poles.

6.2   Problem formulation

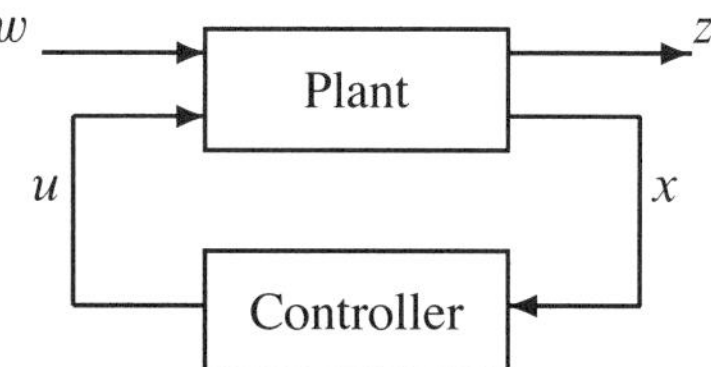

Figure 6.1: Closed-loop system $\Sigma \times \Sigma_{\mathrm{c}}$

We are concerned here with the ADD problem via state feedback with internal $\mathbb{C}_g$-stability (normally coined in the literature as ADDPS) and with the ADD problem via full information feedback with internal $\mathbb{C}_g$-stability. Figure 6.1 depicts a closed-loop system comprising a plant and a state feedback controller. The plant is described by the system Σ:

$$\Sigma : \quad \begin{cases} \sigma x = Ax + Bu + Ew \\ z \;= Cx + Du, \end{cases} \tag{6.1}$$

where, as before, σ is an operator indicating the time derivative $\frac{d}{dt}$ for continuous-time systems and a forward unit time shift for discrete-time systems.

Here as usual $x \in \mathbb{R}^n$ is a state, $u \in \mathbb{R}^m$ is a control input, $w \in \mathbb{R}^\ell$ is an exogenous disturbance input, and $z \in \mathbb{R}^q$ is a controlled output. The given plant or system Σ is characterized by the matrix quintuple (A, B, C, D, E). In what follows, a subsystem of Σ characterized by the matrix quadruple (A, B, C, D) will play a significant role and we will denote it by Σ_{sub}.

The controller Σ_C can be a static or dynamic state feedback controller. We introduce next a family of *dynamic state feedback controllers* parameterized in ε

and given by

$$\Sigma_C^\varepsilon \ : \ \begin{cases} \sigma v = J^\varepsilon v + L^\varepsilon x \\ u \ = M^\varepsilon v + N^\varepsilon x. \end{cases} \tag{6.2}$$

Even though it is a slight abuse of notation, a member of the above family as well as a sequence of controllers are both denoted by Σ_C^ε. The closed-loop system $\Sigma \times \Sigma_C^\varepsilon$, i.e., the interconnection of the given system Σ and a controller Σ_C^ε, can then be described by

$$\Sigma \times \Sigma_C^\varepsilon \ : \ \begin{cases} \sigma x_{cl} = A_{cl}^\varepsilon x_{cl} + E_{cl} w \\ z \ \ \ = C_{cl}^\varepsilon x_{cl}, \end{cases} \tag{6.3}$$

where

$$x_{cl} = \begin{pmatrix} x \\ v \end{pmatrix}, \quad A_{cl}^\varepsilon = \begin{pmatrix} A + BN^\varepsilon & BM^\varepsilon \\ L^\varepsilon & J^\varepsilon \end{pmatrix}, \quad E_{cl} = \begin{pmatrix} E \\ 0 \end{pmatrix},$$

and

$$C_{cl}^\varepsilon = \begin{pmatrix} C + DN^\varepsilon & DM^\varepsilon \end{pmatrix}.$$

Thus, the closed-loop transfer function from the disturbance w to the controlled output z is given by

$$T_{zw}(\Sigma \times \Sigma_C^\varepsilon) = C_{cl}^\varepsilon (\sigma I - A_{cl}^\varepsilon)^{-1} E_{cl}. \tag{6.4}$$

A special case of the controllers given in (6.2) is given by a family of *static state feedback controllers* that is of the form:

$$\Sigma_C^\varepsilon : u = -F^\varepsilon x. \tag{6.5}$$

Whenever a family of static feedback controller is used, $T_{zw}(\Sigma \times \Sigma_C^\varepsilon)$ is given by

$$T_{zw}(\Sigma \times \Sigma_C^\varepsilon) = (C - DF^\varepsilon)(\sigma I - A + BF^\varepsilon)^{-1} E. \tag{6.6}$$

We will also consider another type of family of controllers, namely *static full information feedback controllers*. In static full information feedback controllers, in addition to static state feedback, one also uses static feedback from the disturbance w whenever it is available. That is, static full information feedback controllers are of the form:

$$\Sigma_C^\varepsilon : u = -F^\varepsilon x + G^\varepsilon w. \tag{6.7}$$

The closed-loop transfer function when a family of static full information feedback controllers is used is given by

$$T_{zw}(\Sigma \times \Sigma_C^\varepsilon) = (C - DF^\varepsilon)(\sigma I - A + BF^\varepsilon)^{-1}(BG^\varepsilon + E) + DG^\varepsilon. \tag{6.8}$$

We introduced above three different architectures for controllers, dynamic state feedback, static state feedback, and full information feedback architectures. For any such architecture of controllers, we are now ready to formulate formally the ADD problem as given below.

Problem 6.1 Consider a continuous- or discrete-time system Σ given by (6.1). Also, consider any one of the three different architectures for controllers, dynamic state feedback, static state feedback, or full information feedback architectures. For a given architecture of controllers, the H_2 (H_∞) **almost disturbance decoupling problem with internal $\mathbb{C}_g$-stability** for Σ, denoted by H_2 ADD (H_∞ ADD) problem, is the problem of finding, if it exists, a sequence of controllers Σ_C^ε, parameterized in ε, such that the following hold:

(*i*) The resulting closed-loop system is internally stable for all ε.

(*ii*) The H_2 (H_∞) norm of the corresponding closed-loop transfer function $T_{zw}(\Sigma \times \Sigma_C^\varepsilon)$ converges to zero as $\varepsilon \to 0$.

In what follows, whenever we are discussing H_2 ADD and H_∞ ADD together, we simply call them ADD problems. However, when we need to emphasize only one of them, we quote it explicitly. Also, a sequence of state feedback controllers that solves an ADD problem is called a sequence of ADD controllers.

We study in this chapter several issues associated with the ADD problems. The first issue is the existence of sequences of controllers that solve them. The next issue, whenever a sequence of ADD controllers exists, concerns the limiting behavior of the closed-loop poles as $\varepsilon \to 0$. Expectedly, reminiscent of classic root-loci theory, as $\varepsilon \to 0$, some closed-loop poles go to finite locations in the complex plane, and these can be called the finite asymptotic modes, whereas the other closed-loop poles in the case of continuous-time systems may go to infinity (for discrete-time systems, in view of the stability requirement, all poles must be restricted to the interior of the unit circle and, hence, all asymptotic modes must be finite). This leads us to ask whether the requirements of ADD, namely the closed-loop stability for ε small enough and the H_2 (H_∞) norm of the closed-loop transfer function $T_{zw}(\Sigma \times \Sigma_C^\varepsilon)$ converging to zero as $\varepsilon \to 0$, impose any constraints on the finite asymptotic modes. Although similar questions can be posed on the limitations and constraints imposed on the modes that go to infinity, this issue is not yet completely resolved even though in [45, 46] important progress has been made. Our design will limit the number of modes that go to infinity. However, we will not achieve a minimum number of eigenvalues that go to infinity. Also the rate and directional information of the modes going to infinity (although solid results are available) will not be discussed.

Regarding the finite asymptotic modes, it turns out that some of them are *fixed* whereas others can be freely assigned by an appropriate choice of the sequence of ADD controllers. Here the word *fixed* that qualifies the asymptotic finite modes emphasizes that the location of these finite asymptotic modes is the *same* for all sequences of ADD controllers. Obviously, two attributes of the finite asymptotic fixed modes exist, namely their number and their locations. Both of these attributes are integrated in the set of all finite asymptotic fixed modes, which is formally defined below.

Definition 6.2 (*Finite asymptotic fixed modes of ADD controllers*) *Consider a continuous- or discrete-time system Σ given by (6.1) and characterized by the quintuple (A, B, C, D, E). Also, consider any one of the three different architectures for controllers: dynamic state feedback, static state feedback, and full information feedback architectures as well as the associated ADD problems. Then, for a given architecture of controllers, a scalar $\lambda \in \mathbb{C}^-$ (continuous time) or $\lambda \in \mathbb{C}^\ominus$ (discrete time) is said to be an H_2 (H_∞) ADD finite asymptotic fixed mode with algebraic multiplicity α if for every parameterized sequence of controllers Σ_C^ε that solves the H_2 ADD (H_∞ ADD), poles λ_i^ε, $i = 1, 2, \cdots, \alpha$, of the closed-loop system $\Sigma \times \Sigma_C^\varepsilon$ exist that converge to λ as ε tends to zero. The sets of all ADD finite asymptotic fixed modes (including multiplicities) corresponding to the sequences of dynamic, static, and full information feedback controllers are, respectively, denoted by $\boldsymbol{\Omega}_d^2$, $\boldsymbol{\Omega}_s^2$, and $\boldsymbol{\Omega}_{s,f}^2$ for H_2 ADD, and by $\boldsymbol{\Omega}_d^\infty$, $\boldsymbol{\Omega}_s^\infty$, and $\boldsymbol{\Omega}_{s,f}^\infty$ for H_∞ ADD.*

Sometimes we want to indicate the dependence of the sets $\boldsymbol{\Omega}_s^2$, $\boldsymbol{\Omega}_d^2$, etc., on the quintuple (A, B, C, D, E) explicitly and we use, for instance, $\boldsymbol{\Omega}_s^2(A, B, C, D, E)$ instead of $\boldsymbol{\Omega}_s^2$.

Remark 6.3 *Obviously by enlarging the class of feedback controllers, one can never enlarge the set of fixed modes. Thus, we note that*

$$\boldsymbol{\Omega}_d^2(A, B, C, D, E) \subseteq \boldsymbol{\Omega}_s^2(A, B, C, D, E),$$
$$\boldsymbol{\Omega}_{s,f}^2(A, B, C, D, E) \subseteq \boldsymbol{\Omega}_s^2(A, B, C, D, E),$$
$$\boldsymbol{\Omega}_d^\infty(A, B, C, D, E) \subseteq \boldsymbol{\Omega}_s^\infty(A, B, C, D, E),$$
$$\boldsymbol{\Omega}_{s,f}^\infty(A, B, C, D, E) \subseteq \boldsymbol{\Omega}_s^\infty(A, B, C, D, E).$$

One of our interests in this chapter is to determine the sets defined above. Not surprisingly, yet another of our interests in this chapter is to develop an algorithm of constructing a sequence or sequences of ADD controllers with a feature of assigning those finite asymptotic modes that are not fixed.

We enumerate below all tasks we formulated and outlined so far for H_2 ADD and H_∞ ADD:

(i) to determine the solvability conditions,

(ii) to capture the set of ADD finite asymptotic fixed modes,

(iii) to develop an algorithm for constructing a sequence or sequences of ADD controllers with flexibility in assigning those finite asymptotic modes that are not fixed.

In fact, for each H_2 ADD and H_∞ ADD, the algorithm we develop here to construct a sequence or sequences of ADD controllers also incorporates in itself the determination of the set of ADD finite asymptotic fixed modes.

6.3 Solvability conditions for ADD

In this section, we recall the solvability conditions for H_2 and H_∞ ADD. We do so first for continuous-time systems and then for discrete-time systems.

6.3.1 Solvability conditions for ADD—continuous time

In this subsection, for continuous-time systems, we recall the conditions (see, for instance, [72]) under which ADD can be achieved. These conditions are stated in terms of two subspaces from the geometric theory for linear systems, namely the $\mathbb{C}_g$-stabilizable weakly unobservable subspace $\mathcal{V}_g(\Sigma_*)$ and the $\mathbb{C}_g$-detectable strongly controllable subspace $\mathcal{S}_g(\Sigma_*)$ as defined in Definition 3.28.

The following two theorems consider H_2 ADD.

Theorem 6.4 *Consider the continuous-time system Σ as given by (6.1) and the associated H_2 ADD problem as defined by Problem 6.1. Also, consider dynamic state feedback controllers of the form (6.2) as well as static state feedback controllers of the form (6.5). Then, the following three statements are equivalent:*

(i) The H_2 ADD problem is solvable via static state feedback controllers.

(ii) The H_2 ADD problem is solvable via dynamic state feedback controllers.

(iii) (A, B) is $\mathbb{C}^-$-stabilizable and

$$\operatorname{im} E \subseteq \mathcal{S}^*(\Sigma_{\mathrm{sub}}) + \mathcal{V}^{-0}(\Sigma_{\mathrm{sub}}). \tag{6.9}$$

Theorem 6.5 *Consider the continuous-time system Σ as given in (6.1) and the associated H_2 ADD problem as defined by Problem 6.1, but this time using static full information feedback controllers of the form (6.7). Then, the H_2 ADD problem for Σ is solvable by a sequence of static full information feedback controllers if and only if the pair (A, B) is $\mathbb{C}^-$-stabilizable, and*

$$\operatorname{im} E \subseteq \mathcal{S}^*(\Sigma_{\mathrm{sub}}) + \mathcal{V}^{-0}(\Sigma_{\mathrm{sub}}) + B \ker D. \tag{6.10}$$

Although the above two theorems pertain to H_2 ADD, the following two theorems consider H_∞ ADD.

Theorem 6.6 *Consider the continuous-time system Σ as given by (6.1) and the associated H_∞ ADD problem as defined by Problem 6.1. Also, consider dynamic state feedback controllers of the form (6.2) as well as static state feedback controllers of the form (6.5). Then, the following three statements are equivalent:*

(i) The H_∞ ADD problem is solvable via static state feedback controllers.

(ii) The H_∞ ADD problem is solvable via dynamic state feedback controllers.

(iii) (A, B) is $\mathbb{C}^-$-stabilizable,

$$\operatorname{im} E \subseteq \mathcal{S}^*(\Sigma_{\mathrm{sub}}) + \mathcal{V}^{-0}(\Sigma_{\mathrm{sub}}), \tag{6.11}$$

and for any $\varepsilon > 0$ and any invariant zero s_0 of (A, B, C, D) on the imaginary axis, a matrix F exists such that $s_0 I - A + BF$ is invertible and

$$\|(C - DF)(s_0 I - A + BF)^{-1} E\| < \varepsilon.$$

Remark 6.7 *In [12, 85], a different condition for the solvability of H_∞ ADD has been given. The H_∞ ADD is solvable if and only if*

$$\operatorname{im} E \subseteq \left(\mathcal{S}^*(\Sigma_{\mathrm{sub}}) + \mathcal{V}^{-0}(\Sigma_{\mathrm{sub}})\right) \cap \{\cap_{\lambda \in \mathbb{C}^0} \mathcal{S}_\lambda(\Sigma_{\mathrm{sub}})\}. \tag{6.12}$$

The subspace on the right has been studied in the context of the SCB in Chapter 3; see (3.41) and the surrounding discussion. Obviously this alternative solvability condition can be shown to be equivalent to the condition presented in Theorem 6.6.

Remark 6.8 *The last condition in (iii) can be rewritten, and we can state equivalently that for any invariant zero s_0 of (A, B, C, D) on the imaginary axis, we must have*

$$\operatorname{im} C(s_0 I - A)^{-1} E \subseteq \operatorname{im} \left[C(s_0 I - A)^{-1} B + D \right].$$

However, this latter equivalent condition can only be used when $s_0 I - A$ is invertible. Otherwise, we can first apply a preliminary state feedback, $u = -NX + \bar{u}$, such that $A - BN$ has no eigenvalues on the imaginary axis, and then the condition becomes that for any invariant zero s_0 of (A, B, C, D) on the imaginary axis, we must have

$$\operatorname{im}(C - DN)(s_0 I - A + BN)^{-1} E \subseteq$$
$$\operatorname{im} \left[(C - DN)(s_0 I - A + BN)^{-1} B + D \right].$$

Remark 6.9 *Both Theorems 6.4 and 6.6 point out that the necessary and sufficient conditions for achieving ADD are the same whether static or dynamic state feedback controllers are used.*

Theorem 6.10 *Consider the continuous-time system Σ as given in (6.1) and the associated H_∞ ADD problem as defined by Problem 6.1 but this time using static* *full information feedback controllers of the form (6.7). Then, the H_∞ ADD problem for Σ is solvable by a sequence of static full information feedback controllers if and only if the pair (A, B) is $\mathbb{C}^-$-stabilizable,*

$$\operatorname{im} E \subseteq \mathcal{S}^*(\Sigma_{\text{sub}}) + \mathcal{V}^{-0}(\Sigma_{\text{sub}}) + B \ker D, \tag{6.13}$$

and for any $\varepsilon > 0$ and any invariant zero s_0 of (A, B, C, D) on the imaginary axis, a matrix F exists such that $s_0 I - A + BF$ is invertible and

$$\|(C - DF)(s_0 I - A + BF)^{-1} E\| < \varepsilon.$$

Remark 6.11 *Note that the additional condition regarding the invariant zeros on the imaginary axis is not affected when we use the more general class of full information feedback controllers.*

6.3.2 Solvability conditions for ADD—discrete time

As in the continuous-time case, we recall below the solvability conditions first for H_2 ADD and then for H_∞ ADD (see, for instance, [72]).

Theorem 6.12 *Consider the discrete-time system Σ as given by (6.1) and the associated H_2 ADD problem as defined by Problem 6.1. Also, consider dynamic state feedback controllers of the form (6.2) as well as static state feedback controllers of the form (6.5). Then, the following three statements are equivalent:*

(i) *The H_2 ADD problem is solvable via static state feedback controllers.*

(ii) *The H_2 ADD problem is solvable via dynamic state feedback controllers.*

(iii) *(A, B) is $\mathbb{C}^\ominus$-stabilizable and $\operatorname{im} E \subseteq \mathcal{V}^\otimes(\Sigma_{\text{sub}})$.*

Theorem 6.13 *Consider the discrete-time system Σ as given in (6.1) and the associated H_2 ADD problem as defined by Problem 6.1 but this time using static full information feedback controllers of the form (6.7). Then, the H_2 ADD problem for Σ is solvable by a sequence of static full information feedback controllers if and only if the pair (A, B) is $\mathbb{C}^\ominus$-stabilizable, and that*

$$\operatorname{im} E \subseteq \mathcal{V}^\otimes(\Sigma_{\text{sub}}) + B \ker D. \tag{6.14}$$

Although the above two theorems pertain to H_2 ADD, the following two theorems consider H_∞ ADD.

Theorem 6.14 *Consider the discrete-time system Σ as given by (6.1) and the associated H_∞ ADD problem as defined by Problem 6.1. Also, consider dynamic state feedback controllers of the form (6.2) as well as static state feedback controllers of the form (6.5). Then, the following three statements are equivalent:*

   (i) The H_∞ ADD problem is solvable via static state feedback controllers.

   (ii) The H_∞ ADD problem is solvable via dynamic state feedback controllers.

   (iii) (A, B) is $\mathbb{C}^{\ominus}$-stabilizable, $\operatorname{im} E \subseteq \mathcal{V}^{\otimes}(\Sigma_{\mathrm{sub}})$ and for any $\varepsilon > 0$ and for any invariant zero z_0 of (A, B, C, D) on the unit circle, a matrix F exists such that $z_0 I - A + BF$ is invertible and

$$\|(C - DF)(z_0 I - A + BF)^{-1} E\| < \varepsilon.$$

Remark 6.15 *Note that the last condition in (iii) can also be expressed differently as discussed in Remarks 6.7 and 6.8. The only modification needed compared with these continuous-time conditions is that we have to replace the imaginary axis with the unit circle.*

Theorem 6.16 *Consider the discrete-time system Σ as given in (6.1) and the associated H_∞ ADD problem as defined by Problem 6.1 but this time using static full information feedback controllers of the form (6.7). Then, the H_∞ ADD problem for Σ is solvable by a sequence of static full information feedback controllers if and only if the pair (A, B) is $\mathbb{C}^{\ominus}$-stabilizable,*

$$\operatorname{im} E \subseteq \mathcal{V}^{\otimes}(\Sigma_{\mathrm{sub}}) + B \ker D, \tag{6.15}$$

and for any $\varepsilon > 0$ and any invariant zero z_0 of (A, B, C, D) on the unit circle, a matrix F exists such that $z_0 I - A + BF$ is invertible and

$$\|(C - DF)(z_0 I - A + BF)^{-1} E\| < \varepsilon.$$

As seen from above and as in the continuous-time case, we note that the necessary and sufficient conditions for the solvability of ADD problems are the same whether static or dynamic state feedback controllers are used.

6.4   More on ADD finite asymptotic fixed modes

Our goal in this section is to study the relationship between $\boldsymbol{\Omega}_s^2$ and $\boldsymbol{\Omega}_d^2$ and similarly between $\boldsymbol{\Omega}_s^\infty$ and $\boldsymbol{\Omega}_d^\infty$. As mentioned in Remark 6.3, we have $\boldsymbol{\Omega}_d^2 \subseteq \boldsymbol{\Omega}_s^2$ and $\boldsymbol{\Omega}_d^\infty \subseteq \boldsymbol{\Omega}_s^\infty$. Also, Theorems 6.4 and 6.6 for continuous time and similarly

Theorems 6.12 and 6.14 for discrete time point out that the necessary and sufficient conditions for solvability of ADD are the same whether static or dynamic state feedback controllers are used. This immediately begs the question as to what one would gain by using dynamic state feedback controllers rather than simple static state feedback controllers. For instance, can we indeed decrease the number of finite asymptotic fixed modes by enlarging the class of controllers from static to dynamic state feedback controllers? This cannot be done as shown below. That is, for a given system, the sets of finite asymptotic fixed modes are the same whether static or dynamic state feedback controllers are used. This immediately implies that there is no necessity to enlarge the class of state feedback controllers beyond the static state feedback controllers.

The following theorem pertains to continuous-time systems.

Theorem 6.17 *Let the continuous-time system Σ be given by (6.1). Consider both the H_2 and the H_∞ ADD problems as defined by Problem 6.1.*

- *Assume that the H_2 ADD problem is solvable; i.e.,*

$$\operatorname{im} E \subseteq \mathcal{S}^*(\Sigma_{\text{sub}}) + \mathcal{V}^{-0}(\Sigma_{\text{sub}}). \tag{6.16}$$

  Then we have

$$\Omega_d^2 = \Omega_s^2.$$

- *Assume that the H_∞ ADD problem is solvable; i.e., (6.16) is satisfied, and for any $\varepsilon > 0$ and any invariant zero s_0 of (A, B, C, D) on the imaginary axis, a matrix F exists such that $s_0 I - A + BF$ is invertible and*

$$\|(C - DF)(s_0 I - A + BF)^{-1} E\| < \varepsilon.$$

  Then we have

$$\Omega_d^\infty = \Omega_s^\infty.$$

Proof : As we said earlier, we have $\Omega_d^2 \subseteq \Omega_s^2$ and $\Omega_d^\infty \subseteq \Omega_s^\infty$. Also, along the lines of the proof of Theorem 5.25, we can establish that, for a particular sequence of H_2 (or H_∞) ADD dynamic state feedback controllers, the finite asymptotic modes of the corresponding closed-loop system contains all finite asymptotic fixed modes [i.e., the set Ω_s^2 (or Ω_s^∞)] of the given system. Hence the result. ∎

Similar to the above result, we can establish the following analogous result for discrete-time systems.

Theorem 6.18 *Let the discrete-time system Σ be given by (6.1). Consider both the H_2 and the H_∞ ADD problems as defined by Problem 6.1.*

- *Assume that the H_2 ADD problem is solvable; i.e.,*

$$\operatorname{im} E \subseteq \mathcal{V}^{\otimes}(\Sigma_{\mathrm{sub}}). \tag{6.17}$$

Then we have

$$\mathit{\Omega}_{d}^{2} = \mathit{\Omega}_{s}^{2}.$$

- *Assume that the H_∞ ADD problem is solvable; i.e., (6.17) is satisfied, and for any $\varepsilon > 0$ and any invariant zero z_0 of (A, B, C, D) on the unit circle, a matrix F exists such that $s_0 I - A + BF$ is invertible and*

$$\|(C - DF)(z_0 I - A + BF)^{-1} E\| < \varepsilon.$$

Then we have

$$\mathit{\Omega}_{d}^{\infty} = \mathit{\Omega}_{s}^{\infty}.$$

6.5 H_2 ADD—design

In this section we focus on the issues related to designing the families of controllers that solve the H_2 ADD problem. To be specific, we will describe a mechanism to obtain the H_2 ADD finite asymptotic fixed modes and, at the same time, obtain suitable families of controllers that solve the H_2 ADD problem while pointing out how to exploit some of the flexibility available in assigning certain closed-loop poles. We will do this first for continuous time and then for discrete time.

6.5.1   *Computation of $\mathit{\Omega}_{s}^{2}$ and designing sequences of static H_2 ADD controllers—continuous time*

In this subsection, we develop what we will call the H_2 ADD *algorithm* for continuous-time systems. Obviously the input to the H_2 ADD algorithm is the quintuple (A, B, C, D, E) which characterizes the H_2 ADD problem. The H_2 ADD algorithm first determines whether the solvability conditions for the H_2 ADD problem as given by Theorem 6.4 are satisfied, and if they are satisfied, it determines explicitly the set of H_2 ADD finite asymptotic fixed modes $\mathit{\Omega}_{s}^{2}$. Also, the H_2 ADD algorithm is a complete design algorithm in the sense that it gives a step-by-step design of sequences of static state feedback H_2 ADD controllers that have the flexibility of assigning arbitrarily most of the finite asymptotic modes that are not fixed. We describe below step by step the H_2 ADD algorithm.

The H_2 ADD algorithm—continuous time

Step 1: Representation of Σ_{sub} in SCB.

In this step, as described in Chapter 3, we first construct the SCB of the subsystem Σ_{sub}, which is characterized by the quadruple (A, B, C, D). We will use a more compact form where the special structure of x_d is not made explicit and where x_a^- and x_a^0 are viewed together as x_a^{-0} as given in (3.50)–(3.53).

As in Chapter 3, let Γ_s, Γ_i, and Γ_o be the state, input, and output transformation matrices that take the given system Σ_{sub} to its SCB form. From Property 3.24 of SCB, it is simple then to see that the solvability condition (6.9) for H_2 ADD implies that

$$
\Gamma_s^{-1} E = \begin{pmatrix} E_a^{-0} \\ 0 \\ 0 \\ E_c \\ E_d \end{pmatrix}. \tag{6.18}
$$

If $\Gamma_s^{-1} E$ is not of the form (6.18), the H_2 ADD problem is not solvable and the procedure of H_2 ADD algorithm stops at this point. Otherwise, continue to the next step.

Step 2: Decomposition of A_{aa}^{-0}.

In this step, we decompose A_{aa}^{-0} as given in the SCB compact form (3.50) into two parts: One part is controllable via the disturbance w, and the other part is unaffected by the disturbance. Consider the pair (A_{aa}^{-0}, E_a^{-0}). This pair need not be controllable, i.e., the disturbance w need not affect all the modes of A_{aa}^{-0}. Thus, we can compute a nonsingular transformation T_a such that

$$
T_a^{-1} A_{aa}^{-0} T_a = \begin{pmatrix} A_{aa}^{11} & A_{aa}^{12} \\ 0 & A_{aa}^{22} \end{pmatrix} \quad \text{and} \quad T_a^{-1} E_a^{-0} = \begin{pmatrix} E_a^1 \\ 0 \end{pmatrix}, \tag{6.19}
$$

where the pair (A_{aa}^{11}, E_a^1) is controllable.

Step 3: Decomposition of x_d and y_d, and the definition of a fictitious output y_{d3}.

In this step, depending on how the disturbance w affects the state x_d, we reorder the state x_d and the output y_d and, thus, the input u_d. Such a reordering allows us to partition x_d and y_d as

$$
x_d = \begin{pmatrix} x_{d1} \\ x_{d2} \end{pmatrix}, \qquad y_d = \begin{pmatrix} y_{d1} \\ y_{d2} \end{pmatrix}.
$$

We define also a fictitious output vector y_{d3}.

To proceed with, to explain our method of reordering and partitioning of x_d and y_d, let us first recall the dynamics of x_d. As indicated in Chapter 3, in SCB there is an array of chains of integrators extending from the input u_d to the output y_d. There are m_d number of integrator chains and the outputs of all the integrators of all chains form the state x_d. Let us consider an ith chain of integrators extending from the input u_i to the output y_i as shown in Figure 3.4 where $1 \le i \le m_d$. In our method of decomposing x_d and y_d, starting with the output end, we examine each integrator input for the presence or absence of any components of the disturbance w. In such an examination, we encounter three possible cases.

Case 1: In this case, none of the integrators in the chain contain any components of the disturbance w in their inputs. For this case, all state variables belonging to the chain are allotted as components of the state x_{d1}. Moreover, the output y_i of the ith chain is allotted as a component of the output y_{d1}.

Case 2: In this case, as we examine the chain of integrators starting from the output end, the very first integrator has at least one or more components of the disturbance w in its input. For this case, all state variables belonging to the chain are allotted as components of the state x_{d2}. Moreover, the output y_i of the ith chain is allotted as a component of the output y_{d2}.

Case 3: In this case, as we examine the chain of integrators starting from the output end, the very first integrator in the beginning does not contain any components of the disturbance w in its input. Also, among other integrators in the chain, there exists at least one integrator having an input containing one or more components of the disturbance w in it. For this case, some state variables of the chain are components of x_{d1}, whereas the others are components of x_{d2}. The procedure of how the state variables are allotted as components of x_{d1} or x_{d2} is as follows: As we examine, all outputs of the integrators in the chain from the output end until up to but not including the first integrator that contains one or more components of w in its input are selected as components of the state x_{d1}. The outputs of the rest of the integrators in the chain are selected as components of the state x_{d2}. The output y_i of the ith chain is allotted as a component of the output y_{d1}. For this case, we also define a fictitious output. The state variable of the very first integrator in the chain of integrators that has components of the disturbance in its input is called a fictitious output and is assigned as a component of a fictitious output vector denoted by y_{d3}.

We repeat the above procedure of decomposition for all chains of integrators; i.e., for all i, $1 \le i \le m_d$.

Let us define next matrices C_{d1} and C_{d2} such that

$$y_d = \begin{pmatrix} y_{d1} \\ y_{d2} \end{pmatrix} = \begin{pmatrix} C_{d1} & 0 \\ 0 & C_{d2} \end{pmatrix} \begin{pmatrix} x_{d1} \\ x_{d2} \end{pmatrix}.$$

Also, let us define next a matrix C_{d3} such that

$$y_{d3} = C_{d3}x_{d2}.$$

We observe now that the above reordering and partitioning of x_d and y_d as well as the definition of the fictitious output vector y_{d3} necessitates a compatible reordering and partitioning of the input u_d and the matrices A_{dd}, C_d, B_d, and E_d. When these matrices are reordered and partitioned, they have the following structure (note that the bar over the matrices signifies that the reordering has taken place):

$$\bar{A}_{dd} = \begin{pmatrix} A_{d1} & L_{d1}C_{d23} \\ L_{d2}C_{d1} & A_{d2} \end{pmatrix}, \quad C_{d23} = \begin{pmatrix} C_{d2} \\ C_{d3} \end{pmatrix}, \quad \bar{C}_d = \begin{pmatrix} C_{d1} & 0 \\ 0 & C_{d2} \end{pmatrix}.$$

$$\bar{B}_d = \begin{pmatrix} B_{d1} & 0 \\ 0 & B_{d2} \end{pmatrix}, \quad \bar{E}_d = \begin{pmatrix} 0 \\ E_{d2} \end{pmatrix}.$$

Here $(A_{d2}, B_{d2}, C_{d23}, 0)$ is invertible without finite zeros and $C_{d23}E_{d2}$ is injective.

Step 4: Modified SCB.

The goal of this step is to incorporate the reordering of the SCB that occurred in the previous steps into a new compact form of SCB. Such a reordering implies that the transformation matrices Γ_s, Γ_i, and Γ_o are replaced by $\bar{\Gamma}_s$, $\bar{\Gamma}_i$, and $\bar{\Gamma}_o$ such that a preliminary feedback law

$$u = -F_{\text{pre}}x + v \tag{6.20}$$

exists with the matrix F_{pre} of suitable dimension, which yields a new compact form of SCB with the following structure:

$$\bar{\Gamma}_s^{-1}(A - BF_{\text{pre}})\bar{\Gamma}_s =$$

$$\begin{pmatrix} A_{aa}^{11} & A_{aa}^{12} & 0 & L_{ab}^1 C_b & L_{ad}^{11} C_{d1} & 0 & L_{ad}^{12} C_{d2} \\ 0 & A_{aa}^{22} & 0 & L_{ab}^2 C_b & L_{ad}^{21} C_{d1} & 0 & L_{ad}^{22} C_{d2} \\ 0 & 0 & A_{aa}^+ & L_{ab}^+ C_b & L_{ad}^{+1} C_{d1} & 0 & L_{ad}^{+2} C_{d2} \\ 0 & 0 & 0 & A_{bb} & L_{bd}^1 C_{d1} & 0 & L_{bd}^2 C_{d2} \\ 0 & 0 & 0 & 0 & A_{d1} & 0 & L_{d1} C_{d23} \\ 0 & 0 & 0 & 0 & L_{cd}^1 C_{d1} & A_{cc} & L_{cd}^2 C_{d2} \\ 0 & 0 & 0 & 0 & L_{d2} C_{d1} & 0 & A_{d2} \end{pmatrix},$$

$$\bar{\Gamma}_s^{-1}B\bar{\Gamma}_i = \begin{pmatrix} B_{a0}^1 & 0 & 0 & 0 \\ B_{a0}^2 & 0 & 0 & 0 \\ B_{a0}^+ & 0 & 0 & 0 \\ B_{b0} & 0 & 0 & 0 \\ B_{d0}^1 & B_{d1} & 0 & 0 \\ B_{c0} & 0 & 0 & B_c \\ B_{d0}^2 & 0 & B_{d2} & 0 \end{pmatrix}, \quad \bar{\Gamma}_s^{-1}E = \begin{pmatrix} E_a^1 \\ 0 \\ 0 \\ 0 \\ 0 \\ E_c \\ E_{d2} \end{pmatrix},$$

$$\bar{\Gamma}_o^{-1}(C - DF_{\text{pre}})\bar{\Gamma}_s = \begin{pmatrix} 0 & 0 & 0 & 0 & 0 & 0 & 0 \\ 0 & 0 & 0 & 0 & C_{d1} & 0 & 0 \\ 0 & 0 & 0 & C_b & 0 & 0 & 0 \\ 0 & 0 & 0 & 0 & 0 & 0 & C_{d2} \end{pmatrix},$$

$$\bar{\Gamma}_o^{-1}D\bar{\Gamma}_i = \begin{pmatrix} I_{m0} & 0 & 0 & 0 \\ 0 & 0 & 0 & 0 \\ 0 & 0 & 0 & 0 \\ 0 & 0 & 0 & 0 \end{pmatrix}.$$

Based on the above development, we form the matrices A_z, B_z, and C_z as follows:

$$
A_z := \begin{pmatrix} A_{aa}^{22} & 0 & L_{ab}^2 C_b & L_{ad}^{21} C_{d1} \\ 0 & A_{aa}^+ & L_{ab}^+ C_b & L_{ad}^{+1} C_{d1} \\ 0 & 0 & A_{bb} & L_{bd}^1 C_{d1} \\ 0 & 0 & 0 & A_{d1} \end{pmatrix},
$$

$$
B_z := \begin{pmatrix} L_{ad}^{22} C_{d2} & B_{a0}^2 & 0 \\ L_{ad}^{+2} C_{d2} & B_{a0}^+ & 0 \\ L_{bd}^2 C_{d2} & B_{b0} & 0 \\ L_{d1} C_{d23} & B_{d0}^1 & B_{d1} \end{pmatrix}, \quad C_z := \begin{pmatrix} 0 & 0 & 0 & C_{d1} \\ 0 & 0 & C_b & 0 \end{pmatrix}.
$$

$$(6.21)$$

By using obvious compact notation, we can rewrite B_z as

$$
B_z = \begin{pmatrix} \tilde{L}_{ad}^{22} C_{d23} & B_{a0}^2 & 0 \\ \tilde{L}_{ad}^{+2} C_{d23} & B_{a0}^+ & 0 \\ \tilde{L}_{bd}^2 C_{d23} & B_{b0} & 0 \\ \tilde{L}_{d1} C_{d23} & B_{d0}^1 & B_{d1} \end{pmatrix} = \begin{pmatrix} \tilde{L}_{zd} C_{d23} & B_{z0} & B_{z1} \end{pmatrix}.
$$

From Property 3.3 of SCB, it is simple to verify that the pair (A_z, B_z) is $\mathbb{C}^-$ stabilizable whenever the pair (A, B) is $\mathbb{C}^-$ stabilizable.

Using the above notation of A_z and B_z, and using a compact notation, we can rewrite the given system in another compact form of SCB as

$$
\bar{\Gamma}_s^{-1}(A - BF_{\mathrm{pre}})\bar{\Gamma}_s = \begin{pmatrix} A_{aa}^{11} & A_{az} & 0 & \tilde{L}_{ad}^{12} C_{d23} \\ 0 & A_z & 0 & \tilde{L}_{zd} C_{d23} \\ 0 & A_{cz} & A_{cc} & \tilde{L}_{cd}^2 C_{d23} \\ 0 & A_{dz} & 0 & A_{d2} \end{pmatrix},
$$

$$(6.22)$$

$$
\bar{\Gamma}_s^{-1} B \bar{\Gamma}_i = \begin{pmatrix} B_{a0}^1 & 0 & 0 & 0 \\ B_{z0} & B_{z1} & 0 & 0 \\ B_{c0} & 0 & 0 & B_c \\ B_{d0}^2 & 0 & B_{d2} & 0 \end{pmatrix}, \quad \bar{\Gamma}_s^{-1} E = \begin{pmatrix} E_a^1 \\ 0 \\ E_c \\ E_{d2} \end{pmatrix},
$$

$$(6.23)$$

$$
\bar{\Gamma}_o^{-1}(C - DF_{\mathrm{pre}})\bar{\Gamma}_s = \begin{pmatrix} 0 & 0 & 0 & 0 \\ 0 & C_z & 0 & 0 \\ 0 & 0 & 0 & C_{d2} \end{pmatrix},
$$

$$(6.24)$$

$$
\bar{\Gamma}_o^{-1} D \bar{\Gamma}_i = \begin{pmatrix} I_{m_0} & 0 & 0 & 0 \\ 0 & 0 & 0 & 0 \\ 0 & 0 & 0 & 0 \end{pmatrix}.
$$

$$(6.25)$$

In view of the above structure, the preliminary feedback law (6.20) can be rewritten as

$$\bar{u} = -\bar{\Gamma}_i^{-1} F_{\text{pre}} \bar{\Gamma}_s \bar{x} + \bar{v},$$ (6.26)

where

$$\bar{x} = \bar{\Gamma}_s^{-1} x, \quad \bar{u} = \bar{\Gamma}_i^{-1} u, \quad \text{and} \quad \bar{v} = \bar{\Gamma}_i^{-1} v = \begin{pmatrix} v_1 \\ v_2 \\ v_3 \\ v_4 \end{pmatrix}.$$ (6.27)

Step 5: Computation of Ω_s^2.

We can determine the set of H_2 ADD finite asymptotic fixed modes Ω_s^2 as

$$\Omega_s^2 = \lambda(A_{aa}^{11}) \cup \{\text{input decoupling zeros of } (A_z, B_z)\}.$$ (6.28)

Step 6: Construction of H_2 ADD families of controllers.

The goal of this step is to find sequences of H_2 ADD static state feedback controllers while possibly assigning arbitrarily as desired those finite asymptotic modes that are not fixed. This step is divided into five substeps.

Step 6a: Assigning the first part of finite asymptotic modes.

Choose

$$\bar{F} = \begin{pmatrix} F_1 \\ F_2 \\ F_3 \end{pmatrix}$$

such that $A_z - B_z \bar{F}$ is stable and has the desirable eigenvalues in the open left-half complex plane. Note that $\bar{F}$ is partitioned in conformity with the partitioning of B_z as given in (6.21). Also, note that the desirable eigenvalues must obviously contain the set of input decoupling zeros of (A_z, B_z), which is a subset of Ω_s^2.

Step 6b: Assigning the second part of finite asymptotic modes.

Choose F_4 such that $A_{cc} - B_c F_4$ is stable and has the desirable eigenvalues in the open left-half complex plane. As (A_{cc}, B_c) is always controllable, all eigenvalues of $A_{cc} - B_c F_4$ can be assigned to arbitrary locations.

Step 6c: Low-gain design.

This step involves a low-gain design. To do so, we first define

$$\tilde{B}_{az} = \begin{pmatrix} \tilde{L}_{ad}^{12} C_{d23} & B_{a0}^1 & 0 \\ \tilde{L}_{zd} C_{d23} & B_{z0} & B_{z1} \end{pmatrix}, \quad \tilde{E}_{az} = \begin{pmatrix} E_a^1 \\ 0 \end{pmatrix},$$

and

$$\tilde{A}_{az} = \begin{pmatrix} A_{aa}^{11} & A_{az} \\ 0 & A_z \end{pmatrix} - \tilde{B}_{az} \bar{F} \begin{pmatrix} 0 & I \end{pmatrix}.$$

We construct next a sequence of static state feedback controllers $\{\widetilde{F}^\varepsilon\}_{\varepsilon>0}$ such that the system,

$$\left[\widetilde{A}_{az} - \widetilde{B}_{az}\widetilde{F}^\varepsilon,\ \widetilde{E}_{az},\ \begin{pmatrix} -\widetilde{F}_2^\varepsilon - F_2\begin{pmatrix} 0 & I \end{pmatrix} \\ \begin{pmatrix} 0 & C_z \end{pmatrix} \end{pmatrix},\ 0 \right],$$

is stable for all $\varepsilon > 0$ and its H_2 norm converges to zero as $\varepsilon \downarrow 0$. Moreover, $\widetilde{F}^\varepsilon \to 0$ as $\varepsilon \downarrow 0$. In the above, $\widetilde{F}_2^\varepsilon$ is a component of $\widetilde{F}^\varepsilon$ when it is partitioned to be compatible with the decomposition of $\widetilde{B}_{az}$ as

$$\widetilde{F}^\varepsilon = \begin{pmatrix} \widetilde{F}_1^\varepsilon \\ \widetilde{F}_2^\varepsilon \\ \widetilde{F}_3^\varepsilon \end{pmatrix}.$$

The construction of $\widetilde{F}^\varepsilon$ can be done using a Riccati-based design. For each ε, let $\widetilde{P}^\varepsilon$ be the stabilizing solution of the H_2 CARE,

$$\widetilde{A}'_{az}\widetilde{P}^\varepsilon + \widetilde{P}^\varepsilon\widetilde{A}_{az} - \widetilde{P}^\varepsilon\widetilde{B}_{az}\widetilde{B}'_{az}\widetilde{P}^\varepsilon + \varepsilon^2 I = 0,$$

and $\widetilde{F}^\varepsilon = \widetilde{B}'_{az}\widetilde{P}^\varepsilon$. Note that the eigenvalues of $\widetilde{A}_{az} - \widetilde{B}_{az}\widetilde{F}^\varepsilon$ converge to the eigenvalues of $\widetilde{A}_{az}$ as $\varepsilon \downarrow 0$. We denote by γ^ε the resulting H_2 norm of the above system for a given ε.

Step 6d: High-gain design.

This step involves a high-gain design. To do so, we construct a sequence of state feedback controllers $\{\overline{F}_\delta\}_{\delta>0}$ such that the system,

$$(A_{d2} - B_{d2}\overline{F}_\delta,\ I,\ C_{d23},\ 0),$$

is stable for all $\delta > 0$, and its H_∞ norm converges to zero as $\delta \to 0$. Note that, if we guarantee that the H_∞ norm converges to zero, then the same property is guaranteed to hold for the H_2 norm. One way of finding such a sequence of controllers is through a Riccati-based design. For each δ, let $\overline{P}_\delta$ be the stabilizing solution of the H_∞^1 CARE,

$$A'_{d2}\overline{P}_\delta + \overline{P}_\delta A_{d2} - \delta^{-2}\overline{P}_\delta B_{d2}B'_{d2}\overline{P}_\delta + \rho^{-2}\overline{P}_\delta^2 + C'_{d23}C_{d23} = 0,$$

for some $\rho > 0$ in such a way that $\rho \to 0$ as $\delta \to 0$. Let $\overline{F}_\delta = \delta^{-2}B'_{d2}\overline{P}_\delta$. Note that all eigenvalues of $A_{d2} - B_{d2}\overline{F}_\delta$ converge to infinity as δ converges to zero.

Step 6e: Combining individual components to obtain an H_2 ADD family of controllers.

In this step, we construct finally a sequence of H_2 ADD static state feedback controllers. To do so, we define

$$u = \overline{\Gamma}_i\overline{u} = -F_{\text{pre}}x + \overline{\Gamma}_i\overline{v}, := -F^\varepsilon x \tag{6.29}$$

where

$$\bar{v} = \begin{pmatrix} v_1 \\ v_2 \\ v_3 \\ v_4 \end{pmatrix} = - \begin{pmatrix} \widetilde{F}_2^{\varepsilon} x_{az} + F_2 x_z \\ \widetilde{F}_3^{\varepsilon} x_{az} + F_3 x_z \\ \bar{F}_{\delta} \left(x_{d2} + \widetilde{F}_1^{\varepsilon} x_{az} + F_1 x_z \right) \\ F_4 x_c \end{pmatrix} \quad \text{with } \bar{x} = \bar{\Gamma}_s^{-1} x = \begin{pmatrix} x_{az} \\ x_z \\ x_c \\ x_{d2} \end{pmatrix}.$$

$$(6.30)$$

For each ε, we choose δ small enough, for instance, such that the H_2 norm of the resulting closed-loop system is less than γ^{ε}. Thus, we have created above a sequence of static state feedback controllers $u = -F^{\varepsilon} x$ with $\varepsilon > 0$.

This concludes the description of the H_2 ADD algorithm. The following theorem proves the assertions made in this algorithm.

Theorem 6.19 *Consider the continuous-time system Σ as in (6.1) along with the associated H_2 ADD problem as defined by Problem 6.1. Assume that the solvability conditions for the H_2 ADD problem as given by Theorem 6.4 are satisfied; i.e., assume that the pair (A, B) is $\mathbb{C}^-$-stabilizable and that*

$$\operatorname{im} E \subseteq \mathcal{S}^*(\Sigma_{\text{sub}}) + \mathcal{V}^{-0}(\Sigma_{\text{sub}}).$$

Then the following hold:

(i) As claimed in (6.28), the set of all H_2 ADD finite asymptotic fixed modes Ω_s^2 defined in Definition 6.2 is given by

$$\Omega_s^2 = \lambda(A_{aa}^{11}) \cup \{\textit{input decoupling zeros of } (A_z, B_z)\}.$$

(ii) Any sequences of static state feedback controllers as designed according to the H_2 ADD algorithm [i.e., $u = -F^{\varepsilon} x$ as given by (6.29)] solves the H_2 ADD problem; that is, the closed-loop system is stable for ε small enough, and the transfer matrix of the closed-loop system converges to 0 in H_2 norm as $\varepsilon \downarrow 0$.

We need to state some preliminary results before we prove Theorem 6.19. We have the following two lemmas.

Lemma 6.20 *Consider a stable finite-dimensional linear system characterized by the quadruple $(A, B, C, 0)$ with transfer matrix G. For any $\delta > 0$, we define the shifted H_{∞} norm as*

$$\|G\|_{\delta,\infty} := \sup_{\text{Re } s > \delta} \|G(s)\|.$$

Then we have

$$\|G\|_{\delta,\infty} \leqslant \frac{\|G\|_2}{\sqrt{2\delta}}.$$

Proof : We have $\|G\|_2^2 = \text{trace}\, C'PC$, where P satisfies

$$AP + PA' + BB' = 0$$

and we define

$$\gamma = \frac{\|G\|_2}{\sqrt{2\delta}}.$$

In this case, we have trace $C'PC \leqslant 2\delta\gamma^2$, which implies that $\gamma^{-2}PCC'P \leqslant 2\delta P$, and this in turn yields that

$$(A - \delta I)P + P(A - \delta I) + \gamma^{-2}PCC'P + BB' \leqslant 0.$$

The bounded real lemma (see Lemma 11.45) then immediately yields that $\|\bar{G}\| \leqslant \gamma$ where

$$\bar{G}(s) = C(sI - (A - \delta I))^{-1}B.$$

But this immediately yields $\|G\|_{\delta,\infty} \leqslant \gamma$, which completes the proof. ∎

Lemma 6.21 *Assume that a sequence of rational matrices G^ε of fixed McMillan degree converges (pointwise) to the rational matrix G_0. Then a subset of the poles of G^ε must converge to the poles of G_0 (counting multiplicities).*

Proof : We know that the degree of a pole λ of a rational matrix is equal to the largest degree of the pole in any minor of that rational matrix (see [35], for instance). Note that the minors are defined as the determinant of any submatrix of arbitrary size of the original rational matrix. Consider an arbitrary minor of g^ε of G^ε and the corresponding minor g_0 of G. Factorize $g_0 = n_0/d_0$, where n_0 and d_0 are coprime polynomials. Viewing the poles of g^ε as elements of the Riemann sphere (the one-point compactification of the complex plane), we know that a subsequence exists for which all poles converge to elements of the Riemann sphere. In the standard complex plane, this implies that a certain number of poles (say m) converge to infinity, whereas the remaining poles converge to finite locations in the complex plane. Let $\lambda_1^\varepsilon, \ldots, \lambda_n^\varepsilon$ denote the poles (counting multiplicities) of g^ε. Assume (after possible reordering) that the first m eigenvalues go to infinity. Then we have that

$$d^\varepsilon = [(\lambda_1^\varepsilon)^{-1}s - 1]\cdots[(\lambda_m^\varepsilon)^{-1}s - 1][s - \lambda_{m+1}^\varepsilon]\cdots(s - \lambda_n^\varepsilon)$$

has bounded coefficients. Obviously $n^\varepsilon = d^\varepsilon g^\varepsilon$ is then a polynomial function whose coefficients are bounded because d^ε and g^ε are both bounded. But then we have pointwise that

$$n^\varepsilon(s)d_0(s) \to d^\varepsilon(s)n_0(s).$$

As the coefficients of n^ε and d^ε are bounded, this implies that the zeros of these polynomials must converge to each other. But then this implies that a subset of the zeros of n^ε must converge to the zeros of n_0, and hence, a subset of the poles of this minor converges to the poles of the corresponding minor of G_0.

We have only showed that the poles of G^ε converge to the poles of G_0 for a subsequence whose poles converge (on the Riemann sphere). Assume that G_0 has n_1 poles, whereas G^ε has n poles. Suppose a subsequence of G^ε has $n - n_1 + 1$ poles that remain bounded away from the poles of G_0. Then our earlier proof implies that a subsequence of this subsequence must have a subset of its poles converging to the poles of G_0 (which is then obviously impossible). ■

We proceed now to prove Theorem 6.19.

Proof of Theorem 6.19 : After we choose u according to (6.26) and v according to (6.30), then the closed-loop system can be written in a compact form as

$$\dot{\tilde{x}} = \begin{pmatrix} \bar{A}_{acz} & \tilde{L}_{acz} C_{d23} \\ A_{dacz} & A_{d2} - B_{d2}\bar{F}_8 + L_{d2}^2 C_{d23} \end{pmatrix} \tilde{x} + \begin{pmatrix} \bar{E}_{acz} \\ \bar{E}_{d2} \end{pmatrix} w, \quad \tilde{x} = \begin{pmatrix} \tilde{x}_1 \\ \tilde{x}_2 \end{pmatrix},$$
$$z = \begin{pmatrix} \bar{C}_{acz} & 0 \\ \bar{C}_{dacz} & C_{d2} \end{pmatrix} \tilde{x},$$

$$(6.31)$$

where

$$\tilde{x}_1 = \begin{pmatrix} x_{a1}^{-0} \\ x_z \\ x_c \end{pmatrix}, \quad \text{and} \quad \tilde{x}_2 = x_{d2} + F_1 x_z + \tilde{F}_1^\varepsilon x_{az},$$

and all submatrices in (6.31) are obviously defined, whereas

$$A_{caz} = \begin{pmatrix} 0 & A_{cz} \end{pmatrix} - \begin{pmatrix} \tilde{L}_{cd}^2 C_{d23} & B_{c0} & 0 \end{pmatrix}(\bar{F}\begin{pmatrix} 0 & I \end{pmatrix} + \tilde{F}^\varepsilon),$$

$$\bar{A}_{acz} = \begin{pmatrix} \tilde{A}_{az} - \tilde{B}_{az}\tilde{F}^\varepsilon & 0 \\ A_{caz} & A_{cc} - B_c F_4 \end{pmatrix},$$

$$\tilde{L}_{acz} = \begin{pmatrix} \tilde{L}_{ad}^{12} \\ \tilde{L}_{zd} \\ \tilde{L}_{cd}^2 \end{pmatrix},$$

$$F_{11}^\varepsilon = \begin{pmatrix} F_1\begin{pmatrix} 0 & I \end{pmatrix} + \tilde{F}_1^\varepsilon & 0 \end{pmatrix},$$

$$A_{dacz} = \begin{pmatrix} 0 & A_{dz} & 0 \end{pmatrix} - A_{d2}F_{11}^\varepsilon + F_{11}^\varepsilon \bar{A}_{acz} - B_{d0}^2\begin{pmatrix} F_2\begin{pmatrix} 0 & I \end{pmatrix} + \tilde{F}_2^\varepsilon & 0 \end{pmatrix},$$

$$L_{d2}^2 = F_{11}^\varepsilon \tilde{L}_{acz},$$

$$\bar{E}_{acz} = \begin{pmatrix} E_a^1 \\ 0 \\ E_c \end{pmatrix},$$

$$\overline{E}_{d2} = E_{d2} - F_{11}^{\varepsilon}\overline{E}_{acz},$$

$$\overline{C}_{acz} = \begin{pmatrix} -F_2\begin{pmatrix} 0 & I \end{pmatrix} - \widetilde{F}_2^{\varepsilon} & 0 \\ \begin{pmatrix} 0 & C_z \end{pmatrix} & 0 \end{pmatrix},$$

$$\overline{C}_{dacz} = -C_{d2}F_{11}^{\varepsilon}.$$

Here the subsystem,

$$\left[\overline{A}_{acz}, \overline{E}_{acz}, \begin{pmatrix} \overline{C}_{acz} \\ \overline{C}_{dacz} \end{pmatrix}, 0 \right],$$

has an H_2 norm less than or equal to γ^{ε} and the eigenvalues of $\overline{A}_{acz}$ are close to the desired locations (except of course for the modes that go to infinity). We view the above system as the interconnection of

$$\dot{\widetilde{x}}_1 = \overline{A}_{acz}\widetilde{x}_1 + \overline{E}_{acz}w + L_{acz}\widetilde{w},$$

$$z = \begin{pmatrix} \overline{C}_{acz} \\ \overline{C}_{dacz} \end{pmatrix}\widetilde{x}_1 + \begin{pmatrix} 0 & 0 \\ I & 0 \end{pmatrix}\widetilde{w},$$

$$\widetilde{z} = A_{dacz}\widetilde{x}_1 + \overline{E}_{d2}w + L_{d2}^2\widetilde{w},$$

and

$$\dot{\widetilde{x}}_2 = (A_{d2} - B_{d2}\overline{F}_8)\widetilde{x}_2 + \widetilde{z},$$

$$\widetilde{w} = C_{d23}\widetilde{x}_2.$$

Both systems are asymptotically stable. The transfer matrix of the second subsystem we denote by Δ, and we know that $\|\Delta\|_{\infty} \to 0$ as $\delta \to 0$. Denote the transfer matrices of the first subsystem $G_{\widetilde{w},\widetilde{z}}$, $G_{\widetilde{w},z}$, $G_{w,\widetilde{z}}$, and $G_{w,z}$ using obvious notation. The standard small gain theorem yields that the closed-loop system is asymptotically stable for δ small enough. Moreover, a result from [93] yields the following upper bound for the H_2 norm of the interconnected system:

$$\|G_{w,z}\|_2 + \|\Delta\|_{\infty}\frac{\|G_{w,\widetilde{z}}\|_{\infty}\|G_{\widetilde{w},z}\|_2}{1 - \|G_{\widetilde{w},\widetilde{z}}\|_{\infty}\|\Delta\|_{\infty}}.$$

As $\|\Delta\|_{\infty}$ converges to zero as $\delta \to 0$ while $\|G_{w,z}\|_2 \leqslant \gamma^{\varepsilon}$, this implies that for δ small enough we know that the H_2 norm is less than γ^{ε}.

We still need to evaluate the asymptotic behavior of the eigenvalues of the matrix:

$$A_{cl} = \begin{pmatrix} \overline{A}_{acz} & L_{acz}C_{d23} \\ A_{dacz} & A_{d2} - B_{d2}\overline{F}_8 + L_{d2}^2C_{d23} \end{pmatrix}.$$

Note that

$$G_{\delta}(s) = C_{d23}(sI - A_{d2} + B_{d2}\overline{F}_8)^{-1}$$

has the property that $\|G_\delta\|_\infty \to 0$ as ε goes to zero. It is then easily checked that for any fixed point s in the right-half plane:

$$(sI - A_{cl})^{-1}\begin{pmatrix} I \\ 0 \end{pmatrix} \to \begin{pmatrix} (sI - \bar{A}_{acz})^{-1} \\ 0 \end{pmatrix}.$$

By Lemma 6.21, this implies that the poles of this transfer matrix converge to the poles of $(sI - \bar{A}_{acz})^{-1}$ and, hence, to the eigenvalues of $\bar{A}_{acz}$. Let n_2 be the number of remaining eigenvalues. Again, for any fixed point s in the right-half plane, we can show that

$$(sI - A_{cl})^{-1}\begin{pmatrix} 0 \\ I \end{pmatrix} \to 0.$$

Consider $s = 1$. In this case, we see that $(I - A_{cl})^{-1}$ has at least n_2 eigenvalues going to zero, which implies that A_{cl} has at least n_2 eigenvalues going to infinity.

We established in the above that the ADD finite asymptotic fixed modes are a subset of Ω_s^2. To show equality, we need to establish that any other design does not yield more flexibility in the asymptotic behavior of the poles of the closed-loop system.

Consider the decomposition of the system in Step 4 of our algorithm. Consider the first two components of the state:

$$\begin{pmatrix} \dot{x}_{a1}^{-0} \\ \dot{x}_{a2}^{-0} \end{pmatrix} = \begin{pmatrix} A_{aa}^{11} & A_{aa}^{12} \\ 0 & A_{aa}^{22} \end{pmatrix}\begin{pmatrix} x_{a1}^{-0} \\ x_{a2}^{-0} \end{pmatrix} + \begin{pmatrix} B_{a0}^{1} & L_{ad}^{11} & L_{ab}^{1} & L_{ad}^{12} \\ B_{a0}^{2} & L_{ad}^{21} & L_{ab}^{2} & L_{ad}^{22} \end{pmatrix}\bar{y} + \begin{pmatrix} E_a^1 \\ 0 \end{pmatrix}w.$$

We are applying an arbitrary family of controllers $u = -F^\varepsilon x$, and we know this results into a closed-loop transfer matrix $\bar{G}^\varepsilon$ from w to $\bar{y}$ with the property that $\|\bar{G}^\varepsilon\|_2 \to 0$. By Lemma 6.20, this implies that $\bar{G}^\varepsilon(s)$ converges pointwise to 0 for all s with $\operatorname{Re} s > 0$. But then it is easily checked that the transfer matrix from w to x_{a1}^{-0} converges to $(sI - A_{aa}^{11})^{-1}E_a^1$. As (A_{aa}^{11}, E_a^1) is controllable, the poles of this latter transfer matrix are equal to the eigenvalues of the matrix A_{aa}^{11}. But then Lemma 6.21 implies that a subset of the poles of the transfer matrix from w to x_{a1}^{-0} converge to the eigenvalues of A_{aa}^{11}. Hence, our arbitrarily chosen sequence of state feedback controllers that achieves ADD must result in a closed-loop transfer matrix of which a subset of its modes converges to the eigenvalues of A_{aa}^{11}. Obviously, from the structure in (6.22) and (6.23), it is obvious that any state feedback will result in a closed-loop system whose modes include the input-decoupling zeros of (A_z, B_z). Hence, a subset of the closed-loop poles must converge to the finite asymptotic fixed modes as identified in the H_2 ADD algorithm. ∎

The following theorem pertains to static full information feedback controllers of the form (6.7).

Theorem 6.22 *Consider the continuous-time system Σ as in (6.1) along with the associated H_2 ADD problem as defined by Problem 6.1 but using static full information feedback controllers of the form (6.7). Assume that the solvability conditions for the H_2 ADD problem via full information feedback controllers as described in Theorem 6.5 are satisfied. Then, a matrix G always exists such that*

$$\mathrm{im}(E + BG) \subseteq \mathcal{S}^*(\Sigma_{\mathrm{sub}}) + \mathcal{V}^{-0}(\Sigma_{\mathrm{sub}}),$$

while $DG = 0$. Moreover, with G selected as such the following properties hold:

(i) $\Omega^2_{s,f}(A, B, C, D, E) = \Omega^2_s(A, B, C, D, E + BG)$.

(ii) Any sequence of static state feedback controllers $u = -F^\varepsilon x$ as designed according to the H_2 ADD algorithm with E replaced by $E + BG$ results in a sequence of static full information feedback controllers $u = -F^\varepsilon x + Gw$ that solves the H_2 ADD problem; i.e., the closed-loop system is stable for ε small enough and the transfer matrix converges to 0 in H_2 norm as $\varepsilon \downarrow 0$.

Remark 6.23 *The above theorem has two important consequences. Clearly, the sequence of static full information feedback controllers consists of a static state feedback and a static disturbance feedback. However, it does not help to vary the disturbance feedback part and we can fix it a priori. Even stronger, any static disturbance feedback that can be combined with a sequence of static state feedback controllers to solve the H_2 ADD problem yields the same flexibility in the asymptotic behavior of the closed-loop poles.*

Proof : To start with, from the given system Σ (6.1), let us construct an auxiliary system $\bar{\Sigma}$:

$$\bar{\Sigma} : \begin{cases} \dot{x} = Ax + Bu + Ex_2 \\ \dot{x}_2 = -x_2 + w \\ z = Cx + Du. \end{cases}$$

We can observe easily that the existence conditions for a sequence of H_2 ADD static state feedback controllers for the auxiliary system $\bar{\Sigma}$ coincide precisely with the existence conditions for a sequence of H_2 ADD static full information feedback controllers for the given system Σ, namely the pair (A, B) being $\mathbb{C}^-$-stabilizable and satisfying (6.10). Moreover, if a sequence of static full information feedback controllers $u = -F^\varepsilon x + G^\varepsilon w$ solves the H_2 ADD problem for the given system Σ, then the sequence of static state feedback controllers $u = -F^\varepsilon x + G^\varepsilon x_2$ solves the H_2 ADD problem for the auxiliary system $\bar{\Sigma}$. Also, we can observe easily that the closed-loop eigenvalues of the auxiliary system $\bar{\Sigma}$ with static state feedback $u = -F^\varepsilon x + G^\varepsilon x_2$ indeed contain all closed-loop eigenvalues of the original system Σ with static full information feedback

$u = -F^\varepsilon x + G^\varepsilon w$ as well as additional ℓ eigenvalues in -1, where ℓ is the dimension of w. With the above observations that connect the two systems Σ and $\bar{\Sigma}$, it immediately follows that, when the ℓ eigenvalues in -1 are excluded, the H_2 ADD finite asymptotic fixed modes of $\bar{\Sigma}$ by static state feedback are contained within the corresponding set of H_2 ADD finite asymptotic fixed modes of Σ by static full information feedback.

Next, we consider a special subset of static full information feedback controllers of the form $u = -F^\varepsilon x + Gw$ with G fixed and satisfying the conditions as given in the theorem. Using the static state feedback of Theorem 6.19, we can characterize the fixed modes when applying this type of feedback controllers to the system Σ, which is equivalent to a preliminary disturbance feedback $u = Gw + v$, and then solving a standard static state feedback problem. These finite asymptotic fixed modes turn out to be the same as the finite asymptotic fixed modes of $\bar{\Sigma}$ (when considering static state feedback controllers) except for the additional ℓ fixed modes in -1. But we already argued that the finite asymptotic fixed modes of $\bar{\Sigma}$ (excluding the ℓ in -1) must be part of any design of static full information feedback controllers for Σ. Hence, requiring the full information part G^ε to be fixed does not introduce new fixed modes.

Therefore, using this special class of full information feedback controllers with G fixed, we find that the set of H_2 ADD finite asymptotic fixed modes of Σ (with E replaced by $E + BG$) by static state feedback coincides with the corresponding set of Σ by static full information feedback.   ∎

6.5.2 *Computation of Ω_s^2 and designing sequences of static H_2 ADD controllers—discrete time*

In this subsection, for discrete-time systems, we will describe a mechanism to obtain the finite asymptotic fixed modes and, at the same time, obtain suitable families of controllers that solve the H_2 ADD problem with a certain flexibility in placing the closed-loop poles.

The H_2 ADD algorithm—discrete time

Step 1: Representation of Σ_{sub} in SCB.

In this step, we first construct the SCB of Σ_{sub}, which is characterized by the quadruple (A, B, C, D) as described in Chapter 3. We will use a more compact form where the special structure of x_d is not made explicit and where x_a^- and x_a^0 are viewed together as x_a^{-0}. In this case, we get (3.50)–(3.53).

As in Chapter 3, let Γ_s, Γ_i, and Γ_o be the state, input, and output transformation matrices that take the given system Σ_{sub} to its SCB form. From Property 3.24 of SCB, it is simple to see that

$$\operatorname{im} E \subseteq \mathcal{V}^\otimes(\Sigma_{\text{sub}})$$

implies that

$$\Gamma_s^{-1} E = \begin{pmatrix} E_a^{-0} \\ 0 \\ 0 \\ E_c \\ 0 \end{pmatrix}. \tag{6.32}$$

If $\Gamma_s^{-1} E$ is not of the form (6.32), the H_2 ADD problem is not solvable and the procedure of the H_2 ADD algorithm stops at this point. Otherwise, it continues to the next step.

Step 2: Decomposition of A_{aa}^{-0}.

In this step, we decompose A_{aa}^{-0} into two parts: One part is controllable via the disturbance w, and the other part is unaffected by the disturbance.

Consider the pair (A_{aa}^{-0}, E_a^{-0}). This pair need not be controllable, i.e., the disturbance w need not affect all modes of A_{aa}^{-0}. Thus, we can compute a nonsingular transformation T_a such that

$$T_a^{-1} A_{aa}^{-0} T_a = \begin{pmatrix} A_{aa}^{11} & A_{aa}^{12} \\ 0 & A_{aa}^{22} \end{pmatrix} \quad \text{and} \quad T_a^{-1} E_a^{-0} = \begin{pmatrix} E_a^1 \\ 0 \end{pmatrix}, \tag{6.33}$$

where the pair (A_{aa}^{11}, E_a^1) is controllable.

Step 3: Modified SCB.

The goal of this step is to incorporate the reordering of state, input, and output variables of the SCB that occurred in the previous step into a new compact form of SCB. Such a reordering implies that the transformation matrices Γ_s, Γ_i, and Γ_o that take the given system Σ_{sub} to its SCB form need to be amended as $\bar{\Gamma}_s$, $\bar{\Gamma}_i$, and $\bar{\Gamma}_o$. Then, a matrix F_{pre} of suitable dimension and a preliminary feedback law

$$u = -F_{\text{pre}} x + v$$

exist such that one can obtain a new compact form of SCB having the following structure:

$$\bar{\Gamma}_s^{-1} (A - B F_{\text{pre}}) \bar{\Gamma}_s = \begin{pmatrix} A_{aa}^{11} & A_{aa}^{12} & 0 & L_{ab}^1 C_b & L_{ad}^1 C_d & 0 \\ 0 & A_{aa}^{22} & 0 & L_{ab}^2 C_b & L_{ad}^2 C_d & 0 \\ 0 & 0 & A_{aa}^+ & L_{ab}^+ C_b & L_{ad}^+ C_d & 0 \\ 0 & 0 & 0 & A_{bb} & L_{bd} C_d & 0 \\ 0 & 0 & 0 & 0 & A_{dd} & 0 \\ 0 & 0 & 0 & 0 & L_{cd} C_d & A_{cc} \end{pmatrix},$$

$$\bar{\Gamma}_s^{-1} B \bar{\Gamma}_i = \begin{pmatrix} B_{a0}^1 & 0 & 0 \\ B_{a0}^2 & 0 & 0 \\ B_{a0}^+ & 0 & 0 \\ B_{b0} & 0 & 0 \\ B_{d0} & B_d & 0 \\ B_{c0} & 0 & B_c \end{pmatrix}, \qquad \bar{\Gamma}_s^{-1} E = \begin{pmatrix} E_a^1 \\ 0 \\ 0 \\ 0 \\ 0 \\ E_c \end{pmatrix},$$

$$\bar{\Gamma}_o^{-1}(C - DF_{\mathrm{pre}})\bar{\Gamma}_s = \begin{pmatrix} 0 & 0 & 0 & 0 & 0 & 0 \\ 0 & 0 & 0 & 0 & C_d & 0 \\ 0 & 0 & 0 & C_b & 0 & 0 \end{pmatrix},$$

$$\bar{\Gamma}_o^{-1} D \bar{\Gamma}_i = \begin{pmatrix} I_{m_0} & 0 & 0 \\ 0 & 0 & 0 \\ 0 & 0 & 0 \end{pmatrix}.$$

Based on the above development, by using obvious notation, we form the matrices A_z, B_z, and C_z as

$$A_z := \begin{pmatrix} A_{aa}^{22} & 0 & L_{ab}^2 C_b & L_{ad}^2 C_d \\ 0 & A_{aa}^+ & L_{ab}^+ C_b & L_{ad}^+ C_d \\ 0 & 0 & A_{bb} & L_{bd} C_d \\ 0 & 0 & 0 & A_{dd} \end{pmatrix}, \qquad B_z := \begin{pmatrix} B_{a0}^2 & 0 \\ B_{a0}^+ & 0 \\ B_{b0} & 0 \\ B_{d0} & B_d \end{pmatrix},$$

$$C_z := \begin{pmatrix} 0 & 0 & 0 & 0 \\ 0 & 0 & 0 & C_d \\ 0 & 0 & C_b & 0 \end{pmatrix}. \tag{6.34}$$

We can rewrite B_z as

$$B_z = \begin{pmatrix} B_{z0} & B_{z1} \end{pmatrix}.$$

From Property 3.3 of the SCB, it is simple to verify that the pair (A_z, B_z) is $\mathbb{C}^\ominus$-stabilizable when the pair (A, B) is $\mathbb{C}^\ominus$ stabilizable.

Using the above definition, we can rewrite the system more compactly as

$$\bar{\Gamma}_s^{-1}(A - BF_{\mathrm{pre}})\bar{\Gamma}_s = \begin{pmatrix} A_{aa}^{11} & A_{az} & 0 \\ 0 & A_z & 0 \\ 0 & A_{cz} & A_{cc} \end{pmatrix}, \tag{6.35}$$

$$\bar{\Gamma}_s^{-1} B \bar{\Gamma}_i = \begin{pmatrix} B_{a0}^1 & 0 & 0 \\ B_{z0} & B_{z1} & 0 \\ B_{c0} & 0 & B_c \end{pmatrix}, \qquad \bar{\Gamma}_s^{-1} E = \begin{pmatrix} E_a^1 \\ 0 \\ E_c \end{pmatrix}, \tag{6.36}$$

$$\bar{\Gamma}_o^{-1}(C - DF_{\mathrm{pre}})\bar{\Gamma}_s = \begin{pmatrix} 0 & 0 & 0 \\ 0 & C_z & 0 \end{pmatrix}, \tag{6.37}$$

$$\bar{\Gamma}_o^{-1} D \bar{\Gamma}_i = \begin{pmatrix} I_{m_0} & 0 & 0 \\ 0 & 0 & 0 \end{pmatrix}. \tag{6.38}$$

Step 4: Characterization of Ω_s^2.

We can determine the set of H_2 ADD finite asymptotic fixed modes Ω_s^2 as

$$\Omega_s^2 = \lambda(A_{aa}^{11}) \cup \{\text{input decoupling zeros of } (A_z, B_z)\}. \tag{6.39}$$

Step 5: Construction of H_2 ADD families of controllers

The goal of this step is to find a sequence of H_2 ADD static state feedback controllers while possibly assigning arbitrarily as desired those finite asymptotic modes that are not fixed. This step is divided into four substeps.

Step 5a, Assigning first part of finite asymptotic modes.

Choose

$$\bar{F} = \begin{pmatrix} F_1 \\ F_2 \end{pmatrix}$$

such that $A_z - B_z \bar{F}$ is stable and has the desirable eigenvalues within the unit circle. Note that $\bar{F}$ is partitioned in conformity with the partitioning of B_z as given in (6.34). Also, note that the desirable eigenvalues must obviously contain the set of input decoupling zeros of (A_z, B_z), which is a subset of Ω_s^2.

Step 5b: Assigning second part of finite asymptotic modes.

Choose F_3 such that $A_{cc} - B_c F_3$ is stable and has the desirable eigenvalues within the unit circle. As (A_{cc}, B_c) is always controllable, all eigenvalues of $A_{cc} - B_c F_3$ can be assigned to arbitrary locations.

Step 5c: Low-gain design.

We first define

$$\tilde{B}_{az} = \begin{pmatrix} B_{a0}^1 & 0 \\ B_{z0} & B_{z1} \end{pmatrix}, \quad \tilde{E}_{az} = \begin{pmatrix} E_a^1 \\ 0 \end{pmatrix},$$

and

$$\tilde{A}_{az} = \begin{pmatrix} A_{aa}^{11} & A_{az} \\ 0 & A_z \end{pmatrix} - \tilde{B}_{az} \bar{F} \begin{pmatrix} 0 & I \end{pmatrix}.$$

We construct a sequence of feedback gains $\{\tilde{F}^\varepsilon\}_{\varepsilon > 0}$ such that the system,

$$\left[\tilde{A}_{az} - \tilde{B}_{az} \tilde{F}^\varepsilon, \tilde{E}_{az}, \begin{pmatrix} -\tilde{F}_2^\varepsilon - F_2 \begin{pmatrix} 0 & I \end{pmatrix} \\ \begin{pmatrix} 0 & C_z \end{pmatrix} \end{pmatrix}, 0 \right],$$

is stable for all $\varepsilon > 0$ and the H_2 norm converges to zero as $\varepsilon \downarrow 0$. Moreover, $\widetilde{F}^\varepsilon \to 0$ as $\varepsilon \downarrow 0$. In the above, $\widetilde{F}_2^\varepsilon$ is a component of $\widetilde{F}^\varepsilon$ when it is partitioned to be compatible with the decomposition of $\widetilde{B}_{az}$ as

$$\widetilde{F}^\varepsilon = \begin{pmatrix} \widetilde{F}_1^\varepsilon \\ \widetilde{F}_2^\varepsilon \end{pmatrix}.$$

The construction of $\widetilde{F}^\varepsilon$ can be done using a Riccati-based design. For each ε, let $\widetilde{P}^\varepsilon$ be the stabilizing solution of the H_2 DARE,

$$\widetilde{P}^\varepsilon = \widetilde{A}'_{az} \widetilde{P}^\varepsilon \widetilde{A}_{az} + \varepsilon^2 I - \widetilde{A}'_{az} \widetilde{P}^\varepsilon \widetilde{B}_{az} \left(\widetilde{B}_{az} \widetilde{P}^\varepsilon \widetilde{B}'_{az} + I \right)^{-1} \widetilde{B}'_{az} \widetilde{P}^\varepsilon \widetilde{A}_{az},$$

and

$$\widetilde{F}^\varepsilon = \left(\widetilde{B}_{az} \widetilde{P}^\varepsilon \widetilde{B}'_{az} + I \right)^{-1} \widetilde{B}'_{az} \widetilde{P}^\varepsilon \widetilde{A}_{az}.$$

Note that the eigenvalues of $\widetilde{A}_{az} - \widetilde{B}_{az} \widetilde{F}^\varepsilon$ converge to the eigenvalues of $\widetilde{A}_{az}$ as $\varepsilon \downarrow 0$. We denote by γ^ε the resulting H_2 norm for a given ε.

Step 5d: Combining individual components to obtain an H_2 ADD family of controllers.

To do so, we define

$$u = \bar{\Gamma}_i \bar{u} = -F_{\mathrm{pre}} x + \bar{\Gamma}_i \bar{v} := -F^\varepsilon x, \tag{6.40}$$

where

$$\bar{v} = \begin{pmatrix} v_1 \\ v_2 \\ v_3 \end{pmatrix} = -\begin{pmatrix} \widetilde{F}_1^\varepsilon x_{az} + F_1 x_z \\ \widetilde{F}_2^\varepsilon x_{az} + F_2 x_z \\ F_3 x_c \end{pmatrix} \quad \text{with} \quad \bar{x} = \bar{\Gamma}_s^{-1} x = \begin{pmatrix} x_{az} \\ x_z \\ x_c \end{pmatrix}. \tag{6.41}$$

In terms of the original basis, we have thus created a sequence of feedback controllers $u = -F^\varepsilon x$ with $\varepsilon > 0$. Note that we are using the same notation as for continuous-time systems even though the design and characteristics are obviously very different.

This concludes the description of the H_2 ADD algorithm for discrete-time systems. The following theorem proves the assertions made in this algorithm.

Theorem 6.24 *Consider the discrete-time system Σ as in (6.1) along with the associated H_2 ADD problem as defined by Problem 6.1. Assume that the solvability conditions for the H_2 ADD problem as given by Theorem 6.12 are satisfied; i.e., assume that the pair (A, B) is $\mathbb{C}^{\ominus}$-stabilizable, and that*

$$\mathrm{im}\, E \subseteq \mathcal{V}^{\otimes}(\Sigma_{\mathrm{sub}}).$$

Then the following hold:

(*i*) *As claimed in (6.39), the set of all H_2 ADD finite asymptotic fixed modes Ω_s^2 defined in Definition 6.2 is given by*

$$\Omega_s^2 = \lambda(A_{aa}^{11}) \cup \{input\ decoupling\ zeros\ of\ (A_z, B_z)\}.$$

(*ii*) *Any sequence of state feedback controllers as designed according to the H_2 ADD algorithm [i.e., $u = -F^\varepsilon x$ as given by (6.40)] solves the H_2 ADD problem; that is, the closed-loop system is stable for ε small enough, and the transfer matrix converges to 0 in H_2 norm as $\varepsilon \downarrow 0$.*

Proof : The proof that any controller designed according to the H_2 ADD algorithm solves the H_2 ADD problem can be easily established. After all, the transfer matrix from w to z equals

$$\Gamma_o \left(\begin{pmatrix} -\widetilde{F}^\varepsilon \\ 0 \quad C_z \end{pmatrix} \right) \left(sI - \widetilde{A}_{az} + \widetilde{B}_{az} \widetilde{F}^\varepsilon \right)^{-1} \widetilde{E}_{az},$$

whose H_2 norm converges to 0 by our low-gain design. The closed-loop eigenvalues converge to the eigenvalues of $\widetilde{A}_{az}$ because $\widetilde{F}^\varepsilon \to 0$, and this yields the required asymptotic behavior of the closed-loop eigenvalues.

For sufficiency, the arguments used in the continuous time in the proof of Theorem 6.19 can be converted trivially to the discrete time. ∎

Theorem 6.25 *Consider the discrete-time system Σ as in (6.1) along with the associated H_2 ADD problem as defined by Problem 6.1 but using static full information feedback controllers of the form (6.7). Assume that the existence conditions as in Theorem 6.13 are satisfied, and let G be such that*

$$\mathrm{im}(E + BG) \subseteq \mathcal{V}^\otimes$$

while $DG = 0$. Then we have:

(*i*) $\Omega_{s,f}^2(A, B, C, D, E) = \Omega_s^2(A, B, C, D, E + BG).$

(*ii*) *Any sequence of state feedback controllers $u = -F^\varepsilon x$ as designed according to the H_2 ADD algorithm with E replaced by $E + BG$ results in a sequence of full information feedback controllers $u = -F^\varepsilon x + Gw$, which solves the H_2 ADD problem; i.e., the closed-loop system is stable for ε small enough, and the transfer matrix converges to 0 in H_2 norm as $\varepsilon \downarrow 0$.*

Proof : We can use the same arguments as in the proof of Theorem 6.22. This time, we define the auxiliary system $\bar{\Sigma}$ by

$$
\bar{\Sigma} : \begin{cases} x(k+1) = Ax(k) + Bu(k) + Ex_2(k) \\ x_2(k+1) = w(k) \\ \quad z(k) \quad = Cx(k) + Du(k), \end{cases}
$$

which adds eigenvalues in the origin. The rest of the proof is identical with obvious modifications. $\blacksquare$

6.6 H_∞ ADD—design

In the previous section, we focused on the issues related to designing families of controllers that solve the H_2 ADD problem. In this section we do the same for H_∞ ADD. To be specific, we will describe here a mechanism to obtain the H_∞ ADD finite asymptotic fixed modes and, at the same time, obtain suitable families of controllers that solve the H_∞ ADD problem while pointing out how to exploit some of the flexibility available in assigning certain closed-loop poles. As in the case of H_2 ADD, we will do this first for continuous time and then for discrete time.

6.6.1   *Computation of $\boldsymbol{\Omega}_s^\infty$ and designing sequences of static H_∞ ADD controllers—continuous time*

Paralleling the development in Subsection 6.5.1, in this subsection, we develop what we call the H_∞ ADD *algorithm* for continuous-time systems. Clearly, the input to the H_∞ ADD algorithm is the quintuple (A, B, C, D, E). As before, the H_∞ ADD algorithm at first determines whether the solvability conditions for the H_∞ ADD problem as given by Theorem 6.6 are satisfied, and if they are satisfied, it determines explicitly the set of H_∞ ADD finite asymptotic fixed modes $\boldsymbol{\Omega}_s^\infty$. Also, the H_∞ ADD algorithm is a complete design algorithm in the sense that it gives a step-by-step design of sequences of static state feedback H_∞ ADD controllers that have the flexibility of assigning arbitrarily those finite asymptotic modes that are not fixed.

We describe below step by step the H_∞ ADD algorithm.

The H_∞ ADD algorithm—continuous time

Step 1: Representation of Σ_{sub} in SCB.

In this step, as described in Chapter 3, we first construct the SCB of Σ_{sub} that is characterized by the quadruple (A, B, C, D). We will use a more compact form where the special structure of x_d is not made explicit as given in (3.15)–(3.19).

As in Chapter 3, let Γ_s, Γ_i, and Γ_o be the state, input, and output transformation matrices that take the given system Σ_{sub} to its SCB form. From Property 3.24 of the SCB, it is simple to see that the conditions of Theorem 6.6 imply that

$$\Gamma_s^{-1} E = \begin{pmatrix} E_a^- \\ E_a^0 \\ 0 \\ 0 \\ E_c \\ E_d \end{pmatrix}. \tag{6.42}$$

If $\Gamma_s^{-1} E$ is not of the form (6.42), the H_∞ ADD problem is not solvable and the procedure of H_∞ ADD algorithm stops at this point. Otherwise, it continues to the next step.

Step 2: Decomposition of A_{aa}^-.

In this step, we decompose A_{aa}^- into two parts: One part is controllable via the disturbance w, and the other part is unaffected by the disturbance. Consider the pair (A_{aa}^-, E_a^-). This pair need not be controllable; i.e., the disturbance w need not affect all modes of A_{aa}^-. Thus, we can compute a nonsingular transformation T_a such that

$$T_a^{-1} A_{aa}^- T_a = \begin{pmatrix} A_{aa}^{11} & A_{aa}^{12} \\ 0 & A_{aa}^{22} \end{pmatrix} \quad \text{and} \quad T_a^{-1} E_a^- = \begin{pmatrix} E_a^{1-} \\ 0 \end{pmatrix}, \tag{6.43}$$

where the pair (A_{aa}^{11}, E_a^{1-}) is controllable. We define

$$T_a^{-1} x_a^- = \begin{pmatrix} x_{a1}^- \\ x_{a2}^- \end{pmatrix}.$$

Step 3: Decomposition of A_{aa}^0.

The solvability condition for the H_∞ ADD problem as given in Theorem 6.6 requires more specific structure compared with H_2 ADD. To be more explicit, beyond the special structure given in (6.42), we need additional structure for A_{aa}^0 and E_a^0. This special structure is a consequence of Remark 6.7 and the characterization of the space given in (3.41) in Chapter 3.

We need a modification of that decomposition. In a suitable basis, we can obtain for A_{aa}^0, B_{a0}^0, and E_a^0:

$$\begin{pmatrix} \bar{A}_1 & 0 \\ 0 & A_{a0}^{33} \end{pmatrix}, \quad \begin{pmatrix} \bar{B}_1 \\ B_{a0}^{03} \end{pmatrix}, \quad \begin{pmatrix} \bar{E}_1 \\ 0 \end{pmatrix},$$

respectively where $\bar{A}_1$ contains all Jordan blocks of A_{aa}^0 that are affected by the disturbances through E_a^0, whereas A_{a0}^{33} contains all Jordan blocks of A_{aa}^0 that are not affected by the disturbances through E_a^0.

Next we decompose $\bar{A}_1$ into the following block diagonal form:

$$\bar{A}_1 = \begin{pmatrix} \bar{A}_1^1 & 0 & \cdots & 0 \\ 0 & \bar{A}_1^2 & \ddots & \vdots \\ \vdots & \ddots & \ddots & 0 \\ 0 & \cdots & 0 & \bar{A}_1^k \end{pmatrix},$$

where, for $i = 1, \ldots, k$, block $\bar{A}_1^i$ has either one real eigenvalue λ_i or two complex eigenvalues λ_i and its conjugate λ_i^* and all blocks have distinct eigenvalues. We decompose $\bar{B}_1$ and $\bar{E}_1$ compatibly:

$$\bar{B}_1 = \begin{pmatrix} \bar{B}_1^1 \\ \vdots \\ \bar{B}_1^k \end{pmatrix}, \quad \bar{E}_1 = \begin{pmatrix} \bar{E}_1^1 \\ \vdots \\ \bar{E}_1^k \end{pmatrix}.$$

Next we use the Kalman decomposition such that we have the following extra decomposition:

$$\bar{A}_1^i = \begin{pmatrix} A_{11}^i & A_{12}^i \\ 0 & A_{22}^i \end{pmatrix}, \quad \bar{B}_1^i = \begin{pmatrix} B_{1,i} \\ B_{2,i} \end{pmatrix}, \quad \bar{E}_1^i = \begin{pmatrix} E_{1,i} \\ 0 \end{pmatrix},$$

where $(A_{11}^i, E_{1,i})$ is controllable. If H_∞ ADD problem is solvable, then we have, using the alternative characterization of Scherer as presented in Remark 6.7, that (6.12) is satisfied. Using the characterization of

$$\bigcap_{\lambda \in \mathbb{C}^0} \mathcal{S}_\lambda(\Sigma_{\text{sub}})$$

presented in Chapter 3, this implies that we must have that the geometric multiplicity of λ_i as an eigenvalue of $\bar{A}_1^i$ is equal to the geometric multiplicity of λ_i as an eigenvalue of A_{22}^i. Using a permutation of the above-described basis, we find that a basis transformation T_{a0} exists such that

$$T_{a0}^{-1} A_{aa}^0 T_{a0} = \begin{pmatrix} A_{a0}^{11} & A_{a0}^{12} & 0 \\ 0 & A_{a0}^{22} & 0 \\ 0 & 0 & A_{a0}^{33} \end{pmatrix}, \quad T_{a0}^{-1} E_a^0 = \begin{pmatrix} E_a^{01} \\ 0 \\ 0 \end{pmatrix},$$

$$T_{a0}^{-1} B_{a0}^0 = \begin{pmatrix} B_{a0}^{01} \\ B_{a0}^{02} \\ B_{a0}^{03} \end{pmatrix}, \quad T_{a0}^{-1} x_a^0 = \begin{pmatrix} x_{a1}^0 \\ x_{a2}^0 \\ x_{a3}^0 \end{pmatrix},$$

$$(6.44)$$

with

$$
A_{a0}^{11} = \begin{pmatrix} A_{11}^1 & 0 & \cdots & 0 \\ 0 & A_{11}^2 & \ddots & \vdots \\ \vdots & \ddots & \ddots & 0 \\ 0 & \cdots & 0 & A_{11}^k \end{pmatrix}, \qquad A_{a0}^{22} = \begin{pmatrix} A_{22}^1 & 0 & \cdots & 0 \\ 0 & A_{22}^2 & \ddots & \vdots \\ \vdots & \ddots & \ddots & 0 \\ 0 & \cdots & 0 & A_{22}^k \end{pmatrix},
$$

and

$$
A_{a0}^{12} = \begin{pmatrix} A_{12}^1 & 0 & \cdots & 0 \\ 0 & A_{12}^2 & \ddots & \vdots \\ \vdots & \ddots & \ddots & 0 \\ 0 & \cdots & 0 & A_{12}^k \end{pmatrix}.
$$

Step 4: Decomposition of x_d and y_d, and the definition of a fictitious output y_{d3}.

As in Step 3 of the H_2 ADD algorithm, in this step, depending on how the disturbance w affects the state x_d and the output y_d, we reorder the state x_d and thus the output y_d and the input u_d. Such a reordering allows us to partition x_d and y_d as

$$
x_d = \begin{pmatrix} x_{d1} \\ x_{d2} \end{pmatrix}, \qquad y_d = \begin{pmatrix} y_{d1} \\ y_{d2} \end{pmatrix}.
$$

We define also a fictitious output vector y_{d3}.

We continue here exactly as in Step 3 of the H_2 ADD algorithm. To proceed with, to explain our method of reordering and partitioning of x_d and y_d, let us first recall the dynamics of x_d. As indicated in Chapter 3, in SCB there is an array of chains of integrators extending from the input u_d to the output y_d. There are m_d number of integrator chains, and the outputs of all integrators of all chains form the state x_d. Let us consider an ith chain of integrators extending from the input u_i to the output y_i as shown in Figure 3.4 where $1 \le i \le m_d$. In our method of decomposing x_d and y_d, starting with the output end, we examine each integrator input for the presence or absence of any components of the disturbance w. In such an examination, we encounter three possible cases.

Case 1: In this case, none of the integrators in the chain contain any components of the disturbance w in their inputs. For this case, all state variables belonging to the chain are allotted as components of the state x_{d1}. Moreover, the output y_i of the ith chain is allotted as a component of the output y_{d1}.

Case 2: In this case, as we examine the chain of integrators starting from the output end, the very first integrator has at least one or more components of the disturbance w in its input. For this case, all state variables belonging to the chain are allotted as components of the state x_{d2}. Moreover, the output y_i of the ith chain is allotted as a component of the output y_{d2}.

Case 3: In this case, as we examine the chain of integrators starting from the output end, the very first integrator in the beginning does not contain any components of the disturbance w in its input. Also, among other integrators in the chain, there exists at least one integrator having an input containing one or more components of the disturbance w in it. For this case, some state variables of the chain are components of x_{d1} whereas the others are components of x_{d2}. The procedure of how the state variables are allotted as components of x_{d1} or x_{d2} is as follows: As we examine, all outputs of the integrators in the chain from the output end until up to but not including the first integrator that contains one or more components of w in its input are selected as components of the state x_{d1}. The outputs of the rest of the integrators in the chain are selected as components of the state x_{d2}. The output y_i of the ith chain is allotted as a component of the output y_{d1}. For this case, we also define a fictitious output. The state variable of the very first integrator in the chain of integrators that has components of the disturbance in its input is called a fictitious output and is assigned as a component of a fictitious output vector denoted by y_{d3}.

We repeat the above procedure of decomposition for all chains of integrators, i.e., for all i, $1 \leq i \leq m_d$.

Let us define next matrices C_{d1} and C_{d2} such that

$$y_d = \begin{pmatrix} y_{d1} \\ y_{d2} \end{pmatrix} = \begin{pmatrix} C_{d1} & 0 \\ 0 & C_{d2} \end{pmatrix} \begin{pmatrix} x_{d1} \\ x_{d2} \end{pmatrix}.$$

Also, let us define next a matrix C_{d3} such that

$$y_{d3} = C_{d3} x_{d2}.$$

We observe now that the above reordering and partitioning of x_d and y_d as well as the definition of the fictitious output vector y_{d3} necessitates a compatible reordering and partitioning of the input u_d and the matrices A_{dd}, C_d, B_d, and E_d. When these matrices are reordered and partitioned, they have the following structure (note that the bar over the matrices signifies that the reordering has taken place):

$$\bar{A}_{dd} = \begin{pmatrix} A_{d1} & L_{d1}C_{d23} \\ L_{d2}C_{d1} & A_{d2} \end{pmatrix}, \quad C_{d23} = \begin{pmatrix} C_{d2} \\ C_{d3} \end{pmatrix}, \quad \bar{C}_d = \begin{pmatrix} C_{d1} & 0 \\ 0 & C_{d2} \end{pmatrix},$$

$$\bar{B}_d = \begin{pmatrix} B_{d1} & 0 \\ 0 & B_{d2} \end{pmatrix}, \quad \bar{E}_d = \begin{pmatrix} 0 \\ E_{d2} \end{pmatrix}.$$

Here $(A_{d2}, B_{d2}, C_{d23}, 0)$ is invertible without finite zeros and $C_{d23}E_{d2}$ is injective.

Step 5: Modified SCB.

The goal of this step is to incorporate the reordering of state, input, and output variables of the SCB that occurred in the last two steps into a new compact form of the SCB. Such a reordering implies that the transformation matrices Γ_s, Γ_i, and Γ_o that take the given system Σ_{sub} to its SCB form need to be amended as $\bar{\Gamma}_s$,

$\bar{\Gamma}_i$, and $\bar{\Gamma}_o$. Then, a matrix F_{pre} of suitable dimension and a preliminary feedback law

$$u = -F_{\text{pre}}x + v \tag{6.45}$$

exist such that in a suitable basis, we get the following structure:

$$\bar{\Gamma}_s^{-1}(A - BF_{\text{pre}})\bar{\Gamma}_s =$$

$$\begin{pmatrix}
A_{aa}^{11} & 0 & 0 & 0 & A_{aa}^{12} & 0 & L_{ab}^{1-}C_b & L_{ad}^{11}C_{d1} & 0 & L_{ad}^{12}C_{d2} \\
0 & A_{a0}^{11} & A_{a0}^{12} & 0 & 0 & 0 & L_{ab}^{01}C_b & L_{ad}^{21}C_{d1} & 0 & L_{ad}^{22}C_{d2} \\
0 & 0 & A_{a0}^{22} & 0 & 0 & 0 & L_{ab}^{02}C_b & L_{ad}^{31}C_{d1} & 0 & L_{ad}^{32}C_{d2} \\
0 & 0 & 0 & A_{a0}^{33} & 0 & 0 & L_{ab}^{03}C_b & L_{ad}^{41}C_{d1} & 0 & L_{ad}^{42}C_{d2} \\
0 & 0 & 0 & 0 & A_{aa}^{22} & 0 & L_{ab}^{2-}C_b & L_{ad}^{51}C_{d1} & 0 & L_{ad}^{52}C_{d2} \\
0 & 0 & 0 & 0 & 0 & A_{aa}^{+} & L_{ab}^{+}C_b & L_{ad}^{+1}C_{d1} & 0 & L_{ad}^{+2}C_{d2} \\
0 & 0 & 0 & 0 & 0 & 0 & A_{bb} & L_{bd}^{1}C_{d1} & 0 & L_{bd}^{2}C_{d2} \\
0 & 0 & 0 & 0 & 0 & 0 & 0 & A_{d1} & 0 & L_{d1}C_{d23} \\
0 & 0 & 0 & 0 & 0 & 0 & 0 & L_{cd}^{1}C_{d1} & A_{cc} & L_{cd}^{2}C_{d2} \\
0 & 0 & 0 & 0 & 0 & 0 & 0 & L_{d2}C_{d1} & 0 & A_{d2}
\end{pmatrix},$$

$$\bar{\Gamma}_s^{-1}B\bar{\Gamma}_i = \begin{pmatrix}
B_{a0}^{1-} & 0 & 0 & 0 \\
B_{a0}^{01} & 0 & 0 & 0 \\
B_{a0}^{02} & 0 & 0 & 0 \\
B_{a0}^{03} & 0 & 0 & 0 \\
B_{a0}^{2-} & 0 & 0 & 0 \\
B_{a0}^{+} & 0 & 0 & 0 \\
B_{b0} & 0 & 0 & 0 \\
B_{d0}^{1} & B_{d1} & 0 & 0 \\
B_{c0} & 0 & 0 & B_c \\
B_{d0}^{2} & 0 & B_{d2} & 0
\end{pmatrix}, \qquad \bar{\Gamma}_s^{-1}E = \begin{pmatrix}
E_a^{1-} \\
E_a^{01} \\
0 \\
0 \\
0 \\
0 \\
0 \\
0 \\
E_c \\
E_{d2}
\end{pmatrix},$$

$$\bar{\Gamma}_o^{-1}(C - DF_{\text{pre}})\bar{\Gamma}_s = \begin{pmatrix}
0 & 0 & 0 & 0 & 0 & 0 & 0 & 0 & 0 & 0 \\
0 & 0 & 0 & 0 & 0 & 0 & 0 & C_{d1} & 0 & 0 \\
0 & 0 & 0 & 0 & 0 & 0 & C_b & 0 & 0 & 0 \\
0 & 0 & 0 & 0 & 0 & 0 & 0 & 0 & 0 & C_{d2}
\end{pmatrix},$$

$$\bar{\Gamma}_o^{-1}D\bar{\Gamma}_i = \begin{pmatrix}
I_{m0} & 0 & 0 & 0 \\
0 & 0 & 0 & 0 \\
0 & 0 & 0 & 0 \\
0 & 0 & 0 & 0
\end{pmatrix}.$$

Based on the above development, we can finally form the matrices A_z, B_z, and C_z as follows while using obvious notation:

$$A_z := \begin{pmatrix} A_{a0}^{33} & 0 & 0 & L_{ab}^{03}C_b & L_{ad}^{41}C_{d1} \\ 0 & A_{aa}^{22} & 0 & L_{ab}^{2-}C_b & L_{ad}^{51}C_{d1} \\ 0 & 0 & A_{aa}^{+} & L_{ab}^{+}C_b & L_{ad}^{+1}C_{d1} \\ 0 & 0 & 0 & A_{bb} & L_{bd}^{1}C_{d1} \\ 0 & 0 & 0 & 0 & A_{d1} \end{pmatrix}, \tag{6.46}$$

$$B_z := \begin{pmatrix} L_{ad}^{42}C_{d2} & B_{a0}^{03} & 0 \\ L_{ad}^{52}C_{d2} & B_{a0}^{2-} & 0 \\ L_{ad}^{+2}C_{d2} & B_{a0}^{+} & 0 \\ L_{bd}^{2}C_{d2} & B_{b0} & 0 \\ L_{d1}C_{d23} & B_{d0}^{1} & B_{d1} \end{pmatrix} = \begin{pmatrix} \tilde{L}_{zd}C_{d23} & B_{z0} & B_{z1} \end{pmatrix}, \tag{6.47}$$

and

$$C_z := \begin{pmatrix} 0 & 0 & 0 & 0 & C_{d1} \\ 0 & 0 & 0 & C_b & 0 \end{pmatrix}. \tag{6.48}$$

From Property 3.3 of the SCB, it is simple to verify that the pair (A_z, B_z) is $\mathbb{C}^-$ stabilizable when the pair (A, B) is $\mathbb{C}^-$ stabilizable.

Using the above definition, we can rewrite the system more compactly as

$$\bar{\Gamma}_s^{-1}(A - BF_{\text{pre}})\bar{\Gamma}_s = \begin{pmatrix} A_{aa}^{11} & 0 & 0 & A_{az}^{-} & 0 & \tilde{L}_{ad}^{12}C_{d23} \\ 0 & A_{a0}^{11} & A_{a0}^{12} & A_{az}^{01} & 0 & \tilde{L}_{ad}^{22}C_{d23} \\ 0 & 0 & A_{a0}^{22} & A_{az}^{02} & 0 & \tilde{L}_{ad}^{32}C_{d23} \\ 0 & 0 & 0 & A_z & 0 & \tilde{L}_{zd}C_{d23} \\ 0 & 0 & 0 & A_{cz} & A_{cc} & \tilde{L}_{cd}^{2}C_{d23} \\ 0 & 0 & 0 & A_{dz} & 0 & A_{d2} \end{pmatrix},$$

$$\bar{\Gamma}_s^{-1}B\bar{\Gamma}_i = \begin{pmatrix} B_{a0}^{1-} & 0 & 0 & 0 \\ B_{a0}^{01} & 0 & 0 & 0 \\ B_{a0}^{02} & 0 & 0 & 0 \\ B_{z0} & B_{z1} & 0 & 0 \\ B_{c0} & 0 & 0 & B_c \\ B_{d0}^{2} & 0 & B_{d2} & 0 \end{pmatrix}, \quad \bar{\Gamma}_s^{-1}E = \begin{pmatrix} E_a^{1-} \\ E_a^{01} \\ 0 \\ 0 \\ E_c \\ E_{d2} \end{pmatrix},$$

$$\bar{\Gamma}_o^{-1}(C - DF_{\text{pre}})\bar{\Gamma}_s = \begin{pmatrix} 0 & 0 & 0 & 0 & 0 & 0 \\ 0 & 0 & 0 & C_z & 0 & 0 \\ 0 & 0 & 0 & 0 & 0 & C_{d2} \end{pmatrix},$$

$$\bar{\Gamma}_o^{-1} D \bar{\Gamma}_i = \begin{pmatrix} I_{m_0} & 0 & 0 & 0 \\ 0 & 0 & 0 & 0 \\ 0 & 0 & 0 & 0 \end{pmatrix}.$$

In view of the above structure, the preliminary feedback law (6.45) can be rewritten as

$$\bar{u} = -\bar{\Gamma}_i^{-1} F_{\text{pre}} \bar{\Gamma}_s \bar{x} + \bar{v}, \tag{6.49}$$

where

$$\bar{x} = \bar{\Gamma}_s^{-1} x, \quad \bar{u} = \bar{\Gamma}_i^{-1} u, \quad \text{and} \quad \bar{v} = \bar{\Gamma}_i^{-1} v = \begin{pmatrix} v_1 \\ v_2 \\ v_3 \\ v_4 \end{pmatrix}. \tag{6.50}$$

Step 6: Computation of $\boldsymbol{\Omega}_s^\infty$.

We can determine the set of H_∞ ADD finite asymptotic fixed modes $\boldsymbol{\Omega}_s^\infty$ as

$$\boldsymbol{\Omega}_s^\infty = \lambda(A_{aa}^{11}) \cup \lambda(A_{a0}^{11}) \cup \{\text{input decoupling zeros of } (A_z, B_z)\}. \tag{6.51}$$

Step 7: Construction of H_∞ ADD families of controllers.

The next step is to find a sequence of state feedback controllers that solves H_∞ ADD while possibly assigning the asymptotic behavior of the closed-loop poles arbitrarily as desired beyond the above fixed modes.

Step 7a: Assigning the first part of finite asymptotic modes.

Choose

$$\bar{F} = \begin{pmatrix} F_1 \\ F_2 \\ F_3 \end{pmatrix}$$

such that $A_z - B_z \bar{F}$ is stable and has the desirable eigenvalues in the open left-half complex plane. Note that $\bar{F}$ is partitioned in conformity with the partitioning of B_z as given in (6.47). Also, note that the desirable eigenvalues must obviously contain the set of input decoupling zeros of (A_z, B_z), which is a subset of $\boldsymbol{\Omega}_s^\infty$.

Step 7b: Assigning the second part of finite asymptotic modes.

Choose F_4 such that $A_{cc} - B_c F_4$ is stable and has the desirable eigenvalues in the open left-half complex plane. As (A_{cc}, B_c) is always controllable, all eigenvalues of $A_{cc} - B_c F_4$ can be assigned to arbitrary locations.

Step 7c: Assigning the third part of finite asymptotic modes combined with a low-gain design.

We first define

$$
\tilde{B}_{az} = \begin{pmatrix} \tilde{L}^{22}_{ad}C_{d23} & B^{01}_{a0} & 0 \\ \tilde{L}^{32}_{ad}C_{d23} & B^{02}_{a0} & 0 \\ \tilde{L}_{zd}C_{d23} & B_{z0} & B_{z1} \end{pmatrix} = \begin{pmatrix} B^{11}_{a0} \\ B^{12}_{a0} \\ B_z \end{pmatrix}, \quad \tilde{E}_{az} = \begin{pmatrix} E^{01}_a \\ 0 \\ 0 \end{pmatrix},
$$

and

$$
\tilde{A}_{az} = \begin{pmatrix} A^{11}_{a0} & A^{12}_{a0} & A^{01}_{az} \\ 0 & A^{22}_{a0} & A^{02}_{az} \\ 0 & 0 & A_z \end{pmatrix} - \tilde{B}_{az}\bar{F}\begin{pmatrix} 0 & 0 & I \end{pmatrix} = \begin{pmatrix} A^{11}_{a0} & A^{12}_{a0} & \tilde{A}^{01}_{az} \\ 0 & A^{22}_{a0} & \tilde{A}^{02}_{az} \\ 0 & 0 & A_{zF} \end{pmatrix}.
$$

Our goal here is to construct a sequence of feedback gains $\{\tilde{F}^\varepsilon\}_{\varepsilon>0}$ such that the system,

$$
\left(\tilde{A}_{az} - \tilde{B}_{az}\tilde{F}^\varepsilon, \tilde{E}_{az}, \begin{pmatrix} -\tilde{F}^\varepsilon_2 - F_2\begin{pmatrix} 0 & 0 & I \end{pmatrix} \\ \begin{pmatrix} 0 & 0 & C_z \end{pmatrix} \end{pmatrix}, 0 \right), \tag{6.52}
$$

is stable for all $\varepsilon > 0$; the H_∞ norm of it converges to zero as $\varepsilon \to 0$. Moreover, part of the eigenvalues of $\tilde{A}_{az} - \tilde{B}_{az}\tilde{F}^\varepsilon$ converge to the eigenvalues of A^{11}_{a0} and A_{zF} while the remaining eigenvalues can be assigned arbitrarily. Here we used the decomposition,

$$
\tilde{F}^\varepsilon = \begin{pmatrix} \tilde{F}^\varepsilon_1 \\ \tilde{F}^\varepsilon_2 \\ \tilde{F}^\varepsilon_3 \end{pmatrix},
$$

which is compatible with the decomposition of $\tilde{B}_{az}$. In what follows, we transform the task of designing the feedback gain $\tilde{F}^\varepsilon$ to a similar task of designing a feedback gain for another system that has a suitable structure. To do so, we start with a basis transformation T_1 such that we obtain the appropriate structure we are looking for. Because the matrices

$$
\begin{pmatrix} A^{11}_{a0} & A^{12}_{a0} \\ 0 & A^{22}_{a0} \end{pmatrix}
$$

and A_{zF} have disjoint eigenvalues, a basis transformation of the form

$$
T_1 = \begin{pmatrix} I & 0 & T^{13}_1 \\ 0 & I & T^{23}_1 \\ 0 & 0 & I \end{pmatrix}
$$

exists such that

$$
T_1^{-1}\tilde{A}_{az}T_1 = \begin{pmatrix} A^{11}_{a0} & A^{12}_{a0} & 0 \\ 0 & A^{22}_{a0} & 0 \\ 0 & 0 & A_{zF} \end{pmatrix}, \quad T_1^{-1}\tilde{B}_{az} = \begin{pmatrix} \tilde{B}^{11}_{a0} \\ \tilde{B}^{12}_{a0} \\ B_z \end{pmatrix},
$$

and

$$T_1^{-1}\widetilde{E}_{az} = \begin{pmatrix} E_a^{01} \\ 0 \\ 0 \end{pmatrix}, \quad \begin{pmatrix} x_\ell \\ x_z \end{pmatrix} = T_1^{-1}\begin{pmatrix} x_{a1}^0 \\ x_{a2}^0 \\ x_z \end{pmatrix}.$$

Let us a define a new system Σ_ℓ as

$$\Sigma_\ell \begin{cases} \dot{x}_\ell = \begin{pmatrix} A_{a0}^{11} & A_{a0}^{12} \\ 0 & A_{a0}^{22} \end{pmatrix} x_\ell + \begin{pmatrix} \widetilde{B}_{a0}^{11} \\ \widetilde{B}_{a0}^{12} \end{pmatrix} u_\ell + \begin{pmatrix} E_a^{01} \\ 0 \end{pmatrix} w_\ell \\ z_\ell = u_\ell. \end{cases} \tag{6.53}$$

In view of the above development, it can be easily seen that the task of designing the feedback gain $\widetilde{F}^\varepsilon$ can be converted to the task of designing a feedback gain F_ℓ^ε such that the state feedback $u_\ell = -F_\ell^\varepsilon x_\ell$ when applied to the system Σ_ℓ stabilizes the resulting closed-loop system while the transfer matrix from w_ℓ to z_ℓ has an H_∞ norm converging to zero as $\varepsilon \to 0$. Finally, the closed-loop eigenvalues partially converge to the eigenvalues of A_{a0}^{11} while the remaining asymptotic locations of the eigenvalues can be freely assigned.

After obtaining such a F_ℓ^ε, we can obtain $\widetilde{F}^\varepsilon$ having the required properties we sought earlier by setting $\widetilde{F}^\varepsilon = \begin{pmatrix} F_\ell^\varepsilon & 0 \end{pmatrix} T_1^{-1}$.

We proceed now to design the feedback gain F_ℓ^ε. We are going to do so using a recursive method. We design $k - 1$ preliminary feedback controllers, and in the last step, we obtain a regular feedback controller. These preliminary feedback controllers are partial low-gain feedback controllers designed for a subsystem with a special structure that is preserved in the recursive steps. We initialize by setting $i = 1$.

We have the following system:

$$\Sigma_\ell^i \begin{cases} \dot{x}_\ell^i = \begin{pmatrix} \overline{A}_{11}^i & \overline{A}_{12}^i \\ 0 & \overline{A}_{22}^i \end{pmatrix} x_\ell^i + \begin{pmatrix} \widetilde{B}_1^i \\ \widetilde{B}_2^i \end{pmatrix} v_\ell^i + \begin{pmatrix} \widetilde{E}^i \\ 0 \end{pmatrix} w_\ell, \\ z_\ell^i = v_\ell^i, \end{cases} \tag{6.54}$$

which, for $i = 1$, is equal to Σ_ℓ where we define,

$$x_\ell^1 = x_\ell, v_\ell^1 = u_\ell, z_\ell^1 = z_\ell, \overline{A}_{11}^1 = A_{a0}^{11}, \overline{A}_{12}^1 = A_{a0}^{12}, \overline{A}_{22}^1 = A_{a0}^{22},$$
$$\widetilde{B}_1^1 = \widetilde{B}_{a0}^{11}, \widetilde{B}_2^1 = \widetilde{B}_{a0}^{12}, \widetilde{E}^1 = E_a^{01}.$$

We decompose the state,

$$x_\ell^i = \begin{pmatrix} x_{1,i}^i \\ \vdots \\ x_{1,k}^i \\ x_{2,i}^i \\ \vdots \\ x_{2,k}^i \end{pmatrix}, \tag{6.55}$$

compatible with the special structure of the matrices $\bar{A}^i_{11}$ and $\bar{A}^i_{22}$, which for $i = 1$, follows from the special structure of the matrices A^{11}_{a0}, A^{12}_{a0}, and A^{22}_{a0} as explained earlier in Step 3. We also decompose $\tilde{B}^i_1$, $\tilde{B}^i_2$, and $\tilde{E}^i$ compatibly to the decomposition of $\bar{A}^i_{11}$ and $\bar{A}^i_{22}$,

$$
\tilde{B}^i_1 = \begin{pmatrix} B^i_{1,i} \\ \vdots \\ B^i_{1,k} \end{pmatrix}, \quad
\tilde{B}^i_2 = \begin{pmatrix} B^i_{2,i} \\ \vdots \\ B^i_{2,k} \end{pmatrix}, \quad
\tilde{E}^i = \begin{pmatrix} E^i_{1,i} \\ \vdots \\ E^i_{1,k} \end{pmatrix}.
$$

We next reorder and partition the states of the system Σ^i_ℓ as

$$
x^i_i = \begin{pmatrix} x^i_{1,i} \\ x^i_{2,i} \end{pmatrix}, \quad
\check{x}^i = \begin{pmatrix} x^i_{1,i+1} \\ \vdots \\ x^i_{1,k} \\ x^i_{2,i+1} \\ \vdots \\ x^i_{2,k} \end{pmatrix}.
$$

With the above reordering and partitioning, we can rewrite the system Σ^i_ℓ as

$$
\Sigma^i_\ell : \begin{cases}
\dot{x}^i_i = \begin{pmatrix} A^i_{11} & A^i_{12} \\ 0 & A^i_{22} \end{pmatrix} x^i_i + \begin{pmatrix} B^i_{1,i} \\ B^i_{2,i} \end{pmatrix} v^i_\ell + \begin{pmatrix} E^i_{1,i} \\ 0 \end{pmatrix} w_\ell \\[3mm]
\dot{\check{x}}^i = \begin{pmatrix} \bar{A}^{i+1}_{11} & \bar{A}^{i+1}_{12} \\ 0 & \bar{A}^{i+1}_{22} \end{pmatrix} \check{x}^i + \begin{pmatrix} \check{B}^i_1 \\ \check{B}^i_2 \end{pmatrix} v^i_\ell + \begin{pmatrix} \check{E}^i_{1,i} \\ 0 \end{pmatrix} w_\ell \\[3mm]
z^i_\ell = v^i_\ell,
\end{cases}
$$

where we define

$$
\bar{A}^{i+1}_{11} = \begin{pmatrix} A^{i+1}_{11} & 0 & \cdots & 0 \\ 0 & A^{i+2}_{11} & \ddots & \vdots \\ \vdots & \ddots & \ddots & 0 \\ 0 & \cdots & 0 & A^k_{11} \end{pmatrix}, \quad
\bar{A}^{i+1}_{12} = \begin{pmatrix} A^{i+1}_{12} & 0 & \cdots & 0 \\ 0 & A^{i+2}_{12} & \ddots & \vdots \\ \vdots & \ddots & \ddots & 0 \\ 0 & \cdots & 0 & A^k_{12} \end{pmatrix},
$$

$$
\bar{A}^{i+1}_{22} = \begin{pmatrix} A^{i+1}_{22} & 0 & \cdots & 0 \\ 0 & A^{i+2}_{22} & \ddots & \vdots \\ \vdots & \ddots & \ddots & 0 \\ 0 & \cdots & 0 & A^k_{22} \end{pmatrix},
$$

and

$$
\breve{B}_1^i = \begin{pmatrix} B_{1,i+1}^i \\ \vdots \\ B_{1,k}^i \end{pmatrix}, \quad
\breve{B}_2^i = \begin{pmatrix} B_{2,i+1}^i \\ \vdots \\ B_{2,k}^i \end{pmatrix}, \quad
\breve{E}_{1,i}^i = \begin{pmatrix} E_{1,i+1}^i \\ \vdots \\ E_{1,k}^i \end{pmatrix}.
$$

In our design in step i, we design a preliminary feedback controller

$$
v_\ell^i = -F_{\varepsilon_i}^i x_i^i + v_\ell^{i+1} \tag{6.56}
$$

with suitable properties where for $i = k$, we set $v_\ell^{k+1} = 0$.

We know that the pair,

$$
\left[\begin{pmatrix} A_{11}^i & A_{12}^i \\ 0 & A_{22}^i \end{pmatrix}, \begin{pmatrix} B_{1,i}^i \\ B_{2,i}^i \end{pmatrix} \right] \tag{6.57}
$$

is controllable. Note that the matrix $\overline{A}_1^i$,

$$
\overline{A}_1^i = \begin{pmatrix} A_{11}^i & A_{12}^i \\ 0 & A_{22}^i \end{pmatrix}, \tag{6.58}
$$

has only eigenvalues λ_i and its complex conjugate λ_i^* in case λ_i is complex. Hence, for λ_i, a matrix K_i exists such that

$$
\left[\lambda_i I - \begin{pmatrix} A_{11}^i & A_{12}^i \\ 0 & A_{22}^i \end{pmatrix} \quad \begin{pmatrix} B_{1,i}^i \\ B_{2,i}^i \end{pmatrix} K_i \right]
$$

has full row rank, whereas the number of columns n_i of K_i is equal to the geometric multiplicity of λ_i as an eigenvalue of $\overline{A}_1^i$. By the Hautus test, this implies that the pair

$$
\left[\begin{pmatrix} A_{11}^i & A_{12}^i \\ 0 & A_{22}^i \end{pmatrix}, \begin{pmatrix} B_{1,i}^i \\ B_{2,i}^i \end{pmatrix} K_i \right]
$$

is controllable.

Note that, by the special structure of (6.44), the geometric multiplicity of an eigenvalue of (6.58) equals the geometric order as an eigenvalue of A_{22}^i. Let $\widetilde{F}_{2,2}^i$ be such that $A_{22}^i - B_{2,i}^i K_i \widetilde{F}_{2,2}^i$ is stable and has eigenvalues $\{\mu_1, \ldots, \mu_{m_i}\}$ at the desirable locations. This is possible due to the controllability of the pair $(A_{22}^i, B_{2,i}^i K_i)$. We choose the preliminary feedback controller (6.56) of the form

$$
v_\ell^i = -F_{\varepsilon_i}^i x_i^i + v_\ell^{i+1} = K_i \left[-B_i' P_{\varepsilon_i}^i - \begin{pmatrix} 0 & \widetilde{F}_{2,2}^i \end{pmatrix} \right] x_i^i + v_\ell^{i+1}, \tag{6.59}
$$

where $P_{\varepsilon_i}^i$ is the stabilizing solution of the algebraic Riccati equation H_∞^1 CARE,

$$
A_i' P + P A_i - P B_i B_i' P + \gamma^{-2} P E_i E_i' P + \varepsilon_i^2 I = 0, \tag{6.60}
$$

where γ is fixed (independent of ε_i) and large enough such that this equation has a solution for all ε_i small enough. Moreover,

$$A_i = \begin{pmatrix} A_{11}^i & A_{12}^i - B_{1,i}^i K_i \tilde{F}_{2,2}^i \\ 0 & A_{22}^i - B_{2,i}^i K_i \tilde{F}_{2,2}^i \end{pmatrix}, \quad B_i = \begin{pmatrix} B_{1,i}^i \\ B_{2,i}^i \end{pmatrix} K_i, \quad E_i = \begin{pmatrix} E_{1,i}^i \\ 0 \end{pmatrix}.$$

We apply this preliminary feedback (6.59) to (6.54). Note that this is not a low-gain feedback because the matrix $\tilde{F}_{2,2}^i$ does not depend on ε_i and is hence fixed. Clearly,

$$\begin{pmatrix} \overline{A}_{11}^{i+1} & \overline{A}_{12}^{i+1} \\ 0 & \overline{A}_{22}^{i+1} \end{pmatrix} \tag{6.61}$$

has only eigenvalues on the imaginary axis. Consider the system Σ_ℓ^i after the preliminary feedback (6.59). We define

$$\overline{A}_i = A_i - B_i B_i' P_{\varepsilon_i}^i, \qquad \overline{C}_i = -F_{\varepsilon_i}^i.$$

Note that the matrix $\overline{A}_i$ is asymptotically stable. Hence, (6.61) and $\overline{A}_i$ have disjoint eigenvalues, and it is easily seen that this implies that a matrix X^{ε_i} exists such that the basis transformation $x_\ell^{i+1} = \breve{x}^i - X^{\varepsilon_i} x_i^i$ applied to Σ_ℓ^i, after applying the preliminary feedback (6.59), results in a system having the structure,

$$\tilde{\Sigma}_\ell^i \left\{ \begin{aligned} \begin{pmatrix} \dot{x}_i^i \\ \dot{x}_\ell^{i+1} \end{pmatrix} &= \begin{pmatrix} \overline{A}_i & 0 & 0 \\ 0 & \overline{A}_{11}^{i+1} & \overline{A}_{12}^{i+1} \\ 0 & 0 & \overline{A}_{22}^{i+1} \end{pmatrix} \begin{pmatrix} x_i^i \\ x_\ell^{i+1} \end{pmatrix} + \begin{pmatrix} B_i \\ \tilde{B}_1^{i+1} \\ \tilde{B}_2^{i+1} \end{pmatrix} v_\ell^{i+1} \\ &\quad + \begin{pmatrix} E_i \\ \tilde{E}^{i+1} \\ 0 \end{pmatrix} w_\ell, \\ \tilde{z}_\ell^{i+1} &= \begin{pmatrix} \overline{C}_i & 0 & 0 \end{pmatrix} \begin{pmatrix} x_i^i \\ x_\ell^{i+1} \end{pmatrix} + v_\ell^{i+1}, \end{aligned} \right. \tag{6.62}$$

where various matrices are defined appropriately and the H_∞ norm of the subsystem $(\overline{A}_i, E_i, \overline{C}_i, 0)$ converges to zero as $\varepsilon_i \to 0$, whereas part of the eigenvalues of $\overline{A}_i$ converges to eigenvalues of A_{11}^i while the remaining eigenvalues converge to the arbitrarily chosen locations $\{\mu_1, \ldots, \mu_{m_i}\}$. This will be proven in the proof of Theorem 6.26. Hence, for any ρ, a constant ε_i exists, such that H_∞ norm of the subsystem $(\overline{A}_i, E_i, \overline{C}_i, 0)$ is less than or equal to ρ, whereas the eigenvalues of $\overline{A}_i$ are within distance ρ of their limiting values.

In view of (6.54) and (6.62), the dynamics of x_ℓ^{i+1} with output $z_\ell^{i+1} = v_\ell^{i+1}$ is given by

$$\Sigma_\ell^{i+1} \left\{ \begin{aligned} \dot{x}_\ell^{i+1} &= \begin{pmatrix} \overline{A}_{11}^{i+1} & \overline{A}_{12}^{i+1} \\ 0 & \overline{A}_{22}^{i+1} \end{pmatrix} x_\ell^{i+1} + \begin{pmatrix} \tilde{B}_1^{i+1} \\ \tilde{B}_2^{i+1} \end{pmatrix} v_\ell^{i+1} + \begin{pmatrix} \tilde{E}^{i+1} \\ 0 \end{pmatrix} w_\ell, \\ z_\ell^{i+1} &= v_\ell^{i+1}. \end{aligned} \right.$$

$$\tag{6.63}$$

Assume that we can design a sequence of controllers $v_\ell^{i+1} = -F_{\varepsilon_{i+1}}^{i+1} x_\ell^{i+1}$ parameterized by ε_{i+1} for the subsystem Σ_ℓ^{i+1} such that, as $\varepsilon_{i+1} \to 0$, the H_∞ norm converges to zero, whereas the closed-loop eigenvalues are all stable and converge to asymptotic locations consisting of the eigenvalues of $\bar{A}_{11}^{i+1}$ and some, a priori, chosen arbitrary stable locations. It can then be easily seen that, for ε_{i+1} small enough, the system (6.54) after applying the preliminary feedback (6.59) [which results in the system (6.62)] and the feedback $v_\ell^{i+1} = -F_{\varepsilon_{i+1}}^{i+1} x_\ell^{i+1}$ has H_∞ norm strictly less than 2ρ, whereas part of the closed-loop eigenvalues is within distance ρ of the eigenvalues of $\bar{A}_{11}$ and the remaining closed-loop eigenvalues are within distance ρ of the, a priori, chosen asymptotic locations.

We repeat the above procedure by setting $i = i + 1$. In this regard, we observe that the system Σ_ℓ^{i+1} has the same structure as Σ_ℓ^i and hence the above procedure can be repeated. After $k - 1$ steps we end up with the requirement of the need to design a feedback controller for the system,

$$
\Sigma_\ell^k \left\{
\begin{aligned}
\dot{x}_\ell^k &= \begin{pmatrix} A_{11}^k & A_{12}^k \\ 0 & A_{22}^k \end{pmatrix} x_\ell^k + \begin{pmatrix} \widetilde{B}_{1,1}^k \\ \widetilde{B}_{2,1}^k \end{pmatrix} v_\ell^k + \begin{pmatrix} \widetilde{E}_{1,1}^k \\ 0 \end{pmatrix} w_\ell \\
z_\ell^k &= v_\ell^k,
\end{aligned}
\right.
\tag{6.64}
$$

for which we need a controller $v_\ell^k = -F_{\varepsilon_k}^k x_\ell^k$ such that, as $\varepsilon_k \to 0$, the H_∞ norm converges to zero, whereas the closed-loop eigenvalues are all stable and converge to asymptotic locations consisting of the eigenvalues of A_{11}^k and some, a priori, chosen arbitrary stable locations. Obviously, for this last step, we can then design a feedback controller using the same methodology except that the need for extracting a subsystem is no longer needed.

By the above recursive design procedure, for any given ε, we can choose the constants $\varepsilon_1, \ldots, \varepsilon_k$ small enough such that the designed controller when applied to the system Σ_ℓ stabilizes the resulting closed-loop system, whereas the transfer matrix from w_ℓ to z_ℓ has an H_∞ norm less than ε and the closed-loop poles are within a radius ε of their asymptotic locations. Such a controller leads us to the the needed state feedback gain $\widetilde{F}^\varepsilon$.

Step 7d: High-gain design.

We construct a sequence of feedback gains $\{\bar{F}_\delta\}_{\delta>0}$ such that the system,

$$
(A_{d2} - B_{d2}\bar{F}_\delta, I, C_{d23}, 0),
$$

is stable for all $\delta > 0$ and the H_∞ norm converges to zero as $\delta \to 0$. For each δ, let $\bar{P}_\delta$ be the stabilizing solution of the algebraic Riccati equation H_∞^1 CARE,

$$
A_{d2}'\bar{P}_\delta + \bar{P}_\delta A_{d2} - \delta^{-2}\bar{P}_\delta B_{d2}B_{d2}'\bar{P}_\delta + \rho^{-2}\bar{P}_\delta^2 + C_{d23}'C_{d23} = 0,
$$

for some $\rho > 0$ in such a way that $\rho \to 0$ as $\delta \to 0$. Let $\bar{F}_\delta = -\delta^{-2}B_{d2}'\bar{P}_\delta$. Note that all eigenvalues of $A_{d2} - B_{d2}\bar{F}_\delta$ converge to infinity as δ converges to zero.

Step 7e: Combining individual components to obtain an H_∞ ADD family of controllers.

In this step, we construct finally a sequence of H_∞ ADD static state feedback controllers. To do so, we define

$$u = \bar{\Gamma}_i \bar{u} = -F_{\mathrm{pre}} x + \bar{\Gamma}_i \begin{pmatrix} v_1 \\ v_2 \\ v_3 \\ v_4 \end{pmatrix} := -F^\varepsilon x, \qquad (6.65)$$

where

$$v_1 = -\tilde{F}_2^\varepsilon x_\ell - F_2 x_z, \qquad (6.66)$$

$$v_2 = -\tilde{F}_3^\varepsilon x_\ell - F_3 x_z, \qquad (6.67)$$

$$v_3 = -\bar{F}_\delta \left(x_{d2} + \tilde{F}_1^\varepsilon x_\ell + F_1 x_z \right), \qquad (6.68)$$

$$v_4 = -F_4 x_c. \qquad (6.69)$$

For each ε we choose δ large enough, for instance, such that the H_∞ norm of the complete closed-loop system Σ is less than γ^ε.

This concludes the description of the H_∞ ADD algorithm. The following theorem proves the assertions made in this algorithm.

Theorem 6.26 *Consider the continuous-time system Σ as in (6.1) with the associated H_∞ ADD problem as defined by Problem 6.1. Assume that the solvability conditions for H_∞ ADD problem as given by Theorem 6.6 are satisfied; i.e., assume that (A, B) is $\mathbb{C}^-$-stabilizable,*

$$\mathrm{im}\, E \subseteq \mathcal{S}^*(\Sigma_{\mathrm{sub}}) + \mathcal{V}^{-0}(\Sigma_{\mathrm{sub}}),$$

and for any $\varepsilon > 0$ and any invariant zero s_0 of (A, B, C, D) on the imaginary axis, a matrix F exists such that $s_0 I - A + BF$ is invertible and

$$\|(C - DF)(s_0 I - A + BF)^{-1} E\| < \varepsilon.$$

Then the following hold:

(i) As claimed in (6.51), the set of all H_∞ ADD finite asymptotic fixed modes Ω_s^∞ defined in Definition 6.2 is given by

$$\Omega_s^\infty = \lambda(A_{aa}^{11}) \cup \lambda(A_{a0}^{11}) \cup \{input\ decoupling\ zeros\ of\ (A_z, B_z)\}.$$

(ii) Any sequences of static state feedback controllers as designed according to the H_∞ ADD algorithm, i.e., $u = -F^\varepsilon x$ as given by (6.65), solves the H_∞ ADD problem. In other words, the closed-loop system is stable for ε small enough, and the transfer matrix of the closed-loop system converges to 0 in the H_∞ norm as $\varepsilon \downarrow 0$.

Before we give the proof of this theorem, we present the following lemma:

Lemma 6.27 *Let $A \in \mathbb{R}^{n \times n}$ be a matrix such that*

$$\text{rank}(s_0 I - A) = n - n_k$$

for some $s_0 \in \mathbb{C}$. Let $B \in \mathbb{R}^{n \times n_k}$ and $F \in \mathbb{R}^{n_k \times n}$ be such that

$$\text{rank}(s_0 I - A + BF) = n.$$

Then we have

$$I - F(s_0 I - A + BF)^{-1} B = 0.$$

Proof : The proof of this lemma follows from the following:

$$n + \text{rank}(I - F(s_0 I - A + BF)^{-1} B)$$

$$= \text{rank} \begin{pmatrix} I - F(s_0 I - A + BF)^{-1} B & 0 \\ 0 & s_0 I - A + BF \end{pmatrix}$$

$$= \text{rank} \begin{pmatrix} I & F \\ B & s_0 I - A + BF \end{pmatrix}$$

$$= \text{rank} \begin{pmatrix} I & F \\ 0 & s_0 I - A \end{pmatrix}$$

$$= n_k + (n - n_k) = n,$$

where we use some standard Schur complement arguments. ∎

We proceed now to prove Theorem 6.26.

Proof of Theorem 6.26 : In the design, we claimed that our construction in Step 7c results in a family of feedback controllers such that the H_∞ norm of the system (6.52) converges to zero as $\varepsilon \to 0$. Moreover, we claimed that part of the closed-loop eigenvalues of it converges to the eigenvalues of A_{a0}^{11} and A_{zF}, whereas the remaining eigenvalues can be assigned arbitrarily. This needs to be substantiated.

We claim that the feedback

$$u_\ell = -F_{\varepsilon_i}^i x_i^i = -K_i \left[B_i' P_{\varepsilon_i}^i + \begin{pmatrix} 0 & \tilde{F}_{2,2}^i \end{pmatrix} \right] x_i^i$$

is such that the H_∞ norm of the subsystem $(\bar{A}_i, E_i, \bar{C}_i, 0)$ converges to zero as $\varepsilon_i \to 0$, whereas part of the eigenvalues of $\bar{A}_i$ converge to the eigenvalues of A_{11}^i while the remaining part converges to $\{\mu_1, \ldots, \mu_{m_i}\}$. To substantiate this, we first note that the eigenvalues of A_i are in the closed left-half plane, and hence, the stabilizing solution of the H_∞^1 CARE (6.60) converges to zero as $\varepsilon \to 0$. Therefore, we have that $\bar{A}_i \to A_i$ as $\varepsilon \to 0$, which immediately yields the desired asymptotic behavior of the eigenvalues. Next, we consider the H_∞ norm of this system. The transfer matrix of $(\bar{A}_i, E_i, \bar{C}_i, 0)$ can be rewritten as

$$G^{\varepsilon_i}(s) = G_1^{\varepsilon_i}(s) G_2^{\varepsilon_i}(s), \qquad (6.70)$$

where

$$G_1^{\varepsilon_i}(s) = I - \tilde{F}_{\varepsilon_i,2}^i (sI - A_{22}^i + \bar{B}_{2,i}^i \tilde{F}_{\varepsilon_i,2}^i)^{-1} \bar{B}_{2,i}^i,$$
$$G_2^{\varepsilon_i}(s) = -F_{\varepsilon_i,1}^i (sI - A_{11}^i + \bar{B}_{1,i}^i F_{\varepsilon_i,1}^i + R(s) F_{\varepsilon_i,1}^i)^{-1} E_{1,i}^i,$$
$$R(s) = (A_{12}^i - \bar{B}_{1,i}^i \tilde{F}_{\varepsilon_i,2}^i)(sI - A_{22}^i + \bar{B}_{2,i}^i \tilde{F}_{\varepsilon_i,2}^i)^{-1} \bar{B}_{2,i}^i,$$

whereas

$$F_{\varepsilon_i}^i = K_i \begin{pmatrix} F_{\varepsilon_i,1}^i & \tilde{F}_{\varepsilon_i,2}^i \end{pmatrix} = K_i \begin{pmatrix} F_{\varepsilon_i,1}^i & F_{\varepsilon_i,2}^i + \tilde{F}_{2,2}^i \end{pmatrix}$$

and

$$\bar{B}_{1,i}^i = B_{1,i}^i K_i, \quad \bar{B}_{2,i}^i = B_{2,i}^i K_i.$$

After some straightforward manipulations, we can check that

$$G_2^{\varepsilon_i}(s) = \left[I + F_{\varepsilon_i,2}^i (sI - \tilde{A}_{22}^i)^{-1} \bar{B}_{2,i}^i \right] \bar{G}^{\varepsilon_i}(s),$$

where

$$\tilde{A}_{22}^i = A_{22}^i - \bar{B}_{2,i}^i \tilde{F}_{2,2}^i$$

and

$$\bar{G}^{\varepsilon_i}(s) = -\begin{pmatrix} F_{\varepsilon_i,1}^i & F_{\varepsilon_i,2}^i \end{pmatrix} \left[sI - A_i + B_i \begin{pmatrix} F_{\varepsilon_i,1}^i & F_{\varepsilon_i,2}^i \end{pmatrix} \right]^{-1} E_i.$$

By our design, the H_2 norm of $\bar{G}^{\varepsilon_i}$ converges to zero, whereas its H_∞ norm is less than γ. Then, as $\tilde{A}_{22}^i$ is asymptotically stable, it is easily seen that $G_2^{\varepsilon_i}$ has an H_2 norm converging to zero, whereas its H_∞ norm is less than 2γ for ε_i small enough (recall $F_{\varepsilon_i,2}^i \to 0$ as $\varepsilon_i \to 0$).

On the other hand, for ε_i small enough, $A_{22}^i - \bar{B}_{2,i} \tilde{F}_{\varepsilon_i,2}^i$ is asymptotically stable, and using Lemma 6.27, we get that $G_1^{\varepsilon_i}$ has the property that $G_1^{\varepsilon_i}(s_0) = 0$ for any eigenvalue s_0 of A_{22}^i (which is on the imaginary axis). Here we make use of the fact that K_i was chosen such that its number of columns equals the geometric multiplicity of s_0 as an eigenvalue of A_{22}^i.

Let us next look at the H_∞ norm of the product in (6.70). Our claim is that it converges to zero as $\varepsilon_i \to 0$. We will prove this by contradiction and assume that for small ε, the H_∞ norm is larger than ρ. For an eigenvalue s_0 of A_{22}^i, we have that $G_1^{\varepsilon_i}(s_0) = 0$. As the poles of $G_1^{\varepsilon_i}$ are bounded away from s_0, an open neighborhood $\mathcal{U}_{s_0}$ of s_0 exists such that $\|G_1^{\varepsilon_i}(s)\| < \rho/(2\gamma)$, and hence, $\|G^{\varepsilon_i}(s)\| < \rho$ in this neighborhood $\mathcal{U}_{s_0}$. In this way, we find open neighborhoods around each eigenvalue of A_{22}^i, where $\|G^{\varepsilon_i}(s)\| < \rho$. Let $\mathcal{U}^c$ denote the complement of the union of these open neighborhoods in the closed right-half plane. In $\mathcal{U}^c$ we have that $G_1^{\varepsilon_i}$ is bounded, and hence, if $G_2^{\varepsilon_i}$ converges to zero uniformly on $\mathcal{U}^c$, then this would yield a contradiction with the claim that for small ε_i, the H_∞ norm of G^{ε_i} is larger than ρ. Moreover, we can easily find C such that for $|s| > C$, we have $G_2^{\varepsilon_i}$ converging to zero uniformly. Next we consider the region

$$\widetilde{\mathcal{U}}^c = \mathcal{U}^c \cap \{s \in \mathbb{C} \mid |s| < C\}.$$

If $G_2^{\varepsilon_i}$ converges to zero uniformly in this compact set, then we obtain a contradiction.

For small ε_i, the poles of G_2^ε converge to the eigenvalues of A_{22}^i (note that the eigenvalues of A_{11}^i equal the eigenvalues of A_{22}^i not counting multiplicities). But as $G_2^{\varepsilon_i}$ has no poles in $\widetilde{\mathcal{U}}^c$, we can easily see that the derivative of $G_2^{\varepsilon_i}$ (as a function of s) is locally bounded.

Assume that maximum on $\widetilde{\mathcal{U}}^c$ of $G_2^{\varepsilon_i}$ is larger than η for ε small enough. Because $\widetilde{\mathcal{U}}^c$ is compact, this implies that $G_2^{\varepsilon_i}$ cannot converge to zero locally in $\widetilde{\mathcal{U}}^c$. But this is impossible because, as its derivative is bounded, this would imply that a small region (whose size is independent of ε) inside $\mathcal{U}^c$ exists, where $\|G_2^{\varepsilon_i}\|$ is larger than $\eta/2$. This contradicts with the fact that the H_2 norm of $G_2^{\varepsilon_i}$ converges to zero. Hence, by contradiction, we find that the H_∞ norm of G^{ε_i} converges to zero.

From our recursive design, it is then clear that we find a family of stabilizing feedback controllers $u_\ell = -F_\ell^\varepsilon x_\ell$ for Σ_ℓ, which results in an H_∞ norm converging to zero. After this, it is trivial to verify that the matrix

$$\widetilde{F}^\varepsilon = \begin{pmatrix} F_\ell^\varepsilon & 0 \end{pmatrix} T_1^{-1}$$

is such that (6.52) is stable for all $\varepsilon > 0$, and the H_∞ norm of it converges to zero as $\varepsilon \to 0$. Moreover, part of the eigenvalues of

$$\widetilde{A}_{az} - \widetilde{B}_{az} \widetilde{F}^\varepsilon$$

converges to the eigenvalues of A_{a0}^{11} and A_{zF}, whereas the remaining part converges to the, a priori chosen, arbitrary stable locations.

After we choose u according to (6.65), we can rewrite the closed-loop system for a given ε in the form,

$$\begin{aligned}
\dot{\widetilde{x}} &= \begin{pmatrix} \overline{A}_{acz} & L_{acz} C_{d23} \\ A_{dacz} & A_{d2} - B_{d2}\overline{F}_8 + L_{d2}^2 C_{d23} \end{pmatrix} \widetilde{x} + \begin{pmatrix} \overline{E}_{acz} \\ \overline{E}_{d2} \end{pmatrix} w, \\
z &= \begin{pmatrix} \overline{C}_{acz} & 0 \\ \overline{C}_{dacz} & C_{d2} \end{pmatrix} \widetilde{x}, \quad \widetilde{x} = \begin{pmatrix} \widetilde{x}_1 \\ \widetilde{x}_2 \end{pmatrix},
\end{aligned} \tag{6.71}$$

where

$$\tilde{x}_1 = \begin{pmatrix} x_{a1}^- \\ x_{a1}^0 \\ x_{a2}^0 \\ x_z \\ x_c \end{pmatrix}, \quad \text{and} \quad \tilde{x}_2 = x_{d2} + F_1 x_z + \tilde{F}_1^\varepsilon x_{az},$$

and all submatrices in (6.71) are obviously defined, whereas

$$A_{caz} = \begin{pmatrix} 0 & 0 & A_{cz} \end{pmatrix} - \begin{pmatrix} \tilde{L}_{cd}^2 C_{d23} & B_{c0} & 0 \end{pmatrix} \left[\bar{F} \begin{pmatrix} 0 & 0 & I \end{pmatrix} + \tilde{F}^\varepsilon \right],$$

$$A_{aaz} = \begin{pmatrix} 0 & 0 & A_{az}^- \end{pmatrix} - \begin{pmatrix} \tilde{L}_{ad}^{12} C_{d23} & B_{a0}^{1-} & 0 \end{pmatrix} \left[\bar{F} \begin{pmatrix} 0 & 0 & I \end{pmatrix} + \tilde{F}^\varepsilon \right],$$

$$\bar{A}_{acz} = \begin{pmatrix} A_{aa}^{11} & A_{aaz} & 0 \\ 0 & \tilde{A}_{az} - \tilde{B}_{az}\tilde{F}^\varepsilon & 0 \\ 0 & A_{caz} & A_{cc} - B_c F_4 \end{pmatrix},$$

$$L_{acz} = \begin{pmatrix} \tilde{L}_{ad}^{12} \\ \tilde{L}_{ad}^{22} \\ \tilde{L}_{ad}^{32} \\ \tilde{L}_{zd} \\ \tilde{L}_{cd}^2 \end{pmatrix},$$

$$F_{11}^\varepsilon = \begin{pmatrix} 0 & F_1 \begin{pmatrix} 0 & 0 & I \end{pmatrix} + \tilde{F}_1^\varepsilon & 0 \end{pmatrix},$$

$$A_{dacz} = \begin{pmatrix} 0 & 0 & 0 & A_{dz} & 0 \end{pmatrix} - A_{d2} F_{11}^\varepsilon + F_{11}^\varepsilon \bar{A}_{acz},$$

$$L_{d2}^2 = F_{11}^\varepsilon L_{acz},$$

$$\bar{E}_{acz} = \begin{pmatrix} E_a^{1-} \\ E_a^{01} \\ 0 \\ 0 \\ E_c \end{pmatrix},$$

$$\bar{E}_{d2} = E_{d2} + F_{11}^\varepsilon \bar{E}_{acz},$$

$$\bar{C}_{acz} = \begin{pmatrix} 0 & -F_2 \begin{pmatrix} 0 & 0 & I \end{pmatrix} - \tilde{F}_2^\varepsilon & 0 \\ 0 & \begin{pmatrix} 0 & 0 & C_z \end{pmatrix} & 0 \end{pmatrix},$$

$$\bar{C}_{dacz} = -C_{d2} F_{11}^\varepsilon.$$

Here the subsystem

$$\left[\bar{A}_{acz}, \bar{E}_{acz}, \begin{pmatrix} \bar{C}_{acz} \\ \bar{C}_{dacz} \end{pmatrix}, 0 \right]$$

has an H_∞ norm less than or equal to γ^ε, and the eigenvalues of $\bar{A}_{acz}$ are close to the desired locations. We view the above system as the interconnection of

$$\dot{\tilde{x}}_1 = \bar{A}_{acz}\tilde{x}_1 + \bar{E}_{acz}w + L_{acz}\tilde{w},$$
$$z = \begin{pmatrix} \bar{C}_{acz} \\ \bar{C}_{dacz} \end{pmatrix}\tilde{x}_1 + \begin{pmatrix} 0 & 0 \\ I & 0 \end{pmatrix}\tilde{w},$$
$$\tilde{z} = A_{dacz}\tilde{x}_1 + \bar{E}_{d2}w + L^2_{d2}\tilde{w},$$

and

$$\dot{\tilde{x}}_2 = (A_{d2} - B_{d2}\bar{F}_\delta)\tilde{x}_2 + \tilde{z},$$
$$\tilde{w} = C_{d23}\tilde{x}_2.$$

Both systems are asymptotically stable. The transfer matrix of the second subsystem we denote by Δ, and we know $\|\Delta\|_\infty \to 0$ as $\delta \to 0$. Denote the transfer matrices of the first subsystem as $G_{\tilde{w},\tilde{z}}$, $G_{\tilde{w},z}$, $G_{w,\tilde{z}}$, and $G_{w,z}$ using obvious notation. The standard small gain theorem yields that the closed-loop system is asymptotically stable for δ small enough. Moreover, by using the submultiplicative nature of the H_∞ norm, we get the following upper bound for the H_∞ norm of the interconnected system:

$$\|G_{w,z}\|_\infty + \|\Delta\|_\infty \frac{\|G_{w,\tilde{z}}\|_\infty \|G_{\tilde{w},z}\|_\infty}{1 - \|G_{\tilde{w},\tilde{z}}\|_\infty \|\Delta\|_\infty}.$$

Choose an arbitrary γ. We know that $\|G_{w,z}\|_\infty \to 0$ as $\varepsilon \to 0$ and, hence, can be made less than $\gamma/2$ by choosing an appropriate ε. As $\|\Delta\|_\infty$ converges to zero as $\delta \to 0$ while $\|G_{w,z}\|_\infty \leq \gamma/2$ while $\|G_{w,\tilde{z}}\|_\infty$ and $\|_\infty \|G_{\tilde{w},z}\|_\infty$ are independent of δ, we know that for δ small enough, the H_∞ norm is less than γ. Hence, we can make the H_∞ norm arbitrary small.

We still need to evaluate the asymptotic behavior of the eigenvalues of the matrix:

$$A_{cl} = \begin{pmatrix} \bar{A}_{acz} & L_{acz}C_{d23} \\ A_{dacz} & A_{d2} - B_{d2}\bar{F}_\delta + L^2_{d2}C_{d23} \end{pmatrix}.$$

Note that

$$G_\delta(s) = C_{d23}(sI - A_{d2} + B_{d2}\bar{F}_\delta)^{-1}$$

has the property that $\|G_\delta\|_\infty \to 0$ as ε goes to zero. It is then easily checked that, for any fixed point s in the right-half plane, we have

$$(sI - A_{cl})^{-1}\begin{pmatrix} I \\ 0 \end{pmatrix} \to \begin{pmatrix} (sI - \bar{A}_{acz})^{-1} \\ 0 \end{pmatrix}.$$

By Lemma 6.21, this implies that the poles of this transfer matrix converge to the poles of $(sI - \bar{A}_{acz})^{-1}$ and hence to the eigenvalues of $\bar{A}_{acz}$. Let n_2 be the number of remaining eigenvalues. Again, for any fixed point s in the right-half plane, we can show that

$$(sI - A_{cl})^{-1} \begin{pmatrix} 0 \\ I \end{pmatrix} \to 0.$$

Consider $s = 1$. In this case, we see that $(I - A_{cl})^{-1}$ has at least n_2 eigenvalues going to zero, which implies that A_{cl} has at least n_2 eigenvalues going to infinity.

We have already established in the above that the almost disturbance decoupling finite asymptotic fixed modes are a subset of Ω_s^∞. To show equality, we need to establish that any other design does not yield more flexibility in the asymptotic behavior of the poles of the closed-loop system. But from our construction, it is clear that Ω_s^∞ equals Ω_s^2. But as any family of controllers that solves the H_∞ ADD problem also solves the H_2 ADD problem, it is immediately clear from Theorem 6.19 that part of the eigenvalues must converge to the set Ω_s^2 and hence to Ω_s^∞, which shows that the ADD finite asymptotic fixed modes equal Ω_s^∞. ∎

Theorem 6.28 *Consider the H_∞ ADD problem as defined by Problem 6.1 for the continuous-time system Σ as in (6.1). Assume that the existence conditions as given in Theorem 6.10 are satisfied. Let G be such that*

$$\operatorname{im}(E + BG) \subseteq \mathcal{S}^*(\Sigma_{\text{sub}}) + \mathcal{V}^{-0}(\Sigma_{\text{sub}})$$

while $DG = 0$. Then the following hold:

(i) $\Omega_{s,f}^\infty(A, B, C, D, E) = \Omega_s^\infty(A, B, C, D, E + BG).$

(ii) Any sequence of state feedback controllers designed according to the H_∞ ADD algorithm with E replaced by $E + BG$ results in a sequence of full information feedback controllers $u = -F^\varepsilon x + Gw$, that solves the H_∞ ADD problem; i.e., the closed-loop system is stable for ε small enough and the transfer matrix converges to 0 in H_∞ norm as $\varepsilon \downarrow 0$.

Proof : This follows along the same lines as the proof of the results for the H_2 almost disturbance decoupling problem with full information feedback in Theorem 6.22. ∎

6.6.2 Computation of Ω_s^∞ and designing sequences of static H_∞ ADD controllers—discrete time

In this subsection, for discrete-time systems, we will describe a mechanism to obtain the set of H_∞ ADD finite asymptotic fixed modes and, at the same time, obtain suitable families of controllers that solve the H_∞ ADD problem with a certain flexibility in placing the closed-loop poles.

The H_∞ ADD algorithm—discrete time

Step 1: Representation of Σ_{sub} in SCB.

In this step, we first construct the SCB of Σ_{sub}, which is characterized by the quadruple (A, B, C, D) as described in Chapter 3. We will use a more compact form where the special structure of x_d is not made explicit as given in (3.15)–(3.19).

From Property 3.24 of the SCB, it is simple to see that the conditions of Theorem 6.14 imply that

$$
\Gamma_s^{-1} E = \begin{pmatrix} E_a^- \\ E_a^0 \\ 0 \\ 0 \\ E_c \\ 0 \end{pmatrix}.
\tag{6.72}
$$

As in Chapter 3, let Γ_s, Γ_i, and Γ_o be the state, input, and output transformation matrices that take the given system Σ_{sub} to its SCB form. If $\Gamma_s^{-1} E$ is not of the form (6.72), the H_∞ almost disturbance decoupling problem is not solvable and the procedure of H_∞ ADD algorithm stops at this point. Otherwise, it continues to the next step.

Step 2: Decomposition of A_{aa}^-.

In this step, we decompose A_{aa}^- into two parts: One part is controllable via the disturbance w, and the other part is unaffected by the disturbance. Consider the pair (A_{aa}^-, E_a^-). This pair need not be controllable; i.e., the disturbance w need not affect all the modes of A_{aa}^-. Thus, we can compute a nonsingular transformation T_a such that

$$
T_a^{-1} A_{aa}^- T_a = \begin{pmatrix} A_{aa}^{11} & A_{aa}^{12} \\ 0 & A_{aa}^{22} \end{pmatrix} \quad \text{and} \quad T_a^{-1} E_a^- = \begin{pmatrix} E_a^{1-} \\ 0 \end{pmatrix},
\tag{6.73}
$$

where the pair (A_{aa}^{11}, E_a^{1-}) is controllable. We define

$$
T_a^{-1} x_a^- = \begin{pmatrix} x_{a1}^- \\ x_{a2}^- \end{pmatrix}.
$$

Step 3: Decomposition of A_{aa}^0.

The solvability condition for the H_∞ almost disturbance decoupling problem as given in Theorem 6.14 requires more specific structure compared with H_2 ADD. To be more explicit, beyond the special structure given in (6.72), we need an additional structure for A_{aa}^0 and E_a^0. Using the same arguments as in the continuous time, we find that in a suitable basis, we have

$$
T_{a0}^{-1} A_{aa}^0 T_{a0} = \begin{pmatrix} A_{a0}^{11} & A_{a0}^{12} & 0 \\ 0 & A_{a0}^{22} & 0 \\ 0 & 0 & A_{a0}^{33} \end{pmatrix}, \quad T_{a0}^{-1} E_a^0 = \begin{pmatrix} E_a^{01} \\ 0 \\ 0 \end{pmatrix},
$$

$$
T_{a0}^{-1} B_{a0}^0 = \begin{pmatrix} B_{a0}^{01} \\ B_{a0}^{02} \\ B_{a0}^{03} \end{pmatrix}, \quad T_{a0}^{-1} x_a^0 = \begin{pmatrix} x_{a1}^0 \\ x_{a2}^0 \\ x_{a3}^0 \end{pmatrix},
$$

(6.74)

where (A_{a0}^{11}, E_a^{01}) is controllable, whereas

$$
\begin{pmatrix} A_{a0}^{11} & A_{a0}^{12} \\ 0 & A_{a0}^{22} \end{pmatrix}
$$

has the same eigenvalues as A_{a0}^{22} with the same geometric multiplicities. Moreover,

$$
A_{a0}^{11} = \begin{pmatrix} A_{11}^1 & 0 & \cdots & 0 \\ 0 & A_{11}^2 & \ddots & \vdots \\ \vdots & \ddots & \ddots & 0 \\ 0 & \cdots & 0 & A_{11}^k \end{pmatrix}, \quad A_{a0}^{22} = \begin{pmatrix} A_{22}^1 & 0 & \cdots & 0 \\ 0 & A_{22}^2 & \ddots & \vdots \\ \vdots & \ddots & \ddots & 0 \\ 0 & \cdots & 0 & A_{22}^k \end{pmatrix},
$$

and

$$
A_{a0}^{12} = \begin{pmatrix} A_{12}^1 & 0 & \cdots & 0 \\ 0 & A_{12}^2 & \ddots & \vdots \\ \vdots & \ddots & \ddots & 0 \\ 0 & \cdots & 0 & A_{12}^k \end{pmatrix},
$$

where for $i = 1, \ldots, k$, we have that A_{11}^i and A_{22}^i have both either one real eigenvalue λ_i or two complex eigenvalues λ_i and its conjugate λ_i^*. Finally the diagonal blocks of A_{a0}^{11} (and similarly for A_{a0}^{22}) have all distinct eigenvalues.

Step 4: Modified SCB.

The goal of this step is to incorporate the reordering of state, input, and output variables of the SCB that occurred in the two previous steps into a new compact form of SCB. Such a reordering implies that the transformation matrices Γ_s, Γ_i, and Γ_o that take the given system Σ_{sub} to its SCB form need to be amended as $\bar{\Gamma}_s$, $\bar{\Gamma}_i$, and $\bar{\Gamma}_o$. Then, a matrix F_{pre} of suitable dimension and a preliminary feedback law

$$
u = -F_{\text{pre}} x + v
$$

(6.75)

exist such that in a suitable basis, we get the following structure:

$$\bar{\Gamma}_s^{-1}(A - BF_{\text{pre}})\bar{\Gamma}_s =$$

$$\begin{pmatrix}
A_{aa}^{11} & 0 & 0 & 0 & A_{aa}^{12} & 0 & L_{ab}^{1-}C_b & L_{ad}^{1}C_d & 0 \\
0 & A_{a0}^{11} & A_{a0}^{12} & 0 & 0 & 0 & L_{ab}^{01}C_b & L_{ad}^{2}C_d & 0 \\
0 & 0 & A_{a0}^{22} & 0 & 0 & 0 & L_{ab}^{02}C_b & L_{ad}^{3}C_d & 0 \\
0 & 0 & 0 & A_{a0}^{33} & 0 & 0 & L_{ab}^{03}C_b & L_{ad}^{4}C_d & 0 \\
0 & 0 & 0 & 0 & A_{aa}^{22} & 0 & L_{ab}^{2-}C_b & L_{ad}^{5}C_d & 0 \\
0 & 0 & 0 & 0 & 0 & A_{aa}^{+} & L_{ab}^{+}C_b & L_{ad}^{+}C_d & 0 \\
0 & 0 & 0 & 0 & 0 & 0 & A_{bb} & L_{bd}C_d & 0 \\
0 & 0 & 0 & 0 & 0 & 0 & 0 & A_{dd} & 0 \\
0 & 0 & 0 & 0 & 0 & 0 & 0 & L_{cd}C_d & A_{cc}
\end{pmatrix},$$

$$\bar{\Gamma}_s^{-1}B\bar{\Gamma}_i = \begin{pmatrix}
B_{a0}^{1-} & 0 & 0 \\
B_{a0}^{01} & 0 & 0 \\
B_{a0}^{02} & 0 & 0 \\
B_{a0}^{03} & 0 & 0 \\
B_{a0}^{2-} & 0 & 0 \\
B_{a0}^{+} & 0 & 0 \\
B_{b0} & 0 & 0 \\
B_{d0} & B_d & 0 \\
B_{c0} & 0 & B_c
\end{pmatrix}, \qquad
\bar{\Gamma}_s^{-1}E = \begin{pmatrix}
E_a^{1-} \\
E_a^{01} \\
0 \\
0 \\
0 \\
0 \\
0 \\
0 \\
E_c
\end{pmatrix},$$

$$\bar{\Gamma}_o^{-1}(C - DF_{\text{pre}})\bar{\Gamma}_s = \begin{pmatrix}
0 & 0 & 0 & 0 & 0 & 0 & 0 & 0 & 0 \\
0 & 0 & 0 & 0 & 0 & 0 & 0 & C_d & 0 \\
0 & 0 & 0 & 0 & 0 & 0 & C_b & 0 & 0
\end{pmatrix},$$

$$\bar{\Gamma}_o^{-1}D\bar{\Gamma}_i = \begin{pmatrix}
I_{m_0} & 0 & 0 \\
0 & 0 & 0 \\
0 & 0 & 0
\end{pmatrix}.$$

Based on the above development, we can form the matrices A_z, B_z, and C_z by using obvious notation as

$$A_z := \begin{pmatrix}
A_{a0}^{33} & 0 & 0 & L_{ab}^{03}C_b & L_{ad}^{41}C_d \\
0 & A_{aa}^{22} & 0 & L_{ab}^{2-}C_b & L_{ad}^{51}C_d \\
0 & 0 & A_{aa}^{+} & L_{ab}^{+}C_b & L_{ad}^{+1}C_d \\
0 & 0 & 0 & A_{bb} & L_{bd}^{1}C_d \\
0 & 0 & 0 & 0 & A_{dd}
\end{pmatrix}, \qquad (6.76)$$

$$B_z := \begin{pmatrix} B_{a0}^{03} & 0 \\ B_{a0}^{2-} & 0 \\ B_{a0}^{+} & 0 \\ B_{b0} & 0 \\ B_{d0} & B_d \end{pmatrix} = \begin{pmatrix} B_{z0} & B_{z1} \end{pmatrix},$$

and

$$C_z := \begin{pmatrix} 0 & 0 & 0 & 0 & 0 \\ 0 & 0 & 0 & 0 & C_d \\ 0 & 0 & 0 & C_b & 0 \end{pmatrix}. \tag{6.77}$$

From Property 3.3 of the SCB, it is simple to verify that the pair (A_z, B_z) is $\mathbb{C}^{\ominus}$ stabilizable when the pair (A, B) is $\mathbb{C}^{\ominus}$ stabilizable.

Using the above definition, we can rewrite the system more compactly as

$$\bar{\Gamma}_s^{-1}(A - BF_{\mathrm{pre}})\bar{\Gamma}_s = \begin{pmatrix} A_{aa}^{11} & 0 & 0 & A_{az}^{-} & 0 \\ 0 & A_{a0}^{11} & A_{a0}^{12} & A_{az}^{01} & 0 \\ 0 & 0 & A_{a0}^{22} & A_{az}^{02} & 0 \\ 0 & 0 & 0 & A_z & 0 \\ 0 & 0 & 0 & A_{cz} & A_{cc} \end{pmatrix},$$

$$\bar{\Gamma}_s^{-1} B \bar{\Gamma}_i = \begin{pmatrix} B_{a0}^{1-} & 0 & 0 \\ B_{a0}^{01} & 0 & 0 \\ B_{a0}^{02} & 0 & 0 \\ B_{z0} & B_{z1} & 0 \\ B_{c0} & 0 & B_c \end{pmatrix}, \qquad \bar{\Gamma}_s^{-1} E = \begin{pmatrix} E_a^{1-} \\ E_a^{01} \\ 0 \\ 0 \\ E_c \end{pmatrix},$$

$$\bar{\Gamma}_o^{-1}(C - DF_{\mathrm{pre}})\bar{\Gamma}_s = \begin{pmatrix} 0 & 0 & 0 & 0 & 0 \\ 0 & 0 & 0 & C_z & 0 \end{pmatrix},$$

$$\bar{\Gamma}_o^{-1} D \bar{\Gamma}_i = \begin{pmatrix} I_{m0} & 0 & 0 \\ 0 & 0 & 0 \end{pmatrix}.$$

In view of the above structure, the preliminary feedback law (6.75) can be rewritten as

$$\bar{u} = -\bar{\Gamma}_i^{-1} F_{\mathrm{pre}} \bar{\Gamma}_s \bar{x} + \bar{v}, \tag{6.78}$$

where

$$\bar{x} = \bar{\Gamma}_s^{-1} x, \quad \bar{u} = \bar{\Gamma}_i^{-1} u, \quad \text{and} \quad \bar{v} = \bar{\Gamma}_i^{-1} v = \begin{pmatrix} v_1 \\ v_2 \\ v_3 \end{pmatrix}. \tag{6.79}$$

Step 5: Computation of Ω_s^{∞}.

We can determine the set of H_∞ ADD finite asymptotic fixed modes $\boldsymbol{\Omega}_s^\infty$ as

$$\boldsymbol{\Omega}_s^\infty = \lambda(A_{aa}^{11}) \cup \lambda(A_{a0}^{11}) \cup \{\text{input decoupling zeros of } (A_z, B_z)\}. \qquad (6.80)$$

Step 6: Construction of H_∞ ADD families of controllers.

The next step is to find a sequence of state feedback controllers that achieves H_∞ ADD while possibly assigning the asymptotic behavior of the closed-loop poles arbitrarily as desired beyond the above fixed modes.

Step 6a: Assigning the first part of finite asymptotic modes.

Choose

$$\bar{F} = \begin{pmatrix} F_1 \\ F_2 \end{pmatrix}$$

such that $A_z - B_z \bar{F}$ is stable and has the desirable eigenvalues within the unit circle. Note that $\bar{F}$ is partitioned in conformity with the partitioning of B_z as given in (6.77). Also, note that the desirable eigenvalues must obviously contain the set of input decoupling zeros of (A_z, B_z), which is a subset of $\boldsymbol{\Omega}_s^\infty$.

Step 6b: Assigning the first part of finite asymptotic modes.

Choose F_3 such that $A_{cc} - B_c F_3$ is stable and has the desirable eigenvalues within the unit circle. As (A_{cc}, B_c) is always controllable, all eigenvalues of $A_{cc} - B_c F_3$ can be assigned to arbitrary locations.

Step 6c: Assigning the third part of finite asymptotic modes combined with a low-gain design.

We first define

$$\tilde{B}_{az} = \begin{pmatrix} \tilde{L}_{ad}^{22} C_{d23} & B_{a0}^{01} & 0 \\ \tilde{L}_{ad}^{32} C_{d23} & B_{a0}^{02} & 0 \\ \tilde{L}_{zd} C_{d23} & B_{z0} & B_{z1} \end{pmatrix} = \begin{pmatrix} B_{a0}^{11} \\ B_{a0}^{12} \\ B_z \end{pmatrix}, \quad \tilde{E}_{az} = \begin{pmatrix} E_a^{01} \\ 0 \\ 0 \end{pmatrix},$$

and

$$\tilde{A}_{az} = \begin{pmatrix} A_{a0}^{11} & A_{a0}^{12} & A_{az}^{01} \\ 0 & A_{a0}^{22} & A_{az}^{02} \\ 0 & 0 & A_z \end{pmatrix} - \tilde{B}_{az} \bar{F} \begin{pmatrix} 0 & 0 & I \end{pmatrix} = \begin{pmatrix} A_{a0}^{11} & A_{a0}^{12} & \tilde{A}_{az}^{01} \\ 0 & A_{a0}^{22} & \tilde{A}_{az}^{02} \\ 0 & 0 & A_{zF} \end{pmatrix}.$$

Our goal here is to construct a sequence of feedback gains $\{\tilde{F}^\varepsilon\}_{\varepsilon>0}$ such that the system,

$$\left(\tilde{A}_{az} - \tilde{B}_{az} \tilde{F}^\varepsilon, \ \tilde{E}_{az}, \ \begin{pmatrix} -\tilde{F}_2^\varepsilon - F_2 \begin{pmatrix} 0 & 0 & I \end{pmatrix} \\ \begin{pmatrix} 0 & 0 & C_z \end{pmatrix} \end{pmatrix}, 0 \right), \qquad (6.81)$$

is stable for all $\varepsilon > 0$; the H_∞ norm of it converges to zero as $\varepsilon \to 0$. Moreover, part of the eigenvalues of $\tilde{A}_{az} - \tilde{B}_{az} \tilde{F}^\varepsilon$ converge to the eigenvalues of A_{a0}^{11} and

A_{zF} while the remaining eigenvalues can be assigned arbitrarily. Here we used the decomposition,

$$\widetilde{F}^{\varepsilon} = \begin{pmatrix} \widetilde{F}^{\varepsilon}_1 \\ \widetilde{F}^{\varepsilon}_2 \\ \widetilde{F}^{\varepsilon}_3 \end{pmatrix},$$

which is compatible with the decomposition of $\widetilde{B}_{az}$. In what follows, we transform the task of designing the feedback gain $\widetilde{F}^{\varepsilon}$ to a similar task of designing a feedback gain for another system that has a suitable structure. To do so, we start with a basis transformation T_1 such that we obtain the appropriate structure we are looking for. Because the matrices

$$\begin{pmatrix} A_{a0}^{11} & A_{a0}^{12} \\ 0 & A_{a0}^{22} \end{pmatrix}$$

and A_{zF} have disjoint eigenvalues, a basis transformation of the form

$$T_1 = \begin{pmatrix} I & 0 & T_1^{13} \\ 0 & I & T_1^{23} \\ 0 & 0 & I \end{pmatrix}$$

exists such that

$$T_1^{-1} \widetilde{A}_{az} T_1 = \begin{pmatrix} A_{a0}^{11} & A_{a0}^{12} & 0 \\ 0 & A_{a0}^{22} & 0 \\ 0 & 0 & A_{zF} \end{pmatrix}, \quad T_1^{-1} \widetilde{B}_{az} = \begin{pmatrix} \widetilde{B}_{a0}^{11} \\ \widetilde{B}_{a0}^{12} \\ B_z \end{pmatrix},$$

and

$$T_1^{-1} \widetilde{E}_{az} = \begin{pmatrix} E_a^{01} \\ 0 \\ 0 \end{pmatrix}, \quad \begin{pmatrix} x_{\ell} \\ x_z \end{pmatrix} = T_1^{-1} \begin{pmatrix} x_{a1}^0 \\ x_{a2}^0 \\ x_z \end{pmatrix}.$$

Let us a define a new system Σ_{ℓ} as

$$\Sigma_{\ell} : \begin{cases} \sigma x_{\ell} = \begin{pmatrix} A_{a0}^{11} & A_{a0}^{12} \\ 0 & A_{a0}^{22} \end{pmatrix} x_{\ell} + \begin{pmatrix} \widetilde{B}_{a0}^{11} \\ \widetilde{B}_{a0}^{12} \end{pmatrix} u_{\ell} + \begin{pmatrix} E_a^{01} \\ 0 \end{pmatrix} w_{\ell} \\ z_{\ell} = u_{\ell}. \end{cases} \tag{6.82}$$

In view of the above development, it can be easily seen that the task of designing the feedback gain $\widetilde{F}^{\varepsilon}$ can be converted to the task of designing a feedback gain F_{ℓ}^{ε} such that the state feedback $u_{\ell} = -F_{\ell}^{\varepsilon} x_{\ell}$ when applied to the system Σ_{ℓ} stabilizes the resulting closed-loop system, whereas the transfer matrix from w_{ℓ} to z_{ℓ} has an H_{∞} norm converging to zero as $\varepsilon \to 0$. Finally, the closed-loop eigenvalues partially converge to the eigenvalues of A_{a0}^{11}, whereas the remaining asymptotic locations of the eigenvalues can be freely assigned.

After obtaining such a F_ℓ^ε, we can obtain $\widetilde{F}^\varepsilon$ having the required properties we sought earlier by setting

$$\widetilde{F}^\varepsilon = \left(F_\ell^\varepsilon \quad 0 \right) T_1^{-1}.$$

We proceed now to design the feedback gain F_ℓ^ε. We are going to do so using a recursive method. We design $k - 1$ preliminary feedback controllers, and in the last step, we obtain a regular feedback controller. These preliminary feedback controllers are partial low-gain feedback controllers designed for a subsystem with a special structure that is preserved in the recursive steps. We initialize by setting $i = 1$.

We have the following system:

$$\Sigma_\ell^i : \begin{cases} \sigma x_\ell^i = \begin{pmatrix} \overline{A}_{11}^i & \overline{A}_{12}^i \\ 0 & \overline{A}_{22}^i \end{pmatrix} x_\ell^i + \begin{pmatrix} \widetilde{B}_1^i \\ \widetilde{B}_2^i \end{pmatrix} v_\ell^i + \begin{pmatrix} \widetilde{E}^i \\ 0 \end{pmatrix} w_\ell \\ z_\ell^i = v_\ell^i, \end{cases} \tag{6.83}$$

which, for $i = 1$, is equal to Σ_ℓ where we define

$$x_\ell^1 = x_\ell, v_\ell^1 = u_\ell, z_\ell^1 = z_\ell, \overline{A}_{11}^1 = A_{a0}^{11}, \overline{A}_{12}^1 = A_{a0}^{12}, \overline{A}_{22}^1 = A_{a0}^{22},$$
$$\widetilde{B}_1^1 = \widetilde{B}_{a0}^{11}, \widetilde{B}_2^1 = \widetilde{B}_{a0}^{12}, \widetilde{E}^1 = E_a^{01}.$$

We decompose the state x_ℓ^i as

$$x_\ell^i = \begin{pmatrix} x_{1,i}^i \\ \vdots \\ x_{1,k}^i \\ x_{2,i}^i \\ \vdots \\ x_{2,k}^i \end{pmatrix} \tag{6.84}$$

compatible with the special structure of the matrices $\overline{A}_{11}^i$ and $\overline{A}_{22}^i$, which for $i = 1$ follows from the special structure of the matrices A_{a0}^{11}, A_{a0}^{12}, and A_{a0}^{22} as explained in Step 3. We also decompose $\widetilde{B}_1^i$, $\widetilde{B}_2^i$, and $\widetilde{E}^i$ compatibly to the decomposition of $\overline{A}_{11}^i$ and $\overline{A}_{22}^i$:

$$\widetilde{B}_1^i = \begin{pmatrix} B_{1,i}^i \\ \vdots \\ B_{1,k}^i \end{pmatrix}, \quad \widetilde{B}_2^i = \begin{pmatrix} B_{2,i}^i \\ \vdots \\ B_{2,k}^i \end{pmatrix}, \quad \widetilde{E}^i = \begin{pmatrix} E_{1,i}^i \\ \vdots \\ E_{1,k}^i \end{pmatrix}.$$

We next reorder and partition the states of the system Σ_ℓ^i as

$$x_i^i = \begin{pmatrix} x_{1,i}^i \\ x_{2,i}^i \end{pmatrix}, \quad \check{x}^i = \begin{pmatrix} x_{1,i+1}^i \\ \vdots \\ x_{1,k}^i \\ x_{2,i+1}^i \\ \vdots \\ x_{2,k}^i \end{pmatrix}.$$

With the above reordering and partitioning, we can rewrite the system Σ_ℓ^i as

$$\Sigma_\ell^i : \begin{cases} \sigma x_i^i = \begin{pmatrix} A_{11}^i & A_{12}^i \\ 0 & A_{22}^i \end{pmatrix} x_i^i + \begin{pmatrix} B_{1,i}^i \\ B_{2,i}^i \end{pmatrix} v_\ell^i + \begin{pmatrix} E_{1,i}^i \\ 0 \end{pmatrix} w_\ell \\[4mm] \sigma \check{x}^i = \begin{pmatrix} \overline{A}_{11}^{i+1} & \overline{A}_{12}^{i+1} \\ 0 & \overline{A}_{22}^{i+1} \end{pmatrix} \check{x}^i + \begin{pmatrix} \check{B}_1^i \\ \check{B}_2^i \end{pmatrix} v_\ell^i + \begin{pmatrix} \check{E}_{1,i}^i \\ 0 \end{pmatrix} w_\ell \\[4mm] z_\ell^i = v_\ell^i, \end{cases}$$

where we define

$$\overline{A}_{11}^{i+1} = \begin{pmatrix} A_{11}^{i+1} & 0 & \cdots & 0 \\ 0 & A_{11}^{i+2} & \ddots & \vdots \\ \vdots & \ddots & \ddots & 0 \\ 0 & \cdots & 0 & A_{11}^k \end{pmatrix}, \quad \overline{A}_{12}^{i+1} = \begin{pmatrix} A_{12}^{i+1} & 0 & \cdots & 0 \\ 0 & A_{12}^{i+2} & \ddots & \vdots \\ \vdots & \ddots & \ddots & 0 \\ 0 & \cdots & 0 & A_{12}^k \end{pmatrix},$$

$$\overline{A}_{22}^{i+1} = \begin{pmatrix} A_{22}^{i+1} & 0 & \cdots & 0 \\ 0 & A_{22}^{i+2} & \ddots & \vdots \\ \vdots & \ddots & \ddots & 0 \\ 0 & \cdots & 0 & A_{22}^k \end{pmatrix},$$

and

$$\check{B}_1^i = \begin{pmatrix} B_{1,i+1}^i \\ \vdots \\ B_{1,k}^i \end{pmatrix}, \quad \check{B}_2^i = \begin{pmatrix} B_{2,i+1}^i \\ \vdots \\ B_{2,k}^i \end{pmatrix}, \quad \check{E}_{1,i}^i = \begin{pmatrix} E_{1,i+1}^i \\ \vdots \\ E_{1,k}^i \end{pmatrix}.$$

In our design in step i, we design a preliminary feedback controller

$$v_\ell^i = -F_{\varepsilon_i}^i x_i^i + v_\ell^{i+1} \tag{6.85}$$

with suitable properties, where for $i = k$, we set $v_\ell^{k+1} = 0$.

We know that the pair

$$\left[\begin{pmatrix} A^i_{11} & A^i_{12} \\ 0 & A^i_{22} \end{pmatrix} , \begin{pmatrix} B^i_{1,i} \\ B^i_{2,i} \end{pmatrix} \right] \tag{6.86}$$

is controllable. Note that the matrix $\bar{A}^i_1$,

$$\bar{A}^i_1 = \begin{pmatrix} A^i_{11} & A^i_{12} \\ 0 & A^i_{22} \end{pmatrix}, \tag{6.87}$$

has only eigenvalues λ_i, and its complex conjugate λ_i^* in case λ_i is complex. Hence, for λ_i, a matrix K_i exists such that

$$\left[\lambda_i I - \begin{pmatrix} A^i_{11} & A^i_{12} \\ 0 & A^i_{22} \end{pmatrix} \quad \begin{pmatrix} B^i_{1,i} \\ B^i_{2,i} \end{pmatrix} K_i \right]$$

has full row rank, whereas the number of columns n_i of K_i is equal to the geometric multiplicity of λ_i as an eigenvalue of $\bar{A}^i_1$. By the Hautus test, this implies that the pair

$$\left[\begin{pmatrix} A^i_{11} & A^i_{12} \\ 0 & A^i_{22} \end{pmatrix} , \begin{pmatrix} B^i_{1,i} \\ B^i_{2,i} \end{pmatrix} K_i \right]$$

is controllable.

Note that, by the special structure of (6.74), the geometric multiplicity of an eigenvalue of (6.87) equals the geometric order as an eigenvalue of A^i_{22}. Let $\tilde{F}^i_{2,2}$ be such that $A^i_{22} - B^i_{2,i} K_i \tilde{F}^i_{2,2}$ is stable and has the eigenvalues $\{\mu_1,\ldots,\mu_{m_i}\}$ at the desirable locations. This is possible due to the controllability of the pair $(A^i_{22}, B^i_{2,i} K_i)$. We choose the preliminary feedback controller (6.85) of the form

$$v^i_\ell = -F^i_{\varepsilon_i} x^i_i + v^{i+1}_\ell = -K_i \left[\tilde{F}^i_{\varepsilon,i} + \begin{pmatrix} 0 & \tilde{F}^i_{2,2} \end{pmatrix} \right] x^i_i + v^{i+1}_\ell. \tag{6.88}$$

Let $P^i_{\varepsilon_i}$ be the stabilizing solution of the H^1_∞ DARE,

$$P = A'_i P A_i + \varepsilon^2_i I$$

$$- \begin{pmatrix} B'_i P A_i \\ E'_i P A_i \end{pmatrix}' \begin{pmatrix} B'_i P B_i + I & B'_i P E_i \\ E'_i P B_i & E'_i P E_i - \gamma^2 I \end{pmatrix}^{-1} \begin{pmatrix} B'_i P A_i \\ E'_i P A_i \end{pmatrix},$$

such that

$$\gamma^2 I > E'_i P^i_{\varepsilon_i} E_i,$$

where γ is fixed (independent of ε_i) and large enough such that this equation has a solution for all ε_i small enough. Moreover,

$$A_i = \begin{pmatrix} A^i_{11} & A^i_{12} - B^i_{1,i} K_i \tilde{F}^i_{2,2} \\ 0 & A^i_{22} - B^i_{2,i} K_i \tilde{F}^i_{2,2} \end{pmatrix}, \quad B_i = \begin{pmatrix} B^i_{1,i} \\ B^i_{2,i} \end{pmatrix} K_i, \quad E_i = \begin{pmatrix} E^i_{1,i} \\ 0 \end{pmatrix},$$

and we define

$$\widetilde{F}^i_{\varepsilon_i} = \gamma^2 \left[I + \gamma^2 B'_i P^i_{\varepsilon_i} (\gamma^2 I - E_i E'_i P^i_{\varepsilon_i})^{-1} B_i \right]^{-1}$$
$$\times B'_i P^i_{\varepsilon_i} (\gamma^2 I - E_i E'_i P^i_{\varepsilon_i})^{-1} A_i.$$

We apply this preliminary feedback (6.88) to (6.83). Note that this is not a low-gain feedback because the matrix $\widetilde{F}^i_{2,2}$ does not depend on ε_i and is hence fixed. Clearly,

$$\begin{pmatrix} \overline{A}^{i+1}_{11} & \overline{A}^{i+1}_{12} \\ 0 & \overline{A}^{i+1}_{22} \end{pmatrix} \tag{6.89}$$

has only eigenvalues on the unit circle. Consider the system Σ^i_ℓ after the preliminary feedback (6.88). The matrix

$$\overline{A}_i = A_i - B_i \widetilde{F}^i_{\varepsilon_i}$$

is asymptotically stable. Hence, (6.89) and $\overline{A}_i$ have disjoint eigenvalues, and it is easily seen that this implies that a matrix X^{ε_i} exists such that the basis transformation $x^{i+1}_\ell = \check{x}^i - X^{\varepsilon_i} x^i_i$ applied to Σ^i_ℓ, after applying the preliminary feedback (6.88), results in a system having the structure,

$$\widetilde{\Sigma}^i_\ell : \begin{cases} \sigma \begin{pmatrix} x^i_i \\ x^{i+1}_\ell \end{pmatrix} = \begin{pmatrix} \overline{A}_i & 0 & 0 \\ 0 & \overline{A}^{i+1}_{11} & \overline{A}^{i+1}_{12} \\ 0 & 0 & \overline{A}^{i+1}_{22} \end{pmatrix} \begin{pmatrix} x^i_i \\ x^{i+1}_\ell \end{pmatrix} \\ \qquad\qquad + \begin{pmatrix} B_i \\ \widetilde{B}^{i+1}_1 \\ \widetilde{B}^{i+1}_2 \end{pmatrix} v^{i+1}_\ell + \begin{pmatrix} E_i \\ \widetilde{E}^{i+1} \\ 0 \end{pmatrix} w_\ell \\ \widetilde{z}^{i+1}_\ell = \begin{pmatrix} \overline{C}_i & 0 & 0 \end{pmatrix} \begin{pmatrix} x^i_i \\ x^{i+1}_\ell \end{pmatrix} + v^{i+1}_\ell, \end{cases} \tag{6.90}$$

where various matrices are defined appropriately and the H_∞ norm of the subsystem $(\overline{A}_i, E_i, \overline{C}_i, 0)$ converges to zero as $\varepsilon_i \to 0$, whereas part of the eigenvalues of $\overline{A}_i$ converge to eigenvalues of A^i_{11} and the remaining part converges to the arbitrarily chosen locations $\{\mu_1, \ldots, \mu_{m_i}\}$. This will be proven in the proof of Theorem 6.26. Hence, for any ρ, a constant ε_i exists such that the H_∞ norm of the subsystem $(\overline{A}_i, E_i, \overline{C}_i, 0)$ is less than or equal to ρ, whereas the eigenvalues of $\overline{A}_i$ are within distance ρ of their limiting values.

In view of (6.83) and (6.90), the dynamics of x^{i+1}_ℓ with output $z^{i+1}_\ell = v^{i+1}_\ell$ is given by

$$\Sigma^{i+1}_\ell : \begin{cases} \sigma x^{i+1}_\ell = \begin{pmatrix} \overline{A}^{i+1}_{11} & \overline{A}^{i+1}_{12} \\ 0 & \overline{A}^{i+1}_{22} \end{pmatrix} x^{i+1}_\ell \\ \qquad\qquad + \begin{pmatrix} \widetilde{B}^{i+1}_1 \\ \widetilde{B}^{i+1}_2 \end{pmatrix} v^{i+1}_\ell + \begin{pmatrix} \widetilde{E}^{i+1} \\ 0 \end{pmatrix} w_\ell \\ z^{i+1}_\ell = v^{i+1}_\ell. \end{cases} \tag{6.91}$$

Assume that we can design a sequence of controllers $v_\ell^{i+1} = -F_{\varepsilon_{i+1}}^{i+1} x_\ell^{i+1}$ parameterized by ε_{i+1} for the subsystem Σ_ℓ^{i+1} such that, as $\varepsilon_{i+1} \to 0$, the H_∞ norm converges to zero, whereas the closed-loop eigenvalues are all stable and converge to asymptotic locations consisting of the eigenvalues of $\overline{A}_{11}^{i+1}$ and some, a priori, chosen arbitrary stable locations. It can then be easily seen that, for ε_{i+1} small enough, the system (6.83) after applying the preliminary feedback (6.88) [which results in the system (6.90)] and the feedback $v_\ell^{i+1} = -F_{\varepsilon_{i+1}}^{i+1} x_\ell^{i+1}$ has an H_∞ norm strictly less than 2ρ, whereas part of the closed-loop eigenvalues is within distance ρ of the eigenvalues of $\overline{A}_{11}$ and the remaining part is within distance ρ of the, a priori, chosen asymptotic locations.

We repeat the above procedure by setting $i = i + 1$. In this regard, we observe that the system Σ_ℓ^{i+1} has the same structure as Σ_ℓ^i, and hence, the above procedure can be repeated. After $k - 1$ steps, we end up with the requirement of the need to design a feedback for the system

$$\Sigma_\ell^k : \begin{cases} \sigma x_\ell^k := \begin{pmatrix} A_{11}^k & A_{12}^k \\ 0 & A_{22}^k \end{pmatrix} x_\ell^k + \begin{pmatrix} \widetilde{B}_{1,1}^k \\ \widetilde{B}_{2,1}^k \end{pmatrix} v_\ell^k + \begin{pmatrix} \widetilde{E}_{1,1}^k \\ 0 \end{pmatrix} w_\ell \\ z_\ell^k = v_\ell^k \end{cases} \tag{6.92}$$

for which we need a controller $v_\ell^k = -F_{\varepsilon_k}^k x_\ell^k$ such that, as $\varepsilon_k \to 0$, the H_∞ norm converges to zero, whereas the closed-loop eigenvalues are all stable and converge to asymptotic locations consisting of the eigenvalues of A_{11}^k and some, a priori, chosen arbitrary stable locations. Obviously, for this last step, we can then design a feedback using the same methodology except that the need for extracting a subsystem is no longer needed.

By the above recursive design procedure, for any given ε, we can choose the constants $\varepsilon_1, \ldots, \varepsilon_k$ small enough such that the designed controller when applied to the system Σ_ℓ stabilizes the resulting closed-loop system, whereas the transfer matrix from w_ℓ to z_ℓ has an H_∞ norm less than ε and the closed-loop poles are within a radius ε of their asymptotic locations. Such a controller leads us to the the needed state feedback gain $\widetilde{F}^\varepsilon$.

Step 6d: Combining individual components to obtain an H_∞ ADD family of controllers.

$$u = \overline{\Gamma}_i \overline{u} = -F_{\text{pre}} x + \overline{\Gamma}_i \begin{pmatrix} v_1 \\ v_2 \\ v_3 \end{pmatrix} := -F^\varepsilon x, \tag{6.93}$$

where

$$v_1 = -\widetilde{F}_1^\varepsilon x_a^0 - F_1 x_z, \tag{6.94}$$
$$v_2 = -\widetilde{F}_2^\varepsilon x_a^0 - F_2 x_z, \tag{6.95}$$
$$v_3 = -F_4 x_c. \tag{6.96}$$

In terms of the original basis, we have thus created a sequence of feedback controllers $u = -F^\varepsilon x$ with $\varepsilon > 0$. Note that is the same notation as for continuous-time systems, even though the design and characteristics are obviously very different.

This concludes the description of the H_∞ ADD algorithm. The following theorem proves the assertions made in this algorithm.

Theorem 6.29 *Consider the discrete-time system Σ as in (6.1) with the associated H_∞ ADD problem as defined by Problem 6.1. Assume that the solvability conditions of Theorem 6.14 are satisfied; i.e., assume that the pair (A, B) is $\mathbb{C}^\ominus$-stabilizable, $\mathrm{im}\, E \subseteq \mathcal{V}^\otimes(\Sigma_{\mathrm{sub}})$, and for any $\varepsilon > 0$ and for any invariant zero z_0 of (A, B, C, D) on the unit circle, a matrix F exists such that $z_0 I - A + BF$ is invertible, and*

$$\|(C - DF)(z_0 I - A + BF)^{-1} E\| < \varepsilon.$$

Then we have:

(*i*) *As claimed in (6.80), the set of all H_∞ ADD finite asymptotic fixed modes Ω_s^∞ defined in Definition 6.2 is given by*

$$\Omega_s^\infty = \lambda(A_{aa}^{11}) \cup \lambda(A_{a0}^{11}) \cup \{\textit{input decoupling zeros of }(A_z, B_z)\}.$$

(*ii*) *Any sequence of state feedback controllers $u = -F^\varepsilon x$ (6.93) as designed according to the H_∞ ADD algorithm solves the H_∞ ADD problem; i.e., the closed-loop system is stable for ε small enough, and the transfer matrix converges to 0 in H_∞ norm as $\varepsilon \downarrow 0$.*

Proof : The proof that any family of controllers designed according to the H_∞ ADD algorithm solves the H_∞ ADD problem follows almost directly along the same lines as in the proof of Theorem 6.24. The fact that the low-gain design from Step 6c has the desired properties follows in the same line as the proof of the continuous-time version in Theorem 6.26. Finally the design only shows that the set of all H_∞ ADD finite asymptotic fixed modes is contained in the set Ω_s^∞. To establish equality, we use the same argument as in Theorem 6.26. First, note that we have that Ω_s^∞ equals Ω_s^2 using the H_2 result from Theorem 6.24. Next, any H_∞ ADD family of controllers is also an H_2 ADD family of controllers. Therefore, the H_2 asymptotic fixed modes are a subset of the H_∞ asymptotic fixed modes, and hence, Ω_s^∞ is contained in the the H_∞ asymptotic fixed modes. ∎

Theorem 6.30 *Consider the H_∞ ADD problem as defined by Problem 6.1 for the discrete-time system Σ as in (6.1). Assume that the existence conditions as given in Theorem 6.16 are satisfied. Let G be such that*

$$\mathrm{im}(E + BG) \subseteq \mathcal{V}^{\otimes}(\Sigma_{\mathrm{sub}})$$

while $DG = 0$. Then we have:

(i) $\boldsymbol{\Omega}^{\infty}_{s,f}(A, B, C, D, E) = \boldsymbol{\Omega}^{\infty}_{s}(A, B, C, D, E + BG)$.

(ii) *Any sequence of state feedback controllers designed according to the H_∞ ADD algorithm with E replaced by $E + BG$ results in a sequence of full information feedback controllers $u = -F^{\varepsilon}x + Gw$ that solves the H_∞ ADD problem; i.e., the closed-loop system is stable for ε small enough, and the transfer matrix converges to 0 in H_∞ norm as $\varepsilon \downarrow 0$.*

Proof : This follows along the same lines as the proof of Theorem 6.25. ∎

7
Exact input-decoupling filters

7.1 Introduction

Our goal in this chapter is to estimate a desired output of a linear time-invariant continuous- or discrete-time system by using a measured output of the system but not the inputs or the disturbances that affect the system. Obviously, various estimation or filtering problems emerge depending on the properties sought for the estimation error, i.e., the difference between the actual and the estimated values of the desired output. The problem we would like to study in this chapter is an exact estimation problem. By exact estimation, we mean that the error should tend to zero asymptotically as the time progresses to infinity irrespective of the nature of the unknown inputs into the system, including what can be called persistent inputs. Such a requirement of exact estimation dictates that the transfer function or transfer matrix from the unknown inputs to the estimation error be identically zero. In other words, in this chapter, we seek a filter that estimates the desired output in such a way that the error in the estimation of the desired output is completely decoupled from the unknown input(s). For this reason, we call the problem we study here as the *exact input-decoupling filtering problem*, or for short the *EID filtering problem*, and the filters that solve such a problem as the *exact input-decoupling filters* or EID filters. Clearly, the motivation to study the EID filtering problem and the ensuing EID filters arises from various fields of engineering including loop transfer recovery (see, for instance, [71]) and fault detection and isolation (see, for instance, [53]). The issues associated with fault detection and isolation will be discussed in later chapters.

The EID filtering problem we study here is not a new problem. It is studied in the literature by many, and the resulting filters are called unknown input observers. The necessary and sufficient conditions under which the EID filtering problem can be solved were first formulated in a clear and rigorous way by Hautus in 1983 [30]. However, Hautus did not develop any design methodologies to construct such a filter whenever it exists. The work of Hautus was missed by many. Thus, under a myriad of conditions that may or may not be necessary, several authors developed methods of constructing EID filters, mostly for the special case when the desired output is the state of the system; see [77].

Our intentions in this chapter are two-fold. At first, we would like to develop the necessary and sufficient conditions under which the EID filtering problem is solvable. After a careful study of the solvability conditions, we would like to develop methods of constructing or designing appropriate EID filters. Several issues arise in designing the EID filters. One primary issue in constructing any filter is the shaping of the error dynamics. It is a well-known fact that the dynamics of a system are heavily influenced by its poles. Thus, a fundamental question arises: Can a filter be designed in such a way that it solves the EID filtering problem while simultaneously letting the designer assign its poles arbitrarily as desired? In other words, there is an intrinsic need to study the available flexibility or freedom in assigning the poles of an EID filter. As expected, it turns out that the constraint of EID reduces the available freedom in assigning the poles of an EID filter. That is, the requirement of EID dictates that some poles of the filter are fixed at certain locations, whereas the others can be freely assigned. This leads us to define what will be called the EID filter *fixed modes* and then proceed to investigate the methods of designing EID filters while simultaneously shaping the error dynamics by an appropriate assignment of their poles.

Typically, a filter is designed by first choosing a specific architecture. In this chapter, we study EID filter design by using what is known as a CSS architecture for filters. The CSS architecture we study here includes filters of the same dynamic order as that of the given system as well as reduced-order filters with dynamic order lower than that of the given system by the number equal to the number of measurements that do not contain any inputs in them. It turns out that we also need to distinguish between proper and strictly proper filters.

7.2   Preliminaries

Consider a plant or system model given by

$$\Sigma : \begin{cases} \sigma x = Ax + Bu \\ y \ = Cx + Du \\ z \ = Ex + Fu, \end{cases} \tag{7.1}$$

where σ is an operator indicating the time derivative $\frac{d}{dt}$ for continuous-time systems and a forward unit time shift for discrete-time systems. Also, $x \in \mathbb{R}^n$ denotes the state, $u \in \mathbb{R}^m$ denotes the unknown input or disturbance, $y \in \mathbb{R}^p$ denotes the measured output, and $z \in \mathbb{R}^q$ denotes the desired output signal to be estimated. As known input signals in any estimation or filtering problem can easily be taken into account in a standard and obvious way, without loss of generality, we assume that all input signals are unknown.

As discussed in Section 7.1, our interest lies in estimating the desired output signal z while using only the measured output y but not the input u. Let $\hat{z}$ be the estimate of z as given by a filter. Also, let e_z be the estimation error, $e_z = z - \hat{z}$. Figure 7.1 depicts the setup.

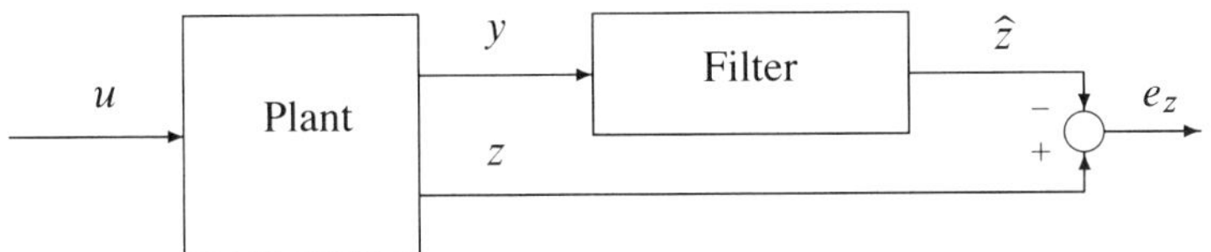

Figure 7.1: General block diagram

We use the following assumption throughout the book. As is well known, it is necessary and natural for any meaningful estimation of z by using y.

Assumption 7.1 *The matrix pair* (C, A) *is* $\mathbb{C}^-$*-detectable for continuous-time systems and* $\mathbb{C}^\ominus$*-detectable for discrete-time systems.*

We consider a general proper filter of the form:

$$\Sigma_f : \begin{cases} \sigma\xi = L\xi + My \\ \widehat{z} \ = N\xi + Py. \end{cases} \tag{7.2}$$

Whenever $P = 0$, the above filter is said to be a strictly proper filter. When the above filter is used as shown in Figure 7.1, the dynamic equations of the error e_z are described by

$$\Sigma^{ue} : \begin{cases} \sigma x = Ax + Bu \\ \sigma\xi = MCx + L\xi + MDu \\ e_z = (E - PC)x - N\xi + (F - PD)u. \end{cases} \tag{7.3}$$

Hence, the transfer matrix from u to e_z can be computed as

$$G^{ue} = \begin{pmatrix} E - PC & -N \end{pmatrix} \begin{pmatrix} \sigma I - A & 0 \\ -MC & \sigma I - L \end{pmatrix}^{-1} \begin{pmatrix} B \\ MD \end{pmatrix} + (F - PD). \tag{7.4}$$

7.3 Statement of EID filtering problem and its solvability conditions

In this section, we first define formally the EID filtering problem while using the class of linear stable unbiased filters, and then study its solvability conditions. We start by defining what we mean by unbiased filters.

Definition 7.2 *(**Unbiasedness**) Consider a continuous- or discrete-time system Σ as in (7.1). We say a linear stable strictly proper (or proper) filter (7.2) is unbiased if, in the absence of the input u, the estimation error e_z decays asymptotically to zero for all possible finite initial values of the system (7.1) and the filter (7.2).*

We have the following formal EID filtering problem statement.

Problem 7.3 The **exact input-decoupling (EID) filtering problem** consists of finding, whenever it exists, a linear stable strictly proper (or proper) filter such that

(**Unbiasedness**) the estimation error e_z, in the absence of the input u, decays asymptotically to zero for all possible finite initial values of the system (7.1) and the filter (7.2), and

(**Performance**) the transfer matrix G^{ue} from u to e_z is zero.

Remark 7.4 *It is easy to see that the above EID filtering problem seeks to find a linear stable strictly proper (or proper) filter, whenever it exists, such that $e_z(t) \rightarrow 0$ as $t \rightarrow \infty$ for any input u and for any finite initial values of the system (7.1) and the filter (7.2).*

The structure of the above-defined EID filtering problem is different for the two cases when $F = 0$ and $F \neq 0$, i.e., whether the unknown input or disturbance directly appears in the to-be-estimated signal. In fact, whenever $F \neq 0$, EID filtering cannot be achieved by a strictly proper filter. As seen from (7.4), the direct feedthrough matrix from u to e_z equals $F - PD$. This implies that a necessary condition to solve the EID filtering problem is that a matrix P exists such that $F - PD = 0$. This immediately implies that the EID filtering problem is solvable only if $\ker D \subseteq \ker F$. In particular, this means that the EID filtering problem via a strictly proper filter (i.e., with $P = 0$) is solvable only if $F = 0$. The following lemma formalizes this.

Lemma 7.5 *Consider a continuous- or discrete-time system as in (7.1). The following results hold:*

  (*i*) *The EID filtering problem is solvable by a strictly proper filter only if $F = 0$.*

(*ii*) *The EID filtering problem is solvable by a proper filter only if*

$$\ker D \subseteq \ker F. \tag{7.5}$$

Proof : The results are obvious in view of the transfer matrix G^{ue} containing the direct feedthrough matrix $F - PD$ from u to e_z as given in (7.4). ∎

In what follows, we study the necessary and sufficient conditions under which the EID filtering problem is solvable. We first consider the case when strictly proper filters are used. Lemma 7.5 then dictates that the EID filtering problem is solvable by a strictly proper filter only if $F = 0$, i.e., only if we consider the special case of the general estimation problem involving the estimation of a linear function of the state. We have the following specific results.

Theorem 7.6 *Consider a continuous- or discrete-time system as in (7.1). Let Assumption 7.1 be satisfied. Then, we have the following:*

(*i*) *For the continuous-time case, the EID filtering problem is solvable via a strictly proper filter if and only if $F = 0$, and*

$$\mathcal{S}^-(A, B, C, D) \subseteq \ker E.$$

(*ii*) *For the discrete-time case, the EID filtering problem is solvable via a strictly proper filter if and only if $F = 0$, and*

$$\mathcal{S}^{\ominus}(A, B, C, D) \subseteq \ker E.$$

Proof : Theorem 7.69 in Appendix 7.A immediately yields that the EID filtering problem is solvable if and only if for a dual system the corresponding disturbance-decoupling problem is solvable by dynamic state feedback. For the latter problem, the solvability conditions can be found in Chapter 5. The theorem then follows by using the duality between the different geometric subspaces. ∎

For the case when $E = I$, i.e., when the entire state is to be estimated, the conditions of Theorem 7.6 simplify because, in this case, $\ker E = \{0\}$. Thus, in this case, we need $\mathcal{S}^-(A, B, C, D) = \{0\}$ for continuous-time systems and $\mathcal{S}^{\ominus}(A, B, C, D) = \{0\}$ for discrete-time systems.

We now consider the EID filtering problem by using proper filters. Our development here is divided into two parts depending on whether $F = 0$ or $F \neq 0$. We have the following results for the case when $F = 0$.

We recall from Section 2.1, the following notation:

$$C^{-1}\{\operatorname{im} D\} = \{\, x \in \mathbb{R}^n \mid Cx \in \operatorname{im} D \,\}$$

which is well defined even when C is neither square nor invertible.

Theorem 7.7 *Consider a continuous- or discrete-time system as in (7.1) with $F = 0$. Let Assumption 7.1 be satisfied. Then, we have the following:*

(i) For the continuous-time case, the EID filtering problem is solvable via a proper filter if and only if

$$\mathcal{S}^-(A, B, C, D) \cap C^{-1}\{\mathrm{im}\, D\} \subseteq \ker E.$$

(ii) For the discrete-time case, the EID filtering problem is solvable via a proper filter if and only if

$$\mathcal{S}^\ominus(A, B, C, D) \cap C^{-1}\{\mathrm{im}\, D\} \subseteq \ker E.$$

Proof : This theorem is a direct consequence of Theorem 7.71 as presented in Appendix 7.A. That is, the solvability conditions for the EID filtering problem can be expressed as the solvability conditions of a dual disturbance-decoupling problem by dynamic state and static disturbance feedback as studied in Section 5.6. The theorem then follows by using the duality between the different geometric subspaces. ∎

Remark 7.8 *It is interesting to compare the solvability conditions when strictly proper filters are used (i.e., those given in Theorem 7.6) with those when proper filters are used (i.e., those given in Theorem 7.7). In fact, for both the continuous- and the discrete-time systems, the solvability conditions are much weaker when proper rather than strictly proper filters are used.*

We proceed now to study the general EID filtering problem when the matrix F is not zero. To do so, we transform the EID filtering problem for a given system for which the matrix F is nonzero to an equivalent EID filtering problem for an auxiliary system where the corresponding F can be taken as zero. The needed auxiliary system is constructed by redefining the desired output z. The new desired output z^* is obtained by applying a preliminary static output injection. That is, let

$$z^* = z - P^* y = (E - P^*C)x + (F - P^*D)u, \tag{7.6}$$

where the matrix P^* is to be selected shortly. The auxiliary system is then defined by

$$\Sigma^* : \begin{cases} \sigma x = Ax + Bu, \\ y \;= Cx + Du, \\ z^* \;= (E - P^*C)x + (F - P^*D)u. \end{cases} \tag{7.7}$$

For the above auxiliary system, let us consider a filter of the form:

$$\begin{aligned}
\sigma\xi &= L\xi + My, \\
\hat{z}^* &= N\xi + \bar{P}y.
\end{aligned} \tag{7.8}$$

Then, it is easy to show that the interconnection of the filter (7.8) and the auxiliary system Σ^* with input u and output $e_z^* = z^* - \hat{z}^*$ yields the same dynamics as the interconnection of the original filter (7.2) when applied to the original Σ with input u and output $e_z = z - \hat{z}$ as long as $P - \bar{P} = P^*$. Therefore, there is a $1-1$ correspondence between filters for the original system and filters for the auxiliary system.

The above simple analysis suggests a method of designing first P^* such that $F - P^*D = 0$ (which is possible by Lemma 7.5) and then designing a filter for the system Σ^* such that the transfer matrix from u to $e_{z*} = z^* - \hat{z}^*$ has the desired value (namely, zero for EID filtering). The designed filter for Σ^* can then easily be translated for the original system Σ by defining $P = \bar{P} + P^*$.

We have the following results.

Theorem 7.9 *Consider a continuous- or discrete-time system as in (7.1). Let Assumption 7.1 be satisfied. Then, we have the following:*

(i) *For the continuous-time case, the EID filtering problem is solvable via a proper filter if and only if*

$$\left(\mathcal{S}^-(A, B, C, D) \oplus \mathbb{R}^m\right) \cap \ker\begin{pmatrix} C & D \end{pmatrix} \subseteq \ker\begin{pmatrix} E & F \end{pmatrix}. \tag{7.9}$$

(ii) *For the discrete-time case, the EID filtering problem is solvable via a proper filter if and only if*

$$\left(\mathcal{S}^\ominus(A, B, C, D) \oplus \mathbb{R}^m\right) \cap \ker\begin{pmatrix} C & D \end{pmatrix} \subseteq \ker\begin{pmatrix} E & F \end{pmatrix}. \tag{7.10}$$

Proof : In view of Lemma 7.5, we first note that a matrix P^* must exist that is a solution of $F - PD = 0$ for P. From the arguments presented when the auxiliary system was introduced, it is obvious that the EID problem is solvable for the original system if and only if the EID filtering problem is solvable for the auxiliary system.

Let us first consider continuous-time systems. Then, in view of Theorem 7.7, an EID filter of the type (7.8) exists for Σ^* if and only if

$$\mathcal{S}^-(A, B, C, D) \cap C^{-1}\{\mathrm{im}\, D\} \subseteq \ker(E - P^*C).$$

This is equivalent to the existence of a matrix G with $GD = 0$ such that

$$\mathcal{S}^-(A, B, C, D) \subseteq \ker(E - P^*C - GC).$$

The condition $GD = 0$ can be included in the subspace inclusion, and thus we obtain the condition that a matrix G must exist such that

$$\mathcal{S}^-(A, B, C, D) \oplus \mathbb{R}^m \subseteq \ker\left(E - P^*C - GC \quad -GD \right).$$

As $F - P^*D = 0$, this can be rewritten as

$$\mathcal{S}^-(A, B, C, D) \oplus \mathbb{R}^m \subseteq \ker\left[\left(E \quad F \right) - (P^* + G)\left(C \quad D \right)\right].$$

The existence of a G satisfying the above is clearly equivalent to (7.9). This proves the result for continuous-time systems. The results for discrete-time systems follow in a similar way. $\blacksquare$

7.4 Uniqueness of EID filters in the sense of transfer function matrix

For a given system, whenever they exist, EID filters are not necessarily unique. Our goal here is to develop the conditions under which EID filters are unique. We observe that the notion of uniqueness of an EID filter can be viewed either in the sense of its transfer function or in the sense of its state-space realization with a fixed architecture. In this section, we view the uniqueness of an EID filter in the sense of its transfer function and not in the sense of its state-space realization. The uniqueness of an EID filter in the sense of its state-space realization with a particular architecture will be discussed in Section 7.5.

We have the following results for the case when strictly proper filters of the form (7.2) with $P = 0$ are used. We recall that strictly proper filters can be used only when $F = 0$.

Theorem 7.10 *Consider a continuous- or discrete-time system as in (7.1). Let Assumption 7.1 be satisfied. Let the solvability conditions for the EID filtering problem via strictly proper filters (as in Theorem 7.6) be satisfied. Then, the transfer function of the EID filter is unique if and only if the subsystem characterized by the quadruple (A, B, C, D) is right-invertible.*

Proof : Consider Figure 7.1. Let the transfer function of the plant from u to y be G^{uy} and from u to z be G^{uz}. Also, let the transfer function of the filter from y to $\hat{z}$ be G^f. Then, the transfer function from u to e_z can be computed as

$$G^{ue} = G^{uz} - G^f G^{uy}. \tag{7.11}$$

Obviously, G^f that renders $G^{ue} = 0$ is unique if and only if G^{uy} is right-invertible, i.e., if and only if the subsystem characterized by the quadruple (A, B, C, D) is right-invertible. ∎

The above theorem pertains to the case when strictly proper filters are used. The following theorem considers proper filters.

Theorem 7.11 *Consider a continuous- or discrete-time system as in (7.1). Let Assumption 7.1 be satisfied. Assume that the solvability conditions for the EID filtering problem via proper filters (as in Theorem 7.9) are satisfied. Then, the transfer function of the EID filter is unique if and only if the subsystem characterized by the quadruple (A, B, C, D) is right-invertible.*

Proof : The proof is exactly the same as that of Theorem 7.10. ∎

7.5   Design of EID filters

In this section, we present explicit algorithms for designing EID filters with a capability to assign the poles as desired while honoring certain conditions imposed by the requirement of EID filtering. In this regard, we observe that typically any filter design is initiated by first assuming a fixed architecture to the filter. The architecture we use for the filters is the CSS architecture that was developed earlier (see, e.g., [15, 71]) in the context of loop transfer recovery. The CSS architecture contains full-order filters as well as reduced-order filters. As usual, full-order filters have the same dynamic order as that of the given system, whereas the reduced-order filters have their dynamic order lower than that of the given system by the number equal to the number of measurements that do not contain any inputs in them. Two versions of full-order CSS filters exist: One is a strictly proper filter and the other is a proper filter. On the other hand, by their nature, the reduced-order filters, except for some special cases, are not strictly proper. For design of filters, throughout this book, we use only filters of CSS architecture, whether they be strictly proper or proper, and whether they be full- or reduced-order type. Thus, henceforth in connection with the design of a filter, the word *filter* without any qualifier is always to be understood as a *full-order filter of CSS architecture*. Whenever we use a reduced-order filter, we always qualify it as such. Whenever we need to emphasize, especially in definitions and theorems, we shall use the qualifiers *full-order* and *CSS architecture*.

As the conditions for the existence of EID filters as developed in Theorems 7.6 and 7.9 do not assume any particular architecture for filters, one fundamental question that arises naturally is as follows: Do EID filters of CSS architecture exist under the same conditions as developed in Theorems 7.6 and 7.9? In this section, we not only answer this question affirmatively, but we also develop systematic methods of designing them. Moreover, we examine here the structure of EID filters regarding their poles. It turns out that the requirement of EID imposes that some of the poles of EID filters be fixed at certain locations in the complex plane, whereas others can be arbitrarily placed as desired. This prompts us to determine the so-called EID filter fixed modes for each class of filters we consider here. Having done so, we then develop methods of designing the EID filters while placing their poles as desired subject to the fixed-mode constraints.

In this section, we do not examine the transient performance of the filters we design; that is, we do not examine the energy of the estimation error signal. This will be done later on in Section 10.17.

In what follows, our presentation is organized into three subsections. The first and second subsections consider, respectively, strictly proper and proper filters, whereas the third subsection considers reduced-order proper filters.

7.5.1 Strictly proper EID filters of CSS architecture

In this subsection, we pursue the design of strictly proper EID filters. As explained, strictly proper filters can be used only when $F = 0$. As such, throughout this subsection we assume that $F = 0$.

The full-order strictly proper CSS architecture of a filter is given by

$$\Sigma_{\text{sp-CSS}} : \begin{cases} \sigma\xi = A\xi + K(y - C\xi) \\ \widehat{z} = E\xi. \end{cases} \tag{7.12}$$

A block diagram representation of the strictly proper filter along with the given plant is shown in Figure 7.2.

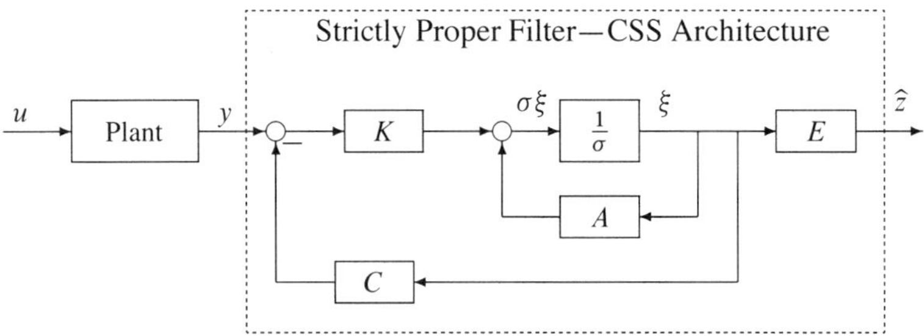

Figure 7.2: Block diagram of a strictly proper filter—CSS architecture

Error Dynamics: Let us define the error $e = x - \xi$; then the error between the actual desired output z and the estimated desired output $\widehat{z}$ is $e_z = E(x-\xi) = Ee$. In view of these definitions and in view of (7.1) and (7.12), the dynamics of error is given by

$$\Sigma^{ue}_{\text{sp-CSS}} : \begin{cases} \sigma e = (A - KC)e + (B - KD)u \\ e_z = Ee. \end{cases} \tag{7.13}$$

Also, the transfer matrix $G^{ue}_{\text{sp-CSS}}$ from u to e_z can obviously be written as

$$G^{ue}_{\text{sp-CSS}} = E(\sigma I - A + KC)^{-1}(B - KD). \tag{7.14}$$

Remark 7.12 *In view of (7.12) and (7.13), it is easy to see that both the filter equation and the error equation have the same poles that are the eigenvalues of* $A - KC$.

We observe that the only unknown in the filter equation (7.12) and consequently in the error equation (7.13) is the matrix K, which is normally referred to as the filter gain. We need to determine or design K in such a way that $G^{ue}_{\text{sp-CSS}} = 0$. In general, the gain K, which renders $G^{ue}_{\text{sp-CSS}} = 0$ is nonunique. This lets us use the available freedom in selecting K to shape the error dynamics appropriately. Whenever $G^{ue}_{\text{sp-CSS}} = 0$, we note that for continuous-time systems,

$$e_z(t) = E e^{(A-KC)t} e(0),$$

whereas for discrete-time systems,

$$e_z(k) = E(A - KC)^k e(0),$$

where the initial condition of e at time $t = 0$ or $k = 0$ is given by $e(0)$. Obviously then, the eigenstructure of $A - KC$ plays a fundamental role in shaping the dynamics of the error e_z. In what follows, we concentrate on developing an algorithm of designing K to render $G^{ue}_{\text{sp-CSS}} = 0$ and to place the eigenvalues of $A - KC$ as desired while honoring certain conditions imposed by the requirement of EID filtering.

We pause to emphasize an important aspect. Theorem 7.6 developed in Section 7.3 gives the conditions under which an EID filter exists among the general class of strictly proper filters of the form (7.2) with $P = 0$. In other words, Theorem 7.6 does not restrict itself to any fixed architecture for a filter such as the one for $\Sigma_{\text{sp-CSS}}$ given in (7.12). Nevertheless, as the theorem that follows shortly shows, whenever the conditions of Theorem 7.6 are satisfied, we can determine the gain parameter K such that the filter $\Sigma_{\text{sp-CSS}}$ is an EID filter. In this regard, let us define next a set of filter gains, denoted henceforth by $\boldsymbol{K}^{\text{eid}}_{\text{sp-CSS}}(A, B, C, D, E, 0)$, which is the set of all EID filter gains, meaning that any gain $K \in \boldsymbol{K}^{\text{eid}}_{\text{sp-CSS}}$ renders $\Sigma_{\text{sp-CSS}}$ an EID filter for (7.1), and conversely, any gain K that renders $\Sigma_{\text{sp-CSS}}$ an EID filter for (7.1) is an element of $\boldsymbol{K}^{\text{eid}}_{\text{sp-CSS}}$.

We have the following result.

Theorem 7.13 *Consider a continuous- or discrete-time system as in (7.1), and assume that the conditions of Theorem 7.6 are satisfied. Then, the set $K^{\text{eid}}_{\text{sp-CSS}}$ $(A, B, C, D, E, 0)$ is nonempty and is given by*

$$K^{\text{eid}}_{\text{sp-CSS}}(A, B, C, D, E, 0) = \{K \mid K' \in F_s(A', C', B', D', E')\}, \qquad (7.15)$$

where the set of EDD static state feedback gains $F_s(A', C', B', D', E')$ is an output of the EDD algorithm (see Section 5.4) with its input as the quintuple (A', C', B', D', E').

Proof: As indicated in Appendix 7.A, there is a $1-1$ correspondence between the stable filters of the form $\Sigma_{\text{sp-CSS}}$ given in (7.12) and the stabilizing state feedbacks $v = -K'\tilde{x}$ for the dual system (7.69). Also, the question of existence of a gain K that renders the filter $\Sigma_{\text{sp-CSS}}$ given in (7.12) an EID filter simply reduces to the existence of a static state feedback that solves the exact disturbance-decoupling problem for the dual system. Moreover, from Chapter 5, we know that the exact disturbance-decoupling problem by dynamic feedback is solvable if and only if the exact disturbance-decoupling problem by static feedback is solvable. This implies that the solvability of the EID filtering problem by arbitrary filters as used in Theorem 7.6 is equivalent to the solvability of the EID filtering problem by a specific filter of the form $\Sigma_{\text{sp-CSS}}$ given in (7.12). Hence, under the conditions of Theorem 7.6, an EID filter of the form $\Sigma_{\text{sp-CSS}}$ exists. This duality concept also immediately yields the characterization of the set $K^{\text{eid}}_{\text{sp-CSS}}(A, B, C, D, E, 0)$ as given in (7.15). ∎

The set $K^{\text{eid}}_{\text{sp-CSS}}$ has an interesting telescopic property with respect to the size of the matrix E. This is formalized below.

Lemma 7.14 *Consider a continuous- or discrete-time system as in (7.1). Consider two different EID filtering problems both as defined in Definition 7.3, however, with one taking a value E_a and the other a value E_b for the matrix E. Then, we have the following telescopic property:*

$$\ker E_a \supseteq \ker E_b \supseteq \mathcal{S}^g(A, B, C, D)$$
$$\implies K^{\text{eid}}_{\text{sp-CSS}}(A, B, C, D, E_a, 0) \supseteq K^{\text{eid}}_{\text{sp-CSS}}(A, B, C, D, E_b, 0),$$

where g denotes $\mathbb{C}^-$ for continuous-time systems and $\mathbb{C}^\ominus$ for discrete-time systems.

Proof: Once again, as indicated in Appendix 7.A, there is a $1-1$ correspondence between the stable filters of the form $\Sigma_{\text{sp-CSS}}$ given in (7.12) and the stabilizing state feedbacks $v = -K'\tilde{x}$ for the dual system (7.69). Then, as in the proof of Theorem 7.13, the proof follows from a similar property of the set $F_s(A', C', B', D', E')$. ∎

In general, whenever they exist, strictly proper EID filters of the form $\Sigma_{\text{sp-CSS}}$ given in (7.12) are not necessarily unique. The following theorem addresses the issue of the uniqueness of such filters.

Theorem 7.15 *Consider a continuous- or discrete-time system as in (7.1). Let Assumption 7.1 be satisfied. Then, a unique strictly proper full-order EID filter of CSS architecture of the form $\Sigma_{\text{sp-CSS}}$ given in (7.12) exists if and only if the following conditions are satisfied:*

    (i) The subsystem characterized by the quadruple (A, B, C, D) does not have any invariant zeros on the imaginary axis for continuous-time systems or on the unit circle for discrete-time systems.

    (ii) The matrix D is invertible.

    (iii) The matrix pair $(E, A - BD^{-1}C)$ is observable.

Whenever the above conditions are satisfied, the unique EID filter is given by

$$\sigma\xi = (A - BD^{-1}C)\xi + BD^{-1}y \quad \text{with} \quad \hat{z} = E\xi.$$

Proof : In view of Theorem 7.13, the uniqueness of the filter $\Sigma_{\text{sp-CSS}}$ is equivalent to the set $K^{\text{eid}}_{\text{sp-CSS}}(A, B, C, D, E, 0)$, which is a singleton set. Then, in view of (7.15), $K^{\text{eid}}_{\text{sp-CSS}}$, is a singleton if and only if $F_s(A', C', B', D', E')$ is a singleton. A close examination of the EDD$_{li}$ algorithm (see Section 5.4) that generates F_s reveals the conditions given in the theorem for F_s to be a singleton. ∎

Remark 7.16 *We observe that the conditions given in Theorem 7.15 imply those given in Theorem 7.10 but not conversely. This is natural and is expected because Theorem 7.15 requires the uniqueness of an EID filter having a particular architecture, namely, a strictly proper full-order CSS architecture.*

In general, whenever it exists, a strictly proper EID filter of the form $\Sigma_{\text{sp-CSS}}$ given in (7.12) is not unique. The nonuniqueness of such a filter can indeed be a blessing as some other specifications can then be imposed to come up with an appropriate filter. As $G^{ue}_{\text{sp-CSS}} = 0$ for all EID filters, all of them have the same steady-state performance. However, the transient behavior of error could indeed be different for different EID filters. It is well known that the dynamics of any system is heavily influenced by its poles. As seen from (7.3), for a filter of an arbitrary architecture, both the poles of the given system or plant as well as those of the filter influence the transient behavior of error. Thus, it is prudent to examine the poles of an EID filter. Also, although for filters of arbitrary architecture, the poles of error dynamics include those of the filter, for the particular case of filters of CSS architecture, the poles of the filter $\Sigma_{\text{sp-CSS}}$ given in (7.12) are exactly the same as those of the error dynamics $\Sigma^{ue}_{\text{sp-CSS}}$ as given in (7.13). Such poles are indeed the eigenvalues of $(A - KC)$. This brings up a fundamental question: Can a gain K be designed in such a way that the resulting filter $\Sigma_{\text{sp-CSS}}$ is an EID filter while simultaneously letting the designer assign the filter poles as arbitrarily as desired? Expectedly, it turns out that the constraint of EID filtering reduces the available freedom in assigning the poles of a filter. In fact, for each given system, a set of complex numbers exists that every EID filter must have among its poles. As the following definition formalizes, such a set of complex numbers can be termed as the EID filter fixed modes.

Definition 7.17 (Fixed modes of strictly proper EID filters of CSS architecture) *Consider a given system as in (7.1) and the EID filtering problem 7.3 characterized by the matrix sextuple $(A, B, C, D, E, 0)$. Assume that the solvability conditions as specified by Theorem 7.6 are satisfied. Then, a scalar $\lambda \in \mathbb{C}^-$ for continuous-time systems or $\lambda \in \mathbb{C}^{\ominus}$ for discrete-time systems is said to be the fixed mode of an EID filter with the strictly proper full-order CSS architecture if λ is a pole (i.e., an eigenvalue of $A - KC$) of every EID filter of such an architecture $\Sigma_{\text{sp-CSS}}$ as given in (7.12). The set of all such EID filter fixed modes is denoted here by $\boldsymbol{\Omega}^{\text{eid}}_{\text{sp-CSS}}(A, B, C, D, E, 0)$.*

We have the following theorem that characterizes the set $\boldsymbol{\Omega}^{\text{eid}}_{\text{sp-CSS}}$.

Theorem 7.18 *Consider a continuous- or discrete-time system as in (7.1). Let Assumption 7.1 be satisfied. Consider strictly proper full-order filters of CSS architecture, and assume that the solvability conditions as specified by Theorem 7.6 are satisfied. Then, we have*

$$\boldsymbol{\Omega}^{\text{eid}}_{\text{sp-CSS}}(A, B, C, D, E, 0) = \boldsymbol{\Omega}_s(A', C', B', D', E'), \qquad (7.16)$$

where $\boldsymbol{\Omega}_s(A', C', B', D', E')$ is obtained by using the EDD algorithm with its input as the quintuple (A', C', B', D', E').

Proof : It follows easily in view of the proof of Theorem 7.13. ∎

Remark 7.19 *Obviously, when the strictly proper EID filter is unique (i.e., when the conditions given in Theorem 7.15 are satisfied), the set $\boldsymbol{\Omega}^{\text{eid}}_{\text{sp-CSS}}$ (A, B, C, D, E, 0) coincides with the set $\lambda(A - BD^{-1}C)$.*

As seen from (7.16), the set $\boldsymbol{\Omega}^{\text{eid}}_{\text{sp-CSS}}$ is obviously characterized by the quintuple (A, B, C, D, E). Also, the set $\boldsymbol{\Omega}^{\text{eid}}_{\text{sp-CSS}}$ is a subset of the invariant zeros of the system characterized by the quadruple (A, B, C, D). It turns out that the exact size of $\boldsymbol{\Omega}^{\text{eid}}_{\text{sp-CSS}}$ is pretty much dictated by the matrix E, which prescribes the variable z that is being estimated. In this respect, as in the case of the set $\boldsymbol{K}^{\text{eid}}_{\text{sp-CSS}}$, a certain interesting telescopic property (nested property) exists as discussed below.

Lemma 7.20 *Consider a continuous- or discrete-time system as in (7.1). Consider two different EID filtering problems both as defined in Definition 7.3, however, with one taking a value E_a and the other a value E_b for the matrix E. Then, we have the following telescopic property:*

$$\ker E_a \supseteq \ker E_b \supseteq \mathscr{S}^g(A, B, C, D)$$
$$\implies \boldsymbol{\Omega}^{\text{eid}}_{\text{sp-CSS}}(A, B, C, D, E_a, 0) \subseteq \boldsymbol{\Omega}^{\text{eid}}_{\text{sp-CSS}}(A, B, C, D, E_b, 0),$$

where g denotes $\mathbb{C}^-$ for continuous-time systems and $\mathbb{C}^{\ominus}$ for discrete-time systems.

Proof : In view of Theorem 7.18, the proof follows from the nested property of the set $\boldsymbol{\Omega}_s$ as described in Chapter 5. $\blacksquare$

Next, we would like to design an EID filter while placing its poles at desired locations. The following theorem shows that we can design a filter gain K such that $\Sigma_{\text{sp-CSS}}$ given in (7.12) is an EID filter while its poles are at the prescribed locations with the restriction that they need to contain $\boldsymbol{\Omega}^{\text{eid}}_{\text{sp-CSS}}$ among them.

Theorem 7.21 *(**Strictly proper EID filter of CSS architecture with pole placement**) Consider a continuous- or discrete-time system as in (7.1). Let the conditions of Theorem 7.6 be satisfied. Also, consider the sets $\boldsymbol{\Omega}^{\text{eid}}_{\text{sp-CSS}}(A, B, C, D, E, 0)$ and $\boldsymbol{K}^{\text{eid}}_{\text{sp-CSS}}(A, B, C, D, E, 0)$ as described, respectively, in Theorems 7.18 and 7.13. Moreover, let Λ be a prescribed set of n self-conjugate elements in the open left-half complex plane $\mathbb{C}^-$ for continuous-time systems or a prescribed set of n self-conjugate elements within the unit disk $\mathbb{C}^{\ominus}$ for discrete-time systems such that Λ includes $\boldsymbol{\Omega}^{\text{eid}}_{\text{sp-CSS}}$. Then, a filter gain $K \in \boldsymbol{K}^{\text{eid}}_{\text{sp-CSS}}(A, B, C, D, E, 0)$ exists such that the strictly proper filter $\Sigma_{\text{sp-CSS}}$ given in (7.12) is an EID filter and, moreover, $\lambda(A - KC) = \Lambda$.*

Proof : Theorem 7.18 already characterized the EID fixed modes. This immediately yields the necessity. To actually establish the existence of an EID filter with the required eigenvalues, we again use the duality between filters of the form $\Sigma_{\text{sp-CSS}}$ given in (7.12) and static state feedbacks for the dual system (7.69). The existence of a static state feedback that achieves exact disturbance-decoupling with the required property that $\lambda(A' - C'K') = \Lambda$ follows from the results in Chapter 5. ∎

An algorithm for designing a strictly proper EID filter with simultaneous filter pole placement:

Step 1: By using the SCB of the system characterized by the quadruple (A, B, C, D), compute $\mathcal{S}^-(A, B, C, D)$ or $\mathcal{S}^\ominus(A, B, C, D)$ depending on whether a continuous- or a discrete-time system is considered. Check the solvability condition $\mathcal{S}^-(A, B, C, D) \subseteq \ker E$ (continuous time) or $\mathcal{S}^\ominus(A, B, C, D) \subseteq \ker E$ (discrete time). If this is not satisfied, an EID filter does not exist.

Step 2: Compute the set $\Omega^{\text{eid}}_{\text{sp-CSS}}(A, B, C, D, E, 0) = \Omega_s(A', C', B', D', E')$ by using the EDD algorithm (see Section 5.4) with its input as the quintuple (A', C', B', D', E').

Step 3: Choose a desired set Λ of n self-conjugate elements in the open left-half complex plane $\mathbb{C}^-$ for continuous-time systems or of n self-conjugate elements within the unit disk $\mathbb{C}^\ominus$ for discrete-time systems such that Λ includes $\Omega^{\text{eid}}_{\text{sp-CSS}}(A, B, C, D, E, 0)$.

Step 4: Design K by using the EDD algorithm. As shown in the block diagram of Figure 7.3, the inputs to the EDD algorithm are the matrix quintuple (A', C', B', D', E') and the matrix Λ. The output of the algorithm is the transpose of the filter gain matrix K.

With the filter gain K as computed above, it is easy to verify that the resulting filter $\Sigma_{\text{sp-CSS}}$ given in (7.12) is an EID filter with its poles at the locations specified by Λ, i.e., $\lambda(A - KC) = \Lambda$. As discussed in EDD algorithm, although the algorithm concentrates only on pole placement, certain freedom exists as well to place the eigenvectors of $A - KC$ but this is not pursued here.

7.5.2 *Proper EID filters of CSS architecture*

As emphasized, use of proper filters rather than strictly proper filters weakens the solvability conditions for EID for both continuous- as well as discrete-time systems. This obviously motivates us to develop and design proper filters. With this motivation, we develop in this subsection a proper full-order filter based on CSS architecture. Indeed, in what follows, to construct a full-order proper EID filter for the given system, we construct a strictly proper EID filter for an auxiliary system $\widetilde{\Sigma}^*$. The required auxiliary system $\widetilde{\Sigma}^*$ is constructed in three layers. In the first

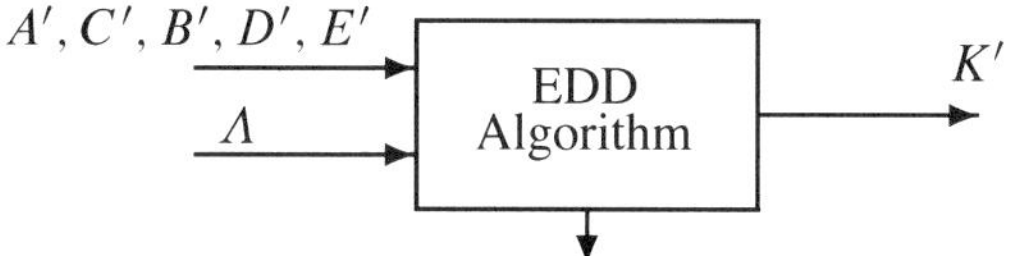

Fails, if conditions of Theorem 7.21 are not satisfied.

Figure 7.3: Strictly proper filter design with pole placement

layer, the given system Σ is rewritten in a suitable form by an appropriate coordinate transformation on the measured output y. Here the output y is decomposed into two parts y_0 and y_1; the first part y_0 contains explicitly the unknown input u, whereas the second part y_1 does not contain any input u. As the matrix F in general is nonzero, in the second layer by an appropriate preliminary injection of the output y into the desired output z, as discussed in Subsection 7.3, we construct another system Σ^* where the corresponding matrix F is zero. The first two layers are common for continuous- and discrete-time systems. However, the third layer differs distinctly for continuous- and discrete-time systems. For continuous-time systems, we augment the measured output y by adding a new component y_2 to it. The added component y_2 equals $\dot{y}_1$, the derivative of that part of y that does not contain explicitly any input in it. On the other hand, for discrete-time systems, we simply replace y_1 with its time-shifted version σy_1. This leads to the construction of the auxiliary system $\widetilde{\Sigma}^*$ for which we construct a strictly proper EID filter. Such a filter when appropriately translated for the given system Σ turns out to be a proper EID filter.

We now proceed to a detailed presentation of our design method. To do so, as alluded to, we first decompose the measured output y into two parts y_0 and y_1 in such a way that y_0 contains explicitly the unknown input u in it, whereas y_1 does not contain any input u in it. That is, we rewrite the given system (7.1) as

$$\Sigma : \begin{cases} \sigma x \quad = Ax + Bu \\ \begin{pmatrix} y_0 \\ y_1 \end{pmatrix} = \begin{pmatrix} C_0 \\ C_1 \end{pmatrix} x + \begin{pmatrix} D_0 \\ 0 \end{pmatrix} u = Cx + Du \\ z \quad = Ex + Fu, \end{cases} \tag{7.17}$$

where rank D = rank $D_0 = m_0$. We note that, without any loss of generality, we can rewrite the given system (7.1) in the form (7.17) by an appropriate coordinate transformation.

In general, the matrix F in (7.17) is nonzero. If so, we can use a preliminary injection of output y into the desired output z as discussed in Section 7.3, and we rewrite the desired output. To this end, let P^* be a solution of $F - PD = 0$ for

P. Then, as in (7.6), let

$$z^* = z - P^* y = (E - P^* C)x + (F - P^* D)u = E^* x, \qquad (7.18)$$

where $E^* = E - P^* C$. In view of (7.17) and (7.18), we can define a new system Σ^* as

$$\Sigma^* : \begin{cases} \sigma x & = Ax + Bu \\[4pt] \begin{pmatrix} y_0 \\ y_1 \end{pmatrix} = \begin{pmatrix} C_0 \\ C_1 \end{pmatrix} x + \begin{pmatrix} D_0 \\ 0 \end{pmatrix} u \\[4pt] z^* & = E^* x. \end{cases} \qquad (7.19)$$

By now we have rewritten the given system in a form suitable for filter development. However, before we proceed further, we need one more amendment of the measured output y. The needed amendment depends on whether we deal with continuous- or discrete-time systems. As such what follows, we divide our presentation into two parts: one pertaining to continuous-time systems and the other pertaining to discrete-time systems.

Proper EID filters of CSS architecture—continuous-time systems:

In what follows, we deal with continuous-time systems. Here we form a new measurement variable $\tilde{y}$ by augmenting y with another part $y_2 = \dot{y}_1$. That is, we let

$$\tilde{y} = \begin{pmatrix} y_0 \\ y_1 \\ y_2 \end{pmatrix} = \begin{pmatrix} y_0 \\ y_1 \\ \dot{y}_1 \end{pmatrix} \quad \text{and} \quad \tilde{C} = \begin{pmatrix} C_0 \\ C_1 \\ C_1 A \end{pmatrix}, \quad \tilde{D} = \begin{pmatrix} D_0 \\ 0 \\ C_1 B \end{pmatrix}. \qquad (7.20)$$

Then, it is easy to see that

$$\tilde{y} = \tilde{C} x + \tilde{D} u.$$

We note that y_2 is not directly available. However, as will be seen shortly it can be eliminated from the filter equation.

Auxiliary System $\tilde{\Sigma}^*$: We define next an auxiliary system with its measured output as $\tilde{y}$:

$$\tilde{\Sigma}^* : \begin{cases} \dot{x} & = Ax + Bu \\ \tilde{y} & = \tilde{C} x + \tilde{D} u \\ z^* & = E^* x. \end{cases} \qquad (7.21)$$

Before we proceed further, it is important that we investigate certain structural properties of $\tilde{\Sigma}^*$ and relate them to those of Σ^*. We have the following results.

Lemma 7.22 *Consider the systems $\tilde{\Sigma}^*$ and Σ^* that are, respectively, characterized by the matrix quintuples $(A, B, \tilde{C}, \tilde{D}, E^*)$ and (A, B, C, D, E^*). Consider two subsystems: one corresponding to $\tilde{\Sigma}^*$ and characterized by the quadruple $(A, B, \tilde{C}, \tilde{D})$, and the other corresponding to Σ^* and characterized by the quadruple (A, B, C, D). Then, the following results hold:*

(*i*) *The invariant zeros of the subsystem characterized by* $(A, B, \tilde{C}, \tilde{D})$ *are the same as those of the subsystem characterized by* (A, B, C, D).

(*ii*) *The matrix pair* $(\tilde{C}, A)$ *is* $\mathbb{C}^-$*-detectable if and only if the matrix pair* (C, A) *is* $\mathbb{C}^-$*-detectable.*

(*iii*) *Orders of infinite zeros of the subsystem characterized by* $(A, B, \tilde{C}, \tilde{D})$ *are reduced by one from those of the subsystem characterized by* (A, B, C, D).

(*iv*) $\mathcal{S}^-(A, B, \tilde{C}, \tilde{D}) = \mathcal{S}^-(A, B, C, D) \cap C^{-1}\{\mathrm{im}\, D\}$.

(*v*) $\mathcal{S}^-(A, B, \tilde{C}, \tilde{D}) = \{0\}$ *if and only if the subsystem characterized by* (A, B, C, D) *is left-invertible and has only invariant zeros in* $\mathbb{C}^-$ *and has no infinite zeros of order higher than one.*

Proof : To establish (*i*), we first premultiply the Rosenbrock system matrix of the subsystem characterized by the quadruple (A, B, C, D) with a matrix that has full rank for all s, and we obtain the Rosenbrock system matrix of the subsystem characterized by the quadruple $(A, B, \tilde{C}, \tilde{D})$:

$$\begin{pmatrix} I & 0 & 0 \\ 0 & I & 0 \\ 0 & 0 & I \\ -C_1 & 0 & sI \end{pmatrix} \begin{pmatrix} sI - A & -B \\ C_0 & D_1 \\ C_1 & 0 \end{pmatrix} = \begin{pmatrix} sI - A & -B \\ \tilde{C} & \tilde{D} \end{pmatrix}.$$

It is now obvious that the two systems have the same invariant zeros.

To establish (*ii*), we note that

$$\begin{pmatrix} I & 0 & 0 \\ 0 & I & 0 \\ 0 & 0 & I \\ -C_1 & 0 & sI \end{pmatrix} \begin{pmatrix} sI - A \\ C_0 \\ C_1 \end{pmatrix} = \begin{pmatrix} sI - A \\ \tilde{C} \end{pmatrix}.$$

Using the Hautus test (see [29]) for detectability, we find that the original system is detectable if and only if the transformed system is detectable.

In a suitable basis for the input space, we can obviously factorize

$$D_0 = \begin{pmatrix} \hat{D}_0 & 0 \end{pmatrix} \quad \text{and} \quad B = \begin{pmatrix} B_0 & B_1 \end{pmatrix}$$

such that $\hat{D}_0$ is invertible. To establish (*iii*), we first note that clearly the infinite zeros of (A, B, C, D) are equal to the infinite zeros of $(A, B_1, C_1, 0)$. Similarly the infinite zeros of $(A, B, \tilde{C}, \tilde{D})$ are equal to the infinite zeros of

$$\left(A, B_1, \begin{pmatrix} C_1 \\ C_1 A \end{pmatrix}, \begin{pmatrix} 0 \\ C_1 B_1 \end{pmatrix} \right).$$

The transfer matrix of the latter system is given by

$$\begin{pmatrix} C_1(\lambda I - A)^{-1} B_1 \\ \lambda C_1(\lambda I - A)^{-1} B_1 \end{pmatrix},$$

and hence, the infinite zeros of this system are determined by the infinite zeros of the transfer matrix $\lambda C_1(\lambda I - A)^{-1} B_1$, whereas the infinite zeros of the original system are determined by the infinite zeros of the transfer matrix $C_1(\lambda I - A)^{-1} B_1$. Property (iii) then follows immediately.

For property (iv), we first establish that

$$\mathcal{S}^*(A, B, \widetilde{C}, \widetilde{D}) = \mathcal{S}^*(A, B, C, D) \cap C^{-1}\{\mathrm{im}\, D\}. \tag{7.22}$$

The strongly controllable subspace $\mathcal{S}^*(A, B, C, D)$ is determined by the following recursion (see [98]):

$$\mathcal{S}_0 = \{0\}, \quad \mathcal{S}_{k+1} = \begin{pmatrix} A & B \end{pmatrix}\left[(\mathcal{S}_k \times \mathbb{R}^m) \cap \ker\begin{pmatrix} C & D \end{pmatrix}\right]$$

with $\mathcal{S}_0 \subseteq \mathcal{S}_1 \subseteq \mathcal{S}_2 \subseteq \cdots$. After at most n steps, we have $\mathcal{S}_{k+1} = \mathcal{S}_k$ and $\mathcal{S}^*(A, B, C, D) = \mathcal{S}_k$. Similarly the space $\mathcal{S}^*(A, B, \widetilde{C}, \widetilde{D})$ is determined by the recursion

$$\widetilde{\mathcal{S}}_0 = \{0\}, \quad \widetilde{\mathcal{S}}_{k+1} = \begin{pmatrix} A & B \end{pmatrix}\left[\left(\widetilde{\mathcal{S}}_k \times \mathbb{R}^m\right) \cap \ker\begin{pmatrix} \widetilde{C} & \widetilde{D} \end{pmatrix}\right].$$

It is then easily verified that

$$\widetilde{\mathcal{S}}_k = \mathcal{S}_k \cap C^{-1}\{\mathrm{im}\, D\}$$

for all k, and hence, (7.22) follows. Note that the differences between $\mathcal{S}^*$ and $\mathcal{S}^-$ are determined by the invariant zeros. But property (i) establishes that the two systems have the same invariant zeros and the zero dynamics are always contained in $C^{-1}\{\mathrm{im}\, D\}$, and hence, property (iv) follows.

For Property (v), it is clear from the characterization in terms of SCB that $\mathcal{S}^-(A, B, \widetilde{C}, \widetilde{D})$ contains the unstable zero dynamics. Hence, a necessary condition for $\mathcal{S}^-(A, B, \widetilde{C}, \widetilde{D})$ to be zero is that all invariant zeros are contained in $\mathcal{S}^-$. In that case, we immediately know that

$$\mathcal{S}^-(A, B, \widetilde{C}, \widetilde{D}) = \mathcal{S}^*(A, B, \widetilde{C}, \widetilde{D}).$$

The latter space is the smallest subspace $\mathcal{S}$ for which a matrix K exists such that $\mathcal{S}$ is $A + K\widetilde{C}$-invariant and contains $B + K\widetilde{D}$. This subspace is therefore equal to zero if and only if a matrix K exists such that $B + K\widetilde{D} = 0$. It is easily seen that this is equivalent to

$$\ker\begin{pmatrix} D \\ (C + DF)B \end{pmatrix} \subseteq \ker B \tag{7.23}$$

for some F. The latter is clearly equivalent to the system being left-invertible without infinite zeros of order larger than 1 because we have the standing assumption that $\begin{pmatrix} B' & D' \end{pmatrix}$ is surjective. In view of this, (7.23) implies that

$$\ker \begin{pmatrix} D \\ (C + DF)B \end{pmatrix} = \{0\}. \qquad \blacksquare$$

As mentioned, we plan to design first a strictly proper filter to solve the EID filtering problem for the auxiliary system $\widetilde{\Sigma}^*$, and then we modify it to obtain a proper filter that solves the EID filtering problem for the original system Σ. This obviously is possible if and only if the conditions for the existence of a strictly proper EID filter for the auxiliary system $\widetilde{\Sigma}^*$ coincide with the conditions for the existence of a proper EID filter for the system Σ. The following lemma that is a consequence of Lemma 7.22 formalizes this.

Lemma 7.23 *Consider the continuous-time systems Σ and $\widetilde{\Sigma}^*$, respectively, as given in (7.1) and (7.21). Then, the following two statements are equivalent:*

(i) A strictly proper EID filter exists for the auxiliary system $\widetilde{\Sigma}^*$.

(ii) A proper EID filter exists for the system Σ.

Moreover, there is a $1-1$ relationship between the strictly proper EID filter of CSS architecture for $\widetilde{\Sigma}^*$ and the proper EID filter of CSS architecture for Σ; that is, one of these filters can be constructed from the other.

Proof : In view of Theorems 7.6 and 7.9, the equivalence between the statements (i) and (ii) follows directly from Lemma 7.22. The $1-1$ relationship between the aforementioned filters can be verified easily. $\qquad \blacksquare$

For the auxiliary system $\widetilde{\Sigma}^*$, we form a strictly proper filter of CSS architecture as

$$\begin{cases} \dot{\widetilde{\xi}} = A\widetilde{\xi} + K(\widetilde{y} - \widetilde{C}\widetilde{\xi}) \\ \hat{z}^* = E^*\widetilde{\xi}, \end{cases} \qquad (7.24)$$

where the matrix K is the filter gain. Then, in view of (7.18), the estimate $\hat{z}$ of z is given by

$$\hat{z} = \hat{z}^* + P^*y = E^*\widetilde{\xi} + P^*y. \qquad (7.25)$$

The above development focuses on developing the strictly proper filter (7.24) that, however, is not directly implementable because $y_2 = \dot{y}_1$ is not available as a measured variable. But we can eliminate the need for $\dot{y}_1$ by defining a new variable:

$$\xi = \tilde{\xi} - K_2 y_1. \tag{7.26}$$

Here K_2 is obtained by partitioning K in conformity with the partitioning of y. That is,

$$K = \begin{pmatrix} K_0 & K_1 & K_2 \end{pmatrix}.$$

With the definition of ξ as in (7.26), we can rewrite the filter equation (7.24) as

$$\begin{cases} \dot{\xi} = (A - K\tilde{C})\xi + \begin{pmatrix} K_0 & K_1 + (A - K\tilde{C})K_2 \end{pmatrix} y \\ \hat{z}^* = E^*(\xi + K_2 y_1). \end{cases} \tag{7.27}$$

Obviously, the filter given above does not use $\dot{y}_1$. Moreover, it is proper rather than strictly proper. The filter given in (7.27) is indeed the proper full-order CSS filter that is to be used for Σ^*.

Proper filter of CSS architecture for Σ: Finally, in view of (7.25), we can rewrite (7.27) as an implementable proper filter for Σ:

$$\Sigma_{\text{p-CSS}} : \begin{cases} \dot{\xi} = (A - K\tilde{C})\xi + \tilde{K}y \\ \hat{z} = E^*\xi + \tilde{P}y, \end{cases} \tag{7.28}$$

where

$$\tilde{K} = \begin{pmatrix} K_0 & K_1 + (A - K\tilde{C})K_2 \end{pmatrix} \quad \text{and} \quad \tilde{P} = \begin{pmatrix} 0 & E^*K_2 \end{pmatrix} + P^*.$$

A block diagram representation of the proper filter of CSS architecture along with the given plant is shown in Figure 7.4.

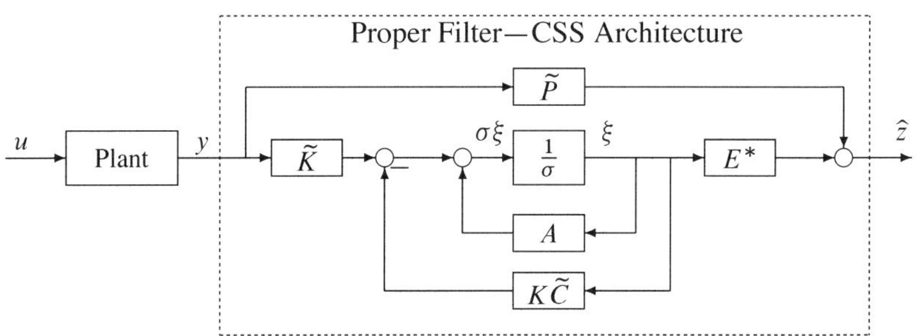

Figure 7.4: Block diagram of a proper filter—CSS architecture

Error Dynamics: By defining the error $e = x - \tilde{\xi}$, the error between the actual desired output $z = E^*x + P^*y$ and the estimated desired output $\hat{z} = E^*\tilde{\xi} + P^*y$ can be written as

$$e_z = z - \hat{z} = E^*e.$$

Then, in view of (7.17) and (7.24), the dynamics of the error signal are given by

$$\Sigma_{\text{p-CSS}}^{ue} : \begin{cases} \dot{e} = (A - K\widetilde{C})e + (B - K\widetilde{D})u \\ e_z = E^*e. \end{cases} \tag{7.29}$$

Also, the transfer matrix $G_{\text{p-CSS}}^{ue}$ from u to e_z can obviously be written as

$$G_{\text{p-CSS}}^{ue} = E^*(sI - A + K\widetilde{C})^{-1}(B - K\widetilde{D}). \tag{7.30}$$

Remark 7.24 *In view of (7.24) and (7.29), it is easy to see that both the filter equation and the error equation have the same poles, which are the eigenvalues of $A - K\widetilde{C}$.*

Everything in the filters (7.24), (7.27), and (7.28) is known except the gain K. Following the notation used in connection with strictly proper filters, let us denote the set of all EID filter gains by $K_{\text{p-CSS}}^{\text{eid}}(A, B, C, D, E, F)$, meaning that any gain $K \in K_{\text{p-CSS}}^{\text{eid}}$ renders $\Sigma_{\text{p-CSS}}$ given in (7.28) an EID filter for (7.1), and conversely, any gain K that renders $\Sigma_{\text{p-CSS}}$ an EID filter for (7.1) is an element of $K_{\text{p-CSS}}^{\text{eid}}$.

We have the following result.

Theorem 7.25 *Consider a continuous-time system as in (7.1). Let the conditions of Theorem 7.9 be satisfied. Then, the set $K_{\text{p-CSS}}^{\text{eid}}(A, B, C, D, E, F)$ is nonempty and is given by*

$$K_{\text{p-CSS}}^{\text{eid}}(A, B, C, D, E, F)$$
$$= \bigcup_{P^* \in \mathscr{P}^*} K_{\text{sp-CSS}}^{\text{eid}}(A, B, \widetilde{C}, \widetilde{D}, E - P^*C, 0), \tag{7.31}$$

where $\mathscr{P}^*$ is the set of all P^* that solve $F - PD = 0$ for P and $K_{\text{sp-CSS}}^{\text{eid}}$ is as defined in Theorem 7.13.

Proof : The proof follows from Lemma 7.23. ∎

Remark 7.26 *We observe that $K_{\text{sp-CSS}}^{\text{eid}}(A, B, \widetilde{C}, \widetilde{D}, E - P^*C, 0)$ is nonempty for any $P^* \in \mathscr{P}^*$.*

As in the case of strictly proper filters, the set $K_{\text{p-CSS}}^{\text{eid}}(A, B, C, D, E, F)$ has the following telescopic property with respect to the matrices $(E \quad F)$.

Lemma 7.27 *Consider a continuous-time system as in (7.1). Consider two different EID filtering problems both as defined in Definition 7.3, however, with one having the matrix pair (E_a, F_a) and the other (E_b, F_b) for the pair (E, F). Then, we have the following telescopic property:*

$$\ker \begin{pmatrix} E_a & F_a \end{pmatrix} \supseteq \ker \begin{pmatrix} E_b & F_b \end{pmatrix}$$
$$\implies K^{\mathrm{eid}}_{\mathrm{p\text{-}CSS}}(A, B, C, D, E_a, F_a) \supseteq K^{\mathrm{eid}}_{\mathrm{p\text{-}CSS}}(A, B, C, D, E_b, F_b),$$

where we assume that the matrix pairs (E_a, F_a) and (E_b, F_b) are such that, for the respective underlying system, the EID filtering problem is solvable (i.e., the conditions of Theorem 7.9 are satisfied).

Proof : The proof follows in view of Lemmas 7.23 and 7.14 and Theorem 7.25. ∎

In general, whenever they exist, proper EID filters of the form $\Sigma_{\mathrm{p\text{-}CSS}}$ given in (7.28) are not necessarily unique. The following theorem addresses the issue of the uniqueness of such filters.

Theorem 7.28 *Consider a continuous-time system as in (7.1). Let Assumption 7.1 be satisfied. Then, a unique proper full-order EID filter of CSS architecture of the form $\Sigma_{\mathrm{p\text{-}CSS}}$ given in (7.28) exists if and only if the following conditions are satisfied:*

(i) The subsystem characterized by the quadruple (A, B, C, D) does not have any invariant zeros on the imaginary axis.

(ii) The matrix D is invertible.

(iii) The matrix pair $(E - FD^{-1}C, \ A - BD^{-1}C)$ is observable.

Whenever the above conditions are satisfied, the unique EID filter is given by

$$\dot{\xi} = (A - BD^{-1}C)\xi + BD^{-1}y \quad \text{with} \quad \hat{z} = (E - FD^{-1}C)\xi + FD^{-1}y.$$

Proof : In view of Theorem 7.25, the uniqueness of the filter $\Sigma_{\mathrm{p\text{-}CSS}}$ is equivalent to the condition that the set

$$K^{\mathrm{eid}}_{\mathrm{p\text{-}CSS}}(A, B, C, D, E, F)$$

is a singleton set. Then, in view of (7.31), the condition that the set $K^{\mathrm{eid}}_{\mathrm{sp\text{-}CSS}}$ is a singleton set is equivalent to the condition that the set $F_s(A', \widetilde{C}', B', \widetilde{D}', (E - P^*C)')$ is a singleton set. A close examination of the EDD_{li} algorithm (see Section 5.4) that generates F_s reveals the conditions given in the theorem for F_s to be a singleton. ∎

Remark 7.29 *We observe that the conditions given in Theorem 7.28 imply those given in Theorem 7.11 but not conversely. This is natural and is expected because Theorem 7.28 requires the uniqueness of an EID filter having a particular architecture, namely, proper full-order CSS architecture.*

In general, whenever it exists, a proper EID filter of the form $\Sigma_{\text{p-CSS}}$ given in (7.28) is not unique. The nonuniqueness of such a filter can indeed be a blessing as some other specifications can then be imposed to come up with an appropriate filter. As in the case of strictly proper filters, we would like to make use of the available freedom to place the poles of an EID filter at desired locations. However, as in the case of strictly proper filters, the requirement of EID dictates that some poles of the filter be fixed at certain locations while the others be free to be assigned. In fact, as before, for each given system, a set of complex numbers exists that every EID filter must have among its poles. As the following definition formalizes, such a set of complex numbers can be termed as the EID filter fixed modes.

Definition 7.30 *(**Fixed modes of proper EID filters of CSS architecture**) Consider a continuous-time system as in (7.1) and the EID filtering problem 7.3 characterized by the matrix sextuple (A, B, C, D, E, F). Assume that the solvability conditions as specified by Theorem 7.9 are satisfied. Then, a scalar $\lambda \in \mathbb{C}^-$ is said to be the fixed mode of an EID filter with the proper full-order CSS architecture if λ is a pole (i.e., an eigenvalue of $A - K\widetilde{C}$) of every EID filter of such an architecture $\Sigma_{\text{p-CSS}}$ as given in (7.28). The set of all such EID filter fixed modes is denoted here by $\mathbf{\Omega}^{\text{eid}}_{\text{p-CSS}}(A, B, C, D, E, F)$.*

The following lemma is needed before we characterize the set $\mathbf{\Omega}^{\text{eid}}_{\text{p-CSS}}$.

Lemma 7.31 *Consider a continuous-time system as in (7.1). Let the conditions of Theorem 7.9 be satisfied. Let P_1^* and P_2^* be two different solutions of $F - PD = 0$ for P. Then, we have*

$$\mathbf{\Omega}^{\text{eid}}_{\text{sp-CSS}}(A, B, \widetilde{C}, \widetilde{D}, E - P_1^*C, 0)$$

$$= \mathbf{\Omega}^{\text{eid}}_{\text{sp-CSS}}(A, B, \widetilde{C}, \widetilde{D}, E - P_2^*C, 0), \quad (7.32)$$

where for $i = 1, 2$, the set $\mathbf{\Omega}^{\text{eid}}_{\text{sp-CSS}}(A, B, \widetilde{C}, \widetilde{D}, E - P_i^*C, 0)$ is as defined in Theorem 7.18.

Proof : This is a direct consequence of Theorems 7.18 and 5.32.   ■

We have the following theorem.

Theorem 7.32 *Consider a continuous-time system as in (7.1). Let Assumption 7.1 be satisfied. Consider proper full-order filters of CSS architecture, and assume that the solvability conditions as specified by Theorem 7.9 are satisfied. Also, let P^* be a solution of $F - PD = 0$ for P. Then, we have*

$$\Omega^{\mathrm{eid}}_{\mathrm{p\text{-}CSS}}(A, B, C, D, E, F) = \Omega^{\mathrm{eid}}_{\mathrm{sp\text{-}CSS}}(A, B, \tilde{C}, \tilde{D}, E - P^*C, 0). \qquad (7.33)$$

Moreover, the set $\Omega^{\mathrm{eid}}_{\mathrm{p\text{-}CSS}}$ is invariant with respect to the choice of P^* that solves $F - PD = 0$ for P.

Proof : The proof is an obvious consequence of Lemmas 7.23 and 7.31.   ∎

Remark 7.33 *In view of (7.33), the set $\Omega^{\mathrm{eid}}_{\mathrm{sp\text{-}CSS}}(A, B, \tilde{C}, \tilde{D}, E - P^*C, 0)$ seems to depend on P^*. However, as Lemma 7.31 shows, it is invariant of the choice of P^*. The seeming dependency of $\Omega^{\mathrm{eid}}_{\mathrm{sp\text{-}CSS}}$ on P^* can be suppressed. To do so, we need to decompose the matrix E into two parts as $E = E_1 + E_2$ such that*

$$\mathcal{S}^-(A, B, C, D) \subseteq \ker E_1, \qquad (7.34)$$

and moreover,

$$\ker \begin{pmatrix} C & D \end{pmatrix} \subseteq \ker \begin{pmatrix} E_2 & F \end{pmatrix}. \qquad (7.35)$$

The decomposition of E indicated above can always be done under the solvability conditions given by Theorem 7.9. Indeed, one can do so by rewriting the system characterized by the quadruple (A, B, C, D) in its SCB and then accordingly rewriting the matrix E. Having decomposed E as given above, one can show easily that a specific matrix P^* exists such that $E^* = E - P^*C = E_1$ with P^* being a solution of $F - PD = 0$ for P. As the matrix E_1 is obtained from the structural property of the given system Σ, the selection of $E^* = E_1$ eliminates the dependency of $\Omega^{\mathrm{eid}}_{\mathrm{sp\text{-}CSS}}$ on P^* and avoids solving the equation $F - PD = 0$ for P.

Remark 7.34 *Obviously, when the proper EID filter is unique (i.e., when the conditions given in Theorem 7.28 are satisfied), the set $\Omega^{\mathrm{eid}}_{\mathrm{p\text{-}CSS}}(A, B, C, D, E, F)$ coincides with the set $\lambda(A - BD^{-1}C)$.*

As in the case of strictly proper filters, the set $\Omega^{\mathrm{eid}}_{\mathrm{p\text{-}CSS}}(A, B, C, D, E, F)$ has the following telescopic property.

Lemma 7.35 *Consider a continuous-time system as in (7.1). Consider two different EID filtering problems both as defined in Definition 7.3, however, with one having the matrix pair (E_a, F_a) and the other (E_b, F_b) for the pair (E, F). Then, we have the following telescopic property:*

$$\ker \begin{pmatrix} E_a & F_a \end{pmatrix} \supseteq \ker \begin{pmatrix} E_b & F_b \end{pmatrix}$$
$$\implies \boldsymbol{\Omega}^{\text{eid}}_{\text{p-CSS}}(A, B, C, D, E_a, F_a) \subseteq \boldsymbol{\Omega}^{\text{eid}}_{\text{p-CSS}}(A, B, C, D, E_b, F_b),$$

where we assume that the matrix pairs (E_a, F_a) and (E_b, F_b) are such that, for the respective underlying system, the EID filtering problem is solvable (i.e., the conditions of Theorem 7.9 are satisfied).

Proof : The proof follows from Lemmas 7.14 and 7.23 and Theorem 7.32. ■

The set of fixed modes obviously presents a constraint that one must contend with in any EID filter design. In this regard, one ponders with the following. Suppose the conditions for the existence of strictly proper EID filters as specified by Theorem 7.6 are satisfied. In this case, both strictly proper as well as proper EID filters exist. The question that arises then is as follows. Will the number and the location of fixed modes be different for strictly proper and proper EID filters? The following lemma answers this question negatively whenever we use filters of CSS architecture.

Lemma 7.36 *Consider a continuous-time system as in (7.1). Let Assumption 7.1 be satisfied. Assume that the solvability conditions for the existence of strictly proper EID filters as specified by Theorem 7.6 are satisfied. Then, we have*

$$\boldsymbol{\Omega}^{\text{eid}}_{\text{p-CSS}}(A, B, C, D, E, F) = \boldsymbol{\Omega}^{\text{eid}}_{\text{sp-CSS}}(A, B, C, D, E, 0). \tag{7.36}$$

Proof : We first note from Theorem 7.32 that

$$\boldsymbol{\Omega}^{\text{eid}}_{\text{p-CSS}}(A, B, C, D, E, F) = \boldsymbol{\Omega}^{\text{eid}}_{\text{sp-CSS}}(A, B, \widetilde{C}, \widetilde{D}, E, 0)$$

because we must have that $F = 0$, and hence, we can choose $P^* = 0$. Next, we note from Theorem 7.18 that

$$\boldsymbol{\Omega}^{\text{eid}}_{\text{sp-CSS}}(A, B, \widetilde{C}, \widetilde{D}, E, 0) = \boldsymbol{\Omega}_s(A', \widetilde{C}', B', \widetilde{D}', E').$$

According to Theorem 5.11 and the associated Remark 5.12, this set $\boldsymbol{\Omega}_s$ consists of a subset of the invariant zeros of $(A', \widetilde{C}', B', \widetilde{D}')$, which are controllable from

E' and the input-decoupling zeros of $(A', (\tilde{C}'\ E'))$. From these characterizations and the structure of $\tilde{C}$ and $\tilde{D}$, it becomes clear that

$$\Omega_s(A', \tilde{C}', B', \tilde{D}', E') = \Omega_s(A', C', B', D', E').$$

The result then follows by noting that

$$\Omega_{\text{sp-CSS}}^{\text{eid}}(A, B, C, D, E, 0) = \Omega_s(A', C', B', D', E'). \qquad \blacksquare$$

Finally, we would like to design an EID filter while placing its poles at desired locations. In this regard, the following theorem shows that we can design a filter gain K such that $\Sigma_{\text{p-CSS}}$ given in (7.28) is an EID filter while its poles are at the prescribed desired locations with the restriction that they need to contain $\Omega_{\text{p-CSS}}^{\text{eid}}(A, B, C, D, E, F)$ among them.

Theorem 7.37 *(EID proper filter using CSS architecture with pole placement)*
Consider a continuous-time system as in (7.1). Let the conditions of Theorem 7.9 be satisfied. Also, let $E^* = E - P^*C$, with P^* being a solution of $F - PD = 0$ for P. Consider the set $\Omega_{\text{p-CSS}}^{\text{eid}}(A, B, C, D, E, F)$, which is the set of EDD fixed modes discussed in Theorem 7.32. Also, consider the nonempty set $K_{\text{p-CSS}}^{\text{eid}}(A, B, C, D, E, F)$ discussed in Theorem 7.25. Moreover, let Λ be a prescribed set of n self-conjugate elements in the open left-half complex plane $\mathbb{C}^-$ such that Λ includes $\Omega_{\text{p-CSS}}^{\text{eid}}$. Then, a filter gain $K \in K_{\text{p-CSS}}^{\text{eid}}(A, B, C, D, E, F)$ exists such that the full-order proper filter of CSS architecture, namely $\Sigma_{\text{p-CSS}}$ as given in (7.28), is an EID filter, and moreover, $\lambda(A - K\tilde{C}) = \Lambda$.

Proof : The proof follows easily in view of Theorems 7.9 and 7.25. $\qquad \blacksquare$

An algorithm for designing a proper EID filter with simultaneous filter pole placement:

Step 1: Given the system Σ as in (7.17) and characterized by the sextuple (A, B, C, D, E, F), form the auxiliary system $\tilde{\Sigma}^*$ characterized by the quintuple $(A, B, \tilde{C}, \tilde{D}, E^*)$, where $\tilde{C}$ and $\tilde{D}$ are as in (7.20), and as usual, $E^* = E - P^*C$, with matrix P^* being any solution of $F - PD = 0$ for P.
Step 2: Choose a desired set Λ of n self-conjugate elements in the open left-half complex plane $\mathbb{C}^-$. [Step 3 given below checks to make sure that Λ includes $\Omega_{\text{p-CSS}}^{\text{eid}}(A, B, C, D, E, F)$].
Step 3: Following the algorithm for designing a strictly proper EID filter with simultaneous filter pole placement as given in Subsection 7.5.1, and using the quintuple $(A, B, \tilde{C}, \tilde{D}, E^*)$ as well as the matrix Λ as inputs to that algorithm, obtain the gain K.

With the filter gain K as computed above, it is easy to verify that the resulting filter $\Sigma_{\text{p-CSS}}$ given in (7.28) is an EID filter with its poles at the locations specified by Λ; i.e., $\lambda(A - K\widetilde{C}) = \Lambda$.

Proper EID filters of CSS architecture—Discrete-time systems:

In what follows, we deal with discrete-time systems. Unlike in continuous-time systems, a new measurement variable $\widetilde{y}$ is formed by replacing y_1 in y with $y_2 = \sigma y_1$. That is, we let

$$\widetilde{y} = \begin{pmatrix} y_0 \\ y_2 \end{pmatrix} = \begin{pmatrix} y_0 \\ \sigma y_1 \end{pmatrix}, \quad \widetilde{C} = \begin{pmatrix} C_0 \\ C_1 A \end{pmatrix}, \quad \text{and} \quad \widetilde{D} = \begin{pmatrix} D_0 \\ C_1 B \end{pmatrix}. \tag{7.37}$$

Then, it is easy to see that

$$\widetilde{y} = \widetilde{C}x + \widetilde{D}u.$$

As in continuous-time systems, y_2 is not directly available. However, as before, it can be eliminated from the filter equation.

Auxiliary System $\widetilde{\Sigma}^*$: We define next an auxiliary system with its measured output as $\widetilde{y}$,

$$\widetilde{\Sigma}^* : \begin{cases} \sigma x = Ax + Bu \\ \widetilde{y} \; = \widetilde{C}x + \widetilde{D}u \\ z^* = E^* x, \end{cases} \tag{7.38}$$

where $E^* = E - P^* C$, with P^* being a solution of $F - PD = 0$ for P. As in continuous-time systems, before we proceed further, it is important that we investigate certain structural properties of $\widetilde{\Sigma}^*$ and relate them to those of Σ^*. We have the following results.

Lemma 7.38 *Consider the systems $\widetilde{\Sigma}^*$ and Σ^*, which are, respectively, characterized by the matrix quintuples $(A, B, \widetilde{C}, \widetilde{D}, E^*)$ and (A, B, C, D, E^*). Consider two subsystems: one corresponding to $\widetilde{\Sigma}^*$ and characterized by the quadruple $(A, B, \widetilde{C}, \widetilde{D})$, and the other corresponding to Σ^* and characterized by the quadruple (A, B, C, D). Then, the following results hold:*

(i) The invariant zeros of the subsystem characterized by $(A, B, \widetilde{C}, \widetilde{D})$ contain the invariant zeros of the subsystem characterized by (A, B, C, D) in addition to (possibly) some invariant zeros at the origin.

(ii) The matrix pair $(\widetilde{C}, A)$ is $\mathbb{C}^{\ominus}$-detectable if and only if the matrix pair (C, A) is $\mathbb{C}^{\ominus}$-detectable.

(iii) The orders of infinite zeros of the subsystem characterized by $(A, B, \widetilde{C}, \widetilde{D})$ are reduced by one from those of the subsystem characterized by (A, B, C, D).

(*iv*) *The subsystem characterized by $(A, B, \widetilde{C}, \widetilde{D})$ is left-invertible if and only if the subsystem characterized by (A, B, C, D) is left-invertible.*

(*v*) $\mathcal{S}^{\ominus}(A, B, \widetilde{C}, \widetilde{D}) = \mathcal{S}^{\ominus}(A, B, C, D) \cap C^{-1}\{\text{im } D\}.$

(*vi*) $\mathcal{S}^{\ominus}(A, B, \widetilde{C}, \widetilde{D}) = \{0\}$ *if and only if the subsystem characterized by (A, B, C, D) is left-invertible, has only invariant zeros in $\mathbb{C}^{\ominus}$, and has no infinite zeros of order higher than one.*

Proof : The proof follows by the same arguments as used in the proof of Lemma 7.22. ∎

Remark 7.39 *Lemma 7.38 for discrete-time systems is similar to Lemma 7.22 for continuous-time systems. In continuous-time systems, the invariant zeros of the subsystem characterized by $(A, B, \widetilde{C}, \widetilde{D})$ are the same as those of the subsystem characterized by (A, B, C, D). However, for discrete-time systems, the subsystem characterized by $(A, B, \widetilde{C}, \widetilde{D})$ has additional invariant zeros at the origin.*

As in continuous-time systems, we plan to design first a strictly proper filter to solve the EID filtering problem for the auxiliary system $\widetilde{\Sigma}^*$, and then we modify it to obtain a proper filter that solves the EID filtering problem for the original system Σ. This obviously is possible if and only if the conditions for the existence of a strictly proper EID filter for the auxiliary system $\widetilde{\Sigma}^*$ coincide with the conditions for the existence of a proper EID filter for the system Σ. The following lemma, which is a consequence of Lemma 7.38, formalizes this.

Lemma 7.40 *Consider the discrete-time systems Σ and $\widetilde{\Sigma}^*$, respectively, as given in (7.1) and (7.38). Then, the following two statements are equivalent:*

(*i*) *A strictly proper EID filter exists for the auxiliary system $\widetilde{\Sigma}^*$.*

(*ii*) *A proper EID filter exists for the system Σ.*

Moreover, there is a $1 - 1$ relationship between the strictly proper EID filter of CSS architecture for $\widetilde{\Sigma}^*$ and the proper EID filter of CSS architecture for Σ; that is, one of these filters can be constructed from the other.

Proof : In view of Theorems 7.6 and 7.9, the equivalence between the statements (*i*) and (*ii*) follows directly from Lemma 7.38. The $1 - 1$ relationship between the aforementioned filters can be verified easily. ∎

For the auxiliary system $\widetilde{\Sigma}^*$, we form a strictly proper filter of CSS architecture as

$$\begin{cases} \sigma\widetilde{\xi} = A\widetilde{\xi} + K(\widetilde{y} - \widetilde{C}\widetilde{\xi}) \\ \widehat{z}^* = E^*\widetilde{\xi}, \end{cases} \tag{7.39}$$

where the matrix K is the filter gain. Then, in view of (7.18), the estimate $\widehat{z}$ of z is given by

$$\widehat{z} = \widehat{z}^* + P^*y = E^*\widetilde{\xi} + P^*y. \tag{7.40}$$

The above development focuses on developing the strictly proper filter (7.24), which, however, is not directly implementable because $y_2 = \sigma y_1$ is not available as a measured variable. But we can eliminate the need for σy_1 by defining a new variable:

$$\xi = \widetilde{\xi} - K_2 y_1. \tag{7.41}$$

Here K_2 is obtained by partitioning K in conformity with the partitioning of y. That is,

$$K = \begin{pmatrix} K_0 & K_2 \end{pmatrix}.$$

With the definition of ξ as in (7.41), we can rewrite the filter equation (7.39) as

$$\begin{cases} \sigma\xi = (A - K\widetilde{C})\xi + \begin{pmatrix} K_0 & (A - K\widetilde{C})K_2 \end{pmatrix}y \\ \widehat{z}^* = E^*(\xi + K_2 y_1). \end{cases} \tag{7.42}$$

Obviously, the filter given above does not use σy_1. Moreover, it is proper rather than strictly proper. The filter given in (7.42) is indeed the proper full-order CSS filter that is to be used for Σ^*.

The structure of (7.42) essentially is similar to the structure of current type of filters well known in discrete-time literature.

Proper filter of CSS architecture for Σ:

Finally, in view of (7.40), we can rewrite (7.42) as an implementable proper filter for Σ:

$$\Sigma_{\text{p-CSS}} : \begin{cases} \sigma\xi = (A - K\widetilde{C})\xi + \widetilde{K}y \\ \widehat{z} = E^*\xi + \widetilde{P}y, \end{cases} \tag{7.43}$$

where

$$\widetilde{K} = \begin{pmatrix} K_0 & (A - K\widetilde{C})K_2 \end{pmatrix} \quad \text{and} \quad \widetilde{P} = \begin{pmatrix} 0 & E^*K_2 \end{pmatrix} + P^*.$$

A block diagram representation of the above proper filter is structurally the same as the one in Figure 7.4.

Error dynamics: By defining the error $e = x - \widetilde{\xi}$, the error between the actual desired output $z = E^*x + P^*y$ and the estimated desired output $\widehat{z} = E^*\widetilde{\xi} + P^*y$ can be written as

$$e_z = z - \widehat{z} = E^*e.$$

Then, in view of (7.17) and (7.39), the dynamics of error is given by

$$\Sigma_{\text{p-CSS}}^{ue} : \begin{cases} \sigma e = (A - K\tilde{C})e + (B - K\tilde{D})u \\ e_z = E^*e. \end{cases} \tag{7.44}$$

Also, the transfer matrix $G_{\text{p-CSS}}^{ue}$ from u to e_z can obviously be written as

$$G_{\text{p-CSS}}^{ue} = E^*(\sigma I - A + K\tilde{C})^{-1}(B - K\tilde{D}). \tag{7.45}$$

Remark 7.41 *In view of (7.39) and (7.44), it is easy to see that both the filter equation and the error equation have the same poles that are the eigenvalues of $A - K\tilde{C}$.*

Everything in the filters (7.39), (7.42), and (7.43) is known except the gain K. Following the notation used in the continuous-time case, let us denote the set of all EID filter gains by $K_{\text{p-CSS}}^{\text{eid}}(A, B, C, D, E, F)$, meaning that any gain $K \in K_{\text{p-CSS}}^{\text{eid}}$ renders $\Sigma_{\text{p-CSS}}$ given in (7.43) an EID filter for (7.1), and conversely, any gain K that renders $\Sigma_{\text{p-CSS}}$ an EID filter for (7.1) is an element of $K_{\text{p-CSS}}^{\text{eid}}$.

We have the following result.

Theorem 7.42 *Consider a discrete-time system as in (7.1). Let the conditions of Theorem 7.9 be satisfied. Then, the set $K_{\text{p-CSS}}^{\text{eid}}(A, B, C, D, E, F)$ is nonempty and is given by*

$$K_{\text{p-CSS}}^{\text{eid}}(A, B, C, D, E, F)$$
$$= \bigcup_{P^* \in \mathscr{P}^*} K_{\text{sp-CSS}}^{\text{eid}}(A, B, \tilde{C}, \tilde{D}, E - P^*C, 0) \tag{7.46}$$

where $\mathscr{P}^*$ is the set of all P^* that solve $F - PD = 0$ for P and $K_{\text{sp-CSS}}^{\text{eid}}$ is as defined in Theorem 7.13.

Proof : The proof follows from Lemma 7.40. ∎

Remark 7.43 *We observe that $K_{\text{sp-CSS}}^{\text{eid}}(A, B, \tilde{C}, \tilde{D}, E - P^*C, 0)$ is nonempty for any $P^* \in \mathscr{P}^*$.*

As in the case of continuous-time systems, the set $K_{\text{p-CSS}}^{\text{eid}}$ has the following telescopic property.

Lemma 7.44 *Consider a discrete-time system as in (7.1). Consider two different EID filtering problems both as defined in Definition 7.3, however, with one having the matrix pair (E_a, F_a) and the other (E_b, F_b) for the pair (E, F). Then, we have the following telescopic property:*

$$\ker \begin{pmatrix} E_a & F_a \end{pmatrix} \supseteq \ker \begin{pmatrix} E_b & F_b \end{pmatrix}$$
$$\implies K_{\text{p-CSS}}^{\text{eid}}(A, B, C, D, E_a, F_a) \supseteq K_{\text{p-CSS}}^{\text{eid}}(A, B, C, D, E_b, F_b),$$

where we assume that the matrix pairs (E_a, F_a) and (E_b, F_b) are such that, for the respective underlying system, the EID filtering problem is solvable (i.e., the conditions of Theorem 7.9 are satisfied).

Proof : The proof follows in view of Lemmas 7.40 and 7.14 and Theorem 7.42.
∎

In general, whenever they exist, EID filters of the form $\Sigma_{\text{p-CSS}}$ given in (7.43) are not necessarily unique. The following theorem addresses the issue of the uniqueness of such filters.

Theorem 7.45 *Consider a discrete-time system as in (7.1). Let Assumption 7.1 be satisfied. Then, a unique proper full-order EID filter of CSS architecture of the form $\Sigma_{\text{p-CSS}}$ given in (7.43) exists if and only if the following conditions are satisfied:*

(i) The subsystem characterized by the quadruple (A, B, C, D) does not have any invariant zeros on the unit circle.

(ii) The matrix D is invertible.

(iii) The matrix pair $(E - FD^{-1}C, \; A - BD^{-1}C)$ is observable.

Whenever the above conditions are satisfied, the unique EID filter is given by

$$\sigma\xi = (A - BD^{-1}C)\xi + BD^{-1}y \quad \text{with} \quad \hat{z} = (E - FD^{-1}C)\xi + FD^{-1}y.$$

Proof : In view of Theorem 7.42, the uniqueness of the filter $\Sigma_{\text{p-CSS}}$ is equivalent to the condition that the set $K_{\text{p-CSS}}^{\text{eid}}(A, B, C, D, E, F)$ is a singleton set. Then, in view of (7.31), the condition that $K_{\text{sp-CSS}}^{\text{eid}}$ is a singleton is equivalent to the condition that the set $F_s(A', \tilde{C}', B', \tilde{D}', (E - P^*C)')$ is a singleton set. A close examination of the EDD_{li} algorithm (see Section 5.4) that generates F_s reveals the conditions given in the theorem for F_s to be a singleton. ∎

Remark 7.46 *We observe that the conditions given in Theorem 7.45 imply those given in Theorem 7.11 but not conversely. This is natural and is expected because Theorem 7.45 requires the uniqueness of an EID filter having a particular architecture, namely, proper full-order CSS architecture.*

In general, whenever it exists, a proper EID filter of the form $\Sigma_{\text{sp-CSS}}$ given in (7.43) is not unique. We can use the nonuniqueness of filters to our advantage. That is, as in continuous-time systems, we would like to design an EID filter while placing its poles at desired locations. As before, it is expected that all proper EID filters share some fixed modes as defined below.

Definition 7.47 *(Fixed modes of proper EID filters of CSS architecture) Consider a discrete-time system as in (7.1) and the EID filtering problem 7.3 characterized by the matrix sextuple (A, B, C, D, E, F). Assume that the solvability conditions as specified by Theorem 7.9 are satisfied. Then, a scalar $\lambda \in \mathbb{C}^{\ominus}$ is said to be the fixed mode of an EID filter with the proper full-order CSS architecture if λ is a pole (i.e., an eigenvalue of $A - K\tilde{C}$) of every EID filter of such an architecture $\Sigma_{\text{p-CSS}}$ as given in (7.43). The set of all such EID filter fixed modes is denoted here by $\boldsymbol{\Omega}_{\text{p-CSS}}^{\text{eid}}(A, B, C, D, E, F)$.*

We have the following theorem.

Theorem 7.48 *Consider a discrete-time system as in (7.1). Let Assumption 7.1 be satisfied. Consider proper full-order filters of CSS architecture, and assume that the solvability conditions as specified by Theorem 7.9 are satisfied. Also, let $E^* = E - P^*C$, with P^* being a solution of $F - PD = 0$ for P. Then, we have*

$$\boldsymbol{\Omega}_{\text{p-CSS}}^{\text{eid}}(A, B, C, D, E, F) = \boldsymbol{\Omega}_{\text{sp-CSS}}^{\text{eid}}(A, B, \tilde{C}, \tilde{D}, E - P^*C, 0). \qquad (7.47)$$

Moreover, the set $\boldsymbol{\Omega}_{\text{p-CSS}}^{\text{eid}}$ is invariant with respect to the choice of P^* that solves $F - PD = 0$ for P.

Proof : The proof is an obvious consequence of Lemmas 7.31 and 7.40. ∎

Remark 7.49 *In view of (7.47), the set $\boldsymbol{\Omega}_{\text{sp-CSS}}^{\text{eid}}(A, B, \tilde{C}, \tilde{D}, E - P^*C)$ seems to depend on P^*. However, it is invariant of the choice of P^*. As in the continuous-time case, the seeming dependency of $\boldsymbol{\Omega}_{\text{sp-CSS}}^{\text{eid}}$ on P^* can be suppressed. To do so, once again we need to decompose the matrix E into two parts as $E = E_1 + E_2$ such that*

$$\mathcal{S}^{\ominus}(A, B, C, D) \subseteq \ker E_1. \qquad (7.48)$$

and moreover,

$$\ker \begin{pmatrix} C & D \end{pmatrix} \subseteq \ker \begin{pmatrix} E_2 & F \end{pmatrix}. \tag{7.49}$$

The decomposition of E indicated above can always be done under the solvability conditions given by Theorem 7.9. Indeed, one can do so by rewriting the system characterized by the quadruple (A, B, C, D) in its SCB and then accordingly rewriting the matrix E. Having decomposed E as given above, one can show easily that a specific P^* exists such that $E^* = E - P^*C = E_1$, with P^* being a solution of $F - PD = 0$ for P. As the matrix E_1 is obtained from the structural property of the given system Σ, the selection of $E^* = E_1$ eliminates the dependency of $\Omega^{\text{eid}}_{\text{sp-CSS}}$ on P^* and avoids solving the equation $F - PD = 0$ for P.

Remark 7.50 *Obviously, when the proper EID filter is unique (i.e., when the conditions given in Theorem 7.45 are satisfied), the set $\Omega^{\text{eid}}_{\text{p-CSS}}(A, B, C, D, E, F)$ coincides with the set $\lambda(A - BD^{-1}C)$.*

As in continuous-time case, the set $\Omega^{\text{eid}}_{\text{p-CSS}}(A, B, C, D, E, F)$ has the following telescopic property.

Lemma 7.51 *Consider a discrete-time system as in (7.1). Consider two different EID filtering problems both as defined in Definition 7.3, however, with one having the matrix pair (E_a, F_a) and the other (E_b, F_b) for the pair (E, F). Then, we have the following telescopic property:*

$$\ker \begin{pmatrix} E_a & F_a \end{pmatrix} \supseteq \ker \begin{pmatrix} E_b & F_b \end{pmatrix}$$
$$\implies \Omega^{\text{eid}}_{\text{p-CSS}}(A, B, C, D, E_a, F_a) \subseteq \Omega^{\text{eid}}_{\text{p-CSS}}(A, B, C, D, E_b, F_b),$$

where we assume that the matrix pairs (E_a, F_a) and (E_b, F_b) are such that, for the respective underlying system, the EID filtering problem is solvable (i.e., conditions of Theorem 7.9 are satisfied).

Proof : The proof follows from Lemmas 7.14 and 7.40 and Theorem 7.48. ∎

The set of fixed modes obviously presents a constraint that one must contend with in any EID filter design. In this regard, one ponders with the following. Suppose the conditions for the existence of strictly proper EID filters as specified by Theorem 7.6 are satisfied. In this case, both strictly proper as well as proper EID filters exist. The question that arises then is as follows: Will the number and the location of fixed modes be different for strictly proper and proper EID filters? In contrast to continuous-time case, the following lemma notes that in discrete time the additional fixed modes in the origin can be introduced whenever we use filters of CSS architecture.

Lemma 7.52 *Consider a discrete-time system as in (7.1). Let Assumption 7.1 be satisfied. Assume that the solvability conditions for the existence of strictly proper EID filters as specified by Theorem 7.6 are satisfied. Then, we have*

$$\Omega^{\text{eid}}_{\text{p-CSS}}(A, B, C, D, E, F) \supset \Omega^{\text{eid}}_{\text{sp-CSS}}(A, B, C, D, E, 0) \tag{7.50}$$

with the only possible difference between the two sets being some additional fixed modes at the origin.

Proof : The proof is similar to that of Lemma 7.36 that pertains to the continuous-time case. The only difference is that we know from Lemma 7.38 that in discrete time, the system $(A, B, \widetilde{C}, \widetilde{D})$ can have additional fixed modes at the origin.   ∎

As in the continuous-time case, the following theorem shows that we can design a filter gain K such that (7.43) is an EID filter with its poles at the prescribed desired locations provided that these locations contain the elements of the set $\Omega^{\text{eid}}_{\text{p-CSS}}(A, B, C, D, E, F)$.

Theorem 7.53 *(EID proper filter using CSS architecture with pole placement)*
Consider a discrete-time system as in (7.1). Let the conditions of Theorem 7.9 be satisfied. Also, let $E^* = E - P^*C$, with P^* being a solution of $F - PD = 0$ for P. Consider the set $\Omega^{\text{eid}}_{\text{p-CSS}}(A,B,C, D,E,F)$, which is the set of EDD fixed modes discussed in Theorem 7.48. Also, consider the nonempty set $K^{\text{eid}}_{\text{p-CSS}}(A, B, C, D, E, F)$ discussed in Theorem 7.42. Moreover, let Λ be a prescribed set of n self-conjugate elements within the unit disk $\mathbb{C}^{\ominus}$, and it includes $\Omega^{\text{eid}}_{\text{p-CSS}}$. Then, a filter gain $K \in K^{\text{eid}}_{\text{p-CSS}}(A, B, C, D, E, F)$ exists such that the proper filter (7.43) is an EID filter, and moreover, $\lambda(A - K\widetilde{C}) = \Lambda$.

Proof : The proof follows easily in view of Theorems 7.9 and 7.42.   ∎

An algorithm for designing a proper EID filter with simultaneous filter pole placement:

Step 1: Given the system Σ as in (7.17) and characterized by the sextuple (A, B, C, D, E, F), form the auxiliary system $\widetilde{\Sigma}^*$ characterized by the quintuple $(A, B, \widetilde{C}, \widetilde{D}, E^*)$, where $\widetilde{C}$ and $\widetilde{D}$ are as in (7.37), and as usual, $E^* = E - P^*C$, with matrix P^* being any solution of $F - PD = 0$ for P.

Step 2: Choose a desired set Λ of n self-conjugate elements within the unit circle $\mathbb{C}^{\ominus}$. [Step 3 given below checks to make sure that Λ includes $\Omega^{\text{eid}}_{\text{p-CSS}}(A, B, C, D, E, F)$].

Step 3: Following the algorithm for designing a strictly proper EID filter with simultaneous filter pole placement as given in Subsection 7.5.1, and using the quintuple $(A, B, \widetilde{C}, \widetilde{D}, E^*)$ as well as the matrix Λ as inputs to that algorithm, obtain the gain K.

With the filter gain K as computed above, it is easy to verify that the resulting filter (7.43) is an EID filter with its poles at the locations specified by Λ; i.e., $\lambda(A - K\widetilde{C}) = \Lambda$.

Remark 7.54 *In developing the proper EID filters, we used*

$$\widetilde{y} = \begin{pmatrix} y_0' & y_1' & \sigma y_1' \end{pmatrix}'$$

for continuous-time systems and

$$\widetilde{y} = \begin{pmatrix} y_0' & \sigma y_1' \end{pmatrix}'$$

for discrete-time systems. The reason for this difference lies in the fact that the use of $\widetilde{y} = (y_0' \quad \sigma y_1')'$ for continuous-time systems will not in general enable the resulting pair $(\widetilde{C}, A)$ $\mathbb{C}^-$-detectable, and hence, one cannot always estimate z by using such a $\widetilde{y}$. However, whenever the matrix A is stable (i.e., has all its eigenvalues in the open left-half complex plane $\mathbb{C}^-$), as $(\widetilde{C}, A)$ is then $\mathbb{C}^-$-detectable, one can use $\widetilde{y} = (y_0' \quad \sigma y_1')'$ to estimate z even for continuous-time systems. That is, whenever A is a stable matrix, the procedure developed for discrete-time systems can equally be used for continuous-time systems. Also, for discrete-time systems, the method developed for continuous-time systems using $\widetilde{y}$ as $(y_0' \quad y_1' \quad \sigma y_1')'$ can always be used without any restrictions.

7.5.3   Reduced-order EID filters of CSS architecture

The previous two subsections construct, respectively, strictly proper and proper full-order filters that solve a designated problem. By full-order filters, as usual, we mean filters having the same dynamic order as that of the given system. Our goal in this subsection is to develop reduced-order filters having their dynamic order lower than that of the given system.

We proceed now to construct an appropriate reduced-order system. To start with, let us rewrite the matrices C and D of (7.1) as

$$C = \begin{pmatrix} 0 & C_{02} \\ I_{p-m_0} & 0 \end{pmatrix}, \quad D = \begin{pmatrix} D_0 \\ 0 \end{pmatrix},$$

where again rank D = rank D_0 = m_0. This can always be done without any loss of generality by appropriate coordinate transformations. In view of the above partitioning of C and D, we can partition the given system Σ as

$$
\Sigma : \begin{cases}
\begin{pmatrix} \sigma x_1 \\ \sigma x_2 \end{pmatrix} = \begin{pmatrix} A_{11} & A_{12} \\ A_{21} & A_{22} \end{pmatrix} \begin{pmatrix} x_1 \\ x_2 \end{pmatrix} + \begin{pmatrix} B_{11} \\ B_{22} \end{pmatrix} u \\[2ex]
y = \begin{pmatrix} y_0 \\ y_1 \end{pmatrix} = \begin{pmatrix} 0 & C_{02} \\ I & 0 \end{pmatrix} \begin{pmatrix} x_1 \\ x_2 \end{pmatrix} + \begin{pmatrix} D_0 \\ 0 \end{pmatrix} u \\[2ex]
z \qquad\qquad = Ex + Fu,
\end{cases}
\tag{7.51}
$$

where different variables have obvious meanings.

In general, the matrix F in (7.51) is nonzero. If so, we can use a preliminary injection of output y into the desired output z as discussed in Section 7.3, and we rewrite the desired output. To this end, as before, let P^* be a solution of $F - PD = 0$ for P. Then, let

$$
z^* = z - P^* y = (E - P^* C)x + (F - P^* D)u = E^* x
\tag{7.52}
$$

where $E^* = E - P^* C$. In view of (7.51) and (7.52), we can define a new system Σ^* as

$$
\Sigma^* : \begin{cases}
\begin{pmatrix} \sigma x_1 \\ \sigma x_2 \end{pmatrix} = \begin{pmatrix} A_{11} & A_{12} \\ A_{21} & A_{22} \end{pmatrix} \begin{pmatrix} x_1 \\ x_2 \end{pmatrix} + \begin{pmatrix} B_{11} \\ B_{22} \end{pmatrix} u \\[2ex]
y = \begin{pmatrix} y_0 \\ y_1 \end{pmatrix} = \begin{pmatrix} 0 & C_{02} \\ I & 0 \end{pmatrix} \begin{pmatrix} x_1 \\ x_2 \end{pmatrix} + \begin{pmatrix} D_0 \\ 0 \end{pmatrix} u \\[2ex]
z^* = E^* x = E_1^* x_1 + E_2^* x_2,
\end{cases}
\tag{7.53}
$$

with E^* being partitioned in conformity with the partitioning of x into x_1 and x_2 as

$$
E^* = \begin{pmatrix} E_1^* & E_2^* \end{pmatrix}.
\tag{7.54}
$$

We note that the y_1 is not contaminated by the input u, and hence, $x_1 = y_1$ is known exactly from the measurement y. Thus, all we need to do next is to estimate the state x_2. To proceed further, let us rewrite the state equation for x_1 in terms of the output y_1 and the state x_2 as

$$
\sigma y_1 = A_{11} y_1 + A_{12} x_2 + B_{11} u.
\tag{7.55}
$$

The above equation can be rewritten as

$$
\sigma y_1 - A_{11} y_1 = A_{12} x_2 + B_{11} u.
$$

Treating σy_1 as known, we can define a new measurement variable y_r:

$$
y_r = \begin{pmatrix} y_0 \\ \sigma y_1 - A_{11} y_1 \end{pmatrix}.
$$

Although σy_1 is not directly available, as we did earlier in the case of proper full-order filters, we can eliminate it from any filter that is constructed using y_r as a measured output. With this in mind, we form the following auxiliary system:

$$\Sigma_r^* : \begin{cases} \sigma x_r = A_r x_r + B_r u + A_{21} y_1 \\ y_r = C_r x_r + D_r u \\ z_r^* = E_r^* x_r, \end{cases} \tag{7.56}$$

where

$$x_r = x_2, \quad A_r = A_{22}, \quad B_r = B_{22}, \quad C_r = \begin{pmatrix} C_{02} \\ A_{12} \end{pmatrix}, \quad D_r = \begin{pmatrix} D_0 \\ B_{11} \end{pmatrix}, \quad E_r^* = E_2^*. \tag{7.57}$$

We note that the dynamic order n_r of the above Σ_r^* is less than the dynamic order n of the given system Σ by a number equal to the dimension of $x_1 = y_1$.

Before we proceed further, it is beneficial to investigate certain structural properties of Σ_r^* and relate them to those of the Σ. We have the following results.

Lemma 7.55 *Consider the systems Σ_r^* and Σ, which are, respectively, characterized by $(A_r, B_r, C_r, D_r, E_r^*)$ and (A, B, C, D, E, F). Consider two subsystems: one corresponding to Σ_r^* and characterized by the quadruple (A_r, B_r, C_r, D_r), and the other corresponding to Σ and characterized by the quadruple (A, B, C, D). Then, the following results hold:*

(i) *The invariant zeros of the subsystem characterized by (A_r, B_r, C_r, D_r) are the same as those of the subsystem characterized by (A, B, C, D).*

(ii) *The matrix pair (C_r, A_r) is $\mathbb{C}^-$-detectable if and only if the matrix pair (C, A) is $\mathbb{C}^-$-detectable.*

(iii) *The orders of infinite zeros of the subsystem characterized by (A_r, B_r, C_r, D_r) are reduced by one from those of the subsystem characterized by (A, B, C, D).*

(iv) *The subsystem characterized by (A_r, B_r, C_r, D_r) is left-invertible if and only if the subsystem characterized by (A, B, C, D) is left-invertible.*

(v) $\begin{pmatrix} 0 \\ I \end{pmatrix} \mathcal{S}^g (A_r, B_r, C_r, D_r) = \mathcal{S}^g (A, B, C, D) \cap C^{-1}\{\mathrm{im}(D)\}$, *where g denotes $\mathbb{C}^-$ for continuous-time systems and $\mathbb{C}^\ominus$ for discrete-time systems.*

(vi) *Let g denote $\mathbb{C}^-$ for continuous-time systems and $\mathbb{C}^\ominus$ for discrete-time systems. Then, $\mathcal{S}^g (A_r, B_r, C_r, D_r) = \{0\}$ if and only if Σ is left-invertible, has invariant zeros only in $\mathbb{C}^g$, and has no infinite zeros of order higher than one.*

Proof : Property (i) follows from the identity:

$$\begin{pmatrix} 0 & I & 0 & 0 \\ 0 & 0 & I & 0 \\ 0 & 0 & 0 & I \\ -I & 0 & 0 & sI - A_{11} \end{pmatrix} \begin{pmatrix} sI - A & -B \\ C & D \end{pmatrix} = \begin{pmatrix} I & 0 & 0 \\ 0 & sI - A_r & -B_r \\ 0 & C_r & D_r \end{pmatrix}.$$

Note the partitioning of the matrices A, B, C, and D as in (7.51). For Property (ii), we consider

$$\begin{pmatrix} 0 & I & 0 & 0 \\ 0 & 0 & I & 0 \\ 0 & 0 & 0 & I \\ -I & 0 & 0 & sI - A_{11} \end{pmatrix} \begin{pmatrix} sI - A \\ C \end{pmatrix} = \begin{pmatrix} I & 0 \\ 0 & sI - A_r \\ 0 & C_r \end{pmatrix},$$

which yields the required result via the Hautus test. Property (iii) can be shown by first noting that D_0 is surjective. Therefore, it is obvious that by considering the transfer matrix,

$$\begin{pmatrix} I & 0 \\ 0 & sI \end{pmatrix} \left(C(sI - A)^{-1}B + D \right),$$

the infinite zeros are reduced in order by exactly one compared with the infinite zeros of Σ. Premultiplying this transfer matrix with

$$\begin{pmatrix} I & -C_{02}(sI - A_{22})^{-1}A_{21} \\ -A_{12}(sI - A_{22})^{-1}s^{-1}A_{21} & I - s^{-1}A_{11} \end{pmatrix}$$

results in the transfer matrix of Σ_r^*. As this matrix converges to the identity matrix when s tends to infinity, it is clear that this last premultiplication does not change the infinite zeros. Property (iii) then follows immediately.

The relationship between the transfer matrices of Σ and Σ_r^*, which was established to show Property (iii), also establishes Property (iv) because it is clear that the transfer matrix of one is left-invertible if and only if the transfer matrix of the other is left-invertible.

The proof that

$$\begin{pmatrix} 0 \\ I \end{pmatrix} \mathcal{S}^*(A_r, B_r, C_r, D_r) = \mathcal{S}^*(A, B, C, D) \cap C^{-1}\{\mathrm{im}(D)\}$$

follows similarly as in the proof of Lemma 7.22 using the recursion that characterizes $\mathcal{S}^*$. The difference between $\mathcal{S}^*$ and $\mathcal{S}^g$ is determined by the zero dynamics. We already established in Property (i) that the two systems have the same zero dynamics. Moreover, it is always guaranteed that the zero dynamics are contained in $C^{-1}\{\mathrm{im}(D)\}$. Property (v) then follows immediately.

For Property (*vi*), we again follow a similar argument as in the proof of Lemma 7.22. First we note that a necessary condition for the $\mathbb{C}_g$-detectable strongly controllable subspace to be zero is that all invariant zeros are contained in $\mathbb{C}_g$. In that case, we automatically have

$$\mathcal{S}^g(A_r, B_r, C_r, D_r) = \mathcal{S}^*(A_r, B_r, C_r, D_r),$$
$$\mathcal{S}^g(A, B, C, D) = \mathcal{S}^*(A, B, C, D).$$

Then we still need to guarantee when $\mathcal{S}^*(A_r, B_r, C_r, D_r) = 0$, which trivially is the case if and only if a matrix K exists such that $B_r + KD_r = 0$. It is easily seen that this is equivalent to

$$\ker \begin{pmatrix} D \\ (C + DF)B \end{pmatrix} \subseteq \ker B$$

for some F. The latter is clearly equivalent to the system being left-invertible without infinite zeros of order larger than 1 because we have the standing assumption that the matrix $(B'\quad D')$ is injective. $\blacksquare$

Next, let us suppose we can construct an EID filter for the auxiliary system Σ_r^* to arrive at an estimate $\hat{z}_r^*$ of z_r^*. (Here y_1 is a known input, and hence, it will show up accordingly in the dynamics of any filter we construct for Σ_r^*.) Then, we can easily obtain an estimate $\hat{z}$ of z as

$$\hat{z} = E_1^* x_1 + \hat{z}_r^* + P^* y. \tag{7.58}$$

This motivates us to construct a full-order strictly proper filter for the auxiliary system Σ_r^* to arrive at an estimate $\hat{z}_r^*$ of z_r^*. A full-order filter constructed as such for Σ_r^* is indeed a reduced-order filter for Σ. However, in view of (7.58), such a reduced-order filter for Σ is a proper (rather than a strictly proper) filter for Σ. A fundamental issue that arises next is under what conditions one can construct a strictly proper EID filter for Σ_r^*. Expectedly, it turns out that one can construct a strictly proper EID filter for the reduced-order system Σ_r^* if and only if one can construct a proper EID filter for the given system Σ. The following lemma formalizes this result.

Lemma 7.56 *Consider continuous- or discrete-time systems Σ and Σ_r^* respectively, as given in (7.1) and (7.56). Then, the following two statements are equivalent.*

(i) A strictly proper EID filter exists for the reduced-order system Σ_r^*.

(ii) A proper EID filter exists for the system Σ.

Moreover, there is a $1-1$ relationship between the strictly proper EID filter of CSS architecture for Σ_r^* and the proper EID filter of CSS architecture for Σ; that is, one of these filters can be constructed from the other.

Proof : In view of Theorems 7.6 and 7.9, the equivalence between the statements (i) and (ii) follows directly from Lemma 7.55. The $1-1$ relationship between the aforementioned filters can be verified easily.   ∎

The above lemmas lead to the following development for the reduced-order filter design for the original system Σ.

Reduced-order proper filter of CSS architecture for Σ:

Assume that we first design a strictly proper filter for the auxiliary system of the form:
$$\begin{cases} \sigma\hat{x}_r = A_r\hat{x}_r + K_r\left[y_r - C_r\hat{x}_r\right] + A_{21}y_1 \\ \quad \hat{z}_r = E_r^*\hat{x}_r. \end{cases}$$

Let us partition $K_r = \begin{pmatrix} K_{r0} & K_{r1} \end{pmatrix}$ so as to be compatible with the partitioning of y_r. Also, let
$$\xi_r = \hat{x}_r - K_{r1}y_1. \tag{7.59}$$

The reduced-order proper filter of CSS Architecture for Σ is given by
$$\Sigma_{\text{r-CSS}} : \begin{cases} \sigma\xi_r = (A_r - K_rC_r)\xi_r + \tilde{K}_r y \\ \hat{z} \quad = E_r^*\xi_r + \tilde{P}_r y, \end{cases} \tag{7.60}$$

where
$$\tilde{K}_r = \begin{pmatrix} K_{r0} & A_{21} - K_{r1}A_{11} + (A_r - K_rC_r)K_{r1} \end{pmatrix}$$

and
$$\tilde{P}_r = \begin{pmatrix} 0 & E_1^* + E_r^*K_{r1} \end{pmatrix} + P^*.$$

A block diagram representation of the reduced-order proper filter of CSS architecture along with the given plant is shown in Figure 7.5.

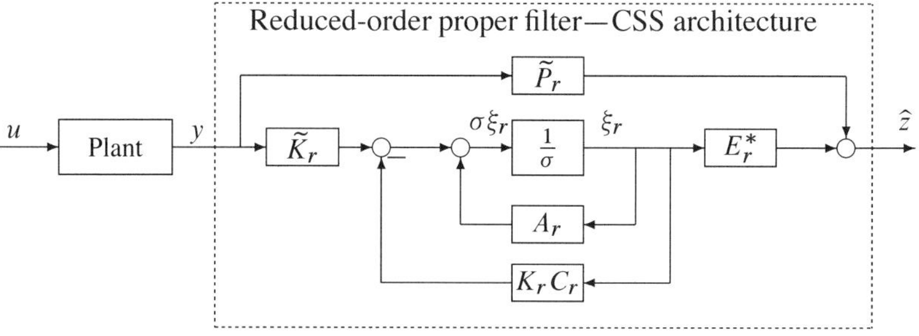

Figure 7.5: Block diagram of a reduced-order filter—CSS architecture

Error Dynamics: By defining the error $e_r = x_r - \tilde{\xi}_r$, the error between the actual desired output $z = E^*x + P^*y$ and the estimated desired output $\hat{z} = E_1^*x_1 + E_r^*\tilde{\xi}_r + P^*y$ can be written as
$$e_z = z - \hat{z} = E_r^*e_r.$$

Then, the dynamics of the error signal is given by

$$\Sigma_{\text{r-CSS}}^{ue} : \begin{cases} \sigma e_r = (A_r - K_r C_r)e_r + (B_r - K_r D_r)u \\ e_z = E_r^* e_r. \end{cases} \tag{7.61}$$

Also, the transfer matrix $G_{\text{r-CSS}}^{ue}$ from u to e_z can obviously be written as

$$G_{\text{r-CSS}}^{ue} = E_r^* (\sigma I - A_r + K_r C_r)^{-1} (B_r - K_r D_r). \tag{7.62}$$

Remark 7.57 *It is easy to see that both the filter equation and the error equation have the same poles, which are the eigenvalues of $A_r - K_r C_r$.*

Everything in the filter $\Sigma_{\text{r-CSS}}$ is known except the gain K_r. Following the notation used earlier, let us denote the set of all EID filter gains by $\boldsymbol{K}_{\text{r-CSS}}^{\text{eid}}(A, B, C, D, E, F)$, meaning that any gain $K_r \in \boldsymbol{K}_{\text{r-CSS}}^{\text{eid}}$ renders $\Sigma_{\text{r-CSS}}$ given in (7.60) an EID filter for (7.1), and conversely, any gain K_r that renders $\Sigma_{\text{r-CSS}}$ an EID filter for (7.1) is an element of $\boldsymbol{K}_{\text{r-CSS}}^{\text{eid}}$.

We have the following result.

Theorem 7.58 *Consider a continuous- or discrete-time system as in (7.1). Let the conditions of Theorem 7.9 be satisfied. Then, the set $\boldsymbol{K}_{\text{r-CSS}}^{\text{eid}}(A, B, C, D, E, F)$ is nonempty, and is given by*

$$\boldsymbol{K}_{\text{r-CSS}}^{\text{eid}}(A, B, C, D, E, F) = \bigcup_{P^* \in \mathcal{P}^*} \boldsymbol{K}_{\text{sp-CSS}}^{\text{eid}}(A_r, B_r, C_r, D_r, E_r^*, 0), \tag{7.63}$$

where the quintuple $(A_r, B_r, C_r, D_r, E_r^*)$ is as defined in (7.57), and $\mathcal{P}^*$ is the set of all P^* that solve $F - PD = 0$ for P. Also, $\boldsymbol{K}_{\text{sp-CSS}}^{\text{eid}}$ is as defined in Theorem 7.13.

Proof : The proof follows from Lemma 7.56 and Theorem 7.13. ∎

Remark 7.59 *We observe that $\boldsymbol{K}_{\text{sp-CSS}}^{\text{eid}}(A_r, B_r, C_r, D_r, E_r^*, 0)$ is nonempty for any $P^* \in \mathcal{P}^*$.*

As in the case of full-order filters, one could easily show the following telescopic property:

$$\ker \begin{pmatrix} E_a & F_a \end{pmatrix} \supseteq \ker \begin{pmatrix} E_b & F_b \end{pmatrix}$$
$$\implies \boldsymbol{K}_{\text{r-CSS}}^{\text{eid}}(A, B, C, D, E_a, F_a) \supseteq \boldsymbol{K}_{\text{r-CSS}}^{\text{eid}}(A, B, C, D, E_b, F_b).$$

Remark 7.60 *It turns out that an EID filter of the form $\Sigma_{r\text{-CSS}}$ given in (7.60) is unique only if the matrix D is invertible. In this case, the so-called reduced-order filter coalesces with the full-order filter.*

Next, as in full-order filters, we would like to design an EID reduced-order filter while placing its poles at desired locations. As before, it is expected that all EID filters of reduced-order CSS architecture share some fixed modes as defined below.

Definition 7.61 *(**Fixed modes of reduced-order proper EID filters of CSS architecture**) Consider a continuous- or discrete-time system as in (7.1) and the EID filtering problem 7.3 characterized by the matrix sextuple (A, B, C, D, E, F). Assume that the solvability conditions as specified by Theorem 7.9 are satisfied. Then, a scalar $\lambda \in \mathbb{C}^-$ for continuous-time systems or a $\lambda \in \mathbb{C}^\ominus$ for discrete-time systems is said to be the fixed mode of an EID filter with the proper reduced-order CSS architecture if λ is a pole (i.e., an eigenvalue of $A_r - K_r C_r$) of every EID filter of such an architecture $\Sigma_{r\text{-CSS}}$ as given in (7.60). The set of all such EID filter fixed modes is denoted here by $\boldsymbol{\Omega}^{\mathrm{eid}}_{r\text{-CSS}}(A, B, C, D, E, F)$.*

We have the following theorem that characterizes the set $\boldsymbol{\Omega}^{\mathrm{eid}}_{r\text{-CSS}}$.

Theorem 7.62 *Consider a continuous- or discrete-time system as in (7.1). Let Assumption 7.1 be satisfied. Consider proper reduced-order filters of CSS architecture, and assume that the solvability conditions as specified by Theorem 7.9 are satisfied. Then, we have*

$$\boldsymbol{\Omega}^{\mathrm{eid}}_{r\text{-CSS}}(A, B, C, D, E, F) = \boldsymbol{\Omega}^{\mathrm{eid}}_{sp\text{-CSS}}(A_r, B_r, C_r, D_r, E_r^*, 0), \qquad (7.64)$$

where $\boldsymbol{\Omega}^{\mathrm{eid}}_{sp\text{-CSS}}(A_r, B_r, C_r, D_r, E_r^*, 0)$ is characterized in Theorem 7.18. Moreover, the set $\boldsymbol{\Omega}^{\mathrm{eid}}_{r\text{-CSS}}$ is invariant with respect to the choice of P^* that solves $F - PD = 0$ for P.

Proof : The proof follows in view of Lemmas 7.56 and 7.31. ∎

As in the case of full-order filters, one could easily show the following telescopic property:

$$\ker\begin{pmatrix} E_a & F_a \end{pmatrix} \supseteq \ker\begin{pmatrix} E_b & F_b \end{pmatrix}$$
$$\implies \boldsymbol{\Omega}^{\mathrm{eid}}_{r\text{-CSS}}(A, B, C, D, E_a, F_a) \subseteq \boldsymbol{\Omega}^{\mathrm{eid}}_{r\text{-CSS}}(A, B, C, D, E_b, F_b).$$

As we discussed, the set of fixed modes obviously presents a constraint that one must contend with in any EID filter design. A question that arises then is whether the EID filter fixed modes of reduced-order filters be different from those of full-order filters. The following lemma addresses this issue and shows that it is not the case in continuous time, whereas in discrete time, the only possible difference is having additional fixed modes at the origin.

Lemma 7.63 *Consider a continuous- or discrete-time system as in (7.1). Let Assumption 7.1 be satisfied. Assume that the solvability conditions for the existence of proper EID filters as specified by Theorem 7.9 are satisfied. Then, we have in continuous time,*

$$\Omega^{\text{eid}}_{\text{p-CSS}}(A, B, C, D, E, F) = \Omega^{\text{eid}}_{\text{r-CSS}}(A, B, C, D, E, F), \qquad (7.65)$$

whereas in discrete time

$$\Omega^{\text{eid}}_{\text{p-CSS}}(A, B, C, D, E, F) \supset \Omega^{\text{eid}}_{\text{r-CSS}}(A, B, C, D, E, F), \qquad (7.66)$$

where the only possible difference between these two sets are additional fixed modes at the origin.

Proof : We first note from Theorem 7.62 that

$$\Omega^{\text{eid}}_{\text{r-CSS}}(A, B, C, D, E, F) = \Omega^{\text{eid}}_{\text{sp-CSS}}(A_r, B_r, C_r, D_r, E_r^*, 0).$$

Next, we note from Theorem 7.18 that

$$\Omega^{\text{eid}}_{\text{sp-CSS}}(A_r, B_r, C_r, D_r, E_r^*, 0) = \Omega_s(A_r', C_r', B_r', D_r', (E_r^*)').$$

According to Theorem 5.11 and the associated Remark 5.12, this set Ω_s consists of a subset of the invariant zeros of (A_r', C_r', B_r', D_r'), which are controllable from E' and the input-decoupling zeros of $(A_r', (C_r'\ E_r'))$. From these characterizations and the structure of A_r, B_r, C_r, D_r and E_r, it becomes clear that

$$\Omega^{\text{eid}}_{\text{r-CSS}}(A, B, C, D, E, F) = \Omega_s(A_r', C_r', B_r', D_r', E_r')$$
$$= \Omega_s(A', C', B', D', (E^*)'). \qquad (7.67)$$

Noting that $E^* = E - P^*C$, whereas $F - P^*D = 0$, the result then follows by noting that for the system parameterized by $(A, B, C, D, E^*, 0)$, we know that EID is solvable by strictly proper filters. But then in continuous time, we obtain from Lemma 7.36 that

$$\Omega^{\text{eid}}_{\text{p-CSS}}(A, B, C, D, E, F) = \Omega^{\text{eid}}_{\text{p-CSS}}(A, B, C, D, E^*, 0)$$
$$= \Omega^{\text{eid}}_{\text{sp-CSS}}(A, B, C, D, E^*),$$

and Theorem 7.18 then results in

$$\Omega\,_{\text{p-CSS}}^{\text{eid}}(A,B,C,D,E,F) = \Omega\,_s(A',C',B',D',(E^*)').$$

Combined with (7.67), this yields (7.65)

In discrete time we use Lemma 7.52 and Theorem 7.18, which yields

$$\Omega\,_{\text{p-CSS}}^{\text{eid}}(A,B,C,D,E,F) \supset \Omega\,_s(A',C',B',D',(E^*)'),$$

with the only possible difference between the two sets additional fixed modes in the origin. This then yields (7.66). ∎

Earlier in Lemmas 7.36 and 7.52, whenever strictly proper EID filters exist, we showed that the sets of fixed modes are the same whether one uses strictly proper or proper EID filters. Following the theme of these lemmas, the following lemma extends this result even further by considering the reduced-order filters as well.

Lemma 7.64 *Consider a continuous- or discrete-time system as in (7.1). Let Assumption 7.1 be satisfied. Assume that the solvability conditions for the existence of strictly proper EID filters as specified by Theorem 7.6 are satisfied. Then, we have*

$$\Omega\,_{\text{sp-CSS}}^{\text{eid}}(A,B,C,D,E,0) = \Omega\,_{\text{r-CSS}}^{\text{eid}}(A,B,C,D,E,F).$$

Additionally, for continuous-time systems, we have

$$\Omega\,_{\text{sp-CSS}}^{\text{eid}}(A,B,C,D,E,0) = \Omega\,_{\text{p-CSS}}^{\text{eid}}(A,B,C,D,E,F),$$

whereas for discrete-time systems, we have

$$\Omega\,_{\text{sp-CSS}}^{\text{eid}}(A,B,C,D,E,0) \subset \Omega\,_{\text{p-CSS}}^{\text{eid}}(A,B,C,D,E,F),$$

with the only possible difference between the two sets being some possible fixed modes at the origin.

Proof : The proof follows in view of Lemmas 7.36, 7.52, and 7.63. ∎

As in the full-order case, the following theorem shows that we can design a filter gain K_r such that the filter $\Sigma_{\text{r-CSS}}$ as given in (7.60) is a reduced-order EID filter while its poles are at the prescribed desired locations except that they need to contain $\Omega\,_{\text{r-CSS}}^{\text{eid}}(A,B,C,D,E,F)$ among them.

Theorem 7.65 *(**EID reduced-order proper filter of CSS architecture with pole placement**) Consider a continuous- or discrete-time system as in (7.1). Let the conditions of Theorem 7.9 be satisfied. Also, let $E^* = E - P^*C$, with P^* being* *a solution of $F - PD = 0$ for P. Consider the set $\boldsymbol{\Omega}^{\mathrm{eid}}_{\mathrm{r\text{-}CSS}}(A, B, C, D, E, F)$, which is the set of EDD fixed modes discussed in Theorem 7.62. Also, consider the nonempty set $\boldsymbol{K}^{\mathrm{eid}}_{\mathrm{r\text{-}CSS}}(A, B, C, D, E, F)$ discussed in Theorem 7.58. Finally, let Λ_r be a prescribed set of n_r self-conjugate elements in the open left-half complex plane $\mathbb{C}^-$ for continuous-time systems or inside the unit disk $\mathbb{C}^\ominus$ for discrete-time systems such that Λ_r includes $\boldsymbol{\Omega}^{\mathrm{eid}}_{\mathrm{r\text{-}CSS}}$. Then, a filter gain $K_r \in \boldsymbol{K}^{\mathrm{eid}}_{\mathrm{r\text{-}CSS}}(A, B, C, D, E, F)$ exists such that the associated proper reduced-order filter $\Sigma_{\mathrm{r\text{-}CSS}}$ given in (7.60) is an EID filter, and moreover, $\lambda(A_r - K_r C_r) = \Lambda_r$.*

Proof : The proof follows easily in view of Theorems 7.9 and 7.58.   ∎

An algorithm for designing a reduced-order proper EID filter with simultaneous filter pole placement:

Step 1: Given the system Σ as in (7.17) and characterized by the sextuple (A, B, C, D, E, F), form the auxiliary system $\widetilde{\Sigma}_r^*$ characterized by the quintuple $(A_r, B_r, C_r, D_r, E_r^*)$ as defined in (7.57).

Step 2: Choose a desired set Λ_r of n_r self-conjugate elements in the open left-half complex plane $\mathbb{C}^-$ for continuous-time systems or within the unit circle $\mathbb{C}^\ominus$ for discrete-time systems [Step 3 given below checks to make sure that Λ_r includes $\boldsymbol{\Omega}^{\mathrm{eid}}_{\mathrm{r\text{-}CSS}}(A, B, C, D, E, F)$].

Step 3: Following the algorithm for designing a strictly proper EID filter with simultaneous filter pole placement as given in Subsection 7.5.1, and using the quintuple $(A_r, B_r, C_r, D_r, E_r^*)$ as well as the matrix Λ_r as inputs to that algorithm, obtain the gain K_r.

With the filter gain K_r as computed above, it is easy to verify that the resulting filter $\Sigma_{\mathrm{r\text{-}CSS}}$ given in (7.60) is an EID filter with its poles at the locations specified by Λ_r; i.e., $\lambda(A_r - K_r C_r) = \Lambda_r$.

7.6 Fixed modes of EID filters with arbitrary architecture

As we discussed, the set of fixed modes obviously presents a constraint that one must contend with in any EID filter design. In this regard, it pays to study fixed modes of EID filters of different types with a hope of finding a desirable set among them. However, when filters of CSS architecture are used and when conditions of Theorem 7.6 for the existence of strictly proper EID filters are satisfied, Lemmas 7.36 and 7.64 show in continuous time that the sets of fixed modes are the same whether one uses full-order strictly proper or full-order proper or even reduced-order proper filters.

Furthermore, in continuous time, when filters of CSS architecture are used and when conditions of Theorem 7.9 for the existence of proper EID filters are satisfied, Lemma 7.63 shows that the sets of fixed modes are the same whether one uses full-order proper or reduced-order proper filters.

In discrete-time we have similar results but with a small difference. If the conditions for the existence of strictly proper EID filters are satisfied, then Lemmas 7.52 and 7.64 show that strictly proper and reduced-order proper filters of CSS architecture have the same fixed modes, but proper filters of CSS architecture can have additional fixed modes at the origin.

A question that arises next is whether any advantages can be gained by using any arbitrary architecture other than CSS architecture for filters. That is, we ask ourselves whether filters of different architecture can result in different sets of EID filter fixed modes. The following two theorems address this issue.

Theorem 7.66 *Consider a continuous- or discrete-time system as in (7.1). Let Assumption 7.1 be satisfied, and assume that the solvability conditions for the existence of strictly proper EID filters as specified by Theorem 7.6 are satisfied. Consider also the set of fixed modes $\Omega_{\text{sp-CSS}}^{\text{eid}}(A, B, C, D, E)$ that pertains to strictly proper EID filters of CSS architecture. Moreover, consider any arbitrary EID filter Σ_f such as the one in (7.2) with $P = 0$ and thus characterized by the triplet (L, M, N). Then, any $\lambda \in \Omega_{\text{sp-CSS}}^{\text{eid}}(A, B, C, D, E, 0)$, which is neither an input decoupling zero nor an output decoupling zero of the subsystem characterized by the quadruple (A, B, C, D) is also a pole of the EID filter (i.e., an eigenvalue of L) characterized by (L, M, N).*

Proof : Using Theorem 7.69, we have a $1 - 1$ correspondence between strictly proper EID filters and dynamic EDD state feedback controllers for a dual system. However, because this is based on transfer matrices, input- and output-decoupling zeros can be removed. Using Theorem 5.29 we know that the fixed modes using dynamic state feedback controllers are the same as the fixed modes using static state feedback controllers. Finally, according to Theorem 7.13, there is a $1 - 1$ correspondence between static EDD state feedback controllers achieving disturbance decoupling and strictly proper EID filters of CSS architecture. ∎

Theorem 7.67 *Consider a continuous- or discrete-time system as in (7.1). Let Assumption 7.1 be satisfied, and assume that the solvability conditions for the existence of proper EID filters as specified by Theorem 7.9 are satisfied. Consider also the set of fixed modes $\Omega_{\text{p-CSS}}^{\text{eid}}(A, B, C, D, E, F)$ that pertains to proper EID filters of CSS architecture. Moreover, consider any arbitrary EID filter Σ_f such as the one in (7.2) and characterized by the quadruple (L, M, N, P).*

In continuous time, any $\lambda \in \Omega_{\text{p-CSS}}^{\text{eid}}(A, B, C, D, E, F)$ that is neither an input-decoupling zero nor an output-decoupling zero of the subsystem characterized by *the quadruple* (A, B, C, D) *is also the pole of the EID filter (eigenvalue of L) characterized by* (L, M, N, P).

In discrete time, any $\lambda \in \boldsymbol{\Omega}^{\mathrm{eid}}_{\mathrm{p\text{-}CSS}}(A, B, C, D, E, F)$ *unequal to zero that is neither an input-decoupling zero nor an output-decoupling zero of the subsystem characterized by the quadruple* (A, B, C, D) *is also the pole of the EID filter (eigenvalue of L) characterized by* (L, M, N, P).

Proof : This uses similar arguments as the proof of Theorem 7.66 except that we connect the proper EID filters with dynamic EDD full-information feedback controllers. Then we use the equivalence of the fixed modes of static and dynamic EDD full-information feedback controllers. Finally the fixed modes of static full-information feedback controllers can be connected to the fixed modes of proper filters of CSS architecture. ∎

7.A   Duality between filtering and control

This section establishes a relationship between a filter for a given system and an appropriately constructed controller for a dual of the given system. To do so, let us consider a continuous- or discrete-time system given in (7.1) and reproduced below as

$$
\Sigma : \begin{cases} \sigma x = Ax + Bu \\ y \;\; = Cx + Du \\ z \;\; = Ex + Fu, \end{cases}
\tag{7.68}
$$

where x is the state, y is the measured output, and z is the desired output that needs to be estimated. For the above system Σ, we can define the dual system as

$$
\Sigma_d : \begin{cases} \sigma \widetilde{x} = A'\widetilde{x} + C'v + E'w \\ z \;\; = B'\widetilde{x} + D'v + F'w, \end{cases}
\tag{7.69}
$$

where $\widetilde{x}$ is the state, v is the control input, w is the disturbance input, and z is the desired output on which the effects of the disturbance w are to be suppressed.

Our goal in this section is to establish a $1 - 1$ relationship between linear stable unbiased filters for Σ and dynamic state feedback controllers for the dual system Σ_d. Before we do so, we need the following preliminary lemma.

We recall from Section 2.1 that $\mathcal{RH}_2$ denotes the set of strictly proper and stable rational matrices and that $\mathcal{RH}_\infty$ denotes the set of proper and stable rational matrices.

Lemma 7.68 *(Parameterization of linear stable unbiased filters) Consider a continuous- or discrete-time system Σ as in (7.68). Let Assumption 7.1 be satisfied. Then, the class of linear stable unbiased proper (or strictly proper) filters for Σ can be parameterized in $P_f(\sigma) \in \mathcal{RH}_\infty$ (or in $P_f(\sigma) \in \mathcal{RH}_2$) as*

$$\Sigma_f : \begin{cases} \sigma\hat{x} = A\hat{x} + K(y - C\hat{x}) \\ \hat{z} = E\hat{x} + P_f(\sigma)(y - C\hat{x}), \end{cases} \tag{7.70}$$

where K is any fixed given matrix such that $A - KC$ is stable. That is,

    (i) For any $P_f \in \mathcal{RH}_\infty$ (or $P_f \in \mathcal{RH}_2$), the filter Σ_f given in (7.70) is unbiased for the system Σ.

    (ii) For any unbiased proper (or strictly proper) filter with transfer function G_f, a transfer function $P_f \in \mathcal{RH}_\infty$ (or $P_f \in \mathcal{RH}_2$) exists such that the filter Σ_f given in (7.70) has the transfer function G_f.

Proof : We prove the lemma for the case of proper filters. The proof is similar for the case of strictly proper filters. To start with, it is trivial to show that the filter Σ_f with $P_f \in \mathcal{RH}_\infty$ is unbiased for the system Σ. Next, let a linear stable unbiased proper filter be given with transfer function G_f. Define a new transfer function $\bar{G}_f$ as

$$\bar{G}_f(\sigma) = -E(\sigma I - A + KC)^{-1}K + G_f(\sigma).$$

Then, choose P_f as

$$P_f(\sigma) = \bar{G}_f(\sigma)\left[I - C(\sigma I - A + KC)^{-1}K\right]^{-1}$$

or equivalently as

$$P_f(\sigma) = \bar{G}_f(\sigma)\left[I + C(\sigma I - A)^{-1}K\right].$$

Then, it is trivial to verify that $P_f \in \mathcal{RH}_\infty$ and that the transfer function of Σ_f in (7.70) with this choice of P_f is indeed G_f.   ■

The following theorem establishes a $1 - 1$ relationship between linear stable unbiased strictly proper filters for Σ given in (7.68) and dynamic state feedback controllers for the dual system Σ_d given in (7.69).

Theorem 7.69 *Consider a continuous- or discrete-time system Σ as in (7.68) with $F = 0$. Let Assumption 7.1 be satisfied. Also, consider the dual system Σ_d given in (7.69). Then the following two statements are equivalent:*

(i) A linear stable unbiased strictly proper filter Σ_f exists for Σ given in (7.68) resulting in a transfer function G from u to $(z - \hat{z})$.

(ii) A stabilizing proper dynamic state feedback controller Σ_c exists for the dual system Σ_d resulting in a closed-loop transfer function G' from w to z.

Moreover, there is a $1 - 1$ relationship between the filter Σ_f in statement (i) and the controller Σ_c in statement (ii); that is, with the knowledge of Σ_f, one can construct the controller Σ_c and vice-versa.

Proof : To start with, without loss of generality, we can assume that E is surjective. Let a linear stable unbiased strictly proper filter Σ_f for Σ be given such that it results in a transfer function from u to $(z - \hat{z})$ as $G(\sigma)$. By Lemma 7.68, Σ_f can be realized in the form given in (7.70) for a certain parameter $P_f(\sigma) \in \mathcal{RH}_\infty$. It is then easy to verify that the feedback controller of the form,

$$v = -K'\tilde{x} + \bar{F}'(\sigma)w, \tag{7.71}$$

when applied to the system (7.69) can result in a closed-loop transfer function from w to z as $G'(\sigma)$ for some strictly proper $\bar{F}'(\sigma)$. Clearly, (7.71) is not yet a state feedback controller. However, the feedback controller (7.71) can be written in a state-space form for some triplet (L, M, N) as

$$\begin{aligned} \sigma\tilde{\xi} &= L'\tilde{\xi} + N'w, \\ v &= -K'\tilde{x} + M'\tilde{\xi}. \end{aligned} \tag{7.72}$$

Then, exploiting that

$$E'w = \sigma\tilde{x} - A'\tilde{x} - C'v,$$

we can rewrite the above feedback controller in the form,

$$\begin{aligned} \sigma\breve{\xi} &= L'\breve{\xi} - N'(EE')^{-1}EC'v - [N'(EE')^{-1}EA' - L'N'(EE')^{-1}E]\tilde{x}, \\ v &= M'\breve{\xi} + \left(M'N'(EE')^{-1}E - K'\right)\tilde{x}, \end{aligned} \tag{7.73}$$

which is clearly a state feedback controller for the dual system (7.69). It is not difficult then to check that the above transformation yields the same poles and the same transfer matrix from w to z.

Conversely, in a similar fashion, given a stabilizing proper dynamic state feedback controller Σ_c for the dual system Σ_d that results in a closed-loop transfer function $G'(\sigma)$ from w to z, we can establish that a transfer function $P_f \in \mathcal{RH}_\infty$ exists such that Σ_f given in (7.70) when applied to the system Σ results in a transfer function G from u to $(z - \hat{z})$. The proof is now complete. ∎

Remark 7.70 *Let us recall a strictly proper filter of CSS architecture $\Sigma_{\text{sp-CSS}}$ given in (7.12) and repeated below as*

$$\Sigma_{\text{sp-CSS}} : \begin{cases} \sigma\hat{x} = A\hat{x} + K(y - C\hat{x}) \\ \hat{z} = E\hat{x}. \end{cases} \tag{7.74}$$

We note that, for any gain K such that $A - KC$ is a stable matrix, the above filter is an unbiased filter for the system Σ. Using this class of filters in statement (i) of Theorem 7.69, it is easy to see that in statement (ii) of Theorem 7.69, one needs only a static state feedback controller given by

$$v = -K'\tilde{x}. \tag{7.75}$$

We note easily that there is a $1 - 1$ relationship between the filter (7.74) and the controller (7.75).

Theorem 7.69 pertains to linear stable unbiased strictly proper filters. The following theorem considers proper filters rather than strictly proper filters.

Theorem 7.71 *Consider a continuous- or discrete-time system Σ as in (7.68). Let Assumption 7.1 be satisfied. Also, consider the dual system Σ_d given in (7.69). Then the following two statements are equivalent:*

   (i) *A linear stable unbiased proper filter Σ_f exists for Σ given in (7.68) resulting in a transfer function G from u to $(z - \hat{z})$.*

   (ii) *A stabilizing proper dynamic state and static disturbance feedback controller Σ_c exists for the dual system Σ_d resulting in a closed-loop transfer function G' from w to z.*

Moreover, there is a $1 - 1$ relationship between the filter Σ_f in statement (i) and the controller Σ_c in statement (ii); that is, with the knowledge of Σ_f, one can construct the controller Σ_c and vice-versa.

Proof : Let a linear stable unbiased proper filter Σ_f for Σ be given such that it results in a transfer function from u to $(z - \hat{z})$ as $G(\sigma)$. By Lemma 7.68, Σ_f can be realized in the form given in (7.70) for a certain parameter $P_f \in \mathcal{RH}_\infty$. It is then easy to verify that the feedback controller of the form,

$$v = -K'\tilde{x} - P'w - \bar{F}'(\sigma)w, \tag{7.76}$$

when applied to the system (7.69) can result in a closed-loop transfer function G' from w to z for some strictly proper $\bar{F}'(\sigma)$. Then using the same type of transformations as in the proof of Theorem 7.69, we can transform this feedback controller for (7.69) to the form,

$$v = -G_c(\sigma)\tilde{x} - P'w.$$

Thus, we obtain a stabilizing dynamic state and static disturbance feedback controller for the system (7.69).

The converse implication starting with a stabilizing dynamic state and static disturbance feedback controller for the system (7.69) and resulting in a stable unbiased proper filter for the system (7.68) can be obtained in a similar way.   ■

8
Almost input-decoupled filtering under white noise input

8.1 Introduction

Chapter 7 considers exact input-decoupled (EID) filtering problems. In that chapter, we seek perfect performance; that is, we try to make the impact of the unknown input on the asymptotic error absolutely zero irrespective of what is the input, whether it is persistent or not. Such a severe performance measure demands that the transfer function or transfer matrix from the input to the estimation error be identically zero. Of course, as discussed in Chapter 7, the EID filtering problem is not always solvable. It is natural then to think of methods of relaxing the performance requirements so that the solvability conditions can possibly be weakened and thus allowing us to deal with a larger class of systems. There are a number of ways by which the performance requirements can be weakened. As we said in the introduction to this book, we plan to relax the performance requirements progressively layer by layer to form a hierarchy of problems. In this chapter, we introduce the first layer of relaxing the performance requirements. We seek here to make the impact of the unknown input on the asymptotic error to be *almost zero* or equivalently *arbitrarily small* or *as small as desired* instead of being identically zero. To be more precise, we try to find a family of filters parameterized by some positive ε such that when applied to the system, the asymptotic error converges to zero as $\varepsilon \downarrow 0$. Thus, in this chapter, we deal with what can be termed as almost-input-decoupled (AID) filtering problems, or for short AID filtering problems. The filters that solve the AID filtering problems are termed not surprisingly as AID filters.

To start with, in AID filtering problems, we need to define an appropriate metric to measure what we mean by the word *small*. To do so, a good framework to adapt is a stochastic framework where the unknown input is modeled as a stochastic process. Whenever the input is a stochastic process, all signals, i.e., the state, the measured output, and the desired output of the given plant or system, are naturally stochastic processes. This lets us use a stochastic norm to measure the error signal. One of the most useful stochastic norms used in engineering is the RMS norm of a signal. Our goal in this chapter is then to construct a family of filters such that the RMS norm of the error signal be as small as desired. This goal leads us to consider to two different and alternative frameworks.

In the first framework, the input to the given system is modeled as a zero mean wide-sense stationary white noise stochastic process of power spectral density (PSD) equal to an identity matrix, or to be called simply as white noise of unit intensity. In fact, we can assume without much loss of generality the input as any zero mean wide-sense stationary stochastic process of known PSD, which is not necessarily a white noise. In this regard, let us recall the well-known fact that one can always generate a wide-sense stationary stochastic process as the output of a linear time-invariant system driven by white noise of unit intensity so that the PSD of such a generated stochastic process approximates arbitrarily closely any known PSD of a wide-sense stationary stochastic process. The needed linear time-invariant system to do so can always be appended to the given system. As such, there is not much loss of generality in assuming the input to be zero mean white noise of unit intensity. In the second framework, however, no kind of statistical information on the input signal is assumed at all.

In this chapter, our study of AID filtering considers the first case where we model the input as a zero mean wide-sense stationary white noise of unit PSD. The next chapter considers the second case where the input is simply unknown either statistically or otherwise. In this chapter, as well as in the next chapter, our study involves first developing the necessary and sufficient conditions under which AID filters exist, and then developing methods of constructing them. As in Chapter 7, we use both full- and reduced-order filters of CSS architecture for AID filter design.

8.2   Preliminaries

Let us reconsider the plant or system model given in (7.1) and rewritten here as

$$\Sigma : \begin{cases} \sigma x = Ax + Bu \\ y \ \ = Cx + Du \\ z \ \ = Ex + Fu, \end{cases} \tag{8.1}$$

where, as before, $u \in \mathbb{R}^m$ is the input, $x \in \mathbb{R}^n$ is the state, $y \in \mathbb{R}^p$ is the measured output, and $z \in \mathbb{R}^q$ is the desired output signal to be estimated.

As before, our interest lies in estimating the desired output signal z while using only the measured output y but not the input u. As usual, let $\hat{z}$ be the estimate of z as given by a filter, and let e_z be the estimation error, $e_z = z - \hat{z}$ as depicted in Figure 8.1.

As before, it is necessary to use the following assumption throughout this chapter as well.

Assumption 8.1 *The matrix pair* (C, A) *is* $\mathbb{C}^-$*-detectable for continuous-time systems and* $\mathbb{C}^\ominus$*-detectable for discrete-time systems.*

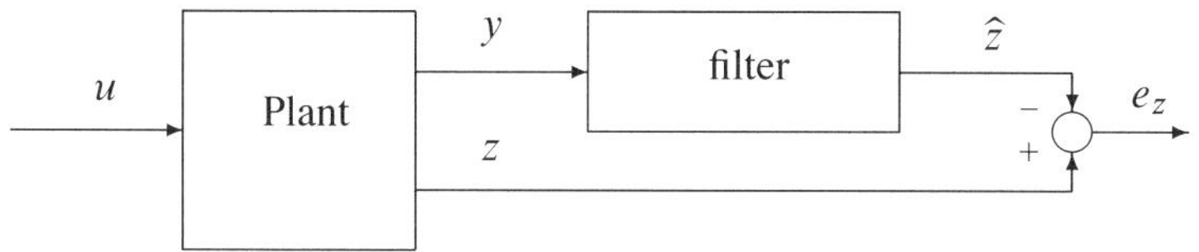

Figure 8.1: General block diagram

In this chapter, we consider a parameterized family of general proper filters of the form,

$$\Sigma_f^\varepsilon : \begin{cases} \sigma\xi = L^\varepsilon\xi + M^\varepsilon y \\ \hat{z} = N^\varepsilon\xi + P^\varepsilon y. \end{cases} \tag{8.2}$$

Here ε is a positive parameter belonging to $(0, \varepsilon^*]$ for some positive ε^*. Whenever $P = 0$, the above filter is said to be a strictly proper filter. When the above filter is used as shown in Figure 8.1, the dynamic equations of the error e_z is described by

$$\Sigma_{ue}^\varepsilon : \begin{cases} \sigma x = Ax + Bu \\ \sigma\xi = M^\varepsilon Cx + L^\varepsilon\xi + M^\varepsilon Du \\ e_z = (E - P^\varepsilon C)x - N^\varepsilon\xi + (F - P^\varepsilon D)u. \end{cases} \tag{8.3}$$

Hence, the transfer matrix from u to e_z can be computed as

$$G_{ue}^\varepsilon = \begin{pmatrix} E - P^\varepsilon C & -N^\varepsilon \end{pmatrix} \begin{pmatrix} \sigma I - A & 0 \\ -M^\varepsilon C & \sigma I - L^\varepsilon \end{pmatrix}^{-1} \begin{pmatrix} B \\ M^\varepsilon D \end{pmatrix} + (F - P^\varepsilon D). \tag{8.4}$$

8.3 Statement of AID filtering problem and its solvability conditions

As explained in Section 8.1, we assume throughout this chapter that the input is a zero mean wide-sense stationary white noise. As in Chapter 7, we consider only the class of unbiased filters; see Definition 7.2 for what "unbiasedness" means. We start by defining formally the following AID filtering problem.

Problem 8.2 Consider the system Σ given in (8.1) where the input $u(t)$ is a zero mean wide-sense stationary white noise with a unit PSD. Then, the **almost input-decoupled (AID) filtering problem under white noise input** is defined as follows: Find, whenever it exists, a family of linear stable strictly proper (or proper) filters of the type Σ_f^ε given in (8.2) and parameterized in positive $\varepsilon \in (0, \varepsilon^*]$ such that

(Unbiasedness) for any $\varepsilon \in (0, \varepsilon^*]$ the filter Σ_f^ε is unbiased, that is, the estimation error e_z, in the absence of the input u, decays asymptotically to zero for all possible finite initial values of the system (8.1) and the filter (8.2), and

(Performance)

$$\|e_z\|_{\mathrm{RMS}} \to 0 \ \text{ as } \ \varepsilon \to 0.$$

The above problem statement can be given a deterministic interpretation. To do so, for any given filter, we proceed first to evaluate the PSD of the error signal e_z and then $\|e_z\|_{\mathrm{RMS}}$. Let $G_{ue}^\varepsilon(\sigma)$ denote the transfer matrix from the input u to the error e_z. Consider first continuous-time systems. Then, $G_{ue}^\varepsilon(j\omega)$ is the Fourier transform of the impulse response of the system from u to e_z. As the input u is a white noise with unit PSD, the PSD of the error signal e_z can easily be determined as $G_{ue}^\varepsilon(j\omega)(G_{ue}^\varepsilon)^{\mathrm{H}}(j\omega)$, where as stated, the superscript H denotes the complex conjugate transpose. Similarly, for discrete-time systems, $G_{ue}^\varepsilon(e^{j\omega})$ is the discrete-time Fourier transform of the impulse response of the system from u to e_z, and thus, the PSD of the error signal e_z can be determined as $G_{ue}^\varepsilon(e^{j\omega})(G_{ue}^\varepsilon)^{\mathrm{H}}(e^{j\omega})$. Then, by the definitions of the RMS norm and H_2 norm as given in Section 2.4, we see that

$$\|e_z\|_{\mathrm{RMS}} = \|G_{ue}^\varepsilon\|_2, \tag{8.5}$$

where $\|G_{ue}^\varepsilon\|_2$ is the H_2 norm of G_{ue}^ε.

In view of (8.5), the AID filtering problem as described in Problem 8.2 is solvable if and only if a family of linear stable strictly proper (or proper) unbiased filters parameterized in positive ε exists such that

$$\|G_{ue}^\varepsilon\|_2 \to 0 \ \text{ as } \ \varepsilon \to 0.$$

This implies that Problem 8.2 can be interpreted in a deterministic setting as an H_2 AID filtering problem as follows.

Problem 8.3 (H_2 almost input-decoupled (AID) filtering problem) Consider the system Σ given in (8.1). Then, find, whenever it exists, a family of linear stable strictly proper (or proper) filters of the type Σ_f^ε given in (8.2) and parameterized in positive $\varepsilon \in (0, \varepsilon^*]$ such that

(Unbiasedness) for any $\varepsilon \in (0, \varepsilon^*]$ the filter Σ_f^ε is unbiased, that is, the estimation error e_z, in the absence of the input u, decays asymptotically to zero for all possible finite initial values of the system (8.1) and the filter (8.2), and

(Performance)

$$\|G_{ue}^\varepsilon\|_2 \to 0 \ \text{ as } \ \varepsilon \to 0.$$

A family of filters that solves the H_2 AID filtering problem is said to be a family of H_2 AID filters.

8.4   Existence conditions — continuous-time case

In what follows, we develop the conditions for the existence of a family of H_2 AID filters for a continuous-time system. Let us first observe that, to render $\|G_{ue}^{\varepsilon}\|_2$ arbitrarily small, the direct feedthrough matrix of G_{ue}^{ε} must necessarily converge to zero as $\varepsilon \to 0$. This implies that a family of matrices P^{ε} must exist such that $F - P^{\varepsilon} D \to 0$ as $\varepsilon \to 0$. It is easily seen then that a matrix P^* must exist such that $F - P^* D = 0$. This in turn implies that the conditions of Lemma 7.5 apply to the H_2 AID filtering problem as well. We have the following lemma formalizing this.

Lemma 8.4 *Consider a continuous-time system as in (8.1). The following results hold:*

   (i) The H_2 AID filtering problem is solvable via a family of proper filters only if

$$\ker D \subseteq \ker F. \tag{8.6}$$

   (ii) The H_2 AID filtering problem is solvable via a family of strictly proper filters only if $F = 0$.

Proof : The results are obvious. ∎

We are now ready to give the solvability conditions for the H_2 AID filtering problem. We first consider a family of strictly proper filters.

Theorem 8.5 *Consider a continuous-time system as in (8.1). Let Assumption 8.1 be satisfied. Then, the H_2 AID filtering problem is solvable via a family of strictly proper filters if and only if $F = 0$ and*

$$\mathcal{S}^{-0}(A, B, C, D) \cap \mathcal{V}^*(A, B, C, D) \subseteq \ker E. \tag{8.7}$$

Proof : The condition that F needs to be zero follows directly from Lemma 8.4. By Theorem 7.69, it is clear that the H_2 AID filtering problem is solvable via a family of strictly proper filters if and only if the H_2 almost disturbance decoupling (ADD) problem is solvable by a family of dynamic state feedback laws for the dual system:

$$\Sigma_d : \begin{cases} \dot{x}_d = A'x_d + C'u_d + E'w_d \\ z_d = B'x_d + D'u_d. \end{cases} \tag{8.8}$$

By Theorem 6.4, this H_2 ADD problem is solvable if and only if

$$\operatorname{im} E' \subseteq \mathcal{S}^*(A', C', B', D') + \mathcal{V}^{-0}(A', C', B', D').$$

Taking the orthogonal complement on both sides in the above inclusion yields that this condition is equivalent to (8.7).   ∎

We consider next the H_2 AID filtering problem by using a family of proper filters. We first consider the case when $F = 0$. Whenever $F \neq 0$, we will soon redefine the desired output to be estimated such that in the new setting, the corresponding F equals zero.

Theorem 8.6 *Consider a continuous-time system as in (8.1) with $F = 0$. Let Assumption 8.1 be satisfied. Then, the H_2 AID filtering problem is solvable via a family of proper filters if and only if*

$$\mathcal{S}^{-0}(A, B, C, D) \cap \mathcal{V}^*(A, B, C, D) \subseteq \ker E. \tag{8.9}$$

Proof : As in the proof of Theorem 8.5, we use a duality argument. By Theorem 7.71, the H_2 AID filtering problem is solvable for the system (8.1) with $F = 0$ via a family of proper filters if and only if the H_2 ADD problem is solvable by a family of dynamic full-information feedback laws for the dual system Σ_d given in (8.8). As outlined in Section 6.4, this H_2 ADD problem is solvable by a family of dynamic full-information feedback laws if and only if the H_2 ADD problem is solvable by a family of static full-information feedback laws. Finally, by Theorem 6.22, the H_2 ADD is solvable by a family of static full-information feedback laws if and only if

$$\operatorname{im} E' \subseteq \mathcal{S}^*(A', C', B', D') + \mathcal{V}^{-0}(A', C', B', D') + C' \ker D'.$$

Taking the orthogonal complement on both sides in the above inclusion yields that this condition is equivalent to (8.9).   ∎

Remark 8.7 *For the case of $F = 0$, and for continuous-time systems that are under consideration, we observe that the solvability conditions are one and the same whether strictly proper or proper filters are used. On the other hand, as we shall see, for discrete-time systems, the solvability conditions when proper filters are used are much weaker than those when strictly proper filters are used. This is due to the fact that, in continuous-time systems, one can use fast filters that use high gain. However, in the case of discrete-time systems, it is not so.*

Remark 8.8 *It is interesting to look at the case when $E = I$, i.e., when the entire state is to be estimated. For this case, because $\ker E = \{0\}$, the conditions of Theorems 8.5 and 8.6 simplify. The simplified conditions are that the system characterized by the quadruple (A, B, C, D) is left-invertible, and moreover, it has no invariant zeros with positive real parts.*

We consider next the case of $F \neq 0$. As in EID filtering, we study here the H_2 AID filtering problem for a given system for which the matrix F is nonzero by transforming it to an equivalent H_2 AID filtering problem for an auxiliary system where the corresponding F can be taken as zero. The needed auxiliary system is constructed exactly the same way as in EID filtering by redefining the desired output z. To do so, we define a new desired output z^* as

$$z^* = z - P^* y = (E - P^* C)x + (F - P^* D)u. \tag{8.10}$$

Next, we consider the following auxiliary system:

$$\Sigma^* : \begin{cases} \dot{x} &= Ax + Bu \\ y &= Cx + Du \\ z^* &= (E - P^* C)x + (F - P^* D)u. \end{cases} \tag{8.11}$$

Also, for the above auxiliary system, let us consider a family of filters of the form

$$\begin{aligned} \dot{\xi} &= L^\varepsilon \xi + M^\varepsilon y, \\ \hat{z}^* &= N^\varepsilon \xi + \bar{P}^\varepsilon y, \end{aligned} \tag{8.12}$$

with ε as a positive parameter. As in Chapter 7, we observe that the transfer matrix from u to $e_{z*} = z^* - \hat{z}^*$ when the new filter (8.12) is applied to the auxiliary system Σ^* is the same as the transfer matrix from u to $e_z = z - \hat{z}$ when the original filter (8.2) is applied to the original Σ as long as $P^\varepsilon = \bar{P}^\varepsilon + P^*$.

The above analysis suggests a method of designing first P^* such that $F - P^* D = 0$ and then designing a family of filters parameterized in ε for the system Σ^* such that the transfer matrix from u to $e_{z*} = z^* - \hat{z}^*$ has the desired value (namely, its H_2 norm tends to zero as $\varepsilon \to 0$). The designed family of filters for Σ^* can then easily be translated for the original system Σ by allowing $P^\varepsilon = \bar{P}^\varepsilon + P^*$.

We have the following results.

Theorem 8.9 *Consider a continuous-time system as in (8.1). Let Assumption 8.1 be satisfied. Then, the H_2 AID filtering problem is solvable via a family of proper filters if and only if*

$$\left(\left[\mathcal{S}^{-0}(A, B, C, D) \cap \mathcal{V}^*(A, B, C, D) \right] \oplus \mathbb{R}^m \right) \cap \ker \begin{pmatrix} C & D \end{pmatrix}$$
$$\subseteq \ker \begin{pmatrix} E & F \end{pmatrix}. \tag{8.13}$$

Proof : By Lemma 8.4, if the H_2 AID filtering problem is solvable via a family of proper filters, then a matrix P^* exists such that $F - P^* D = 0$. Clearly, the H_2 AID filtering problem is solvable via a family of proper filters for the continuous-time system (8.1) if and only if the H_2 AID filtering problem is solvable via a family of proper filters for the following system:

$$\bar{\Sigma} : \begin{cases} \dot{x} & = Ax + Bu, \\ y & = Cx + Du, \\ z^* & = z - P^* y = (E - P^* C)x. \end{cases}$$

Using Theorem 8.6, we then find that for this latter system, the H_2 AID filtering problem is solvable via a family of proper filters if and only if

$$\mathcal{S}^{-0}(A, B, C, D) \cap \mathcal{V}^*(A, B, C, D) \subseteq \ker(E - P^* D).$$

This condition combined with $F - P^* D = 0$ is easily seen to be equivalent to (8.13). ∎

Remark 8.10 *Readers can easily verify that the solvability conditions for H_2 AID filtering are weaker than the solvability conditions for EID filtering (see Theorems 7.6, 7.7, and 7.9), which is of course as it should be. Such a verification can be done by noting that, in general,*

$$\mathcal{S}^{-0}(A, B, C, D) \subseteq \mathcal{S}^-(A, B, C, D),$$
$$\mathcal{V}^*(A, B, C, D) \subseteq C^{-1}\{\operatorname{im} D\}.$$

8.5   Existence conditions—discrete-time case

In this section, we develop the conditions for the existence of a family of H_2 AID filters for a discrete-time system. We observe again that, to render $\|G_{ue}^\varepsilon\|_2$ arbitrarily small, G_{ue}^ε must necessarily tend to be strictly proper as $\varepsilon \to 0$. Thus, the conditions of Lemma 8.4 must also be valid for discrete-time systems. That is, the H_2 AID filtering problem is solvable via a family of proper filters only if $\ker D \subseteq \ker F$, and via a family of strictly proper filters only if $F = 0$.

We have the following results when strictly proper filters are used.

Theorem 8.11 *Consider a discrete-time system as in (8.1). Let Assumption 8.1 be satisfied. Then, the H_2 AID filtering problem is solvable via a family of strictly proper filters if and only if $F = 0$ and*

$$\mathcal{S}^{\otimes}(A, B, C, D) \subseteq \ker E. \tag{8.14}$$

Proof : This proof follows along the same lines as the proof of Theorem 8.5, except that this time we rely on the discrete-time result in Theorem 6.12 instead of the continuous-time result in Theorem 6.4.   ∎

We now consider the H_2 AID filtering problem by using proper filters. As in the continuous-time case, it is easier to give the results first for the case when $F = 0$.

Theorem 8.12 *Consider a discrete-time system as in* (8.1) *with* $F = 0$. *Let Assumption 8.1 be satisfied. Then, the* H_2 *AID filtering problem is solvable via a family of proper filters if and only if*

$$\mathcal{S}^{\otimes}(A, B, C, D) \cap C^{-1}\{\operatorname{im} D\} \subseteq \ker E.$$

Proof : This proof follows along the same lines as the proof of Theorem 8.6, except that this time we rely on the discrete-time result in Theorem 6.25 instead of the continuous-time result in Theorem 6.22.   ∎

Remark 8.13 *As we remarked earlier, for the case of* $F = 0$ *and for continuous-time systems, the solvability conditions are one and the same whether strictly proper or proper filters are used. On the other hand, for discrete-time systems, the solvability conditions when proper filters are used are much weaker than those when strictly proper filters are used.*

Remark 8.14 *Again, it is interesting to consider the case when* $E = I$, *i.e., when the entire state is to be estimated. As in continuous-time systems, because for this case,* $\ker E = \{0\}$, *the conditions of Theorems 8.11 and 8.12 simplify. The simplified conditions are that the system characterized by the quadruple* (A, B, C, D) *is left-invertible, has no invariant zeros outside the unit circle, and the order of infinite zeros must be zero (i.e., the "relative degree" must be zero).*

We consider now the case of $F \neq 0$. Once again, as in the previous section, we study the H_2 AID filtering problem for a given system for which the matrix F is nonzero by transforming it to an equivalent H_2 AID filtering problem for an auxiliary system where the corresponding F can be taken as zero. The needed auxiliary system is exactly the same as (8.11). In fact, we come to the same conclusion as before, namely, that the transfer matrix from u to $e_{z*} = z^* - \hat{z}^*$ when the new filter (8.12) is applied to the auxiliary system Σ^* is the same as the transfer matrix from u to $e_z = z - \hat{z}$ when the original filter (8.2) is applied to the original Σ as long as $P^{\varepsilon} = \bar{P}^{\varepsilon} + P^*$. As before, this suggests a method of designing first P^*

such that $F - P^*D = 0$ and then designing a family of filters parameterized in ε for the system Σ^* such that the transfer matrix from u to $e_{z^*} = z^* - \hat{z}^*$ has the desired value (namely, its H_2 norm tends to zero as $\varepsilon \to 0$). The designed family of filters for Σ^* can then easily be translated for the original system Σ by allowing $P^\varepsilon = \bar{P}^\varepsilon + P^*$.

We have the following results.

Theorem 8.15 *Consider a discrete-time system as in (8.1). Let Assumption 8.1 be satisfied. Then, the H_2 AID filtering problem is solvable via a family of proper filters if and only if*

$$\left(\mathcal{S}^{\otimes}(A, B, C, D) \oplus \mathbb{R}^m\right) \cap \ker \begin{pmatrix} C & D \end{pmatrix} \subseteq \ker \begin{pmatrix} E & F \end{pmatrix}. \tag{8.15}$$

Proof : This proof follows along the same lines as the proof of Theorem 8.9. As noted, it is easy to see that Lemma 8.4 is also valid for discrete-time systems. Moreover, this time we rely on the discrete-time result in Theorem 8.12 instead of the continuous-time result in Theorem 8.6. ∎

Remark 8.16 *Once again, as in continuous-time systems, readers can easily verify that the solvability conditions for H_2 AID filtering are weaker than the solvability conditions for EID filtering (see Theorems 7.6, 7.7, and 7.9), which of course is as it should be. Such a verification can be done by noting that, in general,*

$$\mathcal{S}^{\otimes}(A, B, C, D) \subseteq \mathcal{S}^{\ominus}(A, B, C, D).$$

Remark 8.17 *Once again, it is interesting to consider the case when $E = I$, i.e., when the entire state is to be estimated. As for this case, $\ker E = \{0\}$, the conditions of Theorem 8.15 simplify. The simplified conditions are that the system characterized by the quadruple (A, B, C, D) is left-invertible, has no invariant zeros outside the unit circle, and the order of infinite zeros must be at most one (i.e., the "relative degree" must be at most one).*

8.6   Design of a family of H_2 AID filters of CSS architecture

In this section, we present a method of designing a parameterized family of H_2 AID filters with some flexibility to assign the poles as desired while honoring certain conditions imposed by the requirement of H_2 AID filtering. In this regard, as in the previous chapter, we first observe that typically any filter design is initiated by first assuming a fixed architecture to the filter. The architecture we use for the filters is the CSS architecture that was developed and used in the previous chapter. In most of the following development, the procedure we follow to develop various filters of CSS architecture is similar to what has been given in Section 7.5. However, there are some subtle changes important enough to warrant their redevelopment here.

As the conditions for the existence of a family of H_2 AID filters as developed earlier in Sections 8.4 and 8.5 do not assume any particular architecture for filters, one fundamental question that arises naturally is as follows: Does a family of H_2 AID filters of CSS architecture exist under the same conditions as developed earlier? In this section, we not only answer this question affirmatively, but also we develop systematic methods of designing them. As in the case of EID filters, the requirement of H_2 AID imposes certain constraints on the behavior of poles. As such, we need to examine the structure of H_2 AID filters regarding their poles. That is, clearly, H_2 AID filtering by definition requires a family of filters parameterized in some parameter ε, and we need to examine the asymptotic behavior of H_2 AID filters as the parameter ε tends to a critical value, say, zero. It turns out that the poles of H_2 AID filters have the following asymptotic behavior:

- For both continuous- and discrete-time systems, as $\varepsilon \to 0$, in general some of the poles of a family of H_2 AID filters go asymptotically to some fixed finite locations, whereas some others are free to be assigned.

- For continuous-time systems, as $\varepsilon \to 0$, in general, some of the poles of a family of H_2 AID filters must go asymptotically to infinity. Obviously, for discrete-time systems, such a phenomena does not occur in view of the stability requirement.

Our primary interest in this section is to study and compute finite asymptotic fixed modes while prescribing a method of designing filters of CSS architecture. Interesting questions can also be posed regarding the infinite asymptotic modes, such as 'how many minimum number of infinite asymptotic modes exist?', etc. The answers to such questions remain to be addressed and are beyond the scope of this book.

In what follows, our presentation is organized into three subsections. The first and second subsections consider, respectively, strictly proper and proper filters, whereas the third subsection considers reduced-order proper filters.

8.6.1   *A family of full-order strictly proper H_2 AID filters—CSS architecture*

In this section, we pursue the design of a family of strictly proper H_2 AID filters while simultaneously using the available flexibility to assign the closed-loop eigenvalues as desired. The architecture we use is the full-order CSS architecture. As explained, strictly proper filters can be used only when $F = 0$. As such, throughout this subsection, we assume that $F = 0$. We developed, in Subsection 7.5.1, the full-order strictly proper filter of CSS architecture. It is given by (7.12) and depicted in Figure 7.2.

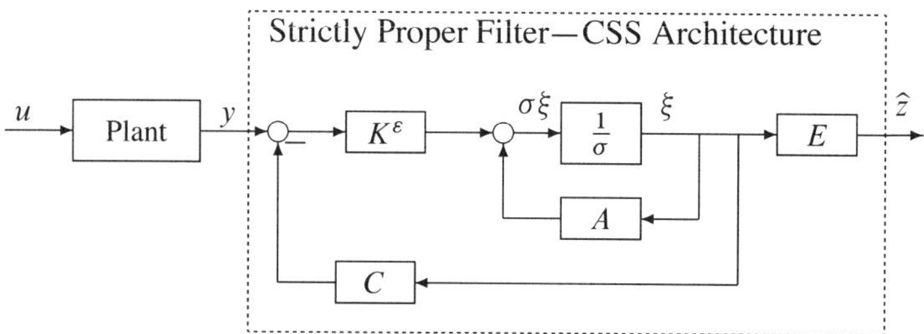

Figure 8.2: Block diagram of a strictly proper filter—CSS architecture

A family of strictly proper filters of CSS architecture parameterized in ε is described by

$$\Sigma^\varepsilon_{\text{sp-CSS}} : \begin{cases} \sigma\xi = A\xi + K^\varepsilon(y - C\xi) \\ \hat{z} = E\xi, \end{cases} \tag{8.16}$$

where the matrix K^ε is parameterized in ε.

Error dynamics: Let us define the error $e = x - \xi$; then the error e_z between the actual desired output z and the estimated desired output $\hat{z}$ is $e_z = E(x-\xi) = Ee$. Also, the dynamics of error is given by

$$\Sigma^{ue,\varepsilon}_{\text{sp-CSS}} : \begin{cases} \sigma e = (A - K^\varepsilon C)e + (B - K^\varepsilon D)u \\ e_z = Ee. \end{cases} \tag{8.17}$$

The transfer matrix $G^{ue,\varepsilon}_{\text{sp-CSS}}$ from u to e_z can obviously be written as

$$G^{ue,\varepsilon}_{\text{sp-CSS}} = E(\sigma I - A + K^\varepsilon C)^{-1}(B - K^\varepsilon D). \tag{8.18}$$

Remark 8.18 *In view of (8.16) and (8.17), it is easy to see that both the filter equation and the error equation have the same poles, which are the eigenvalues of $A - K^\varepsilon C$.*

We observe that the only unknown in the filter equation (8.16) and consequently in the error equation (8.17) is the parameterized matrix K^ε, which is normally referred to as the filter gain. We need to determine or design K^ε in such a way that the H_2 norm of $G^{ue,\varepsilon}_{\text{sp-CSS}}$ is arbitrarily small. In general, such a gain K^ε is nonunique. This lets us use the available freedom in selecting K^ε to shape the error dynamics appropriately, i.e., to shape the eigenstructure of $A - K^\varepsilon C$ as desired while honoring any constraints imposed by the requirement of H_2 AID filtering.

As in the case of EID filtering, we pause to emphasize an important point. Theorems 8.5 and 8.11 developed earlier give the conditions under which a family of H_2 AID filters exists among the general class of strictly proper filters of the form (8.2) with $P^\varepsilon = 0$. In other words, Theorems 8.5 and 8.11 do not restrict themselves to any fixed architecture for a filter such as the one given in (8.16). Nevertheless, as the following theorem shows, whenever the conditions of Theorems 8.5 and 8.11 are satisfied, we can determine the parameterized gain K^ε such that the family of filters given in (8.16) is a family of H_2 AID filters.

Theorem 8.19 *Consider a continuous- or discrete-time system given by (8.1). Let Assumption 8.1 be satisfied. Consider strictly proper filters, and assume that the conditions of Theorem 8.5 for continuous-time systems or Theorem 8.11 for discrete-time systems are satisfied. Then, a family of H_2 AID filters of the type $\Sigma_{\mathrm{sp\text{-}CSS}}^{\varepsilon}$ exists; i.e., a sequence of parameterized gains K^ε exists such that the family of filters given in (8.16) is a family of H_2 AID filters.*

Proof : We use the same duality arguments as in the case of EID filters. The only additional property that is needed is that there is a $1 - 1$ correspondence between the filters of the form (8.16) for the original system and static state feedback controllers of the form $u_d = (K^\varepsilon)' x_d$ solving the H_2 ADD problem for the dual system given in (8.8), where the poles of the filter are equal to the closed-loop poles after applying the static feedback to the dual system. Moreover, the transfer matrix from w to e_z when applying the filter to the original system is equal to the transpose of the transfer matrix from w to z for the dual system after applying the static state feedback. This latter property was discussed earlier in Section 7.A. ∎

As it is well known, the dynamics of any system is heavily influenced by its poles. When we look at the poles of the error dynamics and those of the filter as given in (8.16), we observe that they are the same and equal to the eigenvalues of $A - K^\varepsilon C$. A natural question then becomes whether any constraints in choosing the filter poles that shape the error dynamics exist. As we said in the beginning of this section, the requirement of H_2 AID filtering imposes that certain poles of a family of H_2 AID filters tend as ε tends to zero to certain fixed locations termed as finite asymptotic fixed modes. We have the following formal definition of finite asymptotic fixed modes.

Definition 8.20 (*Finite asymptotic fixed modes of strictly proper H_2 AID filters of CSS architecture*) *Consider the given system (8.1) and the associated H_2 AID filtering problem 8.3. Assume that the solvability conditions as specified by Theorem 8.5 for continuous-time systems or as specified by Theorem 8.11 for discrete-time systems are satisfied. Then, a finite scalar $\lambda \in \mathbb{C}^-$ for the continuous-time case or $\lambda \in \mathbb{C}^\ominus$ for the discrete-time case, is said to be an*

H_2 finite asymptotic fixed mode *with algebraic multiplicity α if for every family of H_2 AID filters parameterized in ε and having the strictly proper full-order CSS architecture, poles λ_i^ε, $i = 1, 2, \cdots, \alpha$, of the family of filters exist such that all λ_i^ε tend to λ as ε tends to zero. The set of all H_2 finite asymptotic fixed modes when strictly proper filters of full-order CSS architecture are used is denoted by $\Omega_{\text{sp-CSS}}^{h2-\text{aid}}(A, B, C, D, E, 0)$.*

We have the following theorem that characterizes the set $\Omega_{\text{sp-CSS}}^{h2-\text{aid}}$.

Theorem 8.21 *Consider a continuous- or discrete-time system as in (8.1). Let Assumption 8.1 be satisfied. Consider strictly proper full-order filters of CSS architecture and assume that the solvability conditions as specified by Theorem 8.5 for continuous-time systems or as specified by Theorem 8.11 for discrete-time systems are satisfied. Then, we have*

$$\Omega_{\text{sp-CSS}}^{h2-\text{aid}}(A, B, C, D, E, 0) = \Omega_s^2(A', C', B', D', E') \tag{8.19}$$

where $\Omega_s^2(A', C', B', D', E')$ is obtained by using the H_2 ADD algorithm (see Subsection 6.5.1 for continuous-time systems and Subsection 6.5.2 for discrete-time systems) with its input as the quintuple (A', C', B', D', E').

Proof : It follows easily in view of the proof of Theorem 8.19. ∎

Obviously, we would like to next explore the development of an algorithm to design H_2 AID filters of full-order strictly proper CSS architecture, which gives us the flexibility to place the finite asymptotic modes of the filters at desired locations, of course, within the constraints imposed by Theorem 8.21. The following theorem considers this issue.

Theorem 8.22 *Consider a continuous- or discrete-time system given by (8.1). Let Assumption 8.1 be satisfied. Consider strictly proper filters, and assume that the conditions of Theorem 8.5 for continuous-time systems or Theorem 8.11 for discrete-time systems are satisfied. Then, the following two statements are equivalent:*

(i) The parameterized gain sequence K^ε renders the family of filters given in (8.16) as a family of H_2 AID filters.

(ii) The parameterized gain sequence K^ε is such that the family of control laws $u_d = -(K^\varepsilon)'x_d$ solves the H_2 ADD problem for the dual system given in (8.8).

Proof : It follows easily in view of the proof of Theorem 8.19.   ∎

Theorem 8.22 provides a roadmap to design the parameterized gain sequence K^ε that renders the family of filters given in (8.16) as a family of H_2 AID filters. All that needs to be done is to use the H_2 ADD Algorithm (see Subsection 6.5.1 for continuous-time systems and Subsection 6.5.2 for discrete-time systems) with the quintuple (A', C', B', D', E') as its input and obtain a parameterized gain sequence F^ε and then transpose it to obtain K^ε. Furthermore, the said H_2 ADD algorithm determines the set of H_2 finite asymptotic fixed modes $\Omega_{\text{sp-CSS}}^{h2-\text{aid}}(A, B, C, D, E, 0)$ for the class of filters given in (8.16). Moreover, the said H_2 ADD algorithm has certain flexibility to place the finite asymptotic modes of the filter as desired but within the constraints imposed by Theorem 8.21.

8.6.2 A family of full-order proper H_2 AID filters—CSS architecture

Our goal in this subsection is to design a family of full-order proper H_2 AID filters of CSS architecture. In connection with the design of EID filters, the proper full-order filter of CSS architecture was developed in Subsection 7.5.2 of Chapter 7. It is given by (7.28) and depicted in Figure 7.4. We follow here basically the same procedure, however, with some subtle but important changes. To construct a family of proper H_2 AID filters of full-order CSS architecture for the given system, we construct here a family of strictly proper H_2 AID filters for an auxiliary system $\widetilde{\Sigma}^*$. To do so, we first decompose the measured output y into two parts y_0 and y_1 in such a way that y_0 contains explicitly the unknown input u in it, whereas y_1 does not contain any input u in it. That is, we rewrite the given system equation (8.1) as

$$\Sigma : \begin{cases} \sigma x & = Ax + Bu \\ \begin{pmatrix} y_0 \\ y_1 \end{pmatrix} = \begin{pmatrix} C_0 \\ C_1 \end{pmatrix} x + \begin{pmatrix} D_0 \\ 0 \end{pmatrix} u = Cx + Du \\ z & = Ex + Fu, \end{cases} \tag{8.20}$$

where rank $D = $ rank $D_0 = m_0$. We note that, without any loss of generality, we can rewrite the given system (8.1) in the form (8.20) by an appropriate coordinate transformation.

We next need to do some preliminary work to consider a nonzero matrix F. We need to decompose the matrix E into two parts: E_1 and E_2. That is, let

$$E = E_1 + E_2 \tag{8.21}$$

such that

$$\mathcal{S}^{-0}(A, B, C, D) \cap \mathcal{V}^*(A, B, C, D) \subseteq \ker E_1 \tag{8.22}$$

for continuous-time systems, or

$$\mathcal{S}^{\otimes}(A, B, C, D) \cap C^{-1}\{\operatorname{im} D\} \subseteq \ker E_1 \tag{8.23}$$

for discrete-time systems, and moreover

$$\ker \begin{pmatrix} C & D \end{pmatrix} \subseteq \ker \begin{pmatrix} E_2 & F \end{pmatrix}. \tag{8.24}$$

The decomposition of E indicated above can always be done under the solvability conditions given by Theorems 8.9 and 8.15. Also, the choice of E_1 and E_2 as chosen above guarantees that a matrix P^* exists such that $E_1 = E - P^*C$ and $F - P^*D = 0$. We can now use a preliminary injection of output y into the desired output z as discussed before in Section 7.3, and we rewrite the desired output. Let

$$z^* = z - P^*y = (E - P^*C)x + (F - P^*D)u = E_1x. \tag{8.25}$$

In view of (8.20) and (8.25), we can define a new system Σ^* as

$$\Sigma^* : \begin{cases} \sigma x & = Ax + Bu \\ \begin{pmatrix} y_0 \\ y_1 \end{pmatrix} = \begin{pmatrix} C_0 \\ C_1 \end{pmatrix} x + \begin{pmatrix} D_0 \\ 0 \end{pmatrix} u \\ z^* & = E_1x. \end{cases} \tag{8.26}$$

The above procedure of developing Σ^* is the same as the one used in developing Σ^* in (7.19).

By now we have rewritten the given system in a form suitable for filter development. However, before we proceed further, we need one more amendment of the measured output y. The needed amendment depends on whether we deal with continuous- or discrete-time systems. As such, in what follows, we divide our presentation into two parts: one pertaining to continuous-time systems and the other pertaining to discrete-time systems.

A family of full-order proper H_2 AID filters—continuous-time systems:

In what follows, we deal with continuous-time systems. We first form a new measurement variable $\tilde{y}$ by augmenting y with another part $y_2 = \dot{y}_1$. That is, we let

$$\tilde{y} = \begin{pmatrix} y_0 \\ y_1 \\ y_2 \end{pmatrix} = \begin{pmatrix} y_0 \\ y_1 \\ \dot{y}_1 \end{pmatrix}, \quad \text{and} \quad \tilde{C} = \begin{pmatrix} C_0 \\ C_1 \\ C_1 A \end{pmatrix}, \quad \tilde{D} = \begin{pmatrix} D_0 \\ 0 \\ C_1 B \end{pmatrix}. \tag{8.27}$$

Then, it is easy to see that

$$\tilde{y} = \tilde{C}x + \tilde{D}u.$$

The dimension of $\tilde{y}$ is $\tilde{p} = 2p - \text{rank } D_0$. We note that y_2 is not directly available. However, as in Chapter 7 and as will be seen shortly, it can be eliminated from the filter equation.

Auxiliary system $\widetilde{\Sigma}^*$: We define next an auxiliary system with its measured output as $\widetilde{y}$:

$$\widetilde{\Sigma}^* : \begin{cases} \dot{x} & = Ax + Bu \\ \widetilde{y} & = \widetilde{C}x + \widetilde{D}u \\ z^* & = E_1 x. \end{cases} \qquad (8.28)$$

Before we proceed, it is important that we examine certain structural properties of $\widetilde{\Sigma}^*$ and relate them to those of Σ^*. Lemma 7.22 examines such properties. It is repeated here (with minor modifications) for completeness.

Lemma 8.23 *Consider the systems $\widetilde{\Sigma}^*$ and Σ^*, which are, respectively, characterized by the matrix quintuples $(A, B, \widetilde{C}, \widetilde{D}, E_1)$ and (A, B, C, D, E_1). Consider two subsystems, one corresponding to $\widetilde{\Sigma}^*$ and characterized by the quadruple $(A, B, \widetilde{C}, \widetilde{D})$, and the other corresponding to Σ^* and characterized by the quadruple (A, B, C, D). Then, the following results hold:*

(i) *The invariant zeros of the system characterized by $(A, B, \widetilde{C}, \widetilde{D})$ are the same as those of the system characterized by (A, B, C, D).*

(ii) *The matrix pair $(\widetilde{C}, A)$ is $\mathbb{C}^-$-detectable if and only if the matrix pair (C, A) is $\mathbb{C}^-$-detectable.*

(iii) *Orders of infinite zeros of the system characterized by $(A, B, \widetilde{C}, \widetilde{D})$ are reduced by one from those of the system characterized by (A, B, C, D).*

(iv) $\mathcal{V}^*(A, B, \widetilde{C}, \widetilde{D}) = \mathcal{V}^*(A, B, C, D)$.

(v) $\mathcal{S}^{-0}(A, B, \widetilde{C}, \widetilde{D}) = \mathcal{S}^{-0}(A, B, C, D) \cap C^{-1}\{\operatorname{im} D\}$.

(vi) $\mathcal{S}^{-0}(A, B, \widetilde{C}, \widetilde{D}) = \{0\}$ *if and only if Σ is left-invertible, has only invariant zeros in $\mathbb{C}^- \cup \mathbb{C}^0$, and has no infinite zeros of order higher than one.*

Proof : Most of these properties were already presented in Lemma 7.22. Property *(iv)* follows from the fact that for any F, an $A + BF$ invariant subspace contained in $\ker(C + DF)$ is contained in $\ker(\widetilde{C} + \widetilde{D}F)$ and, conversely, an $A + BF$ invariant subspace contained in $\ker(\widetilde{C} + \widetilde{D}F)$ is contained in $\ker(C + DF)$. ∎

As in Chapter 7, we plan to design first a family of strictly proper filters to solve the H_2 AID filtering problem for the auxiliary system $\widetilde{\Sigma}^*$ and then modify it to obtain a family of proper filters that solves the H_2 AID filtering problem for the original system Σ. This obviously is possible if and only if the conditions for the existence of a family of strictly proper H_2 AID filters for the auxiliary system $\widetilde{\Sigma}^*$ coincide with the conditions for the existence of a family of proper H_2 AID filters for the system Σ. The following lemma formalizes this.

Lemma 8.24 *Consider the continuous-time systems Σ and $\widetilde{\Sigma}^*$, respectively, as given in (8.1) and (8.28). Then, the following two statements are equivalent:*

(i) A family of strictly proper H_2 AID filters exists for the auxiliary system $\widetilde{\Sigma}^*$.

(ii) A family of proper H_2 AID filters exists for the system Σ.

Proof : The proof follows from Lemma 8.23 in view of Theorems 8.5 and 8.9. ∎

Lemma 8.24 pertains to general strictly proper and proper filters, although our interest here is only on strictly proper and proper filters of CSS architecture. We will relate next strictly proper filters of CSS architecture for $\widetilde{\Sigma}^*$ to proper filters for Σ with a nice desirable structure.

For the auxiliary system $\widetilde{\Sigma}^*$, we form a family of strictly proper filters of CSS architecture as

$$
\widetilde{\Sigma}^{\varepsilon}_{\text{sp-CSS}} : \begin{cases} \dot{\widetilde{\xi}} = A\widetilde{\xi} + K^{\varepsilon}(\widetilde{y} - \widetilde{C}\widetilde{\xi}) \\ \widehat{z}^* = E_1\widetilde{\xi}, \end{cases} \tag{8.29}
$$

where the matrix K^{ε} is a parameterized filter gain. The estimate $\widehat{z}^*$ of z^* is as given by (8.29). Then, in view of (8.25), the estimate $\widehat{z}$ of z is given by

$$
\widehat{z} = \widehat{z}^* + P^* y = E_1\widetilde{\xi} + P^* y. \tag{8.30}
$$

The above development focuses on developing the family of strictly proper filters (8.29). However, the family of filters (8.29) is not directly implementable because $y_2 = \dot{y}_1$ is not available as a measured variable. This implies that some how we need to eliminate y_2 from (8.29). This can be easily done by defining a new variable:

$$
\xi = \widetilde{\xi} - K^{\varepsilon}_2 y_1. \tag{8.31}
$$

Here K^{ε}_2 is obtained by partitioning K^{ε} in conformity with the partitioning of $\widetilde{y}$. That is,

$$
K^{\varepsilon} = \begin{pmatrix} K^{\varepsilon}_0 & K^{\varepsilon}_1 & K^{\varepsilon}_2 \end{pmatrix}. \tag{8.32}
$$

With the definition of ξ as in (8.31), we can rewrite the filter equation (8.29) as

$$
\begin{cases} \dot{\xi} = (A - K^{\varepsilon}\widetilde{C})\xi + \begin{pmatrix} K^{\varepsilon}_0 & K^{\varepsilon}_1 + (A - K^{\varepsilon}\widetilde{C})K^{\varepsilon}_2 \end{pmatrix} y \\ \widetilde{\xi} = \xi + K^{\varepsilon}_2 y_1 \\ \widehat{z}^* = E_1\widetilde{\xi} = E_1(\xi + K^{\varepsilon}_2 y_1). \end{cases} \tag{8.33}
$$

Obviously, the family of filters given above does not use $\dot{y}_1$. Moreover, it is proper rather than strictly proper. The family of filters given in (8.33) is indeed the family of proper full-order CSS filters that is to be used for Σ^*.

Proper filters of CSS architecture for Σ: In view of (8.30), we can rewrite (8.33) as an implementable proper filter for Σ:

$$\Sigma_{\text{p-CSS}}^{\varepsilon} : \begin{cases} \dot{\xi} = (A - K^{\varepsilon}\widetilde{C})\xi + \widetilde{K}^{\varepsilon}y \\ \widetilde{\xi} = \xi + K_2^{\varepsilon}y_1 \\ \hat{z} = E_1\widetilde{\xi} + P^*y = E_1\xi + \widetilde{P}^{\varepsilon}y, \end{cases} \tag{8.34}$$

where

$$\widetilde{K}^{\varepsilon} = \begin{pmatrix} K_0^{\varepsilon} & K_1^{\varepsilon} + (A - K^{\varepsilon}\widetilde{C})K_2^{\varepsilon} \end{pmatrix}, \tag{8.35}$$

and

$$\widetilde{P}^{\varepsilon} = \begin{pmatrix} 0 & E_1 K_2^{\varepsilon} \end{pmatrix} + P^*.$$

Everything in this family of filters is known except the parameterized gain K^{ε}. A block diagram representation of the proper filter along with the given plant is shown in Figure 8.3. Structurally, the block diagram in Figure 8.3 is the same as the one in Figure 7.4.

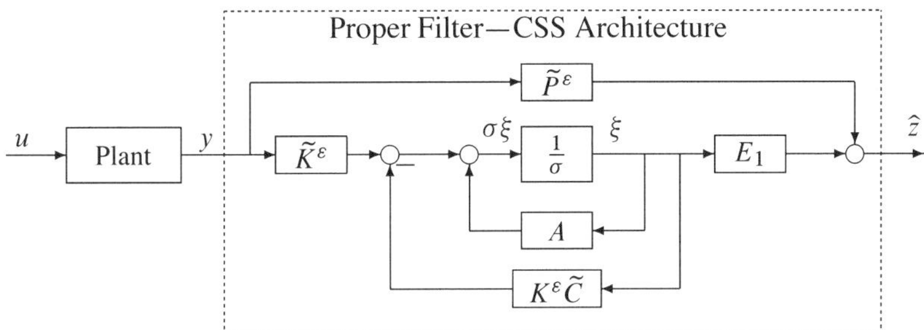

Figure 8.3: Block diagram of a full-order proper filter

Remark 8.25 *The above development shows clearly that there is a $1-1$ relationship between the family of strictly proper filters of CSS architecture for $\widetilde{\Sigma}^*$ and the family of proper filters of CSS architecture for Σ; that is, one of these family of filters can be constructed from the other.*

Error dynamics: By defining the error $e = x - \widetilde{\xi}$, the error e_z between the actual desired output $z = Ex + Fu = E_1x + P^*y$ and the estimated desired output $\hat{z} = E_1\widetilde{\xi} + P^*y$ can be written as

$$e_z = z - \hat{z} = E_1 e.$$

Then, in view of (8.20) and (8.29), the dynamics of the error signal is given by

$$\Sigma_{\text{p-CSS}}^{ue,\varepsilon} : \begin{cases} \dot{e} = (A - K^{\varepsilon}\widetilde{C})e + (B - K^{\varepsilon}\widetilde{D})u \\ e_z = E_1 e. \end{cases} \tag{8.36}$$

Also, the transfer matrix $G^{ue,\varepsilon}$ from u to e_z can obviously be written as

$$G^{ue,\varepsilon}_{\text{p-CSS}} = E_1(\sigma I - A + K^\varepsilon \widetilde{C})^{-1}(B - K^\varepsilon \widetilde{D}). \tag{8.37}$$

Remark 8.26 *In view of* (8.34) *and* (8.36), *it is easy to see that both the filter equation and the error equation have the same poles, which are the eigenvalues of* $A - K^\varepsilon \widetilde{C}$.

As in the case of strictly proper filters, we need to point out an important point. Theorem 8.9 developed earlier gives the conditions under which a family of H_2 AID filters exists among the general class of strictly proper filters of the form (8.2). In other words, Theorem 8.9 does not restrict itself to any fixed architecture for a filter such as the one given in (8.34). Nevertheless, as the following theorem shows, whenever the conditions of Theorem 8.9 are satisfied, a parameterized gain K^ε exists such that the family of filters given in (8.34) is a family of H_2 AID filters.

Theorem 8.27 *Consider a continuous-time system* Σ *given by* (8.1). *Let the conditions of Theorem 8.9 be satisfied. Then, a family of* H_2 *AID filters of the type* $\Sigma^\varepsilon_{\text{p-CSS}}$ *for* Σ *exists; i.e., a parameterized gain sequence* K^ε *exists such that the family of filters given in* (8.34) *is a family of* H_2 *AID filters for* Σ. *Moreover,* K^ε *is the same parameterized gain sequence that renders* $\widetilde{\Sigma}^\varepsilon_{\text{sp-CSS}}$ *of* (8.29) *as a family of* H_2 *AID filters for* $\widetilde{\Sigma}^*$ *of* (8.28).

Proof : Consider the system $\widetilde{\Sigma}^*$ characterized by the quintuple $(A, B, \widetilde{C}, \widetilde{D}, E_1)$, where $E_1 = E - P^*C$ with P^* such that $F - P^*D = 0$, and $\widetilde{C}$ and $\widetilde{D}$ are as defined in (8.27). By Theorem 8.19, a parameterized gain K^ε exists resulting in a family of strictly proper H_2 AID filters for the system $\widetilde{\Sigma}^*$. Before this theorem, it is shown how this family of strictly proper filters can be converted to a family of proper filters (8.34) for the original system Σ with the same error dynamics. Hence the result. ∎

As in the case of strictly proper filters, constraints in choosing the filter poles that shape the error dynamics exist . The requirement of H_2 AID filtering imposes that certain poles of a family of H_2 AID filters tend as ε tends to zero to certain fixed locations termed as finite asymptotic fixed modes. We have the following formal definition of finite asymptotic fixed modes when proper H_2 AID filters of CSS architecture are used.

Definition 8.28 (*Finite asymptotic fixed modes of proper* H_2 *AID filters of CSS architecture*) *Consider the given system* (8.1) *and the associated* H_2 *AID filtering* *problem 8.3. Assume that the solvability conditions as specified by Theorem 8.9 are satisfied. Then, a finite scalar $\lambda \in \mathbb{C}^-$ is said to be an **H_2 finite asymptotic fixed mode** with algebraic multiplicity α if for every family of H_2 AID filters parameterized in ε and having the proper full-order CSS architecture, poles λ_i^ε, $i = 1, 2, \cdots, \alpha$, of the family of filters exist such that all λ_i^ε tend to λ as ε tends to zero. The set of all H_2 finite asymptotic fixed modes when proper filters of full-order CSS architecture are used is denoted by $\mathit{\Omega}_{\text{p-CSS}}^{h2-\text{aid}}(A, B, C, D, E, F)$.*

We have the following theorem that characterizes the set $\mathit{\Omega}_{\text{p-CSS}}^{h2-\text{aid}}$.

Theorem 8.29 *Consider a continuous-time system as in (8.1). Let Assumption 8.1 be satisfied. Consider proper full-order filters of CSS architecture, and assume that the solvability conditions as specified by Theorem 8.9 are satisfied. Also, let E_1 be as discussed in (8.21) to (8.24), and P^* be such that $E_1 = E - P^*C$ and $F - P^*D = 0$. Let $\widetilde{C}$ and $\widetilde{D}$ be as defined in (8.27). Then, we have*

$$\mathit{\Omega}_{\text{p-CSS}}^{h2-\text{aid}}(A, B, C, D, E, F) = \mathit{\Omega}_{\text{sp-CSS}}^{h2-\text{aid}}(A, B, \widetilde{C}, \widetilde{D}, E_1, 0).$$

Proof : The proof is transparent in view of the development of how a family of strictly proper filters for $\widetilde{\Sigma}^*$ can be converted to a family of proper filters (8.34) for the original system Σ with the same error dynamics. ∎

Remark 8.30 *In view of Lemma 8.23 and a detailed examination of H_2 ADD Algorithm (see Subsection 6.5.1) by which $\mathit{\Omega}_{\text{sp-CSS}}^{h2-\text{aid}}(A, B, \widetilde{C}, \widetilde{D}, E_1, 0)$ can be computed, one can easily verify that*

$$\mathit{\Omega}_{\text{sp-CSS}}^{h2-\text{aid}}(A, B, \widetilde{C}, \widetilde{D}, E_1, 0) = \mathit{\Omega}_{\text{sp-CSS}}^{h2-\text{aid}}(A, B, C, D, E_1, 0).$$

As such, in view of Theorem 8.29, we have

$$\mathit{\Omega}_{\text{p-CSS}}^{h2-\text{aid}}(A, B, C, D, E, F) = \mathit{\Omega}_{\text{sp-CSS}}^{h2-\text{aid}}(A, B, C, D, E_1, 0).$$

Remark 8.31 *Assume that the H_2 AID filtering problem by using strictly proper filters is solvable; i.e., the solvability conditions, as specified by Theorem 8.5, are satisfied. Then, one can choose $E = E_1$ and $E_2 = 0$ where E_1 and E_2 are as discussed in (8.21) to (8.24). In this case, because both strictly proper and proper H_2 AID filters exist, an interesting question arises regarding the relationship between $\mathit{\Omega}_{\text{p-CSS}}^{h2-\text{aid}}$ and $\mathit{\Omega}_{\text{sp-CSS}}^{h2-\text{aid}}$. One can verify that in this case,*

$$\mathit{\Omega}_{\text{p-CSS}}^{h2-\text{aid}}(A, B, C, D, E, F) = \mathit{\Omega}_{\text{sp-CSS}}^{h2-\text{aid}}(A, B, C, D, E, 0).$$

Obviously, we would like to explore next the development of an algorithm to design H_2 AID filters of full-order proper CSS architecture, which gives us the flexibility to place the finite asymptotic modes of the filters at desired locations, of course, within the constraints imposed by Theorem 8.29. Such an algorithm can be given by the following steps:

(i) Use the quintuple $(A, B, \widetilde{C}, \widetilde{D}, E_1)$ as the input to the H_2 ADD Algorithm (see Subsection 6.5.1), and obtain a parameterized gain F^ε.

(ii) Let $K^\varepsilon = (F^\varepsilon)'$. Partition K^ε in accordance with (8.32), and then define $\widetilde{K}^\varepsilon$ in accordance with (8.35).

The H_2 ADD algorithm mentioned in the above first step yields the set

$$\boldsymbol{\Omega}_{\text{sp-CSS}}^{h2-\text{aid}}(A, B, \widetilde{C}, \widetilde{D}, E_1, 0) = \boldsymbol{\Omega}_{\text{p-CSS}}^{h2-\text{aid}}(A, B, C, D, E, F).$$

The above equality follows from Theorem 8.29. Also, the H_2 ADD algorithm has certain flexibility to place the finite asymptotic modes of the filter as desired but within the constraints imposed by Theorem 8.29.

A family of full-order proper H_2 AID filters—discrete-time systems:

In what follows, we deal with discrete-time systems. Unlike in continuous-time systems, a new measurement variable $\widetilde{y}$ is formed by replacing y_1 in y with $y_2 = \sigma y_1$. That is, we let

$$\widetilde{y} = \begin{pmatrix} y_0 \\ y_2 \end{pmatrix} = \begin{pmatrix} y_0 \\ \sigma y_1 \end{pmatrix}, \quad \text{and} \quad \widetilde{C} = \begin{pmatrix} C_0 \\ C_1 A \end{pmatrix}, \quad \widetilde{D} = \begin{pmatrix} D_0 \\ C_1 B \end{pmatrix}. \tag{8.38}$$

Then, it is easy to see that

$$\widetilde{y} = \widetilde{C}x + \widetilde{D}u.$$

As in continuous-time systems, y_2 is not directly available. However, as before, it can be eliminated from the filter equation.

Auxiliary system $\widetilde{\Sigma}^*$: We define next an auxiliary system with its measured output as $\widetilde{y}$:

$$\widetilde{\Sigma}^* : \begin{cases} \sigma x = Ax + Bu \\ \widetilde{y} = \widetilde{C}x + \widetilde{D}u \\ z^* = E_1 x, \end{cases} \tag{8.39}$$

where E_1 is as defined in (8.21) and satisfies (8.22), (8.23), and (8.24). As in continuous-time systems, before we proceed further, it is important that we investigate certain structural properties of $\widetilde{\Sigma}^*$ and relate them to those of Σ^*. The following lemma examines such properties. It is the same as Lemma 7.38 (with minor modifications) but reproduced here for convenience.

Lemma 8.32 *Consider the systems $\tilde{\Sigma}^*$ and Σ^*, which are, respectively, characterized by the matrix quintuples $(A, B, \tilde{C}, \tilde{D}, E_1)$ and (A, B, C, D, E_1). Consider two subsystems, one corresponding to $\tilde{\Sigma}^*$ and characterized by the quadruple $(A, B, \tilde{C}, \tilde{D})$, and the other corresponding to Σ^* and characterized by the quadruple (A, B, C, D). Then, the following results hold:*

(i) *The invariant zeros of the system characterized by $(A, B, \tilde{C}, \tilde{D})$ contain the invariant zeros of the system characterized by (A, B, C, D) in addition to the invariant zeros at the origin.*

(ii) *The matrix pair $(\tilde{C}, A)$ is $\mathbb{C}^{\ominus}$-detectable if and only if the matrix pair (C, A) is $\mathbb{C}^{\ominus}$-detectable.*

(iii) *Orders of infinite zeros of the system characterized by $(A, B, \tilde{C}, \tilde{D})$ are reduced by one from those of the system characterized by (A, B, C, D).*

(iv) *The system characterized by $(A, B, \tilde{C}, \tilde{D})$ is left-invertible if and only if the system characterized by (A, B, C, D) is left-invertible.*

(v) $\mathcal{S}^{\otimes}(A, B, \tilde{C}, \tilde{D}) = \mathcal{S}^{\otimes}(A, B, C, D) \cap C^{-1}\{\mathrm{im}\, D\}.$

(vi) $\mathcal{S}^{\otimes}(A, B, \tilde{C}, \tilde{D}) = \{0\}$ *if and only if Σ is left-invertible, has only invariant zeros in $\mathbb{C}^{\otimes}$, and has no infinite zeros of order higher than one.*

Remark 8.33 *Lemma 8.32 for discrete-time systems is similar to Lemma 8.23 for continuous-time systems. In continuous-time systems, the invariant zeros of $\tilde{\Sigma}^*$ are the same as those of Σ^* or equivalently the same as those of Σ. However, for discrete-time systems, $\tilde{\Sigma}^*$ has additional invariant zeros at the origin.*

As in continuous-time systems, we plan to design first a family of strictly proper H_2 AID filters for the auxiliary system $\tilde{\Sigma}^*$ and then modify it to obtain a family of proper H_2 AID filters for the original system Σ. This obviously is possible if and only if the conditions for the existence of a family of strictly proper H_2 AID filters for the auxiliary system $\tilde{\Sigma}^*$ coincide with the conditions for the existence of a family of proper H_2 AID filters for the system Σ. The following lemma formalizes this.

Lemma 8.34 *Consider the discrete-time systems Σ and $\tilde{\Sigma}^*$, respectively, as given in (8.1) and (8.39). Then, the following two statements are equivalent:*

(i) *A family of strictly proper H_2 AID filters exists for the auxiliary system $\tilde{\Sigma}^*$.*

(ii) *A family of proper H_2 AID filters exists for the system Σ.*

Proof : The proof follows from Lemma 8.32 in view of Theorems 8.11 and 8.15.

∎

As in the continuous-time case, the above lemma pertains to general strictly proper and proper filters, although our interest here is only on strictly proper and proper filters of CSS architecture. We will relate next strictly proper filters of CSS architecture for $\widetilde{\Sigma}^*$ to proper filters for Σ with a nice desirable structure.

For the auxiliary system $\widetilde{\Sigma}^*$, we form a family of strictly proper filters of CSS architecture as

$$\widetilde{\Sigma}^{\varepsilon}_{\text{sp-CSS}} : \begin{cases} \sigma\widetilde{\xi} = A\widetilde{\xi} + K^{\varepsilon}(\widetilde{y} - \widetilde{C}\widetilde{\xi}) \\ \widehat{z}^* = E_1\widetilde{\xi}, \end{cases} \tag{8.40}$$

where the matrix K^{ε} is a parameterized filter gain. The estimate $\widehat{z}^*$ of z^* is as given by (8.40). Then, in view of (8.25), the estimate $\widehat{z}$ of z is given by

$$\widehat{z} = \widehat{z}^* + P^*y = E_1\widetilde{\xi} + P^*y. \tag{8.41}$$

The above development focuses on developing the family of strictly proper filters (8.40). However, the family of filters (8.40) is not directly implementable because $y_2 = \sigma y_1$ is not available as a measured variable. This implies that some how we need to eliminate y_2 from the filter equation (8.40). This can be easily done by defining a new variable:

$$\xi = \widetilde{\xi} - K^{\varepsilon}_2 y_1. \tag{8.42}$$

Here K^{ε}_2 is obtained by partitioning K^{ε} in conformity with the partitioning of y. That is,

$$K^{\varepsilon} = \begin{pmatrix} K^{\varepsilon}_0 & K^{\varepsilon}_2 \end{pmatrix}. \tag{8.43}$$

With the definition of ξ as in (8.42), we can rewrite the filter equation (8.40) as

$$\begin{cases} \sigma\xi = (A - K^{\varepsilon}\widetilde{C})\xi + \begin{pmatrix} K^{\varepsilon}_0 & (A - K^{\varepsilon}\widetilde{C})K^{\varepsilon}_2 \end{pmatrix}y \\ \widetilde{\xi} = \xi + K^{\varepsilon}_2 y_1 \\ \widehat{z}^* = E_1\widetilde{\xi} = E_1(\xi + K^{\varepsilon}_2 y_1). \end{cases} \tag{8.44}$$

Obviously, the family of filters given above does not use σy_1. Moreover, it is proper rather than strictly proper. The family of filters given in (8.44) is indeed the family of proper full-order CSS filters that is to be used for $\widetilde{\Sigma}^*$.

The structure of (8.44) essentially is similar to the structure of a current type of filters well known in discrete-time literature.

Proper filters of CSS architecture for Σ: In view of (8.41), we can rewrite (8.44) as an implementable proper filter for Σ:

$$\Sigma^{\varepsilon}_{\text{p-CSS}} : \begin{cases} \sigma\xi = (A - K^{\varepsilon}\widetilde{C})\xi + \widetilde{K}^{\varepsilon}y \\ \widetilde{\xi} = \xi + K^{\varepsilon}_2 y_1 \\ \widehat{z} = E_1\widetilde{\xi} + P^*y = E_1\xi + \widetilde{P}^{\varepsilon}y, \end{cases} \tag{8.45}$$

where

$$\widetilde{K}^\varepsilon = \begin{pmatrix} K_0^\varepsilon & (A - K^\varepsilon \widetilde{C})K_2^\varepsilon \end{pmatrix} \quad \text{and} \quad \widetilde{P}^\varepsilon = \begin{pmatrix} 0 & E_1 K_2^\varepsilon \end{pmatrix} + P^*. \tag{8.46}$$

Everything in this family of filters is known except the parameterized gain K^ε. A block diagram representation of the proper filter along with the given plant is shown in Figure 8.4. Structurally, the block diagram in Figure 8.4 is the same as the one in Figure 8.3.

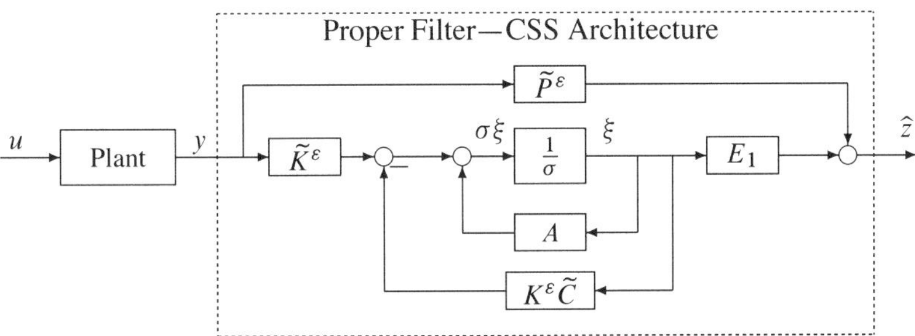

Figure 8.4: Block diagram of a full-order proper filter—CSS architecture

Remark 8.35 *The above development shows clearly that there is a $1-1$ relationship between the family of strictly proper filters of CSS architecture for $\widetilde{\Sigma}^*$ and the family of proper filters of CSS architecture for Σ; that is, one of these family of filters can be constructed from the other.*

Error dynamics: By defining the error $e = x - \widetilde{\xi}$, the error e_z between the actual desired output $z = E_1 x + P^* y$ and the estimated desired output $\widehat{z} = E_1 \widetilde{\xi} + P^* y$ can be written as

$$e_z = z - \widehat{z} = E_1 e.$$

Then, in view of (8.20) and (8.40), the dynamics of error is given by

$$\Sigma_{\text{p-CSS}}^{ue,\varepsilon} : \begin{cases} \sigma e = (A - K^\varepsilon \widetilde{C})e + (B - K^\varepsilon \widetilde{D})u \\ e_z = E_1 e. \end{cases} \tag{8.47}$$

Also, the transfer matrix $G^{ue,\varepsilon}$ from u to e_z can obviously be written as

$$G_{\text{p-CSS}}^{ue,\varepsilon} = E_1(\sigma I - A + K^\varepsilon \widetilde{C})^{-1}(B - K^\varepsilon \widetilde{D}). \tag{8.48}$$

Remark 8.36 *In view of (8.45) and (8.47), it is easy to see that both the filter equation and the error equation have the same poles, which are the eigenvalues of $A - K^\varepsilon \widetilde{C}$.*

As in the continuous-time case, we need to emphasize an important point. Theorem 8.15 developed earlier gives the conditions under which a family of H_2 AID filters exists among the general class of strictly proper filters of the form (8.2). In other words, Theorem 8.15 does not restrict itself to any fixed architecture for a filter such as the one given in (8.45). Nevertheless, as the following theorem shows, whenever the conditions of Theorem 8.15 are satisfied, a parameterized gain K^ε exists such that the family of filters given in (8.45) is a family of H_2 AID filters.

Theorem 8.37 *Consider a discrete-time system Σ given by (8.1). Let the conditions of Theorem 8.15 be satisfied. Then, a family of H_2 AID filters of the type $\Sigma^\varepsilon_{\text{p-CSS}}$ for Σ exists; i.e., a parameterized gain sequence K^ε exists such that the family of filters given in (8.45) is a family of H_2 AID filters for Σ. Moreover, K^ε is the same parameterized gain sequence, which renders $\widetilde{\Sigma}^\varepsilon_{\text{sp-CSS}}$ of (8.40) as a family of H_2 AID filters for $\widetilde{\Sigma}^*$ of (8.39).*

Proof : The proof follows along the lines of the proof of Theorem 8.27.   ■

As in the continuous-time case, constraints exist in choosing the filter poles that shape the error dynamics. The requirement of H_2 AID filtering imposes that certain poles of a family of H_2 AID filters tend as ε tends to zero to certain fixed locations termed again as finite asymptotic fixed modes. We have the following formal definition of finite asymptotic fixed modes when proper H_2 AID filters of CSS architecture are used.

Definition 8.38 (*Finite asymptotic fixed modes of proper H_2 AID filters of CSS architecture*) *Consider the given system (8.1) and the associated H_2 AID filtering problem 8.3. Assume that the solvability conditions as specified by Theorem 8.15 are satisfied. Then, a finite scalar $\lambda \in \mathbb{C}^\ominus$ is said to be an H_2 finite asymptotic fixed mode with algebraic multiplicity α if for every family of H_2 AID filters parameterized in ε and having the proper full-order CSS architecture, poles λ^ε_i, $i = 1, 2, \cdots, \alpha$, of the family of filters exist such that all λ^ε_i tend to λ as ε tends to zero. The set of all H_2 finite asymptotic fixed modes when proper filters of full-order CSS architecture are used is denoted by $\boldsymbol{\Omega}^{h2-\text{aid}}_{\text{p-CSS}}(A, B, C, D, E, F)$.*

We have the following theorem that characterizes the set $\boldsymbol{\Omega}^{h2-\text{aid}}_{\text{p-CSS}}$.

Theorem 8.39 *Consider a discrete-time system given by (8.1). Let the conditions of Theorem 8.15 be satisfied. Also, let E_1 be as discussed in (8.21) to (8.24), and*

P^* be such that $E_1 = E - P^*C$ and $F - P^*D = 0$. Let $\widetilde{C}$ and $\widetilde{D}$ be as defined in (8.38). Then, we have

$$\Omega_{\text{p-CSS}}^{h2-\text{aid}}(A, B, C, D, E, F) = \Omega_{\text{sp-CSS}}^{h2-\text{aid}}(A, B, \widetilde{C}, \widetilde{D}, E_1, 0).$$

Proof : The proof is transparent in view of the development of how a family of strictly proper filters for $\widetilde{\Sigma}^*$ can be converted to a family of proper filters (8.45) for the original system Σ with the same error dynamics. ∎

Remark 8.40 *In continuous-time systems, we remarked that*

$$\Omega_{\text{sp-CSS}}^{h2-\text{aid}}(A, B, \widetilde{C}, \widetilde{D}, E_1, 0) = \Omega_{\text{sp-CSS}}^{h2-\text{aid}}(A, B, C, D, E_1, 0).$$

The above property is not true in the discrete-time case because of possible additional fixed modes at the origin in $\Omega_{\text{sp-CSS}}^{h2-\text{aid}}(A, B, \widetilde{C}, \widetilde{D}, E_1)$. That is, we have

$$\Omega_{\text{sp-CSS}}^{h2-\text{aid}}(A, B, C, D, E_1, 0) \subset \Omega_{\text{sp-CSS}}^{h2-\text{aid}}(A, B, \widetilde{C}, \widetilde{D}, E_1, 0).$$

Remark 8.41 *In continuous-time systems, whenever strictly proper H_2 AID filters exist, we remarked that $\Omega_{\text{p-CSS}}^{h2-\text{aid}} = \Omega_{\text{sp-CSS}}^{h2-\text{aid}}$. Such a property is not true in the discrete-time case because of possible additional fixed modes at the origin as mentioned above.*

Obviously, we would like to explore next the development of an algorithm to design H_2 AID filters of full-order proper CSS architecture, which gives us the flexibility to place the finite asymptotic modes of the filters at desired locations, of course, within the constraints imposed by Theorem 8.39. Such an algorithm can be given by the following steps:

(*i*) Use the quintuple $(A, B, \widetilde{C}, \widetilde{D}, E_1)$ as the input to the H_2 ADD Algorithm (see Subsection 6.5.2), and obtain a parameterized gain F^ε.

(*ii*) Let $K^\varepsilon = (F^\varepsilon)'$. Partition K^ε in accordance with (8.43), and then define $\widetilde{K}^\varepsilon$ in accordance with (8.46).

The H_2 ADD algorithm mentioned in the above first step yields the set

$$\Omega_{\text{sp-CSS}}^{h2-\text{aid}}(A, B, \widetilde{C}, \widetilde{D}, E_1, 0) = \Omega_{\text{p-CSS}}^{h2-\text{aid}}(A, B, C, D, E, F).$$

The above equality follows from Theorem 8.39. Also, the said H_2 ADD algorithm has certain flexibility to place the finite asymptotic modes of the filter as desired but within the constraints imposed by Theorem 8.39.

8.6.3 A family of reduced-order proper H_2 AID filters—CSS architecture

The previous two subsections construct, respectively, strictly proper and proper full-order filters that solve a designated problem. By full-order filters, as usual, we mean filters having the same dynamic order as that of the given system. Our goal in this subsection is to develop reduced-order filters having their dynamic order lower than that of the given system.

The procedure of developing reduced-order filters here follows mostly along the same lines as in Subsection 7.5.3. However, there are some subtle but important differences warranting appropriate justifications, and as such, we need to redevelop them once more here. As before, our method of development transforms the construction of a family of reduced-order filters for a given system to that of a family of full-order filters for a certain reduced-order system.

We proceed now to construct an appropriate reduced-order system. To start with, let us rewrite the matrices C and D of (8.1) as

$$
C = \begin{pmatrix} 0 & C_{02} \\ I_{p-m_0} & 0 \end{pmatrix}, \quad D = \begin{pmatrix} D_0 \\ 0 \end{pmatrix},
$$

where again rank D = rank D_0 = m_0. This can always be done without any loss of generality by appropriate coordinate transformations. In view of the above partitioning of C and D, we can partition the given system Σ as

$$
\Sigma : \begin{cases} \begin{pmatrix} \sigma x_1 \\ \sigma x_2 \end{pmatrix} = \begin{pmatrix} A_{11} & A_{12} \\ A_{21} & A_{22} \end{pmatrix} \begin{pmatrix} x_1 \\ x_2 \end{pmatrix} + \begin{pmatrix} B_{11} \\ B_{22} \end{pmatrix} u \\[2ex] y = \begin{pmatrix} y_0 \\ y_1 \end{pmatrix} = \begin{pmatrix} 0 & C_{02} \\ I & 0 \end{pmatrix} \begin{pmatrix} x_1 \\ x_2 \end{pmatrix} + \begin{pmatrix} D_0 \\ 0 \end{pmatrix} u \\[2ex] z = Ex + Fu, \end{cases} \tag{8.49}
$$

where different variables have obvious meanings.

Next, we use a preliminary injection of output y into the desired output z. To this end, we define a matrix P^* such that $F - P^*D = 0$. Then, let

$$
z^* = z - P^*y = (E - P^*C)x + (F - P^*D)u = E_1 x, \tag{8.50}
$$

where $E_1 = E - P^*C$. We note that E_1 can be defined independent of the choice of any matrix P^*, which renders $F - P^*D = 0$. This can be done by decomposing the matrix E into two parts E_1 and E_2, as discussed in (8.21) to (8.24). Next, in view of (8.49) and (8.50), we can define a new system Σ^* as

$$\Sigma^* : \begin{cases} \begin{pmatrix} \sigma x_1 \\ \sigma x_2 \end{pmatrix} = \begin{pmatrix} A_{11} & A_{12} \\ A_{21} & A_{22} \end{pmatrix} \begin{pmatrix} x_1 \\ x_2 \end{pmatrix} + \begin{pmatrix} B_{11} \\ B_{22} \end{pmatrix} u \\[2ex] y = \begin{pmatrix} y_0 \\ y_1 \end{pmatrix} = \begin{pmatrix} 0 & C_{02} \\ I & 0 \end{pmatrix} \begin{pmatrix} x_1 \\ x_2 \end{pmatrix} + \begin{pmatrix} D_0 \\ 0 \end{pmatrix} u \\[2ex] z^* = E_1 x = E_{11} x_1 + E_{12} x_2, \end{cases} \tag{8.51}$$

with $E_1 = \begin{pmatrix} E_{11} & E_{12} \end{pmatrix}$.

We note that the y_1 is not contaminated by the input u, and hence, $x_1 = y_1$ is known exactly from the measurement y. Thus, all we need to do next is to estimate the state x_2. To proceed further, let us rewrite the state equation for x_1 in terms of the output y_1 and the state x_2 as

$$\sigma y_1 = A_{11} y_1 + A_{12} x_2 + B_{11} u. \tag{8.52}$$

The above equation can be rewritten as

$$\sigma y_1 - A_{11} y_1 = A_{12} x_2 + B_{11} u.$$

Treating σy_1 as known, we can define a new measurement variable y_r:

$$y_r = \begin{pmatrix} y_0 \\ \sigma y_1 - A_{11} y_1 \end{pmatrix}.$$

Although σy_1 is not directly available, as we did earlier in the case of proper full-order CSS filters, we can eliminate it from any filter that is constructed using y_r as a measured output. With this in mind, we form the following auxiliary system:

$$\Sigma_r^* : \begin{cases} \sigma x_r = A_r x_r + B_r u + A_{21} y_1 \\ y_r = C_r x_r + D_r u \\ z_r^* = E_r x_r, \end{cases} \tag{8.53}$$

where $x_r = x_2$ and

$$A_r = A_{22}, \quad B_r = B_{22}, \quad C_r = \begin{pmatrix} C_{02} \\ A_{12} \end{pmatrix}, \quad D_r = \begin{pmatrix} D_0 \\ B_{11} \end{pmatrix}, \quad E_r = E_{12}. \tag{8.54}$$

We note that the dynamic order n_r of the above Σ_r^* is less than the dynamic order n of the given system Σ by a number equal to the dimension of $x_1 = y_1$.

Before we proceed further, it is beneficial to investigate certain structural properties of Σ_r^* and relate them to those of the Σ. We have the following lemma, which intrinsically is the same as Lemma 7.55.

Lemma 8.42 *Consider the systems Σ_r^* and Σ, which are, respectively, characterized by $(A_r,\ B_r,\ C_r,\ D_r,\ E_r)$ and $(A,\ B,\ C,\ D,\ E,\ F)$. Consider two subsystems, one corresponding to Σ_r^* and characterized by the quadruple $(A_r,\ B_r,\ C_r,\ D_r)$, and the other corresponding to Σ and characterized by the quadruple (A, B, C, D). Then, the following results hold:*

(i) *The invariant zeros of the subsystem characterized by (A_r, B_r, C_r, D_r) are the same as those of the subsystem characterized by (A, B, C, D).*

(ii) *The matrix pair (C_r, A_r) is $\mathbb{C}^-$-detectable ($\mathbb{C}^\ominus$-detectable) if and only if the matrix pair (C, A) is $\mathbb{C}^-$-detectable ($\mathbb{C}^\ominus$-detectable).*

(iii) *The orders of infinite zeros of the subsystem characterized by (A_r, B_r, C_r, D_r) are reduced by one from those of the subsystem characterized by (A, B, C, D).*

(iv) *The subsystem characterized by (A_r, B_r, C_r, D_r) is left-invertible if and only if the subsystem characterized by (A, B, C, D) is left-invertible.*

(v) $\begin{pmatrix} 0 \\ I \end{pmatrix} \mathcal{V}^*(A_r, B_r, C_r, D_r) = \mathcal{V}^*(A, B, C, D).$

(vi) $\begin{pmatrix} 0 \\ I \end{pmatrix} \mathcal{S}^g(A_r, B_r, C_r, D_r) = \mathcal{S}^g(A, B, C, D) \cap C^{-1}\{\mathrm{im}(D)\}$, *where g denotes $\mathbb{C}^{-0}$ for continuous-time systems and $\mathbb{C}^\otimes$ for discrete-time systems.*

(vii) *Let g denotes $\mathbb{C}^{-0}$ for continuous-time systems and $\mathbb{C}^\otimes$ for discrete-time systems. Then, $\mathcal{S}^g(A_r, B_r, C_r, D_r) = \{0\}$ if and only if Σ is left-invertible, has invariant zeros only in $\mathbb{C}^g$, and has no infinite zeros of order higher than one.*

Proof : Most claims have been established in the proof of Lemma 7.55. For property (v), we note that obviously $(I \quad 0)\mathcal{V}^*(A, B, C, D) = \{0\}$. Hence, we can define $\overline{\mathcal{V}}^*$ such that

$$\mathcal{V}^*(A, B, C, D) = \left\{ \begin{pmatrix} 0 \\ x \end{pmatrix} \middle|\ x \in \overline{\mathcal{V}}^* \right\}.$$

It is then easily verified that $\overline{\mathcal{V}}^*$ satisfies

$$(A_{12} + B_{11} F_2)\overline{\mathcal{V}}^* = \{0\}, \quad (A_{22} + B_{22} F_2)\overline{\mathcal{V}}^* \subseteq \overline{\mathcal{V}}^*,$$
$$\text{and } (C_{02} + D_0 F_2)\overline{\mathcal{V}}^* = \{0\},$$

where $F = (F_1 \quad F_2)$ satisfies the conditions of Definition 3.28. It follows that $\overline{\mathcal{V}}^* \subseteq \mathcal{V}^*(A_r, B_r, C_r, D_r)$. The reverse implication follows along the same lines.

∎

Next, let us suppose we can construct a family of strictly proper H_2 AID filters for the auxiliary system Σ_r^* in order to arrive at an estimate $\hat{z}_r^*$ of z_r^*. (Here y_1 is a known input and, hence, will show up accordingly in the dynamics of any filter we construct for Σ_r^*.) Then, we can easily obtain an estimate $\hat{z}$ of z as

$$\hat{z} = E_{11}x_1 + \hat{z}_r^* + P^*y. \tag{8.55}$$

This motivates us to construct a family of full-order strictly proper filters for the auxiliary system Σ_r^* to arrive at an estimate $\hat{z}_r^*$ of z_r^*. A family of full-order filters constructed as such for Σ_r^* is indeed a family of reduced-order filters for Σ. However, in view of (8.55), such a family of reduced-order filters for Σ is a family of proper (rather than a family of strictly proper) filters for Σ. A fundamental issue that arises next is under what conditions one can construct a family of strictly proper H_2 AID filters for Σ_r^*. Expectedly, it turns out that one can construct a family of strictly proper H_2 AID filters for the reduced-order system Σ_r^* if and only if one can construct a family of proper H_2 AID filters for the given system Σ as stated in the following lemma.

Lemma 8.43 *Consider the continuous- or discrete-time systems Σ and Σ_r^*, respectively, as given in (8.1) and (8.53). Then, the following two statements are equivalent:*

(i) A family of strictly proper H_2 AID filters exists for the reduced-order system Σ_r^*.

(ii) A family of proper H_2 AID filters exists for the system Σ.

Proof : The proof follows from Lemma 8.42 in view of Theorems 8.5 and 8.9 for continuous time and Theorems 8.11 and 8.15 for discrete time.   ∎

To construct the required family of reduced-order proper H_2 AID filters for Σ, in the spirit of the above development, we first construct a family of strictly proper full-order filters of CSS architecture for the reduced-order system Σ_r^* as

$$\Sigma_{\text{sp-CSS}}^\varepsilon : \begin{cases} \sigma\tilde{\xi}_r = A_r\tilde{\xi}_r + K_r^\varepsilon(y_r - C_r\tilde{\xi}_r) + A_{21}y_1 \\ \hat{z}_r^* = E_r\tilde{\xi}_r, \end{cases} \tag{8.56}$$

where the matrix K_r^ε is a parameterized filter gain. As σy_1 is not available, we need to modify the above family of filters. To this end, let us partition

$$K_r^\varepsilon = (K_{r0}^\varepsilon \quad K_{r1}^\varepsilon) \tag{8.57}$$

so as to be compatible with the partitioning of y_r. Also, let

$$\xi_r = \tilde{\xi}_r - K^\varepsilon_{r1} y_1. \tag{8.58}$$

We can then easily rewrite the family of filters (8.56) as a family of filters for Σ as shown below.

A family of reduced-order proper filters for Σ: The family of reduced-order proper filters of CSS architecture for Σ is given by

$$\Sigma^\varepsilon_{\text{r-CSS}} : \begin{cases} \sigma \xi_r = (A_r - K^\varepsilon_r C_r)\xi_r + \tilde{K}^\varepsilon_r y \\ \hat{z} \quad = E_r \xi_r + \tilde{P}^\varepsilon_r y, \end{cases} \tag{8.59}$$

where

$$\tilde{K}^\varepsilon_r = \begin{pmatrix} K^\varepsilon_{r0} & A_r K^\varepsilon_{r1} + A_{21} - K^\varepsilon_{r1} A_{11} - K^\varepsilon_r C_r K^\varepsilon_{r1} \end{pmatrix} \tag{8.60}$$

and

$$\tilde{P}^\varepsilon_r = \begin{pmatrix} 0 & E_{11} + E_r K^\varepsilon_{r1} \end{pmatrix} + P^*. \tag{8.61}$$

A block diagram representation of the reduced-order proper filter along with the given plant is shown in Figure 8.5, which is structurally the same as the one in Figure 7.5.

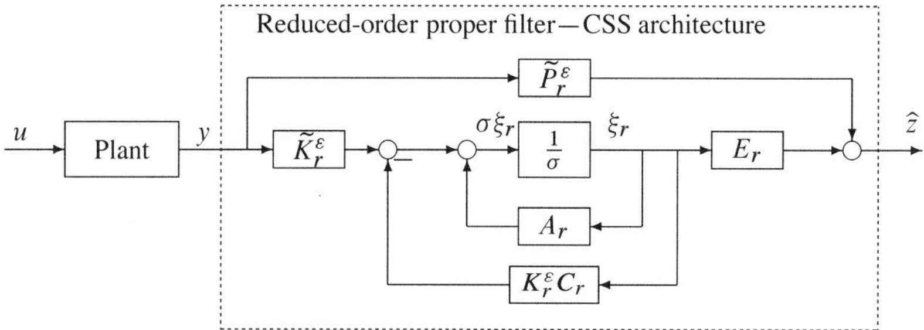

Figure 8.5: Block diagram of a reduced-order proper filter

Error dynamics: By defining the error $e_r = x_r - \tilde{\xi}_r$, the error e_z between the actual desired output $z = E_1 x + P^* y$ and the estimated desired output $\hat{z} = E_{11} x_1 + E_r \tilde{\xi}_r + P^* y$ can be written as

$$e_z = z - \hat{z} = E_r e_r.$$

Then, the dynamics of error is given by

$$\Sigma^{ue,\varepsilon}_{\text{r-CSS}} : \begin{cases} \sigma e_r = (A_r - K^\varepsilon_r C_r)e_r + (B_r - K^\varepsilon_r D_r)u \\ e_z \quad = E_r e_r. \end{cases} \tag{8.62}$$

Also, the transfer matrix $G^{ue,\varepsilon}$ from u to e_z can obviously be written as

$$G^{ue,\varepsilon}_{\text{r-CSS}} = E_r (\sigma I - A_r + K^\varepsilon_r C_r)^{-1}(B_r - K^\varepsilon_r D_r). \tag{8.63}$$

Remark 8.44 *It is easy to see that both the filter equation and the error equation have the same poles, which are the eigenvalues of $A_r - K_r^\varepsilon C_r$.*

Everything in the family of filters (8.59) is known except the parameterized gain K_r^ε. Next, we need to enquire first whether a parameterized gain K_r^ε exists such that the resulting family of proper filters (8.59) solves the H_2 AID filtering problem for the original system Σ given in (8.1) and then, whenever it exists, develop a method of obtaining such a parameterized gain K_r^ε. We have the following results.

Theorem 8.45 *Consider a continuous- or discrete-time system Σ given by (8.1). Let the conditions of Theorem 8.9 for continuous-time systems or the conditions of Theorem 8.15 for discrete-time systems be satisfied. Also, let E_1 be as discussed in (8.21) to (8.24) and P^* be such that $E_1 = E - P^*C$ and $F - P^*D = 0$. Then, a parameterized gain K_r^ε exists such that the family of filters given in (8.59) is a family of H_2 AID filters. Moreover, K_r^ε is the same parameterized gain sequence that renders $\Sigma_{\text{sp-CSS}}^\varepsilon$ of (8.56) as a family of H_2 AID filters for Σ_r^* of (8.53).*

Proof : By Lemma 8.43, we know that for the system Σ_r^*, the H_2 AID filtering problem is solvable. Moreover, any family of filter gains $\{K_r^\varepsilon\}_{\varepsilon>0}$ resulting in a family of strictly proper H_2 AID filters for Σ_r^* results in a family of reduced-order proper filters (8.59) that yield the same error dynamics for the original system Σ and, therefore, achieve H_2 AID for Σ. ∎

As in the case of full-order strictly proper and proper filters, constraints in choosing the filter poles that shape the error dynamics exist. The requirement of H_2 AID filtering imposes that certain poles of a family of H_2 AID filters tend as ε tends to zero to certain fixed locations termed as finite asymptotic fixed modes. We have the following formal definition of finite asymptotic fixed modes when reduced-order proper H_2 AID filters of CSS architecture are used.

Definition 8.46 *(Finite asymptotic fixed modes of reduced-order H_2 AID filters of CSS architecture) Consider the given system (8.1) and the associated H_2 AID filtering problem 8.3. Assume that the solvability conditions as specified by Theorem 8.9 (for continuous time) or Theorem 8.15 (for discrete time) are satisfied. Then, a finite scalar $\lambda \in \mathbb{C}^-$ for the continuous-time case or $\lambda \in \mathbb{C}^\ominus$ for the discrete-time case is said to be an $\boldsymbol{H_2}$ **finite asymptotic fixed mode** with algebraic multiplicity α if for every family of H_2 AID filters parameterized in ε and having the proper reduced-order CSS architecture, poles λ_i^ε, $i = 1, 2, \cdots, \alpha$, of the family of filters exist such that all λ_i^ε tend to λ as ε tends to zero. The set of all H_2 finite asymptotic fixed modes when proper filters of reduced-order CSS architecture are used is denoted by $\Omega_{\text{r-CSS}}^{h2-\text{aid}}(A, B, C, D, E, F)$.*

We have the following theorem that characterizes the set $\boldsymbol{\Omega}^{h2-\mathrm{aid}}_{\mathrm{r\text{-}CSS}}$.

Theorem 8.47 *Consider a continuous- or discrete-time system given by (8.1). Let the conditions of Theorem 8.15 be satisfied. Also, let E_1 be as discussed in equations (8.21) to (8.24), and P^* be such that $E_1 = E - P^*C$ and $F - P^*D = 0$. Let the quintuple $(A_r, B_r, C_r, D_r, E_r)$ be as defined in (8.54). Then, we have*

$$\boldsymbol{\Omega}^{h2-\mathrm{aid}}_{\mathrm{r\text{-}CSS}}(A, B, C, D, E, F) = \boldsymbol{\Omega}^{h2-\mathrm{aid}}_{\mathrm{sp\text{-}CSS}}(A_r, B_r, C_r, D_r, E_r, 0).$$

Proof : The proof is transparent in view of the above development of how a family of full-order strictly proper filters for Σ_r^* can be converted to a family of reduced-order proper filters (8.59) for the original system Σ with the same error dynamics.

∎

For continuous time, the full-order filter (8.34) and the reduced-order filter (8.59) are both proper filters. An interesting question arises as to the relationship between the sets of fixed modes associated with them. The following lemma shows that they are one and the same. Similarly, for discrete time, the full-order filter (8.45) and the reduced-order filter (8.59) are both proper filters. Once again, the sets of fixed modes associated with them are almost the same where the only possible difference between these two sets is some additional fixed modes at the origin.

Lemma 8.48 *Consider a continuous- or discrete-time system given by (8.1). Let the conditions of Theorem 8.9 for continuous time or Theorem 8.15 for discrete time be satisfied. Then the following hold:*

(i) *For continuous time, consider the set of fixed modes $\boldsymbol{\Omega}^{h2-\mathrm{aid}}_{\mathrm{p\text{-}CSS}}$ as in Definition 8.28 and the set of fixed modes $\boldsymbol{\Omega}^{h2-\mathrm{aid}}_{\mathrm{r\text{-}CSS}}$ as in Definition 8.46. Then we have*

$$\boldsymbol{\Omega}^{h2-\mathrm{aid}}_{\mathrm{p\text{-}CSS}}(A, B, C, D, E, F) = \boldsymbol{\Omega}^{h2-\mathrm{aid}}_{\mathrm{r\text{-}CSS}}(A, B, C, D, E, F).$$

(ii) *For discrete time, consider the set of fixed modes $\boldsymbol{\Omega}^{h2-\mathrm{aid}}_{\mathrm{p\text{-}CSS}}$ as in Definition 8.38 and the set of fixed modes $\boldsymbol{\Omega}^{h2-\mathrm{aid}}_{\mathrm{r\text{-}CSS}}$ as in Definition 8.46. Then we have,*

$$\boldsymbol{\Omega}^{h2-\mathrm{aid}}_{\mathrm{p\text{-}CSS}}(A, B, C, D, E, F) \supset \boldsymbol{\Omega}^{h2-\mathrm{aid}}_{\mathrm{r\text{-}CSS}}(A, B, C, D, E, F),$$

where the only possible difference between these two sets is some additional fixed modes at the origin

Proof : We first note that from Theorem 8.47, we obtain in both discrete and continuous time

$$\boldsymbol{\Omega}_{\text{r-CSS}}^{h2-\text{aid}}(A, B, C, D, E, F) = \boldsymbol{\Omega}_{\text{sp-CSS}}^{h2-\text{aid}}(A_r, B_r, C_r, D_r, E_r, 0).$$

Next, using Theorem 8.21, we obtain

$$\boldsymbol{\Omega}_{\text{sp-CSS}}^{h2-\text{aid}}(A_r, B_r, C_r, D_r, E_r, 0) = \boldsymbol{\Omega}_s^2(A_r', C_r', B_r', D_r', E_r').$$

Theorem 6.19 (continuous time) or Theorem 6.24 (discrete time) can then be used to establish that

$$\boldsymbol{\Omega}_s^2(A_r', C_r', B_r', D_r', E_r') = \boldsymbol{\Omega}_s^2(A', C', B', D', E_1').$$

We have

$$\boldsymbol{\Omega}_s^2(A', C', B', D', E_1') = \boldsymbol{\Omega}_{\text{sp-CSS}}^{h2-\text{aid}}(A, B, C, D, E_1, 0),$$

and finally, using Remark 8.30, we obtain for continuous-time systems,

$$\boldsymbol{\Omega}_{\text{sp-CSS}}^{h2-\text{aid}}(A, B, C, D, E_1, 0) = \boldsymbol{\Omega}_{\text{p-CSS}}^{h2-\text{aid}}(A, B, C, D, E, F)$$

given that $E_1 = E - P^*C$ while $F - P^*D = 0$. For discrete-time systems, we have using Remark 8.41,

$$\boldsymbol{\Omega}_{\text{sp-CSS}}^{h2-\text{aid}}(A, B, C, D, E_1, 0) \subset \boldsymbol{\Omega}_{\text{p-CSS}}^{h2-\text{aid}}(A, B, C, D, E, F),$$

where the only possible difference between these two sets is some fixed modes at the origin. ∎

Obviously, we would like to explore next the development of an algorithm to design H_2 AID filters of reduced-order proper CSS architecture, which gives us the flexibility to place the finite asymptotic modes of the filters at desired locations, of course, within the constraints imposed by Theorem 8.47. Such an algorithm can be given by the following steps:

(*i*) Use the quintuple $(A_r, B_r, C_r, D_r, E_r)$ as the input to the H_2 ADD Algorithm (see Subsection 6.5.1 for continuous-time systems or Subsection 6.5.2 for discrete-time systems), and obtain a parameterized gain F^ε.

(*ii*) Let $K_r^\varepsilon = (F^\varepsilon)'$. Partition K_r^ε in accordance with (8.57), and then define $\widetilde{K}_r^\varepsilon$ and $\widetilde{P}_r^\varepsilon$, respectively, in accordance with (8.60) and (8.61).

The H_2 ADD algorithm mentioned in the above first step yields the set

$$\boldsymbol{\Omega}_{\text{sp-CSS}}^{h2-\text{aid}}(A_r, B_r, C_r, D_r, E_r, 0) = \boldsymbol{\Omega}_{\text{r-CSS}}^{h2-\text{aid}}(A, B, C, D, E, F).$$

The above equality follows from Theorem 8.47. Also, the said H_2 ADD algorithm has certain flexibility to place the finite asymptotic modes of the filter as desired but within the constraints imposed by Theorem 8.47.

9

Almost input-decoupled filtering without statistical assumptions on input

9.1   Introduction

Chapter 8 considers almost input-decoupled (AID) filtering problems under white noise input with a known power spectral density (PSD). The objective of Chapter 8 is to check whether it is possible to make the impact of the unknown input on the asymptotic error arbitrarily small. Under white noise input with a known PSD, whenever the performance measure is the RMS norm of the error signal, such an objective translates to an objective of reducing the H_2 norm of the transfer function from the unknown input to the error signal.

In this chapter, unlike in Chapter 8, we assume no information on the input except that it has a finite RMS norm. Under such an unknown input, we seek to render the ratio of RMS value of the error to the RMS value of the input arbitrarily small. The filters that solve such an AID filtering problem are termed once again as AID filters. Our study of AID filtering, in this chapter, as before involves first developing the necessary and sufficient conditions under which such AID filters exist and then developing methods of constructing them. As in Chapter 8, we use both full- and reduced-order filters of CSS architecture for filter design.

9.2   Preliminaries

Let us reconsider the plant or system model given in (7.1) and rewritten here as

$$\Sigma : \begin{cases} \sigma x = Ax + Bu \\ y \;\; = Cx + Du \\ z \;\; = Ex + Fu, \end{cases} \tag{9.1}$$

where, as before, $u \in \mathbb{R}^m$ is the input, $x \in \mathbb{R}^n$ is the state, $y \in \mathbb{R}^p$ is the measured output, and $z \in \mathbb{R}^q$ is the desired output signal to be estimated.

As before, our interest lies in estimating the desired output signal z while using only the measured output y but not the input u. As usual, let $\hat{z}$ be the estimate of z as given by a filter, and let e_z be the estimation error, $e_z = z - \hat{z}$ as depicted in Figure 9.1.

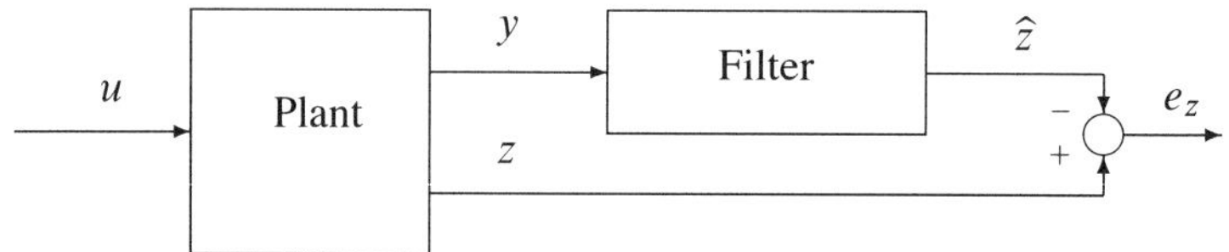

Figure 9.1: General block diagram

As before, it is necessary to use the following assumption throughout this chapter as well.

Assumption 9.1 *The matrix pair (C, A) is $\mathbb{C}^-$-detectable for continuous-time systems and $\mathbb{C}^\ominus$-detectable for discrete-time systems.*

As in Chapter 8, we consider a parameterized family of general proper filters of the form:

$$\Sigma_f^\varepsilon : \begin{cases} \sigma\xi = L^\varepsilon\xi + M^\varepsilon y \\ \hat{z} = N^\varepsilon\xi + P^\varepsilon y, \end{cases} \tag{9.2}$$

where ε is a positive parameter belonging to $(0, \varepsilon^*]$ for some positive ε^*. Whenever $P = 0$, the above filter is said to be a strictly proper filter. When the above filter is used as shown in Figure 9.1, the dynamic equations of the error e_z is described by

$$\Sigma_{ue}^\varepsilon : \begin{cases} \sigma x = Ax + Bu, \\ \sigma\xi = M^\varepsilon Cx + L^\varepsilon\xi + M^\varepsilon Du, \\ e_z = (E - P^\varepsilon C)x - N^\varepsilon\xi + (F - P^\varepsilon D)u. \end{cases} \tag{9.3}$$

Hence, the transfer matrix from u to e_z can be computed as

$$G_{ue}^\varepsilon = \begin{pmatrix} E - P^\varepsilon C & -N^\varepsilon \end{pmatrix} \begin{pmatrix} \sigma I - A & 0 \\ -M^\varepsilon C & \sigma I - L^\varepsilon \end{pmatrix}^{-1} \begin{pmatrix} B \\ M^\varepsilon D \end{pmatrix}$$
$$+ (F - P^\varepsilon D). \tag{9.4}$$

9.3   Statement of AID filtering problem and its solvability conditions

Unlike in Chapter 8, we do not assume here any information regarding the input, statistical or otherwise. As in the previous chapters, we consider only the class of unbiased filters (see Definition 7.2 for what "unbiasedness" means). The performance goal in this chapter is to render the RMS value of the error relative to the RMS value of the input arbitrarily small, i.e., to render the worst-case ratio

$$\frac{\|e_z\|_{\text{RMS}}}{\|u\|_{\text{RMS}}}$$

arbitrarily small. We have the following formal definition of the AID filtering problem under no statistical assumptions on input.

Problem 9.2 (AID filtering problem under no statistical assumptions on input) Consider the system Σ given in (9.1). Let us suppose we do not have any information on the input $u(t)$ except that it has a finite RMS value. Then, the AID filtering problem under no statistical assumptions on input is defined as follows: Find, whenever it exists, a family of linear stable strictly proper (or proper) filters of the type Σ_f^ε given in (9.2) and parameterized in positive $\varepsilon \in (0, \varepsilon^*]$ such that

(Unbiasedness) for any $\varepsilon \in (0, \varepsilon^*]$ the filter Σ_f^ε is unbiased; that is, the estimation error e_z, in the absence of the input u, decays asymptotically to zero for all possible finite initial values of the system (9.1) and the filter (9.2), and

(Performance) the ratio
$$\frac{\|e_z\|_{\text{RMS}}}{\|u\|_{\text{RMS}}} \to 0 \ \text{ as } \ \varepsilon \to 0$$

for any possible input u with bounded and nonzero RMS norm.

Once again, as in Chapter 8, the above problem statement can be given a deterministic interpretation. Let $G_{ue}^\varepsilon(\sigma)$ denote the transfer matrix from the input u to the error e_z. Then, by the definitions and the discussion given in Section 2.6, we know that the H_∞ norm of the transfer function of a system is indeed the ratio of the RMS norm of the output to the RMS norm of the input; i.e.,

$$\|G_{ue}^\varepsilon\|_\infty = \frac{\|e_z\|_{\text{RMS}}}{\|u\|_{\text{RMS}}}, \tag{9.5}$$

where $\|G_{ue}^\varepsilon\|_\infty$ is the H_∞ norm of G_{ue}^ε.

In view of (9.5), the AID filtering problem as described in Problem 9.2 is solvable if and only if a family of linear stable strictly proper (or proper) unbiased filters parameterized in positive ε exists such that

$$\|G_{ue}^\varepsilon\|_\infty \to 0 \ \text{ as } \ \varepsilon \to 0.$$

This implies that Problem 9.2 can be interpreted in a deterministic setting as an H_∞ AID filtering problem as described below.

Problem 9.3 (H_∞ **AID filtering problem**) Consider the system Σ given in (9.1). Then, find, whenever it exists, a family of linear stable strictly proper (or proper) filters of the type Σ_f^ε given in (8.2) and parameterized in positive $\varepsilon \in (0, \varepsilon^*]$ such that

(**Unbiasedness**) for any $\varepsilon \in (0, \varepsilon^*]$, the filter Σ_f^ε is unbiased; that is, the estimation error e_z, in the absence of the input u, decays asymptotically to zero for all possible finite initial values of the system (9.1) and the filter (9.2), and

(**Performance**)

$$\|G_{ue}^\varepsilon\|_\infty \to 0 \ \ \text{as} \ \ \varepsilon \to 0.$$

A family of filters that solves the H_∞ AID filtering problem is said to be a family of H_∞ AID filters.

9.4   Existence conditions for H_∞ AID filters—continuous-time case

We now proceed to develop the solvability conditions for the H_∞ AID filtering problem. Let us consider a general proper filter of the form (9.2). Then, the transfer matrix from u to e_z can be computed as in (9.4). To render $\|G_{ue}^\varepsilon\|_\infty$ arbitrarily small, we necessarily need the direct feedthrough matrix of G_{ue}^ε to be made arbitrarily small. This implies that a family of matrices P^ε must exist such that $F - P^\varepsilon D \to 0$ as $\varepsilon \to 0$. It is easily seen that this implies that a matrix P^* must exist such that $F - P^* D = 0$. This in turn implies that the conditions of Lemmas 7.5 and 8.4 apply to the H_∞ AID filtering problem as well. We have the following lemma formalizing this.

Lemma 9.4 *Consider a continuous -time system as in* (9.1). *The following results hold:*

> (*i*) *The H_∞ AID filtering problem is solvable via a family of proper filters only if*

$$\ker D \subseteq \ker F. \tag{9.6}$$

> (*ii*) *The H_∞ AID filtering problem is solvable via a family of strictly proper filters only if $F = 0$.*

Proof : The results are obvious.   ∎

We are now ready to give the solvability conditions for the H_∞ AID filtering problem. We first consider a family of strictly proper filters.

Theorem 9.5 *Consider a continuous-time system as in (9.1). Let Assumption 9.1 be satisfied. Then, the H_∞ AID filtering problem is solvable via a family of strictly proper filters if and only if $F = 0$ and the following conditions are satisfied:*

(i) $\mathcal{S}^{-0}(A, B, C, D) \cap \mathcal{V}^*(A, B, C, D) \subseteq \ker E.$

(ii) Consider the invariant zeros of the system characterized by the quadruple (A, B, C, D). For any such an invariant zero s_0 on the imaginary axis and for all $\varepsilon > 0$, a matrix K exists such that $s_0 I - A + KC$ is invertible, and

$$\| E(s_0 I - A + KC)^{-1}(B - KD) \| < \varepsilon.$$

Proof : We follow the same arguments as in the proof of Theorem 8.6. The condition that F needs to be zero follows directly from Lemma 9.4. By Theorem 7.69, it is clear that the H_∞ AID filtering problem is solvable via a family of strictly proper filters if and only if the H_∞ almost disturbance decoupling (ADD) problem is solvable via a family of dynamic state feedback controllers for the dual system:

$$\Sigma_d : \begin{cases} \dot{x}_d = A'x_d + C'u_d + E'w_d \\ z_d = B'x_d + D'u_d. \end{cases} \tag{9.7}$$

By Theorem 6.6, this ADD problem is solvable if and only if

$$\operatorname{im} E' \subseteq \mathcal{S}^*(A', C', B', D') + \mathcal{V}^{-0}(A', C', B', D'),$$

and for any $\varepsilon > 0$ and any invariant zero s_0 of (A, B, C, D) on the imaginary axis, a matrix K exists such that $s_0 I - A' + C'K'$ is invertible, and

$$\| (B' - D'F)(s_0 I - A' + C'F)^{-1} E' \| < \varepsilon.$$

Taking the orthogonal complement on both sides in the above yields that these conditions are equivalent to the characterization in the above theorem. ∎

Remark 9.6 *Similar to Remark 6.7 and using techniques from [12,85], a different condition for solvability of the H_∞ AID filter problem via a family of strictly proper filters has been given. The H_∞ AID filtering problem via a family of strictly proper filters is solvable if and only if $F = 0$ and*

$$\left(\mathcal{S}^{-0}(A, B, C, D) \cap \mathcal{V}^*(A, B, C, D)\right) + \sum_{\lambda \in \mathbb{C}^0} \mathcal{V}_\lambda(A, B, C, D) \subseteq \ker E.$$

Remark 9.7 *Remark 6.8 can be used to rewrite the last condition in (ii), and we can state equivalently that for any invariant zero s_0 of (A, B, C, D) on the imaginary axis, we must have that*

$$\ker\left[C(s_0 I - A)^{-1} B + D\right] \subseteq \ker E(s_0 I - A)^{-1} B.$$

However, this latter equivalent condition can only be used when $s_0 I - A$ is invertible. Otherwise, we first need to find a matrix N such that $A - BN$ has no eigenvalues on the imaginary axis and then the condition becomes that we must have for any invariant zero s_0 of (A, B, C, D) on the imaginary axis

$$\ker\left[(C - DN)(s_0 I - A + BN)^{-1} B + D\right] \subseteq \ker E(s_0 I - A + BN)^{-1} B.$$

We consider next the H_∞ AID filtering problem by using a family of proper filters. We first consider the case when $F = 0$. As in the H_2 AID filtering problem, whenever $F \neq 0$, we will soon redefine the desired output to be estimated such that in the new setting the corresponding F equals zero.

Theorem 9.8 *Consider a continuous-time system as in (9.1) with $F = 0$. Let Assumption 9.1 be satisfied. Then, the H_∞ AID filtering problem is solvable via a family of proper filters if and only if the following conditions are satisfied:*

(i) $\mathcal{S}^{-0}(A, B, C, D) \cap \mathcal{V}^*(A, B, C, D) \subseteq \ker E$.

(ii) *Consider the invariant zeros of the system characterized by the quadruple (A, B, C, D). For any such an invariant zero s_0 on the imaginary axis and for all $\varepsilon > 0$, a matrix K exists such that $s_0 I - A + KC$ is invertible, and*

$$\|E(s_0 I - A + KC)^{-1}(B - KD)\| < \varepsilon.$$

Proof : We follow the same arguments as in the proof of Theorem 8.6. By Theorem 7.71, the H_∞ AID filtering problem is solvable for the system (9.1) with $F = 0$ via a family of proper filters if and only if the H_∞ ADD problem is solvable via a family of dynamic full-information feedback laws for the dual system Σ_d given in (9.7). As outlined in Section 6.4, H_∞ ADD problem is solvable via a family of dynamic full-information feedback laws if and only if H_∞ ADD problem is solvable via a family of static full-information feedback laws. Finally, by Theorem 6.10, H_∞ ADD problem is solvable via a family of static full-information feedback laws if and only if

$$\operatorname{im} E' \subseteq \mathcal{S}^*(A', C', B', D') + \mathcal{V}^{-0}(A', C', B', D') + C' \ker D',$$

and for any $\varepsilon > 0$ and any invariant zero s_0 of (A, B, C, D) on the imaginary axis, a matrix K exists such that $s_0 I - A' + C'K'$ is invertible, and

$$\|(B' - D'K')(s_0 I - A' + C'K')^{-1} E'\| < \varepsilon.$$

Taking the orthogonal complement on both sides in the above yields that these conditions are equivalent to the characterization in the above theorem.   ∎

Remark 9.9 *For continuous-time systems when $F = 0$, we note that, as in the case of H_2 AID filtering, the solvability conditions for H_∞ AID filtering are one and the same whether strictly proper or proper filters are used. On the other hand, as we shall see, for discrete-time systems, the solvability conditions when proper filters are used are much weaker than those when strictly proper filters are used. This is because, in continuous-time systems, one can use fast filters that use high gain. However, in the case of discrete-time systems, it is not so.*

Remark 9.10 *It is interesting to look at the case when $E = I$, i.e., when the entire state is to be estimated. For this case, as $\ker E = \{0\}$, the condition (i) of Theorems 9.5 and 9.8 can be rewritten. The rewritten condition is that the system characterized by the quadruple (A, B, C, D) is left-invertible, and moreover, it has no invariant zeros with positive real parts.*

We consider next the case of $F \neq 0$. As in H_2 AID filtering, we study here the H_∞ AID filtering problem for a given system for which the matrix F is nonzero by transforming it to an equivalent H_∞ AID filtering problem for an auxiliary system where the corresponding F can be taken as zero. The needed auxiliary system is constructed in exactly the same way as in H_2 AID filtering by redefining the desired output z. To do so, we define a new desired output z^* as

$$z^* = z - P^* y = (E - P^* C)x + (F - P^* D)u. \tag{9.8}$$

Next, we consider the following auxiliary system:

$$\Sigma^* : \begin{cases} \dot{x} &= Ax + Bu \\ y &= Cx + Du \\ z^* &= (E - P^* C)x + (F - P^* D)u. \end{cases} \tag{9.9}$$

Also, for the above auxiliary system, let us consider a family of filters of the form

$$\begin{aligned} \dot{\xi} &= L^\varepsilon \xi + M^\varepsilon y, \\ \hat{z}^* &= N^\varepsilon \xi + \bar{P}^\varepsilon y, \end{aligned} \tag{9.10}$$

with ε as a positive parameter. As in Chapter 8, we observe that the transfer matrix from u to $e_{z^*} = z^* - \hat{z}^*$ when the new filter (9.10) is applied to the auxiliary system Σ^* is the same as the transfer matrix from u to $e_z = z - \hat{z}$ when the original filter (9.2) is applied to the original Σ as long as $P^\varepsilon = \bar{P}^\varepsilon + P^*$.

The above analysis suggests a method of designing first P^* such that $F - P^*D = 0$ and then designing a family of filters parameterized in ε for the system Σ^* such that the transfer matrix from u to $e_{z^*} = z^* - \hat{z}^*$ has the desired value (namely, its H_∞ norm tends to zero as $\varepsilon \to 0$). The designed family of filters for Σ^* can then easily be translated for the original system Σ by allowing $P^\varepsilon = \bar{P}^\varepsilon + P^*$.

We have the following results.

Theorem 9.11 *Consider a continuous-time system as in* (9.1). *Let Assumption 9.1 be satisfied. Then, the H_∞ AID filtering problem is solvable via a family of proper filters if and only if the following conditions are satisfied:*

(i)

$$\left(\left[\mathcal{S}^{-0}(A, B, C, D) \cap \mathcal{V}^*(A, B, C, D)\right] \oplus \mathbb{R}^m\right) \cap \ker\begin{pmatrix} C & D \end{pmatrix}$$
$$\subseteq \ker\begin{pmatrix} E & F \end{pmatrix}.$$

(ii) Consider the invariant zeros of the system characterized by the quadruple (A, B, C, D). For any such an invariant zero s_0 on the imaginary axis and for all $\varepsilon > 0$, matrices K and L exist such that $s_0 I - A + KC$ is invertible, and

$$\|(E - LC)(s_0 I - A + KC)^{-1}(B - KD) + (F - LD)\| < \varepsilon.$$

Proof : We use the same arguments as in the proof of Theorem 8.9. By Lemma 9.4 if the H_∞ AID filtering problem is solvable via a family of proper filters, then a matrix P^* exists such that $F - P^*D = 0$. Clearly, H_∞ AID filtering problem is solvable via a family of proper filters for the continuous-time system (8.1) if and only if H_∞ AID filtering problem is solvable via a family of proper filters for the system:

$$\bar{\Sigma} : \begin{cases} \dot{x} = Ax + Bu \\ y = Cx + Du \\ z = z - P^*y = (E - P^*C)x. \end{cases}$$

Using Theorem 9.8, we then find that for this latter system the H_∞ AID filtering problem is solvable via a family of proper filters if and only if

$$\mathcal{S}^{-0}(A, B, C, D) \cap \mathcal{V}^*(A, B, C, D) \subseteq \ker(E - P^*C),$$

and for all invariant zeros on the imaginary axis of the system characterized by the quadruple (A, B, C, D) and for all $\varepsilon > 0$, a matrix K exists such that $s_0 I - A + KC$ is invertible, and

$$\|(E - P^*C)(s_0 I - A + KC)^{-1}(B - KD)\| < \varepsilon.$$

These conditions combined with $F - P^*D = 0$ are easily seen to be equivalent to the conditions in the above theorem. ∎

Remark 9.12 *In line with Remark 6.8, the last condition in (ii) is equivalent to the requirement that for any invariant zero s_0 of (A, B, C, D) on the imaginary axis, we must have that*

$$\ker\left[C(s_0 I - A)^{-1}B + D\right] \subseteq \ker E(s_0 I - A)^{-1}B + F.$$

However, as before, this condition can only be used when $s_0 I - A$ is invertible. Otherwise, we first need to find a matrix N such that $A - BN$ has no eigenvalues on the imaginary axis and then the condition becomes that we must have for any invariant zero s_0 of (A, B, C, D) on the imaginary axis

$$\ker\left[(C - DN)(s_0 I - A + BN)^{-1}B + D\right]$$
$$\subseteq \ker(E - FN)(s_0 I - A + BN)^{-1}B + F.$$

Remark 9.13 *Readers can easily verify that the solvability conditions for H_∞ AID filtering are weaker than the solvability conditions for EID filtering (see Theorems 7.6, 7.7, and 7.9), which of course is as it should be. Such a verification can be done by noting that, in general, $\mathcal{S}^{-0}(A, B, C, D) \subseteq \mathcal{S}^{-}(A, B, C, D)$ and $\mathcal{V}^*(A, B, C, D) \subseteq C^{-1}\{\mathrm{im}\, D\}$.*

9.5 Existence conditions for H_∞ AID filters—discrete-time case

In this section, we develop the conditions for the existence of a family of H_∞ AID filters for a discrete-time system. We observe again that, to render $\|G_{ue}^\varepsilon\|_\infty$ arbitrarily small, G_{ue}^ε must necessarily tend to be strictly proper as $\varepsilon \to 0$. Thus, the conditions of Lemma 9.4 must be valid for discrete-time systems as well. That is, the H_∞ AID filtering problem is solvable via a family of proper filters only if $\ker D \subseteq \ker F$, and via a a family of strictly proper filters only if $F = 0$.

We have the following results when strictly proper filters are used.

Theorem 9.14 *Consider a discrete-time system as in (9.1). Let Assumption 9.1 be satisfied. Then, the H_∞ AID filtering problem is solvable via a family of strictly proper filters if and only if $F = 0$, and*

(i) $\mathcal{S}^{\otimes}(A, B, C, D) \subseteq \ker E.$

(ii) *Consider the invariant zeros of the system characterized by the quadruple (A, B, C, D). For any such an invariant zero z_0 on the unit circle and for all $\varepsilon > 0$, a matrix K exists such that $z_0 I - A + KC$ is invertible, and*

$$\|E(z_0 I - A + KC)^{-1}(B - KD)\| < \varepsilon.$$

Proof : This theorem can be proven similar to the continuous-time result in Theorem 9.5, except that this time we need to refer to the discrete-time H_∞ ADD as given in Theorem 6.14. ∎

We now consider the H_∞ AID filtering problem by using proper filters. As in the continuous-time case, it is easier to give the results first for the case when $F = 0$.

Theorem 9.15 *Consider a discrete-time system as in (9.1) with $F = 0$. Let Assumption 9.1 be satisfied. Then, the H_∞ AID filtering problem is solvable via a family of proper filters if and only if*

(i) $\mathcal{S}^{\otimes}(A, B, C, D) \cap C^{-1}\{\operatorname{im} D\} \subseteq \ker E.$

(ii) *Consider the invariant zeros of the system characterized by the quadruple (A, B, C, D). For any such an invariant zero z_0 on the unit circle and for all $\varepsilon > 0$, a matrix K exists such that $z_0 I - A + KC$ is invertible, and*

$$\|E(z_0 I - A + KC)^{-1}(B - KD)\| < \varepsilon.$$

Proof : This theorem can be proven similar to the continuous-time result in Theorem 9.8, except that this time we need to refer to the discrete-time H_∞ ADD by full-information feedback laws as given in Theorem 6.30. ∎

Remark 9.16 *We noted that for the case of $F = 0$ and for continuous-time systems, the solvability conditions are one and the same whether strictly proper or proper filters are used. On the other hand, for discrete-time systems, the solvability conditions when proper filters are used are much weaker than those when strictly proper filters are used.*

Remark 9.17 *Again, it is interesting to consider the case when $E = I$, i.e., when the entire state is to be estimated. As in continuous-time systems, because for this case, $\ker E = \{0\}$, the condition (i) of Theorems 9.14 and 9.15 can be rewritten. The rewritten condition is that the system characterized by the quadruple (A, B, C, D) is left-invertible, has no invariant zeros outside the unit circle, and the order of infinite zeros must be zero (i.e., the "relative degree" must be zero).*

Next, we consider the case when $F \neq 0$. As in the previous section, we study the H_∞ AID filtering problem for a given system for which the matrix F is nonzero by transforming it to an equivalent H_∞ AID filtering problem for an auxiliary system where the corresponding F is zero. The needed auxiliary system is exactly the same as (9.9), where P^* is such that $F - P^*D = 0$. In fact, we come to the same conclusion as before, namely, that the transfer matrix from u to $e_{z^*} = z^* - \hat{z}^*$ when the new filter (9.10) is applied to the auxiliary system Σ^* is the same as the transfer matrix from u to $e_z = z - \hat{z}$ when the original filter (9.2) is applied to the original Σ as long as $P^\varepsilon = \bar{P}^\varepsilon + P^*$. Also, as before, this suggests a method of designing first P^* such that $F - P^*D = 0$ and then designing a family of filters parameterized in ε for the system Σ^* such that the transfer matrix from u to $e_{z^*} = z^* - \hat{z}^*$ has the desired property (namely, its H_∞ norm tends to zero as $\varepsilon \to 0$). The designed family of filters for Σ^* can then easily be translated for the original system Σ by setting $P^\varepsilon = \bar{P}^\varepsilon + P^*$.

We have the following results.

Theorem 9.18 *Consider a discrete-time system as in* (9.1). *Let Assumption 9.1 be satisfied. Then, the H_∞ AID filtering problem is solvable via a family of proper filters if and only if*

(*i*)

$$\left(\mathcal{S}^\otimes(A, B, C, D) \oplus \mathbb{R}^m \right) \cap \ker \begin{pmatrix} C & D \end{pmatrix} \subseteq \ker \begin{pmatrix} E & F \end{pmatrix}.$$

(*ii*) *Consider the invariant zeros of the system characterized by the quadruple* (A, B, C, D). *For any such an invariant zero z_0 on the unit circle and for all $\varepsilon > 0$, matrices K and L exist such that $z_0 I - A + KC$ is invertible, and*

$$\| (E - LC)(z_0 I - A + KC)^{-1}(B - KD) + (F - LD) \| < \varepsilon.$$

Proof : This theorem can be proven similar to the continuous-time result in Theorem 9.11, except that this time we need to refer to the discrete-time H_∞ AID filtering problem as given in Theorem 9.15. ∎

Remark 9.19 *It is useful to note that, as for continuous-time systems, the solvability conditions for H_∞ AID filtering are weaker than the solvability conditions for EID filtering (see Theorems 7.6, 7.7, and 7.9), which of course is as it should be. Such a verification can be done by noting that, in general, $\mathcal{S}^\otimes(A, B, C, D) \subseteq \mathcal{S}^\ominus(A, B, C, D)$ and $\mathcal{V}^*(A, B, C, D) \subseteq C^{-1}\{\text{im } D\}$.*

Remark 9.20 *Once again, it is interesting to consider the case when $E = I$, i.e., when the entire state is to be estimated. As for this case $\ker E = \{0\}$, the condition (i) of Theorem 9.18 can be rewritten. The rewritten condition is that the system characterized by the quadruple (A, B, C, D) is left-invertible, has no invariant zeros outside the unit circle, and the order of infinite zeros must be at most one (i.e., the "relative degree" must be at most one).*

9.6   Design of a family of H_∞ AID filters of CSS architecture

In this section, we present a method of designing a parameterized family of H_∞ AID filters with some flexibility to assign the poles as desired while honoring certain conditions imposed by the requirement of H_∞ AID filtering. In this regard, as in Chapter 8, we typically start by first assuming a fixed architecture to the filter. The architecture we use for the filters is the CSS architecture that was developed and used in earlier chapters. In most of the following development, the procedure we follow to develop the CSS architecture of appropriate filters is similar to what has been given in Section 8.6 of Chapter 8. However, some subtle changes are important enough to warrant their redevelopment here.

As the conditions for the existence of a family of H_∞ AID filters as developed earlier in Sections 9.4 and 9.5 do not assume any particular architecture for filters, once again one of the fundamental questions that arises naturally is as follows: Does a family of H_∞ AID filters of CSS architecture exist under the same conditions as developed earlier? In this section, we not only answer this question affirmatively, but we also develop systematic methods of designing them. As in the case of EID filters and H_2 AID filters, the requirement of H_∞ AID imposes certain constraints on the behavior of poles. As such, we need to examine the structure of H_∞ AID filters regarding their poles. That is, clearly, H_∞ AID filtering by definition requires a family of filters parameterized in some parameter ε, and we need to examine the asymptotic behavior of H_∞ AID filters as the parameter ε tends to a critical value, say zero. It turns out that the poles of H_∞ AID filters have the following asymptotic behavior:

- For both continuous- and discrete-time systems, as $\varepsilon \to 0$, in general some of the poles of a family of H_∞ AID filters go asymptotically to some fixed finite locations, whereas some others are free to be assigned.

- For continuous-time systems, as $\varepsilon \to 0$, in general some of the poles of a family of H_∞ AID filters must go asymptotically to infinity. Obviously, for discrete-time systems, such a phenomena does not occur in view of the stability requirement.

As in Chapter 8, our primary interest in this section is to study and compute finite asymptotic fixed modes while prescribing a method of designing filters of

CSS architecture. Interesting questions can also be posed regarding the infinite asymptotic modes, such as "how many minimum number of infinite asymptotic modes exist?". The answers to such questions remain to be addressed and are beyond the scope of this book.

As in Chapter 8, our presentation here is organized into three subsections. The first and second subsections consider, respectively, strictly proper and proper filters, whereas the third subsection considers reduced-order proper filters.

9.6.1   A family of full-order strictly proper H_∞ AID filters—CSS architecture

We pursue here the design of a family of strictly proper H_∞ AID filters while simultaneously using the available flexibility to place its poles. The architecture we use is the full-order CSS architecture. As explained, strictly proper filters can be used only when $F = 0$. As such, throughout this subsection we assume that $F = 0$. We developed earlier, in Chapters 7 and 8, the full-order strictly proper filter of CSS architecture. It is reproduced below.

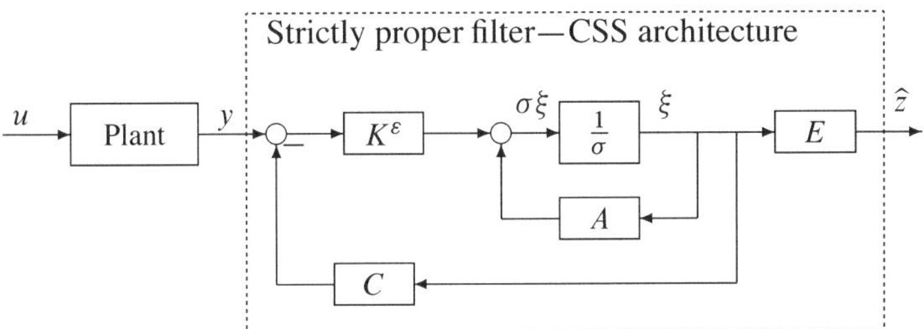

Figure 9.2: Block diagram of a strictly proper filter—CSS architecture

A family of strictly proper filters parameterized in ε is described by

$$\Sigma^\varepsilon_{\text{sp-CSS}} : \begin{cases} \sigma\xi = A\xi + K^\varepsilon(y - C\xi) \\ \widehat{z} = E\xi, \end{cases} \tag{9.11}$$

where the matrix K^ε is parameterized in ε.

Error dynamics: Let us denote the error between the state and its estimate by $e = x - \xi$. The error e_z between the actual desired output z and the estimated desired output $\widehat{z}$ is then given by $e_z = E(x - \xi) = Ee$. Also, the error dynamics is given by

$$\Sigma^{ue,\varepsilon}_{\text{sp-CSS}} : \begin{cases} \sigma e = (A - K^\varepsilon C)e + (B - K^\varepsilon D)u \\ e_z = Ee. \end{cases} \tag{9.12}$$

The transfer matrix $G^{ue,\varepsilon}_{\text{sp-CSS}}$ from u to e_z can obviously be written as

$$G^{ue,\varepsilon}_{\text{sp-CSS}} = E(\sigma I - A + K^\varepsilon C)^{-1}(B - K^\varepsilon D). \tag{9.13}$$

Remark 9.21 *In view of* (9.11) *and* (9.12)*, it is easy to see that both the filter equation and the error equation have the same poles, which are the eigenvalues of* $A - K^\varepsilon C$.

We observe that the only unknown in the filter equation (9.11), and consequently in the error equation (9.12) is the parameterized matrix K^ε, which is normally referred to as the filter gain. We need to determine or design K^ε in such a way that the H_∞ norm of G_{ue}^ε is arbitrarily small. In general, such a gain K^ε is nonunique. This lets us use the available freedom in selecting K^ε to shape the error dynamics appropriately, i.e., to shape the eigenstructure of $A - K^\varepsilon C$ as desired while honoring the constraints imposed by the requirement of H_∞ AID filtering. We pause to emphasize an important aspect. Theorems 9.5 and 9.14 developed earlier give the conditions under which a family of H_∞ AID filters exists among the general class of strictly proper filters of the form (9.2) with $P^\varepsilon = 0$. In other words, Theorems 9.5 and 9.14 do not restrict themselves to any fixed architecture for a filter such as the one given in (9.11). Nevertheless, as the following theorem shows, whenever the conditions of Theorems 9.5 and 9.14 are satisfied, we can determine the parameterized gain K^ε such that the family of filters given in (9.11) is a family of H_∞ AID filters.

Theorem 9.22 *Consider a continuous- or discrete-time system as in (9.1). Let Assumption 9.1 be satisfied. Consider strictly proper filters, and assume that the conditions of Theorem 9.5 for continuous-time systems or Theorem 9.14 for discrete-time systems are satisfied. Then, a family of H_∞ AID filters of the type $\Sigma_{\text{sp-CSS}}^\varepsilon$ exists; i.e., a sequence of parameterized gains K^ε exists such that the family of filters given in (9.11) is a family of H_∞ AID filters.*

Proof : We use the same duality argument as in the case of EID and H_2 AID filters. The only additional property that is needed is that there is a $1 - 1$ correspondence between filters of the form (9.11) for the original system and static state feedbacks of the form $u_d = (K^\varepsilon)' x_d$ solving the H_∞ ADD problem for the dual system given in (9.7), where the poles of the filter are equal to the closed-loop poles after applying the static feedback to the dual system. Moreover, the transfer matrix from w to e_z when applying the filter to the original system is equal to the transpose of the transfer matrix from w to z for the dual system after applying the static state feedback. This latter property was discussed earlier in Section 7.A. ∎

As it is well known, the dynamics of any system is heavily influenced by its poles. When we look at the poles of the error dynamics as given in (9.12) and those of the filter as given in (9.11), we observe that they are the same and equal to the eigenvalues of $A - K^\varepsilon C$. We enquire then whether any constraints exist in choosing the filter poles that shape the error dynamics. As in Chapter 8, the requirement of H_∞ AID filtering imposes that certain poles of a family of H_∞ AID filters tend as ε tends to zero to certain fixed locations termed as finite asymptotic fixed modes. We have the following formal definition of finite asymptotic fixed modes.

Definition 9.23 *(**Finite asymptotic fixed modes of strictly proper H_∞ AID filters of CSS architecture**) Consider the given system (9.1) and the associated H_∞ AID filtering problem 9.3. Assume that the solvability conditions as specified by Theorem 9.5 for continuous-time systems or as specified by Theorem 9.14 for discrete-time systems are satisfied. Then, a finite scalar $\lambda \in \mathbb{C}^-$ for the continuous-time case or $\lambda \in \mathbb{C}^\ominus$ for the discrete-time case is said to be an **H_∞ finite asymptotic fixed mode** with algebraic multiplicity α if for every family of H_∞ AID filters parameterized in ε and having the strictly proper full-order CSS architecture, poles λ_i^ε, $i = 1, 2, \cdots, \alpha$, of the family of filters exist such that all λ_i^ε tend to λ as ε tends to zero. The set of all H_∞ finite asymptotic fixed modes when strictly proper filters of full-order CSS architecture are used is denoted by $\Omega_{\text{sp-CSS}}^{h\infty-\text{aid}}(A, B, C, D, E, 0)$.*

We have the following theorem that characterizes the set $\Omega_{\text{sp-CSS}}^{h\infty-\text{aid}}$.

Theorem 9.24 *Consider a continuous- or discrete-time system as in (9.1). Let Assumption 9.1 be satisfied. Consider strictly proper full-order filters of CSS architecture, and assume that the solvability conditions as specified by Theorem 9.5 for continuous-time systems or as specified by Theorem 9.14 for discrete-time systems are satisfied. Then, we have*

$$\Omega_{\text{sp-CSS}}^{h\infty-\text{aid}}(A, B, C, D, E, 0) = \Omega_s^\infty(A', C', B', D', E'), \qquad (9.14)$$

where $\Omega_s^\infty(A', C', B', D', E')$ is obtained by using the H_∞ ADD algorithm (see Subsection 6.6.1 and Subsection 6.6.2 for continuous- and discrete-time systems, respectively) with its input as the quintuple (A', C', B', D', E').

Proof : It follows easily in view of the proof of Theorem 9.22. ■

Obviously, we would like to explore next the development of an algorithm to design H_∞ AID filters of full-order strictly proper CSS architecture, which gives us the flexibility to place the finite asymptotic modes of the filters at desired locations, of course, within the constraints imposed by Theorem 9.24. The following theorem considers this issue.

Theorem 9.25 *Consider a continuous- or discrete-time system given by (9.1). Let Assumption 9.1 be satisfied. Consider strictly proper filters, and assume that* *the conditions of Theorem 9.5 for continuous-time systems or Theorem 9.14 for discrete-time systems are satisfied. Then, the following two statements are equivalent:*

(i) *The parameterized gain sequence K^ε renders the family of filters given in (9.11) as a family of H_∞ AID filters.*

(ii) *The parameterized gain sequence K^ε is such that the family of control laws $u_d = -(K^\varepsilon)'x_d$ solves the H_∞ ADD problem for the dual system given in (9.7).*

Proof : It follows easily in view of the proof of Theorem 9.22.   ∎

Theorem 9.25 provides a roadmap to design the parameterized gain sequence K^ε that renders the family of filters given in (9.11) as a family of H_∞ AID filters. All that needs to be done is to use the H_∞ ADD Algorithm (see Subsection 6.6.1 for continuous-time systems and Subsection 6.6.2 for discrete-time systems) with the quintuple (A', C', B', D', E') as its input and obtain a parameterized gain sequence F^ε and then transpose it to obtain K^ε. Furthermore, the said H_∞ ADD algorithm determines the set of H_∞ finite asymptotic fixed modes $\boldsymbol{\Omega}_{\text{sp-CSS}}^{h\infty-\text{aid}}(A, B, C, D, E, 0)$ for the class of filters given in (9.11). Moreover, the said H_∞ ADD algorithm has certain flexibility to place the finite asymptotic modes of the filter as desired but within the constraints imposed by Theorem 9.24.

9.6.2 *A family of full-order proper H_∞ AID filters—CSS architecture*

Our goal in this subsection is to design a family of full-order proper H_∞ AID filters of CSS architecture. In connection with the design of H_2 AID filters, the proper full-order filter of CSS architecture was developed in Subsection 8.6.2 of Chapter 8. We follow here basically the same procedure, however, with some subtle but important changes. To construct a family of proper H_∞ AID filters of full-order CSS architecture for the given system, we construct here a family of strictly proper H_∞ AID filters for an auxiliary system $\widetilde{\Sigma}^*$. To do so, we first decompose the measured output y into two parts y_0 and y_1 in such a way that y_0 contains explicitly the unknown input u while y_1 does not contain the input u. That is, we rewrite the given system equation (9.1) as

$$\Sigma : \begin{cases} \sigma x &= Ax + Bu \\ \begin{pmatrix} y_0 \\ y_1 \end{pmatrix} = \begin{pmatrix} C_0 \\ C_1 \end{pmatrix}x + \begin{pmatrix} D_0 \\ 0 \end{pmatrix}u = Cx + Du \\ z &= Ex + Fu, \end{cases} \tag{9.15}$$

where rank $D = \text{rank } D_0 = m_0$ and D_0 surjective. We note that, without any loss of generality, we can rewrite the given system (9.1) in the form (9.15) by an appropriate coordinate transformation.

As in Chapter 8, we need to do some preliminary work to consider a nonzero matrix F. We need to rewrite the matrix E as

$$E = E_1 + E_2 \tag{9.16}$$

such that (8.22), (8.23), and (8.24) are satisfied. Such a decomposition of E can always be done under the solvability conditions given by Theorems 9.11 and 9.18. Also, such a choice of E_1 and E_2 guarantees that a matrix P^* exists such that $E_1 = E - P^*C$ and $F - P^*D = 0$. We can then use a preliminary injection of output y into the desired output z by letting,

$$z^* = z - P^*y = (E - P^*C)x + (F - P^*D)u = E_1 x. \tag{9.17}$$

In view of (9.15) and (9.17), we can define a new system Σ^* as

$$\Sigma^* : \begin{cases} \sigma x = Ax + Bu \\ \begin{pmatrix} y_0 \\ y_1 \end{pmatrix} = \begin{pmatrix} C_0 \\ C_1 \end{pmatrix} x + \begin{pmatrix} D_0 \\ 0 \end{pmatrix} u \\ z^* = E_1 x. \end{cases} \tag{9.18}$$

The above procedure of developing Σ^* is the same one used in developing Σ^* in (8.26).

By now we have rewritten the given system in a form suitable for filter development. However, before we proceed further, we need one more amendment of the measured output y. The needed amendment depends on whether we deal with continuous- or discrete-time systems. As such, in what follows, we divide our presentation into two parts: one pertaining to continuous-time systems and the other pertaining to discrete-time systems.

A family of full-order proper H_∞ AID filters—continuous-time systems:

In what follows, we deal with continuous-time systems. We form first a new measurement variable $\tilde{y}$ by augmenting y with another part $y_2 = \dot{y}_1$. That is, we let

$$\tilde{y} = \begin{pmatrix} y_0 \\ y_1 \\ y_2 \end{pmatrix} = \begin{pmatrix} y_0 \\ y_1 \\ \dot{y}_1 \end{pmatrix}, \quad \text{and} \quad \tilde{C} = \begin{pmatrix} C_0 \\ C_1 \\ C_1 A \end{pmatrix}, \quad \tilde{D} = \begin{pmatrix} D_0 \\ 0 \\ C_1 B \end{pmatrix}. \tag{9.19}$$

Then, it is easy to see that

$$\tilde{y} = \tilde{C}x + \tilde{D}u.$$

The dimension of $\tilde{y}$ is $\tilde{p} = 2p - \text{rank } D_0$. We note that y_2 is not directly available. However, as in Chapter 8 and as will be seen shortly, it can be eliminated from the filter equation.

Auxiliary system $\widetilde{\Sigma}^*$: We define next an auxiliary system with its measured output as $\widetilde{y}$,

$$\widetilde{\Sigma}^* : \begin{cases} \dot{x} &= Ax + Bu \\ \widetilde{y} &= \widetilde{C}x + \widetilde{D}u \\ z^* &= E_1 x. \end{cases} \tag{9.20}$$

The above system Σ^* is the same one as in (8.28).

Before we proceed further, it is important that we examine certain structural properties of $\widetilde{\Sigma}^*$ and relate them to those of Σ^*. Lemma 8.23 examines such properties.

As in Chapter 8, we plan to design first a family of strictly proper filters to solve the H_∞ AID filtering problem for the auxiliary system $\widetilde{\Sigma}^*$ and then modify it to obtain a family of proper filters that solves the H_∞ AID filtering problem for the original system Σ. This obviously is possible if and only if the conditions for the existence of a family of strictly proper H_∞ AID filters for the auxiliary system $\widetilde{\Sigma}^*$ coincide with the conditions for the existence of a family of proper H_∞ AID filters for the system Σ. The following lemma formalizes this.

Lemma 9.26 *Consider the continuous-time systems Σ and $\widetilde{\Sigma}^*$, respectively, as given in (9.1) and (9.20). Then, the following two statements are equivalent:*

(i) A family of strictly proper H_∞ AID filters exists for the auxiliary system $\widetilde{\Sigma}^*$.

(ii) A family of proper H_∞ AID filters exists for the system Σ.

Proof : The proof follows from Lemma 8.23 in view of Theorems 9.5 and 9.11. ∎

Lemma 9.26 pertains to general strictly proper and proper filters, although our interest here is only on strictly proper and proper filters of CSS architecture. We will relate next strictly proper filters of CSS architecture for $\widetilde{\Sigma}^*$ to proper filters for Σ with a nice desirable structure.

For the auxiliary system $\widetilde{\Sigma}^*$, we form a family of strictly proper filters of CSS architecture as

$$\widetilde{\Sigma}^\varepsilon_{\text{sp-CSS}} : \begin{cases} \dot{\widetilde{\xi}} &= A\widetilde{\xi} + K^\varepsilon(\widetilde{y} - \widetilde{C}\widetilde{\xi}) \\ \widehat{z}^* &= E_1\widetilde{\xi}, \end{cases} \tag{9.21}$$

where the matrix K^ε is a parameterized filter gain. The estimate $\widehat{z}^*$ of z^* is as given by (9.21). Then, in view of (9.17), the estimate $\widehat{z}$ of z is given by

$$\widehat{z} = \widehat{z}^* + P^* y = E_1\widetilde{\xi} + P^* y. \tag{9.22}$$

The above development focuses on developing the family of strictly proper filters (9.21). However, the family of filters (9.21) is not directly implementable because $y_2 = \dot{y}_1$ is not available as a measured variable. This implies that somehow we need to eliminate y_2 from (9.21). This can be easily done by defining a new variable:

$$\xi = \widetilde{\xi} - K_2^\varepsilon y_1. \tag{9.23}$$

Here K_2^ε is obtained by partitioning K^ε in conformity with the partitioning of $\widetilde{y}$. That is,

$$K^\varepsilon = \begin{pmatrix} K_0^\varepsilon & K_1^\varepsilon & K_2^\varepsilon \end{pmatrix}. \tag{9.24}$$

With the definition of ξ as in (9.23), we can rewrite the filter equation (9.21) as

$$\begin{cases} \dot{\xi} = (A - K^\varepsilon \widetilde{C})\xi + \begin{pmatrix} K_0^\varepsilon & K_1^\varepsilon + (A - K^\varepsilon \widetilde{C})K_2^\varepsilon \end{pmatrix} y \\ \widetilde{\xi} = \xi + K_2^\varepsilon y_1 \\ \widehat{z}^* = E_1 \widetilde{\xi} = E_1(\xi + K_2^\varepsilon y_1). \end{cases} \tag{9.25}$$

Obviously, the family of filters given above does not use $\dot{y}_1$. Moreover, it is proper rather than strictly proper. The family of filters given in (9.25) is indeed the family of proper full-order CSS filters that is to be used for Σ^*.

Proper filter of CSS architecture for Σ:

In view of (9.22), we can rewrite (9.25) as an implementable proper filter for Σ:

$$\Sigma_{\text{p-CSS}}^\varepsilon : \begin{cases} \dot{\xi} = (A - K^\varepsilon \widetilde{C})\xi + \widetilde{K}^\varepsilon y \\ \widetilde{\xi} = \xi + K_2^\varepsilon y_1 \\ \widehat{z} = E_1 \widetilde{\xi} + P^* y = E_1 \xi + \widetilde{P}^\varepsilon y, \end{cases} \tag{9.26}$$

where

$$\widetilde{K}^\varepsilon = \begin{pmatrix} K_0^\varepsilon & K_1^\varepsilon + (A - K^\varepsilon \widetilde{C})K_2^\varepsilon \end{pmatrix} \tag{9.27}$$

and

$$\widetilde{P}^\varepsilon = \begin{pmatrix} 0 & E_1 K_2^\varepsilon \end{pmatrix} + P^*.$$

A block diagram representation of the proper filter along with the given plant is shown in Figure 9.3. Structurally, the block diagram in Figure 9.3 is the same as the one in Figure 8.3.

Remark 9.27 *The above development shows clearly that there is a $1-1$ relationship between the family of strictly proper filters of CSS architecture for $\widetilde{\Sigma}^*$ and the family of proper filters of CSS architecture for Σ; that is, one of these family of filters can be constructed from the other.*

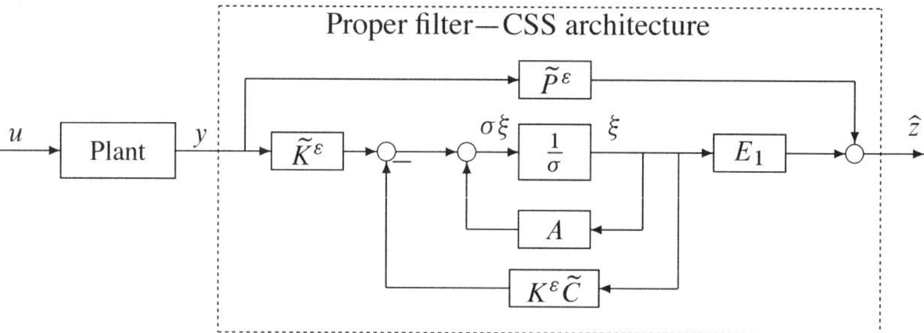

Figure 9.3: Block diagram of a full-order proper filter—CSS architecture

Error dynamics: By defining the error $e = x - \widetilde{\xi}$, the error e_z between the actual desired output $z = Ex + Fu = E_1 x + P^* y$ and the estimated desired output $\widehat{z} = E_1 \widetilde{\xi} + P^* y$ can be written as

$$e_z = z - \widehat{z} = E_1 e.$$

Then, in view of (9.15) and (9.21), the dynamics of error is given by

$$\Sigma^{ue,\varepsilon}_{\text{p-CSS}} : \begin{cases} \dot{e} = (A - K^{\varepsilon}\widetilde{C})e + (B - K^{\varepsilon}\widetilde{D})u \\ e_z = E_1 e. \end{cases} \tag{9.28}$$

Also, the transfer matrix $G^{ue,\varepsilon}$ from u to e_z can obviously be written as

$$G^{ue,\varepsilon}_{\text{p-CSS}} = E_1(\sigma I - A + K^{\varepsilon}\widetilde{C})^{-1}(B - K^{\varepsilon}\widetilde{D}). \tag{9.29}$$

Remark 9.28 *In view of (9.26) and (9.28), it is easy to see that both the filter equation and the error equation have the same poles, which are the eigenvalues of $A - K^{\varepsilon}\widetilde{C}$.*

As in the case of strictly proper filters, we need to point out an important point. Theorem 9.11 developed earlier gives the conditions under which a family of H_{∞} AID filters exists among the general class of strictly proper filters of the form (9.2). In other words, Theorem 9.11 does not restrict itself to any fixed architecture for a filter such as the one given in (9.26). Nevertheless, as the following theorem shows, whenever the conditions of Theorem 9.11 are satisfied, a parameterized gain K^{ε} exists such that the family of filters given in (9.26) is a family of H_{∞} AID filters.

Theorem 9.29 *Consider the continuous-time system Σ as in (9.1). Let the conditions of Theorem 9.11 be satisfied. Then, a family of H_{∞} AID filters of the type $\Sigma^{\varepsilon}_{\text{p-CSS}}$ for Σ exists; i.e., a parameterized gain sequence K^{ε} exists such that the family of filters given in (9.26) is a family of H_{∞} AID filters for Σ. Moreover, K^{ε} is the same parameterized gain sequence, which renders $\widetilde{\Sigma}^{\varepsilon}_{\text{sp-CSS}}$ of (9.21) as a family of H_{∞} AID filters for $\widetilde{\Sigma}^*$ of (9.20).*

Proof : Consider the system $\tilde{\Sigma}^*$, which is characterized by $(A, B, \tilde{C}, \tilde{D}, E_1)$, where $E_1 = E - P^*C$ with P^* such that $F - P^*D = 0$ and $\tilde{C}$ and $\tilde{D}$ are as defined in (9.19). By Theorem 9.22, a parameterized gain K^ε exists resulting in a family of strictly proper H_∞ AID filters for the system $\tilde{\Sigma}^*$. Before this theorem it is shown how this family of strictly proper filters can be converted to a family of proper filters (9.26) for the original system Σ with the same error dynamics. Hence the result. ∎

As in the case of strictly proper filters, constraints exist in choosing the filter poles that shape the error dynamics. The requirement of H_∞ AID filtering imposes that certain poles of a family of H_∞ AID filters tend as ε tends to zero to certain fixed locations termed as finite asymptotic fixed modes. We have the following formal definition of finite asymptotic fixed modes when proper H_∞ AID filters of CSS architecture are used.

Definition 9.30 *(Finite asymptotic fixed modes of proper H_∞ AID filters of CSS architecture) Consider the given system (9.1) and the associated H_∞ AID filtering problem 9.3. Assume that the solvability conditions as specified by Theorem 9.11 are satisfied. Then, a finite scalar $\lambda \in \mathbb{C}^-$ is said to be an H_∞ finite asymptotic fixed mode with algebraic multiplicity α if for every family of H_∞ AID filters parameterized in ε and having the proper full-order CSS architecture, poles λ_i^ε, $i = 1, 2, \cdots, \alpha$ of the family of filters exist such that all λ_i^ε tend to λ as ε tends to zero. The set of all H_∞ finite asymptotic fixed modes when proper filters of full-order CSS architecture are used is denoted by $\Omega_{\text{p-CSS}}^{h\infty-\text{aid}}(A, B, C, D, E, F)$.*

We have the following theorem that characterizes the set $\Omega_{\text{p-CSS}}^{h\infty-\text{aid}}$.

Theorem 9.31 *Consider a continuous-time system as in (9.1). Let Assumption 9.1 be satisfied. Consider proper full-order filters of CSS architecture, and assume that the solvability conditions as specified by Theorem 9.11 are satisfied. Also, let E_1 be as discussed in (9.16), (8.22), (8.23), and (8.24), and P^* be such that $E_1 = E - P^*C$ and $F - P^*D = 0$. Let $\tilde{C}$ and $\tilde{D}$ be as defined in (9.19). Then, we have*

$$\Omega_{\text{p-CSS}}^{h\infty-\text{aid}}(A, B, C, D, E, F) = \Omega_{\text{sp-CSS}}^{h\infty-\text{aid}}(A, B, \tilde{C}, \tilde{D}, E_1, 0).$$

Proof : The proof is transparent in view of the development of how a family of strictly proper filters for $\tilde{\Sigma}^*$ can be converted to a family of proper filters (9.26) for the original system Σ with the same error dynamics. ∎

Remark 9.32 *In view of Lemma 8.23 and a detailed examination of H_∞ ADD Algorithm (see Subsection 6.6.1) by which $\Omega_{\text{sp-CSS}}^{h\infty-\text{aid}}(A, B, \widetilde{C}, \widetilde{D}, E_1, 0)$ can be computed, one can easily verify that*

$$\Omega_{\text{sp-CSS}}^{h\infty-\text{aid}}(A, B, \widetilde{C}, \widetilde{D}, E_1, 0) = \Omega_{\text{sp-CSS}}^{h\infty-\text{aid}}(A, B, C, D, E_1, 0).$$

As such, in view of Theorem 9.31, we have

$$\Omega_{\text{p-CSS}}^{h\infty-\text{aid}}(A, B, C, D, E, F) = \Omega_{\text{sp-CSS}}^{h\infty-\text{aid}}(A, B, C, D, E_1, 0).$$

Remark 9.33 *Assume that the H_∞ AID filtering problem by using strictly proper filters is solvable, i.e., the solvability conditions, as specified by Theorem 9.5 are satisfied. Then, one can choose $E = E_1$ and $E_2 = 0$, where E_1 is as defined in (9.16) and satisfies (8.22), (8.23), and (8.24). In this case, as both strictly proper and proper H_∞ AID filters exist, an interesting question arises regarding the relationship between $\Omega_{\text{p-CSS}}^{h\infty-\text{aid}}$ and $\Omega_{\text{sp-CSS}}^{h\infty-\text{aid}}$. One can verify that in this case,*

$$\Omega_{\text{p-CSS}}^{h\infty-\text{aid}}(A, B, C, D, E, F) = \Omega_{\text{sp-CSS}}^{h\infty-\text{aid}}(A, B, C, D, E, 0).$$

Obviously, we would like to explore next the development of an algorithm to design H_∞ AID filters of full-order proper CSS architecture, which gives us the flexibility to place the finite asymptotic modes of the filters at desired locations, of course, within the constraints imposed by Theorem 9.31. Such an algorithm can be given by the following steps:

(*i*) Use the quintuple $(A', \widetilde{C}', B', \widetilde{D}', E_1')$ as the input to the H_∞ ADD Algorithm (see Subsection 6.6.1), and obtain a parameterized gain F^ε.

(*ii*) Let $K^\varepsilon = (F^\varepsilon)'$. Partition K^ε in accordance with (9.24), and then define $\widetilde{K}^\varepsilon$ in accordance with (9.27).

The H_∞ ADD algorithm mentioned in the above first step yields the set

$$\Omega_{\text{sp-CSS}}^{h\infty-\text{aid}}(A, B, \widetilde{C}, \widetilde{D}, E_1, 0) = \Omega_{\text{p-CSS}}^{h\infty-\text{aid}}(A, B, C, D, E, F).$$

The above equality follows from Theorem 9.31. Also, the said H_∞ ADD algorithm has certain flexibility to place the finite asymptotic modes of the filter as desired but within the constraints imposed by Theorem 9.31.

A family of full-order proper H_∞ AID filters—discrete-time systems:

In what follows, we deal with discrete-time systems. Unlike in continuous-time systems, a new measurement variable $\widetilde{y}$ is formed by replacing y_1 in y with $y_2 = \sigma y_1$. That is, we let

$$\widetilde{y} = \begin{pmatrix} y_0 \\ y_2 \end{pmatrix} = \begin{pmatrix} y_0 \\ \sigma y_1 \end{pmatrix}, \quad \text{and} \quad \widetilde{C} = \begin{pmatrix} C_0 \\ C_1 A \end{pmatrix}, \quad \widetilde{D} = \begin{pmatrix} D_0 \\ C_1 B \end{pmatrix}. \tag{9.30}$$

Then, it is easy to see that

$$\widetilde{y} = \widetilde{C}x + \widetilde{D}u.$$

As in continuous-time systems, y_2 is not directly available. However, as before, it can be eliminated from the filter equation.

Auxiliary system $\widetilde{\Sigma}^*$: We define next an auxiliary system with its measured output as $\widetilde{y}$:

$$\widetilde{\Sigma}^* : \begin{cases} \sigma x = Ax + Bu \\ \widetilde{y} \ = \widetilde{C}x + \widetilde{D}u \\ z^* = E_1 x, \end{cases} \tag{9.31}$$

where E_1 is as defined in (8.21) and satisfies (8.22), (8.23), and (8.24). As in continuous-time systems, before we proceed further, it is important that we get familiar with certain structural properties of $\widetilde{\Sigma}^*$ and relate them to those of Σ^*. Such properties have already been examined in Lemma 8.32.

As in continuous-time systems, we plan to design first a family of strictly proper H_∞ AID filters for the auxiliary system $\widetilde{\Sigma}^*$ and then modify it to obtain a family of proper H_∞ AID filters for the original system Σ. This obviously is possible if and only if the conditions for the existence of a family of strictly proper H_∞ AID filters for the auxiliary system $\widetilde{\Sigma}^*$ coincide with the conditions for the existence of a family of proper H_∞ AID filters for the system Σ. The following lemma formalizes this.

Lemma 9.34 *Consider the discrete-time systems Σ and $\widetilde{\Sigma}^*$, respectively, as given in (9.1) and (9.31). Then, the following two statements are equivalent:*

(i) A family of strictly proper H_∞ AID filters exists for the auxiliary system $\widetilde{\Sigma}^*$.

(ii) A family of proper H_∞ AID filters exists for the system Σ.

Proof : The proof follows from Lemma 8.32 in view of Theorems 9.14 and 9.18.
∎

As in the continuous-time case, the above lemma pertains to general strictly proper and proper filters, although our interest here is only on strictly proper and proper filters of CSS architecture. We will relate next strictly proper filters of CSS architecture for $\widetilde{\Sigma}^*$ to proper filters for Σ with a nice desirable structure.

For the auxiliary system $\widetilde{\Sigma}^*$, we form a family of strictly proper filters of CSS architecture as

$$\widetilde{\Sigma}^\varepsilon_{\text{sp-CSS}} : \begin{cases} \sigma\widetilde{\xi} = A\widetilde{\xi} + K^\varepsilon(\widetilde{y} - \widetilde{C}\widetilde{\xi}) \\ \widehat{z}^* = E_1\widetilde{\xi}, \end{cases} \tag{9.32}$$

where the matrix K^ε is a parameterized filter gain. The estimate $\hat{z}^*$ of z^* is as given by (9.32). Then, in view of (9.17), the estimate $\hat{z}$ of z is given by

$$\hat{z} = \hat{z}^* + P^* y = E_1 \tilde{\xi} + P^* y. \tag{9.33}$$

The above development focuses on developing the family of strictly proper filters (9.32). However, the family of filters (9.32) is not directly implementable because $y_2 = \sigma y_1$ is not available as a measured variable. This implies that somehow we need to eliminate y_2 from the filter equation (9.32). This can be easily done by defining a new variable:

$$\xi = \tilde{\xi} - K_2^\varepsilon y_1. \tag{9.34}$$

Here K_2^ε is obtained by partitioning K^ε in conformity with the partitioning of y. That is,

$$K^\varepsilon = \begin{pmatrix} K_0^\varepsilon & K_2^\varepsilon \end{pmatrix}. \tag{9.35}$$

With the definition of ξ as in (9.34), we can rewrite the filter equation (9.32) as

$$\begin{cases} \sigma\xi = (A - K^\varepsilon \tilde{C})\xi + \begin{pmatrix} K_0^\varepsilon & (A - K^\varepsilon \tilde{C})K_2^\varepsilon \end{pmatrix} y \\ \tilde{\xi} = \xi + K_2^\varepsilon y_1 \\ \hat{z}^* = E_1 \tilde{\xi} = E_1(\xi + K_2^\varepsilon y_1). \end{cases} \tag{9.36}$$

Obviously, the family of filters given above does not use σy_1. Moreover, it is proper rather than strictly proper. The family of filters given in (9.36) is indeed the family of proper full-order CSS filters that is to be used for Σ^*.

The structure of (9.36) essentially is similar to the structure of current type of filters well known in discrete-time literature.

Proper filter of CSS architecture for Σ:

Finally, in view of (9.33), we can rewrite (9.36) as an implementable proper filter for Σ:

$$\Sigma_{\text{p-CSS}}^\varepsilon : \begin{cases} \sigma\xi = (A - K^\varepsilon \tilde{C})\xi + \tilde{K}^\varepsilon y \\ \tilde{\xi} = \xi + K_2^\varepsilon y_1 \\ \hat{z} = E_1 \tilde{\xi} + P^* y = E_1 \xi + \tilde{P}^\varepsilon y, \end{cases} \tag{9.37}$$

where

$$\tilde{K}^\varepsilon = \begin{pmatrix} K_0^\varepsilon & (A - K^\varepsilon \tilde{C})K_2^\varepsilon \end{pmatrix} \quad \text{and} \quad \tilde{P}^\varepsilon = \begin{pmatrix} 0 & E_1 K_2^\varepsilon \end{pmatrix} + P^*. \tag{9.38}$$

A block diagram representation of the proper filter along with the given plant is shown in Figure 9.4. Structurally, the block diagram in Figure 9.4 is the same as the one in Figure 8.3.

Remark 9.35 *The above development shows clearly that there is a $1-1$ relationship between the family of strictly proper filters of CSS architecture for $\tilde{\Sigma}^*$ and the family of proper filters of CSS architecture for Σ; that is, one of these family of filters can be constructed from the other.*

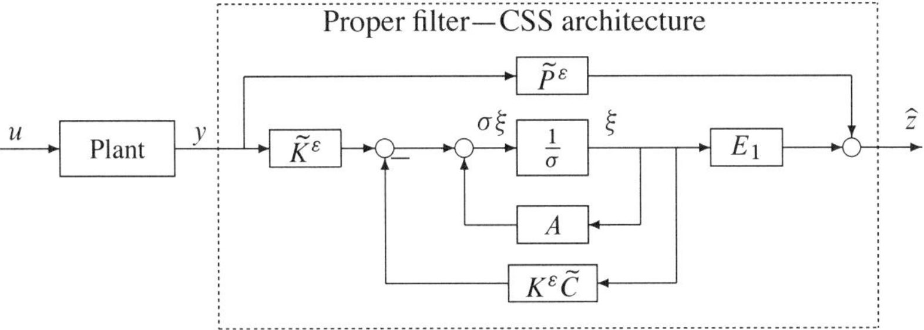

Figure 9.4: Block diagram of a full-order proper filter—CSS architecture

Error dynamics: By defining the error $e = x - \widetilde{\xi}$, the error e_z between the actual desired output $z = E_1 x + P^* y$ and the estimated desired output $\widehat{z} = E_1\widetilde{\xi} + P^* y$ can be written as

$$e_z = z - \widehat{z} = E_1 e.$$

Then, in view of (9.15) and (9.32), the dynamics of error is given by

$$\Sigma_{\text{p-CSS}}^{ue,\varepsilon} : \begin{cases} \sigma e = (A - K^\varepsilon\widetilde{C})e + (B - K^\varepsilon\widetilde{D})u \\ e_z = E_1 e. \end{cases} \tag{9.39}$$

Also, the transfer matrix $G^{ue,\varepsilon}$ from u to e_z can obviously be written as

$$G_{\text{p-CSS}}^{ue,\varepsilon} = E_1(\sigma I - A + K^\varepsilon\widetilde{C})^{-1}(B - K^\varepsilon\widetilde{D}). \tag{9.40}$$

Remark 9.36 *In view of (9.37) and (9.39), it is easy to see that both the filter equation and the error equation have the same poles, which are the eigenvalues of $A - K^\varepsilon\widetilde{C}$.*

As in the continuous-time case, we need to emphasize an important point. Theorem 9.18 developed earlier gives the conditions under which a family of H_∞ AID filters exists among the general class of strictly proper filters of the form (9.2). In other words, Theorem 9.18 does not restrict itself to any fixed architecture for a filter such as the one given in (9.37). Nevertheless, as the following theorem shows, whenever the conditions of Theorem 9.18 are satisfied, a parameterized gain K^ε exists such that the family of filters given in (9.37) is a family of H_∞ AID filters.

Theorem 9.37 *Consider a discrete-time system Σ given by (9.1). Let the conditions of Theorem 9.18 be satisfied. Then, a family of H_∞ AID filters of the type $\Sigma_{\text{p-CSS}}^\varepsilon$ for Σ exists; i.e., a parameterized gain sequence K^ε exists such that the family of filters given in (9.37) is a family of H_∞ AID filters for Σ. Moreover, K^ε is the same parameterized gain sequence, which renders $\widetilde{\Sigma}_{\text{sp-CSS}}^\varepsilon$ of (9.32) as a family of H_∞ AID filters for $\widetilde{\Sigma}^*$ of (9.31).*

Proof : The proof follows along the lines of the proof of Theorem 9.29.   ∎

As in the continuous-time case, constraints exist in choosing the filter poles that shape the error dynamics. The requirement of H_∞ AID filtering imposes that certain poles of a family of H_∞ AID filters tend as ε tends to zero to certain fixed locations termed again as finite asymptotic fixed modes. We have the following formal definition of finite asymptotic fixed modes when proper H_∞ AID filters of CSS architecture are used.

Definition 9.38 *(Finite asymptotic fixed modes of proper H_∞ AID filters of CSS architecture) Consider the given system (9.1) and the associated H_∞ AID filtering problem 9.3. Assume that the solvability conditions as specified by Theorem 9.18 are satisfied. Then, a finite scalar $\lambda \in \mathbb{C}^\ominus$ is said to be an H_∞ finite asymptotic fixed mode with algebraic multiplicity α if for every family of H_∞ AID filters parameterized in ε and having the proper full-order CSS architecture, poles λ_i^ε, $i = 1, 2, \cdots, \alpha$ of the family of filters exist such that all λ_i^ε tends to λ as ε tends to zero. The set of all H_∞ finite asymptotic fixed modes when proper filters of full-order CSS architecture are used is denoted by $\Omega_{\mathrm{p\text{-}CSS}}^{h\infty-\mathrm{aid}}(A, B, C, D, E, F)$.*

We have the following theorem that characterizes the set $\Omega_{\mathrm{p\text{-}CSS}}^{h\infty-\mathrm{aid}}$.

Theorem 9.39 *Consider a discrete-time system given by (9.1). Let the conditions of Theorem 9.18 be satisfied. Also, let E_1 be as discussed in (9.16), (8.22), (8.23), and (8.24), and P^* be such that $E_1 = E - P^*C$ and $F - P^*D = 0$. Let $\tilde{C}$ and $\tilde{D}$ be as defined in (9.30). Then, we have*

$$\Omega_{\mathrm{p\text{-}CSS}}^{h\infty-\mathrm{aid}}(A, B, C, D, E, F) = \Omega_{\mathrm{sp\text{-}CSS}}^{h\infty-\mathrm{aid}}(A, B, \tilde{C}, \tilde{D}, E_1, 0).$$

Proof : The proof is transparent in view of the development of how a family of strictly proper filters for $\tilde{\Sigma}^*$ can be converted to a family of proper filters (9.37) for the original system Σ with the same error dynamics.   ∎

Remark 9.40 *In continuous-time systems, we remarked that*

$$\Omega_{\mathrm{sp\text{-}CSS}}^{h\infty-\mathrm{aid}}(A, B, \tilde{C}, \tilde{D}, E_1, 0) = \Omega_{\mathrm{sp\text{-}CSS}}^{h\infty-\mathrm{aid}}(A, B, C, D, E_1, 0).$$

The above property is not true in the discrete-time case because of possible additional fixed modes at the origin in $\Omega_{\mathrm{sp\text{-}CSS}}^{h\infty-\mathrm{aid}}(A, B, \tilde{C}, \tilde{D}, E_1, 0)$. That is, we have

$$\Omega_{\mathrm{sp\text{-}CSS}}^{h\infty-\mathrm{aid}}(A, B, C, D, E_1, 0) \subset \Omega_{\mathrm{sp\text{-}CSS}}^{h\infty-\mathrm{aid}}(A, B, \tilde{C}, \tilde{D}, E_1, 0).$$

Remark 9.41 *In continuous-time systems, whenever strictly proper H_∞ AID filters exist, we remarked that $\Omega^{h\infty-\text{aid}}_{\text{p-CSS}} = \Omega^{h\infty-\text{aid}}_{\text{sp-CSS}}$. Such a property is not true in the discrete-time case because of possible additional fixed modes at the origin as mentioned above.*

Obviously, we would like to explore next the development of an algorithm to design H_∞ AID filters of full-order proper CSS architecture, which gives us the flexibility to place the finite asymptotic modes of the filters at desired locations, of course, within the constraints imposed by Theorem 9.39. Such an algorithm can be given by the following steps:

(i) Use the quintuple $(A', \widetilde{C}', B', \widetilde{D}', E_1')$ as the input to the H_∞ ADD Algorithm (see Subsection 6.6.2), and obtain a parameterized gain F^ε.

(ii) Let $K^\varepsilon = (F^\varepsilon)'$. Partition K^ε in accordance with (9.35), and then define $\widetilde{K}^\varepsilon$ in accordance with (9.38).

The H_∞ ADD algorithm mentioned in the above first step yields the set:

$$\Omega^{h\infty-\text{aid}}_{\text{sp-CSS}}(A, B, \widetilde{C}, \widetilde{D}, E_1, 0) = \Omega^{h\infty-\text{aid}}_{\text{p-CSS}}(A, B, C, D, E, F),$$

where the equality follows from Theorem 9.39. Also, the said H_∞ ADD algorithm has certain flexibility to place the finite asymptotic modes of the filter as desired but within the constraints imposed by Theorem 9.39.

9.6.3 A family of reduced-order proper H_∞ AID filters—CSS architecture

The previous two subsections construct, respectively, strictly proper and proper full-order filters that solve a designated problem. By full-order filters, as usual, we mean filters having the same dynamic order as that of the given system. Our goal in this subsection is to develop reduced-order filters having their dynamic order lower than that of the given system.

The procedure of developing reduced-order filters here follows mostly along the same lines as in Subsection 8.6.3. However, there are some subtle but important differences. As before, our method of development transforms the construction of a family of reduced-order filters for a given system to that of a family of full-order filters for a certain reduced-order system.

We proceed now to construct an appropriate reduced-order system. To start with, let us rewrite the matrices C and D of (9.1) as

$$C = \begin{pmatrix} 0 & C_{02} \\ I_{p-m_0} & 0 \end{pmatrix}, \quad D = \begin{pmatrix} D_0 \\ 0 \end{pmatrix},$$

where again rank $D = $ rank $D_0 = m_0$. This can always be done without any loss of generality by appropriate coordinate transformations. In view of the above partitioning of C and D, we can partition the given system Σ as

$$
\Sigma : \begin{cases}
\begin{pmatrix} \sigma x_1 \\ \sigma x_2 \end{pmatrix} = \begin{pmatrix} A_{11} & A_{12} \\ A_{21} & A_{22} \end{pmatrix} \begin{pmatrix} x_1 \\ x_2 \end{pmatrix} + \begin{pmatrix} B_{11} \\ B_{22} \end{pmatrix} u \\[4mm]
y = \begin{pmatrix} y_0 \\ y_1 \end{pmatrix} = \begin{pmatrix} 0 & C_{02} \\ I & 0 \end{pmatrix} \begin{pmatrix} x_1 \\ x_2 \end{pmatrix} + \begin{pmatrix} D_0 \\ 0 \end{pmatrix} u \\[4mm]
z = Ex + Fu,
\end{cases}
\tag{9.41}
$$

where different variables have obvious meanings.

Next, we use a preliminary injection of output y into the desired output z. To this end, we define a matrix P^* such that $F - P^*D = 0$. Then, let

$$
z^* = z - P^*y = (E - P^*C)x + (F - P^*D)u = E_1 x
\tag{9.42}
$$

where $E_1 = E - P^*C$. We note that E_1 can be defined independent of the choice of any matrix P^*, which renders $F - P^*D = 0$. This can be done by decomposing the matrix E into two parts E_1 and E_2 as discussed in (9.16) [also, see (8.22), (8.23), and (8.24)]. In view of (9.41) and (9.42), we can define a new system Σ^* as

$$
\Sigma^* : \begin{cases}
\begin{pmatrix} \sigma x_1 \\ \sigma x_2 \end{pmatrix} = \begin{pmatrix} A_{11} & A_{12} \\ A_{21} & A_{22} \end{pmatrix} \begin{pmatrix} x_1 \\ x_2 \end{pmatrix} + \begin{pmatrix} B_{11} \\ B_{22} \end{pmatrix} u \\[4mm]
y = \begin{pmatrix} y_0 \\ y_1 \end{pmatrix} = \begin{pmatrix} 0 & C_{02} \\ I & 0 \end{pmatrix} \begin{pmatrix} x_1 \\ x_2 \end{pmatrix} + \begin{pmatrix} D_0 \\ 0 \end{pmatrix} u \\[4mm]
z^* = E_1 x = E_{11} x_1 + E_{12} x_2,
\end{cases}
\tag{9.43}
$$

with $E_1 = \begin{pmatrix} E_{11} & E_{12} \end{pmatrix}$.

We note that the y_1 is not contaminated by the input u, and hence, $x_1 = y_1$ is known exactly from the measurement y. Thus, all we need to do is to estimate the state x_2. To proceed further, let us rewrite the state equation for x_1 in terms of the output y_1 and the state x_2 as

$$
\sigma y_1 = A_{11} y_1 + A_{12} x_2 + B_{11} u.
\tag{9.44}
$$

The above equation can be rewritten as

$$
\sigma y_1 - A_{11} y_1 = A_{12} x_2 + B_{11} u.
$$

Treating σy_1 as known, we can define a new measurement variable y_r:

$$
y_r = \begin{pmatrix} y_0 \\ \sigma y_1 - A_{11} y_1 \end{pmatrix}.
$$

Although σy_1 is not directly available, as we did earlier in the case of proper full-order CSS filters, we can eliminate it from any filter that is constructed using y_r as a measured output. With this in mind, we form the following auxiliary system:

$$\Sigma_r^* : \begin{cases} \sigma x_r = A_r x_r + B_r u + A_{21} y_1 \\ y_r \;\;= C_r x_r + D_r u \\ z_r^* \;\;= E_r x_r, \end{cases} \tag{9.45}$$

where $x_r = x_2$ and

$$A_r = A_{22}, \quad B_r = B_{22}, \quad C_r = \begin{pmatrix} C_{02} \\ A_{12} \end{pmatrix}, \quad D_r = \begin{pmatrix} D_0 \\ B_{11} \end{pmatrix}, \quad E_r = E_{22}. \tag{9.46}$$

We note that the dynamic order n_r of the above Σ_r^* is less than the dynamic order n of the given system Σ by a number equal to the dimension of $x_1 = y_1$.

Before we proceed further, it is beneficial to be familiar with certain structural properties of Σ_r^* and relate them to those of the Σ. Such properties have been studied earlier in Lemma 8.42.

Next, let us suppose we can construct a family of strictly proper H_∞ AID filters for the auxiliary system Σ_r^* to arrive at an estimate $\hat{z}_r^*$ of z_r^*. (Here y_1 is a known input and hence will show up accordingly in the dynamics of any filter we construct for Σ_r^*.) Then, we can easily obtain an estimate $\hat{z}$ of z as

$$\hat{z} = E_{11} x_1 + \hat{z}_r^* + P^* y. \tag{9.47}$$

This motivates us to construct a family of full-order strictly proper filters for the auxiliary system Σ_r^* to arrive at an estimate $\hat{z}_r^*$ of z_r^*. A family of full-order filters constructed as such for Σ_r^* is indeed a family of reduced-order filters for Σ. However, in view of (9.47), such a family of reduced-order filters for Σ is a family of proper (rather than a family of strictly proper) filters for Σ. A fundamental issue that arises next is under what conditions one can construct a family of strictly proper H_∞ AID filters for Σ_r^*. Expectedly, it turns out that one can construct a family of strictly proper H_∞ AID filters for the reduced-order system Σ_r^* if and only if one can construct a family of proper H_∞ AID filters for the given system Σ as stated in the following lemma.

Lemma 9.42 *Consider the continuous- or discrete-time systems Σ and Σ_r^*, respectively, as given in (9.1) and (9.45). Then, the following two statements are equivalent:*

(i) *A family of strictly proper H_∞ AID filters exists for the reduced-order system Σ_r^*.*

(ii) *A family of proper H_∞ AID filters exists for the system Σ.*

Proof : The proof follows from Lemma 8.42 in view of Theorems 9.5 and 9.11 for continuous time, and Theorems 9.14 and 9.18 for discrete time. ■

To construct the required family of reduced-order proper H_∞ AID filters for Σ, in the spirit of the above development, we first construct a family of strictly proper full-order filters of CSS architecture for the reduced-order system Σ_r^* as

$$\widetilde{\Sigma}_{\text{sp-CSS}}^{\varepsilon} : \begin{cases} \sigma\widetilde{\xi}_r = A_r\widetilde{\xi}_r + A_{21}y_1 + K_r^{\varepsilon}(y_r - C_r\widetilde{\xi}_r) \\ \widehat{z}_r^* = E_r\widetilde{\xi}_r, \end{cases} \tag{9.48}$$

where the matrix K_r^{ε} is a parameterized filter gain. As σy_1 is not available, we need to modify the above family of filters. To this end, let us partition

$$K_r^{\varepsilon} = \begin{pmatrix} K_{r0}^{\varepsilon} & K_{r1}^{\varepsilon} \end{pmatrix} \tag{9.49}$$

so as to be compatible with the partitioning of y_r. Also, let

$$\xi_r = \widetilde{\xi}_r - K_{r1}^{\varepsilon}y_1. \tag{9.50}$$

We can then easily rewrite the family of filters (9.48) as a family of filters for Σ as shown below.

A family of reduced-order proper filters of CSS architecture for Σ:

The family of reduced-order proper filters of CSS architecture for Σ is given by

$$\Sigma_{\text{r-CSS}}^{\varepsilon} : \begin{cases} \sigma\xi_r = (A_r - K_r^{\varepsilon}C_r)\xi_r + \widetilde{K}_r^{\varepsilon}y \\ \widetilde{\xi}_r = \xi_r + K_{r1}^{\varepsilon}y_1 \\ \widehat{z} = E_{11}x_1 + E_r\widetilde{\xi}_r + P^*y = E_r\xi_r + \widetilde{P}_r^{\varepsilon}y, \end{cases} \tag{9.51}$$

where

$$\widetilde{K}_r^{\varepsilon} = \begin{pmatrix} K_{r0}^{\varepsilon} & A_{21} - K_{r1}^{\varepsilon}A_{11} + (A_r - K_r^{\varepsilon}C_r)K_{r1}^{\varepsilon} \end{pmatrix} \tag{9.52}$$

and

$$\widetilde{P}_r^{\varepsilon} = \begin{pmatrix} 0 & E_{11} + E_r K_{r1}^{\varepsilon} \end{pmatrix} + P^*. \tag{9.53}$$

A block diagram representation of the reduced-order proper filter along with the given plant is shown in Figure 9.5, which is structurally the same as the one in Figure 8.5.

Error dynamics: By defining the error $e_r = x_r - \widetilde{\xi}_r$, the error e_z between the actual desired output $z = E_1x + P^*y$ and the estimated desired output $\widehat{z} = E_{11}x_1 + E_r\widetilde{\xi}_r + P^*y$ can be written as

$$e_z = z - \widehat{z} = E_r e_r.$$

Then, the dynamics of error is given by

$$\Sigma_{\text{r-CSS}}^{ue,\varepsilon} : \begin{cases} \sigma e_r = (A_r - K_r^{\varepsilon}C_r)e_r + (B_r - K_r^{\varepsilon}D_r)u \\ e_z = E_r e_r. \end{cases} \tag{9.54}$$

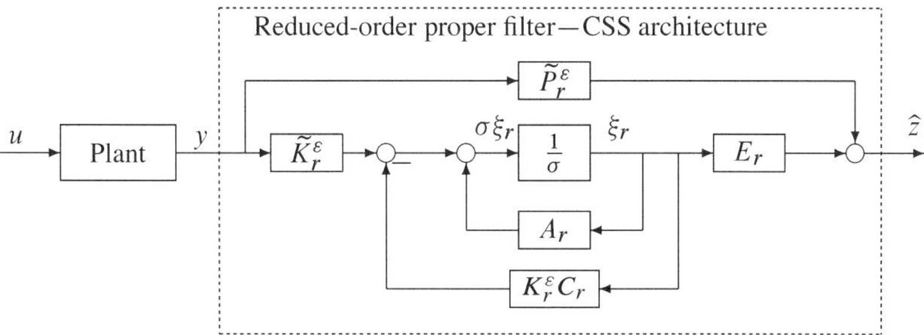

Figure 9.5: Block diagram of a reduced-order proper filter—CSS architecture

Also, the transfer matrix $G^{ue,\varepsilon}$ from u to e_z can obviously be written as

$$G^{ue,\varepsilon}_{\text{r-CSS}} = E_r(\sigma I - A_r + K_r^\varepsilon C_r)^{-1}(B_r - K_r^\varepsilon D_r). \qquad (9.55)$$

Remark 9.43 *It is easy to see that both the filter equation and the error equation have the same poles, which are the eigenvalues of $A_r - K_r^\varepsilon C_r$.*

Everything in the family of filters (9.51) is known except the parameterized gain K_r^ε. Next, we need to enquire first whether a parameterized gain K_r^ε exists such that the resulting family of proper filters (9.51) solves the H_∞ AID filtering problem for the original system Σ given in (9.1) and then, whenever it exists, develop a method of obtaining such a parameterized gain K_r^ε. We have the following results.

Theorem 9.44 *Consider a continuous- or discrete-time system Σ given by (9.1). Let the conditions of Theorem 9.11 for continuous-time systems or the conditions of Theorem 9.18 for discrete-time systems be satisfied. Also, let $E_1 = E - P^*C$ with P^* such that $F - P^*D = 0$. Then, a parameterized gain K_r^ε exists such that the family of filters given in (9.51) is a family of H_∞ AID filters for Σ. Moreover, K^ε is the same parameterized gain sequence, which renders $\widetilde{\Sigma}^\varepsilon_{\text{sp-CSS}}$ of (9.48) as a family of H_∞ AID filters for Σ_r^* of (9.45).*

Proof : By Lemma 9.42, we know that for the system $\widetilde{\Sigma}_r^*$, the H_∞ AID filtering problem is solvable. Moreover, any family of filter gains $\{K_r^\varepsilon\}_{\varepsilon>0}$ resulting in a family of strictly proper H_∞ AID filters for Σ_r^* results in a family of reduced-order proper filters (9.51), which yield the same error dynamics for the original system Σ and therefore achieve H_∞ AID for Σ. $\blacksquare$

As in the case of full-order strictly proper and proper filters, constraints exist in choosing the filter poles that shape the error dynamics. The requirement of H_∞

AID filtering imposes that certain poles of a family of H_∞ AID filters tend as ε tends to zero to certain fixed locations termed as finite asymptotic fixed modes. We have the following formal definition of finite asymptotic fixed modes when reduced-order proper H_∞ AID filters of CSS architecture are used.

Definition 9.45 (*Finite asymptotic fixed modes of reduced-order H_∞ AID filters of CSS architecture*) *Consider the given system* (9.1) *and the associated H_∞ AID filtering problem 9.3. Assume that the solvability conditions as specified by Theorem 9.11 (for continuous-time) or Theorem 9.18 (for discrete-time) are satisfied. Then, a finite scalar $\lambda \in \mathbb{C}^-$ for the continuous-time case or $\lambda \in \mathbb{C}^\ominus$ for the discrete-time case is said to be an H_∞ finite asymptotic fixed mode with algebraic multiplicity α if for every family of H_∞ AID filters parameterized in ε and having the proper reduced-order CSS architecture, poles λ_i^ε, $i = 1, 2, \cdots, \alpha$ of the family of filters exist such that all λ_i^ε tend to λ as ε tends to zero. The set of all H_∞ finite asymptotic fixed modes when proper filters of reduced-order CSS architecture are used is denoted by $\Omega_{\text{r-CSS}}^{h\infty-\text{aid}}(A, B, C, D, E, F)$.*

We have the following theorem that characterizes the set $\Omega_{\text{r-CSS}}^{h\infty-\text{aid}}$.

Theorem 9.46 *Consider a continuous- or discrete-time system given by* (9.1). *Let the conditions of Theorem 9.18 be satisfied. Also, let E_1 be as discussed in* (9.16), (8.22), (8.23), *and* (8.24), *and P^* be such that $E_1 = E - P^*C$ and $F - P^*D = 0$. Let the quintuple $(A_r, B_r, C_r, D_r, E_r)$ be as defined in* (9.46). *Then, we have*

$$\Omega_{\text{r-CSS}}^{h\infty-\text{aid}}(A, B, C, D, E, F) = \Omega_{\text{sp-CSS}}^{h\infty-\text{aid}}(A_r, B_r, C_r, D_r, E_r, 0).$$

Proof: The proof is transparent in view of the above development of how a family of full-order strictly proper filters for Σ_r^* can be converted to a family of reduced-order proper filters (9.51) for the original system Σ with the same error dynamics.

∎

For continuous time, the full-order filter (9.26) and the reduced-order filter (9.51) are both proper filters. An interesting question arises as to the relationship between the sets of fixed modes associated with them. The following lemma shows that they are one and the same. Similarly, for discrete time, the full-order filter (9.37) and the reduced-order filter (9.51) are both proper filters. Once again, the sets of fixed modes associated with them are almost the same.

Lemma 9.47 *Consider a continuous- or discrete-time system given by* (9.1). *Let the conditions of Theorem 9.11 for continuous time or Theorem 9.18 for discrete time be satisfied. Then the following hold:*

(*i*) *For continuous time, consider the set of fixed modes $\boldsymbol{\Omega}_{\text{p-CSS}}^{h\infty-\text{aid}}$ as in Definition 9.30 and the set of fixed modes $\boldsymbol{\Omega}_{\text{r-CSS}}^{h\infty-\text{aid}}$ as in Definition 9.45. Then we have*

$$\boldsymbol{\Omega}_{\text{p-CSS}}^{h\infty-\text{aid}}(A, B, C, D, E, F) = \boldsymbol{\Omega}_{\text{r-CSS}}^{h\infty-\text{aid}}(A, B, C, D, E, F).$$

(*ii*) *For discrete time, consider the set of fixed modes $\boldsymbol{\Omega}_{\text{p-CSS}}^{h\infty-\text{aid}}$ as in Definition 9.38 and the set of fixed modes $\boldsymbol{\Omega}_{\text{r-CSS}}^{h\infty-\text{aid}}$ as in Definition 9.45. Then we have*

$$\boldsymbol{\Omega}_{\text{p-CSS}}^{h\infty-\text{aid}}(A, B, C, D, E, F) \supset \boldsymbol{\Omega}_{\text{r-CSS}}^{h\infty-\text{aid}}(A, B, C, D, E, F),$$

where some additional fixed modes at the origin is the only possible difference between these two sets.

Proof : We first note that from Theorem 9.46 we obtain in both discrete and continuous time:

$$\boldsymbol{\Omega}_{\text{r-CSS}}^{h\infty-\text{aid}}(A, B, C, D, E, F) = \boldsymbol{\Omega}_{\text{sp-CSS}}^{h\infty-\text{aid}}(A_r, B_r, C_r, D_r, E_r, 0).$$

Next, using Theorem 9.24, we obtain

$$\boldsymbol{\Omega}_{\text{sp-CSS}}^{h\infty-\text{aid}}(A_r, B_r, C_r, D_r, E_r, 0) = \boldsymbol{\Omega}_s^\infty(A_r', C_r', B_r', D_r', E_r').$$

Theorem 6.26 (continuous time) or Theorem 6.29 (discrete time) can then be used to establish that

$$\boldsymbol{\Omega}_s^\infty(A_r', C_r', B_r', D_r', E_r') = \boldsymbol{\Omega}_s^\infty(A', C', B', D', E_1').$$

We have

$$\boldsymbol{\Omega}_s^\infty(A', C', B', D', E_1') = \boldsymbol{\Omega}_{\text{sp-CSS}}^{h\infty-\text{aid}}(A, B, C, D, E_1, 0),$$

and finally, using Remark 9.33, we obtain for continuous-time systems

$$\boldsymbol{\Omega}_{\text{sp-CSS}}^{h\infty-\text{aid}}(A, B, C, D, E_1, 0) = \boldsymbol{\Omega}_{\text{p-CSS}}^{h\infty-\text{aid}}(A, B, C, D, E, F)$$

given that $E_1 = E - P^*C$, whereas $F - P^*D = 0$. For discrete-time systems, using Remark 9.41, we have

$$\boldsymbol{\Omega}_{\text{sp-CSS}}^{h2-\text{aid}}(A, B, C, D, E_1, 0) \subset \boldsymbol{\Omega}_{\text{p-CSS}}^{h2-\text{aid}}(A, B, C, D, E, F),$$

where some fixed modes at the origin is the only possible difference between these two sets. ∎

Obviously, we would like to explore next the development of an algorithm to design H_∞ AID filters of reduced-order proper CSS architecture, which gives us the flexibility to place the finite asymptotic modes of the filters at desired locations, of course, within the constraints imposed by Theorem 9.46. Such an algorithm can be given by the following steps:

(*i*) Use the quintuple $(A'_r, C'_r, B'_r, D'_r, E'_r)$ as the input to the H_∞ ADD Algorithm (see Subsection 6.6.1 for continuous-time systems and Subsection 6.6.2 for discrete-time systems), and obtain a parameterized gain F^ε.

(*ii*) Let $K^\varepsilon_r = (F^\varepsilon)'$. Partition K^ε_r in accordance with (9.49), and then define $\widetilde{K}^\varepsilon_r$ and $\widetilde{P}^\varepsilon_r$, respectively, in accordance with (9.52) and (9.53).

The H_∞ ADD algorithm mentioned in the above first step yields the set

$$\Omega^{h\infty-\mathrm{aid}}_{\mathrm{sp\text{-}CSS}}(A_r, B_r, C_r, D_r, E_r, 0) = \Omega^{h\infty-\mathrm{aid}}_{\mathrm{r\text{-}CSS}}(A, B, C, D, E, F).$$

The above equality follows from Theorem 9.46. Also, the said H_∞ ADD algorithm has certain flexibility to place the finite asymptotic modes of the filter as desired but within the constraints imposed by Theorem 9.46.

10

Optimally (suboptimally) input-decoupling filtering under white noise input — H_2 filtering

10.1   Introduction

Chapter 7 considers the exact input-decoupled (EID) filtering problem, whereas Chapters 8 and 9 consider almost input-decoupled (AID) filtering problems. EID filtering seeks perfect performance; i.e., it tries to make the impact of the input on the estimation error signal absolutely zero. AID filtering relaxes this requirement by trying to find conditions under which the impact of the input on the error signal can be made arbitrarily small. In particular, Chapter 8 pertains to AID filtering under white noise input, whereas Chapter 9 pertains to AID filtering without any statistical information on the input. AID filtering under white noise input seeks conditions such that one can render the RMS norm of the error signal as small as *desired*. On the other hand, AID filtering under no statistical information on the input seeks conditions such that one can render the ratio of RMS norm of the error signal to the RMS norm of the input as small as *desired*. In this chapter, we relax the requirement even further by seeking the impact of the input on the error signal be as small as *possible* rather than as small as desired. In particular, we follow here the direction set by AID filtering under white noise input. That is, we assume here that the input to the given system is a white noise of unit intensity and seek to make the RMS norm of the error signal as small as *possible*. The problems we deal with here are called optimally input-decoupling (OID) filtering problems under white noise input. The corresponding filters are of course termed as OID filters under white noise input. In a later chapter, we will consider OID filtering problems without any statistical information on the input.

Another concept highly tied to OID filtering is suboptimally input-decoupling (SOID) filtering. In the absence of a formal definition of suboptimality, any filter that is not optimal can be construed as a suboptimal filter. A good definition of suboptimality can be given through the notion of attaining an RMS norm of the error signal arbitrarily close to its infimum. In this regard, a sequence or a family of filters can be called suboptimal filters if one can select a filter from the family such that the resulting RMS norm of the error signal is within an arbitrarily speci-fied value from its infimum. A question that arises next is why one would entertain SOID filtering rather than OID filtering. Obviously, a given system may not sat-isfy the conditions under which OID filters exist. By relaxing the requirement of optimality to suboptimality, one hopes to weaken such existence conditions.

As we said in Chapter 1, the origins of filtering theory can be traced to ancient times and took root in the work of Karl Friederich Gauss, who is generally acknowledged to be the forefather of what is now referred to as estimation theory or filtering theory. The pioneering works of Kolmogorov and Wiener laid foundations to filtering theory. The seminal work of Kalman solidified the field under the umbrella of what is now generally referred to as Kalman filtering. Kalman filtering, which falls in the domain of what we now call here OID filtering under white noise input, has been pursued in a variety of fronts, and an unimaginable number of applications are available covering a vast and diverse scientific disciplines. It is not our intention here to survey the field. Our interest here is to formulate general OID and SOID filtering problems under white noise input and then to develop the solvability conditions for such problems. Under the assumption that the solvability conditions are satisfied, we then develop methods of designing filters that solve the formulated problems.

10.2   Preliminaries

Let us reconsider the plant or system model given in (7.1) and rewritten here as

$$\Sigma : \begin{cases} \sigma x = Ax + Bu, \\ y \ = Cx + Du, \\ z \ = Ex + Fu, \end{cases} \tag{10.1}$$

where, as before, $u \in \mathbb{R}^m$ is the input. We assume that u is a zero mean widesense stationary white noise stochastic process of unit intensity. Also, as before, $x \in \mathbb{R}^n$ is the state, $y \in \mathbb{R}^p$ is the measured output, and $z \in \mathbb{R}^q$ is the desired output signal to be estimated.

As usual, our interest lies in estimating the desired output signal z while using only the measured output y but not the input u. Let $\hat{z}$ be the estimate of z as given by a filter, and let e_z be the estimation error, $e_z = z - \hat{z}$ as depicted in Figure 10.1.

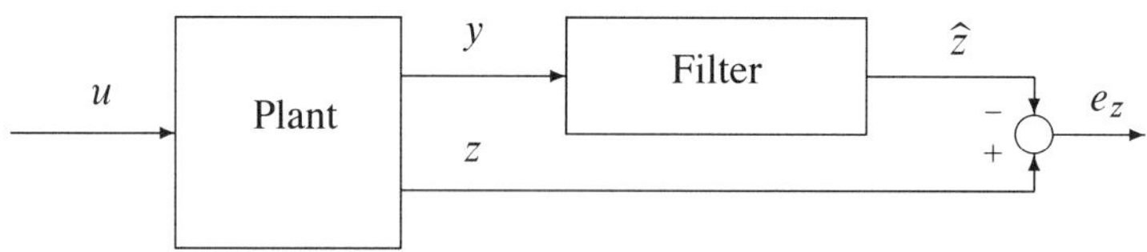

Figure 10.1: General block diagram

As before, it is natural to use the following assumption throughout this chapter as well.

Assumption 10.1 *The matrix pair* (C, A) *is* $\mathbb{C}^-$*-detectable for continuous-time systems and* $\mathbb{C}^\ominus$*-detectable for discrete-time systems.*

Again, as before, we consider a general proper filter of the form (7.2), which is repeated below:

$$\Sigma_f : \begin{cases} \sigma\xi = L\xi + My \\ \hat{z} \ = N\xi + Py. \end{cases} \tag{10.2}$$

Whenever $P = 0$, the above filter is said to be a strictly proper filter. When the above filter is used as shown in Figure 10.1, the dynamic equations of the error e_z are described by

$$\Sigma^{ue} : \begin{cases} \sigma x = Ax + Bu \\ \sigma\xi = MCx + L\xi + MDu \\ e_z \ = (E - PC)x - N\xi + (F - PD)u. \end{cases} \tag{10.3}$$

Hence, the transfer matrix from u to e_z can be computed as

$$G^{ue}(s) = \begin{pmatrix} E - PC & -N \end{pmatrix} \begin{pmatrix} sI - A & 0 \\ -MC & sI - L \end{pmatrix}^{-1} \begin{pmatrix} B \\ MD \end{pmatrix} + (F - PD). \tag{10.4}$$

10.3   OID and SOID filtering problems with white noise input

In this section, we formally define optimal and suboptimal input-decoupling (OID and SOID) filtering problems, while using the class of linear stable unbiased filters. We start this by recalling from Definition 7.2 what we mean by unbiased filters.

Definition 10.2 *Consider a continuous- or discrete-time system* Σ *as in* (10.1). *We say a linear stable strictly proper (or proper) filter* (10.2) *is* **unbiased** *if, in the absence of the input* u, *the estimation error* e_z *decays asymptotically to zero for all possible finite initial values of the system* (10.1) *and the filter* (10.2).

Next, we define the optimal performance that one can achieve under white noise input.

Definition 10.3 *Consider a continuous- or discrete-time system* Σ *of the form* (10.1) *where the input* u *is a zero-mean wide-sense stationary white noise stochastic process of unit intensity. The infimum of the RMS norm of the error signal* e_z

over the set of all linear stable strictly proper (or proper) unbiased filters is called the **optimal input-decoupling (OID) filtering performance under white noise input** *via linear stable strictly proper (or proper) unbiased filters and is denoted by* γ_{sp}^* *(or* γ_p^**).*

We are now ready to define the OID filtering problem under white noise input.

Problem 10.4 Consider a continuous- or discrete-time system Σ as in (10.1) where the input u is a zero mean wide-sense stationary white noise stochastic process of unit intensity. Then, the **OID filtering problem under white noise input** is defined as follows: Find, whenever it exists, a linear stable strictly proper (or proper) filter such that

(*i*) (**Unbiasedness**) the estimation error e_z, in the absence of the input u, decays asymptotically to zero for all possible finite initial values of the system (10.1) and the filter (10.2), and

(*ii*) (**Performance**) the RMS norm of the error signal, namely $\|e_z\|_{\mathrm{RMS}}$, is as small as possible; i.e., $\|e_z\|_{\mathrm{RMS}} = \gamma_{sp}^*$ (or $\|e_z\|_{\mathrm{RMS}} = \gamma_p^*$).

Given Definition 10.3, the minimal achievable RMS norm is equal to γ_{sp}^* or γ_p^* depending on whether we use strictly proper or proper filters. The above problem tries to find a filter that attains this infimum. However, depending on the given system Σ, one may not be able to attain such an infimum. In such a case, one can try to find a family of suboptimal strictly proper or proper filters having the characteristic that one can always select a filter from the family such that the resulting RMS norm of the error signal is within an arbitrarily specified value from the infimum γ_{sp}^* or γ_p^*. This leads to the following problem formulation.

Problem 10.5 Consider a continuous- or discrete-time system Σ as in (10.1) where the input u is a zero-mean wide-sense stationary white noise stochastic process of unit intensity. Then, the **SOID filtering problem under white noise input** is defined as follows: Find, whenever it exists, a family of linear stable strictly proper (or proper) filters parameterized in positive ε such that

(*i*) (**Unbiasedness**) for any given filter in the family, the estimation error e_z, in the absence of the input u, decays asymptotically to zero for all possible finite initial values of the system (10.1) and the filter (10.2), and

(*ii*) (**Performance**) the RMS norm of the error signal, namely $\|e_z\|_{\mathrm{RMS}}$, approaches γ_{sp}^* (or γ_p^*) as ε tends to zero.

The above OID and SOID filtering problems under white noise input can be given deterministic interpretations as in the case of the AID filtering problem. To do so, for any given filter, we proceed first to evaluate the PSD of the error signal e_z and then $\|e_z\|_{\mathrm{RMS}}$. Let G^{ue} denote the transfer matrix from the input u to the error e_z. Consider first continuous-time systems. Then, $G^{ue}(j\omega)$ is the Fourier transform of the impulse response of the system from u to e_z. As the input $u(t)$ is a white noise of unit intensity, the PSD of the error signal $e_z(t)$ can easily be determined as $G^{ue}(j\omega)(G^{ue})^H(j\omega)$, where as usual the superscript H denotes the complex conjugate transpose. Similarly, for discrete-time systems, $G^{ue}(e^{j\omega})$ is the discrete-time Fourier transform of the impulse response of the system from u to e_z, and thus, the PSD of the error signal $e_z(k)$ can be determined as $G^{ue}(e^{j\omega})(G^{ue})^H(e^{j\omega})$. Then, by the definitions of RMS norm and H_2 norm as in Section 2.6, we see that

$$\|e_z\|_{\mathrm{RMS}} = \|G^{ue}\|_2, \qquad (10.5)$$

where $\|G^{ue}\|_2$ is the H_2 norm of G^{ue}.

In view of (10.5), the OID and SOID filtering problems under white noise input as stated in Problems 10.4 and 10.5 can be given deterministic interpretations as described below.

Problem 10.6 Consider a continuous- or discrete-time system Σ as in (10.1) with white noise input. Let G^{ue} denote the transfer matrix from the input u to the error e_z. Then the $\boldsymbol{H_2}$ **OID filtering problem** is to find, whenever it exists, a linear stable strictly proper (or proper) filter such that

(Unbiasedness) the estimation error e_z, in the absence of the input u, decays asymptotically to zero for all possible finite initial values of the system (10.1) and the filter (10.2), and

(Performance) $\|G^{ue}\|_2$ is as small as possible.

A filter that achieves the above two objectives is called an $\boldsymbol{H_2}$ **OID filter**.

On the other hand, the $\boldsymbol{H_2}$ **SOID filtering problem under white noise input** is defined as follows: Find a family of linear stable strictly proper (or proper) filters parameterized in positive ε such that

(Unbiasedness) for any given filter in the family, the estimation error e_z, in the absence of the input u, decays asymptotically to zero for all possible finite initial values of the system (10.1) and the filter (10.2), and

(Performance) as ε tends to zero, $\|G^{ue}\|_2$ approaches its infimum over all linear stable strictly proper (or proper) unbiased filters.

A member of the family of filters that achieves the above two objectives is called an $\boldsymbol{H_2}$ **SOID filter**.

In view of the above definition, we can interpret γ^*_{sp} (or γ^*_p) as the infimum of the H_2 norm of the transfer matrix G^{ue} from u to e_z over the set of all unbiased linear strictly proper (or proper) filters. In other words, γ^*_{sp} (or γ^*_p) can be called the **H_2 OID filtering performance** via linear strictly proper (or proper) filters.

As we said in Section 10.1, the celebrated Kalman filtering falls in the domain of what we now call here OID filtering under white noise input. Most of the available literature on the Kalman filtering problem deals with what is known as a *regular* filtering problem. If it is not a *regular* filtering problem, it is said to be a *singular* filtering problem. We have the following formal definitions.

Definition 10.7 *Consider a continuous- or discrete-time system Σ as in (10.1) with white noise input.*

> *(i) For a continuous-time system Σ, a **regular H_2 OID filtering problem** refers to an H_2 OID filtering problem in which the matrix D is surjective, and the subsystem characterized by the quadruple (A, B, C, D) has no invariant zeros on the imaginary axis.*

> *(ii) For a discrete-time system Σ, a **regular H_2 OID filtering problem** refers to an H_2 OID filtering problem in which the subsystem characterized by the quadruple (A, B, C, D) is right-invertible and has no invariant zeros on the unit circle.*

Definition 10.8 *Consider a continuous- or discrete-time system Σ as in (10.1) with white noise input. An H_2 OID filtering problem is said to be a **singular H_2 OID filtering problem** if it is not a regular H_2 OID filtering problem.*

We will see that an H_2 OID filter always exists for the regular case. The regular case is the one that is always featured predominantly in many textbooks and hence creates the impression that an H_2 OID filter always exists even for the singular case. However, as will be seen subsequently, this is not the case!

10.4   Connection between H_2 OID (H_2 SOID) and EID (H_2 AID) filtering problems—continuous-time case

In this section, we develop two fundamental theorems that enable us to connect the H_2 OID (SOID) filtering problems for a given system to that of EID (H_2 AID) filtering problems for an auxiliary system Σ_Q. Also, in subsequent chapters, we will see that other types of filtering problems for one system can be connected to either EID or AID filtering problems for another system. Such a development is significant as it shows that EID and AID filtering problems play crucial and central roles in filtering theory.

The above-mentioned results provide us a *roadmap* for a variety of filtering issues such as

(*i*) computing the H_2 OID filtering performance via linear stable strictly proper (or proper) unbiased filters, namely, γ_{sp}^* (or γ_p^*),

(*ii*) determining performance limitations of H_2 OID filtering owing to the structural properties of a given system,

(*iii*) developing the existence and uniqueness conditions for H_2 OID (SOID) filters,

(*iv*) developing the design methodologies for H_2 OID (SOID) filters, and

(*v*) studying the structural properties of H_2 OID (SOID) filters.

Some of the above issues are dealt with in separate sections that follow this section. Also, owing to the technical nature of our development, continuous- and discrete-time systems are treated separately. Our task in this section is to develop for continuous-time systems the aforementioned fundamental theorems and their immediate consequences.

Our first task is to define the auxiliary system Σ_Q. To do so, certain preliminaries are necessary. To start with, let us consider a linear matrix inequality (CLMI):

$$G(Q) \geq 0, \tag{10.6}$$

where

$$G(Q) := \begin{pmatrix} AQ + QA' + BB' & QC' + BD' \\ CQ + DB' & DD' \end{pmatrix}.$$

In the expression for $G(Q)$, the matrix Q is unknown, whereas the matrix quadruple (A, B, C, D) corresponds to the data of the given system Σ as in (10.1). We are interested in a semi-stabilizing solution Q of the CLMI (10.6). By Assumption 10.1, the pair (C, A) is $\mathbb{C}^-$-detectable, and hence by applying Theorem 4.116, we can conclude that a unique semi-stabilizing solution Q of the CLMI (10.6) exists. Moreover, such a solution Q is positive semi-definite, rank minimizing, and is the largest among all symmetric solutions. We can determine such a solution by following the procedure described in Section 4.3.1.

Remark 10.9 *Whenever the matrix D is surjective, one can equivalently determine Q by solving for the unique semi-stabilizing solution Q of an H_2-CARE given by*

$$QA' + AQ + BB' - (QC' + BD')(DD')^{-1}(CQ + DB') = 0. \tag{10.7}$$

We note that, for a regular H_2 OID filtering problem, the unique semi-stabilizing solution of the H_2-CARE (10.7) is indeed stabilizing rather than merely semi-stabilizing. Thus, for a regular H_2 OID filtering problem, one can obtain the matrix Q equivalently by solving for the unique stabilizing solution of H_2-CARE (10.7) rather than by solving the CLMI (10.6).

Next, once the matrix Q is determined as discussed above, we can define matrices $B_Q \in \mathbb{R}^{n \times \rho}$ and $D_Q \in \mathbb{R}^{p \times \rho}$ with $\rho = \operatorname{rank} G(Q)$ such that

$$G(Q) = \begin{pmatrix} B_Q \\ D_Q \end{pmatrix} \begin{pmatrix} B'_Q & D'_Q \end{pmatrix}. \tag{10.8}$$

The condition that Q is semi-stabilizing, as defined in Definition 4.109, can then be rewritten as

$$\operatorname{rank} \begin{pmatrix} A - sI & B_Q \\ C & D_Q \end{pmatrix} = n + \operatorname{rank} \begin{pmatrix} B_Q \\ D_Q \end{pmatrix} \tag{10.9}$$

for all $s \in \mathbb{C}^+$.

Next, we know that to achieve a finite RMS norm a matrix P^* must exist such that $F - P^* D = 0$. We can then define the system:

$$\Sigma_Q : \begin{cases} \dot{\tilde{x}} = A\tilde{x} + B_Q \tilde{u} \\ \tilde{y} = C\tilde{x} + D_Q \tilde{u} \\ \tilde{z} = E^* \tilde{x}, \end{cases} \tag{10.10}$$

with

$$E^* = E - P^* C. \tag{10.11}$$

For the above auxiliary system Σ_Q, we consider filters of the form:

$$\tilde{\Sigma}_f : \begin{cases} \dot{\xi} = L\xi + M\tilde{y} \\ \hat{\tilde{z}} = N\xi + (P - P^*)\tilde{y}. \end{cases} \tag{10.12}$$

We next present two main results. The first one relates the system Σ_Q to the system Σ. The second one shows that AID filtering is always possible for the auxiliary system Σ_Q.

Theorem 10.10 *Consider a continuous-time system Σ given by (10.1) where (C, A) is $\mathbb{C}^-$-detectable. Let Q be the unique semi-stabilizing solution of the CLMI (10.6). Also, assume that $F - PD = 0$ has a solution for P, and let P^* be any such solution; then define the auxiliary system Σ_Q given by (10.10). Moreover, given any proper filter of the form (10.2), construct the filter (10.12). Assume that the filter (10.2), applied to the system Σ, results in an error signal*

e_z, whereas the filter (10.12), applied to the system Σ_ϱ, results in an error signal $\tilde{e}_z$. Then, the filter (10.2) applied to the system Σ is unbiased and achieves a finite RMS norm of the error signal e_z if and only if the filter (10.12) applied to the system Σ_ϱ is unbiased and achieves a finite RMS norm of the error signal $\tilde{e}_z$. Moreover, in this case, the resulting RMS norms of the error signals are related by

$$\|e_z\|_{\mathrm{RMS}}^2 = \|\tilde{e}_z\|_{\mathrm{RMS}}^2 + \operatorname{trace}((E - P^*C)Q(E - P^*C)').$$

Proof : We first note that the structure of both systems Σ and Σ_ϱ are very closely related, and that it is trivial to check that the filter (10.2) applied to the system Σ is unbiased if and only if the filter (10.12) applied to the system Σ_ϱ is unbiased. Given that the filter in both cases is unbiased, we can transform the filters into the form (7.70) as presented in Lemma 7.68. Note that this transformation is independent of whether we are working with the original system Σ or the transformed system Σ_ϱ.

A filter of the form (10.2) when applied to Σ is equivalent to applying a filter with stable transfer matrix $\hat{G}_f$ to the system:

$$\bar{\Sigma} : \begin{cases} \dot{\bar{x}} = (A - KC)\bar{x} + (B - KD)u \\ \bar{y} = C\bar{x} + Du \\ \bar{z} = E\bar{x} + Fu. \end{cases}$$

On the other hand, applying the filter (10.12) to Σ_ϱ is equivalent to applying a filter with transfer matrix $\hat{G}_f - P^*$ to the system:

$$\bar{\Sigma}_\varrho : \begin{cases} \dot{\check{x}} = (A - KC)\check{x} + (B_\varrho - KD_\varrho)\tilde{u} \\ \check{y} = C\check{x} + D_\varrho\tilde{u} \\ \check{z} = E^*\check{x}. \end{cases}$$

Note that with respect to these two transformed systems that we created, we still have the relationship:

$$\begin{pmatrix} A_K Q + Q A_K' + B_K B_K' & QC' + B_K D' \\ CQ + DB_K' & DD' \end{pmatrix} = \begin{pmatrix} B_\varrho - KD_\varrho \\ D_\varrho \end{pmatrix} \begin{pmatrix} B_\varrho - KD_\varrho \\ D_\varrho \end{pmatrix}',$$

where $A_K = A - KC$ and $B_K = B - KC$. Note that the transfer matrix of the error dynamics of the filter given by (10.2) when applied to Σ can hence be rewritten as

$$G_1 + (\hat{G}_f - P)G_2,$$

with G_1 and G_2 given by

$$G_1(s) = (E - PC)(sI - A + KC)^{-1}(B - KD) + (F - PD),$$
$$G_2(s) = C(sI - A + KC)^{-1}(B - KD) + D.$$

Similarly, the error dynamics of the filter given by (10.12) when applied to Σ_Q can be rewritten as

$$G_{Q,1} + (\hat{G}_f - P)G_{Q,2},$$

with $G_{Q,1}$ and $G_{Q,2}$ given by

$$G_{Q,1}(s) = (E - PC)(sI - A + KC)^{-1}(B_Q - KD_Q) + (P^* - P)D_Q,$$
$$G_{Q,2}(s) = C(sI - A + KC)^{-1}(B_Q - KD_Q) + D_Q.$$

Note that we have $F - PD = 0$ if and only if $(P^* - P)D_Q = 0$, and it is clear that for a finite RMS norm of the error dynamics, we need to impose $F - PD = 0$. Also, we have

$$\|e_z\|_{\mathrm{RMS}}^2 = \langle G_{cl}, G_{cl}\rangle,$$

where $\langle \cdot, \cdot \rangle$ is the inner product associated with the H_2 norm as clarified in Section 2.6 and G_{cl} is the transfer matrix from the noise u to the error signal e_z. Note that one property of this inner product is important to us, namely,

$$\langle X, Y \rangle = 0$$

when X and Y are stable and antistable transfer matrices, respectively. Using some basic algebraic manipulations, it is easy to show that

$$G_2 G_2^* = G_{Q,2}G_{Q,2}^*,$$
$$G_1 G_2^* - G_{Q,1}G_{Q,2}^* \quad \text{is stable},$$
$$\langle G_1, G_1 \rangle = \langle G_{Q,1}, G_{Q,1} \rangle + \mathrm{trace}((E - PC)Q(E - PC)'),$$

where we used that $F - PD = 0$. The last property is a consequence of the characterization of the H_2 norm in terms of the solution of a Lyapunov equation as given in Section 2.6. As Q satisfies the CLMI (10.6), it can be easily verified that $\mathrm{im}\, Q \subset C^{-1} \mathrm{im}\, D$, which then implies that $\mathrm{im}\, CQ \subset \mathrm{im}\, D$. As $F - PD = 0$ and $F - P^*D = 0$, we find $(P - P^*)D = 0$, which then implies that $(P - P^*)CQ = 0$. But then, it implies that the $\mathrm{trace}((E - PC)Q(E - PC)')$ remains the same for all solutions P of $F - PD = 0$. Hence, we have

$$\langle G_1, G_1 \rangle = \langle G_{Q,1}, G_{Q,1} \rangle + \mathrm{trace}((E - P^*C)Q(E - P^*C)').$$

From these three properties, the theorem follows directly.   ∎

The above theorem will be crucial in our further development because in trying to design an H_2 OID filter or a family of H_2 SOID filters for the original system Σ, according to the above theorem, we can then equally well do these designs for the auxiliary system Σ_Q. This is, of course, useful only if the auxiliary system has desirable properties, and this is the topic of the following theorem.

Theorem 10.11 *Consider a continuous-time system Σ given by (10.1) where (C, A) is $\mathbb{C}^-$-detectable. Let Q be the unique semi-stabilizing solution of the CLMI (10.6). Also, assume that $F - PD = 0$ has a solution for P, and let P^* be any such solution; then define the auxiliary system Σ_Q by (10.10). Then, the H_2 AID filtering problem is solvable for Σ_Q by using strictly proper filters.*

Proof : The system Σ_Q satisfies the property (10.9), and hence, it is clear that the system characterized by the quadruple (A, B_Q, C, D_Q) is left-invertible. But then by Remark 3.34, we have

$$\mathcal{V}^*(A, B_Q, C, D_Q) \cap \mathcal{S}^*(A, B_Q, C, D_Q) = \{0\}.$$

But (10.9) also implies that the system has no invariant zeros in the open right-half plane, and hence,

$$\mathcal{S}^*(A, B_Q, C, D_Q) = \mathcal{S}^{-0}(A, B_Q, C, D_Q).$$

The last two properties together imply that

$$\mathcal{V}^*(A, B_Q, C, D_Q) \cap \mathcal{S}^{-0}(A, B_Q, C, D_Q) = \{0\}.$$

By Theorem 8.5, this implies that the H_2 almost input-decoupled filtering problem for Σ_Q is solvable via a family of strictly proper filters. ∎

Theorems 10.10 and 10.11 have many consequences. We provide below two such consequences that relate the H_2 OID and H_2 SOID filtering problems for the given system Σ to EID and H_2 AID filtering problems for the auxiliary system Σ_Q. To do so, let us first consider strictly proper filters of the type Σ_f given in (10.2) with $P = 0$, and redraw Figure 10.1 for the given system Σ and for Σ_Q as shown in Figure 10.2. Also, observe that H_2 OID and SOID filtering problems are solvable by strictly proper filters only if $F = 0$.

Corollary 10.12 *Consider Figure 10.2, where Σ is a continuous-time system as in (10.1) with (C, A) $\mathbb{C}^-$-detectable and $F = 0$, Σ_Q is the auxiliary system given by (10.10) with $P^* = 0$ (and thus $E^* = E$), and Σ_f is a strictly proper filter of the type given in (10.2) with $P = 0$. Then, the following two statements are equivalent:*

(i) Σ_f is a strictly proper H_2 OID filter for the system Σ.

(ii) Σ_f is a strictly proper EID filter for the auxiliary system Σ_Q.

Let us next consider a proper filter of the type Σ_f given in (10.2) for Σ. Also, for Σ_Q, consider a proper filter of the type $\widetilde{\Sigma}_f$ given in (10.12) where the matrix

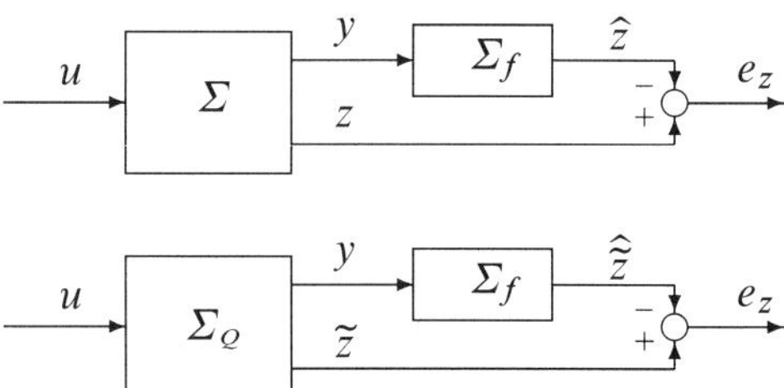

Figure 10.2: Block diagram representation of filtering for two related systems—strictly proper filters

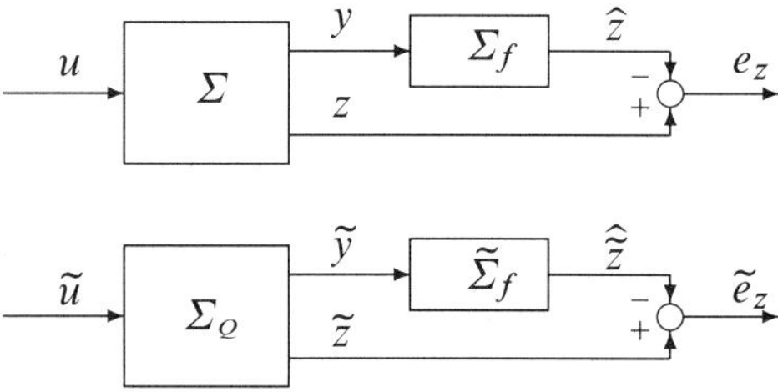

Figure 10.3: Block diagram representation of filtering for two related systems—proper filters

P^* is such that $F - P^*D = 0$. Moreover, let us redraw Figure 10.1 for the given system Σ and for the auxiliary system Σ_Q as shown in Figure 10.3. Also, observe that H_2 OID and H_2 SOID filtering problems can perhaps be solvable by proper filters (unlike in the case of strictly proper filters) even if $F \neq 0$.

Corollary 10.13 *Consider a continuous-time system Σ as in (10.1) with (C, A) $\mathbb{C}^-$-detectable. Assume that $F - PD = 0$ has a solution for P, and let P^* be any such solution; then define the auxiliary system Σ_Q given by (10.10). Consider Figure 10.3, where Σ_f is a proper filter of the type given in (10.2) and $\tilde{\Sigma}_f$ is a proper filter of the type given in (10.12). Then, the following two statements are equivalent:*

(i) *Σ_f is a proper H_2 OID filter for the system Σ.*

(ii) *$\tilde{\Sigma}_f$ is a proper EID filter for the auxiliary system Σ_Q.*

Corollaries 10.12 and 10.13 show that H_2 OID filtering problems for the given system Σ can be related to EID filtering problems for the auxiliary system Σ_Q. Along the same lines, one can show that H_2 SOID filtering problems for the given system Σ can be related to H_2 AID filtering problems for the auxiliary system Σ_Q. To do so, let us consider a family of parameterized filters of the form:

$$\Sigma_f^\varepsilon : \begin{cases} \dot{\xi} = L^\varepsilon \xi + M^\varepsilon y \\ \hat{z} = N^\varepsilon \xi + P^\varepsilon y, \end{cases} \tag{10.13}$$

where L^ε, M^ε, N^ε, and P^ε are matrices parameterized in a positive parameter ε.

We have the following corollary when the class of strictly proper filters are used. We observe that H_2 SOID filtering problems are solvable by strictly proper filters only if $F = 0$.

Corollary 10.14 *Consider a continuous-time system Σ as in (10.1) with (C, A) $\mathbb{C}^-$-detectable. Consider Figure 10.2, where Σ_f is to be replaced by Σ_f^ε as given in (10.13) but with $P^\varepsilon = 0$. Also, let Σ_Q be the auxiliary system given in (10.10) with $P^* = 0$ and thus $E^* = E$. Then, the following two statements are equivalent:*

(i) *The family of filters Σ_f^ε is a family of strictly proper H_2 SOID filters for the system Σ.*

(ii) *The family of filters Σ_f^ε is a family of strictly proper H_2 AID filters for the auxiliary system Σ_Q.*

A result similar to the above corollary can be obtained when the class of proper filters are used. Before we state the result, consider the family of parameterized filters:

$$\widetilde{\Sigma}_f^\varepsilon : \begin{cases} \dot{\xi} = L^\varepsilon \xi + M^\varepsilon \tilde{y} \\ \hat{\tilde{z}} = N^\varepsilon \xi + (P^\varepsilon - P^*)\tilde{y}. \end{cases} \tag{10.14}$$

We have the following corollary when the class of proper filters is used. We observe that H_2 SOID filtering problems can perhaps be solvable by proper filters (unlike in the case of strictly proper filters) even if $F \neq 0$.

Corollary 10.15 *Consider a continuous-time system Σ as in (10.1) with (C, A) $\mathbb{C}^-$-detectable. Assume that $F - PD = 0$ has a solution for P, and let P^* be any such solution; then define the auxiliary system Σ_Q given by (10.10). Consider Figure 10.3, where Σ_f is to be replaced by Σ_f^ε as given in (10.13) and $\widetilde{\Sigma}_f$ is to be replaced by $\widetilde{\Sigma}_f^\varepsilon$ as given in (10.14). Then, the following two statements are equivalent:*

(i) *The family of filters Σ_f^ε is a family of proper H_2 SOID filters for the system Σ.*

(ii) *The family of filters $\widetilde{\Sigma}_f^\varepsilon$ is a family of proper H_2 AID filters for the auxiliary system Σ_Q.*

10.5   Computation of γ_{sp}^* and γ_p^* — continuous-time case

In this section, with the help of Theorems 10.10 and 10.11, we compute the H_2 OID filtering performance via linear stable strictly proper filters (namely, γ_{sp}^*) and via linear stable proper filters (namely, γ_p^*).

We have the following corollaries of Theorems 10.10 and 10.11.

Corollary 10.16 *Consider a continuous-time system as in (10.1). Let Assumption 10.1 be satisfied. Let Q be the unique semi-stabilizing solution of the CLMI (10.6). Then, the H_2 OID filtering performance via linear stable strictly proper filters (denoted by γ_{sp}^* as formulated in Definition 10.3) is finite if and only if $F = 0$, and it is given by*

$$\gamma_{sp}^* = \left(\mathrm{trace}(EQE')\right)^{1/2}. \tag{10.15}$$

Proof : Clearly, if we use only strictly proper filters, $F = 0$ is necessary for a finite RMS norm. The auxiliary system Σ_Q given in (10.10) is then defined by setting $P^* = 0$. By Theorem 10.11, the H_2 OID filtering performance for the system Σ_Q is equal to 0. But then, it is clear by Theorem 10.10 that the H_2 OID filtering performance for the system Σ is given by (10.15).   ∎

Remark 10.17 *From the results of Chapter 4, it is trivial to see that $\gamma_{sp}^* = 0$ if and only if*

$$\mathcal{S}^{-0}(A, B, C, D) \cap \mathcal{V}^*(A, B, C, D) \subseteq \ker E.$$

Obviously, whenever $\gamma_{sp}^* = 0$, the definition of H_2 OID filtering as given in 10.4 coincides with that of H_2 AID filtering as given in 8.2. In fact, as given in Theorem 8.5, the conditions for the solvability of H_2 AID filtering problem via strictly proper filters are the above condition and $F = 0$.

The above corollary considers strictly proper filters. Next, we would like to present a similar result when proper filters are used. We observe that whenever proper filters are considered, one must have $\ker D \subseteq \ker F$ (otherwise, $\|e_z\|_{\mathrm{RMS}}$ is not finite). In this case, the linear equation $F - PD = 0$ has at least one solution.

We have the following result.

Corollary 10.18 *Consider a continuous-time system as in* (10.1). *Let Assumption 10.1 be satisfied. Let Q be the unique semi-stabilizing solution of the CLMI* (10.6). *Then, the H_2 OID performance via linear stable proper filters (namely γ_p^* as defined in Definition 10.3) is finite if and only if* $\ker D \subseteq \ker F$, *and it is given by*

$$\gamma_p^* = \left(\operatorname{trace}((E - P^*C)Q(E - P^*C)')\right)^{1/2}, \qquad (10.16)$$

where P^* is any solution of the equation $F - PD = 0$ for P.

Proof : The condition $\ker D \subseteq \ker F$ is equivalent to the condition that a matrix P^* exits such that $F - P^*D = 0$. This is the condition necessary to guarantee that the closed-loop transfer matrix is strictly proper, which is obviously a necessary condition for a finite RMS norm. The auxiliary system Σ_Q given in (10.10) is then defined by using such a P^*. By Theorem 10.11, we can solve the H_2 AID filtering problem for Σ_Q by strictly proper filters, i.e., filters of the form (10.12) with $P = P^*$. We note that this family of filters is such that $\|\tilde{e}_z\|_{\text{RMS}} \to 0$ as $\varepsilon \downarrow 0$. But then, from Theorem 10.10, it is obvious that this family of filters yields in the limit the infimum of $\|e_z\|_{\text{RMS}}$. Hence, the infimum of the RMS norm of the error signal is equal to (10.16). ∎

Remark 10.19 *The expression for γ_p^* given in* (10.16) *appears to be dependent on the choice of P^*. However, in the proof of Theorem 10.10, it was already established that a different choice of P^*, which still satisfies $F - P^*D = 0$, does not affect the expression for γ_p^* given in* (10.16).

Remark 10.20 *Once again, from the results of Chapter 4, it is trivial to see that $\gamma_p^* = 0$ if and only if*

$$\mathcal{S}^{-0}(A, B, C, D) \cap \mathcal{V}^*(A, B, C, D) \subseteq \ker(E - P^*C).$$

Obviously, whenever $\gamma_p^* = 0$, the definition of H_2 OID filtering as given in 10.4 coincides with that of H_2 AID filtering as given in 8.2. As such, as given in Theorem 8.9, the above condition is the condition for the solvability of the H_2 AID filtering problem via proper filters.

10.5.1 Relationship between γ_{sp}^* and γ_p^* and the structural properties of Σ

We have already developed the expressions (10.15) and (10.16) to determine the H_2 OID filtering performance indices γ_{sp}^* and γ_p^* in terms of the matrix E and the unique semi-stabilizing solution Q of the CLMI (10.6). However, we need to study these expressions to gain some insight as to how γ_{sp}^* and γ_p^* are related to the structural constraints imposed by the subsystem characterized by the quadruple (A, B, C, D). The purpose of this subsection is to relate γ_{sp}^* and γ_p^* to such structural constraints. To do so, we first consider the study of γ_{sp}^* via (10.15). The study of γ_p^* via (10.16) is similar except the matrix E then is replaced by E^*.

As we said, strictly proper filters can be used only when $F = 0$. For this case, the system Σ given in (10.1) is characterized by the quintuple (A, B, C, D, E). As in Chapter 7, the study of a filtering problem for a given system Σ can be transformed to the study of a control problem for a dual system, represented here by Σ_d and characterized by the quintuple $(\bar{A}, \bar{B}, \bar{C}, \bar{D}, \bar{E})$, where

$$\bar{A} = A', \quad \bar{B} = C', \quad \bar{C} = B', \quad \bar{D} = D', \quad \text{and } \bar{E} = E'. \tag{10.17}$$

To view the structural details of the dual system Σ_d, we can transform its subsystem characterized by the quadruple $(\bar{A}, \bar{B}, \bar{C}, \bar{D})$ to SCB. Let $(\bar{\Gamma}_s, \bar{\Gamma}_i, \bar{\Gamma}_o)$ be the related state, input, and output transformation matrices. Let

$$\bar{\Gamma}_s^{-1}\bar{E} = \left((\bar{E}_a^-)' \quad (\bar{E}_a^0)' \quad (\bar{E}_a^+)' \quad (\bar{E}_b)' \quad (\bar{E}_c)' \quad (\bar{E}_d)' \right)'. \tag{10.18}$$

As in Chapter 4, we intend to write the unique semi-stabilizing solution Q of the CLMI (10.6) in terms of a certain CARE. To do so, we extract the following matrices from the SCB constructed above:

$$\bar{A}_s := \begin{pmatrix} \bar{A}_{aa}^+ & \bar{L}_{ab}^+\bar{C}_b \\ 0 & \bar{A}_{bb} \end{pmatrix}, \qquad \bar{B}_s := \begin{pmatrix} \bar{B}_{a0}^+ & \bar{L}_{ad}^+ \\ \bar{B}_{b0} & \bar{L}_{bd} \end{pmatrix}, \tag{10.19}$$

$$\bar{C}_s := \bar{\Gamma}_o \begin{pmatrix} 0 & 0 \\ 0 & 0 \\ 0 & \bar{C}_b \end{pmatrix}, \qquad \bar{D}_s := \bar{\Gamma}_o \begin{pmatrix} I_{\bar{m}_0} & 0 \\ 0 & \bar{C}_d\bar{C}_d' \\ 0 & 0 \end{pmatrix}, \tag{10.20}$$

and

$$\bar{E}_s = \left((\bar{E}_a^+)' \quad (\bar{E}_b)' \right)'. \tag{10.21}$$

We remark that various submatrices in the above definitions come from the SCB as applied to the subsystem characterized by the quadruple $(\bar{A}, \bar{B}, \bar{C}, \bar{D})$(for details, see Chapter 3). Then, in view of Chapter 4, the unique semi-stabilizing solution Q of the CLMI (10.6) is given by

$$Q = (\bar{\Gamma}_s^{-1})' \begin{pmatrix} 0 & 0 & 0 & 0 & 0 \\ 0 & 0 & 0 & 0 & 0 \\ 0 & 0 & Q_s & 0 & 0 \\ 0 & 0 & 0 & 0 & 0 \\ 0 & 0 & 0 & 0 & 0 \end{pmatrix} \bar{\Gamma}_s^{-1}, \tag{10.22}$$

where Q_s is the solution of H_2 CARE:

$$Q_s\bar{A}_s + \bar{A}_s'Q_s + \bar{C}_s'\bar{C}_s$$
$$- (Q_s\bar{B}_s + \bar{C}_s'\bar{D}_s)(\bar{D}_s'\bar{D}_s)^{-1}(\bar{B}_s'Q_s + \bar{D}_s'\bar{C}_s) = 0. \tag{10.23}$$

The above H_2 CARE can be seen to arise from a standard linear quadratic optimization problem. To see this, let us define a state $\bar{x}_s$ as

$$\bar{x}_s = \begin{pmatrix} \bar{x}_a^+ \\ \bar{x}_b \end{pmatrix},$$

and a subsystem $\bar{\Sigma}_{\text{sub}}$ as,

$$\bar{\Sigma}_{\text{sub}} : \quad \begin{cases} \dot{\bar{x}}_s = \bar{A}_s \bar{x}_s + \bar{B}_s \bar{u}_s \\ \bar{z}_s = \bar{C}_s \bar{x}_s + \bar{D}_s \bar{u}_s, \end{cases} \tag{10.24}$$

with $\bar{x}_s(0) = \bar{E}_{si}$, where $\bar{E}_{si}$ is the ith column of $\bar{E}_s$. Clearly, in $\bar{\Sigma}_{\text{sub}}$, $\bar{x}_s$ is the state, $\bar{z}_s$ is the output, and $\bar{u}_s$ is the controlling input. One would like to minimize the performance measure J_{sub} by using static state feedback controllers, where

$$J_{\text{sub}} = \int_0^\infty \bar{z}_s' \bar{z}_s \, dt. \tag{10.25}$$

Then, it is easy to see that the infimum of J_{sub} over all possible static state feedback controllers is given by

$$J_{\text{sub}}^*(x_s(0)) = J_{\text{sub}}^*(\bar{E}_{si}) = \bar{E}_{si}' Q_s \bar{E}_{si}. \tag{10.26}$$

In view of the above development, and in particular in view of (10.15), (10.22), and (10.26), we have the following lemma that relates the H_2 OID filtering performance index γ_{sp}^* to the structural properties of the given system.

Lemma 10.21 *Consider a continuous-time system as in (10.1) with $F = 0$. Let Assumption 10.1 be satisfied. Also, let $J_{\text{sub}}^*(\bar{E}_{si})$ be as in (10.26). Then, the H_2 OID filtering performance via linear stable strictly proper filters, namely γ_{sp}^*, is given by*

$$(\gamma_{sp}^*)^2 = \sum_{i=1}^q J_{\text{sub}}^*(\bar{E}_{si}) = \sum_{i=1}^q \bar{E}_{si}' Q_s \bar{E}_{si}, \tag{10.27}$$

where q is the dimension of the output z in (10.1).

Before we comment on the importance of the above lemma, let us develop a result similar to the above, however, for γ_p^*. This can be done by slightly modifying the development given above. The required modification is to be made in (10.17) by replacing the matrix E by $E - P^*C$, with P^* being any solution of the equation $F - PD = 0$ for P. We have the following result.

Lemma 10.22 *Consider a continuous-time system as in* (10.1). *Let Assumption 10.1 be satisfied. Assume that* $\ker D \subseteq \ker F$. *Moreover, in* (10.17), *let E be replaced by $E - P^*C$, with P^* being any solution of the equation $F - PD = 0$ for P. Also, following the development subsequent to* (10.17) *and culminating in* (10.26), *obtain $J_{\mathrm{sub}}^*(\bar{E}_{si}^*)$ as in* (10.26) *(Note that in this case, $\bar{E}_{si}^*$ replaces $\bar{E}_{si}$). Then, the H_2 OID filtering performance via linear stable proper filters, namely γ_p^*, is given by*

$$(\gamma_p^*)^2 = \sum_{i=1}^{q} J_{\mathrm{sub}}^*(\bar{E}_{si}^*) = \sum_{i=1}^{q} (\bar{E}_{si}^*)' Q_s \bar{E}_{si}^*, \tag{10.28}$$

where q is the dimension of the output z in (10.1).

The above lemmas clearly show the roles played by $\bar{\Sigma}_{\mathrm{sub}}$ and $\bar{E}_s$ or $\bar{E}_s^*$ in dictating the values of γ_{sp}^* and γ_p^*. The subsystem $\bar{\Sigma}_{\mathrm{sub}}$ has two types of dynamics. The first type of dynamics is represented by the state $\bar{x}_a^+$, which is often called the *unstable zero dynamics*. It is present only when the subsystem characterized by (A, B, C, D) [and hence the dual subsystem characterized by $(\bar{A}, \bar{B}, \bar{C}, \bar{D})$] has invariant zeros in the open right-half plane. Such invariant zeros are given by the eigenvalues of $\bar{A}_{aa}^+$. The second type of dynamics is represented by the state $\bar{x}_b$, and it is present only when the subsystem characterized by (A, B, C, D) is non-left-invertible, and hence, the dual subsystem characterized by $(\bar{A}, \bar{B}, \bar{C}, \bar{D})$ is non-right-invertible.

Remark 10.23 *We observe that whenever the subsystem characterized by (A, B, C, D) is left-invertible and is at most weakly non-minimum phase (i.e., it has no invariant zeros in the open right-half plane $\mathbb{C}^+$), then both the states $\bar{x}_a^+$ and $\bar{x}_b$ are nonexistent. As such, both γ_{sp}^* and γ_p^* will then equal zero.*

Let us next assume that the subsystem characterized by (A, B, C, D) is left-invertible. Consequently, $\bar{x}_b$ does not exist. In this case, the subsystem $\bar{\Sigma}_{\mathrm{sub}}$ of (10.24) simplifies to

$$\bar{\Sigma}_{\mathrm{sub}} : \begin{cases} \dot{\bar{x}}_a^+ = \bar{A}_{aa}^+ \bar{x}_a^+ + \bar{B}_s \bar{u}_s, \\ \bar{z}_s = \bar{\Gamma}_o \bar{u}_s. \end{cases} \tag{10.29}$$

Also, in this case, as $\bar{z}_s = \bar{\Gamma}_o \bar{u}_s$, the performance measure J_{sub} has the interpretation of being the energy of control input.

Remark 10.24 *(Energy interpretation) Whenever the subsystem characterized by (A, B, C, D) is left-invertible, $(\gamma_{sp}^*)^2$ (or $(\gamma_p^*)^2$) equals the sum of minimum* *energies required to stabilize the unstable zero dynamics of the subsystem characterized by $(\bar{A}, \bar{B}, \bar{C}, \bar{D})$ for all initial conditions $\bar{x}_a^+(0) = \bar{E}_{si}, i = 1$ to q (or $\bar{x}_a^+(0) = \bar{E}_{si}^*, i = 1$ to q). Moreover, the impact of unstable zero dynamics on γ_{sp}^* and γ_p^* vanishes asymptotically as the unstable zeros (non-minimum-phase zeros) tend toward the imaginary axis.*

Remark 10.25 *The above study can be viewed from a different angle. This time by looking at the roles played by the matrices E and F, which dictate $\bar{E}_{si}$ and $\bar{E}_{si}^*$ in (10.27) and (10.28). We note that the matrices E and F are defined by the output z that is to be estimated. When viewed from the view point of $\bar{E}_{si}$ and $\bar{E}_{si}^*$ that appear in (10.27) and (10.28), the fundamental limitations to the H_2 OID filtering performance arise from the inclusion of two types of dynamics in the output z that is to be estimated, one is the unstable zero dynamics of the subsystem characterized by (A, B, C, D) (as represented here by the dynamics of the state $\bar{x}_a^+$), and the other is the non-left-invertible dynamics of the subsystem characterized by (A, B, C, D) (as represented here by the dynamics of the state $\bar{x}_b$). In the absence of both of these dynamics in the output z, the H_2 OID filtering performance indices γ_{sp}^* and γ_p^* simply equal zero.*

Whenever the subsystem characterized by (A, B, C, D) is left-invertible, we know already how the non-minimum-phase zeros of it affect the H_2 OID filtering performance indices γ_{sp}^* and γ_p^* (see Remark 10.24). We would like to investigate next how the location of the non-minimum-phase zeros of the same subsystem affect γ_{sp}^* and γ_p^* if it is not left-invertible. To do so, we decompose Q_s below into two parts, one arising due to unstable zero dynamics (i.e., the dynamics dictated by the open right-half plane invariant zeros) and the other due to non-left-invertible dynamics. This is done by assuming, without loss of generality, that the output transformation matrix $\bar{\Gamma}_o$ when the subsystem characterized by $(\bar{A}, \bar{B}, \bar{C}, \bar{D})$ is brought to its SCB representation is an identity matrix. Let

$$Q_s = \begin{pmatrix} 0 & 0 \\ 0 & Q_b \end{pmatrix} + Q_r. \tag{10.30}$$

Here Q_b is the stabilizing solution of the following H_2-CARE:

$$\bar{A}_{bb}'Q_b + Q_b\bar{A}_{bb} + \bar{C}_b'\bar{C}_b - Q_b\begin{pmatrix} \bar{B}_{b0} & \bar{L}_{bd} \end{pmatrix}(\bar{D}_s'\bar{D}_s)^{-1}\begin{pmatrix} \bar{B}_{b0}' \\ \bar{L}_{bd}' \end{pmatrix}Q_b = 0, \tag{10.31}$$

which exists because

$$(\bar{A}_{bb}, (\bar{B}_{b0} \quad \bar{L}_{bd}))$$

is controllable. In view of (10.30) and (10.31), it is easy to show that Q_r satisfies the CARE:

$$Q_r V + V'Q_r - Q_r\bar{B}_s(\bar{D}_s'\bar{D}_s)^{-1}\bar{B}_s'Q_r = 0, \tag{10.32}$$

with V given by

$$V = \begin{pmatrix} \bar{A}_{aa}^+ & \bar{L}_{ab}^+ \bar{C}_b - \bar{B}_{a0}^+ \bar{B}_{b0}' Q_b - \bar{L}_{ad}^+ (\bar{C}_d \bar{C}_d')^{-1} \bar{L}_{bd}' Q_b \\ 0 & \bar{A}_{bb} - \bar{L}_{bd}(\bar{C}_d \bar{C}_d')^{-1} \bar{L}_{bd}' Q_b \end{pmatrix} = \begin{pmatrix} V_{11} & V_{12} \\ 0 & V_{22} \end{pmatrix}.$$

As Q_b is a stabilizing solution of (10.31), we know that V_{22} is asymptotically stable. Next, we would like to rewrite Q_r in terms of another matrix $\bar{Q}_r$. To do so, let R be the solution of

$$\bar{A}_{aa}^+ R - R V_{22} + V_{12} = 0.$$

Note that R is bounded because V_{12} is bounded and the eigenvalues of $\bar{A}_{aa}^+$ and V_{22} are in the open right- and left-half plane, respectively, and hence bounded away from each other. We then find that

$$Q_r = \begin{pmatrix} I \\ R \end{pmatrix} \bar{Q}_r \begin{pmatrix} I & R' \end{pmatrix},$$

with $\bar{Q}_r$ being a stabilizing solution of the CARE:

$$\bar{Q}_r \bar{A}_{aa}^+ + (\bar{A}_{aa}^+)' \bar{Q}_r - \bar{Q}_r \bar{B}_{s1} (\bar{D}_s' \bar{D}_s)^{-1} \bar{B}_{s1}' \bar{Q}_r = 0, \tag{10.33}$$

where

$$\begin{pmatrix} I & -R \\ 0 & 0 \end{pmatrix} \bar{B}_s = \begin{pmatrix} \bar{B}_{s1} \\ \bar{B}_{s2} \end{pmatrix}.$$

For the stabilizing solution $\bar{Q}_r$ of (10.33) to exist, the pair $(\bar{A}_{aa}^+, \bar{B}_{s1})$ must be stabilizable. This is so because, after all, the original system was detectable and hence the dual system that we are currently working with is stabilizable. We observe that, if we move the open right-half plane invariant zeros of the subsystem (A, B, C, D), i.e., the eigenvalues of $\bar{A}_{aa}^+$, closer to the imaginary axis, then $\bar{Q}_r$ converges to zero provided Assumption 10.1 is still satisfied.

The above analysis implies that

$$Q_s = \begin{pmatrix} 0 & 0 \\ 0 & Q_b \end{pmatrix} + \begin{pmatrix} I \\ R \end{pmatrix} \bar{Q}_r \begin{pmatrix} I & R' \end{pmatrix}, \tag{10.34}$$

where matrices Q_b and $\bar{Q}_r$ are, respectively, the stabilizing solutions of (10.31) and (10.33). We note that Q_b depends only on the non-left-invertible dynamics, whereas $\bar{Q}_r$ depends only on the unstable zero dynamics (i.e., the dynamics dictated by the open right-half plane invariant zeros) of the subsystem characterized by (A, B, C, D). Having decomposed Q_s as in (10.34), it is easy to see that the H_2 OID filtering performance behaving as

$$\text{trace } E Q E' = \text{trace } W_1 Q_b W_1' + \text{trace } W_2 \bar{Q}_r W_2' \tag{10.35}$$

for some suitably defined matrices W_1 and W_2. Note that W_1 and W_2 are both bounded with respect to changes in the unstable invariant zeros of the system. The first term of (10.35) is completely determined by the non-left-invertible dynamics of the system, and the second term is determined by the unstable zero dynamics. The first term is not affected by the locations of the open right-half plane invariant zeros, i.e., independent of the matrix $\bar{A}_{aa}^{+}$, whereas the second term converges to zero if all the open right-half plane invariant zeros move to the imaginary axis provided Assumption 10.1 is still satisfied.

Let us summarize the above development. *In view of* (10.15) *and* (10.16) *for* γ_{sp}^{*} *and* γ_{p}^{*}, (10.35) *pinpoints that both* γ_{sp}^{*} *and* γ_{p}^{*} *can be decomposed into two parts, one arising due to unstable zero dynamics and the other due to non-left-invertible dynamics of the subsystem characterized by* (A, B, C, D). *The part that is contributed by the unstable zero dynamics to* γ_{sp}^{*} *and* γ_{p}^{*} *tends to zero as the open right-half plane invariant zeros move toward the imaginary axis, and in the same way, it increases as the open right-half plane invariant zeros move away from the imaginary axis.*

10.6   Existence of H_2 OID and SOID filters — continuous-time case

In what follows, we develop the conditions for the existence of H_2 OID filters for continuous-time systems. Note that we have already characterized γ_{sp}^{*} and γ_{p}^{*} in Corollaries 10.16 and 10.18.

We have the following theorem regarding the existence of a strictly proper H_2 OID filter.

Theorem 10.26 *Consider a continuous-time system Σ as in* (10.1). *Let Assumption 10.1 be satisfied. Then, a linear stable strictly proper filter of the form* (10.2) *exists with $P = 0$, which solves the H_2 OID filtering problem 10.6 for Σ if and only if $F = 0$ and $\mathcal{S}^{-}(A, B_Q, C, D_Q) \subseteq \ker E$, where the matrices B_Q and D_Q are as in* (10.8).

Proof : By Corollary 10.12, a strictly proper filter (10.2) (with $P = 0$) solves the H_2 OID filtering problem 10.6 for Σ if and only if $F = 0$, and moreover, such a strictly proper filter (10.2) solves the EID filtering problem for Σ_Q. From Theorem 7.6, we know that the EID filtering problem is solvable if and only if $\mathcal{S}^{-}(A, B_Q, C, D_Q) \subseteq \ker E$. This proves the theorem.   ∎

The above theorem considers strictly proper filters. Next, we would like to present results where proper filters are used. We observe that whenever proper filters are considered, one must have $\ker D \subseteq \ker F$ (otherwise, $\|G^{ue}\|_2$ is not finite). We have the following result.

Theorem 10.27 *Consider a continuous-time system Σ as in (10.1). Let Assumption 10.1 be satisfied. Then, a linear stable proper filter of the form (10.2) exists that solves the H_2 OID filtering problem 10.6 for Σ if and only if* $\ker D \subseteq \ker F$ *and*

$$\mathcal{S}^-(A, B_\varrho, C, D_\varrho) \cap C^{-1}\{\mathrm{im}\, D_\varrho\} \subseteq \ker(E - P^*C), \qquad (10.36)$$

where the matrices B_ϱ and D_ϱ are as in (10.8) and P^* is any solution of $F - PD = 0$ for P.

Proof : By Corollary 10.13, a filter (10.2) solves the H_2 OID filtering problem 10.6 for Σ if and only if the filter (10.12) achieves EID filtering for Σ_ϱ. By Theorem 7.7, the EID filtering problem for Σ_ϱ is solvable if and only if (10.36) is satisfied. The proof is now complete. ∎

Remark 10.28 *Under the condition that $\gamma_{sp}^* = 0$ (respectively, $\gamma_p^* = 0$), the conditions for the solvability of the H_2 OID filtering problem 10.6 for Σ via strictly proper (respectively, proper) filters coincide with the conditions for the solvability of the EID filtering problem 7.3 for Σ via strictly proper (respectively, proper) filters.*

For general singular H_2 OID filtering problems, the solvability conditions are indeed as asserted in Theorems 10.26 and 10.27. However, regular H_2 OID filtering problems are always solvable under the simple assumption that the pair (C, A) is $\mathbb{C}^-$-detectable, i.e., as long as Assumption 10.1 is satisfied. This is formalized in the following corollary of Theorems 10.26 and 10.27.

Corollary 10.29 *Consider a regular H_2 OID filtering problem as in Definition 10.7 for the continuous-time system Σ given in (10.1). Let Assumption 10.1 be satisfied. Then, the regular H_2 OID filtering problem is solvable via a linear strictly proper filter if $F = 0$ and via a proper stable filter if $\ker D \subseteq \ker F$.*

Proof : Note that for a regular problem, we have no invariant zeros on the imaginary axis for the subsystem (A, B, C, D). By Proposition 4.116, this implies that Q is a stabilizing solution of the CLMI (10.6), which implies that (10.9) is satisfied for all $s \in \mathbb{C}^+ \cup \mathbb{C}^0$. But then $(A, B_\varrho, C, D_\varrho)$ is minimum phase. This implies that

$$\mathcal{S}^*(A, B_\varrho, C, D_\varrho) = \mathcal{S}^-(A, B_\varrho, C, D_\varrho).$$

On the other hand, D is surjective implies that D_ϱ is also surjective and this is easily seen to imply that $\mathcal{S}^*(A, B_\varrho, C, D_\varrho) = \{0\}$. But then $\mathcal{S}^-(A, B_\varrho, C, D_\varrho) =$

$\{0\}$. Then, Theorem 10.26 implies that the H_2 OID filtering problem is solvable via a linear strictly proper filter provided $F = 0$, whereas Theorem 10.27 implies that the H_2 OID filtering problem is solvable via linear proper filter provided that $\ker D \subseteq \ker F$. ∎

The above discussion pertains to the existence of H_2 OID filters. We have the following result regarding the H_2 SOID filters.

Theorem 10.30 *Consider a continuous-time system as in* (10.1). *Let Assumption 10.1 be satisfied. Then, the following results hold:*

(*i*) *The H_2 SOID filtering problem via a family of linear stable strictly proper filters is solvable if and only if $F = 0$.*

(*ii*) *The H_2 SOID filtering problem via a family of linear stable proper filters is solvable if and only if $\ker D \subseteq \ker F$.*

Proof : The results follow trivially in view of Corollaries 10.14 and 10.15. ∎

10.7 Connection between H_2 OID (H_2 SOID) and EID (H_2 AID) filtering problems—discrete-time case

In Section 10.4, for continuous-time systems, we developed some fundamental results that provide a *roadmap* to a variety of filtering issues. In particular, the results there enabled us to relate the H_2 OID (SOID) filtering problems for a given system to that of EID (H_2 AID) filtering problems for an auxiliary system. The development in this section pertaining to discrete-time systems follows conceptually along the same lines as that of Section 10.4.

Once again, certain preliminaries are necessary before we introduce our main results of this section. To start with, let us consider a discrete-time linear matrix inequality (DLMI):

$$G(Q) \geqslant 0, \tag{10.37}$$

where

$$G(Q) := \begin{pmatrix} -Q + AQA' + BB' & AQC' + BD' \\ CQA' + DB' & CQC' + DD' \end{pmatrix}.$$

In the expression for $G(Q)$, the matrix Q is unknown, whereas the matrix quadruple (A, B, C, D) corresponds to the data of the given system Σ as in (10.1). We are interested in a semi-stabilizing solution Q of the DLMI (10.37). By Assumption 10.1, the pair (C, A) is $\mathbb{C}^{\ominus}$-detectable, and hence, by applying Theorem 4.147, we can conclude that a unique semi-stabilizing solution Q of the DLMI (10.37) exists. Moreover, such a solution Q is positive semi-definite, strongly rank minimizing, and is the largest among all symmetric solutions. Let us recall how to determine such a solution. According to Theorem 4.141, the semi-stabilizing solution of the DLMI (10.37) is indeed a semi-stabilizing solution of an associated H_2 DARE. The computation of the semi-stabilizing solution of a DARE is discussed in Subsection 4.2.5.

Remark 10.31 *Whenever the system characterized by the quadruple (A, B, C, D) is right-invertible (hence, $CQC' + DD'$ is invertible), one can equivalently compute Q as the unique semi-stabilizing solution Q of the H_2-DARE described by*

$$
\begin{aligned}
Q = AQA' + BB' \\
-(AQC' + BD')(CQC' + DD')^{-1}(CQA' + DB') \quad (10.38) \\
CQC' + DD' > 0.
\end{aligned}
$$

We note that, for a regular H_2 OID filtering problem, the unique semi-stabilizing solution of the H_2-DARE (10.38) is indeed stabilizing rather than merely semi-stabilizing. Thus, for a regular H_2 OID filtering problem, one can obtain the matrix Q equivalently by solving for the unique stabilizing solution of H_2-DARE (10.38) rather than by solving the DLMI (10.37).

Next, once the matrix Q is determined as discussed above, we can define matrices $B_Q \in \mathbb{R}^{n \times \rho}$ and $D_Q \in \mathbb{R}^{p \times \rho}$ with $\rho = \operatorname{rank} G(Q)$ such that

$$
G(Q) = \begin{pmatrix} B_Q \\ D_Q \end{pmatrix} \begin{pmatrix} B'_Q & D'_Q \end{pmatrix}. \quad (10.39)
$$

The condition that Q is semi-stabilizing, as defined in Definition 4.129, can then be rewritten as

$$
\operatorname{rank} \begin{pmatrix} A - zI & B_Q \\ C & D_Q \end{pmatrix} = n + \operatorname{rank} \begin{pmatrix} B_Q \\ D_Q \end{pmatrix} \quad (10.40)
$$

for all $z \in \mathbb{C}^{\oplus}$.

We can now define the following system Σ_Q:

$$
\Sigma_Q : \begin{cases} \sigma \tilde{x} = A\tilde{x} + B_Q \tilde{u} \\ \tilde{y} = C\tilde{x} + D_Q \tilde{u} \\ \tilde{z} = E^* \tilde{x}, \end{cases} \quad (10.41)
$$

with

$$E^* = E - P^*C, \tag{10.42}$$

where P^* minimizes

$$\text{trace}((E - PC)Q(E - PC)' + (F - PD)(F - PD)'). \tag{10.43}$$

Equivalently, we can impose that P^* be a solution of the following equation:

$$EQC' + FD' = P(CQC' + DD'). \tag{10.44}$$

Let us remark that the semi-stabilizing solution Q of the DLMI (10.37) satisfies

$$\ker(CQC' + DD') \subset \ker(EQC' + FD'),$$

and this implies that (10.44) has a solution. More specifically, we can choose

$$P^* = (EQC' + FD')(CQC' + DD')^\dagger,$$

where $\dagger$ denotes the Moore–Penrose generalized inverse. Throughout this chapter, whenever we refer to P^* in connection with discrete-time systems, it is as defined above.

For the auxiliary system Σ_Q, we consider filters of the form

$$\widetilde{\Sigma}_f : \begin{cases} \sigma \xi = L\xi + M\widetilde{y} \\ \widehat{\widetilde{z}} = N\xi + (P - P^*)\widetilde{y}. \end{cases} \tag{10.45}$$

We present next our main results. As in the continuous-time case, the first one relates the auxiliary system Σ_Q to the given system Σ. The second one shows that AID filtering is always possible for the auxiliary system.

Theorem 10.32 *Consider a discrete-time system Σ as in* (10.1) *with* (C, A) $\mathbb{C}^\ominus$-*detectable. Let Q be the unique semi-stabilizing solution of the DLMI* (10.37). *Also, let P^* be a matrix that minimizes the expression in* (10.43). *Define the auxiliary system Σ_Q given by* (10.41). *Moreover, given any proper filter of the form* (10.2), *construct the filter* (10.45). *Assume that the filter* (10.2), *applied to the system Σ, results in an error signal e_z, whereas the filter* (10.45), *applied to the system Σ_Q, results in an error signal $\widetilde{e}_z$. Then, the error dynamics resulting from the filter* (10.2) *applied to the system Σ is stable and achieves a finite RMS norm of the error signal e_z if and only if the error dynamics resulting from the filter* (10.45) *applied to the system Σ_Q is stable and achieves a finite RMS norm of the error signal $\widetilde{e}_z$. Moreover, in this case, the resulting RMS norms of the error signals are related by*

$$\|e_z\|_{\text{RMS}}^2 = \|\widetilde{e}_z\|_{\text{RMS}}^2 + \text{trace}(E - P^*C)Q(E - P^*C)'$$
$$+ \text{trace}(F - P^*D)(F - P^*D)'.$$

Proof : We can use the same arguments as in the proof of Theorem 10.10 for continuous-time systems. It is trivial to check that the filter (10.2) applied to Σ results in stable error dynamics if and only if the filter (10.45) applied to Σ_Q results in stable error dynamics. Under this condition, we note that the filters can be transformed into the form (7.70) as presented in Lemma 7.68.

A filter of the form (10.2) when applied to Σ is equivalent to applying a filter with stable transfer matrix $\hat{G}_f$ to the system:

$$\bar{\Sigma} : \begin{cases} \sigma\bar{x} = (A - KC)\bar{x} + (B - KD)u \\ \bar{y} \;\; = C\bar{x} + Du \\ \bar{z} \;\; = E\bar{x} + Fu. \end{cases}$$

Similarly, application of the filter given by (10.45) to Σ_Q is equivalent to the application of a filter with transfer matrix $\hat{G}_f - P^*$ to the system:

$$\bar{\Sigma}_Q : \begin{cases} \sigma\check{x} = (A - KC)\check{x} + (B_Q - KD_Q)\tilde{u} \\ \check{y} \;\; = C\check{x} + D_Q\tilde{u} \\ \check{z} \;\; = E^*\check{x}. \end{cases}$$

Note that with respect to these two transformed systems that we created, we still have the relationship:

$$\begin{pmatrix} -Q + A_K Q A'_K + B_K B'_K & A_K Q C' + B_K D' \\ CQA'_K + DB'_K & CQC' + DD' \end{pmatrix}$$
$$= \begin{pmatrix} B_Q - KD_Q \\ D_Q \end{pmatrix}\begin{pmatrix} B_Q - KD_Q \\ D'_Q \end{pmatrix}',$$

where $A_K = A - KC$ and $B_K = B - KC$. Note that the error dynamics of the filter given by (10.2) when applied to Σ can hence be rewritten as

$$G_1 + (\hat{G}_f - P^*)G_2,$$

with G_1 and G_2 given by

$$G_1(z) = (E - P^*C)(zI - A + KC)^{-1}(B - KD) + (F - P^*D),$$
$$G_2(z) = C(zI - A + KC)^{-1}(B - KD) + D.$$

Similarly, the error dynamics of the filter given by (10.2) when applied to Σ_Q can be rewritten as

$$G_{Q,1} + \hat{G}_f G_{Q,2},$$

with $G_{Q,1}$ and $G_{Q,2}$ given by

$$G_{Q,1}(z) = (E - P^*C)(zI - A + KC)^{-1}(B_Q - KD_Q),$$
$$G_{Q,2}(z) = C(zI - A + KC)^{-1}(B_Q - KD_Q) + D_Q.$$

Note that in discrete time,
$$\langle X, Y \rangle = 0$$
when X and Y are stable and antistable transfer matrices, respectively, and at least one of them is strictly proper.

Using some basic algebraic manipulations, it is also easy to show that for discrete-time systems, we have

$$G_2 G_2^* = G_{\varrho,2} G_{\varrho,2}^*,$$
$$G_1 G_2^* - G_{\varrho,1} G_{\varrho,2}^* \quad \text{is stable and strictly proper,}$$
$$\langle G_1, G_1 \rangle = \langle G_{\varrho,1}, G_{\varrho,1} \rangle + \text{trace}((E - P^*C)Q(E - P^*C)')$$
$$+(F - P^*D)(F - P^*D)'.$$

The fact that $G_1 G_2^* - G_{\varrho,1} G_{\varrho,2}^*$ is strictly proper is a consequence of our special choice of P^* given in (10.44). The last property is slightly different for discrete-time systems because we no longer need strictly proper error dynamics to achieve a finite RMS norm of the error signal.

From these three properties, the theorem follows directly. ∎

When we consider strictly proper filters, the auxiliary system Σ_ϱ defined earlier in (10.41) is not quite appropriate, and hence, we consider instead the system:

$$\Sigma_\varrho^{sp} : \begin{cases} \sigma \tilde{x} = A\tilde{x} + B_\varrho \tilde{u} \\ \tilde{y} \;\;= C\tilde{x} + D_\varrho \tilde{u} \\ \tilde{z} \;\;= E\tilde{x}. \end{cases} \tag{10.46}$$

In this case, we have the following modified version of Theorem 10.32, which can be established in a similar fashion.

Theorem 10.33 *Consider a discrete-time system Σ as in (10.1) with (C, A) $\mathbb{C}^\ominus$-detectable. Also, with Q being the unique semi-stabilizing solution of the DLMI (10.37), define the auxiliary system Σ_ϱ^{sp} given by (10.46). Moreover, assume that a strictly proper filter of the form (10.2) with $P = 0$, which when applied to the system Σ, results in an error signal e_z and, when applied to Σ_ϱ^{sp}, results in an error signal $\tilde{e}_z$. Then, the filter (10.2) results in a stable error dynamics when applied to Σ if and only if the same filter when applied to Σ_ϱ^{sp} results in stable error dynamics. In this case, the resulting RMS norms of the error signals are related by*
$$\|e_z\|_{\text{RMS}}^2 = \|\tilde{e}_z\|_{\text{RMS}}^2 + \text{trace}(EQE' + FF').$$

As in the continuous-time case, Theorems 10.32 and 10.33 are crucial in our further development because in trying to design an H_2 OID filter or a family of H_2 SOID filters for the original system Σ, according to the above theorems, we can then do all these designs equally well for the auxiliary system Σ_ϱ or Σ_ϱ^{sp}

depending on whether we are using proper or strictly proper filters. This is of course useful only if the auxiliary system has desirable properties, and this is the topic of the following theorem.

Theorem 10.34 *Consider a discrete-time system Σ as in (10.1) with (C, A) $\mathbb{C}^{\ominus}$-detectable. Let Q be the unique semi-stabilizing solution of the DLMI (10.37). Also, let P^* be a matrix that minimizes the expression in (10.43). For both the system Σ_Q defined by (10.41) and the system Σ_Q^{sp} defined by (10.46), the H_2 AID filtering problem is solvable using strictly proper filters.*

Proof : The system Σ_Q satisfies the property (10.40), and hence, the system has no invariant zeros outside the unit circle; therefore,

$$\mathcal{S}^*(A, B_Q, C, D_Q) = \mathcal{S}^{\otimes}(A, B_Q, C, D_Q).$$

On the other hand, as Q is a strongly rank-minimizing solution of the DLMI (10.37), we have

$$\operatorname{rank} \begin{pmatrix} B_Q \\ D_Q \end{pmatrix} = \operatorname{rank} G(Q) = \operatorname{rank}(DD' + CQC') = \operatorname{rank} D_Q.$$

This implies that a matrix K exists such that $B_Q - KD_Q = 0$, which in turn implies that $\mathcal{S}^*(A, B_Q, C, D_Q) = \{0\}$.

The last two properties together imply that

$$\mathcal{S}^{\otimes}(A, B_Q, C, D_Q) = \{0\}. \tag{10.47}$$

By Theorem 8.5, this implies that the H_2 AID filtering problem for Σ_Q is solvable via a family of strictly proper filters. Clearly, by the same arguments, for the system Σ_Q^{sp} given in (10.46), the H_2 AID filtering problem is solvable via a family of strictly proper filters because this system has the same structural property.   ∎

As in the continuous-time case, Theorems 10.32, 10.33, and 10.34 have many consequences. We provide below two such consequences that relate the H_2 OID and H_2 SOID filtering problems for the given system Σ to EID and H_2 AID filtering problems for the auxiliary system (10.41) or (10.46). To do so, let us first consider strictly proper filters of the type Σ_f given in (10.2) with $P = 0$, and redraw Figure 10.1 for the given system Σ and for the auxiliary system Σ_Q^{sp} as shown in Figure 10.4.

Corollary 10.35 *Consider a discrete-time system Σ as in (10.1) with (C, A) $\mathbb{C}^{\ominus}$-detectable. Consider Figure 10.4, where Σ_Q^{sp} is the auxiliary system given by (10.46) and Σ_f is a strictly proper filter of the type given in (10.2) with $P = 0$. Then, the following two statements are equivalent:*

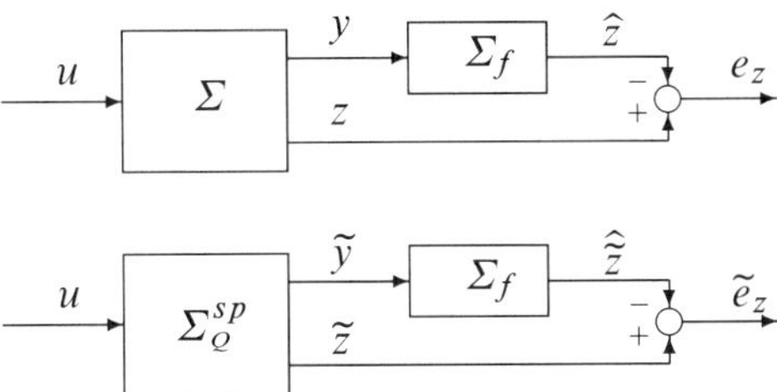

Figure 10.4: Block diagram representation of filtering for two related systems —
strictly proper filters

(*i*) Σ_f *is a strictly proper H_2 OID filter for the system Σ.*

(*ii*) Σ_f *is a strictly proper EID filter for the auxiliary system Σ_Q^{sp}.*

Let us next consider a proper filter of the type Σ_f given in (10.2) for Σ. Also, for Σ_Q, consider a proper filter of the type $\widetilde{\Sigma}_f$ given in (10.45). Moreover, let us redraw Figure 10.1 for the given system Σ and for the auxiliary system Σ_Q as shown in Figure 10.5.

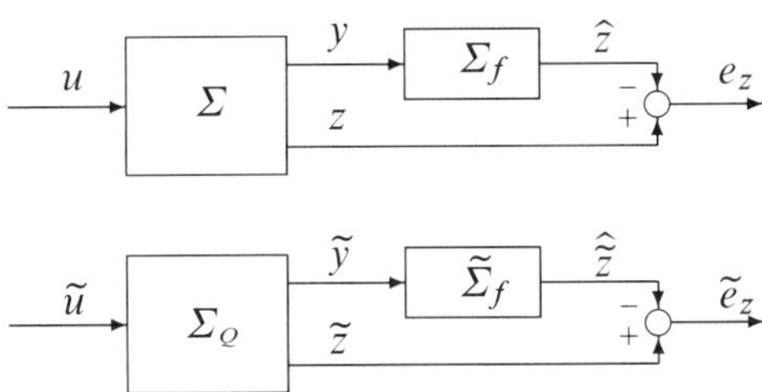

Figure 10.5: Block diagram representation of filtering for two related systems —
proper filters

Corollary 10.36 *Consider a discrete-time system Σ as in (10.1) with (C, A) $\mathbb{C}^\ominus$-detectable. Let P^* be a matrix that minimizes the expression in (10.43). Also, consider Figure 10.5, where Σ_Q is the auxiliary system given by (10.41), Σ_f is a proper filter of the type given in (10.2), and $\widetilde{\Sigma}_f$ is a proper filter of the type given in (10.45). Then, the following two statements are equivalent:*

(i) Σ_f is a proper H_2 OID filter for the system Σ.

(ii) $\tilde{\Sigma}_f$ is a proper EID filter for the auxiliary system Σ_Q.

Corollaries 10.35 and 10.36 show that H_2 OID filtering problems for the given system Σ can be related to EID filtering problems for the auxiliary system Σ_Q^{sp} or Σ_Q. Along the same lines, one can show that H_2 SOID filtering problems for the given system Σ can be related to H_2 AID filtering problems for the auxiliary system Σ_Q^{sp} or Σ_Q. Before doing so, let

$$\Sigma_f^\varepsilon : \begin{cases} \sigma\xi = L^\varepsilon\xi + M^\varepsilon y \\ \hat{z} = N^\varepsilon\xi + P^\varepsilon y \end{cases} \tag{10.48}$$

where L^ε, M^ε, N^ε, and P^ε are matrices parameterized in a positive parameter ε. We have the following corollary when the class of strictly proper filters are used.

Corollary 10.37 *Consider a discrete-time system Σ as in (10.1) with (C, A) $\mathbb{C}^\ominus$-detectable. Also, let Σ_Q^{sp} be the auxiliary system given by (10.46). Consider Figure 10.4, where Σ_f is to be replaced by Σ_f^ε as given in (10.48) but with $P^\varepsilon = 0$. Then, the following two statements are equivalent:*

(i) *The family of filters Σ_f^ε is a family of strictly proper H_2 SOID filters for the system Σ.*

(ii) *The family of filters Σ_f^ε is a family of strictly proper H_2 AID filters for the auxiliary system Σ_Q^{sp}.*

A result similar to the above corollary can be obtained when the class of proper filters are used. Before we state the result, let us consider the family of parameterized filters:

$$\tilde{\Sigma}_f^\varepsilon : \begin{cases} \sigma\xi = L^\varepsilon\xi + M^\varepsilon\tilde{y} \\ \hat{\tilde{z}} = N^\varepsilon\xi + (P^\varepsilon - P^*)\tilde{y}, \end{cases} \tag{10.49}$$

where P^* is a matrix that minimizes the expression in (10.43). We have the following corollary when the class of proper filters are used.

Corollary 10.38 *Consider a discrete-time system Σ as in (10.1) with (C, A) $\mathbb{C}^\ominus$-detectable. Let P^* be a matrix that minimizes the expression in (10.43). Also, let Σ_Q be the auxiliary system given by (10.41). Consider Figure 10.5, where Σ_f is to be replaced by Σ_f^ε as given in (10.48) and $\tilde{\Sigma}_f$ is to be replaced by $\tilde{\Sigma}_f^\varepsilon$ as given in (10.49). Then, the following two statements are equivalent:*

(i) *The family of filters Σ_f^ε is a family of proper H_2 SOID filters for the system Σ.*

(ii) *The family of filters $\tilde{\Sigma}_f^\varepsilon$ is a family of proper H_2 AID filters for the auxiliary system Σ_Q.*

10.8   Computation of γ_{sp}^* and γ_p^* — discrete-time case

In this section, with the help of Theorems 10.32, 10.33, and 10.34, we compute the H_2 OID filtering performance via linear stable strictly proper filters (denoted by γ_{sp}^*) and via linear stable proper filters (denoted by γ_p^*).

We have the following corollaries of Theorems 10.32, 10.33, and 10.34.

Corollary 10.39 *Consider a discrete-time system as in (10.1). Let Assumption 10.1 be satisfied. Let Q be the unique semi-stabilizing solution of the DLMI (10.37). Then, the H_2 OID filtering performance via linear stable strictly proper filters (denoted by γ_{sp}^* as formulated in Definition 10.3) is given by*

$$\gamma_{sp}^* = \left(\mathrm{trace}(EQE' + FF')\right)^{1/2}. \tag{10.50}$$

Proof : We define the auxiliary system Σ_Q^{sp} by (10.46). By Theorem 10.34, the H_2 OID filtering performance for the auxiliary system Σ_Q^{sp} is equal to 0. But then, by Theorem 10.33, it is clear that the H_2 OID filtering performance for the system Σ is given by (10.50). ∎

Remark 10.40 *From the results of Chapter 4, it is trivial to see that $\gamma_{sp}^* = 0$ if and only if $F = 0$ and*

$$\mathcal{S}^{\otimes}(A, B, C, D) \subseteq \ker E.$$

Obviously, whenever $\gamma_{sp}^* = 0$, the definition of H_2 OID filtering as given in 10.4 coincides with that of H_2 AID filtering as given in 8.2. In fact, as given in Theorem 8.11, the conditions for the solvability of H_2 AID filtering problem via strictly proper filters is equal to the above condition.

The above corollary considers strictly proper filters. Next, we would like to present results where proper filters are used. We have the following result.

Corollary 10.41 *Consider a discrete-time system as in (10.1). Let Assumption 10.1 be satisfied. Let Q be the unique semi-stabilizing solution of the DLMI (10.37). Then, the H_2 OID filtering performance via linear stable proper filters (denoted by γ_p^* as formulated in Definition 10.3) is given by*

$$\gamma_p^* = \left[\mathrm{trace}\left(E^* Q(E^*)' + F^*(F^*)'\right)\right]^{1/2}, \tag{10.51}$$

with $E^* = E - P^*C$ and $F^* = F - P^*D$, where P^* is a matrix that minimizes the expression in (10.43).

Proof : We define the auxiliary system Σ_Q by (10.10). By Theorem 10.34, we can solve the H_2 AID filtering problem for Σ_Q. But then, from Theorem 10.32, it is immediate that the infimum of the RMS norm of the error signal is given by (10.51).   ∎

Remark 10.42 *Once again, from the results of Chapter 4, it is trivial to see that $\gamma_p^* = 0$ if and only if a matrix P^* exists such that $F - P^*D = 0$ and*

$$\mathcal{S}^{\otimes}(A, B, C, D) \cap C^{-1}\{\text{im } D\} \subseteq \ker(E - P^*C).$$

Obviously, whenever $\gamma_p^* = 0$, the definition of H_2 OID filtering as given in 10.4 coincides with that of H_2 AID filtering as given in 8.2. As such, as given in Theorem 8.15, the above condition is the condition for the solvability of H_2 AID filtering problem via proper filters.

10.8.1 *Relationship between γ_{sp}^* and γ_p^* and the structural properties of Σ*

We have already developed (10.50) and (10.51) to determine the H_2 OID filtering performance indices γ_{sp}^* and γ_p^* in terms of the matrices E and F and the unique semi-stabilizing solution Q of the DLMI (10.37). However, as in the continuous-time case, we need to study these expressions to gain some insight as to how γ_{sp}^* and γ_p^* are related to the structural constraints imposed by the subsystem characterized by the quadruple (A, B, C, D). The purpose of this subsection is to relate γ_{sp}^* and γ_p^* to such structural constraints. To do so, we first consider the study of γ_{sp}^* via (10.50). The study of γ_p^* via (10.51) is similar, except the matrix E then is replaced by E^*. To start with, we observe that the terms trace(FF') and trace$(F^*(F^*)')$ that appear, respectively, in (10.50) and (10.51) are simply certain bias terms; as such, our study simply involves the examination of EQE' and $E^*Q(E^*)'$.

As in the continuous-time case, the study of the filtering problem for a given system Σ can be transformed to the study of a control problem for a dual system, represented here by Σ_d and characterized by the quintuple $(\bar{A}, \bar{B}, \bar{C}, \bar{D}, \bar{E})$, where

$$\bar{A} = A', \quad \bar{B} = C', \quad \bar{C} = B', \quad \bar{D} = D', \quad \text{and} \quad \bar{E} = E'. \tag{10.52}$$

To view the structural details of the dual system Σ_d, we can transform its subsystem characterized by the quadruple $(\bar{A}, \bar{B}, \bar{C}, \bar{D})$ to SCB. Let $(\bar{\Gamma}_s, \bar{\Gamma}_i, \bar{\Gamma}_o)$ be the related state, input, and output transformation matrices. Also, for convenience, we rearrange the order of the state variables in SCB as

$$\left((\bar{x}_a^-)' \quad (\bar{x}_a^0)' \quad (\bar{x}_c)' \quad (\bar{x}_a^+)' \quad (\bar{x}_b)' \quad (\bar{x}_d)' \right)'.$$

Then, let

$$\bar{\Gamma}_s^{-1}\bar{E} = \left((\bar{E}_a^-)' \quad (\bar{E}_a^0)' \quad (\bar{E}_c)' \quad (\bar{E}_a^+)' \quad (\bar{E}_b)' \quad (\bar{E}_d)' \right)'. \qquad (10.53)$$

As in Chapter 4, we intend to write the unique semi-stabilizing solution Q of the DLMI (10.37) in terms of a certain DARE. To do so, we extract the following matrices from the SCB constructed above:

$$\bar{A}_s = \begin{pmatrix} \bar{A}_{aa}^+ & \bar{L}_{ab}^+\bar{C}_b & \bar{L}_{ad}^+\bar{C}_d \\ 0 & \bar{A}_{bb} & \bar{L}_{bd}\bar{C}_d \\ \bar{B}_d\bar{E}_{da}^+ & \bar{B}_d\bar{E}_{db} & \bar{A}_d \end{pmatrix}, \quad \bar{B}_s = \begin{pmatrix} \bar{B}_{a0}^+ & 0 \\ \bar{B}_{b0} & 0 \\ \bar{B}_{d0} & \bar{B}_d \end{pmatrix}, \qquad (10.54)$$

$$\bar{C}_s = \bar{\Gamma}_o \begin{pmatrix} 0 & 0 & 0 \\ 0 & 0 & \bar{C}_d \\ 0 & \bar{C}_b & 0 \end{pmatrix}, \qquad\qquad \bar{D}_s = \bar{\Gamma}_o \begin{pmatrix} I_{\bar{m}_0} & 0 \\ 0 & 0 \\ 0 & 0 \end{pmatrix}, \qquad (10.55)$$

and

$$\bar{E}_s = \left((\bar{E}_a^+)' \quad (\bar{E}_b)' \quad (\bar{E}_d)' \right)'. \qquad (10.56)$$

We remark that various submatrices in the above definitions come from the SCB as applied to the subsystem characterized by the quadruple $(\bar{A}, \bar{B}, \bar{C}, \bar{D})$ (see Chapter 3).

Then, in view of Chapter 4, the unique semi-stabilizing solution Q of the DLMI (10.37) is given by

$$Q = (\bar{\Gamma}_s^{-1})' \begin{pmatrix} 0 & 0 & 0 & 0 \\ 0 & 0 & 0 & 0 \\ 0 & 0 & 0 & 0 \\ 0 & 0 & 0 & Q_s \end{pmatrix} \bar{\Gamma}_s^{-1}, \qquad (10.57)$$

where the symmetric positive semi-definite matrix Q_s is the solution of the H_2 DARE:

$$\bar{A}_s' Q_s \bar{A}_s - Q_s + \bar{C}_s' \bar{C}_s$$
$$- (\bar{A}_s' Q_s \bar{B}_s + \bar{C}_s' \bar{D}_s)(\bar{B}_s' Q_s \bar{B}_s + \bar{D}_s' \bar{D}_s)^{-1}(\bar{B}_s' Q_s \bar{A}_s + \bar{D}_s' \bar{C}_s) = 0, \quad (10.58)$$

along with the condition that $\bar{B}_s' Q_s \bar{B}_s + \bar{D}_s' \bar{D}_s$ is positive definite. The above H_2 DARE can be seen to arise from a standard linear quadratic optimization problem. To see this, let us define a state $\bar{x}_s$ as

$$\bar{x}_s = \begin{pmatrix} \bar{x}_a^+ \\ \bar{x}_b \\ \bar{x}_d \end{pmatrix},$$

and a subsystem $\bar{\Sigma}_{\text{sub}}$ as

$$\bar{\Sigma}_{\text{sub}} : \begin{cases} \sigma \bar{x}_s = \bar{A}_s \bar{x}_s + \bar{B}_s \bar{u}_s \\ \bar{z}_s \;\; = \bar{C}_s \bar{x}_s + \bar{D}_s \bar{u}_s, \end{cases} \tag{10.59}$$

with $\bar{x}_s(0) = \bar{E}_{si}$, where $\bar{E}_{si}$ is the ith column of $\bar{E}_s$. Clearly, in $\bar{\Sigma}_{\text{sub}}$, $\bar{x}_s$ is the state, $\bar{z}_s$ is the output, and $\bar{u}_s$ is considered as the controlling input. One would like to minimize the performance measure J_{sub} by using static state feedback controllers, where

$$J_{\text{sub}} = \sum_{i=0}^{\infty} \bar{z}_s'(i) \bar{z}_s(i) . \tag{10.60}$$

Then, it is easy to see that the infimum of J_{sub} over all possible static state feedback controllers is given by

$$J_{\text{sub}}^*(x_s(0)) = J_{\text{sub}}^*(\bar{E}_{si}) = \bar{E}_{si}' Q_s \bar{E}_{si} . \tag{10.61}$$

In view of the above development, and in particular in view of (10.50), (10.57), and (10.61), we have the following lemma that relates the H_2 OID filtering performance index γ_{sp}^* to the structural properties of the given system.

Lemma 10.43 *Consider a discrete-time system as in (10.1). Let Assumption 10.1 be satisfied. Also, let $J_{\text{sub}}^*(\bar{E}_{si})$ be as in (10.61). Then, the H_2 OID filtering performance via linear stable strictly proper filters, namely γ_{sp}^*, is given by*

$$(\gamma_{sp}^*)^2 = \sum_{i=1}^{q} J_{\text{sub}}^*(\bar{E}_{si}) + \text{trace } FF' = \sum_{i=1}^{q} \bar{E}_{si}' Q_s \bar{E}_{si} + \text{trace } FF', \tag{10.62}$$

where q is the dimension of the output z in (10.1).

Before we comment on the importance of the above lemma, let us develop a result similar to the above, however, for γ_p^*. This can be done by slightly modifying the development given above. The required modification is to be made in (10.52) by replacing the matrix E by $E - P^*C$, where P^* minimizes the expression in (10.43). We have the following result.

Lemma 10.44 *Consider a discrete-time system as in (10.1). Let Assumption 10.1 be satisfied. Let P^* be a matrix that minimizes the expression in (10.43). Moreover, in (10.52), let E be replaced by $\bar{E}^* = E - P^*C$. Also, following the development subsequent to (10.52) and culminating in (10.61), obtain $J_{\text{sub}}^*(\bar{E}_{si}^*)$ as* *in* (10.61) *(Note that, in this case, $\bar{E}^*_{si}$ replaces $\bar{E}_{si}$). Then, the H_2 OID filtering performance via linear stable proper filters, namely γ^*_p, is given by*

$$(\gamma^*_p)^2 = \sum_{i=1}^{q} J^*_{\text{sub}}(\bar{E}^*_{si}) + \text{trace } F^*(F^*)' = \sum_{i=1}^{q} (\bar{E}^*_{si})' Q_s \bar{E}^*_{si} + \text{trace } F^*(F^*)',$$

$$(10.63)$$

where q is the dimension of the output z in (10.1) and $F^* = F - P^*D$.

The above lemmas clearly show the roles played by $\bar{\Sigma}_{\text{sub}}$ and $\bar{E}_s$ or $\bar{E}^*_s$ in dictating the values of γ^*_{sp} and γ^*_p. The subsystem $\bar{\Sigma}_{\text{sub}}$ has three types of dynamics. The first type of dynamics is represented by the state $\bar{x}^+_a$, which is often called the *unstable zero dynamics*. It is present only when the subsystem characterized by (A, B, C, D) [and hence the dual subsystem characterized by $(\bar{A}, \bar{B}, \bar{C}, \bar{D})$] has invariant zeros outside the unit circle. Such invariant zeros are given by the eigenvalues of $\bar{A}^+_{aa}$. The second type of dynamics is represented by the state $\bar{x}_b$, and it is present only when the subsystem characterized by (A, B, C, D) is non-left-invertible, and hence the dual subsystem characterized by $(\bar{A}, \bar{B}, \bar{C}, \bar{D})$ is non-right-invertible. Finally, the third type of dynamics in $\bar{\Sigma}_{\text{sub}}$ is represented by the state $\bar{x}_d$, and it is present only when the subsystem characterized by (A, B, C, D) has infinite zeros of order greater than or equal to one.

Let us first assume that the subsystem characterized by (A, B, C, D) is left-invertible and has no infinite zeros of order greater than or equal to one. Consequently, the states $\bar{x}_b$ and $\bar{x}_d$ do not exist. In this case, the subsystem $\bar{\Sigma}_{\text{sub}}$ of (10.59) simplifies to

$$\bar{\Sigma}_{\text{sub}} : \begin{cases} \sigma \bar{x}^+_a = \bar{A}^+_{aa} \bar{x}^+_a + \bar{B}_s \bar{u}_s \\ \bar{z}_s \quad = \qquad\quad \bar{\Gamma}_o \bar{u}_s. \end{cases}$$

$$(10.64)$$

Also, in this case, as $\bar{z}_s = \bar{\Gamma}_o \bar{u}_s$, the performance measure J_{sub} has the interpretation of being the energy of control input.

The above simplification, and the results of Lemmas 10.43 and 10.44, enable us to interpret γ^*_{sp} and γ^*_p as explained in the following remarks.

Remark 10.45 *(Energy interpretation) Whenever the subsystem characterized by (A, B, C, D) is left-invertible and has no infinite zeros of order greater than or equal to one, $(\gamma^*_{sp})^2 - \text{trace } FF'$ [or $(\gamma^*_p)^2 - \text{trace } F^*(F^*)'$] equals the sum of minimum energies required to stabilize the unstable zero dynamics of the subsystem characterized by $(\bar{A}, \bar{B}, \bar{C}, \bar{D})$ for all initial conditions $\bar{x}^+_a(0) = \bar{E}_{si}, i = 1$ to q (or $\bar{x}^+_a(0) = \bar{E}^*_{si}, i = 1$ to q).*

Remark 10.46 *The above study can be viewed from a different angle. This time by looking at the roles played by the matrices E and F, which dictate $\bar{E}_{si}$ and*

$\bar{E}_{si}^{*}$ *in* (10.62) *and* (10.63). *We note that the matrices E and F are defined by the output z that is to be estimated. When viewed from the viewpoint of $\bar{E}_{si}$ and $\bar{E}_{si}^{*}$ that appear in* (10.62) *and* (10.63), *the fundamental limitations to the H_2 OID filtering performance arise from the inclusion of three types of dynamics in the output z that is to be estimated, the first one is the unstable zero dynamics of the subsystem characterized by (A, B, C, D) (as represented here by the dynamics of the state $\bar{x}_a^{+}$), the second one is the non-left-invertible dynamics of the subsystem characterized by (A, B, C, D) (as represented here by the dynamics of the state $\bar{x}_b$), and the third one is represented by the state $\bar{x}_d$ that is present only when the subsystem characterized by (A, B, C, D) has infinite zeros of order greater than or equal to one.*

We would like to investigate next how the location of the invariant zeros outside the unit circle (non-minimum-phase zeros) of the subsystem characterized by (A, B, C, D) affect the H_2 OID filtering performance indices γ_{sp}^{*} and γ_{p}^{*}. The investigation in this regard parallels that given in Subsection 10.5.1 for continuous-time systems and enables us to state the following: *Both γ_{sp}^{*} and γ_{p}^{*} can be decomposed into two parts, one arising due to unstable zero dynamics (i.e., the dynamics dictated by the invariant zeros outside the unit circle) and the other due to the dynamics of $\bar{x}_b$ and $\bar{x}_d$ of $\bar{\Sigma}_{\mathrm{sub}}$. The part that is contributed by the unstable zero dynamics to γ_{sp}^{*} and γ_{p}^{*} tends to zero as the invariant zeros outside the unit circle move toward the unit circle, and in the same way, it increases as the invariant zeros outside the unit circle move away from the unit circle.*

10.9   Existence of H_2 OID and SOID filters—discrete-time case

In Section 10.6, we developed the conditions for the existence of H_2 OID filters for continuous-time systems. We do so here for discrete-time systems. We have the following result.

Theorem 10.47 *Consider a discrete-time system as in* (10.1). *Let Assumption 10.1 be satisfied. Then, a linear strictly proper stable filter of the form* (10.2) *exists with $P = 0$ such that it solves the H_2 OID filtering problem 10.6 if and only if $\mathcal{S}^{\ominus}(A, B_Q, C, D_Q) \subseteq \ker E$, where the matrices B_Q and D_Q are as in* (10.39).

Proof : By Theorem 10.33 and Corollary 10.39, a strictly proper filter (10.2) (with $P = 0$) solves the H_2 OID filtering problem 10.6 if and only if the strictly proper filter (10.2) solves the EID filtering problem for Σ_Q^{sp}. From Theorem 7.6, we know that the EID filtering problem is solvable if and only if

$$\mathcal{S}^{\ominus}(A, B_Q, C, D_Q) \subseteq \ker E.$$

This proves the theorem.   ∎

The above theorem considers strictly proper filters. Next, we present results when proper filters are used.

Theorem 10.48 *Consider a discrete-time system as in* (10.1). *Let Assumption 10.1 be satisfied. Let* P^* *be a matrix that minimizes the expression in* (10.43). *Then, a linear proper stable filter of the form* (10.2) *exists that solves the H_2 OID filtering problem 10.6 if and only if*

$$\mathcal{S}^{\ominus}(A, B_Q, C, D_Q) \cap C^{-1}\{\operatorname{im} D_Q\} \subseteq \ker(E - P^*C), \qquad (10.65)$$

where the matrices B_Q and D_Q are as in (10.39).

Proof : By Corollary 10.36, a filter (10.2) solves the H_2 OID filtering problem 10.6 if and only if the filter (10.45) achieves EID filtering for Σ_Q. By Theorem 7.7, EID filtering problem for Σ_Q is solvable if and only if (10.65) is satisfied, and thus, the proof is complete.   ∎

Remark 10.49 *We already know from the proof of Theorem 10.34 that $\mathcal{S}^*(A, B_Q, C, D_Q) = \{0\}$. The structure of SCB combined with Property 3.33 can then be used to establish that*

$$\mathcal{S}^{\ominus}(A, B_Q, C, D_Q) \subset C^{-1}\{\operatorname{im} D_Q\},$$

which sheds more light on (10.65).

Remark 10.50 *Under the condition that $\gamma_{sp}^* = 0$ (respectively, $\gamma_p^* = 0$), the conditions for the solvability of the H_2 OID filtering problem 10.6 for Σ via strictly proper (respectively, proper) filters coincide with the conditions for the solvability of the EID filtering problem 7.3 for Σ via strictly proper (respectively, proper) filters.*

For general singular H_2 OID filtering problems, the solvability conditions are as asserted in Theorems 10.47 and 10.48. However, regular H_2 OID filtering problems are always solvable under the simple assumption that the pair (C, A) is $\mathbb{C}^{\ominus}$-detectable, i.e., as long as Assumption 10.1 is satisfied. This is formalized in the following corollary of Theorems 10.47 and 10.48.

Corollary 10.51 *Consider a regular H_2 OID filtering problem as in Definition 10.7 for the discrete-time system given in* (10.1). *Let Assumption 10.1 be satisfied. Then, the regular H_2 OID filtering problem is always solvable via a linear strictly proper or proper stable filter.*

Proof : Clearly, Q as a strongly rank-minimizing solution of the DLMI (10.37) is indeed a solution of the associated GDARE. Note that for a regular problem, we have no invariant zeros on the unit circle for the subsystem (A, B, C, D). By Theorem 4.95, this implies that Q is a stabilizing solution of the GDARE, which implies that (10.40) is satisfied for all $s \in \mathbb{C}^{\oplus} \cup \mathbb{C}^{\circ}$. But then $(A, B_\varrho, C, D_\varrho)$ is minimum phase. This implies that

$$\mathcal{S}^*(A, B_\varrho, C, D_\varrho) = \mathcal{S}^{\ominus}(A, B_\varrho, C, D_\varrho).$$

We already know from the proof of Theorem 10.34 that $\mathcal{S}^*(A, B_\varrho, C, D_\varrho) = \{0\}$. But then $\mathcal{S}^-(A, B_\varrho, C, D_\varrho) = \{0\}$. Then, Theorem 10.47 implies that the H_2 OID filtering problem is solvable via a linear strictly proper filter, whereas Theorem 10.48 implies that H_2 OID filtering problem is solvable via a linear proper filter. ∎

Note that the above proof does not use the fact that (A, B, C, D) is right-invertible. Only the exclusion of invariant zeros on the unit circle is sufficient to guarantee the existence of H_2 OID filters.

The above discussion pertains to the existence of H_2 OID filters. We have the following result regarding the H_2 SOID filters.

Theorem 10.52 *Consider a discrete-time system as in* (10.1). *Let Assumption 10.1 be satisfied. Then, the following results hold:*

(*i*) *The H_2 SOID filtering problem via a family of linear stable strictly proper filters is always solvable.*

(*ii*) *The H_2 SOID filtering problem via a family of linear stable proper filters is always solvable.*

Proof : The results follow trivially in view of Corollaries 10.37 and 10.38. ∎

10.10   Uniqueness of H_2 OID filters

For a given system, multiple H_2 OID filters might exist. Our goal here is to develop the conditions under which H_2 OID filters are unique. We observe that the notion of uniqueness of an H_2 OID filter can be viewed either in the sense of its transfer matrix or in the sense of its state-space realization with a fixed architecture. In this section, we view the uniqueness of an H_2 OID filter in the sense of its transfer matrix and not in the sense of its state-space realization.

We have the following results for the case when strictly proper filters of the form (10.2) with $P = 0$ are used. We recall that, for continuous-time systems, strictly proper filters can be used only when $F = 0$.

Theorem 10.53 *Consider a continuous- or discrete-time system as in (10.1). Let Assumption 10.1 be satisfied. Let the solvability conditions of the H_2 OID filtering problem via strictly proper filters (as in Theorem 10.26 or 10.47) be satisfied. Then, the solution of the H_2 OID filtering problem via strictly proper filters of the form (10.2) with $P = 0$ is unique, in the sense of transfer matrix of a filter, if and only if the subsystem characterized by the quadruple (A, B, C, D) is right-invertible.*

Proof : For continuous-time systems, as established in Theorem 10.10, the optimization can be written in the form of minimizing the H_2 norm of

$$G_1 + \hat{G}_f G_2$$

over strictly proper transfer matrices $\hat{G}_f$ with G_1 and G_2 given by

$$G_1(s) = E(sI - A + KC)^{-1}(B - KD),$$
$$G_2(s) = C(sI - A + KC)^{-1}(B - KD) + D,$$

where K is such that $A - KC$ is asymptotically stable. It is then obvious that the optimum, if it exists, is unique if and only if G_2 is right-invertible, and thus, the theorem follows. The discrete-time result is established similarly.   ∎

The above theorem pertains to the case when strictly proper filters are used. The following theorem considers proper filters.

Theorem 10.54 *Consider a continuous- or discrete-time system as in (10.1). Let Assumption 10.1 be satisfied. Assume that the solvability conditions for H_2 OID filtering problem via proper filters (as in Theorem 10.27 or 10.48) are satisfied. Then, the solution of the H_2 OID filtering problem via proper filters of the form (10.2) is unique, in the sense of transfer matrix of a filter, if and only if the subsystem characterized by the quadruple (A, B, C, D) is right-invertible.*

Proof : This follows along the same lines as the proof of Theorem 10.53.   ■

10.11 Uniqueness of the transfer matrix of H_2 OID error dynamics

As discussed in the previous section, the transfer matrix of an H_2 OID filter is in general nonunique. An interesting question is then whether different optimal filters result in different error dynamics. In this regard, we know already by definition that all H_2 OID filters must result in the same H_2 norm of the transfer matrix from the input u to the error signal $e_z = z - \hat{z}$, namely G^{ue}. Indeed, as shown below, not merely the H_2 norm of G^{ue} but G^{ue} itself is the same whatever is the H_2 OID filter that one uses.

Theorem 10.55 *Consider a continuous- or discrete-time system as in* (10.1). *Let Assumption 10.1 be satisfied. Let the solvability conditions of the H_2 OID filtering problem via strictly proper filters (as in Theorem 10.26 or 10.47) are satisfied. Then, the solutions of the H_2 OID filtering problem result in the same transfer matrix from the input u to the error signal $e_z = z - \hat{z}$, namely G^{ue}.*

Similarly, assume that the solvability conditions for H_2 OID filtering problem via proper filters (as in Theorem 10.27 or 10.48) be satisfied. Then, the solutions of the H_2 OID filtering problem result in the same transfer matrix from the input u to the error signal $e_z = z - \hat{z}$, namely G^{ue}.

Proof : Similar to the proofs of Theorems 10.53 and 10.54, we note that the transfer matrix of the error dynamics can be rewritten as $G_1 + \tilde{Q} G_2$, where we optimize either over strictly proper or proper stable transfer matrices $\tilde{Q}$. In other words, we try to find the element of minimal H_2 norm of the affine set (assuming a strictly proper $\tilde{Q}$):

$$\left\{ G_1 + \tilde{Q} G_2 \mid \tilde{Q} \in \mathcal{RH}_2 \right\}.$$

As H_2 is a Hilbert space, it is well known that the element of minimal H_2 norm (if it exists) is unique. Note that the existence is not guaranteed because the affine set is not necessarily closed. Using proper $\tilde{Q}$ (corresponding to proper filters), the argument follows along the same lines.   ■

10.12   Design of H_2 OID filters—continuous-time case

For continuous-time systems, we present in this section explicit algorithms of designing H_2 OID filters with a capability to assign its poles as desired while honoring certain conditions imposed by H_2 optimality. In this regard, as in the previous chapters, we first observe that typically any filter design is initiated by first assuming a fixed architecture to the filter. The architecture we use for the filters is the CSS architecture that was developed and used in earlier chapters.

In view of Corollaries 10.12 and 10.13, the design of H_2 OID filters for the given system Σ as in (10.1) can indeed be transformed to the design of EID filters for the auxiliary system Σ_Q given in (10.10). As we have already discussed in Chapter 7 the design of EID filters for any given system, one might wonder about the need for the development that follows in this section. The reason is very simple. The design of H_2 OID filters for Σ through the design of EID filters for the auxiliary system Σ_Q reflects certain important structural properties of H_2 OID filters for Σ through similar structural properties of EID filters for Σ_Q. Our intention by presenting the detailed design of H_2 OID filters in this section is to bring forth the insight and transparency of the structural properties of H_2 OID filters for Σ *directly* in terms of the data that characterize Σ rather than *indirectly* in terms of the data that characterize Σ_Q.

In the following three subsections, we consider the design of full-order strictly proper, full-order proper, and reduced-order proper H_2 OID filters one at a time.

10.12.1   Strictly proper H_2 OID filters of CSS architecture

We pursue here the design of strictly proper H_2 OID filters. As explained earlier, strictly proper filters can be used only when $F = 0$. As such, throughout this subsection, we assume that $F = 0$. The architecture we use for the filter is the full-order CSS architecture that was developed earlier in Subsection 7.5.1 of Chapter 7. It is given by (7.12) and depicted in Figure 7.2. It is reproduced here as

$$\Sigma_{\text{sp-CSS}} : \quad \left\{ \dot{\xi} = (A - KC)\xi + Ky \quad \text{with} \quad \hat{z} = E\xi. \right. \tag{10.66}$$

Error Dynamics: Let us define the error $e = x - \xi$ and then the error between the actual desired output z and the estimated desired output $\hat{z}$ is $e_z = E(x-\xi) = Ee$. Also, the dynamics of error is given by

$$\Sigma_{\text{sp-CSS}}^{ue} : \left\{ \begin{array}{l} \dot{e} = (A - KC)e + (B - KD)u \\ e_z = Ee. \end{array} \right. \tag{10.67}$$

The transfer matrix $G_{\text{sp-CSS}}^{ue}$ from u to e_z can obviously be written as

$$G_{\text{sp-CSS}}^{ue}(s) = E(sI - A + KC)^{-1}(B - KD). \tag{10.68}$$

Remark 10.56 *In view of* (10.66) *and* (10.67), *it is easy to see that both the filter equation and the error equation have the same modes, which are the eigenvalues of* $A - KC$.

We observe that the only unknown in the filter equation (10.66) and consequently in the error equation (10.67) is the matrix K, which is normally referred to as the filter gain. We need to determine or design K in such a way that the H_2 norm of $G^{ue}_{\text{sp-CSS}}$ is as small as possible. Before we do so, we pause to emphasize an important aspect. Theorem 10.26 developed earlier gives the conditions under which an H_2 OID filter exists among the general class of strictly proper filters of the form (10.2) with $P = 0$. In other words, Theorem 10.26 does not restrict itself to any fixed architecture for a filter such as the one $\Sigma_{\text{sp-CSS}}$ given in (10.66). Nevertheless, as the theorem that follows shortly shows, whenever the conditions of Theorem 10.26 are satisfied, we can determine the gain parameter K such that the filter $\Sigma_{\text{sp-CSS}}$ is an H_2 OID filter. In this regard, following the notation of Chapter 7, we define next a set of filter gains $\boldsymbol{K}^{h2-\text{oid}}_{\text{sp-CSS}}(A, B, C, D, E, 0)$, which is the set of all H_2 OID filter gains, meaning that any gain $K \in \boldsymbol{K}^{h2-\text{oid}}_{\text{sp-CSS}}$ renders $\Sigma_{\text{sp-CSS}}$ given in (10.66) an H_2 OID filter for (10.1), and conversely any gain K that renders $\Sigma_{\text{sp-CSS}}$ an H_2 OID filter for (10.1) is an element of $\boldsymbol{K}^{h2-\text{oid}}_{\text{sp-CSS}}$.

We have the following result.

Theorem 10.57 *Consider a continuous-time system as in* (10.1), *and let Assumption 10.1 be satisfied. Also, let the matrices* B_Q *and* D_Q *be as defined by* (10.8), *and consider the system* Σ_Q *defined in* (10.10) *with* $P^* = 0$. *Consider strictly proper filters, and assume that the solvability conditions of the* H_2 *OID filtering problem (as in Theorem 10.26) are satisfied. Then, the set* $\boldsymbol{K}^{h2-\text{oid}}_{\text{sp-CSS}}$ *(A, B, C, D, E, 0) is nonempty and is given by*

$$\boldsymbol{K}^{h2-\text{oid}}_{\text{sp-CSS}}(A, B, C, D, E, F) = \{ K \mid \boldsymbol{K}^{\text{eid}}_{\text{sp-CSS}}(A, B_\varrho, C, D_\varrho, E, 0)\}, \quad (10.69)$$

where the set $\boldsymbol{K}^{\text{eid}}_{\text{sp-CSS}}$ *is given by Theorem 7.13.*

Proof : In Corollary 10.12, we already noted that a filter is an H_2 OID filter for Σ if and only if it is an EID filter for the system Σ_Q defined in (10.10) with $P^* = 0$ (which is possible because the existence of a strictly proper filter achieving finite RMS norm requires that $F = 0$). The result then follows from Theorem 7.13. $\blacksquare$

As in the case of the set $\boldsymbol{K}^{\text{eid}}_{\text{sp-CSS}}$, the set $\boldsymbol{K}^{h2-\text{oid}}_{\text{sp-CSS}}$ has an interesting telescopic property with respect to the size of the matrix E. This is formalized below.

Lemma 10.58 *Consider a continuous-time system as in* (10.1). *Consider two different H_2 OID filtering problems both as defined in Definition 10.6, however, with one taking a value E_a and the other a value E_b for the matrix E. Then, we have the following telescopic property:*

$$\ker E_a \supseteq \ker E_b \supseteq \mathcal{S}^-(A, B_Q, C, D_Q)$$
$$\implies K^{h2-\mathrm{oid}}_{\mathrm{sp\text{-}CSS}}(A, B, C, D, E_a, 0) \supseteq K^{h2-\mathrm{oid}}_{\mathrm{sp\text{-}CSS}}(A, B, C, D, E_b, 0).$$

Proof : In view of Corollary 10.12, the result follows from Theorem 7.14, which discusses a telescopic property for $K^{\mathrm{eid}}_{\mathrm{sp\text{-}CSS}}$. ∎

In general, whenever it exists, a strictly proper H_2 OID filter $\Sigma_{\mathrm{sp\text{-}CSS}}$ of the form given in (10.66) is not unique. The nonuniqueness of such an H_2 OID filter can indeed be a blessing as some other specifications can then be imposed to come up with an appropriate filter. Theorem 10.55 shows that all H_2 OID filters have the same transfer matrix. Consequently, all H_2 OID filters generate the same error dynamics *in steady state*. However, the transient behavior of error could indeed be different for different H_2 OID filters. It is well known that the dynamics of any system is heavily influenced by its poles. As seen from (10.3), for a filter of an arbitrary architecture, both the poles of the given system or plant as well as those of the filter influence the transient behavior of error. Thus, it is prudent to examine the poles of an H_2 OID filter. Also, although for filters of arbitrary architecture, the poles of error dynamics include those of the filter, for the particular case of filters of CSS architecture, the poles of the filter $\Sigma_{\mathrm{sp\text{-}CSS}}$ given in (10.66) are exactly the same as those of the error dynamics $\Sigma^{ue}_{\mathrm{sp\text{-}CSS}}$ as given in (10.67). Such poles are indeed the eigenvalues of $(A - KC)$. This brings up a fundamental question: Can a gain K be designed in such a way that the resulting filter $\Sigma_{\mathrm{sp\text{-}CSS}}$ is an H_2 OID filter while simultaneously letting the designer assign the filter poles as arbitrarily as desired? Expectedly, it turns out that the constraint of H_2 OID filtering reduces the available freedom in assigning the poles of a filter. In fact, as in the case of EID filtering, for each given system, a set of complex numbers exists which every H_2 OID filter must have among its poles. As the following definition formalizes, such a set of complex numbers can be termed as the H_2 OID filter fixed modes.

Definition 10.59 *(Fixed modes of strictly proper H_2 OID filters of CSS architecture) Consider a continuous-time system as given* (10.1) *along with the H_2 OID filtering problem 10.6 characterized by the matrix quintuple (A, B, C, D, E). Assume that the solvability conditions of the H_2 OID filtering problem as in Theorem 10.26 are satisfied. Then, a scalar $\lambda \in \mathbb{C}^-$ is said to be the fixed mode of an H_2 OID filter with the strictly proper full-order CSS architecture if λ is a pole (i.e., an eigenvalue of $A - KC$) of every H_2 OID filter of such an architecture $\Sigma_{\mathrm{sp\text{-}CSS}}$ as given in (10.66). The set of all such H_2 OID filter fixed modes is denoted here by $\Omega^{h2-\mathrm{oid}}_{\mathrm{sp\text{-}CSS}}(A, B, C, D, E, 0)$.*

We have the following theorem that characterizes the set $\Omega_{\text{sp-CSS}}^{h2-\text{oid}}$.

Theorem 10.60 *Consider a continuous-time system as in* (10.1), *and let Assumption 10.1 be satisfied. Also, let the matrices B_Q and D_Q be as defined by* (10.8). *Consider strictly proper filters, and assume that the solvability conditions of the H_2 OID filtering problem (as in Theorem 10.26) are satisfied. Then, we have*

$$\Omega_{\text{sp-CSS}}^{h2-\text{oid}}(A, B, C, D, E, 0) = \Omega_{\text{sp-CSS}}^{\text{eid}}(A, B_Q, C, D_Q, E, 0),$$

where $\Omega_{\text{sp-CSS}}^{\text{eid}}$ is characterized in Theorem 7.18.

Proof : In view of Corollary 10.12, the result follows from the flexibility in EID filters as studied in Chapter 7 and in particular from Theorem 7.18. ∎

Theorem 10.60 prescribes a procedure of obtaining $\Omega_{\text{sp-CSS}}^{h2-\text{oid}}$. However, to gain insight regarding the nature of these fixed modes, it is worthwhile to note that $\Omega_{\text{sp-CSS}}^{h2-\text{oid}}(A, B, C, D, E, 0)$ is a subset of the stable invariant zeros of the system characterized by the quadruple (A, B_Q, C, D_Q). This implies that we need to study certain structural properties of the system characterized by (A, B_Q, C, D_Q). Expectedly, a definite relationship exists between the structural properties of the system characterized by (A, B_Q, C, D_Q) and the system characterized by the given quadruple (A, B, C, D). The following lemma describes such a relationship.

Lemma 10.61 *Consider a continuous-time system characterized by the quadruple (A, B, C, D). Let Q be a semi-stabilizing solution of the CLMI* (10.6), *and define B_Q and D_Q by* (10.8) *resulting in a second system characterized by the quadruple (A, B_Q, C, D_Q). We have the following properties:*

(i) *The system characterized by (A, B_Q, C, D_Q) is left-invertible.*

(ii) *The system characterized by (A, B_Q, C, D_Q) is invertible if and only if the system characterized by (A, B, C, D) is right-invertible.*

(iii) *The invariant zeros of the system characterized by (A, B_Q, C, D_Q) are classified by*

   (a) *the stable (i.e., those in $\mathbb{C}^-$) invariant zeros of the system characterized by (A, B, C, D),*

   (b) *the invariant zeros that are on the imaginary axis of the system characterized by (A, B, C, D),*

   (c) *the mirror images with respect to the imaginary axis of all unstable (i.e., those in $\mathbb{C}^+$) invariant zeros of the system characterized by (A, B, C, D), and*

> (d) *what are termed as compromise zeros, which are some fixed locations in the open left-half complex plane containing those stable output decoupling zeros of the system characterized by (A, B, C, D), which are not invariant zeros.*

(iv) *The system characterized by (A, B_Q, C, D_Q) has the same infinite zero structure as that of the system characterized by (A, B, C, D).*

Proof : From (10.9), it is immediate that the system characterized by (A, B_Q, C, D_Q) is left-invertible and has no invariant zeros in the open right-half plane.

Secondly, it is easy to verify that

$$\begin{pmatrix} x & y \end{pmatrix} \begin{pmatrix} A - sI & B \\ C & D \end{pmatrix} = 0 \text{ if and only if } \begin{pmatrix} x & y \end{pmatrix} \begin{pmatrix} A - sI & B_Q \\ C & D_Q \end{pmatrix} = 0$$

for all s on the imaginary axis. But this clearly implies the property (ii) [exploiting property (i)] and moreover establishes that the zeros on the imaginary axis of both systems are the same.

If D is surjective, then we know that Q is a solution of an H_2 CARE. In this case, it is easily established that all invariant zeros are eigenvalues of the associated Hamiltonian. As the eigenvalues of the Hamiltonian are symmetric with respect to the imaginary axis, this implies that the stable eigenvalues of the Hamiltonian contain the stable invariant zeros of (A, B, C, D) and the mirror images of the invariant zeros of (A, B, C, D) in the open right-half plane. But it is then immediate that, if Q is a semi-stabilizing solution, then the stable eigenvalues of $A - (QC' + BD')(DD')^{-1}C$ are equal to the stable eigenvalues of the Hamiltonian and, hence, include the stable invariant zeros of (A, B, C, D) and the mirror images of the invariant zeros of (A, B, C, D) in the open right-half plane. This immediately yields that the set of invariant zeros of (A, B_Q, C, D_Q) contains the sets (a) and (c). The general case when D is not surjective follows from Theorem 4.114, which connects in the general case, the semi-stabilizing solutions of the CLMI (10.6) to the semi-stabilizing solutions of an associated reduced H_2 CARE.

The last property (iv) follows from the fact established in Lemma 4.112 (where we use the duality between $\mathcal{V}^*$ and $\mathcal{S}^*$) that Q as a solution of the CLMI satisfies $\operatorname{im} Q \subseteq \mathcal{V}^*(A, B, C, D)$, which implies that the change from (A, B, C, D) to (A, B_Q, C, D_Q) does not affect the infinite zero structure.   ∎

As we said, $\Omega^{h2-\mathrm{oid}}_{\mathrm{sp\text{-}CSS}}(A, B, C, D, E, 0)$ is a subset of the stable invariant zeros of the system characterized by the quadruple (A, B_Q, C, D_Q). Then, in view of Lemma 10.61, $\Omega^{h2-\mathrm{oid}}_{\mathrm{sp\text{-}CSS}}(A, B, C, D, E, 0)$ consists of some stable invariant zeros, some mirror images of unstable invariant zeros, and what are known as compromise zeros of the system characterized by (A, B, C, D). The exact size of $\Omega^{h2-\mathrm{oid}}_{\mathrm{sp\text{-}CSS}}(A, B, C, D, E, 0)$ is dictated by the size of matrix E, which prescribes the variable z that is being estimated. In this respect, as in the case of the set $K^{h2-\text{oid}}_{\text{sp-CSS}}$, a certain interesting telescopic property (nested property) exists, namely, *as* $\ker E$ *decreases, the size of* $\boldsymbol{\Omega}^{h2-\text{oid}}_{\text{sp-CSS}}(A, B, C, D, E, 0)$ *increases.* The smallest set $\boldsymbol{\Omega}^{h2-\text{oid}}_{\text{sp-CSS}}(A, B, C, D, E, 0)$ is obtained when $\ker E = \mathbb{R}^n$, whereas it is the largest when $\ker E = \mathcal{S}^-(A, B_Q, C, D_Q)$. The following lemma describes the telescopic property of $\boldsymbol{\Omega}^{h2-\text{oid}}_{\text{sp-CSS}}$.

Lemma 10.62 *Consider a continuous-time system as in* (10.1). *Consider two different H_2 OID filtering problems both as defined in Definition 10.6, however, with one taking a value E_a and the other a value E_b for the matrix E. Then, we have the telescopic property:*

$$\ker E_a \supseteq \ker E_b \supseteq \mathcal{S}^-(A, B_Q, C, D_Q)$$
$$\implies \boldsymbol{\Omega}^{h2-\text{oid}}_{\text{sp-CSS}}(A, B, C, D, E_a, 0) \subseteq \boldsymbol{\Omega}^{h2-\text{oid}}_{\text{sp-CSS}}(A, B, C, D, E_b, 0).$$

Proof : In Proposition 5.13, this nested property was established for the set $\boldsymbol{\Omega}_s$. Using Theorem 7.18, this results in a similar property for the set $\boldsymbol{\Omega}^{\text{eid}}_{\text{sp-CSS}}$, which in turn through Theorem 10.60 results in the property described in the above theorem. $\blacksquare$

Remark 10.63 *The above theorem simply states that, by increasing the number of states to be estimated (i.e., by increasing the dimension of the estimated variable z), the number of fixed modes increases telescopically.*

Having noted the result of Lemma 10.62, we now move on to study $\boldsymbol{\Omega}^{h2-\text{oid}}_{\text{sp-CSS}}$ for two extreme cases for E, namely for

$$\ker E = \mathbb{R}^n \text{ or } \ker E = \mathcal{S}^-(A, B_Q, C, D_Q).$$

For the case when $\ker E = \mathcal{S}^-(A, B_Q, C, D_Q)$, we note that the set $\boldsymbol{\Omega}^{h2-\text{oid}}_{\text{sp-CSS}}(A, B, C, D, E, 0)$ consists of

(i) all stable invariant zeros (i.e., those in $\mathbb{C}^-$) of the system characterized by (A, B, C, D),

(ii) all mirror images of the unstable invariant zeros (i.e., those in $\mathbb{C}^+$) of the system characterized by (A, B, C, D),

(iii) some fixed locations called compromising zeros in $\mathbb{C}^-$.

On the other hand, for the other extreme case when $\ker E = \mathbb{R}^n$, the set $\boldsymbol{\Omega}^{h2-\text{oid}}_{\text{sp-CSS}}$ is empty because the static estimator $\hat{z} = 0$ will be optimal in this case. Next, it is easy to show that, in general, when $\ker E$ is between these two extreme cases, the set $\boldsymbol{\Omega}^{h2-\text{oid}}_{\text{sp-CSS}}(A, B, C, D, E, 0)$ *consists of only some, but not necessarily all, of the stable invariant zeros of* (A, B_Q, C, D_Q).

Besides the presence of H_2 OID filter fixed modes as explained above, another important structural issue in filter design exists. It turns out that in the system $\Sigma^{ue}_{\text{sp-CSS}}$, which represents the error dynamics, certain pole/zero cancellations exist; see (10.67). To be specific, for a given H_2 OID filtering problem, indeed what can be termed as a set of fixed decoupling zeros of $\Sigma^{ue}_{\text{sp-CSS}}$ exist, resulting from H_2 OID filters of CSS architecture. Such a set shows the minimum absolutely necessary number and locations of pole/zero cancellations within the system $\Sigma^{ue}_{\text{sp-CSS}}$. We have the following formal definition.

Definition 10.64 (*Fixed decoupling zeros of $\Sigma^{ue}_{\text{sp-CSS}}$ resulting from strictly proper H_2 OID filters of CSS architecture*) *Consider a continuous-time system as in* (10.1) *along with the H_2 OID filtering problem 10.6 characterized by the matrix quintuple (A, B, C, D, E). Assume that the solvability conditions of the H_2 OID filtering problem as in Theorem 10.26 are satisfied. Then, a scalar $\lambda \in \mathbb{C}^-$ is said to be a fixed decoupling zero of $\Sigma^{ue}_{\text{sp-CSS}}$ resulting from strictly proper H_2 OID filters of CSS architecture if λ is either an input or an output decoupling zero (or both) of the system $\Sigma^{ue}_{\text{sp-CSS}}$ as given in* (10.67) *for any strictly proper H_2 OID filter $\Sigma_{\text{sp-CSS}}$ that one can use. The set of all such fixed decoupling zeros is denoted by $\Lambda^{h2-\text{oid}}_{\text{sp-CSS}}(A, B, C, D, E, 0)$.*

We prescribe below an algorithm of computing $\Lambda^{h2-\text{oid}}_{\text{sp-CSS}}(A, B, C, D, E, 0)$.

An algorithm for computing the set $\Lambda^{h2-\text{oid}}_{\text{sp-CSS}}(A, B, C, D, E, 0)$:

Step 1a: (*Representation of Σ_{sub} in SCB*) We first construct in this step the SCB of Σ_{sub} characterized by (A, B, C, D). In SCB, we have the structure as presented in (3.42)–(3.45). Clearly we can also decompose E with respect to this basis:

$$E\Gamma_s^{-1} = \begin{pmatrix} E^-_{0a} & E^{0+}_{0a} & E_{0b} & E_{0c} & E_{0d} \end{pmatrix}. \tag{10.70}$$

Step 1b: The pair (E^-_{0a}, A^-_{aa}) need not be observable. We apply another transformation:

$$T_a^{-1} A^-_{aa} T_a = \begin{pmatrix} A^{11}_{aa} & A^{12}_{aa} \\ 0 & A^{22}_{aa} \end{pmatrix} \quad \text{and} \quad T_a^{-1} E^-_a = \begin{pmatrix} E^1_a & 0 \end{pmatrix},$$

such that (E^1_a, A^{11}_{aa}) is observable.

Step 2: (*Application of EDD algorithm*) We apply the EDD algorithm to the quintuple (A, B_Q, C, D_Q, E), and we obtain the pair (A_z, B_z) but denote it by

(A_z^Q, B_z^Q) to signify that the input to the EDD algorithm is (A, B_Q, C, D_Q, E). Form the set $\Lambda_{\text{sp-CSS}}^{h2-\text{oid}}(A, B, C, D, E, 0)$ as

$$\Lambda_{\text{sp-CSS}}^{h2-\text{oid}}(A, B, C, D, E, 0) = \lambda(A_{aa}^{11}) \cup \{\text{Input decoupling zeros of } (A_z^Q, B_z^Q)\}.$$
(10.71)

We have the following theorem.

Theorem 10.65 *Consider a continuous-time system as in (10.1). Let Assumption 10.1 be satisfied. Consider strictly proper filters of CSS architecture as given in (10.66), and assume that the solvability conditions of the H_2 OID filtering problem (as in Theorem 10.26) are satisfied. Then, the set $\Lambda_{\text{sp-CSS}}^{h2-\text{oid}}(A, B, C, D, E, 0)$ as given in (10.71) is indeed the set of fixed decoupling zeros of $\Sigma_{\text{sp-CSS}}^{ue}$ resulting from strictly proper H_2 OID filters of CSS architecture as defined in Definition 10.64.*

Proof : It is trivial to show that the set of input decoupling zeros of (A_z^Q, B_z^Q) coincides with the the set of output decoupling zeros of error dynamics $\Sigma_{\text{sp-CSS}}^{ue}$, and, hence, is included in $\Lambda_{\text{sp-CSS}}^{h2-\text{oid}}$. On the other hand, it follows from the properties of EDD algorithm and the properties of SCB that the set $\lambda(A_{aa}^{11})$ contains the input-decoupling zeros of error dynamics $\Sigma_{\text{sp-CSS}}^{ue}$ and, therefore, is included in $\Lambda_{\text{sp-CSS}}^{h2-\text{oid}}$. The proof is now complete. ∎

Lemma 10.62 shows that the set of H_2 OID filter fixed modes $\Omega_{\text{sp-CSS}}^{h2-\text{oid}}$ has a certain telescopic property. We enquire here whether the set of fixed decoupling zeros $\Lambda_{\text{sp-CSS}}^{h2-\text{oid}}$ has a similar property. It turns out that, under the general assumption of the pair (C, A) being detectable, the set $\Lambda_{\text{sp-CSS}}^{h2-\text{oid}}$ does not have any such property. However, whenever the pair (C, A) is observable, it does have such a property as the following lemma shows.

Lemma 10.66 *Consider a continuous-time system as in (10.1) where the pair (C, A) is observable. Consider two different H_2 OID filtering problems both as defined in Definition 10.6, however, with one taking a value E_a and the other a value E_b for the matrix E. Then, we have the telescopic property:*

$$\ker E_a \supseteq \ker E_b \supseteq \mathcal{S}^-(A, B_Q, C, D_Q)$$
$$\implies \Lambda_{\text{sp-CSS}}^{h2-\text{oid}}(A, B, C, D, E_a, 0) \subseteq \Lambda_{\text{sp-CSS}}^{h2-\text{oid}}(A, B, C, D, E_b, 0).$$

Proof : In Proposition 5.13, this nested property was established for the set Λ_s. Using Theorem 10.65, this results in a similar property for the set $\Lambda_{\text{sp-CSS}}^{h2-\text{oid}}$. ∎

The poles of $G_{\text{sp-CSS}}^{ue}$ are given by the following lemma whose proof is obvious.

Lemma 10.67 *Consider a continuous-time system as in (10.1). Let Assumption 10.1 be satisfied. Also, let the solvability conditions of the H_2 OID filtering problem as in Theorem 10.26 be satisfied. Then, the poles of the resulting transfer matrix $G^{ue}_{\text{sp-CSS}}$ as given in (10.68) when it is in its reduced form* are given by*

$$\Omega^{h2-\text{oid}}_{\text{sp-CSS}}(A, B, C, D, E, 0) \, / \, \Lambda^{h2-\text{oid}}_{\text{sp-CSS}}(A, B, C, D, E, 0).$$

In Section 10.10, we discussed the conditions under which the transfer matrix of any H_2 OID filter is unique. We considered there general strictly proper filters of the form (10.2) with $P = 0$ or proper filters of the form (10.2). The following theorem considers a class of filters with a specified architecture, namely the strictly proper CSS architecture, and discusses an H_2 OID filter being unique among them.

Theorem 10.68 *Consider a continuous-time system as in (10.1), and let Assumption 10.1 be satisfied. Also, let the matrices B_Q and D_Q be as defined by (10.8). Consider strictly proper filters, and assume that the solvability conditions of the H_2 OID filtering problem (as in Theorem 10.26) are satisfied. Then, a unique strictly proper full-order H_2 OID filter exists of the form $\Sigma_{\text{sp-CSS}}$ given (10.66) if and only if the following conditions are satisfied:*

   (i) The subsystem characterized by the quadruple (A, B, C, D) is right-invertible.

   (ii) The H_2 OID filtering problem is regular (i.e., the regularity conditions given in Definition 10.7 are satisfied).

   (iii) The matrix pair $(E, \; A - B_Q D_Q^{-1} C)$ is observable.

Moreover, under the above conditions, we have

   (i) $K^{h2-\text{oid}}_{\text{sp-CSS}}(A, B, C, D, E, 0) = \{B_Q D_Q^{-1}\}$, which is a singleton,

   (ii) the set $\Omega^{h2-\text{oid}}_{\text{sp-CSS}}(A, B, C, D, E, 0)$ is the same as the set $\lambda(A - B_Q D_Q^{-1} C)$, which is the union of all stable invariant zeros and the mirror images of unstable invariant zeros and the compromising zeros of the subsystem characterized by the quadruple (A, B, C, D),

   (iii) the set $\Lambda^{h2-\text{oid}}_{\text{sp-CSS}}(A, B, C, D, E, 0)$ is the same as the set of all stable invariant zeros of the subsystem characterized by the quadruple (A, B, C, D).

Also, the unique H_2 OID filter $\Sigma_{\text{sp-CSS}}$ is given by

$$\dot{\xi} = (A - B_Q D_Q^{-1} C)\xi + B_Q D_Q^{-1} y \quad \text{with} \quad \hat{z} = E\xi.$$

---

*A transfer matrix is said to be in its reduced form if all its pole/zero cancellations have been performed.

Proof : From Corollary 10.12, we know that a filter is an H_2 OID filter for Σ if and only if it is an EID filter for Σ_Q. The results then follow from the conditions for the uniqueness of an EID filter for Σ_Q as implied by Theorem 7.15.   ∎

Remark 10.69 *We observe that the conditions given in Theorem 10.68 imply those given in Theorem 10.53 but not conversely. The reason for the more stringent conditions is that here we define the uniqueness of an H_2 OID filter at the state-space level (using strictly proper architecture), whereas earlier we considered the uniqueness in terms of a transfer matrix.*

Finally, we are in a position to design an H_2 OID filter while placing its modes at desired locations subject to the constraint that the set $\Omega_{\text{sp-CSS}}^{h2-\text{oid}}$ must be among the modes of any H_2 OID filter. We have the following result.

Theorem 10.70 (*Strictly proper H_2 OID filter of CSS architecture with pole placement*) *Consider a continuous-time system as in (10.1), and let Assumption 10.1 be satisfied. Also, let the matrices B_Q and D_Q be as defined by (10.8). Consider strictly proper filters, and assume that the solvability conditions of the H_2 OID filtering problem (as in Theorem 10.26) are satisfied. Also, consider the sets $\Omega_{\text{sp-CSS}}^{h2-\text{oid}}(A, B, C, D, E, 0)$ and $K_{\text{sp-CSS}}^{h2-\text{oid}}(A, B, C, D, E, 0)$ as described, respectively, in Theorems 10.60 and 10.57. Moreover, let Λ be a prescribed set of n self-conjugate elements in the open left-half complex plane $\mathbb{C}^-$ such that Λ includes $\Omega_{\text{sp-CSS}}^{h2-\text{oid}}(A, B, C, D, E, 0)$. Then, a filter gain $K \in K_{\text{sp-CSS}}^{h2-\text{oid}}(A, B, C, D, E, 0)$ exists such that the strictly proper filter $\Sigma_{\text{sp-CSS}}$ given in (10.66) is an H_2 OID filter and moreover $\lambda(A - KC) = \Lambda$.*

Proof : As noted in the proofs of the previous theorems, the results simply follow from Corollary 10.12, which implies that a filter achieves H_2 OID filtering for Σ if and only if it achieves EID filtering for the system Σ_Q.   ∎

An algorithm for designing a strictly proper H_2 OID filter with simultaneous filter pole placement:

Throughout this algorithm, the matrices B_Q and D_Q are as defined by (10.8).

Step 1: By using the SCB of the system characterized by the quadruple (A, B_Q, C, D_Q), compute $\mathcal{S}^-(A, B_Q, C, D_Q)$. Check the solvability condition

$$\mathcal{S}^-(A, B_Q, C, D_Q) \subseteq \ker E.$$

If this is not satisfied, an H_2 OID strictly proper filter does not exist.

Step 2: Compute the set

$$\boldsymbol{\Omega}^{h2-\text{oid}}_{\text{sp-CSS}}(A, B, C, D, E, 0) = \boldsymbol{\Omega}^{\text{eid}}_{\text{sp-CSS}}(A, B_Q, C, D_Q, E, 0)$$
$$= \boldsymbol{\Omega}_s(A', C', B'_Q, D'_Q, E')$$

by using the EDD algorithm with the quintuple (A', C', B'_Q, D'_Q, E') as its input.
Step 3: Choose a desired set Λ of n self-conjugate elements in the open left-half complex plane $\mathbb{C}^-$ such that Λ includes $\boldsymbol{\Omega}^{h2-\text{oid}}_{\text{sp-CSS}}(A, B, C, D, E, 0)$.
Step 4: Design K by using the (EDDSPP) algorithm discussed in Section 5.4.3. The inputs to the EDDSPP algorithm are the matrix quintuple (A', C', B'_Q, D'_Q, E') and the matrix Λ. The output of the algorithm is the filter gain matrix K'.

With the filter gain K as computed above, it is easy to verify that the resulting filter $\Sigma_{\text{sp-CSS}}$ given in (10.66) is an H_2 OID filter with its modes at the locations specified by Λ; i.e., $\lambda(A - KC) = \Lambda$. As discussed in the EDDSPP algorithm, although the algorithm concentrates only on pole placement, certain freedom exists as well to place the eigenvectors of $A - KC$, but this is not pursued here.

10.12.2 Proper H_2 OID filters of CSS architecture

We pursue here the design of proper H_2 OID filters while using filters of CSS architecture. In this regard, Corollary 10.13 lays a roadmap to our development here. To start with, for ease of presentation and without loss of generality, we decompose the measured output y into two parts, y_0 and y_1, in such a way that y_0 contains explicitly the unknown input u in it, whereas y_1 does not contain any input u in it. That is, we write

$$y = \begin{pmatrix} y_0 \\ y_1 \end{pmatrix} = Cx + Du \quad \text{and} \quad C = \begin{pmatrix} C_0 \\ C_1 \end{pmatrix}, \quad D = \begin{pmatrix} D_0 \\ 0 \end{pmatrix}, \qquad (10.72)$$

where the matrix D_0 has rank m_0. We can then rewrite the given system equation (10.1) as

$$\Sigma : \begin{cases} \dot{x} &= Ax + Bu \\ \begin{pmatrix} y_0 \\ y_1 \end{pmatrix} = \begin{pmatrix} C_0 \\ C_1 \end{pmatrix} x + \begin{pmatrix} D_0 \\ 0 \end{pmatrix} u \\ z &= Ex + Fu. \end{cases} \qquad (10.73)$$

In view of Corollary 10.13, a proper H_2 OID filter for Σ can be designed through the design of a proper EID filter for the auxiliary system Σ_Q given in (10.10) and characterized by the quintuple (A, B_Q, C, D_Q, E^*), where $E^* = E - P^*C$, with matrix P^* being a solution of $F - PD = 0$ for P and where B_Q and D_Q are as defined in (10.8). A simple examination of proper EID filter design (as given in Chapter 7) for Σ_Q calls for a strictly proper EID filter design for another auxiliary system that is referred to here as $\widetilde{\Sigma}^*_Q$, which is essentially the same as Σ_Q except for some additional measurements. Such a $\widetilde{\Sigma}^*_Q$ is characterized by the quintuple

$(A, B_Q, \tilde{C}, \tilde{D}_Q, E^*)$, where

$$\tilde{C} = \begin{pmatrix} C_0 \\ C_1 \\ C_1 A \end{pmatrix} = \begin{pmatrix} C \\ C_1 A \end{pmatrix} \quad \text{and} \quad \tilde{D}_Q = \begin{pmatrix} D_Q \\ C_1 B_Q \end{pmatrix}. \tag{10.74}$$

For the reasons that will become obvious soon, in what follows, we present an alternative method of obtaining the auxiliary system $\tilde{\Sigma}^*_Q$. In this alternative method, as given below, we first construct an auxiliary system $\tilde{\Sigma}^*$ and then another auxiliary system referred to here as $\tilde{\Sigma}^*_{\tilde{Q}}$, which turns out to be the same as $\tilde{\Sigma}^*_Q$.

Auxiliary System $\tilde{\Sigma}^*$: As in Chapter 7, we define the auxiliary system $\tilde{\Sigma}^*$ as

$$\tilde{\Sigma}^* : \begin{cases} \dot{x} = Ax + Bu \\ \tilde{y} = \begin{pmatrix} y_0 \\ y_1 \\ \dot{y}_1 \end{pmatrix} = \tilde{C}x + \tilde{D}u \\ z^* = E^*x, \end{cases} \tag{10.75}$$

where

$$\tilde{D} = \begin{pmatrix} D_0 \\ 0 \\ C_1 B \end{pmatrix}. \tag{10.76}$$

Auxiliary System $\tilde{\Sigma}^*_{\tilde{Q}}$: To define $\tilde{\Sigma}^*_{\tilde{Q}}$, we need to define the following CLMI (10.77), which is similar to the one developed in (10.6) except that the matrices C and D are replaced now by $\tilde{C}$ and $\tilde{D}$. Consider

$$\tilde{G}(\tilde{Q}) := \begin{pmatrix} A\tilde{Q} + \tilde{Q}A' + BB' & \tilde{Q}\tilde{C}' + B\tilde{D}' \\ \tilde{C}\tilde{Q} + \tilde{D}B' & \tilde{D}\tilde{D}' \end{pmatrix} \geq 0. \tag{10.77}$$

Using Assumption 10.1, it can be verified that $(\tilde{C}, A)$ is $\mathbb{C}^-$-detectable. Hence, one can easily determine the unique semi-stabilizing solution $\tilde{Q}$ of the CLMI (10.77). Section 4.3.1 describes a procedure of determining such a solution $\tilde{Q}$. Once the matrix $\tilde{Q}$ is determined, we define the matrices $\tilde{B}_{\tilde{Q}} \in \mathbb{R}^{n \times \tilde{\rho}}$ and $\tilde{D}_{\tilde{Q}} \in \mathbb{R}^{\tilde{p} \times \tilde{\rho}}$ with $\tilde{p} = 2p - \text{rank } D_0$ and $\tilde{\rho} = \text{rank } \tilde{G}(\tilde{Q})$ such that

$$G(\tilde{Q}) = \begin{pmatrix} \tilde{B}_{\tilde{Q}} \\ \tilde{D}_{\tilde{Q}} \end{pmatrix} \begin{pmatrix} \tilde{B}'_{\tilde{Q}} & \tilde{D}'_{\tilde{Q}} \end{pmatrix}. \tag{10.78}$$

We can now define the auxiliary system:

$$\tilde{\Sigma}^*_{\tilde{Q}} : \begin{cases} \dot{\tilde{x}} = A\tilde{x} + \tilde{B}_{\tilde{Q}}\tilde{u} \\ \tilde{y} = \tilde{C}\tilde{x} + \tilde{D}_{\tilde{Q}}\tilde{u} \\ \tilde{z} = E^*\tilde{x}. \end{cases} \tag{10.79}$$

We observe that the auxiliary system $\widetilde{\Sigma}_{\widetilde{Q}}^*$ is characterized by the quintuple $(A, \widetilde{B}_{\widetilde{Q}}, \widetilde{C}, \widetilde{D}_{\widetilde{Q}}, E^*)$. An important observation we make at this time is that the auxiliary system $\widetilde{\Sigma}_{\widetilde{Q}}^*$ is in fact the same as $\widetilde{\Sigma}_{Q}^*$, which is characterized by the quintuple $(A, B_Q, \widetilde{C}, \widetilde{D}_Q, E^*)$. In other words, as shown in the following lemma, we claim that $\widetilde{B}_{\widetilde{Q}} = B_Q$ and $\widetilde{D}_{\widetilde{Q}} = \widetilde{D}_Q$.

Lemma 10.71 *Consider the continuous-time systems Σ and $\widetilde{\Sigma}^*$, respectively, as given in* (10.73) *and* (10.75). *Let Q be the semi-stabilizing solution of the CLMI* (10.6), *and define B_Q and D_Q according to* (10.8). *Also, let $\widetilde{D}_Q$ be as in* (10.74). *Similarly, let $\widetilde{Q}$ be the semi-stabilizing solution of the CLMI* (10.77), *and define $\widetilde{B}_{\widetilde{Q}}$ and $\widetilde{D}_{\widetilde{Q}}$ according to* (10.78). *Then we have*

$$\widetilde{B}_{\widetilde{Q}} = B_Q \quad and \quad \widetilde{D}_{\widetilde{Q}} = \widetilde{D}_Q. \tag{10.80}$$

Proof : This follows from some straightforward algebraic manipulations.   ■

The above lemma implies that the auxiliary system $\widetilde{\Sigma}_{\widetilde{Q}}^*$ is the same as the auxiliary system $\widetilde{\Sigma}_{Q}^*$. In simple words, this is as follows: First constructing the system $\widetilde{\Sigma}^*$ as in (10.75) and then the system $\widetilde{\Sigma}_{\widetilde{Q}}^*$ as in (10.79) is equivalent to first constructing the system Σ_Q as in (10.10) and then the system $\widetilde{\Sigma}_{Q}^*$.

As mentioned, in view of Corollary 10.13, a proper H_2 OID filter for the given system Σ as in (10.73) can be constructed through the construction of a proper EID filter for Σ_Q. This in turn via the results of Chapter 7 implies that a proper EID filter for Σ_Q can be constructed through the construction of a strictly proper EID filter, however, for $\widetilde{\Sigma}_{Q}^*$. Then, the result of Lemma 10.71 along with the basic result of Corollary 10.12 implies that the construction of a strictly proper EID filter for $\widetilde{\Sigma}_{\widetilde{Q}}^*$ is equivalent to the construction of a strictly proper H_2 OID filter for the system $\widetilde{\Sigma}^*$. This simply leads us to the following lemma.

Lemma 10.72 *Consider continuous-time systems Σ and $\widetilde{\Sigma}^*$, respectively, as in* (10.73) *and* (10.75). *Then, the following two statements are equivalent:*

(i) A strictly proper H_2 OID filter exists for the auxiliary system $\widetilde{\Sigma}^*$.

(ii) A proper H_2 OID filter exists for the system Σ.

Moreover, there is a $1-1$ relationship between the strictly proper H_2 OID filter of CSS architecture for $\widetilde{\Sigma}^*$ and the proper H_2 OID filter of CSS architecture for Σ; that is, one of these filters can be constructed from the other.

All of the above development leads to the fact that instead of designing a proper H_2 OID filter for Σ, one can indeed design a strictly proper H_2 OID filter for $\widetilde{\Sigma}^*$ as emphasized in Lemma 10.72. This helps our quest to understand the structural properties of H_2 OID filters for Σ directly in terms of the data of Σ because all the data that characterize $\widetilde{\Sigma}^*$ is obtained by a simple modification of the data of Σ, and the structural properties of strictly proper H_2 OID filters have already been developed in Subsection 10.12.1.

We move on now to construct a strictly proper filter of CSS architecture for $\widetilde{\Sigma}^*$ as

$$\begin{cases} \dot{\widetilde{\xi}} = A\widetilde{\xi} + K(\widetilde{y} - \widetilde{C}\widetilde{\xi}) \\ \widehat{z}^* = E^*\widetilde{\xi}, \end{cases} \tag{10.81}$$

where the matrix K is a filter gain. The estimate $\widehat{z}$ of z is given by

$$\widehat{z} = \widehat{z}^* + P^*y = E^*\widetilde{\xi} + P^*y. \tag{10.82}$$

We note that the filter (10.81) is not directly implementable because $y_2 = \dot{y}_1$ is not available as a measured variable. As in Chapter 7, we can eliminate the need for $\dot{y}_1$ by defining a new variable:

$$\xi = \widetilde{\xi} - K_2 y_1. \tag{10.83}$$

Here K_2 is obtained by partitioning K in conformity with the partitioning of $\widetilde{y}$. That is,

$$K = \begin{pmatrix} K_0 & K_1 & K_2 \end{pmatrix}.$$

With the definition of ξ as in (10.83), we can rewrite the filter equation (10.81) as

$$\begin{cases} \dot{\xi} = (A - K\widetilde{C})\xi + \begin{pmatrix} K_0 & K_1 + (A - K\widetilde{C})K_2 \end{pmatrix}y \\ \widetilde{\xi} = \xi + K_2 y_1 \\ \widehat{z}^* = E^*\widetilde{\xi} = E^*(\xi + K_2 y_1). \end{cases} \tag{10.84}$$

Obviously, the filter given above does not use $\dot{y}_1$. Moreover, it is proper rather than strictly proper. The filter given in (10.84) is indeed the proper full-order CSS filter that is to be used for $\widetilde{\Sigma}^*$.

Proper filter of CSS architecture for Σ: Finally, in view of (10.82), we can rewrite (10.84) as an implementable proper filter for Σ:

$$\Sigma_{\text{p-CSS}} : \begin{cases} \dot{\xi} = (A - K\widetilde{C})\xi + \widetilde{K}y \\ \widetilde{\xi} = \xi + K_2 y_1 \\ \widehat{z} = E^*\widetilde{\xi} + P^*y = E^*\xi + \widetilde{P}y, \end{cases} \tag{10.85}$$

where

$$\widetilde{K} = \begin{pmatrix} K_0 & K_1 + (A - K\widetilde{C})K_2 \end{pmatrix} \quad \text{and} \quad \widetilde{P} = \begin{pmatrix} 0 & E^*K_2 \end{pmatrix} + P^*.$$

A block diagram representation of the proper filter along with the given plant is structurally the same as the one in Figure 7.4.

Error Dynamics: By defining the error $e = x - \tilde{\xi}$, the error between the actual desired output $z = Ex + Fu = E^*x + P^*y$ and the estimated desired output $\hat{z} = E^*\tilde{\xi} + P^*y$ can be written as

$$e_z = z - \hat{z} = E^*e.$$

Then, in view of (10.73) and (10.81), the dynamics of error is given by

$$\Sigma_{\text{p-CSS}}^{ue} : \begin{cases} \dot{e} = (A - K\tilde{C})e + (B - K\tilde{D})u \\ e_z = E^*e. \end{cases} \tag{10.86}$$

Also, the transfer matrix $G_{\text{p-CSS}}^{ue}$ from u to e_z can obviously be written as

$$G_{\text{p-CSS}}^{ue}(s) = E^*(sI - A + K\tilde{C})^{-1}(B - K\tilde{D}). \tag{10.87}$$

Remark 10.73 *In view of* (10.85) *and* (10.86), *it is easy to see that both the filter equation and the error equation have the same modes, which are the eigenvalues of* $A - K\tilde{C}$.

Everything in the filters (10.81), (10.84), and (10.85) is known except the gain K. Following the notation used in connection with strictly proper filters, we denote the set of all H_2 OID filter gains by $K_{\text{p-CSS}}^{h2-\text{oid}}(A, B, C, D, E, F)$, meaning that any gain $K \in K_{\text{p-CSS}}^{h2-\text{oid}}$ renders $\Sigma_{\text{p-CSS}}$ given in (10.85) an H_2 OID filter for (10.1), and conversely any gain K that renders $\Sigma_{\text{p-CSS}}$ an H_2 OID filter for (10.1) is an element of $K_{\text{p-CSS}}^{h2-\text{oid}}$.

We have the following results.

Theorem 10.74 *Consider a continuous-time system as in* (10.1). *Let the conditions of Theorem 10.27 be satisfied. Then, the set* $K_{\text{p-CSS}}^{h2-\text{oid}}(A, B, C, D, E, F)$ *is nonempty and is given by*

$$K_{\text{p-CSS}}^{h2-\text{oid}}(A, B, C, D, E, F) = \bigcup_{P^* \in \mathcal{P}^*} K_{\text{sp-CSS}}^{h2-\text{oid}}(A, B, \tilde{C}, \tilde{D}, E - P^*C, 0),$$

$$\tag{10.88}$$

where $\mathcal{P}^*$ *is the set of all* P^* *that solve* $F - PD = 0$ *for* P *and* $K_{\text{sp-CSS}}^{h2-\text{oid}}$ *is as defined in Theorem 10.57.*

Proof : The proof follows from Lemma 10.72. ∎

Remark 10.75 *The set $K^{h2-\text{oid}}_{\text{sp-CSS}}(A, B, \widetilde{C}, \widetilde{D}, E - P^*C, 0)$ is nonempty for any P^* that solves $F - PD = 0$ for P.*

As in the case of strictly proper filters, the set $K^{h2-\text{oid}}_{\text{p-CSS}}(A, B, C, D, E, F)$ has the following telescopic property with respect to the matrices $\begin{pmatrix} E & F \end{pmatrix}$.

Lemma 10.76 *Consider a continuous-time system as in* (10.1). *Consider two different H_2 OID filtering problems both as defined in Definition 10.6, however, with one having the matrix pair (E_a, F_a) and the other (E_b, F_b) for the pair (E, F). Then, we have the telescopic property:*

$$\ker \begin{pmatrix} E_a & F_a \end{pmatrix} \supseteq \ker \begin{pmatrix} E_b & F_b \end{pmatrix}$$
$$\implies K^{h2-\text{oid}}_{\text{p-CSS}}(A, B, C, D, E_a, F_a) \supseteq K^{h2-\text{oid}}_{\text{p-CSS}}(A, B, C, D, E_b, F_b),$$

where we assume that the matrix pair (E_a, F_a) and (E_b, F_b) are such that, for the respective underlying system, the H_2 OID filtering problem is solvable (i.e., the conditions of Theorem 10.27 are satisfied).

Proof : The proof follows in view of Lemmas 10.72 and 10.58 and Theorem 10.74. ∎

In general, whenever it exists, a proper H_2 OID filter of the form $\Sigma_{\text{p-CSS}}$ given in (10.85) is not unique. The nonuniqueness of such a filter can indeed be a blessing as some other specifications can then be imposed to come up with an appropriate filter. As in the case of strictly proper filters, we would like to make use of the available freedom to place the poles of error dynamics $\Sigma^{ue}_{\text{p-CSS}}$ to shape its transient behavior. Once again, we observe that the the poles of error dynamics $\Sigma^{ue}_{\text{p-CSS}}$ are the same as those of H_2 OID filter $\Sigma_{\text{p-CSS}}$. Thus, we would like to place the poles of an H_2 OID filter at desired locations. However, as in the case of strictly proper filters, the requirement of H_2 OID filtering dictates that some poles of the filter be fixed at certain locations, whereas the others be free to be assigned. In fact, as before, for each given system, a set of complex numbers exists that every H_2 OID filter must have among its poles. As the following definition formalizes, such a set of complex numbers can be termed as the H_2 OID filter fixed modes.

Definition 10.77 (*Fixed modes of proper H_2 OID filters of CSS architecture*) *Consider a continuous-time system as in* (10.1) *and the H_2 OID filtering problem 10.6 characterized by the matrix sextuple (A, B, C, D, E, F). Assume that the solvability conditions as specified by Theorem 10.27 are satisfied. Then, a scalar*

$\lambda \in \mathbb{C}^-$ *is said to be the fixed mode of an H_2 OID filter with the proper full-order CSS architecture if λ is a pole (i.e., an eigenvalue of $A - K\widetilde{C}$) of every H_2 OID filter of such an architecture $\Sigma_{\text{p-CSS}}$ as given in (10.85). The set of all such H_2 OID filter fixed modes is denoted here by* $\boldsymbol{\Omega}^{h2-\text{oid}}_{\text{p-CSS}}(A, B, C, D, E, F)$.

The following lemma is needed before we characterize the set $\boldsymbol{\Omega}^{h2-\text{oid}}_{\text{p-CSS}}$.

Lemma 10.78 *Consider a continuous-time system as in (10.1). Let the conditions of Theorem 10.26 be satisfied. Let P_1^* and P_2^* be two different solutions of $F - PD = 0$ for P. Then, we have*

$$\boldsymbol{\Omega}^{h2-\text{oid}}_{\text{sp-CSS}}(A, B, \widetilde{C}, \widetilde{D}, E - P_1^*C, 0) = \boldsymbol{\Omega}^{h2-\text{oid}}_{\text{sp-CSS}}(A, B, \widetilde{C}, \widetilde{D}, E - P_2^*C, 0),$$
$$(10.89)$$

where, for $i = 1, 2$, the set $\boldsymbol{\Omega}^{h2-\text{oid}}_{\text{sp-CSS}}(A, B, \widetilde{C}, \widetilde{D}, E - P_i^*C, 0)$ is as characterized in Theorem 10.60.

Proof : This lemma follows directly by combining Theorem 10.60 and Lemma 7.31. ∎

We have the following theorem that characterizes the set $\boldsymbol{\Omega}^{h2-\text{oid}}_{\text{p-CSS}}$.

Theorem 10.79 *Consider a continuous-time system as in (10.1). Let Assumption 10.1 be satisfied. Consider proper filters, and assume that the solvability conditions as specified by Theorem 10.27 are satisfied. Then, we have*

$$\boldsymbol{\Omega}^{h2-\text{oid}}_{\text{p-CSS}}(A, B, C, D, E, F) = \boldsymbol{\Omega}^{h2-\text{oid}}_{\text{sp-CSS}}(A, B, \widetilde{C}, \widetilde{D}, E^*, 0),$$

where $E^* = E - P^*C$, with P^* being any solution of $F - PD = 0$ for P.

Proof : The proof follows from Lemmas 10.72 and 10.78. ∎

As in Lemma 10.62, the set $\boldsymbol{\Omega}^{h2-\text{oid}}_{\text{p-CSS}}(A, B, C, D, E, F)$ has a certain telescopic property as given below.

Lemma 10.80 *Consider a continuous-time system as in (10.1). Consider two different H_2 OID filtering problems both as defined in Definition 10.6, however, with* *one having the matrix pair (E_a, F_a) and the other (E_b, F_b) for the pair (E, F). Then, we have the telescopic property:*

$$\ker \begin{pmatrix} E_a & F_a \end{pmatrix} \supseteq \ker \begin{pmatrix} E_b & F_b \end{pmatrix}$$
$$\implies \Omega_{\text{p-CSS}}^{h2-\text{oid}}(A, B, C, D, E_a, F_a) \subseteq \Omega_{\text{p-CSS}}^{h2-\text{oid}}(A, B, C, D, E_b, F_b),$$

where we assume that the matrix pair (E_a, F_a) and (E_b, F_b) are such that, for the respective underlying system, the H_2 OID filtering problem is solvable (i.e., the conditions of Theorem 10.27 are satisfied).

Proof : The proof follows from Lemmas 10.72 and 10.62, and Theorem 10.79. ■

The set of fixed modes obviously presents a constraint that one must contend with in any H_2 OID filter design. In this regard, one ponders with the following. Suppose the conditions for the existence of strictly proper H_2 OID filters as specified by Theorem 10.26 are satisfied. In this case, both strictly proper as well as proper H_2 OID filters exist. The question that arises then is as follows: Will the number and the location of fixed modes be different for strictly proper and proper H_2 OID filters? The following lemma answers this question negatively whenever we use filters of CSS architecture.

Lemma 10.81 *Consider a continuous-time system as in (10.1). Let Assumption 10.1 be satisfied. Assume that the solvability conditions for the existence of strictly proper H_2 OID filters as specified by Theorem 10.26 are satisfied. Then, we have*

$$\Omega_{\text{p-CSS}}^{h2-\text{oid}}(A, B, C, D, E, F) = \Omega_{\text{sp-CSS}}^{h2-\text{oid}}(A, B, C, D, E, 0). \qquad (10.90)$$

Proof : The proof follows in view of Corollary 10.13 and Lemma 7.36.   ■

Besides the presence of H_2 OID filter fixed modes as explained above, another important structural issue exists in filter design. As in the case of strictly proper filters, in the system $\Sigma_{\text{p-CSS}}^{ue}$ that represents the error dynamics, certain pole/zero cancellations exist; see (10.86). To be specific, for a given H_2 OID filtering problem, there exist indeed what can be termed as a set of fixed decoupling zeros of $\Sigma_{\text{p-CSS}}^{ue}$ resulting from H_2 OID filters of CSS architecture. Such a set shows the minimum absolutely necessary number and locations of pole/zero cancellations within the system $\Sigma_{\text{p-CSS}}^{ue}$. We have the following formal definition.

Definition 10.82 (Fixed decoupling zeros of $\Sigma_{\text{p-CSS}}^{ue}$ resulting from proper H_2 OID filters of CSS architecture) *Consider a continuous-time system as in (10.1)* *and the H_2 OID filtering problem 10.6 characterized by the matrix sextuple (A, B, C, D, E, F). Assume that the solvability conditions of the H_2 OID filtering problem, as in Theorem 10.27, are satisfied. Then, a scalar $\lambda \in \mathbb{C}^-$ is said to be a fixed decoupling zero of $\Sigma^{ue}_{\text{p-CSS}}$ resulting from proper H_2 OID filters of CSS architecture if λ is either an input or an output decoupling zero (or both) of the system $\Sigma^{ue}_{\text{p-CSS}}$ as given in (10.86) for any proper H_2 OID filter $\Sigma_{\text{p-CSS}}$ that one can use. The set of all such fixed decoupling zeros is denoted by $\Lambda^{h2-\text{oid}}_{\text{p-CSS}}$ (A, B, C, D, E, F).*

The following lemma is needed before we characterize the set $\Lambda^{h2-\text{oid}}_{\text{p-CSS}}$.

Lemma 10.83 *Consider a continuous-time system as in (10.1). Let the conditions of Theorem 10.26 be satisfied. Let P_1^* and P_2^* be two different solutions of $F - PD = 0$ for P. Then, we have*

$$\Lambda^{h2-\text{oid}}_{\text{sp-CSS}}(A, B, \widetilde{C}, \widetilde{D}, E - P_1^*C, 0) = \Lambda^{h2-\text{oid}}_{\text{sp-CSS}}(A, B, \widetilde{C}, \widetilde{D}, E - P_2^*C, 0),$$
$$(10.91)$$

where, for $i = 1, 2$, the set $\Lambda^{h2-\text{oid}}_{\text{sp-CSS}}(A, B, \widetilde{C}, \widetilde{D}, E - P_i^*C, 0)$ is as characterized in Theorem 10.65.

Proof : The proof follows along the same lines as that of Lemma 7.31.   ∎

We have the following theorem that characterize the set $\Lambda^{h2-\text{oid}}_{\text{p-CSS}}$.

Theorem 10.84 *Consider a continuous-time system as in (10.1). Let Assumption 10.1 be satisfied. Consider proper filters of CSS architecture as given in (10.85), and assume that the solvability conditions of the H_2 OID filtering problem (as in Theorem 10.27) are satisfied. Then, we have*

$$\Lambda^{h2-\text{oid}}_{\text{p-CSS}}(A, B, C, D, E, F) = \Lambda^{h2-\text{oid}}_{\text{sp-CSS}}(A, B, \widetilde{C}, \widetilde{D}, E^*, 0),$$

where $E^* = E - P^*C$ with P^* being any solution of $F - PD = 0$ for P.

Proof : The proof follows from Lemmas 10.72 and 10.83.   ∎

As in the case of strictly proper filters, it turns out that, under the general assumption of the pair (C, A) being detectable, the set $\Lambda^{h2-\text{oid}}_{\text{sp-CSS}}$ does not have any telescopic property. However, as before, whenever the pair (C, A) is observable, it does have such a property as the following lemma shows.

Lemma 10.85 *Consider a continuous-time system as in* (10.1), *where the pair* (C, A) *is observable. Consider two different* H_2 *OID filtering problems both as defined in Definition 10.6, however, with one having the matrix pair* (E_a, F_a) *and the other* (E_b, F_b) *for the pair* (E, F). *Then, we have the telescopic property:*

$$\ker \begin{pmatrix} E_a & F_a \end{pmatrix} \supseteq \ker \begin{pmatrix} E_b & F_b \end{pmatrix}$$
$$\implies \Lambda_{\text{p-CSS}}^{h2-\text{oid}}(A, B, C, D, E_a, F_a) \subseteq \Lambda_{\text{p-CSS}}^{h2-\text{oid}}(A, B, C, D, E_b, F_b),$$

where we assume that the matrix pair (E_a, F_a) *and* (E_b, F_b) *are such that, for the respective underlying system, the* H_2 *OID filtering problem is solvable (i.e., the conditions of Theorem 10.27 are satisfied).*

Proof : The proof follows from Lemmas 10.72 and 10.66 and Theorem 10.84. ∎

Earlier in Section 10.10, we discussed the conditions under which the transfer matrix of any H_2 OID filter is unique. We considered there general proper filters of the form (10.2). The following theorem considers a class of filters with a specified architecture, namely, the proper CSS architecture, and discusses an H_2 OID filter being unique among them.

Theorem 10.86 *Consider a continuous-time system as in* (10.1), *and let Assumption 10.1 be satisfied. Also, let the matrices* B_Q *and* D_Q *be as defined by* (10.8), *and let* Σ_Q *be as in* (10.10). *Consider proper filters, and assume that the solvability conditions of the* H_2 *OID filtering problem (as in Theorem 10.27) are satisfied. Then, a unique proper full-order* H_2 *OID filter of the form* $\Sigma_{\text{p-CSS}}$ *given* (10.85) *exists if and only if the following conditions are satisfied:*

    (i) The subsystem characterized by the quadruple (A, B, C, D) *is right-invertible.*

    (ii) The H_2 *OID filtering problem is regular (i.e., the regularity conditions given in Definition 10.7 are satisfied).*

    (iii) The matrix pair $(E - FD_Q^{-1}C, \ A - B_Q D_Q^{-1}C)$ *is observable.*

Moreover, under the above conditions, we have

    (i) $\boldsymbol{K}_{\text{p-CSS}}^{h2-\text{oid}}(A, B, C, D, E, F) = \{B_Q D_Q^{-1}\}$, *which is a singleton.*

    (ii) We have

$$\Omega_{\text{p-CSS}}^{h2-\text{oid}}(A, B, C, D, E, F) = \lambda(A - B_Q D_Q^{-1}C),$$

    which is the union of all stable invariant zeros and the mirror images of unstable invariant zeros and the compromising zeros of the subsystem characterized by the quadruple (A, B, C, D).

(iii) The set $\Lambda^{h2-\mathrm{oid}}_{\mathrm{sp\text{-}CSS}}(A, B, C, D, E, 0)$ is the same as the set of all stable invariant zeros of the subsystem characterized by the quadruple (A, B, C, D).

Also, the unique H_2 OID filter $\Sigma_{\mathrm{p\text{-}CSS}}$ is given by

$$\dot{\xi} = (A - B_{\varrho} D_{\varrho}^{-1} C)\xi + B_{\varrho} D_{\varrho}^{-1} y,$$

with

$$\hat{z} = (E - F D_{\varrho}^{-1} C)\xi + F D_{\varrho}^{-1} y.$$

Proof : From Corollary 10.13, we know that a filter is an H_2 OID filter for Σ if and only if it is an EID filter for Σ_{ϱ}. The results then follow from the conditions for the uniqueness of an EID filter for Σ_{ϱ} as implied by Theorem 7.28.   ■

Remark 10.87 *We observe that the conditions given in Theorem 10.86 imply those given in Theorem 10.54 but not conversely. The reason for the more stringent conditions is that here we define the uniqueness of an H_2 OID filter at the state-space level (using proper architecture), whereas earlier we considered the uniqueness in terms of a transfer matrix.*

Finally, we would like to design an H_2 OID filter while placing its modes at desired locations. The following theorem shows that we can design a filter gain K such that $\Sigma_{\mathrm{p\text{-}CSS}}$ given in (10.85) is an H_2 OID filter, whereas its modes are at the prescribed desired locations except that they need to contain $\boldsymbol{\Omega}^{h2-\mathrm{oid}}_{\mathrm{p\text{-}CSS}}(A, B, C, D, E, F)$ among them.

Theorem 10.88 (Proper H_2 OID filter of CSS architecture with pole placement) *Consider a continuous-time system as in (10.1). Let Assumption 10.1 be satisfied. Consider proper filters, and assume that the conditions of Theorem 10.27 are satisfied. Also, let P^* be any solution of $F - PD = 0$ for P. Consider the set $\boldsymbol{\Omega}^{h2-\mathrm{oid}}_{\mathrm{p\text{-}CSS}}(A, B, C, D, E, F)$, which is the set of H_2 OID fixed modes discussed in Theorem 10.79. Also, consider the nonempty set $\boldsymbol{K}^{h2-\mathrm{oid}}_{\mathrm{p\text{-}CSS}}(A, B, C, D, E, F)$ discussed in Theorem 10.74. Moreover, let Λ be a prescribed set of n self-conjugate elements in the open left-half complex plane $\mathbb{C}^-$ such that Λ includes $\boldsymbol{\Omega}^{h2-\mathrm{oid}}_{\mathrm{p\text{-}CSS}}(A, B, C, D, E, F)$. Then, a filter gain $K \in \boldsymbol{K}^{h2-\mathrm{oid}}_{\mathrm{p\text{-}CSS}}(A, B, C, D, E, F)$ exists such that the proper filter $\Sigma_{\mathrm{p\text{-}CSS}}$ given in (10.85) is an H_2 OID filter, and moreover, $\lambda(A - K\widetilde{C}) = \Lambda$.*

Proof : The proof follows from Lemma 10.72 and Theorem 10.70.   ■

An algorithm for designing a proper H_2 OID filter with simultaneous filter pole placement:

Step 1: Given the system Σ as in (10.73) and characterized by the sextuple (A, B, C, D, E, F) form the auxiliary system $\tilde{\Sigma}^*$ characterized by the quintuple $(A, B, \tilde{C}, \tilde{D}, E^*)$, where $\tilde{C}$ and $\tilde{D}$ are, respectively, as in (10.74) and (10.76), and as usual, $E^* = E - P^*C$, with matrix P^* being any solution of $F - PD = 0$ for P.

Step 2: Choose a desired set Λ of n self-conjugate elements in the open left-half complex plane $\mathbb{C}^-$. [Step 3 given below checks to make sure that Λ includes $\Omega_{\text{p-CSS}}^{h2-\text{oid}}(A, B, C, D, E, F)$].

Step 3: Following the algorithm for designing a strictly proper H_2 OID filter with simultaneous filter pole placement as given in Subsection 10.12.1, and using the quintuple $(A, B, \tilde{C}, \tilde{D}, E^*)$ as well as the matrix Λ as inputs to that algorithm, we obtain the gain K.

With the filter gain K as computed above, it is easy to verify that the resulting filter $\Sigma_{\text{p-CSS}}$ given in (10.85) is an H_2 OID filter with its modes at the locations specified by Λ; i.e., $\lambda(A - K\tilde{C}) = \Lambda$.

10.12.3 Reduced-order H_2 OID filters of CSS architecture

The two previous subsections respectively construct full-order strictly proper and proper filters of CSS architecture that solve the H_2 OID filtering problem for continuous-time systems. By full-order filters, as usual, we mean filters having the same dynamic order as that of the given system. Our goal in this subsection is to develop reduced-order filters of CSS architecture having their dynamic order lower than that of the given system.

The procedure of developing reduced-order filters here follows mostly along the same lines as in Subsection 7.5.3, which pertains to EID filters. However, there are some subtle but important differences warranting pertinent justifications, and as such, we need to redevelop them once more here in connection with H_2 OID filters. As before, in essence, our method of development transforms the construction of a reduced-order filter for a given system to that of a full-order filter for a certain reduced-order system.

We proceed now to construct an appropriate reduced-order system. To start with, let us rewrite the matrices C and D of (10.1) as

$$C = \begin{pmatrix} 0 & C_{02} \\ I_{p-m_0} & 0 \end{pmatrix}, \quad D = \begin{pmatrix} D_0 \\ 0 \end{pmatrix},$$

where again rank D = rank D_0 = m_0. This can always be done without any loss of generality by appropriate coordinate transformations. In view of the above partitioning of C and D, we can partition the given system Σ as

$$\Sigma : \begin{cases} \begin{pmatrix} \dot{x}_1 \\ \dot{x}_2 \end{pmatrix} = \begin{pmatrix} A_{11} & A_{12} \\ A_{21} & A_{22} \end{pmatrix} \begin{pmatrix} x_1 \\ x_2 \end{pmatrix} + \begin{pmatrix} B_{11} \\ B_{22} \end{pmatrix} u \\[4mm] y = \begin{pmatrix} y_0 \\ y_1 \end{pmatrix} = \begin{pmatrix} 0 & C_{02} \\ I & 0 \end{pmatrix} \begin{pmatrix} x_1 \\ x_2 \end{pmatrix} + \begin{pmatrix} D_0 \\ 0 \end{pmatrix} u \\[4mm] z = Ex + Fu, \end{cases} \tag{10.92}$$

where different variables have obvious meanings.

Next, we use a preliminary injection of output y into the desired output z. To this end, we define a matrix P^*, which is a solution of $F - PD = 0$ for P. We define a new desired output z^* as

$$z^* = z - P^*y = (E - P^*C)x + (F - P^*D)u = E^*x, \tag{10.93}$$

where $E^* = E - P^*C$. In view of (10.92) and (10.93), we can define a new system Σ^* as

$$\Sigma^* : \begin{cases} \begin{pmatrix} \dot{x}_1 \\ \dot{x}_2 \end{pmatrix} = \begin{pmatrix} A_{11} & A_{12} \\ A_{21} & A_{22} \end{pmatrix} \begin{pmatrix} x_1 \\ x_2 \end{pmatrix} + \begin{pmatrix} B_{11} \\ B_{22} \end{pmatrix} u \\[4mm] y = \begin{pmatrix} y_0 \\ y_1 \end{pmatrix} = \begin{pmatrix} 0 & C_{02} \\ I & 0 \end{pmatrix} \begin{pmatrix} x_1 \\ x_2 \end{pmatrix} + \begin{pmatrix} D_0 \\ 0 \end{pmatrix} u \\[4mm] z^* = E^*x = E_1^*x_1 + E_2^*x_2, \end{cases} \tag{10.94}$$

with $E^* = \begin{pmatrix} E_1^* & E_2^* \end{pmatrix}$.

We note that the y_1 is not contaminated by the input u, and hence, $x_1 = y_1$ is known exactly from the measurement y. Thus, all we need to do next is to estimate the state x_2. To proceed further, let us rewrite the state equation for x_1 in terms of the output y_1 and the state x_2 as,

$$\dot{y}_1 = A_{11}y_1 + A_{12}x_2 + B_{11}u. \tag{10.95}$$

The above equation can be rewritten as

$$\dot{y}_1 - A_{11}y_1 = A_{12}x_2 + B_{11}u.$$

Treating $\dot{y}_1$ as known, we can define a new measurement variable y_r,

$$y_r = \begin{pmatrix} y_0 \\ \dot{y}_1 - A_{11}y_1 \end{pmatrix}.$$

Although $\dot{y}_1$ is not directly available, as we did earlier in the case of full-order proper CSS filters, we can eliminate it from any filter that is constructed using y_r

as a measured output. With this in mind, we form the auxiliary system:

$$\Sigma_r^* : \begin{cases} \dot{x}_r = A_r x_r + B_r u + A_{21} y_1 \\ y_r = C_r x_r + D_r u \\ z_r^* = E_r^* x_r, \end{cases} \qquad (10.96)$$

where $x_r = x_2$ and

$$A_r = A_{22}, \quad B_r = B_{22}, \quad C_r = \begin{pmatrix} C_{02} \\ A_{12} \end{pmatrix}, \quad D_r = \begin{pmatrix} D_0 \\ B_{11} \end{pmatrix}, \quad E_r^* = E_2^*.$$
$$(10.97)$$

We note that the dynamic order n_r of the above Σ_r^* is less than the dynamic order n of the given system Σ by a number equal to the dimension of $x_1 = y_1$.

Before we proceed further, we state that certain structural properties of Σ_r^* are related to those of the given system Σ, and they are developed in Lemma 7.55.

Next, let us suppose we can construct a strictly proper H_2 OID filter for the auxiliary system Σ_r^* in order to arrive at an estimate $\hat{z}_r^*$ of z_r^*. (Here y_1 is a known input and hence will show up accordingly in the dynamics of any filter we construct for Σ_r^*.) Then, we plan to obtain an estimate $\hat{z}$ of z as

$$\hat{z} = E_1^* x_1 + \hat{z}_r^* + P^* y. \qquad (10.98)$$

This motivates us to construct a full-order strictly proper filter for the auxiliary system Σ_r^* to arrive at an estimate $\hat{z}_r^*$ of z_r^*. A full-order filter constructed as such for Σ_r^* is indeed a reduced-order filter for Σ. However, in view of (10.98), such a reduced-order filter for Σ is a proper (rather than a strictly proper) filter for Σ. A fundamental issue that arises next is under what conditions one can construct a strictly proper H_2 OID filter for Σ_r^*. It turns out that one can construct a strictly proper H_2 OID filter for the reduced-order system Σ_r^* if and only if one can construct a proper H_2 OID filter for the given system Σ. The following lemma formalizes this result.

Lemma 10.89 *Consider continuous-time systems Σ and Σ_r^* as given in* (10.1) *and* (10.96), *respectively. Then, the following two statements are equivalent:*

(i) A strictly proper H_2 OID filter exists for the reduced-order system Σ_r^*.

(ii) A proper reduced-order H_2 OID filter exists for the system Σ.

Moreover, there is a $1-1$ relationship between the strictly proper H_2 OID filter of CSS architecture for Σ_r^* and the proper reduced-order H_2 OID filter of CSS architecture for Σ; that is, one of these filters can be constructed from the other.

Proof : The equivalence of statements (*i*) and (*ii*) follows from Lemma 7.55. The $1-1$ relationship between the aforementioned filters can be verified easily.   ∎

Next, to construct the required filter for Σ, in the spirit of the above development, we first construct a strictly proper filter of CSS architecture for the reduced-order system Σ_r^* as

$$\begin{cases} \dot{\tilde{\xi}}_r = A_r\tilde{\xi}_r + A_{21}y_1 + K_r(y_r - C_r\tilde{\xi}_r) \\ \hat{z}_r^* = E_r^*\tilde{\xi}_r, \end{cases} \tag{10.99}$$

where K_r is termed as the reduced-order filter gain. As $\dot{y}_1$ is not available, we need to modify the above filter. To this end, let us partition $K_r = (K_{r0} \quad K_{r1})$ so as to be compatible with the partitioning of y_r. Also, let

$$\xi_r = \tilde{\xi}_r - K_{r1}y_1. \tag{10.100}$$

We can then easily rewrite the filter (10.132) as a proper filter for Σ as shown below.

Reduced-order proper filter of CSS architecture for Σ: The reduced-order proper filter of CSS architecture for Σ is given by

$$\Sigma_{\text{r-CSS}} : \begin{cases} \dot{\xi}_r = (A_r - K_rC_r)\xi_r + \tilde{K}_r y \\ \tilde{\xi}_r = \xi_r + K_{r1}y_1 \\ \hat{z} = E_1^*x_1 + E_r^*\tilde{\xi}_r + P^*y = E_r^*\xi_r + \tilde{P}_r y, \end{cases} \tag{10.101}$$

where

$$\tilde{K}_r = \left(K_{r0} \qquad A_{21} - K_{r1}A_{11} + (A_r - K_rC_r)K_{r1} \right)$$

and

$$\tilde{P}_r = \left(0 \qquad E_1^* + E_r^*K_{r1} \right) + P^*.$$

A block diagram representation of the reduced-order proper filter along with the given plant is structurally the same as in Figure 7.5.

Error dynamics: By defining the error $e_r = x_r - \tilde{\xi}_r$, the error between the actual desired output $z = Ex + Fu = E^*x + P^*y + F^*u$ and the estimated desired output $\hat{z} = E_1^*x_1 + E_r^*\tilde{\xi}_r + P^*y$ can be written as

$$e_z = z - \hat{z} = E_r^*e_r.$$

Then, the dynamics of error is given by

$$\Sigma_{\text{r-CSS}}^{ue} : \begin{cases} \dot{e}_r = (A_r - K_rC_r)e_r + (B_r - K_rD_r)u \\ e_z = E_r^*e_r. \end{cases} \tag{10.102}$$

Also, the transfer matrix $G_{\text{r-CSS}}^{ue}$ from u to e_z can obviously be written as

$$G_{\text{r-CSS}}^{ue}(s) = E_r^*(sI - A_r + K_rC_r)^{-1}(B_r - K_rD_r). \tag{10.103}$$

Remark 10.90 *It is easy to see that both the filter equation (10.101) and the error equation (10.102) have the same modes that are the eigenvalues of $A_r - K_rC_r$.*

Everything in the filter $\Sigma_{\text{r-CSS}}$ is known except the gain K_r. Following the notation used in earlier sections, we denote the set of all H_2 OID filter gains by $K_{\text{r-CSS}}^{h2-\text{oid}}(A, B, C, D, E, F)$, meaning that any gain $K_r \in K_{\text{r-CSS}}^{h2-\text{oid}}$ renders $\Sigma_{\text{r-CSS}}$ given in (10.101) an H_2 OID filter for (10.1), and conversely, any gain K_r that renders $\Sigma_{\text{r-CSS}}$ an H_2 OID filter for (10.1) is an element of $K_{\text{r-CSS}}^{h2-\text{oid}}$.

We have the following results.

Theorem 10.91 *Consider a continuous-time system as in (10.1). Let the conditions of Theorem 10.27 be satisfied. Also, let P^* be a solution of $F - PD = 0$ for P. Then, the set $K_{\text{r-CSS}}^{h2-\text{oid}}(A, B, C, D, E, F)$ is nonempty and is given by*

$$K_{\text{r-CSS}}^{h2-\text{oid}}(A, B, C, D, E, F) = \bigcup_{P^* \in \mathcal{P}^*} K_{\text{sp-CSS}}^{h2-\text{oid}}(A_r, B_r, C_r, D_r, E_r^*, 0),$$

$$(10.104)$$

where E_r^* is obtained by partitioning $E - P^*C$ in conformity with the partitioning of x into x_1 and x_2 as

$$E - P^*C = \begin{pmatrix} E_1^* & E_r^* \end{pmatrix},$$
$$(10.105)$$

and $\mathcal{P}^*$ is the set of all P^* that solve $F - PD = 0$ for P. Also, $K_{\text{sp-CSS}}^{h2-\text{oid}}$ is as defined in Theorem 10.57.

Proof : The proof follows from Lemma 10.89 and Theorem 10.57.   ■

Remark 10.92 *The set $K_{\text{sp-CSS}}^{h2-\text{oid}}(A_r, B_r, C_r, D_r, E_r^*, 0)$ is nonempty for any P^* that solves $F - PD = 0$ for P.*

In general, whenever they exist, reduced-order H_2 OID filters of the form $\Sigma_{\text{r-CSS}}$ given in (10.101) are not necessarily unique. The following remark addresses the issue of uniqueness of such a filter.

Remark 10.93 *A reduced-order H_2 OID filter of CSS architecture for the system Σ given in (10.1) is unique if and only if a full-order proper H_2 OID filter of CSS architecture for the same system Σ is unique. The conditions for the uniqueness of such filters are given in Theorem 10.86. Under such conditions, the so-called reduced-order filter of CSS architecture coalesces with the full-order filter of CSS architecture.*

As in the previous subsections, the set $K_{\text{r-CSS}}^{h2-\text{oid}}(A, B, C, D, E, F)$ has the following telescopic property with respect to the matrices $(E \quad F)$.

Lemma 10.94 *For the continuous-time system given in (10.1), consider two different H_2 OID filtering problems both as defined in Definition 10.6, however, with one having the matrix pair $(E_a,\ F_a)$ and the other $(E_b,\ F_b)$ for the pair $(E,\ F)$. Then, we have the telescopic property:*

$$\ker\begin{pmatrix} E_a & F_a \end{pmatrix} \supseteq \ker\begin{pmatrix} E_b & F_b \end{pmatrix}$$
$$\Longrightarrow K_{\text{r-CSS}}^{h2-\text{oid}}(A, B, C, D, E_a, F_a) \supseteq K_{\text{r-CSS}}^{h2-\text{oid}}(A, B, C, D, E_b, F_b),$$

where we assume that the matrix pair $(E_a,\ F_a)$ and $(E_b,\ F_b)$ are such that, for the respective underlying system, the H_2 OID filtering problem is solvable (i.e., the conditions of Theorem 10.27 are satisfied).

Proof : The proof follows in view of Lemmas 10.89 and 10.58 and Theorem 10.91. ∎

Next, as in full-order filters, we would like to design a reduced-order H_2 OID filter of CSS architecture while placing its poles at desired locations in order to shape the transient behavior of error dynamics. As before, it is expected that all H_2 OID filters of reduced-order CSS architecture share some fixed modes as defined below.

Definition 10.95 *(**Fixed modes of reduced-order proper H_2 OID filters of CSS architecture**) Consider a continuous-time system as in (10.1) and the H_2 OID filtering problem 10.6 characterized by the matrix sextuple (A, B, C, D, E, F). Assume that the solvability conditions as specified by Theorem 10.27 are satisfied. Then, a scalar $\lambda \in \mathbb{C}^-$ is said to be the fixed mode of an H_2 OID filter with the proper reduced-order CSS architecture if λ is a pole (i.e., an eigenvalue of $A_r - K_r C_r$) of every H_2 OID filter of such an architecture $\Sigma_{\text{r-CSS}}$ as given in (10.101). The set of all such H_2 OID filter fixed modes is denoted here by $\Omega_{\text{r-CSS}}^{h2-\text{oid}}(A, B, C, D, E, F)$.*

The following lemma is needed before we characterize the set $\Omega_{\text{r-CSS}}^{h2-\text{oid}}$.

Lemma 10.96 *Consider a continuous-time system as in (10.1). Let the conditions of Theorem 10.26 be satisfied. Let P_1^* and P_2^* be two different solutions of $F - PD = 0$ for P. Then, we have*

$$\Omega_{\text{sp-CSS}}^{h2-\text{oid}}(A_r, B_r, C_r, D_r, E_{r1}^*, 0) = \Omega_{\text{sp-CSS}}^{h2-\text{oid}}(A_r, B_r, C_r, D_r, E_{r2}^*, 0), \quad (10.106)$$

where, for $i = 1, 2$, E_{ri}^* is obtained by partitioning $E - P_i^* C$ in conformity with the partitioning of x into x_1 and x_2 as

$$E - P_i^* C = \begin{pmatrix} E_1^* & E_{ri}^* \end{pmatrix}, \quad (10.107)$$

and $\Omega_{\text{sp-CSS}}^{h2-\text{oid}}$ is characterized as in Theorem 10.60.

Proof : The proof follows along the same lines as that of Lemma 10.78.   ∎

We have the following theorem that characterizes the set $\Omega_{\text{r-CSS}}^{h2-\text{oid}}$.

Theorem 10.97 *Consider a continuous-time system as in (10.1). Let Assumption 10.1 be satisfied. Consider proper filters, and assume that the solvability conditions as specified by Theorem 10.27 are satisfied. Then, we have*

$$\Omega_{\text{r-CSS}}^{h2-\text{oid}}(A, B, C, D, E, F) = \Omega_{\text{sp-CSS}}^{h2-\text{oid}}(A_r, B_r, C_r, D_r, E_r^*, 0),$$

where $\Omega_{\text{sp-CSS}}^{h2-\text{oid}}(A_r, B_r, C_r, D_r, E_r^*, 0)$ *is characterized in Theorem 10.60.*

Proof : The proof follows in view of Lemmas 10.89 and 10.96.   ∎

As in Lemma 10.80, the set $\Omega_{\text{r-CSS}}^{h2-\text{oid}}(A, B, C, D, E, F)$ has a certain telescopic property as given below.

Lemma 10.98 *Consider a continuous-time system as in (10.1). Consider two different H_2 OID filtering problems both as defined in Definition 10.6, however, with one having the matrix pair (E_a, F_a) and the other (E_b, F_b) for the pair (E, F). Then, we have the telescopic property:*

$$\ker \begin{pmatrix} E_a & F_a \end{pmatrix} \supseteq \ker \begin{pmatrix} E_b & F_b \end{pmatrix}$$
$$\implies \Omega_{\text{r-CSS}}^{h2-\text{oid}}(A, B, C, D, E_a, F_a) \subseteq \Omega_{\text{p-CSS}}^{h2-\text{oid}}(A, B, C, D, E_b, F_b),$$

where we assume that the matrix pair (E_a, F_a) and (E_b, F_b) are such that, for the respective underlying system, the H_2 OID filtering problem is solvable (i.e., the conditions of Theorem 10.27 are satisfied).

Proof : The proof follows in view of Lemmas 10.89 and 10.62 and Theorem 10.97.   ∎

As we discussed, the set of fixed modes obviously presents a constraint that one must contend with in any H_2 OID filter design. A question that arises then is whether the H_2 OID filter fixed modes of reduced-order filters are different from those of full-order filters. The following lemma addresses this issue and shows that it is not the case.

Lemma 10.99 *Consider a continuous-time system as in* (10.1). *Let Assumption 10.1 be satisfied. Assume that the solvability conditions for the existence of proper H_2 OID filters as specified by Theorem 10.27 are satisfied. Then, we have*

$$\Omega_{\text{r-CSS}}^{h2-\text{oid}}(A, B, C, D, E, F) = \Omega_{\text{p-CSS}}^{h2-\text{oid}}(A, B, C, D, E, F). \tag{10.108}$$

Proof : The proof follows along the same lines as in the proof of Lemma 7.63. ∎

In Lemma 10.81, we showed that, whenever strictly proper H_2 OID filters exist, the sets of fixed modes are the same whether one uses strictly proper or proper H_2 OID filters. Following the theme of this lemma, the following lemma extends this result even further by considering the reduced-order filters as well.

Lemma 10.100 *Consider a continuous-time system as in* (10.1). *Let Assumption 10.1 be satisfied. Assume that the solvability conditions for the existence of strictly proper H_2 OID filters as specified by Theorem 10.26 are satisfied. Then, we have*

$$\Omega_{\text{p-CSS}}^{h2-\text{oid}}(A, B, C, D, E, F) = \Omega_{\text{sp-CSS}}^{h2-\text{oid}}(A, B, C, D, E, 0)$$
$$= \Omega_{\text{r-CSS}}^{h2-\text{oid}}(A, B, C, D, E, F). \tag{10.109}$$

Proof : The proof follows along the same lines as in the proof of Lemma 7.64. ∎

As in the case of strictly proper and proper filters, besides the presence of H_2 OID filter fixed modes as explained above, another important structural issue exists in filter design. That is, in the system $\Sigma_{\text{r-CSS}}^{ue}$ that represents the error dynamics, certain pole/zero cancellations exist; see (10.102). To be specific, for a given H_2 OID filtering problem, there exist indeed what can be termed as a set of fixed decoupling zeros of $\Sigma_{\text{r-CSS}}^{ue}$ resulting from H_2 OID filters of CSS architecture. Such a set shows the minimum absolutely necessary number and locations of pole/zero cancellations within the system $\Sigma_{\text{r-CSS}}^{ue}$. We have the following formal definition.

Definition 10.101 *(Fixed decoupling zeros of $\Sigma_{\text{r-CSS}}^{ue}$ resulting from proper H_2 OID filters of CSS architecture) Consider a continuous-time system as in* (10.1) *and the H_2 OID filtering problem 10.6 characterized by the matrix sextuple (A, B, C, D, E, F). Assume that the solvability conditions of the H_2 OID filtering problem, as in Theorem 10.27, are satisfied. Then, a scalar $\lambda \in \mathbb{C}^-$ is said to be a fixed decoupling zero of $\Sigma_{\text{r-CSS}}^{ue}$ resulting from reduced-order H_2 OID filters of CSS architecture if λ is either an input or an output decoupling zero (or both) of the system $\Sigma_{\text{r-CSS}}^{ue}$ as given in* (10.102) *for any proper H_2 OID filter $\Sigma_{\text{r-CSS}}$ that one can use. The set of all such fixed decoupling zeros is denoted by $\Lambda_{\text{r-CSS}}^{h2-\text{oid}}(A, B, C, D, E, F)$.*

The following lemma is needed before we characterize the set $\Lambda_{\text{r-CSS}}^{h2-\text{oid}}$.

Lemma 10.102 *Consider a continuous-time system as in* (10.1). *Let the conditions of Theorem 10.26 be satisfied. Let P_1^* and P_2^* be two different solutions of $F - PD = 0$ for P. Then, we have*

$$\Lambda_{\text{sp-CSS}}^{h2-\text{oid}}(A_r, B_r, C_r, D_r, E_{r1}^*, 0) = \Lambda_{\text{sp-CSS}}^{h2-\text{oid}}(A_r, B_r, C_r, D_r, E_{r2}^*, 0), \quad (10.110)$$

where, for $i = 1, 2$, E_{ri}^* is as defined in (10.107) *and $\Lambda_{\text{sp-CSS}}^{h2-\text{oid}}(A_r, B_r, C_r, D_r, E_{ri}^*, 0)$ is characterized as in Theorem 10.65.*

Proof : The proof follows along the same lines as that of Lemma 7.31.   ∎

We have the following theorem that characterize the set $\Lambda_{\text{r-CSS}}^{h2-\text{oid}}$.

Theorem 10.103 *Consider a continuous-time system as in* (10.1). *Let Assumption 10.1 be satisfied. Consider reduced-order filters of CSS architecture as given in* (10.101), *and assume that the solvability conditions of the H_2 OID filtering problem (as in Theorem 10.27) are satisfied. Then, we have*

$$\Lambda_{\text{r-CSS}}^{h2-\text{oid}}(A, B, C, D, E, F) = \Lambda_{\text{sp-CSS}}^{h2-\text{oid}}(A_r, B_r, C_r, D_r, E_r^*, 0).$$

Proof : The proof follows in view of Lemmas 10.89 and 10.102.   ∎

As in the case of strictly proper and proper filters, it turns out that, under the general assumption of the pair (C, A) being detectable, the set $\Lambda_{\text{r-CSS}}^{h2-\text{oid}}$ does not have any telescopic property. However, as before, whenever the pair (C, A) is observable, it does have such a property as the following lemma shows.

Lemma 10.104 *Consider a continuous-time system as in* (10.1) *with the pair (C, A) observable. Consider two different H_2 OID filtering problems both as defined in Definition 10.6, however, with one having the matrix pair (E_a, F_a) and the other (E_b, F_b) for the pair (E, F). Then, we have the telescopic property:*

$$\ker \begin{pmatrix} E_a & F_a \end{pmatrix} \supseteq \ker \begin{pmatrix} E_b & F_b \end{pmatrix}$$

$$\implies \Lambda_{\text{r-CSS}}^{h2-\text{oid}}(A, B, C, D, E_a, F_a) \subseteq \Lambda_{\text{r-CSS}}^{h2-\text{oid}}(A, B, C, D, E_b, F_b),$$

where we assume that the matrix pair (E_a, F_a) and (E_b, F_b) are such that, for the respective underlying system, the H_2 OID filtering problem is solvable (i.e., the conditions of Theorem 10.27 are satisfied).

Proof : The proof follows from Lemmas 10.89 and 10.66 and Theorem 10.103.

■

Finally, the following theorem shows that we can design a filter gain K_r such that $\Sigma_{\text{r-CSS}}$ given in (10.101) is a reduced-order H_2 OID filter, whereas its poles are at the prescribed desired locations, except that they need to contain $\Omega_{\text{r-CSS}}^{h2-\text{oid}}$ (A, B, C, D, E, F) among them.

Theorem 10.105 *(H_2 OID reduced-order proper filter of CSS architecture with pole placement) Consider a continuous-time system as in (10.1). Let the condi- tions of Theorem 10.27 be satisfied. Also, let P^* be a solution of $F - PD = 0$ for P. Consider the set of H_2 OID filter fixed modes $\Omega_{\text{r-CSS}}^{h2-\text{oid}}(A, B, C, D, E, F)$ discussed in Theorem 10.97. Also, consider the nonempty set $K_{\text{r-CSS}}^{h2-\text{oid}}$ (A, B, C, D, E, F) discussed in Theorem 10.91. Moreover, let Λ_r be a given set of n_r self-conjugate elements in the open left-half complex plane $\mathbb{C}^-$ such that Λ_r in- cludes $\Omega_{\text{r-CSS}}^{h2-\text{oid}}$. Then, a filter gain $K_r \in K_{\text{r-CSS}}^{h2-\text{oid}}(A, B, C, D, E, F)$ exists such that the proper filter $\Sigma_{\text{r-CSS}}$ given in (10.101) is an H_2 OID filter, and moreover, $\lambda(A_r - K_r C_r) = \Lambda_r$.*

Proof : The proof follows from Lemma 10.89 and Theorem 10.70. ■

An algorithm for designing a reduced-order proper H_2 OID fil- ter with simultaneous filter pole placement:

Step 1: Given the system Σ as in (10.73) and characterized by the sextuple (A, B, C, D, E, F), form the auxiliary system $\widetilde{\Sigma}_r^*$ characterized by the quintuple $(A_r, B_r, C_r, D_r, E_r^*)$ where these matrices are defined in (10.97).

Step 2: Choose a desired set Λ_r of n_r self-conjugate elements in the open left- half complex plane $\mathbb{C}^-$. [Step 3 given below checks to make sure that Λ_r includes $\Omega_{\text{r-CSS}}^{h2-\text{oid}}(A, B, C, D, E, F)$.]

Step 3: Following the algorithm for designing a strictly proper H_2 OID filter with simultaneous filter pole placement as given in Subsection 10.12.1, and using the quintuple $(A_r, B_r, C_r, D_r, E_r^*)$ as well as the matrix Λ_r as inputs to that algorithm, we obtain the gain K_r.

With the filter gain K_r as computed above, it is easy to verify that the resulting filter $\Sigma_{\text{r-CSS}}$ given in (10.101) is an H_2 OID filter with its modes at the locations specified by Λ_r; i.e., $\lambda(A_r - K_r C_r) = \Lambda_r$.

10.13   Design of H_2 SOID filters—continuous-time case

For continuous-time systems, we present in this section explicit algorithms of designing families of H_2 SOID filters with a capability to assign poles while honoring certain conditions imposed by H_2 suboptimality by connecting the given design objectives to those of H_2 AID filtering design. In this regard, as in previous chapters, we first observe that typically any filter design is initiated by first assuming a fixed architecture to the filter. The architecture we use for the filters is the CSS architecture that was developed and used in the previous section.

In view of Corollaries 10.14 and 10.15, the design of H_2 SOID filters for the given system Σ as in (10.1) can indeed be transformed to the design of AID filters for the auxiliary system Σ_Q given in (10.10). As we have already discussed the design of H_2 AID filters for any given system in Chapter 8, the design of H_2 SOID filters follows from that chapter. Here we briefly illuminate some issues in the translation from H_2 AID filters to H_2 SOID filters.

In the following three subsections, we consider the design of full-order strictly proper, full-order proper, and the reduced-order proper H_2 SOID filters one at a time.

10.13.1 Strictly proper H_2 SOID filters of CSS architecture

We pursue here the design of strictly proper H_2 SOID filters. As explained, strictly proper filters can be used only when $F = 0$. As such, throughout this subsection we assume that $F = 0$. The architecture we use for the filter is the strictly proper full-order CSS architecture. It is given by (8.16) and depicted in Figure 8.2. It is reproduced here as

$$\Sigma^\varepsilon_{\text{sp-CSS}} : \begin{cases} \dot{\xi} = A\xi + K^\varepsilon(y - C\xi) \\ \hat{z} = E\xi, \end{cases} \tag{10.111}$$

where the matrix K^ε is parameterized in ε.

Error Dynamics: Let us define the error $e = x - \xi$; then the error between the actual desired output z and the estimated desired output $\hat{z}$ is $e_z = E(x-\xi) = Ee$. Also, the dynamics of error is given by

$$\Sigma^{ue}_{\text{sp-CSS}} : \begin{cases} \dot{e} = (A - K^\varepsilon C)e + (B - K^\varepsilon D)u \\ e_z = Ee. \end{cases} \tag{10.112}$$

The transfer matrix $G^{ue}_{\text{sp-CSS}}$ from u to e_z can obviously be written as

$$G^{ue}_{\text{sp-CSS}}(s) = E(sI - A + K^\varepsilon C)^{-1}(B - K^\varepsilon D). \tag{10.113}$$

Remark 10.106 *In view of (10.111) and (10.112), it is easy to see that both the filter equation and the error equation have the same modes that are the eigenvalues of $A - K^\varepsilon C$.*

We observe that the only unknown in the filter equation (10.111) and consequently in the error equation (10.112) is the family of matrices K^ε parameterized by ε, which is normally referred to as a family of filter gains. We need to determine a family of filter gains in such a way that the H_2 norm of $G^{ue}_{\text{sp-CSS}}$ converges to its infimum. Before we do so, we pause to emphasize an important aspect. Theorem 10.30 developed earlier gives the conditions under which a sequence of H_2 SOID filters exists among the general class of strictly proper filters of the form (10.13) with $P^\varepsilon = 0$. In other words, Theorem 10.30 does not restrict itself to any fixed architecture for a filter such as the one $\Sigma_{\text{sp-CSS}}$ given in (10.111). Nevertheless, as the following theorem shows, whenever the conditions of Theorem 10.30 are satisfied, we can determine a family of filter gains such that the family of filters $\Sigma_{\text{sp-CSS}}$ is indeed a family of H_2 SOID filters, and such a family of filter gains can be designed via designing a family of H_2 AID filters for the system Σ_Q defined in (10.10) with $P^* = 0$.

Theorem 10.107 *Consider the continuous-time system Σ given in (10.1). Let Assumption 10.1 be satisfied and that $F = 0$, or equivalently, let the conditions of Theorem 10.30 for the strictly proper case be satisfied. Also, let the matrices B_Q and D_Q be as defined by (10.8), and consider the system Σ_Q defined in (10.10) with $P^* = 0$. Then, the following hold:*

(i) *A family of strictly proper H_2 SOID filters of CSS architecture $\Sigma^\varepsilon_{\text{sp-CSS}}$ exists for the given system Σ; i.e., a sequence of parameterized gains K^ε exists such that the family of filters $\Sigma^\varepsilon_{\text{sp-CSS}}$ given in (10.111) is a family of H_2 SOID filters for the given system Σ.*

(ii) *A family of strictly proper filters of CSS architecture $\Sigma^\varepsilon_{\text{sp-CSS}}$ as given in (10.111) is an H_2 SOID family of filters for the given system Σ if and only if this family of filters is a family of H_2 AID filters for the system Σ_Q of (10.10) with $P^* = 0$.*

Proof : This is a direct consequence of Corollary 10.14. ■

As we said so often, the dynamics of any system is heavily influenced by its poles. When we look at the poles of the error dynamics and those of the filter as given in (10.111), we observe that they are the same and equal to the eigenvalues of $A - K^\varepsilon C$. As in the case of H_2 AID filtering, the requirement of H_2 SOID filtering imposes that certain poles of a family of H_2 SOID filters tend as ε tends to zero to certain fixed locations termed as finite asymptotic fixed modes. We have the following formal definition of finite asymptotic fixed modes.

Definition 10.108 (*Finite asymptotic fixed modes of strictly proper H_2 SOID filters of CSS architecture*) *Consider the given system (10.1) and the associated H_2 SOID filtering problem 10.6. Assume that the solvability conditions as specified by Theorem 10.30 are satisfied. Then, a finite scalar $\lambda \in \mathbb{C}^-$ is said to be* *an **H_2 SOID filtering finite asymptotic fixed mode** with algebraic multiplicity α if for every family of H_2 SOID filters parameterized in ε and having the strictly proper full-order CSS architecture, poles λ_i^ε, $i = 1, 2, \cdots, \alpha$, of the family of filters exist such that all λ_i^ε tend to λ as ε tends to zero. The set of all H_2 SOID filtering finite asymptotic fixed modes when strictly proper filters of full-order CSS architecture are used is denoted by $\Omega_{\text{sp-CSS}}^{h2-soid}(A, B, C, D, E, 0)$.*

Note that in Chapter 8 we explicitly characterized the finite asymptotic fixed modes of H_2 AID filters. That characterization captures for a large part the freedom we have in choosing the asymptotic behavior of the modes of the AID H_2 filters as given by the eigenvalues of $A - K^\varepsilon C$ of Chapter 8. That is, we can use the results of Theorem 8.21 of Chapter 8 to compute

$$\Omega_{\text{sp-CSS}}^{h2-soid}(A, B, C, D, E, 0) = \Omega_{\text{sp-CSS}}^{h2-aid}(A, B_\varrho, C, D_\varrho, E, 0). \qquad (10.114)$$

10.13.2 Proper H_2 SOID filters of CSS architecture

We pursue here the design of proper H_2 SOID filters while using filters of CSS architecture. In this regard, Corollary 10.15 lays a roadmap to our development here. To start with, for ease of presentation and without loss of generality, we decompose the measured output y into two parts y_0 and y_1 in such a way that y_0 contains explicitly the unknown input u in it, whereas y_1 does not contain any input u in it. That is, we write

$$y = \begin{pmatrix} y_0 \\ y_1 \end{pmatrix} = Cx + Du \quad \text{and} \quad C = \begin{pmatrix} C_0 \\ C_1 \end{pmatrix}, \quad D = \begin{pmatrix} D_0 \\ 0 \end{pmatrix}, \qquad (10.115)$$

where the matrix D_0 has rank m_0. We can then rewrite the given system equation (10.1) as

$$\Sigma : \begin{cases} \dot{x} & = Ax + Bu \\ \begin{pmatrix} y_0 \\ y_1 \end{pmatrix} = \begin{pmatrix} C_0 \\ C_1 \end{pmatrix}x + \begin{pmatrix} D_0 \\ 0 \end{pmatrix}u \\ z & = Ex + Fu. \end{cases} \qquad (10.116)$$

Proper filter design is based on the design of strictly proper filter design for an auxiliary system $\widetilde{\Sigma}^*$ defined by

$$\widetilde{\Sigma}^* : \begin{cases} \dot{x} & = Ax + Bu \\ \widetilde{y} = \begin{pmatrix} y_0 \\ y_1 \\ \dot{y}_1 \end{pmatrix} = \widetilde{C}x + \widetilde{D}u \\ z^* = E^*x, \end{cases} \qquad (10.117)$$

where $E^* = E - P^*C$, with matrix P^* being a solution of $F - PD = 0$ for P:

$$\tilde{C} = \begin{pmatrix} C_0 \\ C_1 \\ C_1 A \end{pmatrix}, \quad \text{and} \quad \tilde{D} = \begin{pmatrix} D_0 \\ 0 \\ C_1 B \end{pmatrix}. \tag{10.118}$$

The essence of our design philosophy is to translate the design of a family of proper H_2 SOID filters for the given system Σ to the design of a family of strictly proper H_2 SOID filters for the auxiliary system $\tilde{\Sigma}^*$. This is formalized below.

Lemma 10.109 *Consider continuous-time systems Σ and $\tilde{\Sigma}^*$, respectively, as in (10.116) and (10.117). Then, the following two statements are equivalent:*

   (i) A family of strictly proper H_2 SOID filters exists for the auxiliary system $\tilde{\Sigma}^*$.

   (ii) A family of proper H_2 SOID filters exists for the system Σ.

Moreover, there is a $1 - 1$ relationship between the family of strictly proper H_2 SOID filters of CSS architecture for $\tilde{\Sigma}^*$ and the family of proper H_2 SOID filters of CSS architecture for Σ; that is, one of these family of filters can be constructed from the other.

Proof : The first part is obvious. The second part (i.e., the $1 - 1$ relationship) is described in the following development. ∎

We move on to consider a family of strictly proper filters of CSS architecture for $\tilde{\Sigma}^*$ as

$$\tilde{\Sigma}^\varepsilon_{\text{sp-CSS}} : \begin{cases} \dot{\tilde{\xi}} = A\tilde{\xi} + K^\varepsilon(\tilde{y} - \tilde{C}\tilde{\xi}) \\ \hat{z}^* = E^*\tilde{\xi}, \end{cases} \tag{10.119}$$

where the matrix K^ε is a filter gain parameterized by ε. We partition K^ε in conformity with the partitioning of $\tilde{y}$. That is,

$$K^\varepsilon = \begin{pmatrix} K^\varepsilon_0 & K^\varepsilon_1 & K^\varepsilon_2 \end{pmatrix}. \tag{10.120}$$

This family of filters for $\tilde{\Sigma}^*$ can be converted to a family of filters for the system Σ as was done previously for optimal H_2 OID filters in Subsection 10.12.2. That is, we obtain an estimate $\hat{z}$ of z as given by

$$\hat{z} = \hat{z}^* + P^*y = E^*\tilde{\xi} + P^*y. \tag{10.121}$$

We note that the family of filters (10.119) is not directly implementable because $y_2 = \dot{y}_1$ is not available as a measured variable. We eliminate the need for $\dot{y}_1$ by defining a new variable:

$$\xi = \tilde{\xi} - K^\varepsilon_2 y_1. \tag{10.122}$$

With the definition of ξ as in (10.122), we can rewrite the filter equation (10.119) as

$$\begin{cases} \dot{\xi} = (A - K^\varepsilon \tilde{C})\xi + \left(K_0^\varepsilon \quad K_1^\varepsilon + (A - K^\varepsilon \tilde{C})K_2^\varepsilon \right) y \\ \tilde{\xi} = \xi + K_2^\varepsilon y_1 \\ \hat{z}^* = E^* \tilde{\xi} = E^*(\xi + K_2^\varepsilon y_1). \end{cases} \tag{10.123}$$

Obviously, the family of filters given above does not use $\dot{y}_1$. Moreover, it is proper rather than strictly proper. The family of filters given in (10.123) is indeed the family of proper full-order CSS filters that is to be used for Σ^*.

Family of proper filters of CSS architecture for Σ:

Finally, in view of (10.121), we can rewrite (10.123) as an implementable family of proper filters for Σ:

$$\Sigma^\varepsilon_{\text{p-CSS}} : \begin{cases} \dot{\xi} = (A - K^\varepsilon \tilde{C})\xi + \tilde{K}^\varepsilon y \\ \tilde{\xi} = \xi + K_2^\varepsilon y_1 \\ \hat{z} = E^* \tilde{\xi} + P^* y = E^* \xi + \tilde{P}^\varepsilon y, \end{cases} \tag{10.124}$$

where

$$\tilde{K}^\varepsilon = \left(K_0^\varepsilon \quad K_1^\varepsilon + (A - K^\varepsilon \tilde{C})K_2^\varepsilon \right) \quad \text{and} \quad \tilde{P}^\varepsilon = \left(0 \quad E^* K_2^\varepsilon \right) + P^*.$$

We have the following theorem.

Theorem 10.110 *Consider the continuous-time system Σ given in (10.1). Let Assumption 10.1 be satisfied and that $\ker D \subseteq \ker F$, or equivalently, assume that the conditions of Theorem 10.30 for the proper case are satisfied. Define $\tilde{\Sigma}^*$ as in (10.117). Then, the following hold:*

(i) *A family of strictly proper H_2 SOID filters of CSS architecture $\tilde{\Sigma}^\varepsilon_{\text{sp-CSS}}$ for the system $\tilde{\Sigma}^*$ exists; i.e., a sequence of parameterized gains K^ε exists such that the family of filters $\tilde{\Sigma}^\varepsilon_{\text{sp-CSS}}$ given in (10.119) is a family of H_2 SOID filters for $\tilde{\Sigma}^*$.*

(ii) *Any family of strictly proper H_2 SOID filters of CSS architecture $\tilde{\Sigma}^\varepsilon_{\text{sp-CSS}}$ as given in (10.119) for the auxiliary system $\tilde{\Sigma}^*$ results in a family of proper H_2 SOID filters of CSS architecture $\Sigma^\varepsilon_{\text{p-CSS}}$ as given in (10.124) for Σ.*

Proof : Before we prove the two parts of the theorem, we first derive a preliminary result. We define the following system:

$$\Sigma^* : \begin{cases} \dot{x} = Ax + Bu \\ \begin{pmatrix} y_0 \\ y_1 \end{pmatrix} = \begin{pmatrix} C_0 \\ C_1 \end{pmatrix} x + \begin{pmatrix} D_0 \\ 0 \end{pmatrix} u \\ z = E^* x. \end{cases}$$

Let $\gamma_{sp}^*(\Sigma^*)$ and $\gamma_p^*(\Sigma)$, respectively, denote γ_{sp}^* pertaining to Σ^* and γ_p^* pertaining to Σ. We have

$$\gamma_p^*(\Sigma) = \gamma_{sp}^*(\Sigma^*),$$

and there is a $1-1$ relationship between strictly proper H_2 SOID filters for Σ^* and proper H_2 SOID filters for Σ. Next we note that the H_2 OID performance $\gamma_{sp}^*(\widetilde{\Sigma}^*)$ pertaining to $\widetilde{\Sigma}^*$ while using strictly proper filters can only be better than the performance using strictly proper filters for the system Σ^*. This follows from the fact that the only difference between these two systems is an additional measurement. In conclusion, we find that

$$\gamma_{sp}^*(\widetilde{\Sigma}^*) \leqslant \gamma_{sp}^*(\Sigma^*).$$

To establish part (i) of the theorem, we need to show that a sequence of parameterized gains K^ε exists such that the family of filters $\widetilde{\Sigma}_{\text{sp-CSS}}^\varepsilon$ given in (10.119) is a family of H_2 SOID filters for $\widetilde{\Sigma}^*$. This follows easily from Theorem 10.107.

To establish part (ii), we first note that the family of strictly proper H_2 SOID filters $\widetilde{\Sigma}_{\text{sp-CSS}}^\varepsilon$ of CSS architecture as given in (10.119) yields a strictly proper closed-loop dynamics from u to $z^* - \hat{z}^*$ with transfer matrix H_ε^*. Next, we consider the family of proper filters of the form (10.124) constructed from the family of strictly proper filters with a suitable gain K^ε. Such a family of filters when applied to the system Σ results in proper closed-loop dynamics from u to $z - \hat{z}$ with transfer matrix H_ε. It can be verified that $H_\varepsilon = H_\varepsilon^*$, and this clearly implies that

$$\|H_\varepsilon\|_2 = \|H_\varepsilon^*\|_2.$$

As $\varepsilon \to 0$, the right-hand side of the above equation goes to $\gamma_{sp}^*(\widetilde{\Sigma}^*)$. Then, in view of the preliminary result, we get

$$\gamma_{sp}^*(\widetilde{\Sigma}^*) \leqslant \gamma_p^*(\Sigma).$$

As $\gamma_p^*(\Sigma)$ was defined as the H_2 optimal performance via proper filters for the system Σ, this implies that the family of filters of the form (10.124) applied to the system Σ is a family of proper H_2 SOID filters for the system Σ.   ∎

We have converted the design of family of proper H_2 SOID filters for Σ to the design of family of strictly proper H_2 SOID filters for the auxiliary system $\widetilde{\Sigma}^*$. We can then use the tools of the previous subsection to obtain the appropriate filters and the associated finite asymptotic fixed modes. To be explicit, we have the following formal definition of finite asymptotic fixed modes when proper filters are used.

Definition 10.111 (*Finite asymptotic fixed modes of proper H_2 SOID filters of CSS architecture*) *Consider the given system (10.1) and the associated H_2 SOID filtering problem 10.6. Assume that the solvability conditions as specified by Theorem 10.30 are satisfied. Then, a finite scalar $\lambda \in \mathbb{C}^-$ is said to be an H_2 SOID*

filtering finite asymptotic fixed mode *with algebraic multiplicity* α *if for every family of* H_2 *SOID filters parameterized in* ε *and having the proper full-order CSS architecture, poles* λ_i^ε, $i = 1, 2, \cdots, \alpha$, *of the family of filters exist such that all* λ_i^ε *tend to* λ *as* ε *tends to zero. The set of all* H_2 *SOID filtering finite asymptotic fixed modes when proper filters of full-order CSS architecture are used is denoted by* $\boldsymbol{\Omega}_{\text{p-CSS}}^{h2-soid}(A, B, C, D, E, F)$.

Obviously, we can conclude that

$$\boldsymbol{\Omega}_{\text{p-CSS}}^{h2-soid}(A, B, C, D, E, F) = \boldsymbol{\Omega}_{\text{sp-CSS}}^{h2-soid}(A, B, \tilde{C}, \tilde{D}, E - P^*C, 0).$$

We hasten to add that $\boldsymbol{\Omega}_{\text{sp-CSS}}^{h2-soid}(A, B, \tilde{C}, \tilde{D}, E - P^*C, 0)$ is the same whatever may be the matrix P^* that solves $F - PD = 0$. The latter follows from (10.114), Theorem 8.21, and a detailed analysis of the characterization of $\boldsymbol{\Omega}_s^2$ obtained in Theorem 6.19.

10.13.3   Reduced-order H_2 SOID filters of CSS architecture

The two previous subsections respectively construct families of full-order strictly proper and proper filters of CSS architecture that solve the H_2 SOID filtering problem for continuous-time systems. By full-order filters, as usual, we mean filters having the same dynamic order as that of the given system. Our goal in this subsection is to develop families of reduced-order filters of CSS architecture having their dynamic order lower than that of the given system.

The procedure of developing reduced-order filters here follows mostly along the same lines as in Subsection 10.12.3, which pertains to H_2 OID filters. As before, in essence, our method of development transforms the construction of a reduced-order filter for a given system to that of a full-order filter for a certain reduced-order system.

We proceed now to construct an appropriate reduced-order system. To start with, let us rewrite the matrices C and D of (10.1) as

$$C = \begin{pmatrix} 0 & C_{02} \\ I_{p-m_0} & 0 \end{pmatrix}, \quad D = \begin{pmatrix} D_0 \\ 0 \end{pmatrix},$$

where again rank $D = $ rank $D_0 = m_0$. This can always be done without any loss of generality by appropriate coordinate transformations. In view of the above partitioning of C and D, we can partition the given system Σ as

$$\Sigma : \begin{cases} \begin{pmatrix} \dot{x}_1 \\ \dot{x}_2 \end{pmatrix} &= \begin{pmatrix} A_{11} & A_{12} \\ A_{21} & A_{22} \end{pmatrix} \begin{pmatrix} x_1 \\ x_2 \end{pmatrix} + \begin{pmatrix} B_{11} \\ B_{22} \end{pmatrix} u \\ y = \begin{pmatrix} y_0 \\ y_1 \end{pmatrix} &= \begin{pmatrix} 0 & C_{02} \\ I & 0 \end{pmatrix} \begin{pmatrix} x_1 \\ x_2 \end{pmatrix} + \begin{pmatrix} D_0 \\ 0 \end{pmatrix} u \\ z = Ex + Fu, \end{cases} \quad (10.125)$$

where different variables have obvious meanings.

Next, we use a preliminary injection of output y into the desired output z. To this end, we define a matrix P^*, which is a solution of $F - PD = 0$ for P. We define a new desired output z^* as

$$z^* = z - P^*y = (E - P^*C)x + (F - P^*D)u = E^*x, \qquad (10.126)$$

where $E^* = E - P^*C$. In view of (10.125) and (10.126), we can define a new system Σ^* as

$$\Sigma^* : \begin{cases} \begin{pmatrix} \dot{x}_1 \\ \dot{x}_2 \end{pmatrix} = \begin{pmatrix} A_{11} & A_{12} \\ A_{21} & A_{22} \end{pmatrix} \begin{pmatrix} x_1 \\ x_2 \end{pmatrix} + \begin{pmatrix} B_{11} \\ B_{22} \end{pmatrix} u \\[12pt] y = \begin{pmatrix} y_0 \\ y_1 \end{pmatrix} = \begin{pmatrix} 0 & C_{02} \\ I & 0 \end{pmatrix} \begin{pmatrix} x_1 \\ x_2 \end{pmatrix} + \begin{pmatrix} D_0 \\ 0 \end{pmatrix} u \\[12pt] z^* = E^*x = E_1^*x_1 + E_2^*x_2, \end{cases} \qquad (10.127)$$

with $E^* = \begin{pmatrix} E_1^* & E_2^* \end{pmatrix}$.

We note that the measured output y_1 is not contaminated by the input u, and hence, $x_1 = y_1$ is known exactly from the measurement y. Thus, all we need to do next is to estimate the state x_2. To proceed further, let us rewrite the state equation for x_1 in terms of the output y_1 and the state x_2 as

$$\dot{y}_1 = A_{11}y_1 + A_{12}x_2 + B_{11}u. \qquad (10.128)$$

The above equation can be rewritten as

$$\dot{y}_1 - A_{11}y_1 = A_{12}x_2 + B_{11}u.$$

Treating $\dot{y}_1$ as known, we can define a new measurement variable y_r:

$$y_r = \begin{pmatrix} y_0 \\ \dot{y}_1 - A_{11}y_1 \end{pmatrix}.$$

Although $\dot{y}_1$ is not directly available, as we did earlier in the case of full-order proper CSS filters, we can eliminate it from any filter that is constructed using y_r as a measured output. With this in mind, we form the auxiliary system:

$$\Sigma_r^* : \begin{cases} \dot{x}_r = A_r x_r + B_r u + A_{21}y_1 \\ y_r = C_r x_r + D_r u \\ z_r^* = E_r^* x_r, \end{cases} \qquad (10.129)$$

where $x_r = x_2$ and

$$A_r = A_{22}, \quad B_r = B_{22}, \quad C_r = \begin{pmatrix} C_{02} \\ A_{12} \end{pmatrix}, \quad D_r = \begin{pmatrix} D_0 \\ B_{11} \end{pmatrix}, \quad E_r^* = E_2^*. \qquad (10.130)$$

We note that the dynamic order n_r of the above Σ_r^* is less than the dynamic order n of the given system Σ by a number equal to the dimension of $x_1 = y_1$.

Next, let us suppose we can construct a family of strictly proper H_2 SOID filters for the auxiliary system Σ_r^* in order to arrive at an estimate $\hat{z}_r^*$ of z_r^*. We then plan to obtain an estimate $\hat{z}$ of z as

$$\hat{z} = E_1^* x_1 + \hat{z}_r^* + P^* y. \tag{10.131}$$

This motivates us to construct a family of full-order strictly proper filters for the auxiliary system Σ_r^* to arrive at an estimate $\hat{z}_r^*$ of z_r^*. A full-order family of filters constructed as such for Σ_r^* is indeed a family of reduced-order filters for Σ. However, in view of (10.131), such a family of reduced-order filters for Σ is proper (rather than strictly proper) for Σ. Thus, the essence of our design philosophy is to translate the design of a family of reduced-order H_2 SOID filters for the given system Σ to the design of a family of strictly proper H_2 SOID filters for the reduced-order system Σ_r^*. This is formalized below.

Lemma 10.112 *Consider continuous-time systems Σ and Σ_r^* as given in (10.1) and (10.129), respectively. Then, the following two statements are equivalent:*

(i) A family of strictly proper H_2 SOID filters exists for the reduced-order system Σ_r^*.

(ii) A family of proper reduced-order H_2 SOID filters exists for the system Σ.

Moreover, there is a $1-1$ relationship between the family of strictly proper H_2 SOID filters of CSS architecture for Σ_r^* and the family of proper reduced-order H_2 SOID filters of CSS architecture for Σ; that is, one of these filters can be constructed from the other.

Proof : The first part is obvious. The second part (i.e., the $1-1$ relationship) is described in the following development.   ∎

It is obvious from our previous results that one can construct a family of strictly proper H_2 SOID filters for Σ_r^*. Next, to construct the required family of filters for Σ, in the spirit of the above development, we first construct a family of strictly proper filters of CSS architecture for the reduced-order system Σ_r^* as

$$\tilde{\Sigma}_{\text{sp-CSS}}^{\varepsilon} : \begin{cases} \dot{\tilde{\xi}}_r = A_r \tilde{\xi}_r + A_{21} y_1 + K_r^{\varepsilon}(y_r - C_r \tilde{\xi}_r) \\ \hat{z}_r^* = E_r^* \tilde{\xi}_r, \end{cases} \tag{10.132}$$

where K_r^{ε} is a parameterized reduced-order filter gain. To convert this to a filter for Σ, we note that as $\dot{y}_1$ is not available, we need to modify the above family of filters. To this end, let us partition $K_r^\varepsilon = (K_{r0}^\varepsilon \quad K_{r1}^\varepsilon)$ so as to be compatible with the partitioning of y_r. Also, let

$$\xi_r = \tilde{\xi}_r - K_{r1}^\varepsilon y_1. \tag{10.133}$$

We can then easily rewrite the family of filters (10.132) as a family of proper filters for Σ:

$$\Sigma_{\text{r-CSS}}^\varepsilon : \begin{cases} \dot{\xi}_r = (A_r - K_r^\varepsilon C_r)\xi_r + \tilde{K}_r^\varepsilon y \\ \tilde{\xi}_r = \xi_r + K_{r1}^\varepsilon y_1 \\ \hat{z} = E_1^* x_1 + E_r^* \tilde{\xi}_r + P^* y = E_r^* \xi_r + \tilde{P}_r y, \end{cases} \tag{10.134}$$

where

$$\tilde{K}_r^\varepsilon = \left(K_{r0}^\varepsilon \quad A_{21} - K_{r1}^\varepsilon A_{11} + (A_r - K_r^\varepsilon C_r) K_{r1}^\varepsilon \right)$$

and

$$\tilde{P}_r = \left(0 \quad E_1^* + E_r^* K_{r1}^\varepsilon \right) + P^*.$$

The following theorem formalizes the situation.

Theorem 10.113 *Consider the continuous-time system Σ given in (10.1). Let Assumption 10.1 be satisfied and that $\ker D \subseteq \ker F$, or equivalently, assume that the conditions of Theorem 10.30 for the proper case are satisfied. Define the reduced-order system Σ_r^* as in (10.129). Then, the following hold:*

    (i) *A family of strictly proper H_2 SOID filters of CSS architecture $\tilde{\Sigma}_{\text{sp-CSS}}^\varepsilon$ exists for the reduced-order system Σ_r^*; i.e., a sequence of parameterized gains K_r^ε exists such that the family of filters $\tilde{\Sigma}_{\text{sp-CSS}}^\varepsilon$ given in (10.132) is a family of H_2 SOID filters for Σ_r^*.*

    (ii) *Any family of strictly proper H_2 SOID filters of CSS architecture $\tilde{\Sigma}_{\text{sp-CSS}}^\varepsilon$ as given in (10.132) for the reduced-order system Σ_r^* results in a family of reduced-order proper H_2 SOID filters of CSS architecture $\Sigma_{\text{r-CSS}}^\varepsilon$ as given in (10.134) for Σ.*

Proof : Before we prove the two parts of the theorem, we first derive a preliminary result. Consider Σ^* defined by (10.127). Let $\gamma_{sp}^*(\Sigma^*)$ and $\gamma_p^*(\Sigma)$, respectively, denote γ_{sp}^* pertaining to Σ^* and γ_p^* pertaining to Σ. We have

$$\gamma_p^*(\Sigma) = \gamma_{sp}^*(\Sigma^*),$$

and there is a $1 - 1$ relationship between strictly proper H_2 SOID filters for Σ^* and proper H_2 SOID filters for Σ^*. Next we note that the H_2 OID performance $\gamma_{sp}^*(\Sigma_r^*)$ pertaining to Σ_r^* while using strictly proper filters can only be better than the performance using strictly proper filters for the system Σ. This follows from the fact that the only part we eliminated is the known part of the state. In conclusion, we find that

$$\gamma_{sp}^{*}(\Sigma_{r}^{*}) \leqslant \gamma_{sp}^{*}(\Sigma^{*}).$$

To establish part (i) of the theorem, we need to show that a sequence of parameterized gains K_{r}^{ε} exists such that the family of filters $\widetilde{\Sigma}_{\text{sp-CSS}}^{\varepsilon}$ given in (10.132) is a family of H_2 SOID filters for $\widetilde{\Sigma}^{*}$. This follows easily from Theorem 10.107.

To establish part (ii), we first note that the family of strictly proper H_2 SOID filters $\widetilde{\Sigma}_{\text{sp-CSS}}^{\varepsilon}$ of CSS architecture as given in (10.132) yields a strictly proper closed-loop dynamics from u to $z_{r}^{*} - \hat{z}_{r}^{*}$ with transfer matrix H_{ε}^{*}. Next, we consider the family of reduced-order filters of the form (10.134) constructed from the family of strictly proper filters with a suitable gain K_{r}^{ε}. Such a family of filters when applied to the system Σ results in proper closed-loop dynamics from u to $z - \hat{z}$ with transfer matrix H_{ε}. It can be verified that $H_{\varepsilon} = H_{\varepsilon}^{*}$, and this clearly implies that

$$\|H_{\varepsilon}\|_2 = \|H_{\varepsilon}^{*}\|_2.$$

As $\varepsilon \to 0$, the right-hand side of the above equation goes to $\gamma_{sp}^{*}(\Sigma_{r}^{*})$. Then, in view of the preliminary result, we get

$$\gamma_{sp}^{*}(\Sigma_{r}^{*}) \leqslant \gamma_{p}^{*}(\Sigma).$$

As $\gamma_{p}^{*}(\Sigma)$ was defined as the H_2 optimal performance via proper filters for the system Σ, this implies that the family of filters of the form (10.134) applied to the system Σ is a family of proper H_2 SOID filters for the system Σ. ∎

We have converted the design of reduced-order H_2 SOID filters for Σ to the design of strictly proper H_2 SOID filter for the auxiliary system Σ_{r}^{*}. We can then use the tools of Subsection 10.13.1 to obtain the appropriate filters and the associated finite asymptotic fixed modes. To be explicit, we have the following formal definition of finite asymptotic fixed modes when reduced-order filters are used.

Definition 10.114 (Finite asymptotic fixed modes of reduced-order H_2 SOID filters of CSS architecture) *Consider the given system (10.1) and the associated H_2 SOID filtering problem 10.6. Assume that the solvability conditions as specified by Theorem 10.30 are satisfied. Then, a finite scalar $\lambda \in \mathbb{C}^{-}$ is said to be an H_2 SOID filtering finite asymptotic fixed mode with algebraic multiplicity α if for every family of H_2 SOID filters parameterized in ε and having the reduced-order CSS architecture, poles $\lambda_{i}^{\varepsilon}$, $i = 1, 2, \cdots, \alpha$, of the family of filters exist such that all $\lambda_{i}^{\varepsilon}$ tend to λ as ε tends to zero. The set of all H_2 SOID filtering finite asymptotic fixed modes when reduced-order filters of CSS architecture are used is denoted by $\mathbf{\Omega}_{\text{r-CSS}}^{h2-soid}(A, B, C, D, E, F)$.*

Obviously, we can conclude that

$$\boldsymbol{\Omega}_{\text{r-CSS}}^{h2-soid}(A, B, C, D, E, F) = \boldsymbol{\Omega}_{\text{sp-CSS}}^{h2-soid}(A_r, B_r, C_r, D_r, E_r^*, 0).$$

We hasten to add that $\boldsymbol{\Omega}_{\text{sp-CSS}}^{h2-soid}(A_r, B_r, C_r, D_r, E_r^*, 0)$ is the same whatever may be the matrix P^* that solves $F - PD = 0$. The latter follows from (10.114), Theorem 8.21, and a detailed analysis of the characterization of $\boldsymbol{\Omega}_s^2$ obtained in Theorem 6.19.

10.14   Design of H_2 OID filters—discrete-time case

For discrete-time systems, Section 10.9 explores the conditions for the existence of an H_2 OID filter whether it is proper or strictly proper. In this section, we consider the design issues associated with discrete-time H_2 OID filters. As in the continuous-time case, the architecture we use for the filters is the CSS architecture. In fact, in most of the following development, the procedure we use to design appropriate filters follows along the same lines as in the continuous-time case except that now Corollaries 10.35 and 10.36 lay the roadmap instead of Corollaries 10.12 and 10.13. As in the continuous-time case, our development here not only develops the design of appropriate H_2 OID filters, but it also studies their structural properties. In the following three subsections, we consider the design of full-order strictly proper, full-order proper, and reduced-order proper H_2 OID filters one at a time.

10.14.1   Strictly proper H_2 OID filters of CSS architecture

We pursue here the design of strictly proper H_2 OID filters. In view of Corollary 10.35, one can first design a strictly proper EID filter for the auxiliary system Σ_Q^{sp} given in (10.46) and then modify it to obtain a strictly proper H_2 OID filter for the given system Σ. As in continuous-time systems, the full-order strictly proper filter of CSS architecture is given by

$$\Sigma_{\text{sp-CSS}} : \{\sigma\xi = (A - KC)\xi + Ky \quad \text{with} \quad \hat{z} = E\xi. \tag{10.135}$$

Error dynamics: Let us define the error $e = x - \xi$; then the error between the actual desired output z and the estimated desired output $\hat{z}$ is $e_z = E(x - \xi) = Ee + Fu$. Also, the dynamics of error is given by

$$\Sigma_{\text{sp-CSS}}^{ue} : \begin{cases} \sigma e = (A - KC)e + (B - KD)u \\ e_z = Ee + Fu. \end{cases} \tag{10.136}$$

Then, the transfer matrix $G_{\text{sp-CSS}}^{ue}$ from u to e_z can obviously be written as

$$G_{\text{sp-CSS}}^{ue}(z) = E(zI - A + KC)^{-1}(B - KD) + F. \tag{10.137}$$

We observe that, unlike in the continuous-time case, strictly proper filters can be used for discrete-time systems even if $F \neq 0$ provided the conditions of Theorem 10.47 are satisfied.

Remark 10.115 *In view of* (10.135) *and* (10.136), *it is easy to see that both the filter equation and the error equation have the same modes, which are the eigenvalues of $A - KC$.*

We observe that the only unknown in the filter equation (10.135) and, consequently, in the error equation (10.136) is the matrix K, which is normally referred to as the filter gain. We need to determine or design K in such a way that the H_2 norm of $G^{ue}_{\text{sp-CSS}}$ is as small as possible. Before we do so, we pause to emphasize an important aspect. Theorem 10.47 developed earlier gives the conditions under which an H_2 OID filter exists among the general class of strictly proper filters of the form (10.2) with $P = 0$. In other words, Theorem 10.47 does not restrict itself to any fixed architecture for a filter such as the one $\Sigma_{\text{sp-CSS}}$ given in (10.135). Nevertheless, as the theorem that follows shortly shows, whenever the conditions of Theorem 10.47 are satisfied, we can determine the gain parameter K such that the filter given in (10.135) is an H_2 OID filter. In this regard, as in the continuous-time case, we define next a set of filter gains $\boldsymbol{K}^{h2-\text{oid}}_{\text{sp-CSS}}(A, B, C, D, E, F)$, which is the set of all H_2 OID filter gains, meaning that any gain $K \in \boldsymbol{K}^{h2-\text{oid}}_{\text{sp-CSS}}$ renders $\Sigma_{\text{sp-CSS}}$ an H_2 OID filter for (10.1), and conversely, any gain K that renders $\Sigma_{\text{sp-CSS}}$ an H_2 OID filter for (10.1) is an element of $\boldsymbol{K}^{h2-\text{oid}}_{\text{sp-CSS}}$.

We have the following result.

Theorem 10.116 *Consider a discrete-time system as in* (10.1), *and let Assumption 10.1 be satisfied. Also, let the matrices B_Q and D_Q be as defined by* (10.39), *and consider the system Σ^{sp}_Q defined in* (10.46). *Consider strictly proper filters, and assume that the conditions of Theorem 10.47 are satisfied. Then, the set $\boldsymbol{K}^{h2-\text{oid}}_{\text{sp-CSS}}(A, B, C, D, E, F)$ is nonempty and is given by*

$$\boldsymbol{K}^{h2-\text{oid}}_{\text{sp-CSS}}(A, B, C, D, E, F) = \{ K \mid \boldsymbol{K}^{\text{eid}}_{\text{sp-CSS}}(A, B_Q, C, D_Q, E, 0)\}, \quad (10.138)$$

where the set $\boldsymbol{K}^{\text{eid}}_{\text{sp-CSS}}$ is given by Theorem 7.13.

Proof : In Corollary 10.35, we noted that a filter is an H_2 OID filter for Σ if and only if it is an EID filter for the system Σ^{sp}_Q defined in (10.46). The result then follows from Theorem 7.13. ∎

Remark 10.117 *It is easy to see that the set $K^{h2-\text{oid}}_{\text{sp-CSS}}(A, B, C, D, E, F)$ is invariant with respect to the matrix F.*

As in the case of the set $K^{\text{eid}}_{\text{sp-CSS}}$, the set $K^{h2-\text{oid}}_{\text{sp-CSS}}$ has an interesting telescopic property with respect to the size of the matrix E. This is formalized below.

Lemma 10.118 *Consider a discrete-time system as in (10.1). Consider two different H_2 OID filtering problems both as defined in Definition 10.6, however, with one taking a value E_a and the other a value E_b for the matrix E. Then, we have the telescopic property:*

$$\ker E_a \supseteq \ker E_b \supseteq \mathcal{S}^{\ominus}(A, B_Q, C, D_Q)$$
$$\implies K^{h2-\text{oid}}_{\text{sp-CSS}}(A, B, C, D, E_a, F) \supseteq K^{h2-\text{oid}}_{\text{sp-CSS}}(A, B, C, D, E_b, F).$$

Proof : In view of Corollary 10.35, the result follows from Theorem 7.14, which discusses a telescopic property for $K^{\text{eid}}_{\text{sp-CSS}}$. ∎

In general, a strictly proper H_2 OID filter of the form $\Sigma_{\text{sp-CSS}}$ as given in (10.135), whenever it exists, need not be unique. We can use the fact that the filters are not unique to our advantage. That is, as in continuous-time systems, we would like to design an H_2 OID filter so as to shape the transient behavior of error dynamics $\Sigma^{ue}_{\text{sp-CSS}}$ given in (10.136). (Note that all H_2 OID filters result in steady state the same error dynamics.) This can be done by assigning appropriately the poles of $\Sigma^{ue}_{\text{sp-CSS}}$. In this regard, as before, we observe that the poles of $\Sigma^{ue}_{\text{sp-CSS}}$ are exactly the same as those of the filter $\Sigma_{\text{sp-CSS}}$ given in (10.135). Thus, one would like to place the poles of an H_2 OID filter as desired in order to shape the transient behavior of error dynamics. However, as in continuous-time systems, the requirement of H_2 OID filtering forces all H_2 OID filters to share some fixed modes as defined below.

Definition 10.119 (*Fixed modes of strictly proper H_2 OID filters of CSS architecture*) *Consider a discrete-time system as in (10.1) and the H_2 OID filtering problem 10.6 characterized by the matrix quintuple (A, B, C, D, E, F). Assume that the solvability conditions of the H_2 OID filtering problem as in Theorem 10.47 are satisfied. Then, a scalar $\lambda \in \mathbb{C}^{\ominus}$ is said to be the fixed mode of an H_2 OID filter with the strictly proper full-order CSS architecture if λ is a pole (i.e., an eigenvalue of $A - KC$) of every H_2 OID filter of such an architecture $\Sigma_{\text{sp-CSS}}$ as given in (10.135). The set of all such H_2 OID filter fixed modes is denoted here by $\Omega^{h2-\text{oid}}_{\text{sp-CSS}}(A, B, C, D, E, F)$.*

We have the following theorem that characterizes the set $\Omega^{h2-\text{oid}}_{\text{sp-CSS}}$.

Theorem 10.120 *Consider a discrete-time system as in* (10.1), *and let Assumption 10.1 be satisfied. Also, let the matrices B_Q and D_Q be as defined by* (10.39). *Consider strictly proper filters, and assume that the conditions of Theorem 10.47 are satisfied. Then, we have*

$$\Omega_{\text{sp-CSS}}^{h2-\text{oid}}(A, B, C, D, E, F) = \Omega_{\text{sp-CSS}}^{\text{eid}}(A, B_Q, C, D_Q, E, 0),$$

where $\Omega_{\text{sp-CSS}}^{\text{eid}}$ is characterized in Theorem 7.18.

Proof : In Corollary 10.35, we noted that a filter is an H_2 OID filter for Σ if and only if it is an EID filter for the system Σ_Q^{sp} defined in (10.46). The result then follows from the flexibility in EID filters as studied in Chapter 7 and in particular in Theorem 7.18.   ∎

Remark 10.121 *Once again, it is easy to see that the set $\Omega_{\text{sp-CSS}}^{h2-\text{oid}}$ (A, B, C, D, E, F) is invariant with respect to the matrix F.*

Theorem 10.120 prescribes a procedure of obtaining $\Omega_{\text{sp-CSS}}^{h2-\text{oid}}$. However, to gain insight regarding the nature of these fixed modes, it is worthwhile to note that $\Omega_{\text{sp-CSS}}^{h2-\text{oid}}(A, B, C, D, E, F)$ is a subset of the stable invariant zeros of the system characterized by the quadruple (A, B_Q, C, D_Q). This implies that we need to study certain structural properties of the system characterized by (A, B_Q, C, D_Q). Expectedly, a definite relationship exists between the structural properties of the system characterized by (A, B_Q, C, D_Q) and the system characterized by the given quadruple (A, B, C, D). The following lemma describes such a relationship, and it is similar to Lemma 10.61 that pertains to continuous-time systems.

Lemma 10.122 *Consider a discrete-time system characterized by the quadruple (A, B, C, D). Let Q be a semi-stabilizing solution of the DLMI* (10.37), *and define B_Q and D_Q by* (10.39) *resulting in a second system characterized by the quadruple (A, B_Q, C, D_Q). We have the following properties:*

  (i) *The system characterized by (A, B_Q, C, D_Q) is left-invertible.*

  (ii) *The system characterized by (A, B_Q, C, D_Q) is invertible if and only if the system characterized by (A, B, C, D) is right-invertible.*

  (iii) *The invariant zeros of the system characterized by (A, B_Q, C, D_Q) are classified by*

    (a) *the stable (i.e., those in $\mathbb{C}^{\ominus}$) invariant zeros of the system characterized by (A, B, C, D),*

    (b) *the invariant zeros that are on the unit circle of the system characterized by (A, B, C, D),*

    (c) *the mirror images[†] with respect to the unit circle of all unstable (i.e., those in $\mathbb{C}^{\oplus}$) invariant zeros of the system characterized by (A, B, C, D), and*

    (d) *what are termed as compromise zeros that are some fixed locations within the unit circle containing those stable output decoupling zeros of the system characterized by (A, B, C, D), which are not invariant zeros.*

(iv) *The system characterized by (A, B_Q, C, D_Q) has no infinite zeros of order greater than or equal to one.*

Proof: The arguments are very similar to the continuous-time case. From (10.40), it is immediate that the system characterized by (A, B_Q, C, D_Q) is left-invertible and has no invariant zeros outside the unit circle.

Secondly, it is easy to verify that

$$\begin{pmatrix} x & y \end{pmatrix} \begin{pmatrix} A - zI & B \\ C & D \end{pmatrix} = 0 \text{ if and only if } \begin{pmatrix} x & y \end{pmatrix} \begin{pmatrix} A - zI & B_Q \\ C & D_Q \end{pmatrix} = 0$$

for all z on the unit circle. But this clearly implies the property (ii) [exploiting property (i)] and moreover establishes that the zeros on the unit circle of both systems are the same.

If (A, B, C, D) is right-invertible, then we know that Q is a solution of a DARE. In this case, it is easily established that all invariant zeros are eigenvalues of the associated symplectic pencil. As the eigenvalues of the symplectic pencil are symmetric with respect to the unit circle, this implies that the stable zeros of the symplectic pencil contain the stable invariant zeros of (A, B, C, D) and the mirror images of the invariant zeros of (A, B, C, D) outside the unit circle. But it is then immediate that, if Q is a semi-stabilizing solution, then the stable eigenvalues of $A - (AQC' + BD')(CQC' + DD')^{-1}C$ are equal to the stable zeros of the symplectic pencil and, hence, include the stable invariant zeros of (A, B, C, D) and the mirror images of the invariant zeros of (A, B, C, D) outside the unit circle. This immediately yields that the set of invariant zeros of (A, B_Q, C, D_Q) contains the sets (a) and (c). The general case when (A, B, C, D) is not right-invertible follows from Theorem 4.141, which connects in the general case the semi-stabilizing solutions of the DLMI to the semi-stabilizing solutions of an associated reduced H_2 DARE.

[†] For continuous-time systems, the mirror image of a complex number $\alpha + j\beta$ is defined as $-\alpha + j\beta$, whereas in discrete-time systems, the mirror image of a complex number $re^{j\theta}$ is defined as $\frac{1}{r}e^{j\theta}$.

As Q is a semi-stabilizing solution of the DLMI $G(Q) \geq 0$ and therefore also strongly rank minimizing, we have

$$\operatorname{rank} \begin{pmatrix} B_Q \\ D_Q \end{pmatrix} = \operatorname{rank} G(Q) = \operatorname{rank} D_Q.$$

In other words, $B_Q \ker D_Q = \{0\}$. This immediately yields (iv). ∎

As we said, $\Omega_{\text{sp-CSS}}^{h2-\text{oid}}(A, B, C, D, E, F)$ is a subset of the stable invariant zeros of the system characterized by the quadruple (A, B_Q, C, D_Q). Then, in view of Lemma 10.122, $\Omega_{\text{sp-CSS}}^{h2-\text{oid}}(A, B, C, D, E, F)$ consists of some stable invariant zeros, some mirror images of unstable invariant zeros, and what are known as compromise zeros of the system characterized by (A, B, C, D). As in continuous-time systems, the exact size of $\Omega_{\text{sp-CSS}}^{h2-\text{oid}}(A, B, C, D, E, F)$ is dictated by the matrix E, which prescribes the variable z that is being estimated. In this respect, as in the case of the set $K_{\text{sp-CSS}}^{h2-\text{oid}}$, a telescopic property (nested property) exists, namely, *as* $\ker E$ *decreases, the size of* $\Omega_{\text{sp-CSS}}^{h2-\text{oid}}(A, B, C, D, E, F)$ *increases.* The smallest set $\Omega_{\text{sp-CSS}}^{h2-\text{oid}}(A, B, C, D, E, F)$ is obtained when $\ker E = \mathcal{S}^{\ominus}(A, B_Q, C, D_Q)$, whereas it is the largest when $\ker E = \{0\}$. The following lemma describes the telescopic property of $\Omega_{\text{sp-CSS}}^{h2-\text{oid}}$.

Lemma 10.123 *Consider a discrete-time system as in* (10.1). *Consider two different* H_2 *OID filtering problems both as defined in Definition 10.6, however, with one taking a value* E_a *and the other a value* E_b *for the matrix* E. *Then, we have the telescopic property:*

$$\ker E_a \supseteq \ker E_b \supseteq \mathcal{S}^{\ominus}(A, B_Q, C, D_Q)$$
$$\Longrightarrow \Omega_{\text{sp-CSS}}^{h2-\text{oid}}(A, B, C, D, E_a, F) \subseteq \Omega_{\text{sp-CSS}}^{h2-\text{oid}}(A, B, C, D, E_b, F).$$

Proof : In Proposition 5.13, this nested property was established for the set Ω_s. Using Theorem 7.18, this results in a similar property for the set $\Omega_{\text{sp-CSS}}^{\text{eid}}$, which in turn through Theorem 10.120 results in the property described in the above theorem. ∎

Remark 10.124 *The above theorem simply states that, by increasing the number of states to be estimated (i.e., by increasing the dimension of the estimated variable* z*), the number of fixed modes increase telescopically.*

Having noted the result of Lemma 10.123, we now move on to study $\boldsymbol{\Omega}_{\text{sp-CSS}}^{h2-\text{oid}}$ for two extreme cases for E, namely for

$$\ker E = \mathbb{R}^n \text{ or } \ker E = \mathcal{S}^{\ominus}(A, B_Q, C, D_Q).$$

For the case when $\ker E = \mathcal{S}^{\ominus}(A, B_Q, C, D_Q)$, we note that the set $\boldsymbol{\Omega}_{\text{sp-CSS}}^{h2-\text{oid}}$ (A, B, C, D, E, F) consists of

(i) all stable invariant zeros (i.e., those in $\mathbb{C}^{\ominus}$) of the system characterized by (A, B, C, D),

(ii) all mirror images of the unstable invariant zeros (i.e., those in $\mathbb{C}^{\oplus}$) of the system characterized by (A, B, C, D),

(iii) some fixed locations called compromising zeros in $\mathbb{C}^{\ominus}$.

On the other hand, for the other extreme case when $\ker E = \mathbb{R}^n$, the set $\boldsymbol{\Omega}_{\text{sp-CSS}}^{h2-\text{oid}}$ is empty because the static estimator $\hat{z} = 0$ will be optimal in this case. Next, it is easy to show that, in general, when $\ker E$ is between these two extreme cases, the set $\boldsymbol{\Omega}_{\text{sp-CSS}}^{h2-\text{oid}}(A, B, C, D, E, F)$ *consists of only some, but not necessarily all, of the stable invariant zeros, and only some, but not necessarily all, of the mirror images of the unstable invariant zeros of* the system characterized by (A, B, C, D).

As in the continuous-time case, besides the presence of H_2 OID filter fixed modes as explained above, another important structural issue exists in filter design. It turns out that, in the system $\Sigma_{\text{sp-CSS}}^{ue}$ that represents the error dynamics, certain pole/zero cancellations exist; see (10.136). To be specific, for a given H_2 OID filtering problem, there exist indeed what can be termed as a set of fixed decoupling zeros of $\Sigma_{\text{sp-CSS}}^{ue}$ resulting from H_2 OID filters of CSS architecture. Such a set shows the minimum absolutely necessary number and locations of pole/zero cancellations within the system $\Sigma_{\text{sp-CSS}}^{ue}$. We have the following formal definition.

Definition 10.125 (*Fixed decoupling zeros of* $\Sigma_{\text{sp-CSS}}^{ue}$ *resulting from strictly proper H_2 OID filters of CSS architecture*) *Consider a discrete-time system as given (10.1) and the H_2 OID filtering problem 10.6 characterized by the matrix quintuple (A, B, C, D, E, F). Assume that the solvability conditions of the H_2 OID filtering problem as in Theorem 10.47 are satisfied. Then, a scalar $\lambda \in \mathbb{C}^{\ominus}$ is said to be a fixed decoupling zero of $\Sigma_{\text{sp-CSS}}^{ue}$ resulting from strictly proper H_2 OID filters of CSS architecture if λ is either an input or an output decoupling zero (or both) of the system $\Sigma_{\text{sp-CSS}}^{ue}$ as given in (10.136) for any strictly proper H_2 OID filter $\Sigma_{\text{sp-CSS}}$ that one can use. The set of all such fixed decoupling zeros is denoted by $\boldsymbol{\Lambda}_{\text{sp-CSS}}^{h2-\text{oid}}(A, B, C, D, E, F)$.*

We prescribe below an algorithm of computing $\boldsymbol{\Lambda}_{\text{sp-CSS}}^{h2-\text{oid}}(A, B, C, D, E, F)$.

An Algorithm of Computing the Set $\boldsymbol{\Lambda}_{\text{sp-CSS}}^{h2-\text{oid}}(A, B, C, D, E, F)$:

Step 1a: (*Representation of Σ_{sub} in SCB*) As in the continuous-time case, we first construct in this step the SCB of Σ_{sub} characterized by (A, B, C, D). In SCB we have the structure as presented in (3.42)–(3.45). Clearly we can also decompose E with respect to this basis:

$$E\Gamma_s^{-1} = \begin{pmatrix} E_{0a}^- & E_{0a}^{0+} & E_{0b} & E_{0c} & E_{0d} \end{pmatrix}. \tag{10.139}$$

Step 1b: The pair (E_{0a}^-, A_{aa}^-) need not be observable. We apply another transformation:

$$T_a^{-1} A_{aa}^- T_a = \begin{pmatrix} A_{aa}^{11} & A_{aa}^{12} \\ 0 & A_{aa}^{22} \end{pmatrix} \quad \text{and} \quad T_a^{-1} E_a^- = \begin{pmatrix} E_a^1 & 0 \end{pmatrix},$$

such that (E_a^1, A_{aa}^{11}) is observable.

Step 2: (*Application of EDD algorithm*) We apply the EDD algorithm to the quintuple (A, B_Q, C, D_Q, E) and obtain the pair (A_z, B_z) but denote it by (A_z^Q, B_z^Q) to signify that the input to the EDD algorithm is (A, B_Q, C, D_Q, E). Form the set $\Lambda_{\text{sp-CSS}}^{h2-\text{oid}}(A, B, C, D, E, F)$ as

$$\Lambda_{\text{sp-CSS}}^{h2-\text{oid}}(A, B, C, D, E, F) = \lambda(A_{aa}^{11}) \cup \{\text{Input decoupling zeros of } (A_z^Q, B_z^Q)\}. \tag{10.140}$$

We have the following theorem.

Theorem 10.126 *Consider a discrete-time system as in* (10.1). *Let Assumption 10.1 be satisfied. Consider strictly proper filters of CSS architecture as given in* (10.135), *and assume that the solvability conditions of the H_2 OID filtering problem (as in Theorem 10.47) are satisfied. Then, the set $\Lambda_{\text{sp-CSS}}^{h2-\text{oid}}(A, B, C, D, E, F)$ as given in* (10.140) *is indeed the set of fixed decoupling zeros of $\Sigma_{\text{sp-CSS}}^{ue}$ resulting from strictly proper H_2 OID filters of CSS architecture as defined in Definition 10.125.*

Proof : The proof is the same as that in Theorem 10.65 that pertains to continuous-time systems. ∎

Remark 10.127 *Once again, it is easy to see that the set $\Lambda_{\text{sp-CSS}}^{h2-\text{oid}}(A, B, C, D, E, F)$ is invariant with respect to the matrix F.*

As in the continuous-time case, the set of fixed decoupling zeros $\Lambda_{\text{sp-CSS}}^{h2-\text{oid}}$, in general, does not have any telescopic property if the pair (C, A) is merely detectable but not observable. However, whenever the pair (C, A) is observable, it does have such a property as the following lemma shows.

Lemma 10.128 *Consider a discrete-time system as in* (10.1) *with the pair* (C, A) *observable. Consider two different H_2 OID filtering problems both as defined in Definition 10.6, however, with one taking a value E_a and the other a value E_b for the matrix E. Then, we have the telescopic property:*

$$\ker E_a \supseteq \ker E_b \supseteq \mathcal{S}^{\ominus}(A, B_Q, C, D_Q)$$
$$\implies \Lambda_{\text{sp-CSS}}^{h2-\text{oid}}(A, B, C, D, E_a, F) \subseteq \Lambda_{\text{sp-CSS}}^{h2-\text{oid}}(A, B, C, D, E_b, F).$$

Proof : In Proposition 5.13, this nested property was established for the set Λ_s. Using Theorem 10.126, this results in a similar property for the set $\Lambda_{\text{sp-CSS}}^{h2-\text{oid}}$. ∎

The poles of $G_{\text{sp-CSS}}^{ue}$ are given by the following lemma whose proof is obvious.

Lemma 10.129 *Consider a discrete-time system as in* (10.1). *Let Assumption 10.1 be satisfied. Also, let the solvability conditions of the H_2 OID filtering problem as in Theorem 10.47 be satisfied. Then, the poles of the resulting transfer matrix $G_{\text{sp-CSS}}^{ue}$ as given in* (10.137) *when it is in its reduced form are given by*

$$\Omega_{\text{sp-CSS}}^{h2-\text{oid}}(A, B, C, D, E, F) \,/\, \Lambda_{\text{sp-CSS}}^{h2-\text{oid}}(A, B, C, D, E, F).$$

In Section 10.10, we discussed the conditions under which the transfer matrix of any H_2 OID filter is unique. We considered there general strictly proper filters of the form (10.2) with $P = 0$ or proper filters of the form (10.2). The following theorem considers a class of filters with a specified architecture, namely the strictly proper CSS architecture, and discusses an H_2 OID filter being unique among them.

Theorem 10.130 *Consider a discrete-time system as in* (10.1), *and let Assumption 10.1 be satisfied. Also, let the matrices B_Q and D_Q be as defined by* (10.39). *Consider strictly proper filters, and assume that the conditions of Theorem 10.47 are satisfied. Then, a unique strictly proper full-order H_2 OID filter of the form* (10.135) *exists if and only if the following conditions are satisfied:*

   (i) The subsystem characterized by the quadruple (A, B, C, D) is right-invertible.

   (ii) The H_2 OID filtering problem is regular (i.e., the regularity conditions given in Definition 10.7 are satisfied).

   (iii) The matrix pair $(E, A - B_Q D_Q^{-1} C)$ is observable.

Moreover, under the above conditions, we have

(i) $K^{h2-\text{oid}}_{\text{sp-CSS}}(A, B, C, D, E, F) = \{B_\varrho D_\varrho^{-1}\}$, which is a singleton.

(ii) We have

$$\Omega^{h2-\text{oid}}_{\text{sp-CSS}}(A, B, C, D, E, F) = \lambda(A - B_\varrho D_\varrho^{-1}C),$$

which is the union of all stable invariant zeros and the mirror images of unstable invariant zeros and the compromising zeros of the subsystem characterized by the quadruple (A, B, C, D).

(iii) The set $\Lambda^{h2-\text{oid}}_{\text{sp-CSS}}(A, B, C, D, E, F)$ is the same as the set of all stable invariant zeros of the subsystem characterized by the quadruple (A, B, C, D).

Also, the unique H_2 OID filter $\Sigma_{\text{sp-CSS}}$ is given by

$$\sigma\xi = (A - B_\varrho D_\varrho^{-1}C)\xi + B_\varrho D_\varrho^{-1}y \quad \text{with} \quad \hat{z} = E\xi.$$

Proof : From Corollary 10.35, we know that a filter is an H_2 OID filter for Σ if and only if it is an EID filter for Σ_ϱ^{sp}. The results then follow from the conditions for the uniqueness of an EID filter for Σ_ϱ^{sp} as implied by Theorem 7.15. ∎

Remark 10.131 *We observe that the conditions given in Theorem 10.130 imply those given in Theorem 10.53 but not conversely. The reason for the more stringent conditions is that here we define the uniqueness of an H_2 OID filter at the state-space level (using strictly proper architecture), whereas earlier we considered the uniqueness in terms of a transfer matrix.*

Finally, we are in a position to design an H_2 OID filter while placing its modes at desired locations subject to the constraint that the set $\Omega^{h2-\text{oid}}_{\text{sp-CSS}}$ must be among the modes of any H_2 OID filter. We have the following result.

Theorem 10.132 (Strictly proper H_2 OID filter of CSS architecture with pole placement) *Consider a discrete-time system as in (10.1). Let Assumption 10.1 be satisfied. Consider strictly proper filters, and assume that the conditions of Theorem 10.47 are satisfied. Let the matrices B_ϱ and D_ϱ be as defined by (10.39). Also, consider the sets $\Omega^{h2-\text{oid}}_{\text{sp-CSS}}(A, B, C, D, E, F)$ and $K^{h2-\text{oid}}_{\text{sp-CSS}}(A, B, C, D, E, F)$ as described, respectively, in Theorems 10.120 and 10.116. Moreover, let Λ be a prescribed set of n self-conjugate elements within the unit circle $\mathbb{C}^\ominus$ such that Λ includes $\Omega^{h2-\text{oid}}_{\text{sp-CSS}}(A, B, C, D, E, F)$. Then, a filter gain $K \in K^{h2-\text{oid}}_{\text{sp-CSS}}(A, B, C, D, E, F)$ exists such that the strictly proper filter $\Sigma_{\text{sp-CSS}}$ given in (10.135) is an H_2 OID filter, and moreover, $\lambda(A - KC) = \Lambda$.*

Proof : In view of Theorems 10.47 and 10.116, the proof follows easily by using once again the duality principle. ∎

An algorithm for designing a strictly proper H_2 OID filter with simultaneous filter pole placement:

Throughout this algorithm, the matrices B_Q and D_Q are as defined by (10.39).
Step 1: By using the SCB of the system characterized by the quadruple (A, B_Q, C, D_Q), compute $\mathcal{S}^\ominus(A, B_Q, C, D_Q)$. Check the solvability condition

$$\mathcal{S}^\ominus(A, B_Q, C, D_Q) \subseteq \ker E.$$

If this is not satisfied, an H_2 OID strictly proper filter does not exist.
Step 2: Compute the set

$$\mathbf{\Omega}^{h2-\mathrm{oid}}_{\mathrm{sp\text{-}CSS}}(A, B, C, D, E, F) = \mathbf{\Omega}^{*}_{s}(A', C', B'_Q, D'_Q, E')$$

by using the EDD algorithm with its input as the quintuple (A', C', B'_Q, D'_Q, E').
Step 3: Choose a desired set Λ of n self-conjugate elements within the unit circle $\mathbb{C}^\ominus$ such that Λ includes $\mathbf{\Omega}^{h2-\mathrm{oid}}_{\mathrm{sp\text{-}CSS}}(A, B, C, D, E, F)$.
Step 4: Design K by using the (EDDSPP) algorithm discussed in Section 5.4.3. The inputs to the EDDSPP algorithm are the matrix quintuple (A', C', B'_Q, D'_Q, E') and the matrix Λ. The output of the algorithm is the transpose of the filter gain matrix K.

With the filter gain K as computed above, it is easy to verify that the resulting filter $\Sigma_{\mathrm{sp\text{-}CSS}}$ given in (10.135) is an H_2 OID filter with its modes at the locations specified by Λ; i.e., $\lambda(A - KC) = \Lambda$. As discussed in EDDSPP algorithm, although the algorithm concentrates only on pole placement, certain freedom exists as well to place the eigenvectors of $A - KC$, but this is not pursued here.

10.14.2   Proper H_2 OID filters of CSS architecture

We pursue here the design of proper H_2 OID filters while using filters of CSS architecture. We follow basically the same procedure as we did in the continuous-time case, however, with some subtle but important changes. For discrete-time systems, Corollary 10.36 plays the role of Corollary 10.13 and lays a roadmap to our development here. As before, to construct a proper H_2 OID filter of CSS architecture for the given system, we construct here a strictly proper H_2 OID filter for an auxiliary system $\widetilde{\Sigma}^*$.

We start with the system:

$$\Sigma : \begin{cases} \sigma x = Ax + Bu \\ y \;\;= Cx + Du \\ z \;\;= Ex + Fu. \end{cases} \tag{10.141}$$

Let Q be a semi-stabilizing solution of the DLMI (10.37), and define B_Q and D_Q by (10.39). We define $E^* = E - P^*C$ and $F^* = F - P^*D$ where the matrix

P^* minimizes the expression in (10.43). We obtain the system:

$$\Sigma_Q : \begin{cases} \sigma\tilde{x} = A\tilde{x} + B_Q\tilde{u} \\ y_Q = \begin{pmatrix} \tilde{C}_0 \\ \tilde{C}_1 \end{pmatrix}\tilde{x} + \begin{pmatrix} D_{Q0} \\ 0 \end{pmatrix}u = C\tilde{x} + D_Q\tilde{u} \\ z_Q^* = E^*\tilde{x}, \end{cases} \qquad (10.142)$$

where $\operatorname{rank} D_Q = \operatorname{rank} D_{Q0}$ and D_{Q0} surjective. Corollary 10.36 implies that a proper EID filter for the above system Σ_Q is a proper H_2 OID filter for the given system Σ. A proper EID filter for the above system Σ_Q can be obtained via designing a strictly proper EID filter for another auxiliary system $\widetilde{\Sigma}_Q^*$, which has some additional measurements compared with those of Σ_Q. The auxiliary system $\widetilde{\Sigma}_Q^*$ is defined as

$$\widetilde{\Sigma}_Q^* : \begin{cases} \sigma\tilde{x} = A\tilde{x} + B_Q\tilde{u} \\ \tilde{y}_Q = \begin{pmatrix} \tilde{y}_0 \\ \sigma\tilde{y}_1 \end{pmatrix} = \tilde{C}_Q\tilde{x} + \tilde{D}_Q\tilde{u} \\ \tilde{z}_Q^* = E^*\tilde{x}, \end{cases} \qquad (10.143)$$

where

$$\tilde{C}_Q = \begin{pmatrix} \tilde{C}_0 \\ \tilde{C}_1 A \end{pmatrix} \quad \text{and} \quad \tilde{D}_Q = \begin{pmatrix} D_{Q0} \\ \tilde{C}_1 B_Q \end{pmatrix}. \qquad (10.144)$$

As said above, a strictly proper EID filter for $\widetilde{\Sigma}_Q^*$ of (10.143) can be translated as a proper EID filter for Σ_Q given by (10.142), which in turn can be converted to a proper H_2 OID filter for Σ given by (10.141).

To summarize the above, the design of a proper H_2 OID filter for the given system Σ is at first transformed to the design of a proper EID filter for a system Σ_Q given by (10.142). In turn, the design of a proper EID filter Σ_Q can be transformed to the design of a strictly proper EID filter for an auxiliary system $\widetilde{\Sigma}_Q^*$. The auxiliary system $\widetilde{\Sigma}_Q^*$ has some additional measurements compared with those of Σ_Q.

As in the continuous-time case, for discrete-time systems an alternative method exists as well. In this alternative method, the design of a proper H_2 OID filter for the given system Σ is transformed to the design of a strictly proper H_2 OID filter for an auxiliary system $\widetilde{\Sigma}^*$, which has some additional measurements compared with the given system Σ. In turn, the design of a strictly proper H_2 OID filter for the auxiliary system $\widetilde{\Sigma}^*$ can be transformed to the design of a strictly proper EID filter for an auxiliary system denoted by $\widetilde{\Sigma}_{\tilde{Q}}^*$. Unlike in the continuous-time case where the auxiliary system $\widetilde{\Sigma}_Q^*$ of the first method is exactly the same as the auxiliary system $\widetilde{\Sigma}_{\tilde{Q}}^*$ of the alternative method, in the discrete-time case, the two auxiliary systems $\widetilde{\Sigma}_Q^*$ and $\widetilde{\Sigma}_{\tilde{Q}}^*$ are *not* one and the same. Nevertheless, both methods eventually lead to the design of a proper H_2 OID filter for the given system Σ.

In what follows, we present in detail the alternative method of design. As in the continuous-time case, the alternative method helps our quest to understand the structural properties of H_2 OID filters for Σ directly in terms of the data of Σ.

As explained above, in the alternative method, we first need to construct an auxiliary system $\widetilde{\Sigma}^*$ from the given system Σ. To do so, without loss of generality, we first decompose the measured output y into two parts y_0 and y_1 in such a way that y_0 contains explicitly the unknown input u in it, whereas y_1 does not contain any input u in it. That is, as in the case of continuous-time systems, we rewrite the given system equation (10.1) as

$$\Sigma : \begin{cases} \sigma x &= Ax + Bu \\ \begin{pmatrix} y_0 \\ y_1 \end{pmatrix} = \begin{pmatrix} C_0 \\ C_1 \end{pmatrix} x + \begin{pmatrix} D_0 \\ 0 \end{pmatrix} u = Cx + Du \\ z &= Ex + Fu, \end{cases} \tag{10.145}$$

where rank $D = $ rank $D_0 = m_0$.

Auxiliary System $\widetilde{\Sigma}^*$:
As in Chapter 7, we define the auxiliary system $\widetilde{\Sigma}^*$ as

$$\widetilde{\Sigma}^* : \begin{cases} \sigma x = Ax + Bu \\ \widetilde{y} = \begin{pmatrix} y_0 \\ \sigma y_1 \end{pmatrix} = \widetilde{C}x + \widetilde{D}u \\ z^* = E^*x, \end{cases} \tag{10.146}$$

where

$$\widetilde{C} = \begin{pmatrix} C_0 \\ C_1 A \end{pmatrix} \quad \text{and} \quad \widetilde{D} = \begin{pmatrix} D_0 \\ C_1 B \end{pmatrix}, \tag{10.147}$$

and $E^* = E - P^*C$ and P^* is a matrix that minimizes the expression in (10.43). We have the following lemma.

Lemma 10.133 *Consider the discrete-time system Σ as given in (10.141). Let P^* be such that it minimizes (10.43). We define $E^* = E - P^*C$ and $F^* = F - P^*D$, whereas $\widetilde{\Sigma}^*$ is defined as given in (10.146). Then, the following two statements are equivalent:*

(i) A strictly proper H_2 OID filter exists for the auxiliary system $\widetilde{\Sigma}^*$.

(ii) A proper H_2 OID filter exists for the system Σ.

Moreover, there is a $1 - 1$ correspondence between strictly proper H_2 OID filters for $\widetilde{\Sigma}^*$ and proper H_2 OID filters for Σ; that is, one of these filters can be constructed from the other.

Proof : The fact that a direct feedthrough matrix of an H_2 OID filter must minimize (10.43) is a direct consequence of Theorem 10.32. Consider any proper H_2 OID filter for Σ, namely

$$
\Sigma_f : \begin{cases} \sigma\xi = L\xi + My \\ \hat{z} = N\xi + Py, \end{cases}
$$

where the direct feedthrough matrix P must minimize (10.43). As P and P^* both minimize (10.43), we find that

$$
(P - P^*)(CQC' + DD') = 0.
$$

This implies in particular that $(P - P^*)D = 0$, and hence, using a decomposition compatible with the decomposition of C and D, we find that

$$
P - P^* = \begin{pmatrix} 0 & P_1 \end{pmatrix}.
$$

We define the following strictly proper filter $\tilde{\Sigma}_f^*$ for $\tilde{\Sigma}^*$:

$$
\tilde{\Sigma}_f^* : \begin{cases} \sigma\xi = L\xi + M_1\xi_1 + \begin{pmatrix} M_0 & 0 \end{pmatrix}\tilde{y} \\ \sigma\xi_1 = \begin{pmatrix} 0 & I \end{pmatrix}\tilde{y} \\ \hat{z}^* = N\xi + P_1\xi_1, \end{cases}
$$

where we decompose $M = (M_0 \quad M_1)$ compatible with the decomposition of y into y_0 and y_1. This new strictly proper filter $\tilde{\Sigma}_f^*$ yields the same error dynamics when applied to $\tilde{\Sigma}^*$ as the filter Σ_f applied to Σ and hence the same H_2 OID performance.

Conversely, given a strictly proper H_2 OID filter $\bar{\Sigma}_f^*$:

$$
\bar{\Sigma}_f^* : \begin{cases} \sigma\xi = L\xi + M\tilde{y} \\ \hat{z}^* = N\xi \end{cases}
$$

for $\tilde{\Sigma}^*$, we can construct a proper filter $\bar{\Sigma}_f$:

$$
\bar{\Sigma}_f : \begin{cases} \sigma\tilde{\xi} = L\tilde{\xi} + \begin{pmatrix} M_0 & LM_1 \end{pmatrix}y \\ \hat{z} = N\tilde{\xi} + (P^* + NM_1)y. \end{cases}
$$

The above proper filter $\bar{\Sigma}_f$ yields the same error dynamics when applied to Σ as the filter $\bar{\Sigma}_f^*$ applied to $\tilde{\Sigma}^*$ and hence the same H_2 OID performance.

The above two steps immediately show that the H_2 OID performance of Σ using proper filters and the H_2 OID performance of $\tilde{\Sigma}^*$ using strictly proper filters are the same.

We have shown that if for the system Σ the H_2 OID filter problem is solvable, strictly proper H_2 OID filters exist for the system $\widetilde{\Sigma}^*$. Finally, we note that strictly proper filters of CSS architecture that yield H_2 OID performance for $\widetilde{\Sigma}^*$ can then be converted, using the above arguments, to proper filters of CSS architecture for the original system, which yield H_2 OID performance, and conversely. ∎

As in the continuous-time case, all of the above development leads to the fact that instead of designing a proper H_2 OID filter for Σ, one can indeed design a strictly proper H_2 OID filter for $\widetilde{\Sigma}^*$ as emphasized in Lemma 10.133. This helps our quest to understand the structural properties of H_2 OID filters for Σ directly in terms of the data of Σ because all data that characterizes $\widetilde{\Sigma}^*$ is obtained by a simple modification of the data of Σ, and moreover, the structural properties of strictly proper H_2 OID filters have already been developed in Subsection 10.14.1.

We move on now to construct a strictly proper filter of CSS architecture for $\widetilde{\Sigma}^*$ as

$$\begin{cases} \sigma\widetilde{\xi} = A\widetilde{\xi} + K(\widetilde{y} - \widetilde{C}\widetilde{\xi}) \\ \widehat{z}^* = E^*\widetilde{\xi}, \end{cases} \tag{10.148}$$

where the matrix K is the filter gain. The estimate $\widehat{z}$ of z is given by

$$\widehat{z} = \widehat{z}^* + P^*y = E^*\widetilde{\xi} + P^*y. \tag{10.149}$$

The above development focuses on developing a strictly proper filter (10.148). However, the filter (10.148) is not directly implementable because $y_2 = \sigma y_1$ is not available as a measured variable. As in the continuous-time case, we can eliminate the need for σy_1 by defining a new variable:

$$\xi = \widetilde{\xi} - K_2 y_1. \tag{10.150}$$

Here K_2 is obtained by partitioning K in conformity with the partitioning of $\widetilde{y}$. That is,

$$K = \begin{pmatrix} K_0 & K_2 \end{pmatrix}.$$

With the definition of ξ as in (10.150), we can rewrite the filter equation (10.148) as

$$\begin{cases} \sigma\xi = (A - K\widetilde{C})\xi + \begin{pmatrix} K_0 & (A - K\widetilde{C})K_2 \end{pmatrix}y \\ \widetilde{\xi} = \xi + K_2 y_1 \\ \widehat{z}^* = E^*\widetilde{\xi} = E^*(\xi + K_2 y_1). \end{cases} \tag{10.151}$$

Obviously, the filter given above does not use σy_1. Moreover, it is proper rather than strictly proper. The filter given in (10.151) is indeed the proper full-order CSS filter that is to be used for Σ^*.

Proper filter of CSS architecture for Σ: Finally, in view of (10.149), we can rewrite (10.151) as an implementable proper filter for Σ:

$$\Sigma_{\text{p-CSS}} : \begin{cases} \sigma\xi = (A - K\widetilde{C})\xi + \widetilde{K}y \\ \widetilde{\xi} = \xi + K_2 y_1 \\ \widehat{z} = E^*\widetilde{\xi} + P^*y = E^*\xi + \widetilde{P}y, \end{cases} \tag{10.152}$$

where

$$\widetilde{K} = \begin{pmatrix} K_0 & (A - K\widetilde{C})K_2 \end{pmatrix} \quad \text{and} \quad \widetilde{P} = \begin{pmatrix} 0 & E^*K_2 \end{pmatrix} + P^*.$$

A block diagram representation of the proper filter along with the given plant is structurally the same as in Figure 7.4.

Error dynamics: By defining the error $e = x - \widetilde{\xi}$, the error between the actual desired output $z = Ex + Fu = E^*x + P^*y + F^*u$ and the estimated desired output $\widehat{z} = E^*\widetilde{\xi} + P^*y$ can be written as

$$e_z = z - \widehat{z} = E^*e + F^*u,$$

where $E^* = E - P^*C$ and $F^* = F - P^*D$. Then, in view of (10.145) and (10.148), the dynamics of error is given by

$$\Sigma^{ue}_{\text{p-CSS}} : \begin{cases} \sigma e = (A - K\widetilde{C})e + (B - K\widetilde{D})u \\ e_z = E^*e + F^*u. \end{cases} \tag{10.153}$$

Also, the transfer matrix $G^{ue}_{\text{p-CSS}}$ from u to e_z can obviously be written as

$$G^{ue}_{\text{p-CSS}}(z) = E^*(zI - A + K\widetilde{C})^{-1}(B - K\widetilde{D}) + F^*. \tag{10.154}$$

Remark 10.134 *In view of (10.152) and (10.153), it is easy to see that both the filter equation and the error equation have the same modes, which are the eigenvalues of $A - K\widetilde{C}$.*

Everything in the filters (10.148), (10.151), and (10.152) is known except the gain K. Following the notation used earlier, we denote the set of all H_2 OID filter gains by $\boldsymbol{K}^{h2-\text{oid}}_{\text{p-CSS}}(A, B, C, D, E, F)$, meaning that any gain $K \in \boldsymbol{K}^{h2-\text{oid}}_{\text{p-CSS}}$ renders $\Sigma_{\text{p-CSS}}$ given in (10.152) an H_2 OID filter for (10.1), and conversely, any gain K that renders $\Sigma_{\text{p-CSS}}$ an H_2 OID filter for (10.1) is an element of $\boldsymbol{K}^{h2-\text{oid}}_{\text{p-CSS}}$.

We have the following results.

Theorem 10.135 *Consider a discrete-time system as in (10.1). Let the conditions of Theorem 10.48 be satisfied. Then, the set $\boldsymbol{K}^{h2-\text{oid}}_{\text{p-CSS}}(A, B, C, D, E, F)$ is nonempty and is given by*

$$\boldsymbol{K}^{h2-\text{oid}}_{\text{p-CSS}}(A, B, C, D, E, F) = \bigcup_{P^* \in \mathscr{P}^*} \boldsymbol{K}^{h2-\text{oid}}_{\text{sp-CSS}}(A, B, \widetilde{C}, \widetilde{D}, E - P^*C, 0),$$

$$\tag{10.155}$$

where $\mathscr{P}^*$ is the set of all P^* that minimizes the expression in (10.43) and $\boldsymbol{K}^{h2-\text{oid}}_{\text{sp-CSS}}$ is as defined in Theorem 10.116.

Proof : The proof follows from Lemma 10.133.   ∎

Remark 10.136 *The set $K_{\text{sp-CSS}}^{h2-\text{oid}}(A, B, \widetilde{C}, \widetilde{D}, E - P^*C, 0)$ is nonempty for any P^* that minimizes the expression in (10.43).*

As in the case of strictly proper filters, the set $K_{\text{p-CSS}}^{h2-\text{oid}}(A, B, C, D, E, F)$ has the following telescopic property with respect to the matrices $\begin{pmatrix} E & F \end{pmatrix}$.

Lemma 10.137 *Consider a discrete-time system as in (10.1). Consider two different H_2 OID filtering problems both as defined in Definition 10.6, however, with one having the matrix pair (E_a, F_a) and the other (E_b, F_b) for the pair (E, F). Then, we have the telescopic property:*

$$\ker\begin{pmatrix} E_a & F_a \end{pmatrix} \supseteq \ker\begin{pmatrix} E_b & F_b \end{pmatrix}$$
$$\implies K_{\text{p-CSS}}^{h2-\text{oid}}(A, B, C, D, E_a, F_a) \supseteq K_{\text{p-CSS}}^{h2-\text{oid}}(A, B, C, D, E_b, F_b),$$

where we assume that the matrix pair (E_a, F_a) and (E_b, F_b) are such that, for the respective underlying system, the H_2 OID filtering problem is solvable (i.e., the conditions of Theorem 10.48 are satisfied).

Proof : The proof follows in view of Lemmas 10.133 and 10.118 and Theorem 10.135.   ∎

In general, whenever it exists, a proper H_2 OID filter of the form $\Sigma_{\text{p-CSS}}$ given in (10.152) is not unique. The nonuniqueness of such a filter can indeed be a blessing as some other specifications can then be imposed to come up with an appropriate filter. As in the case of strictly proper filters, we would like to make use of the available freedom to place the poles of error dynamics $\Sigma_{\text{p-CSS}}^{ue}$ to shape its transient behavior. Once again, we observe that the the poles of error dynamics $\Sigma_{\text{p-CSS}}^{ue}$ are the same as those of H_2 OID filter $\Sigma_{\text{p-CSS}}$. Thus, we would like to place the poles of an H_2 OID filter at desired locations. However, as in the case of strictly proper filters, the requirement of H_2 OID filtering dictates that some poles of the filter be fixed at certain locations while the others be free to be assigned. In fact, as before, for each given system, a set of complex numbers exists, which every H_2 OID filter must have among its poles. As the following definition formalizes, such a set of complex numbers can be termed as the H_2 OID filter fixed modes.

Definition 10.138 (*Fixed modes of proper H_2 OID filters of CSS architecture*)
*Consider a discrete-time system as in (10.1) and the H_2 OID filtering problem
10.6 characterized by the matrix sextuple $(A,\ B,\ C,\ D,\ E,\ F)$. Assume that the
solvability conditions as specified by Theorem 10.48 are satisfied. Then, a scalar
$\lambda \in \mathbb{C}^{\ominus}$ is said to be the fixed mode of an H_2 OID filter with the proper full-order
CSS architecture if λ is a pole (i.e., an eigenvalue of $A - K\tilde{C}$) of every H_2 OID
filter of such an architecture $\Sigma_{\text{p-CSS}}$ as given in (10.152). The set of all such H_2
OID filter fixed modes is denoted here by $\Omega^{h2-\text{oid}}_{\text{p-CSS}}(A, B, C, D, E, F)$.*

The following lemma is needed before we characterize the set $\Omega^{h2-\text{oid}}_{\text{p-CSS}}$.

Lemma 10.139 *Consider a discrete-time system as in (10.1). Let the conditions
of Theorem 10.47 be satisfied. Let P_1^* and P_2^* be two different matrices, both of
which minimize the expression in (10.43). Then, we have*

$$\Omega^{h2-\text{oid}}_{\text{sp-CSS}}(A, B, \tilde{C}, \tilde{D}, E - P_1^*C, 0) = \Omega^{h2-\text{oid}}_{\text{sp-CSS}}(A, B, \tilde{C}, \tilde{D}, E - P_2^*C, 0),$$

$$(10.156)$$

*where, for $i = 1,\ 2$, the set $\Omega^{h2-\text{oid}}_{\text{sp-CSS}}(A, B, \tilde{C}, \tilde{D}, E - P_i^*C, 0)$ is as characterized
in Theorem 10.120.*

Proof : The proof follows along the same lines as that of Lemma 10.78.   ∎

We have the following theorem that characterizes the set $\Omega^{h2-\text{oid}}_{\text{p-CSS}}$.

Theorem 10.140 *Consider a discrete-time system as in (10.1). Let Assumption
10.1 be satisfied. Consider proper filters, and assume that the solvability conditions as specified by Theorem 10.48 are satisfied. Also, let the matrix $E^* =
E - P^*C$, with P^* minimizing the expression in (10.43). Then, we have*

$$\Omega^{h2-\text{oid}}_{\text{p-CSS}}(A, B, C, D, E, F) = \Omega^{h2-\text{oid}}_{\text{sp-CSS}}(A, B, \tilde{C}, \tilde{D}, E^*, 0),$$

where $\Omega^{h2-\text{oid}}_{\text{sp-CSS}}$ is characterized in Theorem 10.120.

Proof : The proof follows from Lemmas 10.133 and 10.139.   ∎

As in Lemma 10.123, the set $\Omega^{h2-\text{oid}}_{\text{p-CSS}}(A, B, C, D, E, F)$ has a certain telescopic property as given below.

Lemma 10.141 *Consider a discrete-time system as in (10.1). Consider two different H_2 OID filtering problems both as defined in Definition 10.6, however, with one having the matrix pair $(E_a,\ F_a)$ and the other $(E_b,\ F_b)$ for the pair $(E,\ F)$. Then, we have the telescopic property:*

$$\ker\begin{pmatrix} E_a & F_a \end{pmatrix} \supseteq \ker\begin{pmatrix} E_b & F_b \end{pmatrix}$$
$$\implies \varOmega_{\text{p-CSS}}^{h2-\text{oid}}(A, B, C, D, E_a, F_a) \subseteq \varOmega_{\text{p-CSS}}^{h2-\text{oid}}(A, B, C, D, E_b, F_b),$$

where we assume that the matrix pair $(E_a,\ F_a)$ and $(E_b,\ F_b)$ are such that, for the respective underlying system, the H_2 OID filtering problem is solvable (i.e., the conditions of Theorem 10.48 are satisfied).

Proof : The proof follows from Lemma 10.133, Lemma 10.123, and Theorem 10.140. ∎

As before, the set of fixed modes obviously presents a constraint that one must contend with in any H_2 OID filter design. In this regard, one ponders with the following. Suppose the conditions for the existence of strictly proper H_2 OID filters as specified by Theorem 10.47 are satisfied. In this case, both strictly proper as well as proper H_2 OID filters exist. The question that arises then is as follows: Will the number and the location of fixed modes be different for strictly proper and proper H_2 OID filters? The following lemma answers this question.

Lemma 10.142 *Consider a discrete-time system as in (10.1). Let Assumption 10.1 be satisfied. Assume that the solvability conditions for the existence of strictly proper H_2 OID filters as specified by Theorem 10.47 are satisfied. Then, we have*

$$\varOmega_{\text{p-CSS}}^{h2-\text{oid}}(A, B, C, D, E, F) \supset \varOmega_{\text{sp-CSS}}^{h2-\text{oid}}(A, B, C, D, E, F). \qquad (10.157)$$

Proof : The proof follows in view of Corollary 10.36 and Lemma 7.52. ∎

As in the continuous-time case, besides the presence of H_2 OID filter fixed modes as explained above, another important structural issue in filter design exists, namely, in the system $\Sigma_{\text{p-CSS}}^{ue}$ which represents the error dynamics, the certain pole/zero cancellations are possible; see (10.153). We have the following formal definition.

Definition 10.143 (*Fixed decoupling zeros of $\Sigma_{\text{p-CSS}}^{ue}$ resulting from proper H_2 OID filters of CSS architecture*) *Consider a discrete-time system as in (10.1) and the H_2 OID filtering problem 10.6 characterized by the matrix sextuple (A,*

B, C, D, E, F). Assume that the solvability conditions of the H_2 OID filtering problem as in Theorem 10.48 are satisfied. Then, a scalar $\lambda \in \mathbb{C}^{\ominus}$ is said to be a fixed decoupling zero of $\Sigma_{\text{p-CSS}}^{ue}$ resulting from proper H_2 OID filters of CSS architecture if λ is either an input or an output decoupling zero (or both) of the system $\Sigma_{\text{p-CSS}}^{ue}$ as given in (10.153) for any proper H_2 OID filter $\Sigma_{\text{p-CSS}}$ that one can use. The set of all such fixed decoupling zeros is denoted by $\Lambda_{\text{p-CSS}}^{h2-\text{oid}}$ (A, B, C, D, E, F).

The following lemma is needed before we characterize the set $\Lambda_{\text{p-CSS}}^{h2-\text{oid}}$.

Lemma 10.144 *Consider a discrete-time system as in (10.1). Let the conditions of Theorem 10.47 be satisfied. Let P_1^* and P_2^* be two different matrices, both of which minimize the expression in (10.43). Then, we have*

$$\Lambda_{\text{sp-CSS}}^{h2-\text{oid}}(A, B, \tilde{C}, \tilde{D}, E - P_1^*C, 0) = \Lambda_{\text{sp-CSS}}^{h2-\text{oid}}(A, B, \tilde{C}, \tilde{D}, E - P_2^*C, 0),$$

$$(10.158)$$

where, for $i = 1, 2$, the set $\Lambda_{\text{sp-CSS}}^{h2-\text{oid}}(A, B, \tilde{C}, \tilde{D}, E - P_i^*C, 0)$ is as characterized in Theorem 10.126.

Proof : The proof follows along the same lines as that of Lemma 7.31. ∎

We have the following theorem that characterize the set $\Lambda_{\text{p-CSS}}^{h2-\text{oid}}$.

Theorem 10.145 *Consider a discrete-time system as in (10.1). Let Assumption 10.1 be satisfied. Consider proper filters of CSS architecture as given in (10.152), and assume that the solvability conditions of the H_2 OID filtering problem (as in Theorem 10.48) are satisfied. Then, we have*

$$\Lambda_{\text{p-CSS}}^{h2-\text{oid}}(A, B, C, D, E, F) = \Lambda_{\text{sp-CSS}}^{h2-\text{oid}}(A, B, \tilde{C}, \tilde{D}, E^*, 0),$$

where $E^* = E - P^*C$, with matrix P^* minimizing the expression in (10.43).

Proof : The proof follows from Lemmas 10.133 and 10.144. ∎

As in the continuous-time case, it turns out that, under the general assumption of the pair (C, A) being detectable, the set $\Lambda_{\text{sp-CSS}}^{h2-\text{oid}}$ does not have any telescopic property. However, as before, whenever the pair (C, A) is observable, it does have such a property as the following lemma shows.

Lemma 10.146 *Consider a discrete-time system as in (10.1) with the pair (C, A) observable. Consider two different H_2 OID filtering problems both as defined in Definition 10.6, however, with one having the matrix pair (E_a, F_a) and the other (E_b, F_b) for the pair (E, F). Then, we have the telescopic property:*

$$\ker \begin{pmatrix} E_a & F_a \end{pmatrix} \supseteq \ker \begin{pmatrix} E_b & F_b \end{pmatrix}$$
$$\Longrightarrow \Lambda_{\text{p-CSS}}^{h2-\text{oid}}(A, B, C, D, E_a, F_a) \subseteq \Lambda_{\text{p-CSS}}^{h2-\text{oid}}(A, B, C, D, E_b, F_b),$$

where we assume that the matrix pair (E_a, F_a) and (E_b, F_b) are such that, for the respective underlying system, the H_2 OID filtering problem is solvable (i.e., the conditions of Theorem 10.48 are satisfied).

Proof: The proof follows from Lemmas 10.133 and 10.128 and Theorem 10.145.
∎

In Section 10.10, we discussed the conditions under which the transfer matrix of any H_2 OID filter is unique. We considered there general proper filters of the form (10.2). The following theorem considers a class of filters with a specified architecture, namely the proper CSS architecture, and discusses an H_2 OID filter being unique among them.

Theorem 10.147 *Consider a discrete-time system as in (10.1), and let Assumption 10.1 be satisfied. Also, let the matrices B_Q and D_Q be as defined by (10.39), and let Σ_Q be as in (10.41). Consider proper filters, and assume that the solvability conditions of the H_2 OID filtering problem (as in Theorem 10.48) are satisfied. Then, a unique proper full-order H_2 OID filter of the form $\Sigma_{\text{sp-CSS}}$ given (10.152) exists if and only if the following conditions are satisfied:*

   (i) The subsystem characterized by the quadruple (A, B, C, D) is right-invertible.

   (ii) The H_2 OID filtering problem is regular (i.e., the regularity conditions given in Definition 10.7 are satisfied).

   (iii) The matrix pair $(E - FD_Q^{-1}C, \ A - B_Q D_Q^{-1}C)$ is observable.

Moreover, under the above conditions, we have

   (i) $K_{\text{p-CSS}}^{h2-\text{oid}}(A, B, C, D, E, F) = \{B_Q D_Q^{-1}\}$, which is a singleton.

   (ii) The set $\Omega_{\text{p-CSS}}^{h2-\text{oid}}(A, B, C, D, E, F)$ is the same as the set $\lambda(A - B_Q D_Q^{-1}C)$, which is the union of all stable invariant zeros and the mirror images of unstable invariant zeros and the compromising zeros of the subsystem characterized by the quadruple (A, B, C, D).

(iii) The set $\Lambda_{\text{sp-CSS}}^{h2-\text{oid}}(A, B, C, D, E, F)$ is the same as the set of all stable invariant zeros of the subsystem characterized by the quadruple (A, B, C, D).

Also, the unique H_2 OID filter $\Sigma_{\text{p-CSS}}$ is given by

$$\sigma\xi = (A - B_\varrho D_\varrho^{-1} C)\xi + B_\varrho D_\varrho^{-1} y \quad \text{with} \quad \hat{z} = (E - F D_\varrho^{-1} C)\xi + F D_\varrho^{-1} y.$$

Proof : From Corollary 10.36, we know that a filter is an H_2 OID filter for Σ if and only if it is an EID filter for Σ_ϱ. The results then follow from the conditions for the uniqueness of an EID filter for Σ_ϱ as implied by Theorem 7.45.   ∎

Remark 10.148 *We observe that the conditions given in Theorem 10.147 imply those given in Theorem 10.54 but not conversely. The reason for the more stringent conditions is that here we define the uniqueness of an H_2 OID filter at the state-space level (using proper architecture), whereas earlier we considered the uniqueness in terms of a transfer matrix.*

Finally, we would like to design an H_2 OID filter while placing its modes at desired locations. The following theorem shows that we can design a filter gain K such that $\Sigma_{\text{p-CSS}}$ given in (10.152) is an H_2 OID filter while its modes are at the prescribed desired locations except that they need to contain $\Omega_{\text{p-CSS}}^{h2-\text{oid}}(A, B, C, D, E, F)$ among them.

Theorem 10.149 (Proper H_2 OID filter of CSS architecture with pole placement) *Consider a discrete-time system as in (10.1). Let Assumption 10.1 be satisfied. Consider proper filters, and assume that the conditions of Theorem 10.48 are satisfied. Also, let $E^* = E - P^* C$, where P^* minimizes the expression in (10.43). Consider the set $\Omega_{\text{p-CSS}}^{h2-\text{oid}}(A, B, C, D, E, F)$, which is the set of fixed modes discussed in Theorem 10.140. Also, consider the nonempty set $K_{\text{p-CSS}}^{h2-\text{oid}}(A, B, C, D, E, F)$ discussed in Theorem 10.135. Moreover, let Λ be a prescribed set of n self-conjugate elements within the unit circle $\mathbb{C}^\ominus$ such that Λ includes $\Omega_{\text{p-CSS}}^{h2-\text{oid}}(A, B, C, D, E, F)$. Then, a filter gain $K \in K_{\text{p-CSS}}^{h2-\text{oid}}(A, B, C, D, E, F)$ exists such that the proper filter (10.152) is an H_2 OID filter, and moreover, $\lambda(A - K\tilde{C}) = \Lambda$.*

Proof : The proof follows in view of Lemma 10.133 and Theorem 10.132.   ∎

An algorithm of designing a proper H_2 OID filter with simultaneous filter pole placement:

Step 1: Given the system Σ as in (10.73) and characterized by the sextuple (A, B, C, D, E, F), form the auxiliary system $\widetilde{\Sigma}^*$ characterized by the quintuple $(A, B, \widetilde{C}, \widetilde{D}, E^*)$, where $\widetilde{C}$ and $\widetilde{D}$ are, respectively, as in (10.147), and as usual, $E^* = E - P^*C$, with matrix P^* minimizing the expression in (10.43).

Step 2: Choose a desired set Λ of n self-conjugate elements within the unit circle $\mathbb{C}^{\ominus}$. [Step 3 given below checks to make sure that Λ includes $\mathbf{\Omega}_{\text{p-CSS}}^{h2-\text{oid}}(A, B, C, D, E, F)$].

Step 3: Following the algorithm for designing a strictly proper H_2 OID filter with simultaneous filter pole placement as given in Subsection 10.14.1, and using the quintuple $(A, B, \widetilde{C}, \widetilde{D}, E^*)$ as well as the matrix Λ as inputs to that algorithm, we obtain the gain K.

With the filter gain K as computed above, it is easy to verify that the resulting filter $\Sigma_{\text{p-CSS}}$ given in (10.152) is an H_2 OID filter with its modes at the locations specified by Λ; i.e., $\lambda(A - K\widetilde{C}) = \Lambda$.

10.14.3 Reduced-order H_2 OID filters of CSS architecture

The two previous subsections respectively construct full-order strictly proper and proper filters of CSS architecture that solve the H_2 OID filtering problem for discrete-time systems. By full-order filters, as usual, we mean filters having the same dynamic order as that of the given system. Our goal in this subsection is to develop reduced-order filters of CSS architecture having their dynamic order lower than that of the given system.

The procedure of developing reduced-order filters here for discrete-time systems follows mostly along the same lines as in Subsection 10.12.3, which pertains to the continuous-time case. However, there are some subtle but important differences warranting appropriate justifications, and as such, we need to redevelop them once more here. As before, our method of development transforms the construction of a reduced-order filter for a given system to that of a full-order filter for a certain reduced-order system.

We proceed now to construct an appropriate reduced-order system. To start with, let us rewrite the matrices C and D of (10.1) as

$$C = \begin{pmatrix} 0 & C_{02} \\ I_{p-m_0} & 0 \end{pmatrix}, \quad D = \begin{pmatrix} D_0 \\ 0 \end{pmatrix},$$

where again rank D = rank D_0 = m_0. This can always be done without any loss of generality by appropriate coordinate transformations. In view of the above partitioning of C and D, we can partition the given system Σ as

$$\Sigma : \begin{cases} \begin{pmatrix} \sigma x_1 \\ \sigma x_2 \end{pmatrix} = \begin{pmatrix} A_{11} & A_{12} \\ A_{21} & A_{22} \end{pmatrix} \begin{pmatrix} x_1 \\ x_2 \end{pmatrix} + \begin{pmatrix} B_{11} \\ B_{22} \end{pmatrix} u \\[2ex] y = \begin{pmatrix} y_0 \\ y_1 \end{pmatrix} = \begin{pmatrix} 0 & C_{02} \\ I & 0 \end{pmatrix} \begin{pmatrix} x_1 \\ x_2 \end{pmatrix} + \begin{pmatrix} D_0 \\ 0 \end{pmatrix} u \\[2ex] z = Ex + Fu, \end{cases} \qquad (10.159)$$

where different variables have obvious meanings.

Next, we use a preliminary injection of output y into the desired output z. To this end, we choose a matrix P^* that minimizes the expression in (10.43). This aspect of selecting the P^* matrix is one of the fundamental differences between the EID and the H_2 OID filters, and between constructing the continuous-time H_2 OID filters and the current case of constructing the discrete-time H_2 OID filters. Having selected P^*, we define a new desired output z^* as

$$z^* = z - P^* y = (E - P^* C)x + (F - P^* D)u = E^* x + F^* u, \qquad (10.160)$$

where $E^* = E - P^* C$ and $F^* = F - P^* D$. We note that $F^* = 0$ for continuous-time case; however, for discrete-time systems being considered here, F^* is not necessarily zero.

In view of (10.159) and (10.160), we can define a new system Σ^* as

$$\Sigma^* : \begin{cases} \begin{pmatrix} \sigma x_1 \\ \sigma x_2 \end{pmatrix} = \begin{pmatrix} A_{11} & A_{12} \\ A_{21} & A_{22} \end{pmatrix} \begin{pmatrix} x_1 \\ x_2 \end{pmatrix} + \begin{pmatrix} B_{11} \\ B_{22} \end{pmatrix} u \\[2ex] y = \begin{pmatrix} y_0 \\ y_1 \end{pmatrix} = \begin{pmatrix} 0 & C_{02} \\ I & 0 \end{pmatrix} \begin{pmatrix} x_1 \\ x_2 \end{pmatrix} + \begin{pmatrix} D_0 \\ 0 \end{pmatrix} u \\[2ex] z^* = E^* x = E_1^* x_1 + E_2^* x_2 + F^* u, \end{cases} \qquad (10.161)$$

with $E^* = \begin{pmatrix} E_1^* & E_2^* \end{pmatrix}$.

We note that the y_1 is not contaminated by the input u, and hence, $x_1 = y_1$ is known exactly from the measurement y. Thus, all we need to do next is to estimate the state x_2. To proceed further, let us rewrite the state equation for x_1 in terms of the output y_1 and the state x_2 as

$$\sigma y_1 = A_{11} y_1 + A_{12} x_2 + B_{11} u. \qquad (10.162)$$

The above equation can be rewritten as

$$\sigma y_1 - A_{11} y_1 = A_{12} x_2 + B_{11} u.$$

Treating σy_1 as known, we can define a new measurement variable y_r:

$$y_r = \begin{pmatrix} y_0 \\ \sigma y_1 - A_{11} y_1 \end{pmatrix}.$$

Although σy_1 is not directly available, as we did earlier in the case of full-order proper CSS filters, we can eliminate it from any filter that is constructed using y_r as a measured output. With this in mind, we form the auxiliary system:

$$\Sigma_r^* : \begin{cases} \sigma x_r = A_r x_r + B_r u + A_{21} y_1 \\ y_r = C_r x_r + D_r u \\ z_r^* = E_r^* x_r + F^* u, \end{cases} \tag{10.163}$$

where

$$x_r = x_2, \; A_r = A_{22}, \; B_r = B_{22}, \; C_r = \begin{pmatrix} C_{02} \\ A_{12} \end{pmatrix}, \; D_r = \begin{pmatrix} D_0 \\ B_{11} \end{pmatrix}, \; E_r^* = E_2^*. \tag{10.164}$$

We note that the dynamic order n_r of the above Σ_r^* is less than the dynamic order n of the given system Σ by a number equal to the dimension of $x_1 = y_1$.

Before we proceed further, certain structural properties of Σ_r^* as related to those of the given system Σ as developed in Lemma 7.55 are brought to the reader's attention.

Next, let us suppose we can construct a strictly proper H_2 OID filter for the auxiliary system Σ_r^* in order to arrive at an estimate $\hat{z}_r^*$ of z_r^*. Then, we plan to obtain an estimate $\hat{z}$ of z as

$$\hat{z} = E_1^* x_1 + \hat{z}_r^* + P^* y. \tag{10.165}$$

This motivates us to construct a full-order strictly proper filter for the auxiliary system Σ_r^* to arrive at an estimate $\hat{z}_r^*$ of z_r^*. A full-order filter constructed as such for Σ_r^* is indeed a reduced-order filter for Σ. However, in view of (10.165), such a reduced-order filter for Σ is a proper (rather than a strictly proper) filter for Σ. A fundamental issue that arises next is under what conditions one can construct a strictly proper H_2 OID filter for Σ_r^*. Expectedly, it turns out that one can construct a strictly proper H_2 OID filter for the reduced-order system Σ_r^* if and only if one can construct a proper H_2 OID filter for the given system Σ. The following corollary formalizes this result.

Lemma 10.150 *Consider discrete-time systems Σ and Σ_r^*, respectively, as given in (10.1) and (10.163). Then, the following two statements are equivalent:*

(i) A strictly proper H_2 OID filter exists for the reduced-order system Σ_r^*.

(ii) A proper H_2 OID filter exists for the system Σ.

Moreover, there is a $1 - 1$ relationship between the strictly proper H_2 OID filter of CSS architecture for Σ_r^* and the proper reduced-order H_2 OID filter of CSS architecture for Σ; that is, one of these filters can be constructed from the other.

Proof : The equivalence of statements (i) and (ii) follows from Lemma 7.55. The $1 - 1$ relationship between the aforementioned filters can be verified easily.   ∎

Next, to construct the required filter for Σ, in the spirit of the above development, we first construct a strictly proper filter of CSS architecture for the reduced-order system Σ_r^* as

$$\begin{cases} \sigma\tilde{\xi}_r = A_r\tilde{\xi}_r + A_{21}y_1 + K_r(y_r - C_r\tilde{\xi}_r) \\ \hat{z}_r^* = E_r^*\tilde{\xi}_r, \end{cases} \tag{10.166}$$

where K_r is termed as the reduced-order filter gain. As σy_1 is not available, we need to modify the above filter. To this end, let us partition $K_r = \begin{pmatrix} K_{r0} & K_{r1} \end{pmatrix}$ so as to be compatible with the partitioning of y_r. Also, let

$$\xi_r = \tilde{\xi}_r - K_{r1}y_1. \tag{10.167}$$

We can then easily rewrite the filter (10.166) as a proper filter for Σ as shown below.

Reduced-order proper filter for Σ: The reduced-order proper filter of CSS architecture for Σ is given by

$$\Sigma_{\text{r-CSS}} : \begin{cases} \sigma\xi_r = (A_r - K_rC_r)\xi_r + \tilde{K}_r y \\ \tilde{\xi}_r = \xi_r + K_{r1}y_1 \\ \hat{z} = E_1^*x_1 + E_r^*\tilde{\xi}_r + P^*y = E_r^*\xi_r + \tilde{P}_r y, \end{cases} \tag{10.168}$$

where

$$\tilde{K}_r = \begin{pmatrix} K_{r0} & A_{21} - K_{r1}A_{11} + (A_r - K_rC_r)K_{r1} \end{pmatrix}$$

and

$$\tilde{P}_r = \begin{pmatrix} 0 & E_1^* + E_r^*K_{r1} \end{pmatrix} + P^*.$$

A block diagram representation of the reduced-order proper filter along with the given plant is structurally the same as in Figure 7.5.

Error dynamics: By defining the error $e_r = x_r - \tilde{\xi}_r$, the error between the actual desired output $z = Ex + Fu = E^*x + P^*y + F^*u$ and the estimated desired output $\hat{z} = E_1^*x_1 + E_r^*\tilde{\xi}_r + P^*y$ can be written as

$$e_z = z - \hat{z} = E_r^*e_r + F^*u.$$

Then, the dynamics of error is given by

$$\Sigma_{\text{r-CSS}}^{ue} : \begin{cases} \sigma e_r = (A_r - K_rC_r)e_r + (B_r - K_rD_r)u \\ e_z = E_r^*e_r + F^*u. \end{cases} \tag{10.169}$$

Also, the transfer matrix $G_{\text{r-CSS}}^{ue}$ from u to e_z can obviously be written as

$$G_{\text{r-CSS}}^{ue}(z) = E_r^*(zI - A_r + K_rC_r)^{-1}(B_r - K_rD_r) + F^*. \tag{10.170}$$

Remark 10.151 *It is easy to see that both the filter equation* (10.168) *and the error equation* (10.169) *have the same modes, which are the eigenvalues of $A_r - K_r C_r$.*

Everything in the filter $\Sigma_{\text{r-CSS}}$ is known except the gain K_r. Following the notation used in earlier sections, we denote the set of all H_2 OID filter gains by $\boldsymbol{K}_{\text{r-CSS}}^{h2-\text{oid}}(A, B, C, D, E, F)$, meaning that any gain $K_r \in \boldsymbol{K}_{\text{r-CSS}}^{h2-\text{oid}}$ renders $\Sigma_{\text{r-CSS}}$ given in (10.168) an H_2 OID filter for (10.1), and conversely, any gain K_r that renders $\Sigma_{\text{r-CSS}}$ an H_2 OID filter for (10.1) is an element of $\boldsymbol{K}_{\text{r-CSS}}^{h2-\text{oid}}$.

We have the following results.

Theorem 10.152 *Consider a discrete-time system as in* (10.1). *Let the conditions of Theorem 10.48 be satisfied. Also, let $E^* = E - P^*C$ and $F^* = F - P^*D$, where P^* is a matrix that minimizes the expression in* (10.43). *Then, the set $\boldsymbol{K}_{\text{r-CSS}}^{h2-\text{oid}}(A, B, C, D, E, F)$ is nonempty and is given by*

$$\boldsymbol{K}_{\text{r-CSS}}^{h2-\text{oid}}(A, B, C, D, E, F) = \bigcup_{P^* \in \mathscr{P}^*} \boldsymbol{K}_{\text{sp-CSS}}^{h2-\text{oid}}(A_r, B_r, C_r, D_r, E_r^*, 0),$$

$$(10.171)$$

where E_r^* is obtained by partitioning $E - P^*C$ in conformity with the partitioning of x into x_1 and x_2 as

$$E - P^*C = \begin{pmatrix} E_1^* & E_r^* \end{pmatrix}, \tag{10.172}$$

and $\mathscr{P}^*$ is the set of all P^* that minimize the expression in (10.43). *Also, $\boldsymbol{K}_{\text{sp-CSS}}^{h2-\text{oid}}$ is as defined in Theorem 10.116.*

Proof : The proof follows from Lemma 10.150. ∎

Remark 10.153 *The set $\boldsymbol{K}_{\text{sp-CSS}}^{h2-\text{oid}}(A_r, B_r, C_r, D_r, E_r^*, 0)$ is nonempty for any P^* that minimizes the expression in* (10.43).

In general, whenever they exist, reduced-order H_2 OID filters of the form $\Sigma_{\text{r-CSS}}$ given in (10.168) are not necessarily unique. The following remark addresses the issue of the uniqueness of such a filter.

Remark 10.154 *A reduced-order H_2 OID filter of CSS architecture for the system Σ given in* (10.1) *is unique if and only if a full-order proper H_2 OID filter of CSS architecture for the same system Σ is unique. The conditions for the uniqueness of such filters are given in Theorem 10.147. Under such conditions, the so-called reduced-order filter of CSS architecture coalesces with the full-order filter of CSS architecture.*

As in the previous subsections, the set $K_{\text{r-CSS}}^{h2-\text{oid}}(A, B, C, D, E, F)$ has the following telescopic property with respect to the matrices $\begin{pmatrix} E & F \end{pmatrix}$.

Lemma 10.155 *Consider a discrete-time system as in* (10.1). *Consider two different H_2 OID filtering problems both as defined in Definition 10.6, however, with one having the matrix pair (E_a, F_a) and the other (E_b, F_b) for the pair (E, F). Then, we have the telescopic property:*

$$\ker\begin{pmatrix} E_a & F_a \end{pmatrix} \supseteq \ker\begin{pmatrix} E_b & F_b \end{pmatrix}$$
$$\implies K_{\text{r-CSS}}^{h2-\text{oid}}(A, B, C, D, E_a, F_a) \supseteq K_{\text{r-CSS}}^{h2-\text{oid}}(A, B, C, D, E_b, F_b),$$

where we assume that the matrix pairs (E_a, F_a) and (E_b, F_b) are such that, for the respective underlying system, the H_2 OID filtering problem is solvable (i.e., the conditions of Theorem 10.48 are satisfied).

Proof : The proof follows in view of Lemmas 10.150 and 10.118 and Theorem 10.152. ∎

Next, as in full-order filters, we would like to design a reduced-order H_2 OID filter of CSS architecture while placing its poles at desired locations in order to shape the transient behavior of error dynamics. As before, it is expected that all H_2 OID filters of reduced-order CSS architecture share some fixed modes as defined below.

Definition 10.156 *(**Fixed modes of reduced-order proper H_2 OID filters of CSS architecture**) Consider a discrete-time system as in* (10.1) *and the H_2 OID filtering problem 10.6 characterized by the matrix sextuple (A, B, C, D, E, F). Assume that the solvability conditions as specified by Theorem 10.48 are satisfied. Then, a scalar $\lambda \in \mathbb{C}^{\ominus}$ is said to be the fixed mode of an H_2 OID filter with the proper reduced-order CSS architecture if λ is a pole (i.e., an eigenvalue of $A_r - K_r C_r$) of every H_2 OID filter of such an architecture $\Sigma_{\text{r-CSS}}$ as given in* (10.168). *The set of all such H_2 OID filter fixed modes is denoted here by $\Omega_{\text{r-CSS}}^{h2-\text{oid}}(A, B, C, D, E, F)$.*

The following lemma is needed before we characterize the set $\Omega_{\text{r-CSS}}^{h2-\text{oid}}$.

Lemma 10.157 *Consider a discrete-time system as in* (10.1). *Let the conditions of Theorem 10.26 be satisfied. Let P_1^* and P_2^* be two different matrices, both of which minimize the expression in* (10.43). *Then, we have*

$$\Omega_{\text{sp-CSS}}^{h2-\text{oid}}(A_r, B_r, C_r, D_r, E_{r1}^*, 0) = \Omega_{\text{sp-CSS}}^{h2-\text{oid}}(A_r, B_r, C_r, D_r, E_{r2}^*, 0), \quad (10.173)$$

where, for $i = 1, 2$, E_{ri}^* is obtained by partitioning $E - P_i^* C$ in conformity with the partitioning of x into x_1 and x_2 as

$$E - P_i^* C = \left(E_1^* \quad E_{ri}^* \right), \tag{10.174}$$

and $\boldsymbol{\Omega}_{\text{sp-CSS}}^{h2-\text{oid}}(A_r, B_r, C_r, D_r, E_{ri}^*, 0)$ is characterized as in Theorem 10.120.

Proof : The proof follows along the same lines as that of Lemma 10.78. ∎

We have the following theorem that characterizes the set $\boldsymbol{\Omega}_{\text{r-CSS}}^{h2-\text{oid}}$.

Theorem 10.158 *Consider a discrete-time system as in (10.1). Let Assumption 10.1 be satisfied. Consider proper filters, and assume that the solvability conditions as specified by Theorem 10.48 are satisfied. Then, we have*

$$\boldsymbol{\Omega}_{\text{r-CSS}}^{h2-\text{oid}}(A, B, C, D, E, F) = \boldsymbol{\Omega}_{\text{sp-CSS}}^{h2-\text{oid}}(A_r, B_r, C_r, D_r, E_r^*, 0),$$

where $\boldsymbol{\Omega}_{\text{sp-CSS}}^{h2-\text{oid}}(A_r, B_r, C_r, D_r, E_r^*, 0)$ is characterized in Theorem 10.120.

Proof : The proof follows in view of Lemmas 10.150 and 10.157. ∎

As in Lemma 10.141, the set $\boldsymbol{\Omega}_{\text{r-CSS}}^{h2-\text{oid}}(A, B, C, D, E, F)$ has a certain telescopic property as given below.

Lemma 10.159 *Consider a discrete-time system as in (10.1). Consider two different H_2 OID filtering problems both as defined in Definition 10.6, however, with one having the matrix pair (E_a, F_a) and the other (E_b, F_b) for the pair (E, F). Then, we have the telescopic property:*

$$\ker \left(E_a \quad F_a \right) \supseteq \ker \left(E_b \quad F_b \right)$$
$$\implies \boldsymbol{\Omega}_{\text{r-CSS}}^{h2-\text{oid}}(A, B, C, D, E_a, F_a) \subseteq \boldsymbol{\Omega}_{\text{p-CSS}}^{h2-\text{oid}}(A, B, C, D, E_b, F_b),$$

where we assume that the matrix pair (E_a, F_a) and (E_b, F_b) are such that, for the respective underlying system, the H_2 OID filtering problem is solvable (i.e., the conditions of Theorem 10.48 are satisfied).

Proof : The proof follows in view of Lemmas 10.150 and 10.123 and Theorem 10.158. ∎

As we discussed, the set of fixed modes obviously presents a constraint that one must contend with in any H_2 OID filter design. A question that arises then is whether the H_2 OID filter fixed modes of reduced-order filters are different from those of full-order filters. The following lemma addresses this issue and shows that it is not the case.

Lemma 10.160 *Consider a discrete-time system as in (10.1). Let Assumption 10.1 be satisfied. Assume that the solvability conditions for the existence of proper H_2 OID filters as specified by Theorem 10.48 are satisfied. Then, we have*

$$\Omega_{\text{r-CSS}}^{h2-\text{oid}}(A, B, C, D, E, F) = \Omega_{\text{p-CSS}}^{h2-\text{oid}}(A, B, C, D, E, F). \tag{10.175}$$

Proof : The proof follows along the same lines as in the proof of Lemma 7.63. ∎

In Lemmas 10.81 and 10.142, whenever strictly proper H_2 OID filters existed, we showed that the sets of fixed modes are the same whether one uses strictly proper or proper H_2 OID filters. Following the theme of these lemmas, the following lemma extends this result even further by considering the reduced-order filters as well.

Lemma 10.161 *Consider a discrete-time system as in (10.1). Let Assumption 10.1 be satisfied. Assume that the solvability conditions for the existence of strictly proper H_2 OID filters as specified by Theorem 10.47 are satisfied. Then, we have*

$$\Omega_{\text{p-CSS}}^{h2-\text{oid}}(A, B, C, D, E, F) = \Omega_{\text{sp-CSS}}^{h2-\text{oid}}(A, B, C, D, E, F)$$
$$\supset \Omega_{\text{r-CSS}}^{h2-\text{oid}}(A, B, C, D, E, F). \tag{10.176}$$

Proof : The proof follows along the same lines as in the proof of Lemma 7.64. ∎

As in the case of strictly proper and proper filters, besides the presence of H_2 OID filter fixed modes as explained above, another important structural issue in filter design regarding pole/zero cancellations in the system $\Sigma_{\text{r-CSS}}^{ue}$ exists that represents the error dynamics; see (10.169). To be specific, for a given H_2 OID filtering problem, there exist indeed what can be termed as a set of fixed decoupling zeros of $\Sigma_{\text{r-CSS}}^{ue}$ as defined below formally.

Definition 10.162 (Fixed decoupling zeros of $\Sigma_{\text{r-CSS}}^{ue}$ resulting from proper H_2 OID filters of CSS architecture) *Consider a discrete-time system as in (10.1) and the H_2 OID filtering problem 10.6 characterized by the matrix sextuple $(A,$

B, C, D, E, F). *Assume that the solvability conditions of the* H_2 *OID filtering problem as in Theorem 10.48 are satisfied. Then, a scalar* $\lambda \in \mathbb{C}^{\ominus}$ *is said to be a fixed decoupling zero of* $\Sigma^{ue}_{\text{r-CSS}}$ *resulting from reduced-order* H_2 *OID filters of CSS architecture if* λ *is either an input or an output decoupling zero (or both) of the system* $\Sigma^{ue}_{\text{r-CSS}}$ *as given in* (10.169) *for any proper* H_2 *OID filter* $\Sigma_{\text{r-CSS}}$ *that one can use. The set of all such fixed decoupling zeros is denoted by* $\Lambda^{h2-\text{oid}}_{\text{r-CSS}}(A, B, C, D, E, F)$.

The following lemma is needed before we characterize the set $\Lambda^{h2-\text{oid}}_{\text{r-CSS}}$.

Lemma 10.163 *Consider a discrete-time system as in* (10.1). *Let the conditions of Theorem 10.47 be satisfied. Let* P_1^* *and* P_2^* *be two different matrices, both of which minimize the expression in* (10.43). *Then, we have*

$$\Lambda^{h2-\text{oid}}_{\text{sp-CSS}}(A_r, B_r, C_r, D_r, E_{r1}^*, 0) = \Lambda^{h2-\text{oid}}_{\text{sp-CSS}}(A_r, B_r, C_r, D_r, E_{r2}^*, 0), \quad (10.177)$$

where, for $i = 1, 2$, E_{ri}^* *is as defined in* (10.174), *and* $\Lambda^{h2-\text{oid}}_{\text{sp-CSS}}(A_r, B_r, C_r, D_r, E_{ri}^*, 0)$ *is characterized as in Theorem 10.126.*

Proof : The proof follows along the same lines as that of Lemma 7.31. ∎

We have the following theorem that characterize the set $\Lambda^{h2-\text{oid}}_{\text{r-CSS}}$.

Theorem 10.164 *Consider a discrete-time system as in* (10.1). *Let Assumption 10.1 be satisfied. Consider reduced-order filters of CSS architecture as given in* (10.168), *and assume that the solvability conditions of the* H_2 *OID filtering problem (as in Theorem 10.48) are satisfied. Then, we have*

$$\Lambda^{h2-\text{oid}}_{\text{r-CSS}}(A, B, C, D, E, F) = \Lambda^{h2-\text{oid}}_{\text{sp-CSS}}(A_r, B_r, C_r, D_r, E_r^*, 0).$$

Proof : The proof follows in view of Lemmas 10.150 and 10.163. ∎

As in the case of strictly proper and proper filters, it turns out that, under the general assumption of the pair (C, A) being detectable, the set $\Lambda^{h2-\text{oid}}_{\text{r-CSS}}$ does not have any telescopic property. However, as before, whenever the pair (C, A) is observable, it does have such a property as the following lemma shows.

Lemma 10.165 *Consider a discrete-time system as in* (10.1) *with the pair* (C, A) *observable. Consider two different* H_2 *OID filtering problems both as defined in*

Definition 10.6, however, with one having the matrix pair (E_a, F_a) and the other (E_b, F_b) for the pair (E, F). Then, we have the telescopic property:

$$\mathrm{ker}\begin{pmatrix} E_a & F_a \end{pmatrix} \supseteq \mathrm{ker}\begin{pmatrix} E_b & F_b \end{pmatrix}$$
$$\Longrightarrow \Lambda_{\mathrm{r\text{-}CSS}}^{h2-\mathrm{oid}}(A, B, C, D, E_a, F_a) \subseteq \Lambda_{\mathrm{r\text{-}CSS}}^{h2-\mathrm{oid}}(A, B, C, D, E_b, F_b),$$

where we assume that the matrix pairs (E_a, F_a) and (E_b, F_b) are such that, for the respective underlying system, the H_2 OID filtering problem is solvable (i.e., the conditions of Theorem 10.48 are satisfied).

Proof : The proof follows from Lemmas 10.150 and 10.128 and Theorem 10.164.
■

Finally, the following theorem shows that we can design a filter gain K_r such that $\Sigma_{\mathrm{r\text{-}CSS}}$ given in (10.168) is a reduced-order H_2 OID filter while its poles are at the prescribed desired locations except that they need to contain $\boldsymbol{\Omega}_{\mathrm{r\text{-}CSS}}^{h2-\mathrm{oid}}(A, B, C, D, E, F)$ among them.

Theorem 10.166 (*H_2 OID reduced-order proper filter with pole placement*) *Consider a discrete-time system as in (10.1). Let the conditions of Theorem 10.48 be satisfied. Also, let $E^* = E - P^* C$ and $F^* = F - P^* D$, where P^* minimizes the expression in (10.43). Consider the set of H_2 OID filter fixed modes $\boldsymbol{\Omega}_{\mathrm{r\text{-}CSS}}^{h2-\mathrm{oid}}$ discussed in Theorem 10.158. Also, consider the nonempty set $\boldsymbol{K}_{\mathrm{r\text{-}CSS}}^{h2-\mathrm{oid}}$ discussed in Theorem 10.152. Moreover, let Λ_r be a prescribed set of n_r self-conjugate elements inside the unit disk $\mathbb{C}^{\ominus}$ such that Λ_r includes*

$$\boldsymbol{\Omega}_{\mathrm{r\text{-}CSS}}^{h2-\mathrm{oid}}(A, B, C, D, E, F).$$

Then, a filter gain

$$K_r \in \boldsymbol{K}_{\mathrm{r\text{-}CSS}}^{h2-\mathrm{oid}}(A, B, C, D, E, F)$$

exists such that the proper filter (10.168) is an H_2 OID filter, and moreover, $\lambda(A_r - K_r C_r) = \Lambda_r$.

Proof : The proof follows from Lemma 10.150 and Theorem 10.132.   ■

An algorithm for designing a reduced-order proper H_2 OID filter with simultaneous filter pole placement:

Step 1: Given the system Σ as in (10.145) and characterized by the sextuple (A, B, C, D, E, F), form the auxiliary system $\widetilde{\Sigma}_r^*$ characterized by the quintuple (A_r, B_r, C_r, D_r, E_r^*) where these matrices are defined in (10.164).

Step 2: Choose a desired set Λ_r of n_r self-conjugate elements in the open left-half complex plane $\mathbb{C}^\ominus$. [Step 3 given below checks to make sure that Λ_r includes $\Omega_{\text{r-CSS}}^{h2-\text{oid}}(A, B, C, D, E, F)$].

Step 3: Following the algorithm for designing a strictly proper H_2 OID filter with simultaneous filter pole placement as given in Subsection 10.14.1, and using the quintuple (A_r, B_r, C_r, D_r, E_r^*) as well as the matrix Λ_r as inputs to that algorithm, we obtain the gain K_r.

With the filter gain K_r as computed above, it is easy to verify that the resulting filter $\Sigma_{\text{r-CSS}}$ given in (10.168) is an H_2 OID filter with its modes at the locations specified by Λ_r; i.e., $\lambda(A_r - K_r C_r) = \Lambda_r$.

10.15   Design of H_2 SOID filters—discrete-time case

For discrete-time systems, we present in this section explicit algorithms of designing families of H_2 SOID filters with a capability to assign poles while honoring certain conditions imposed by H_2 suboptimality by connecting the given design objectives to those of H_2 AID filtering design. In this regard, as in previous chapters, we first observe that typically any filter design is initiated by first assuming a fixed architecture to the filter. The architecture we use for the filters is the CSS architecture that was developed and used in Sections 10.12, 10.13, and 10.14.

In view of Corollaries 10.37 and 10.38, the design of H_2 SOID filters for the given system Σ as in (10.1) can indeed be transformed to the design of AID filters for the auxiliary system Σ_Q given in (10.41). As we have already discussed the design of H_2 AID filters for any given system in Chapter 8, the design of H_2 SOID filters follows from that chapter. Here we briefly illuminate some issues in the translation from H_2 SOID filters to H_2 AID filters.

In the following three subsections, we consider the design of full-order strictly proper, full-order proper, and reduced-order proper H_2 SOID filters one at a time.

10.15.1   *Strictly proper H_2 SOID filters of CSS architecture*

We pursue here the design of strictly proper H_2 SOID filters. The architecture we use for the filter is the full-order CSS architecture. It is given by (8.16) and depicted in Figure 8.2. It is reproduced here as

$$\Sigma_{\text{sp-CSS}}^{\varepsilon} : \begin{cases} \sigma\xi = A\xi + K^\varepsilon(y - C\xi) \\ \hat{z} = E\xi, \end{cases} \tag{10.178}$$

where the matrix K^ε is parameterized in ε.

Error Dynamics: Let us define the error $e = x - \xi$; then the error between the actual desired output z and the estimated desired output $\hat{z}$ is $e_z = E(x-\xi) = Ee$. Also, the dynamics of error is given by

$$\Sigma_{\text{sp-CSS}}^{ue} : \begin{cases} \sigma e = (A - K^\varepsilon C)e + (B - K^\varepsilon D)u \\ e_z = Ee + Fu. \end{cases} \qquad (10.179)$$

The transfer matrix $G_{\text{sp-CSS}}^{ue}$ from u to e_z can obviously be written as

$$G_{\text{sp-CSS}}^{ue}(z) = E(zI - A + K^\varepsilon C)^{-1}(B - K^\varepsilon D) + F. \qquad (10.180)$$

We observe that, unlike in the continuous-time case, strictly proper filters can be used for discrete-time systems even if $F \neq 0$.

Remark 10.167 *In view of* (10.178) *and* (10.179), *it is easy to see that both the filter equation and the error equation have the same modes, which are the eigenvalues of* $A - K^\varepsilon C$.

We observe that the only unknown in the filter equation (10.178) and consequently in the error equation (10.179) is the family of matrices K^ε parameterized by ε, which is normally referred to as a family of filter gains. We need to determine a family of filter gains in such a way that the H_2 norm of $G_{\text{sp-CSS}}^{ue}$ converges to its infimum. Before we do so, we pause to emphasize an important aspect. Theorem 10.52 developed earlier shows that a sequence of H_2 SOID filters always exists among the general class of strictly proper filters of the form (10.48) with $P^\varepsilon = 0$. In other words, Theorem 10.52 does not restrict itself to any fixed architecture for a filter such as the one $\Sigma_{\text{sp-CSS}}$ given in (10.178). Nevertheless, as the following theorem shows, whenever the conditions of Theorem 10.52 are satisfied, we can determine a family of filter gains such that the family of filters $\Sigma_{\text{sp-CSS}}$ is indeed a family of H_2 SOID filters, and such a family of filter gains can be designed via designing a family of H_2 AID filters for the system Σ_Q^{sp} defined in (10.46).

Theorem 10.168 *Consider a discrete-time system Σ as in* (10.1), *and let Assumption 10.1 be satisfied. Also, let the matrices B_Q and D_Q be as defined by* (10.39), *and consider the system Σ_Q^{sp} defined in* (10.46). *Then, the following hold:*

(i) *A family of strictly proper H_2 SOID filters of CSS architecture $\Sigma_{\text{sp-CSS}}^\varepsilon$ for the given system Σ exists; i.e., a sequence of parameterized gains K^ε exists such that the family of filters $\Sigma_{\text{sp-CSS}}^\varepsilon$ given in* (10.178) *is a family of H_2 SOID filters for the given system Σ.*

(ii) *A family of strictly proper filters of CSS architecture $\Sigma_{\text{sp-CSS}}^\varepsilon$ as given in* (10.178) *is an H_2 SOID family of filters for the given system Σ if and only if this family of filters is a family of H_2 AID filters for the system Σ_Q^{sp} of* (10.46).

Proof : This is a direct consequence of Corollary 10.37.   ∎

Note that in Chapter 8, we explicitly characterized the finite asymptotic fixed modes. This captures for a large part the freedom we have in choosing the asymptotic behavior of the modes of the AID H_2 filters as given by the eigenvalues of $A - K^\varepsilon C$. The above theorem immediately relates this to the freedom in the asymptotic behavior of the modes of H_2 SOID filters. To be explicit, we have the following formal definition of finite asymptotic fixed modes.

Definition 10.169 (*Finite asymptotic fixed modes of strictly proper H_2 SOID filters of CSS architecture*) *Consider the given system (10.1) and the associated H_2 SOID filtering problem 10.6. Assume that the solvability conditions as specified by Theorem 10.52 are satisfied. Then, a finite scalar $\lambda \in \mathbb{C}^\ominus$ is said to be an **H_2 SOID filtering finite asymptotic fixed mode** with algebraic multiplicity α if for every family of H_2 SOID filters parameterized in ε and having the strictly proper full-order CSS architecture, poles λ_i^ε, $i = 1, 2, \cdots, \alpha$, of the family of filters exist such that λ_i^ε tends to λ as ε tends to zero. The set of all H_2 SOID filtering finite asymptotic fixed modes when strictly proper filters of full-order CSS architecture are used is denoted by $\Omega_{\text{sp-CSS}}^{h2-soid}(A, B, C, D, E, F)$.*

We can use the results of Theorem 8.21 of Chapter 8 to compute

$$\Omega_{\text{sp-CSS}}^{h2-soid}(A, B, C, D, E, F) = \Omega_{\text{sp-CSS}}^{h2-aid}(A, B_\varrho, C, D_\varrho, E, 0). \qquad (10.181)$$

10.15.2   Proper H_2 SOID filters of CSS architecture

We pursue here the design of proper H_2 SOID filters while using filters of CSS architecture. In this regard, Corollary 10.38 lays a roadmap to our development here. To start with, for ease of presentation and without loss of generality, we decompose the measured output y into two parts y_0 and y_1 in such a way that y_0 contains explicitly the unknown input u in it, whereas y_1 does not contain any input u in it. That is, we write

$$y = \begin{pmatrix} y_0 \\ y_1 \end{pmatrix} = Cx + Du \quad \text{and} \quad C = \begin{pmatrix} C_0 \\ C_1 \end{pmatrix}, \quad D = \begin{pmatrix} D_0 \\ 0 \end{pmatrix}, \qquad (10.182)$$

where the matrix D_0 has rank m_0. We can then rewrite the given system equation (10.1) as

$$\Sigma : \begin{cases} \sigma x & = Ax + Bu \\ \begin{pmatrix} y_0 \\ y_1 \end{pmatrix} = \begin{pmatrix} C_0 \\ C_1 \end{pmatrix} x + \begin{pmatrix} D_0 \\ 0 \end{pmatrix} u \\ z & = Ex + Fu. \end{cases} \qquad (10.183)$$

Proper filter design is based on the design of strictly proper filter design for an auxiliary system $\widetilde{\Sigma}^*$ defined by

$$\widetilde{\Sigma}^* : \begin{cases} \sigma x = Ax + Bu \\[2mm] \tilde{y} = \begin{pmatrix} y_0 \\ \sigma y_1 \end{pmatrix} = \tilde{C}x + \tilde{D}u \\[2mm] z^* = E^* x, \end{cases}$$

(10.184)

where $E^* = E - P^* C$, with matrix P^* being the solution of the equation:

$$EQC' + FD' = P(CQC' + DD')$$

(10.185)

and

$$\tilde{C} = \begin{pmatrix} C_0 \\ C_1 A \end{pmatrix}, \quad \tilde{D} = \begin{pmatrix} D_0 \\ C_1 B \end{pmatrix}.$$

(10.186)

The essence of our design philosophy is to translate the design of a family of proper H_2 SOID filters for the given system Σ to the design of a family of strictly proper H_2 SOID filters for the auxiliary system $\widetilde{\Sigma}^*$. This is formalized below.

Lemma 10.170 *Consider discrete-time systems Σ and $\widetilde{\Sigma}^*$, respectively, as given in* (10.183) *and* (10.184)*. Then, the following two statements are equivalent:*

(i) A family of strictly proper H_2 SOID filters exists for the auxiliary system $\widetilde{\Sigma}^*$.

(ii) A family of proper H_2 SOID filters exists for the system Σ.

Moreover, there is a $1-1$ relationship between a family of strictly proper H_2 OID filters of CSS architecture for $\widetilde{\Sigma}^*$ and a family of proper H_2 SOID filters of CSS architecture for Σ, that is, one of these families filters can be constructed from the other.

Proof : The first part is obvious. The second part (i.e., the $1 - 1$ relationship) is described in the following development.   ■

We move on to consider a family of strictly proper filters of CSS architecture for $\widetilde{\Sigma}^*$ as

$$\widetilde{\Sigma}^\varepsilon_{\text{sp-CSS}} : \begin{cases} \sigma \tilde{\xi} = A\tilde{\xi} + K^\varepsilon(\tilde{y} - \tilde{C}\tilde{\xi}) \\[2mm] \hat{z}^* = E^* \tilde{\xi}, \end{cases}$$

(10.187)

where the matrix K^ε is a filter gain parameterized by ε.

We partition K^ε in conformity with the partitioning of $\tilde{y}$. That is,

$$K^\varepsilon = \begin{pmatrix} K_0^\varepsilon & K_2^\varepsilon \end{pmatrix}.$$

(10.188)

The above filter for $\widetilde{\Sigma}^*$ can be converted to a filter for Σ as was done previously for optimal H_2 OID filters in Subsection 10.14.2.

We obtain an estimate $\hat{z}$ of z as

$$\hat{z} = \hat{z}^* + P^* y = E^* \widetilde{\xi} + P^* y. \tag{10.189}$$

We note that the filter (10.187) is not directly implementable because $y_2 = \sigma y_1$ is not available as a measured variable. We eliminate the need for σy_1 by defining a new variable:

$$\xi = \widetilde{\xi} - K_2^\varepsilon y_1. \tag{10.190}$$

With the definition of ξ as in (10.190), we can rewrite the filter equation (10.187) as

$$\begin{cases} \sigma\xi = (A - K\widetilde{C})\xi + \begin{pmatrix} K_0^\varepsilon & (A - K^\varepsilon\widetilde{C})K_2^\varepsilon \end{pmatrix} y \\ \widetilde{\xi} = \xi + K_2^\varepsilon y_1 \\ \hat{z}^* = E^*\widetilde{\xi} = E^*(\xi + K_2^\varepsilon y_1). \end{cases} \tag{10.191}$$

Obviously, the filter given above does not use σy_1. Moreover, it is proper rather than strictly proper. The filter given in (10.191) is indeed the proper full-order CSS filter that is to be used for Σ^*.

A family of proper filters of CSS architecture for Σ.

Finally, in view of (10.189), we can rewrite (10.191) as an implementable proper filter for Σ:

$$\Sigma_{\text{p-CSS}}^\varepsilon : \begin{cases} \sigma\xi = (A - K^\varepsilon\widetilde{C})\xi + \widetilde{K}^\varepsilon y \\ \widetilde{\xi} = \xi + K_2^\varepsilon y_1 \\ \hat{z} = E^*\widetilde{\xi} + P^* y = E^*\xi + \widetilde{P}^\varepsilon y, \end{cases} \tag{10.192}$$

where

$$\widetilde{K}^\varepsilon = \begin{pmatrix} K_0^\varepsilon & (A - K^\varepsilon\widetilde{C})K_2^\varepsilon \end{pmatrix} \quad \text{and} \quad \widetilde{P}^\varepsilon = \begin{pmatrix} 0 & E^*K_2^\varepsilon \end{pmatrix} + P^*.$$

We have the following theorem.

Theorem 10.171 *Consider the discrete-time system Σ given in (10.1), and let Assumption 10.1 be satisfied. Define $\widetilde{\Sigma}^*$ as in (10.184). Then, the following hold:*

(i) *A family of strictly proper H_2 SOID filters of CSS architecture $\widetilde{\Sigma}_{\text{sp-CSS}}^\varepsilon$ for the system $\widetilde{\Sigma}^*$ exists; i.e., a sequence of parameterized gains K^ε exists such that the family of filters $\widetilde{\Sigma}_{\text{sp-CSS}}^\varepsilon$ given in (10.187) is a family of H_2 SOID filters for $\widetilde{\Sigma}^*$.*

(ii) *Any family of strictly proper H_2 SOID filters of CSS architecture $\widetilde{\Sigma}_{\text{sp-CSS}}^\varepsilon$ as given in (10.187) for the auxiliary system $\widetilde{\Sigma}^*$ results in a family of proper H_2 SOID filters of CSS architecture $\Sigma_{\text{p-CSS}}^\varepsilon$ as given in (10.192) for Σ.*

Proof: Before we prove the two parts of the theorem, we first derive a preliminary result. We define the following system:

$$\Sigma^* : \begin{cases} \sigma x & = Ax + Bu \\ \begin{pmatrix} y_0 \\ y_1 \end{pmatrix} = \begin{pmatrix} C_0 \\ C_1 \end{pmatrix} x + \begin{pmatrix} D_0 \\ 0 \end{pmatrix} u \\ z & = E^* x. \end{cases}$$

Let $\gamma_{sp}^*(\Sigma^*)$ and $\gamma_p^*(\Sigma)$, respectively, denote γ_{sp}^* pertaining to Σ^* and γ_p^* pertaining to Σ. We have

$$\gamma_p^*(\Sigma)^2 = \gamma_{sp}^*(\Sigma^*)^2 + \|F^*\|^2,$$

and there is a $1-1$ relationship between strictly proper H_2 SOID filters for Σ^* and proper H_2 SOID filters for Σ. Next we note that the H_2 OID performance $\gamma_{sp}^*(\widetilde{\Sigma}^*)$ pertaining to $\widetilde{\Sigma}^*$ while using strictly proper filters can only be better than the performance using strictly proper filters for the system Σ^*. This follows from the fact that $y_1(k)$ is no longer part of the measurement $\widetilde{y}(k)$ but is available because the second component of $\widetilde{y}(k-1)$ equals $y_1(k)$. The only difference is that it is available earlier, which can only improve performance. In conclusion, we find that

$$\gamma_{sp}^*(\widetilde{\Sigma}^*) \leqslant \gamma_{sp}^*(\Sigma^*).$$

To establish part (i) of the theorem, we need to show that a sequence of parameterized gains K^ε exists such that the family of filters $\widetilde{\Sigma}_{\text{sp-CSS}}^\varepsilon$ given in (10.187) is a family of H_2 SOID filters for $\widetilde{\Sigma}^*$. This follows easily from Theorem 10.168.

To establish part (ii), we first note that the family of strictly proper H_2 SOID filters $\widetilde{\Sigma}_{\text{sp-CSS}}^\varepsilon$ of CSS architecture as given in (10.187) yields a strictly proper closed-loop dynamics from u to $z^* - \widehat{z}^*$ with transfer matrix H_ε^*. Next, we consider the family of proper filters of the form (10.192) constructed from the family of strictly proper filters with a suitable gain K^ε. Such a family of filters when applied to the system Σ results in proper closed-loop dynamics from u to $z - \widehat{z}$ with transfer matrix H_ε. It can be verified that

$$H_\varepsilon = H_\varepsilon^* + F^*,$$

and because H_ε^* is strictly proper, this implies that

$$\|H_\varepsilon\|_2^2 = \|H_\varepsilon^*\|_2^2 + \|F^*\|^2.$$

As $\varepsilon \to 0$, the right-hand side of the above equation goes to $\gamma_{sp}^*(\widetilde{\Sigma}^*)^2 + \|F^*\|^2$. Then, in view of the preliminary result, we get

$$\gamma_{sp}^*(\widetilde{\Sigma}^*)^2 + \|F^*\|^2 \leqslant \gamma_p^*(\Sigma)^2.$$

As $\gamma_p^*(\Sigma)$ was defined as the H_2 optimal performance via proper filters for the system Σ, this implies that the family of filters of the form (10.192) applied to the system Σ is a family of proper H_2 SOID filters for the system Σ. ∎

We have converted the design of proper H_2 SOID filters for Σ to the design of strictly proper H_2 SOID filters for the auxiliary system $\widetilde{\Sigma}^*$. We can then use the tools of the previous subsection to obtain the appropriate filters and the associated finite asymptotic fixed modes. To be explicit, we have the following formal definition of finite asymptotic fixed modes when proper filters are used.

Definition 10.172 (*Finite asymptotic fixed modes of proper H_2 SOID filters of CSS architecture*) *Consider the given system* (10.1) *and the associated H_2 SOID filtering problem 10.6. Then, a finite scalar* $\lambda \in \mathbb{C}^{\ominus}$ *is said to be an **H_2 SOID filtering finite asymptotic fixed mode** with algebraic multiplicity* α *if for every family of H_2 SOID filters parameterized in* ε *and having the proper full-order CSS architecture, poles* λ_i^{ε}, $i = 1, 2, \cdots, \alpha$, *of the family of filters exist such that all* λ_i^{ε} *tend to* λ *as* ε *tends to zero. The set of all H_2 SOID filtering finite asymptotic fixed modes when proper filters of full-order CSS architecture are used is denoted by* $\boldsymbol{\Omega}_{\text{p-CSS}}^{h2-soid}(A, B, C, D, E, F)$.

We can easily conclude that

$$\boldsymbol{\Omega}_{\text{p-CSS}}^{h2-soid}(A, B, C, D, E, F) = \boldsymbol{\Omega}_{\text{sp-CSS}}^{h2-soid}(A, B, \widetilde{C}, \widetilde{D}, E - P^*C, 0).$$

We hasten to add that $\boldsymbol{\Omega}_{\text{sp-CSS}}^{h2-soid}(A, B, \widetilde{C}, \widetilde{D}, E - P^*C, 0)$ is the same whatever may be the matrix P^* that solves (10.185). The latter follows from (10.181), Theorem 8.21, and a detailed analysis of the characterization of $\boldsymbol{\Omega}_s^2$ obtained in Theorem 6.24.

10.15.3 Reduced-order H_2 SOID filters of CSS architecture

The two previous subsections respectively construct families of full-order strictly proper and proper filters of CSS architecture that solve the H_2 SOID filtering problem for discrete-time systems. By full-order filters, as usual, we mean filters having the same dynamic order as that of the given system. Our goal in this subsection is to develop families of reduced-order filters of CSS architecture having their dynamic order lower than that of the given system.

The procedure of developing reduced-order filters here follows mostly along the same lines as in Subsection 10.14.3, which pertains to H_2 OID filters. As before, in essence, our method of development transforms the construction of a reduced-order filter for a given system to that of a full-order filter for a certain reduced-order system.

We proceed now to construct an appropriate reduced-order system. To start with, let us rewrite the matrices C and D of (10.1) as

$$C = \begin{pmatrix} 0 & C_{02} \\ I_{p-m_0} & 0 \end{pmatrix}, \quad D = \begin{pmatrix} D_0 \\ 0 \end{pmatrix},$$

where again rank $D = \text{rank } D_0 = m_0$. This can always be done without any loss of generality by appropriate coordinate transformations. In view of the above partitioning of C and D, we can partition the given system Σ as

$$
\Sigma : \begin{cases}
\sigma \begin{pmatrix} x_1 \\ x_2 \end{pmatrix} = \begin{pmatrix} A_{11} & A_{12} \\ A_{21} & A_{22} \end{pmatrix} \begin{pmatrix} x_1 \\ x_2 \end{pmatrix} + \begin{pmatrix} B_{11} \\ B_{22} \end{pmatrix} u \\[2ex]
y = \begin{pmatrix} y_0 \\ y_1 \end{pmatrix} = \begin{pmatrix} 0 & C_{02} \\ I & 0 \end{pmatrix} \begin{pmatrix} x_1 \\ x_2 \end{pmatrix} + \begin{pmatrix} D_0 \\ 0 \end{pmatrix} u \\[2ex]
z = Ex + Fu,
\end{cases}
\tag{10.193}
$$

where different variables have obvious meanings.

Next, we use a preliminary injection of output y into the desired output z. To this end, we define a matrix P^*, which is a solution of (10.185). We define a new desired output z^* as

$$
z^* = z - P^*y = (E - P^*C)x + (F - P^*D)u = E^*x + F^*u, \tag{10.194}
$$

where $E^* = E - P^*C$ and $F^* = F - P^*D$. In view of (10.193) and (10.194), we can define a new system Σ^* as

$$
\Sigma^* : \begin{cases}
\begin{pmatrix} \sigma x_1 \\ \sigma x_2 \end{pmatrix} = \begin{pmatrix} A_{11} & A_{12} \\ A_{21} & A_{22} \end{pmatrix} \begin{pmatrix} x_1 \\ x_2 \end{pmatrix} + \begin{pmatrix} B_{11} \\ B_{22} \end{pmatrix} u \\[2ex]
y = \begin{pmatrix} y_0 \\ y_1 \end{pmatrix} = \begin{pmatrix} 0 & C_{02} \\ I & 0 \end{pmatrix} \begin{pmatrix} x_1 \\ x_2 \end{pmatrix} + \begin{pmatrix} D_0 \\ 0 \end{pmatrix} u \\[2ex]
z^* = E^*x = E_1^*x_1 + E_2^*x_2,
\end{cases}
\tag{10.195}
$$

with $E^* = \begin{pmatrix} E_1^* & E_2^* \end{pmatrix}$.

We note that the measured output y_1 is not contaminated by the input u, and hence, $x_1 = y_1$ is known exactly from the measurement y. Thus, all we need to do next is to estimate the state x_2. To proceed further, let us rewrite the state equation for x_1 in terms of the output y_1 and the state x_2 as

$$
\sigma y_1 = A_{11}y_1 + A_{12}x_2 + B_{11}u. \tag{10.196}
$$

The above equation can be rewritten as

$$
\sigma y_1 - A_{11}y_1 = A_{12}x_2 + B_{11}u.
$$

Treating σy_1 as known, we can define a new measurement variable y_r:

$$
y_r = \begin{pmatrix} y_0 \\ \sigma y_1 - A_{11}y_1 \end{pmatrix}.
$$

Although σy_1 is not directly available, as we did earlier in the case of full-order proper CSS filters, we can eliminate it from any filter that is constructed using y_r as a measured output. With this in mind, we form the auxiliary system:

$$\Sigma_r^* : \begin{cases} \sigma x_r = A_r x_r + B_r u + A_{21} y_1 \\ y_r \;\;= C_r x_r + D_r u \\ z_r^* \;\;= E_r^* x_r, \end{cases} \tag{10.197}$$

where $x_r = x_2$ and

$$A_r = A_{22}, \quad B_r = B_{22}, \quad C_r = \begin{pmatrix} C_{02} \\ A_{12} \end{pmatrix}, \quad D_r = \begin{pmatrix} D_0 \\ B_{11} \end{pmatrix}, \quad E_r^* = E_2^*. \tag{10.198}$$

We note that the dynamic order n_r of the above Σ_r^* is less than the dynamic order n of the given system Σ by a number equal to the dimension of $x_1 = y_1$.

Next, let us suppose we can construct a family of strictly proper H_2 SOID filters for the auxiliary system Σ_r^* in order to arrive at an estimate $\hat{z}_r^*$ of z_r^*. We then plan to obtain an estimate $\hat{z}$ of z as

$$\hat{z} = E_1^* x_1 + \hat{z}_r^* + P^* y. \tag{10.199}$$

This motivates us to construct a full-order strictly proper filter for the auxiliary system Σ_r^* to arrive at an estimate $\hat{z}_r^*$ of z_r^*. A family of full-order filters constructed as such for Σ_r^* is indeed a family of reduced-order filters for Σ. However, in view of (10.199), such a family of reduced-order filters for Σ is proper (rather than strictly proper) for Σ. Thus, the essence of our design philosophy is to translate the design of a family of reduced-order H_2 SOID filters for the given system Σ to the design of a family of strictly proper H_2 SOID filters for the reduced-order system Σ_r^*. This is formalized below.

Lemma 10.173 *Consider discrete-time systems Σ and Σ_r^* as given in* (10.1) *and* (10.197), *respectively. Then, the following two statements are equivalent:*

(i) A family of strictly proper H_2 SOID filters exists for the reduced-order system Σ_r^*.

(ii) A family of proper reduced-order H_2 SOID filters exists for the system Σ.

Moreover, there is a $1 - 1$ relationship between the family of strictly proper H_2 SOID filters of CSS architecture for Σ_r^* and the family of proper reduced-order H_2 SOID filters of CSS architecture for Σ; that is, one of these filters can be constructed from the other.

Proof : The first part is obvious. The second part (i.e., the $1 - 1$ relationship) is described in the following development. ∎

It is obvious from our previous results that one can construct a family of strictly proper H_2 SOID filters for Σ_r^*. Next, to construct the required family of filters for Σ, in the spirit of the above development, we first construct a family of strictly proper filters of CSS architecture for the reduced-order system Σ_r^* as

$$\tilde{\Sigma}^\varepsilon_{\text{sp-CSS}} : \begin{cases} \sigma\tilde{\xi}_r = A_r\tilde{\xi}_r + A_{21}y_1 + K_r^\varepsilon(y_r - C_r\tilde{\xi}_r) \\ \hat{z}_r^* = E_r^*\tilde{\xi}_r, \end{cases} \tag{10.200}$$

where K_r^ε is a parameterized reduced-order filter gain. To convert this to a filter for Σ, because σy_1 is not available, we need to modify the above family of filters. To this end, let us partition $K_r^\varepsilon = (K_{r0}^\varepsilon \quad K_{r1}^\varepsilon)$ so as to be compatible with the partitioning of y_r. Also, let

$$\xi_r = \tilde{\xi}_r - K_{r1}^\varepsilon y_1. \tag{10.201}$$

We can then easily rewrite the family of strictly proper filters (10.200) as a family of proper filters for Σ given by

$$\Sigma^\varepsilon_{\text{r-CSS}} : \begin{cases} \sigma\xi_r = (A_r - K_r^\varepsilon C_r)\xi_r + \tilde{K}_r^\varepsilon y \\ \tilde{\xi}_r = \xi_r + K_{r1}^\varepsilon y_1 \\ \hat{z} = E_1^* x_1 + E_r^* \tilde{\xi}_r + P^* y = E_r^* \xi_r + \tilde{P}_r y, \end{cases} \tag{10.202}$$

where

$$\tilde{K}_r^\varepsilon = \left(K_{r0}^\varepsilon \quad A_{21} - K_{r1}^\varepsilon A_{11} + (A_r - K_r^\varepsilon C_r)K_{r1}^\varepsilon \right)$$

and

$$\tilde{P}_r = \left(0 \quad E_1^* + E_r^* K_{r1}^\varepsilon \right) + P^*.$$

The following theorem formalizes the situation.

Theorem 10.174 *Consider the discrete-time system Σ given in (10.1), and let Assumption 10.1 be satisfied. Define the reduced-order system Σ_r^* as in (10.197). Then, the following hold:*

(i) *A family of strictly proper H_2 SOID filters of CSS architecture $\tilde{\Sigma}^\varepsilon_{\text{sp-CSS}}$ for the reduced-order system Σ_r^* exists; i.e., a sequence of parameterized gains K_r^ε exists such that the family of filters $\tilde{\Sigma}^\varepsilon_{\text{sp-CSS}}$ given in (10.200) is a family of H_2 SOID filters for the system Σ_r^*.*

(ii) *Any family of strictly proper H_2 SOID filters of CSS architecture $\tilde{\Sigma}^\varepsilon_{\text{sp-CSS}}$ as given in (10.132) for the reduced-order system Σ_r^* results in a family of reduced-order proper H_2 SOID filters of CSS architecture $\Sigma^\varepsilon_{\text{r-CSS}}$ as given in (10.202) for Σ.*

Proof : Before we prove the two parts of the theorem, we first derive a preliminary result. Consider Σ^* defined by (10.195). Let $\gamma_{sp}^*(\Sigma^*)$ and $\gamma_p^*(\Sigma)$, respectively, denote γ_{sp}^* pertaining to Σ^* and γ_p^* pertaining to Σ. We have

$$\gamma_p^*(\Sigma)^2 = \gamma_{sp}^*(\Sigma^*)^2 + \|F^*\|^2,$$

and there is a $1-1$ relationship between strictly proper H_2 SOID filters for Σ^* and proper H_2 SOID filters for Σ. Next we note that the H_2 OID performance $\gamma_{sp}^*(\Sigma_r^*)$ pertaining to Σ_r^* while using strictly proper filters can only be better than the performance using strictly proper filters for the system Σ^*. This follows from the fact that the only part we eliminated is the known part of the state. In conclusion, we find that

$$\gamma_{sp}^*(\Sigma_r^*) \leqslant \gamma_{sp}^*(\Sigma^*).$$

To establish part (i) of the theorem, we need to show that a sequence of parameterized gains K_r^ε exists such that the family of filters $\widetilde{\Sigma}_{\text{sp-CSS}}^\varepsilon$ given in (10.200) is a family of H_2 SOID filters for $\widetilde{\Sigma}^*$. This follows easily from Theorem 10.168.

To establish part (ii), we first note that the family of strictly proper H_2 SOID filters $\widetilde{\Sigma}_{\text{sp-CSS}}^\varepsilon$ of CSS architecture as given in (10.200) yields a strictly proper closed-loop dynamics from u to $z_r^* - \hat{z}_r^*$ with transfer matrix H_ε^*. Next, we consider the family of reduced-order filters of the form (10.202) constructed from the family of strictly proper filters with a suitable gain K_r^ε. Such a family of filters when applied to the system Σ results in proper closed-loop dynamics from u to $z - \hat{z}$ with transfer matrix H_ε. It can be verified that

$$H_\varepsilon = H_\varepsilon^* + F^*,$$

and because H_ε^* is strictly proper, this implies that

$$\|H_\varepsilon\|_2^2 = \|H_\varepsilon^*\|_2^2 + \|F^*\|^2.$$

As $\varepsilon \to 0$, the right-hand side of the above equation goes to $\gamma_{sp}^*(\Sigma_r^*)^2 + \|F^*\|^2$. Then, in view of the preliminary result, we get

$$\gamma_{sp}^*(\Sigma_r^*)^2 + \|F^*\|^2 \leqslant \gamma_p^*(\Sigma)^2.$$

As $\gamma_p^*(\Sigma)$ was defined as the H_2 optimal performance via proper filters for the system Σ, this implies that the family of filters of the form (10.202) applied to the system Σ is a family of proper H_2 SOID filters for the system Σ.   ∎

We have converted the design of reduced-order H_2 SOID filters for Σ to the design of strictly proper H_2 SOID filters for the auxiliary system Σ_r^*. We can then use the tools of Subsection 10.15.1 to obtain the appropriate filters and the associated finite asymptotic fixed modes. To be explicit, we have the following formal definition of finite asymptotic fixed modes when reduced-order filters are used.

Definition 10.175 (*Finite asymptotic fixed modes of reduced-order H_2 SOID filters of CSS architecture*) *Consider the given system* (10.1) *and the associated H_2 SOID filtering problem 10.6. Then, a finite scalar $\lambda \in \mathbb{C}^\ominus$ is said to be an H_2 SOID filtering finite asymptotic fixed mode with algebraic multiplicity α if for every family of H_2 SOID filters parameterized in ε and having the reduced-order CSS architecture, poles λ_i^ε, $i = 1, 2, \cdots, \alpha$, of the family of filters exist such that all λ_i^ε tend to λ as ε tends to zero. The set of all H_2 SOID filtering finite asymptotic fixed modes when reduced-order filters of CSS architecture are used is denoted by $\Omega_{\text{r-CSS}}^{h2-soid}(A, B, C, D, E, F)$.*

It is easy to conclude that

$$\Omega_{\text{r-CSS}}^{h2-soid}(A, B, C, D, E, F) = \Omega_{\text{sp-CSS}}^{h2-soid}(A_r, B_r, C_r, D_r, E_r^*, 0).$$

We hasten to add that $\Omega_{\text{sp-CSS}}^{h2-soid}(A_r, B_r, C_r, D_r, E_r^*, 0)$ is the same whatever may be the matrix P^* that solves (10.185). The latter follows from (10.181), Theorem 8.21, and a detailed analysis of the characterization of Ω_s^2 obtained in Theorem 6.24.

10.16 Fixed modes of H_2 OID filters with arbitrary architecture

As we discussed earlier, the set of fixed modes obviously presents a constraint that one must contend with in any H_2 OID filter design. In this regard, it pays to study fixed modes of H_2 OID filters of different types with a hope of finding a desirable set among them. However, when filters of CSS architecture are used and when conditions of Theorem 10.26 or Theorem 10.47 for the existence of *strictly proper* H_2 OID filters are satisfied, Lemmas 10.81, 10.100 , 10.142, and 10.161 show that the sets of fixed modes are the same whether one uses full-order strictly proper or full order proper or even reduced-order proper filters. Furthermore, when filters of CSS architecture are used and when conditions of Theorem 10.27 or Theorem 10.48 for the existence of *proper H_2 OID filters* are satisfied, Lemmas 10.99 and 10.160 show that the sets of fixed modes are the same whether one uses full-order proper or reduced-order proper filters. A question that arises next is whether any advantages can be gained by using any arbitrary architecture other than CSS architecture for filters. That is, we ask ourselves whether filters of different architecture can result in different sets of H_2 OID filter fixed modes. In this regard, as in Chapter 7, it can be shown that any $\lambda \in \Omega_{\text{sp-CSS}}^{h2-oid}(A, B, C, D, E, F)$ [or $\lambda \in \Omega_{\text{p-CSS}}^{h2-oid}(A, B, C, D, E, F)$], which is neither an input decoupling zero nor an output decoupling zero of the subsystem characterized by the quadruple (A, B, C, D) is also the pole of any strictly proper (or proper) H_2 OID filter of any arbitrary architecture.

10.17   Performance measure for unbiasedness of filters with CSS architecture

As in the case of the EID filtering problem, the H_2 OID filtering problem 10.6 has two requirements. One is the unbiased requirement where, when the white noise input $u = 0$, the error signal needs to converge to zero asymptotically for all possible initial conditions of the given system and the filter. The second one is to minimize the RMS norm of the error signal due to the input u. Also, as in the case of EID filtering, both of these requirements are on the asymptotic behavior of the system. Note that, regarding the second requirement, the H_2 OID filtering performance measure, namely γ_{sp}^* or γ_p^*, measures simply asymptotic performance. Such a performance measure is blind to the transient behavior of the filter. A good way of bringing the transient performance into picture is to consider the performance associated with the unbiased requirement. In this case, the effect of the input u is an energy signal, and as such, one can formulate the performance measure due to the unbiasedness as the energy of the estimation error signal under the condition that u is zero.

To start with, consider the filtering block diagram of Figure 10.1 for the system Σ given in (10.1) and any fixed filter such as Σ_f given in (10.2). Whenever the unbiased requirement is satisfied by the filter Σ_f and when $u = 0$, we can define the unbiasedness performance measure J as follows: For continuous-time systems,

$$J(x_0, \xi_0, \Sigma_f) = \int_0^\infty e_z(t)' e_z(t)\, dt,$$

and for discrete-time systems,

$$J(x_0, \xi_0, \Sigma_f) = \sum_{i=0}^\infty e_z(i)' e_z(i).$$

Clearly, the performance measure due to the unbiasedness depends on the filter used. Our aim here is to study $J(x_0, \xi_0, \Sigma_f)$ when Σ_f has the CSS architecture, either strictly proper, proper, or reduced-order type.

10.17.1   Strictly proper filter of CSS architecture

Consider the system Σ given in (10.1) with $u = 0$; i.e., let

$$\begin{aligned}
\sigma x &= Ax, & x(0) &= x_0 \in \mathbb{R}^n, \\
y &= Cx, & & \\
z &= Ex.
\end{aligned} \tag{10.203}$$

Consider a strictly proper filter of CSS architecture [see (10.66) and (10.135)]:

$$\Sigma_{\text{sp-CSS}} : \left\{ \begin{aligned}
\sigma\xi &= (A - KC)\xi + Ky, & \xi(0) &= \xi_0 \in \mathbb{R}^n, \\
\hat{z} &= E\xi,
\end{aligned} \right. \tag{10.204}$$

where K is the filter gain. The error dynamics, in the absence of input signal u (as we assumed), is given by (10.67) or (10.136); i.e.,

$$\sigma e = (A - KC)e, \qquad e(0) = e_0 = x_0 - \xi_0 \in \mathbb{R}^n,$$
$$e_z = Ee. \tag{10.205}$$

We are interested in determining the minimal possible performance measure

$$J(x_0, \xi_0, \Sigma_{\text{sp-CSS}})$$

with respect to the gain K subject to the constraint (10.205). Then, the optimal performance measure, denoted by $J^*(x_0, \xi_0, \Sigma_{\text{sp-CSS}})$, shows the limitation associated with the unbiasedness requirement.

We proceed now to show that $J^*(x_0, \xi_0, \Sigma_{\text{sp-CSS}})$ is indeed related to the H_2 OID performance of an auxiliary system defined by the data of the original system Σ. Consider an auxiliary system:

$$\Sigma_{aux} : \begin{cases} \sigma x_{aux} = A x_{aux} + B_{aux} v \\ y_{aux} = C x_{aux} \\ z_{aux} = E x_{aux}, \end{cases} \tag{10.206}$$

where $B_{aux} = x_0 - \xi_0$ and v is an unknown white noise input. We have the following result whose proof can easily be written.

Lemma 10.176 *Consider the H_2 OID filtering problem defined in Problem 10.6 for a continuous- or discrete-time system Σ given in (10.1). Let Assumption 10.1 be satisfied. Also, consider the strictly proper filter $\Sigma_{\text{sp-CSS}}$ given in (10.204). Let Σ_{aux} be as in (10.206). Then the infimum of the performance measure due to the unbiasedness is given by*

$$J^*(x_0, \xi_0, \Sigma_{\text{sp-CSS}}) = (\gamma_{sp}^*)^2 \ \textit{pertaining to} \ \Sigma_{aux}.$$

We note that the initial condition x_0 of the given system is usually unknown, whereas one can set the initial condition ξ_0 of the filter as one likes. This suggests that one can generate an average performance measure for the unbiasedness requirement. Let $e_i, i = 1, \cdots, n$, form a basis for the state space. Also, assume that one always sets the initial condition ξ_0 of the filter to zero. Then we can define a new performance measure for the unbiasedness requirement while using a strictly proper filter of CSS architecture as

$$\tilde{J}(\Sigma, \Sigma_{\text{sp-CSS}}) = \sum_{i=1}^{n} J(e_i, 0, \Sigma_{\text{sp-CSS}}).$$

We can then define $\tilde{J}^*(\Sigma, \Sigma_{\text{sp-CSS}})$ as the infimum of $\tilde{J}(\Sigma, \Sigma_{\text{sp-CSS}})$ over all possible filter gains K. Once again it is straightforward to show that $\tilde{J}^*(\Sigma, \Sigma_{\text{sp-CSS}})$

is also related to the H_2 OID performance of an appropriately defined auxiliary system. Let

$$\widetilde{\Sigma}_{aux} : \begin{cases} \sigma \widetilde{x}_{aux} = A\widetilde{x}_{aux} + I v \\ \widetilde{y}_{aux} \;\;= C\widetilde{x}_{aux} \\ \widetilde{z}_{aux} \;\;= E\widetilde{x}_{aux}, \end{cases} \qquad (10.207)$$

where v is an unknown white noise input. We have the following result whose proof can easily be written.

Lemma 10.177 *Consider the H_2 OID filtering problem defined in Problem 10.6 for a continuous- or discrete-time system Σ given in (10.1). Let Assumption 10.1 be satisfied. Also, consider the strictly proper filter $\Sigma_{\text{sp-CSS}}$ given in (10.204). Let $\widetilde{\Sigma}_{aux}$ be as in (10.207). Then the infimum of the average performance measure due to the unbiasedness is given by*

$$\widetilde{J}^*(\Sigma, \Sigma_{\text{sp-CSS}}) = (\gamma^*_{sp})^2 \;\; \text{pertaining to} \;\; \widetilde{\Sigma}_{aux}.$$

Remark 10.178 *Following the results of Subsection 10.5.1 for continuous-time systems, we note that only the unstable zero dynamics and the non-left-invertible dynamics of the subsystem characterized by $(A, I, C, 0)$ contribute to the value of γ^*_{sp}. However, it is easy to see that the said subsystem does not have any zero dynamics, and moreover, it is left-invertible only if $\text{rank } C = n$. As almost always $\text{rank } C \neq n$, the said subsystem is almost always non-left-invertible. This implies that γ^*_{sp} pertaining to $\widetilde{\Sigma}_{aux}$ is almost always nonzero. In other words, the average performance measure due to the unbiasedness of a strictly proper filter of CSS architecture is almost always nonzero. Following similar analysis, one can deduce the same result for discrete-time systems.*

10.17.2 Proper filter of CSS architecture

In this subsection, we examine the performance measure due to the unbiasedness of a proper filter of CSS architecture. Consider the system Σ given in (10.1) with $u = 0$, i.e., the system given in (10.203). For such a system, in the case of continuous-time systems, consider a proper filter $\Sigma_{\text{p-CSS}}$ of CSS architecture as given in (10.85) in which the matrix $\widetilde{C}$ is as given in (10.74) and $E^* = E - P^*C$, with matrix P^* being a solution of $F - PD = 0$ for P. Similarly, for discrete-time systems, consider a proper filter $\Sigma_{\text{p-CSS}}$ of CSS architecture as given in (10.152) in which the matrix $\widetilde{C}$ is as given in (10.147) and $E^* = E - P^*C$, with matrix P^* minimizing the expression in (10.43). The error dynamics, in the absence of input signal u (as we assumed), is given by (10.86) with $u = 0$ for continuous-time systems or by (10.153) with $u = 0$ for discrete-time systems. We are interested in determining the minimal possible performance measure $J(x_0, \xi_0, \Sigma_{\text{p-CSS}})$ with respect to the gain K subject to the constraint (10.86) with $u = 0$ (continuous time) or (10.153) with $u = 0$ (discrete time). Then, the optimal performance measure, denoted by $J^*(x_0, \xi_0, \Sigma_{\text{p-CSS}})$, shows the limitation associated with the unbiasedness requirement.

We proceed now to show that $J^*(x_0, \xi_0, \Sigma_{\text{p-CSS}})$ is indeed related to the H_2 OID performance of an auxiliary system defined by the data of the original system Σ. Consider an auxiliary system:

$$\Sigma_{aux} : \begin{cases} \sigma x_{aux} = A x_{aux} + B_{aux} v \\ y_{aux} = \tilde{C} x_{aux} \\ z_{aux} = E^* x_{aux}, \end{cases} \tag{10.208}$$

where $B_{aux} = x_0 - \xi_0$ and v is an unknown white noise input. We have the following result whose proof can easily be written.

Lemma 10.179 *Consider the H_2 OID filtering problem defined in Problem 10.6 for a continuous- or discrete-time system Σ given in (10.1). Let Assumption 10.1 be satisfied. Also, consider the proper filter $\Sigma_{\text{p-CSS}}$ given in (10.85) for a continuous-time system and in (10.152) for a discrete-time system. Let Σ_{aux} be as in (10.208). Then the infimum of the performance measure due to the unbiasedness is given by*

$$J^*(x_0, \xi_0, \Sigma_{\text{p-CSS}}) = (\gamma_p^*)^2 \text{ pertaining to } \Sigma_{aux}.$$

We note that the initial condition x_0 of the given system is usually unknown, whereas one can set the initial condition ξ_0 of the filter as one likes. This suggests as before that one can generate an average performance measure for the unbiasedness requirement. Let $e_i, i = 1, \cdots, n$, form a basis for the state space. Also, assume that one always sets the initial condition ξ_0 of the filter to zero. Then we can define a new performance measure for the unbiasedness requirement while using a proper filter of CSS architecture as

$$\tilde{J}(\Sigma, \Sigma_{\text{p-CSS}}) = \sum_{i=1}^{n} J(e_i, 0, \Sigma_{\text{p-CSS}}).$$

We can then define $\tilde{J}^*(\Sigma, \Sigma_{\text{p-CSS}})$ as the infimum of $\tilde{J}(\Sigma, \Sigma_{\text{p-CSS}})$ over all possible filter gains. Once again it is straightforward to show that $\tilde{J}^*(\Sigma, \Sigma_{\text{p-CSS}})$ is also related to the H_2 OID performance of an appropriately defined auxiliary system. Let

$$\tilde{\Sigma}_{aux} : \begin{cases} \sigma \tilde{x}_{aux} = A \tilde{x}_{aux} + I v \\ \tilde{y}_{aux} = \tilde{C} \tilde{x}_{aux} \\ \tilde{z}_{aux} = E^* \tilde{x}_{aux}, \end{cases} \tag{10.209}$$

where v is an unknown white noise input. We have the following result whose proof can easily be written.

Lemma 10.180 *Consider the H_2 OID filtering problem defined in Problem 10.6 for a continuous- or discrete-time system Σ given in (10.1). Let Assumption 10.1 be satisfied. Also, consider the proper filter $\Sigma_{\text{p-CSS}}$ given in (10.85) for a continuous-time system and in (10.152) for a discrete-time system. Let $\widetilde{\Sigma}_{aux}$ be as in (10.209). Then the infimum of the average performance measure due to the unbiasedness is given by*

$$\widetilde{J}^*(\Sigma, \Sigma_{\text{p-CSS}}) = (\gamma_p^*)^2 \ \text{pertaining to} \ \widetilde{\Sigma}_{aux}.$$

Remark 10.181 *Following the lines of discussion given in Remark 10.178, we can deduce that the average performance measure due to the unbiasedness of a proper filter of CSS architecture is almost always nonzero.*

10.17.3 Reduced-order filter of CSS architecture

In this subsection, we examine the performance measure due to the unbiasedness of a reduced-order filter of CSS architecture. Consider the system Σ given in (10.1) with $u = 0$, i.e., the system given in (10.203). For such a system, in the case of continuous-time systems, consider a reduced-order filter $\Sigma_{\text{r-CSS}}$ of CSS architecture as given in (10.101) in which the matrices A_r, C_r, E_r^* are as given in (10.97) and $E^* = E - P^*C$, with matrix P^* being a solution of $F - PD = 0$ for P. Similarly, for discrete-time systems, consider a reduced-order filter $\Sigma_{\text{r-CSS}}$ of CSS architecture as given in (10.168) in which the matrices A_r, C_r, E_r^* are as given in (10.164) and $E^* = E - P^*C$, with matrix P^* minimizing the expression in (10.43). The error dynamics, in the absence of input signal u (as we assumed), is given by (10.102) with $u = 0$ for continuous-time systems or by (10.169) with $u = 0$ for discrete-time systems. The unbiased requirement demands that the steady-state estimation error $e_z = z - \hat{z}$ be zero. As before, we can define the unbiasedness performance measure J. For continuous-time systems, we have

$$J(x_{r0}, \xi_{r0}, \Sigma_{\text{r-CSS}}) = \int_0^\infty e_z(t)' e_z(t)\, dt,$$

whereas, for discrete-time systems, we have

$$J(x_{r0}, \xi_{r0}, \Sigma_{\text{r-CSS}}) = \sum_{i=0}^\infty e_z(i)' e_z(i).$$

In the above two equations, x_{r0} is the initial condition of the state x_r of the system given in (10.96) with $u = 0$ (continuous time) or in (10.163) with $u = 0$ (discrete time), whereas ξ_{r0} is the initial condition of the state ξ_r of the filter $\Sigma_{\text{r-CSS}}$ given in (10.101) (continuous time) or in (10.168) (discrete time). We are interested in the infimum of $J(x_{r0}, \xi_{r0}, \Sigma_{\text{r-CSS}})$ with respect to the filter gain K_r subject to the constraint (10.102) with $u = 0$ (continuous time) or (10.169) with $u = 0$ (discrete time). Such an infimum performance measure, denoted by $J^*(x_{r0}, \xi_{r0}, \Sigma_{\text{r-CSS}})$, shows the limitation associated with the unbiasedness requirement.

We proceed now to show that $J^*(x_{r0}, \xi_{r0}, \Sigma_{\text{r-CSS}})$ is indeed related to the H_2 OID performance of an auxiliary system defined by the data of the original system Σ. Consider an auxiliary system:

$$\Sigma_{aux} : \begin{cases} \sigma x_{aux} = A_r x_{aux} + B_{aux} v \\ y_{aux} = C_r x_{aux} \\ z_{aux} = E_r^* x_{aux}, \end{cases} \tag{10.210}$$

where $B_{aux} = x_{r0} - \xi_{r0}$ and v is an unknown white noise input. We have the following result whose proof can easily be written.

Lemma 10.182 *Consider the H_2 OID filtering problem defined in Problem 10.6 for a continuous- or discrete-time system Σ given in (10.1). Let Assumption 10.1 be satisfied. Also, consider the reduced-order filter $\Sigma_{\text{r-CSS}}$ given in (10.101) for a continuous-time system and in (10.168) for a discrete-time system. Let Σ_{aux} be as in (10.210). Then the infimum of performance measure due to the unbiasedness is given by*

$$J^*(x_{r0}, \xi_{r0}, \Sigma_{\text{r-CSS}}) = (\gamma_p^*)^2 \ \text{pertaining to} \ \Sigma_{aux}.$$

We note that the initial condition x_{r0} of the given system is usually unknown, whereas one can set the initial condition ξ_{r0} of the filter as one likes. This suggests as before that one can generate an average performance measure for the unbiasedness requirement. Let e_i, $i = 1, \cdots, n_r$, form a basis for the state space. Also, assume that one always sets the initial condition ξ_{r0} of the filter to zero. Then we can define a new performance measure for the unbiasedness requirement while using a reduced-order filter of CSS architecture as

$$\tilde{J}(\Sigma, \Sigma_{\text{r-CSS}}) = \sum_{i=1}^{n_r} J(e_i, 0, \Sigma_{\text{r-CSS}}).$$

We can then define $\tilde{J}^*(\Sigma, \Sigma_{\text{r-CSS}})$ as the infimum of $\tilde{J}(\Sigma, \Sigma_{\text{r-CSS}})$ over all possible filter gains. Once again it is straightforward to show that $\tilde{J}^*(\Sigma, \Sigma_{\text{r-CSS}})$ is also related to the H_2 OID performance of an appropriately defined auxiliary system. Let

$$\tilde{\Sigma}_{aux} : \begin{cases} \sigma \tilde{x}_{aux} = A_r \tilde{x}_{aux} + I v \\ \tilde{y}_{aux} = C_r \tilde{x}_{aux} \\ \tilde{z}_{aux} = E_r^* \tilde{x}_{aux}, \end{cases} \tag{10.211}$$

where v is an unknown white noise input. We have the following result whose proof can easily be written.

Lemma 10.183 *Consider the H_2 OID filtering problem defined in Problem 10.6 for a continuous- or discrete-time system Σ given in (10.1). Let Assumption 10.1 be satisfied. Also, consider the reduced-order filter $\Sigma_{\text{r-CSS}}$ given in (10.101) for a continuous-time system and in (10.168) for a discrete-time system. Let $\widetilde{\Sigma}_{aux}$ be as in (10.211). Then the infimum of the average performance measure due to the unbiasedness is given by*

$$\widetilde{J}^*(\Sigma, \Sigma_{\text{r-CSS}}) = (\gamma_p^*)^2 \ \ pertaining\ to\ \ \widetilde{\Sigma}_{aux}.$$

Remark 10.184 *Following the lines of discussion given in Remark 10.178, we can deduce that the average performance measure due to the unbiasedness of a reduced-order filter of CSS architecture is almost always nonzero.*

11
Optimally (suboptimally) input-decoupled filtering without statistical information on the input — H_∞ filtering

11.1   Introduction

We have been relaxing the performance requirements progressively and successively in previous chapters. Chapter 7 considers exact-input-decoupled (EID) filtering problems. When this is not possible, Chapters 8 and 9 consider almost-input-decoupled (AID) filtering problems, respectively, for the two cases, when the input is white noise and when no statistical information about the input is known. In Chapter 10, we look at optimally-input-decoupled (OID) filtering problems for the case when the input is white noise. As a natural continuation, this chapter looks at the optimally-input-decoupled (OID) filtering problems for the case when no information about the input is available except that it has a finite RMS value.

Recall that EID filtering seeks perfect performance; i.e., it tries to make the impact of the unknown input on the estimation error signal zero. On the other hand, AID filtering relaxes this requirement by testing whether we can make the impact of the unknown input on the error signal arbitrarily small. Next, OID filtering relaxes the requirement even further by trying to make the impact of the unknown input on the error signal to be as small as *possible* rather than arbitrarily small.

Here in this chapter, we continue with the theme of optimally-input-decoupled (OID) filtering and express the filtering performance by the smallest γ for which the RMS norm of the error signal for any input is always less than or equal to γ times the RMS norm of the input. The infimum of the filtering performance over the set of all linear stable strictly proper (or proper) unbiased filters is denoted by γ_{sp}^* (or γ_p^*).

As in the case of Chapter 10, one can traditionally seek two kinds of filters: optimal and suboptimal filters. Clearly, an optimal filter achieves the minimum possible performance, namely γ_{sp}^* or γ_p^*. In the case of a suboptimal filter, for a specified number $\gamma > \gamma_{sp}^*$ or $\gamma > \gamma_p^*$, one seeks a filter that achieves the RMS norm of the error signal for any input less than or equal to γ times the RMS norm of the input. Such a suboptimal filter can be termed as a γ-level suboptimal filter. One can of course seek in general to design an optimal filter or a γ-level suboptimal filter. However, essentially most of the available literature bypasses seeking optimal filters and focuses only on γ-level suboptimal filters. A primary reason to do so is that for general systems, no elegant analytic formula exists that enables one to compute the infimum performance measure γ_{sp}^* or γ_p^*. Only numerical approximations of computing γ_{sp}^* or γ_p^* exist, and this obviously implies that seeking γ-level suboptimal filters is very natural. Also, there is a secondary reason to seek such suboptimal filters, namely that the existence conditions for an optimal filter are prohibitively complex. Moreover, for engineering applications, often suboptimal filters suffice.

As usual, we first develop the solvability conditions for the problems posed. Under the assumption that the solvability conditions are satisfied, we then develop methods of designing filters that solve the posed problems. As before, we use both full- and reduced-order filters of CSS architecture for filter design.

11.2 Preliminaries

Let us reconsider the plant or system model given in (7.1) and rewritten here as

$$\Sigma : \begin{cases} \sigma x = Ax + Bu \\ y \;\; = Cx + Du \\ z \;\; = Ex + Fu, \end{cases} \tag{11.1}$$

where, as before, $u \in \mathbb{R}^m$ is the input. Unlike in Chapter 10, we assume here that no information about the input u is available except that it has a finite RMS value. Also, as before, $x \in \mathbb{R}^n$ is the state, $y \in \mathbb{R}^p$ is the measured output, and $z \in \mathbb{R}^q$ is the desired output signal to be estimated.

As usual, our interest lies in estimating the desired output signal z while using only the measured output y but not the input u. As usual, let $\hat{z}$ be the estimate of z as given by a filter, and let e_z be the estimation error, $e_z = z - \hat{z}$ as depicted in Figure 11.1.

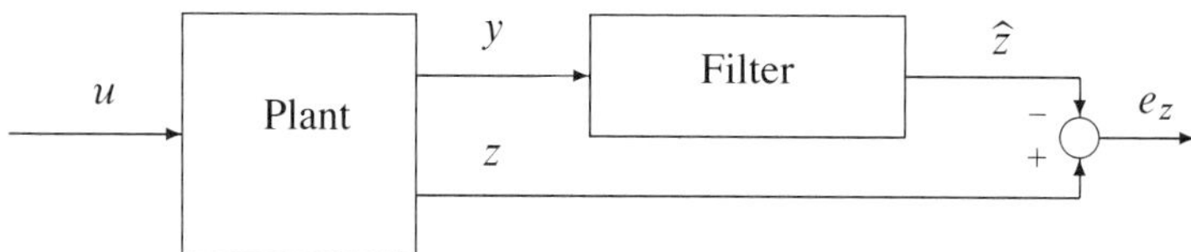

Figure 11.1: General block diagram

As before, we will make the following assumption throughout this chapter. Note that this condition is obviously necessary for the existence of an unbiased filter.

Assumption 11.1 *The matrix pair (C, A) is $\mathbb{C}^{-}$-detectable for continuous-time systems and $\mathbb{C}^{\ominus}$-detectable for discrete-time systems.*

Again, as before, we consider a general proper filter of the form (7.2), which is repeated below:

$$\Sigma_f : \begin{cases} \sigma \xi = L\xi + My \\ \hat{z} = N\xi + Py. \end{cases} \tag{11.2}$$

Whenever $P = 0$, the above filter is said to be a strictly proper filter. When the above filter is used, as shown in Figure 11.1, the dynamic equations of the error e_z are described by

$$\Sigma_{ue} : \begin{cases} \sigma x = Ax + Bu \\ \sigma \xi = MCx + L\xi + MDu \\ e_z = (E - PC)x - N\xi + (F - PD)u. \end{cases} \tag{11.3}$$

Hence, the transfer matrix from u to e_z can be computed as

$$G_{ue}(\sigma) = \begin{pmatrix} E - PC & -N \end{pmatrix} \begin{pmatrix} \sigma I - A & 0 \\ -MC & \sigma I - L \end{pmatrix}^{-1} \begin{pmatrix} B \\ MD \end{pmatrix} + F - PD. \tag{11.4}$$

11.3 OID and SOID filtering problems without statistical information on the input

Unlike in Chapter 10, as said earlier in this chapter, we do not assume any information about the input except that it has a finite RMS value. Obviously, as before, we only consider unbiased filters. Once again, as in Chapter 10, we start our development by recalling from Definition 7.2 what we mean by unbiased filters.

Definition 11.2 *Consider a continuous- or discrete-time system Σ as in (11.1). We say a linear stable strictly proper (or proper) filter (11.2) is **unbiased** if, in the absence of the input u, the estimation error e_z decays asymptotically to zero for all possible finite initial values of the system (11.1) and the filter (11.2).*

Let us begin by formally defining the performance criterion that we want to minimize in this chapter. For any linear unbiased filter Σ_f, we can define a performance measure associated with it as

$$J(\Sigma_f) = \sup_{\substack{u \\ \|u\|_{\mathrm{RMS}} \neq 0}} \frac{\|e_z\|_{\mathrm{RMS}}}{\|u\|_{\mathrm{RMS}}} = \inf\{\, \gamma \mid \|e_z\|_{\mathrm{RMS}} \leqslant \gamma \|u\|_{\mathrm{RMS}} \text{ for all inputs } u \,\}.$$

Note that the fact that the linear filter is unbiased guarantees that the performance level γ is well defined and we will refer to it as the filtering performance.

Definition 11.3 *Consider a continuous- or discrete-time system Σ of the form (11.1) without knowing any statistical information about the input. The infimum of the filtering performance $J(\Sigma_f)$ over the set of all linear stable strictly proper (or proper) unbiased filters is called the* **optimal input-decoupling (OID) filtering performance without any statistical information about the input** *via linear stable strictly proper (or proper) unbiased filters, and it is denoted by γ_{sp}^* (or γ_p^*).*

We are now ready to define the OID filtering problem when no information is available about the input u except that it has a finite RMS value.

Problem 11.4 Consider a continuous- or discrete-time system Σ given by (11.1). Assume that the input u has a finite RMS value. Then, the **OID filtering problem without any statistical information about the input** is defined as follows:

Find, whenever it exists, a linear strictly proper (or proper) filter Σ_f of the form (11.2) such that the following two conditions are satisfied:

(Unbiasedness) the estimation error e_z, in the absence of the input u, decays asymptotically to zero for all possible finite initial values of the system (11.1) and the filter (11.2), and

(Performance) the filtering performance $J(\Sigma_f)$ achieved by the filter Σ_f is as small as possible; i.e., $J(\Sigma_f) = \gamma_{sp}^*$ (or $J(\Sigma_f) = \gamma_p^*$).

Given Definition 11.3, the minimal achievable performance measure $J(\Sigma_f)$ by any filter Σ_f is γ_{sp}^* or γ_p^* depending on whether we use a strictly proper or proper filter Σ_f. The above problem tries to find a filter Σ_f that attains this infimum. However, depending on the given system Σ, one may not be able to attain such an infimum. In this case, as in Chapter 10, it is natural to resort to suboptimal filters. Also, it turns out that in the case of the OID filtering problem when no information is available about the input u as considered in this chapter, the infimum performance measures γ_{sp}^* or γ_p^* are not easily computable. This also necessitates seeking suboptimal filters. Moreover, in many engineering applications, suboptimal filters might suffice. In view of these reasons, most of the literature focuses on suboptimal filters when no information is available about the input u. As such, our development here focuses also on suboptimal filters. Suboptimal filtering is formally defined in the following problem.

Problem 11.5 Consider a continuous- or discrete-time system Σ given by (11.1). Assume that the input u has a finite RMS value. Then, the **γ-level SOID filtering problem without any statistical information about the input** is defined as follows:

Let $\gamma > \gamma_{sp}^*$ (or $\gamma > \gamma_p^*$) be fixed. Find, a linear strictly proper (or proper) filter Σ_f of the form (11.2) such that the following two conditions are satisfied:

(Unbiasedness) the estimation error e_z, in the absence of the input u, decays asymptotically to zero for all possible finite initial values of the system (11.1) and the filter (11.2), and

(Performance) the filtering performance $J(\Sigma_f)$ achieved by the filter is strictly smaller than γ.

Note that the above suboptimal filtering problem indeed is of a different formulation than in the previous chapters where we obtain a sequence of filters whose performance approaches γ_{sp}^* (or γ_p^*) instead of finding one filter whose performance is better than γ. The reason for formulating it in the above way is related to the design methodology as we will see later on. Clearly, finding a sequence of filters whose performance converges to γ_{sp}^* can be easily found based on the above problem formulation because we can solve the above problem for a sequence of γ values that converge to γ_{sp}^* (or γ_p^*).

As in Section 10.3, the above OID filtering problem as well as SOID filtering problem statements can be given a deterministic interpretation. Let G_{ue} denote the transfer matrix from the input u to the error e_z. Then, by the definitions and the discussion given in Section 2.6, we know that

$$\|G_{ue}\|_\infty = \sup_{\substack{u \\ \|u\|_{\mathrm{RMS}}\neq 0}} \frac{\|e_z\|_{\mathrm{RMS}}}{\|u\|_{\mathrm{RMS}}}.$$

This means that the OID filtering performance requirement in Problem 11.4 is that $\|G_{ue}\|_\infty$ associated with the filter Σ_f equal γ_{sp}^* or γ_p^*, and the SOID filtering performance requirement in Problem 11.5 is that $\|G_{ue}\|_\infty$ associated with the filter Σ_f satisfy the inequality

$$\|G_{ue}\|_\infty \leq \gamma.$$

This implies that Problems 11.4 and 11.5 can be interpreted in a deterministic setting as an H_∞ OID filtering problem and a γ-level H_∞ SOID filtering problem as described below.

Problem 11.6 Consider a continuous- or discrete-time system Σ given in (11.1). The objective of the $\boldsymbol{H_\infty}$ **OID filtering problem** is to find, whenever it exists, a stable strictly proper (or proper) filter Σ_f such that the following two conditions are satisfied:

(Unbiasedness) the estimation error e_z, in the absence of the input u, decays asymptotically to zero for all possible finite initial values of the system (11.1) and the filter (11.2), and

(Performance) the H_∞ norm of the resulting G_{ue} equals γ_{sp}^* (or γ_p^*).

A filter that achieves the above two objectives is called an $\boldsymbol{H_\infty}$ **OID filter**.

Problem 11.7 Consider the continuous- or discrete-time system Σ as given in (11.1). Let $\gamma > \gamma_{sp}^*$ (or $\gamma > \gamma_p^*$) be fixed. The objective of the **γ-level H_∞ SOID filtering problem** is to find a linear strictly proper (or proper) filter Σ_f of the form (11.2) such that the following two conditions are satisfied:

(Unbiasedness) the estimation error e_z, in the absence of the input u, decays asymptotically to zero for all possible finite initial values of the system (11.1) and the filter (11.2), and

(Performance) the H_∞ norm of the resulting G_{ue} is less than or equal to γ.

A filter that achieves the above two objectives is called a **γ-level H_∞ SOID filter**.

In view of the above definition, we can interpret γ_{sp}^* (or γ_p^*) as the infimum of the H_∞ norm of the transfer function G^{ue} from u to e_z over the set of all unbiased linear strictly proper (or proper) filters. In other words, γ_{sp}^* (or γ_p^*) can be called the **H_∞ OID filtering performance** via linear strictly proper (or proper) filters.

Remark 11.8 *Throughout this chapter and elsewhere, to avoid notational clutter, whenever it is not explicitly necessary, we judiciously omit the qualifier γ-level. Thus, the γ-level H_∞ SOID filtering problem is simply called the H_∞ SOID filtering problem.*

Remark 11.9 *Although we formally defined above the H_∞ OID filtering problem as well as the γ-level H_∞ SOID filtering problem, as we said earlier, our focus in this chapter is only on the γ-level H_∞ SOID filtering problem.*

We emphasize here that most of the literature also focuses primarily on the γ-level H_∞ SOID filtering problem. In fact, in many cases, such a problem is referred to in the literature as the H_∞ optimal filtering problem even though this is obviously a bad notation because this has nothing to do with optimality.

We note also that most of the literature in fact focuses only on a subclass of H_∞ SOID filtering problems known as *regular* problems. The level of complexity associated with such regular problems is far much less than that associated with general problems. The subclass of regular problems are defined below. If a filtering problem is not a *regular* problem, it is said to be a *singular* problem. We emphasize once again that, as will be seen in this chapter soon, regular problems are much easier to solve than the singular problems. Structurally, a huge level of complexity in singular problems exists both with respect to the existence conditions for a γ-level H_∞ SOID filter as well as with respect to the design of such a filter.

Definition 11.10 *Consider a continuous- or discrete-time system Σ as given in (11.1).*

(*i*) *For a continuous-time system Σ, a **γ-level H_∞ SOID regular filtering problem** refers to a filtering problem in which the matrix D is surjective, and the system characterized by the quadruple (A, B, C, D) has no invariant zeros on the imaginary axis.*

(*ii*) *For a discrete-time system Σ, a **γ-level H_∞ SOID regular filtering problem** refers to a filtering problem in which the subsystem characterized by the quadruple (A, B, C, D) is right-invertible and has no invariant zeros on the unit circle.*

*If a γ-level H_∞ SOID filtering problem is not regular, then it is said to be a **γ-level H_∞ SOID singular filtering problem**.*

11.4   Computation of γ_{sp}^* and γ_p^*

Clearly a γ-level H_∞ SOID filter exists for any given γ if and only if $\gamma > \gamma_{sp}^*$ (or $\gamma > \gamma_p^*$). Thus, determining the H_∞ OID filtering performance measures γ_{sp}^* and γ_p^* is essential. Our goal in this section is to compute them. As we said, for a general class of systems, we do not have explicit analytical expressions for γ_{sp}^* and γ_p^*, and thus, we can compute them only numerically. However, for a fairly large class of systems, we can develop explicit analytical expressions for γ_{sp}^* and γ_p^*. As such, we divide our development here into four subsections: (1) explicit computation of γ_{sp}^* and γ_p^* for a class of continuous-time systems, (2) numerical computation of γ_{sp}^* and γ_p^* for the general class of continuous-time systems, (3) explicit computation of γ_{sp}^* and γ_p^* for a class of discrete-time systems, and (4) numerical computation of γ_{sp}^* and γ_p^* for the general class of discrete-time systems. When we consider the numerical computation of γ_{sp}^* and γ_p^*, we first consider the regular H_∞ OID filtering problems and then the singular H_∞ OID filtering problems. Moreover, in the singular case, we consider first a subclass of singular problems where the given system does not have invariant zeros on the imaginary axis (continuous-time case) or on the unit circle (discrete-time case), and then we consider the case where the given system has invariant zeros on the imaginary axis (continuous-time case) or on the unit circle (discrete-time case).

11.4.1   Explicit computation of γ_{sp}^* and γ_p^*—continuous-time systems

In this subsection, as we said above, we consider the explicit computation of γ_{sp}^* and γ_p^* for a class of continuous-time systems. For the computation of γ_{sp}^*, the class of systems we consider here are of the form (11.1), where $F = 0$ and

$$\mathcal{V}^-(\Sigma_*) \cap \mathcal{S}^-(\Sigma_*) \subset \ker E. \tag{11.5}$$

Here Σ_* is the system characterized by the quadruple (A, B, C, D). The above geometric subspaces have been defined in Section 3.2.5. Similarly, for the computation of γ_p^*, the class of systems we consider here is again of the form (11.1) for which we assume that a matrix K exists such that $F - KD = 0$ and

$$\mathcal{V}^-(\Sigma_*) \cap \mathcal{S}^-(\Sigma_*) \subset \ker(E - KC). \tag{11.6}$$

It is interesting to note that if indeed a matrix K exists such that (11.6) is satisfied while $F - KD = 0$, then we can choose $K = FD'(DD')^\dagger D$.

For the computation of both γ_{sp}^* and γ_p^*, we transform the system Σ_* into its special coordinate basis (SCB) and obtain a decomposition as presented in (3.15)–(3.19). Then, for the strictly proper case, the condition (11.5) implies that

$$E\Gamma_s^{-1} = \begin{pmatrix} E_a^- & E_a^0 & E_a^+ & E_b & 0 & E_d \end{pmatrix}, \tag{11.7}$$

whereas for the proper case, the condition (11.6) implies that

$$(E - KC)\Gamma_s^{-1} = \begin{pmatrix} E_a^- & E_a^0 & E_a^+ & E_b & 0 & E_d \end{pmatrix}. \tag{11.8}$$

As it becomes apparent soon, the part of the dynamics of the given system that contributes to the nonzero values of γ_{sp}^* and γ_p^* is that of x_a^+ and x_c. Because of this, we look at the following system (after applying an output injection of y_0):

$$\Sigma_1 : \begin{cases} \begin{pmatrix} \dot{x}_a^+ \\ \dot{x}_c \end{pmatrix} = \begin{pmatrix} A_{aa}^+ & 0 \\ B_c E_{ca}^+ & A_{cc} \end{pmatrix} \begin{pmatrix} x_a^+ \\ x_c \end{pmatrix} + \begin{pmatrix} 0 & 0 & 0 \\ 0 & 0 & B_c \end{pmatrix} \begin{pmatrix} u_0 \\ u_d \\ u_c \end{pmatrix}, \\[2ex] \tilde{y} = \begin{pmatrix} C_{0a}^+ & C_{0c} \\ E_{da}^+ & E_{dc} \end{pmatrix} \begin{pmatrix} x_a^+ \\ x_c \end{pmatrix} + \begin{pmatrix} I & 0 & 0 \\ 0 & I & 0 \end{pmatrix} \begin{pmatrix} u_0 \\ u_d \\ u_c \end{pmatrix}, \\[2ex] \tilde{z} = \begin{pmatrix} E_a^+ & 0 \end{pmatrix} \begin{pmatrix} x_a^+ \\ x_c \end{pmatrix} + \begin{pmatrix} 0 & 0 & 0 \end{pmatrix} \begin{pmatrix} u_0 \\ u_d \\ u_c \end{pmatrix}. \end{cases}$$

Let

$$\tilde{A} = \begin{pmatrix} A_{aa}^+ & 0 \\ B_c E_{ca}^+ & A_{cc} \end{pmatrix}, \quad \tilde{B} = \begin{pmatrix} 0 & 0 & 0 \\ 0 & 0 & B_c \end{pmatrix},$$

$$\tilde{C} = \begin{pmatrix} C_{0a}^+ & C_{0c} \\ E_{da}^+ & E_{dc} \end{pmatrix}, \quad \tilde{E} = \begin{pmatrix} E_a^+ & 0 \end{pmatrix}.$$

We note that (C, A) being $\mathbb{C}^-$-detectable guarantees that $(\tilde{C}, \tilde{A})$ is $\mathbb{C}^-$-detectable. Also, we note that $(-\tilde{A}, \tilde{B})$ is stabilizable because the structure of the SCB guarantees that $-A_{aa}^+$ is Hurwitz-stable and (A_{cc}, B_{cc}) is controllable. This guarantees according to Theorem 4.40 that the following H_2 Continuous-time Algebraic Riccati Equation (H_2 CARE) associated with Σ_1 has a unique positive definite solution:

$$0 = \tilde{S}\tilde{A} + \tilde{A}'\tilde{S} + \tilde{S}\tilde{B}\tilde{B}'\tilde{S} - \tilde{C}'\tilde{C}. \tag{11.9}$$

It is easy to show that this will be an anti-stabilizing solution (eigenvalues in the open right-half plane). Next, we define $\widetilde{T}$ as the solution of the Lyapunov equation:

$$\widetilde{T}\widetilde{A} + \widetilde{A}'\widetilde{T} = \widetilde{E}'\widetilde{E}. \tag{11.10}$$

The above equation has a solution $\widetilde{T} \geqslant 0$ such that $\widetilde{T}\widetilde{B} = 0$. The following lemma will play a crucial role in the development that is to follow.

Lemma 11.11 *Let s_i be an invariant zero on the imaginary axis of the subsystem characterized by (A, B, C, D). Define*

$$\gamma_{0,i} = \inf_{K} \left\{ \|E(s_i I - A + KC)^{-1}(B - KD)\| \mid s_i I - A + KC \text{ is invertible} \right\}. \tag{11.11}$$

Then, we have

$$\gamma_{0,i} = \|L_{i,1}L_{i,2}^{+}\|, \tag{11.12}$$

where

$$L_{i,1} = E(s_i - A + K_1 C)^{-1}(B - K_1 D),$$
$$L_{i,2} = C(s_i - A + K_1 C)^{-1}(B - K_1 D) + D,$$

with K_1 an arbitrary matrix such that $A - K_1 C$ is Hurwitz-stable while $L_{i,2}^{+}$ is such that

$$(L_{i,2}^{+})'L_{i,2}^{+} = I, \quad \operatorname{im} L_{i,2}^{+} = \ker L_{i,2}.$$

Proof : Clearly

$$\gamma_{0,i} = \inf_{\widetilde{K}} \left\{ \|E(V_i + \widetilde{K}C)^{-1}(B_i - \widetilde{K}D)\| \mid V_i + \widetilde{K}C \text{ is invertible} \right\},$$

where

$$V_i = s_i - A + K_1 C, \qquad B_i = B - K_1 D$$

with V_i invertible. But then

$$E(V_i + \widetilde{K}C)^{-1}(B_i - \widetilde{K}D) = L_{i,1} - EV_i^{-1}\widetilde{K}(I + CV_i^{-1}\widetilde{K})^{-1}L_{i,2}.$$

It is relatively easy to check that we can equivalently minimize over

$$\overline{K} = V_i^{-1}\widetilde{K}(I + CV_i^{-1}\widetilde{K})^{-1}$$

without the constraint that $V_i + \widetilde{K}C$ must be invertible because there is a $1 - 1$ relationship between $\widetilde{K}$ and $\overline{K}$ except for a set of singular points that do not affect the infimum due to continuity. But then we get

$$\gamma_{0,i} = \inf_{\overline{K}} \|L_{i,1} - E\overline{K}L_{i,2}\|.$$

As im $L_{i,1} \subset$ im E, it is easily seen that we get

$$\gamma_{0,i} = \inf_{\widehat{K}} \| L_{i,1} - \widehat{K} L_{i,2} \|.$$

It is then a standard projection result to obtain (11.12).   ∎

We have the following result that pertains to strictly proper case.

Theorem 11.12 *Consider the continuous-time system* (11.1) *and the associated OID filtering performance via strictly proper filters, i.e., γ_{sp}^*. Assume that* (11.5) *is satisfied while $F = 0$. Let $\tilde{S}$ and $\tilde{T}$ be the solution of the H_2 CARE* (11.9) *and the Lyapunov equation* (11.10), *respectively. In this case, we define*

$$\gamma_{sp,1} = \sqrt{\lambda_{\max}(\tilde{T}\tilde{S}^{-1})}.$$

Then, the following hold:

  (i) *For the case when there are no invariant zeros of the system characterized by the quadruple (A, B, C, D) on the imaginary axis, we have $\gamma_{sp}^* = \gamma_{sp,1}$.*

  (ii) *Otherwise, let $s_1, \ldots, s_v$ be the invariant zeros on the imaginary axis of the subsystem characterized by (A, B, C, D) and we define $\gamma_{0,i}$ by* (11.11). *Then, we have*

$$\gamma_{sp}^* = \sup\{\gamma_{sp,1}, \gamma_{0,1}, \ldots, \gamma_{0,v}\}.$$

Proof : In this proof, we will make use of Theorem 11.22 (which will be presented and proven later). We focus first on the case when no invariant zeros on the imaginary axis exist for the system characterized by the quadruple (A, B, C, D). To check whether $\gamma > \gamma_{sp}^*$, we need to check for the existence of a positive semi-definite and semi-stabilizing solution Q of the Continuous-time Quadratic Matrix Inequality (CQMI) (11.20). Using Theorem 4.159 and Proposition 4.29, we note that Q is a positive semi-definite and semi-stabilizing solution of the CQMI (11.20) if and only if Q has the form (using the SCB):

$$Q = (\Gamma_s^{-1})' \begin{pmatrix} 0 & 0 & 0 & 0 & 0 & 0 \\ 0 & 0 & 0 & 0 & 0 & 0 \\ 0 & 0 & V_{11} & 0 & V_{12} & 0 \\ 0 & 0 & 0 & 0 & 0 & 0 \\ 0 & 0 & V_{21} & 0 & V_{22} & 0 \\ 0 & 0 & 0 & 0 & 0 & 0 \end{pmatrix} \Gamma_s^{-1}, \qquad (11.13)$$

where

$$\begin{pmatrix} V_{11} & V_{12} \\ V_{21} & V_{22} \end{pmatrix} = \widehat{Q},$$

and $\tilde{Q}$ is a positive semi-definite and stabilizing solution of the H_∞^1 Continuous-time Algebraic Riccati Equation (H_∞^1 CARE):

$$0 = \tilde{A}\tilde{Q} + \tilde{Q}\tilde{A}' + \tilde{B}\tilde{B}' - \tilde{Q}\tilde{C}'\tilde{C}\tilde{Q} + \gamma^{-2}\tilde{Q}\tilde{E}\tilde{E}\tilde{Q}. \tag{11.14}$$

Therefore, $\gamma > \gamma_{sp}^*$ if and only if a positive semi-definite stabilizing solution $\tilde{Q}$ exists of the H_∞^1 CARE (11.14).

Using the development given before Theorem 11.12, we define

$$\tilde{V} = \tilde{S} - \tfrac{1}{\gamma^2}\tilde{T}. \tag{11.15}$$

We recall that $\tilde{S} > 0$ and $\tilde{T} \geq 0$. As such we observe that $\tilde{V}$ is invertible for all but finitely many γ. Moreover, it is easily verified that $\tilde{V}$ is an antistabilizing solution of the H_∞^1 CARE:

$$0 = \tilde{V}\tilde{A} + \tilde{A}'\tilde{V} + \tilde{V}\tilde{B}\tilde{B}'\tilde{V} - \tilde{C}'\tilde{C} + \gamma^{-2}\tilde{E}\tilde{E}.$$

But then, for all but finitely many γ, the matrix $\tilde{Q} = \tilde{V}^{-1}$ will be a stabilizing solution of the H_∞^1 CARE (11.14). Note that the stabilizing solution of an H_∞^1 CARE is unique. But to guarantee $\gamma > \gamma_{sp}^*$, we need $\tilde{Q} \geq 0$.

Hence, for all but finitely many γ, we have that $\gamma > \gamma_{sp}^*$ if and only if $\tilde{Q} \geq 0$, which is equivalent to $\tilde{V} \geq 0$. Using (11.15), we find that this yields the requirement that

$$\lambda_{\max}(\tilde{T}\tilde{S}^{-1}) < \gamma^2.$$

Obviously the finite number of exceptions is irrelevant because we can use an arbitrarily small perturbation $\tilde{\gamma}$ of γ and then verify whether $\tilde{\gamma} > \gamma_{sp}^*$. The infimum over all γ such that $\gamma > \gamma_{sp}^*$ is then clearly equal to

$$\lambda_{\max}(\tilde{T}\tilde{S}^{-1}),$$

and the part (i) of theorem follows. The proof of part (ii), i.e., for the case when the system characterized by the quadruple (A, B, C, D) has invariant zeros on the imaginary axis, follows in view of the proof of part (i) and Theorem 11.22. ∎

Remark 11.13 *For the case when (A, B, C, D) is left-invertible, we do not need to require that $F = 0$ [we still need to impose (11.5)] and we find that*

$$\gamma_{sp}^* = \sup\{\gamma_{sp,1}, \gamma_{0,1}, \ldots, \gamma_{0,v}, \|F\|\}.$$

The following theorem pertains to the proper case.

Theorem 11.14 *Consider the continuous-time system* (11.1) *and the associated OID filtering performance via proper filters, i.e.,* γ_p^*. *Assume that a matrix* K *exists such that* $F - KD = 0$ *while* (11.6) *is satisfied. Choose* K *such that* $F - KD = 0$ *while* $E - KC$ *satisfies* (11.8). *Let* $\widetilde{S}$ *and* $\widetilde{T}$ *be the solution of the* H_2 *CARE* (11.9) *and the Lyapunov equation* (11.10), *respectively. In this case, we define*

$$\gamma_{p,1} = \sqrt{\lambda_{\max}(\widetilde{T}\widetilde{S}^{-1})}.$$

Then, the following hold:

(i) *For the case when no invariant zeros exist on the imaginary axis for the system characterized by the quadruple* (A, B, C, D), *we have* $\gamma_p^* = \gamma_{p,1}$.

(ii) *Otherwise define* $\gamma_{0,1}, \ldots, \gamma_{0,v}$ *as in Theorem 11.12, with* E *replaced by* $E - KC$. *In this case, we obtain*

$$\gamma_p^* = \sup\{\gamma_{p,1}, \gamma_{0,1}, \ldots, \gamma_{0,v}\}.$$

Proof : This is a direct consequence of Theorem 11.26, which we will present and prove later, and the arguments as used before in deriving Theorem 11.12. ∎

Remark 11.15 *For the case when the system characterized by* (A, B, C, D) *is left-invertible, we do not need to require the existence of a matrix* K *such that* $F - KD = 0$ *[we still need to impose* (11.5)*] and we find that*

$$\gamma_p^* = \sup\{\gamma_{p,1}, \gamma_{0,1}, \ldots, \gamma_{0,v}, \gamma_\infty\},$$

where

$$\gamma_\infty = \sqrt{\|FF' - FD'(DD')^{-1}D'F\|}.$$

11.4.2 Numerical computation of γ_{sp}^* and γ_p^*—continuous-time systems

In the previous subsection we computed explicitly γ_{sp}^* and γ_p^* for a subclass of systems satisfying certain conditions. When such conditions are not satisfied, the existing methods to compute γ_{sp}^* and γ_p^* are numerical in nature. In what follows, we present below such numerical methods. It turns out that such numerical methods are much more complex for the general singular case of γ-level H_∞ SOID problems than for the regular case. In view of this, for clarity as well as transparency of the results, we first present the regular case and then the singular case. Such a presentation also highlights the differences between the regular and singular cases while pointing out the complexity involved in the singular case.

Numerical computation of γ_{sp}^* and γ_p^*—regular case of continuous-time systems.

We consider first the regular γ-level H_∞ SOID problems. The following theorem deals with the strictly proper case.

Theorem 11.16 *Consider the continuous-time system* (11.1) *and the associated OID filtering performance via strictly proper filters, i.e., γ_{sp}^*. Assume that we have a regular filtering problem where D is surjective and the system characterized by (A, B, C, D) has no zeros on the imaginary axis. In this case, $\gamma > \gamma_{sp}^*$ if and only if*

$$\| F \| < \gamma \tag{11.16}$$

and a stabilizing solution $Q \geqslant 0$ exists of the H_∞^1 CARE:

$$0 = AQ + QA' + BB'$$
$$- \begin{pmatrix} CQ + DB' \\ EQ + FB' \end{pmatrix}' \begin{pmatrix} DD' & DF' \\ FD' & FF' - \gamma^2 I \end{pmatrix}^{-1} \begin{pmatrix} CQ + DB' \\ EQ + FB' \end{pmatrix}. \tag{11.17}$$

Proof : The proof will be presented in Subsection 11.5.1. ∎

Remark 11.17 *The Riccati equation* (11.17), *which we claimed as H_∞^1 CARE, looks different from the form of H_∞^1 CARE studied in Chapter 4. However, it can be rewritten in the form of an H_∞^1 CARE when we exploit* (11.16).

Remark 11.18 *The above theorem can be viewed as an existing condition for a γ-level H_∞ SOID filter. This is because it yields a test for any given γ whether $\gamma < \gamma_{sp}^*$ or $\gamma \geqslant \gamma_{sp}^*$. The above theorem also provides a simple bisection algorithm that enables us to compute γ_{sp}^* numerically to any degree of numerical accuracy. In the following algorithm, the first three steps represent initialization, whereas the last step represents the bisection principle:*

(i) *Set $i = 0$, and set $\gamma_\ell = \| F \|$.*

(ii) *Choose any $\gamma_0 > \gamma_\ell$.*

(iii) *Use Theorem 11.16 to test whether $\gamma_i < \gamma_{sp}^*$. If so, set $\gamma_{i+1} = 2\gamma_i$ and $i = i + 1$ and continue with step (iii). Otherwise set $\gamma_u = \gamma_i$ and continue with step (iv).*

(iv) *Set $\gamma = (\gamma_u + \gamma_\ell)/2$. Use Theorem 11.16 to test whether $\gamma < \gamma_{sp}^*$. If so, set $\gamma_\ell = \gamma$, and otherwise, set $\gamma_u = \gamma$, and repeat step (iv).*

In step (iv), we always know that $\gamma_{sp}^* \in [\gamma_\ell, \gamma_u]$ and after each repetition of that step, the size of the interval is divided in half. In [84], the above algorithm has been refined and quadratic convergence has been established.

The following theorem deals with proper case.

Theorem 11.19 *Consider the continuous-time system (11.1) and the associated OID filtering performance via proper filters, i.e., γ_p^*. Assume that we have a regular filtering problem where D is surjective and the system characterized by (A, B, C, D) has no zeros on the imaginary axis. In this case, $\gamma > \gamma_p^*$ if and only if*

$$\|FF' - FD'(DD')^{-1}DF\| < \gamma^2 \tag{11.18}$$

and a stabilizing solution $Q \geq 0$ of the H_∞^1 CARE (11.17) exists.

Proof : The proof will be presented in Subsection 11.5.1.   ■

Remark 11.20 *Once again, the Riccati equation (11.17), which we claimed as H_∞^1 CARE, looks different from the form of H_∞^1 CARE studied in Chapter 4. However, it can be rewritten in the form of an H_∞^1 CARE when we exploit (11.18).*

Remark 11.21 *As before, the above theorem can be viewed as an existing condition for a γ-level H_∞ SOID filter. This is because it yields a test for any given γ whether $\gamma < \gamma_p^*$ or $\gamma \geq \gamma_p^*$. As before, the above theorem also provides a simple bisection algorithm that enables us to compute γ_p^* numerically to any degree of numerical accuracy.*

Numerical computation of γ_{sp}^* and γ_p^*—singular case of continuous-time systems.

In the previous subsubsection, we presented a numerical algorithm of computing γ_{sp}^* and γ_p^* for the regular case of γ-level H_∞ SOID problems. We address here the same issue, however, for the singular case.

The following theorem deals with the strictly proper case.

Theorem 11.22 *Consider the continuous-time system (11.1) and the associated OID filtering performance via strictly proper filters, i.e., γ_{sp}^*. In this case, $\gamma > \gamma_{sp}^*$ if and only if*

- *We have*

$$\|F\| < \gamma. \tag{11.19}$$

- *A semi-stabilizing solution $Q \geqslant 0$ exists of the CQMI:*

$$\begin{pmatrix} AQ + QA' + BB' & QC' + BD' \\ CQ + DB' & DD' \end{pmatrix}$$
$$+ \begin{pmatrix} QE' + BF' \\ DF' \end{pmatrix}(\gamma^2 I - FF')^{-1}\begin{pmatrix} QE' + BF' \\ DF' \end{pmatrix}' \geqslant 0. \tag{11.20}$$

- *For each invariant zero s_0 on the imaginary axis, a matrix K exists such that*

$$\|E(s_0 I - A + KC)^{-1}(B - KD) + F\| < \gamma, \tag{11.21}$$

where K is such that $s_0 I - A + KC$ is invertible.

Proof : The proof will be presented in Subsection 11.5.2.   ∎

Remark 11.23 *In the above, we claimed (11.20) as a CQMI. However, it looks quite different from the form of CQMI studied in Chapter 4. We note that it is straightforward to rewrite (11.20) in the form of a CQMI as studied in Chapter 4 when we exploit (11.16).*

Remark 11.24 *In view of Theorem 11.12, the condition (11.21) of Theorem 11.22 can be rewritten as $\gamma_{0,i} < \gamma$ for $i = 1, \ldots v$.*

Remark 11.25 *As before, the above theorem can be viewed as an existing condition for a γ-level H_∞ SOID filter. This is because it yields a test for any given γ whether $\gamma < \gamma_{sp}^*$ or $\gamma \geqslant \gamma_{sp}^*$. As before, the above theorem also provides a simple bisection algorithm that enables us to compute γ_{sp}^* numerically to any degree of numerical accuracy as described earlier in Remark 11.18.*

The following theorem deals with the proper case.

Theorem 11.26 *Consider the continuous-time system (11.1) and the associated OID filtering performance via proper filters, i.e., γ_p^*. In this case, $\gamma > \gamma_p^*$ if and only if*

- *We have*

$$\| FF' - FD'(DD')^\dagger DF \| < \gamma^2. \tag{11.22}$$

- *A semi-stabilizing solution $Q \geq 0$ exists of the CQMI:*

$$\begin{pmatrix} AQ + QA' + BB' & QC' + BD' \\ CQ + DB' & DD' \end{pmatrix}$$
$$+ \begin{pmatrix} Q\widetilde{E}' + B\widetilde{F}' \\ D\widetilde{F}' \end{pmatrix} (\gamma^2 I - \widetilde{F}\widetilde{F}')^{-1} \begin{pmatrix} Q\widetilde{E}' + B\widetilde{F}' \\ D\widetilde{F}' \end{pmatrix}' \geq 0, \tag{11.23}$$

where

$$\widetilde{E} = E - FD'(DD')^\dagger C, \qquad \widetilde{F} = F - FD'(DD')^\dagger D.$$

- *For each invariant zero s_0 on the imaginary axis, matrices K and P exist such that*

$$\| (E - PC)(s_0 I - A + KC)^{-1}(B - KD) + (F - PD) \| < \gamma,$$

where K is such that $s_0 I - A + KC$ is invertible.

Proof : The proof will be presented in Subsection 11.5.2. ■

Remark 11.27 *In the above, we claimed (11.23) as a CQMI. However, it looks quite different from the CQMI studied in Chapter 4. Once again we note that it is straightforward to rewrite (11.23) in the form of a CQMI as studied in Chapter 4 when we exploit (11.22).*

Remark 11.28 *As before, the above theorem can be viewed as an existing condition for a γ-level H_∞ SOID filter. This is because it yields a test for any given γ whether $\gamma < \gamma_p^*$ or $\gamma \geq \gamma_p^*$. As noted, the above theorem also provides a simple bisection algorithm that enables us to compute γ_p^* numerically to any degree of numerical accuracy .*

Remark 11.29 *Note that there is a strong connection between the two CQMIs (11.20) and (11.23). In particular, when (11.19) is satisfied, then these two CQMIs are the same and hence have the same semi-stabilizing solution. The different format in which we present this CQMI is only needed because the inverse of $\gamma^2 I - FF'$ might not exist when we only know that (11.22) is satisfied.*

11.4.3   Explicit computation of γ_{sp}^* and γ_p^* — discrete-time systems

In this subsection, we consider the explicit computation of γ_{sp}^* and γ_p^* for a class of discrete-time systems. Such an explicit computation can be pursued via two methods of approach. In this regard, we note that the H_∞ norm is invariant under the bilinear transform. Hence, in the first method of approach, the bilinear transform can be used straightforwardly to obtain discrete-time results for the computation of γ_{sp}^* and γ_p^*. The second method of approach is to deal with discrete-time systems directly. For obvious reasons, the direct method is more elegant and more revealing structurally, and hence, we proceed next with the development of direct method.

For the computation of γ_{sp}^*, the class of systems we consider here is of the form (11.1) where $F = 0$ and

$$\mathcal{V}^\ominus(\Sigma_*) \cap \mathcal{S}^\ominus(\Sigma_*) \subset \ker E, \tag{11.24}$$

where Σ_* is the system characterized by the quadruple (A, B, C, D). These geometric subspaces have been defined in Section 3.2.5. Similarly, for the computation of γ_p^*, the class of systems we consider here is again of the form (11.1) for which we assume that a matrix K exists such that $F - KD = 0$ and

$$\mathcal{V}^\ominus(\Sigma_*) \cap \mathcal{S}^\ominus(\Sigma_*) \subset \ker(E - KC). \tag{11.25}$$

We first focus on the computation of γ_{sp}^*. As in the continuous-time case, we transform the system Σ_* into its SCB and obtain a decomposition as presented in (3.15)–(3.19). Then, the above condition (11.24) implies that

$$E\Gamma_s^{-1} = \begin{pmatrix} E_a^- & E_a^0 & E_a^+ & E_b & 0 & E_d \end{pmatrix}. \tag{11.26}$$

As will become apparent soon, the part of the dynamics of the given system that contributes to the nonzero values of γ_{sp}^* and γ_p^* is that of x_a^+, x_c, and x_d. Because of this, we look at the following subsystem (after applying an output injection of y_0):

$$\Sigma_1 : \begin{cases} \sigma \begin{pmatrix} x_a^+ \\ x_c \\ x_d \end{pmatrix} = \begin{pmatrix} A_{aa}^+ & 0 & 0 \\ B_c E_{ca}^+ & A_{cc} & 0 \\ B_d E_{da}^+ & B_d E_{dc} & A_{dd} \end{pmatrix} \begin{pmatrix} x_a^+ \\ x_c \\ x_d \end{pmatrix} \\ \qquad\qquad + \begin{pmatrix} 0 & 0 & 0 \\ 0 & 0 & B_c \\ 0 & B_d & 0 \end{pmatrix} \begin{pmatrix} u_0 \\ u_d \\ u_c \end{pmatrix} \\[2ex] \tilde{y} = \begin{pmatrix} C_{0a}^+ & C_{0c} & C_{0d} \\ 0 & 0 & C_d \end{pmatrix} \begin{pmatrix} x_a^+ \\ x_c \\ x_d \end{pmatrix} + \begin{pmatrix} I & 0 & 0 \\ 0 & 0 & 0 \end{pmatrix} \begin{pmatrix} u_0 \\ u_d \\ u_c \end{pmatrix} \\[2ex] \tilde{z} = \begin{pmatrix} E_a^+ & 0 & E_d \end{pmatrix} \begin{pmatrix} x_a^+ \\ x_c \\ x_d \end{pmatrix} + \begin{pmatrix} 0 & 0 & 0 \end{pmatrix} \begin{pmatrix} u_0 \\ u_d \\ u_c \end{pmatrix}. \end{cases}$$

Let

$$
\tilde{A} = \begin{pmatrix} A_{aa}^+ & 0 & 0 \\ B_c E_{ca}^+ & A_{cc} & 0 \\ B_d E_{da}^+ & B_d E_{dc} & A_{dd} \end{pmatrix}, \quad \tilde{B} = \begin{pmatrix} 0 & 0 & 0 \\ 0 & 0 & B_c \\ 0 & B_d & 0 \end{pmatrix},
$$

$$
\tilde{C} = \begin{pmatrix} C_{0a}^+ & C_{0c} & C_{0d} \\ 0 & 0 & C_d \end{pmatrix}, \quad \tilde{D} = \begin{pmatrix} I & 0 & 0 \\ 0 & 0 & 0 \end{pmatrix}, \quad \tilde{E} = \begin{pmatrix} E_a^+ & 0 & E_d \end{pmatrix}.
$$

We define next the following H_∞^1 DARE associated with Σ_1:

$$
\tilde{Q} = \tilde{A}\tilde{Q}\tilde{A}' + \tilde{B}\tilde{B}'
$$
$$
- \tilde{A}\tilde{Q}\begin{pmatrix} \tilde{C} \\ \tilde{E} \end{pmatrix}' \begin{pmatrix} \tilde{D}\tilde{D}' + \tilde{C}\tilde{Q}\tilde{C}' & \tilde{C}\tilde{Q}\tilde{E}' \\ \tilde{E}\tilde{Q}\tilde{C}' & \tilde{E}\tilde{Q}\tilde{E}' - \gamma^2 I \end{pmatrix}^{-1} \begin{pmatrix} \tilde{C} \\ \tilde{E} \end{pmatrix} \tilde{Q}\tilde{A}'. \quad (11.27)
$$

To establish an exact formula for γ_{sp}^*, we will use Theorem 11.39, which will be presented and proven later. In view of Theorem 11.39, to check whether $\gamma > \gamma_{sp}^*$, we need to check for the existence of a positive semi-definite and semi-stabilizing solution Q of the Generalized Discrete-time Algebraic Riccati Equation (H_∞^1 GDARE) (11.40). However, using the arguments at the end of Subsection 4.2.7, we find that Q is a positive semi-definite and semi-stabilizing solution of (11.40) if and only if Q has the form (using the SCB):

$$
Q = (\Gamma_s^{-1})' \begin{pmatrix} 0 & 0 & 0 & 0 & 0 & 0 \\ 0 & 0 & 0 & 0 & 0 & 0 \\ 0 & 0 & V_{11} & 0 & V_{12} & V_{13} \\ 0 & 0 & 0 & 0 & 0 & 0 \\ 0 & 0 & V_{21} & 0 & V_{22} & V_{23} \\ 0 & 0 & V_{31} & 0 & V_{32} & V_{33} \end{pmatrix} \Gamma_s^{-1}, \quad (11.28)
$$

where

$$
\begin{pmatrix} V_{11} & V_{12} & V_{13} \\ V_{21} & V_{22} & V_{23} \\ V_{31} & V_{32} & V_{33} \end{pmatrix} = \tilde{Q},
$$

and $\tilde{Q}$ is a positive semi-definite and stabilizing solution of the H_∞^1 DARE associated with Σ_1, namely (11.27). Then, Theorem 11.39 implies that $\gamma > \gamma_{sp}^*$ if and only if the positive semi-definite and stabilizing solution of the above H_∞^1 DARE exists and the additional condition (11.41) is satisfied. This is the first step toward explicit computation of γ_{sp}^*. The next step is to determine yet another auxiliary system Σ_2 extracted from Σ_1. Such an extraction is obtained by assuming that the variable x_d is known, which allows us to focus on the estimation of x_a^+ and

x_c while providing additional information (i.e., measurement) about x_a^+ and x_c. The auxiliary system Σ_2 is given by

$$
\Sigma_2 : \begin{cases}
\sigma\begin{pmatrix} x_a^+ \\ x_c \end{pmatrix} = \begin{pmatrix} A_{aa}^+ & 0 \\ B_c E_{ca}^+ & A_{cc} \end{pmatrix}\begin{pmatrix} x_a^+ \\ x_c \end{pmatrix} + \begin{pmatrix} 0 & 0 & 0 \\ 0 & 0 & B_c \end{pmatrix}\begin{pmatrix} u_0 \\ u_d \\ u_c \end{pmatrix} \\[2em]
\bar{y} = \begin{pmatrix} C_{0a}^+ & C_{0c} \\ E_{da}^+ & E_{dc} \end{pmatrix}\begin{pmatrix} x_a^+ \\ x_c \end{pmatrix} + \begin{pmatrix} I & 0 & 0 \\ 0 & I & 0 \end{pmatrix}\begin{pmatrix} u_0 \\ u_d \\ u_c \end{pmatrix} \\[2em]
\bar{z} = \begin{pmatrix} E_a^+ & 0 \end{pmatrix}\begin{pmatrix} x_a^+ \\ x_c \end{pmatrix} + \begin{pmatrix} 0 & 0 & 0 \end{pmatrix}\begin{pmatrix} u_0 \\ u_d \\ u_c \end{pmatrix}.
\end{cases}
$$

To proceed with, let

$$
\bar{A} = \begin{pmatrix} A_{aa}^+ & 0 \\ B_c E_{ca}^+ & A_{cc} \end{pmatrix}, \quad \bar{B} = \begin{pmatrix} 0 & 0 & 0 \\ 0 & 0 & B_c \end{pmatrix},
$$

$$
\bar{C} = \begin{pmatrix} C_{0a}^+ & C_{0c} \\ E_{da}^+ & E_{dc} \end{pmatrix}, \quad \bar{D} = \begin{pmatrix} I & 0 & 0 \\ 0 & I & 0 \end{pmatrix}, \quad \bar{E} = \begin{pmatrix} E_a^+ & 0 \end{pmatrix}.
$$

Next, we consider the H_∞^1 DARE associated with Σ_2:

$$
\bar{Q} = \bar{A}\bar{Q}\bar{A}' + \bar{B}\bar{B}' - \bar{A}\bar{Q}\begin{pmatrix} \bar{C}\,\bar{Q}\,\bar{A}' \\ \bar{E} \end{pmatrix}' \bar{G}(\bar{Q})^{-1}\begin{pmatrix} \bar{C} \\ \bar{E} \end{pmatrix}\bar{Q}\bar{A}', \tag{11.29}
$$

where

$$
\bar{G}(\bar{Q}) = \begin{pmatrix} \bar{D}\,\bar{D}' + \bar{C}\,\bar{Q}\,\bar{C}' & \bar{C}\,\bar{Q}\,\bar{E}' \\ \bar{E}\,\bar{Q}\,\bar{C}' & \bar{E}\,\bar{Q}\,\bar{E}' - \gamma^2 I \end{pmatrix}.
$$

It is clear by the introduction of Σ_2 that for γ larger than the H_∞ OID filtering performance of Σ_1, we need γ to be larger than the H_∞ OID filtering performance of Σ_2. Hence, according to Theorem 11.39, for a chosen γ, the existence of a positive semi-definite stabilizing solution of the H_∞^1 DARE (11.29) implies the existence of a positive semi-definite semi-stabilizing solution of the H_∞^1 DARE (11.27). Moreover, if the H_∞^1 DARE (11.29) has a positive semi-definite stabilizing solution $\bar{Q}$, then some algebra shows that

$$
\tilde{Q} = \begin{pmatrix} \bar{Q} & 0 \\ 0 & 0 \end{pmatrix}
$$

is a positive semi-definite stabilizing solution of the H_∞^1 DARE (11.27). Therefore, to prove the existence of a semi-definite and semi-stabilizing solution Q

of the H_∞^1 GDARE (11.40), it is sufficient to establish the existence of a semi-definite stabilizing solution $\bar{Q}$ of (11.29).

Note that (C, A) being $\mathbb{C}^\ominus$-detectable guarantees that $(\bar{C}, \bar{A})$ is $\mathbb{C}^\ominus$-detectable. Next, we note that $(-\bar{A}, \bar{B})$ is stabilizable because the structure of the SCB guarantees that $-A_{aa}^+$ is Schur-stable and (A_{cc}, B_{cc}) is controllable. This guarantees according to Theorem 4.80 that the following H_2 DARE has a positive definite solution:

$$\bar{L} = \bar{A}\bar{L}\bar{A}' + \bar{B}\bar{B}' - \bar{A}\bar{L}\bar{C}'(I + \bar{C}\bar{L}\bar{C}')^{-1}\bar{C}\bar{L}\bar{A}'. \tag{11.30}$$

Next, we define $\bar{T}$ as the solution of the Lyapunov equation:

$$\bar{T} + \bar{E}'\bar{E} = \bar{A}'\bar{T}\bar{A}, \tag{11.31}$$

which has a solution $\bar{T} \geq 0$ such that $\bar{T}\bar{B} = 0$. Then we can show using some algebraic manipulations that for all but finitely many γ,

$$\bar{Q} = (\bar{L}^{-1} - \gamma^{-2}\bar{T})^{-1} \tag{11.32}$$

is the unique positive semi-definite stabilizing solution of the H_∞^1 DARE (11.29). Then, the explicit positive semi-definite and semi-stabilizing solution Q of H_∞^1 GDARE (11.40) is given by

$$Q = (\Gamma_s^{-1})' \begin{pmatrix} 0 & 0 & 0 & 0 & 0 & 0 \\ 0 & 0 & 0 & 0 & 0 & 0 \\ 0 & 0 & V_{11} & 0 & V_{12} & 0 \\ 0 & 0 & 0 & 0 & 0 & 0 \\ 0 & 0 & V_{21} & 0 & V_{22} & 0 \\ 0 & 0 & 0 & 0 & 0 & 0 \end{pmatrix} \Gamma_s^{-1}, \tag{11.33}$$

where

$$\begin{pmatrix} V_{11} & V_{12} \\ V_{21} & V_{22} \end{pmatrix} = \bar{Q},$$

with $\bar{Q}$ is as in (11.32). Note that the additional condition (11.41) of Theorem 11.39 for $F = 0$ is $EQE' \leq \gamma^2 I$. This along with positive semi-definiteness of Q implies that

$$\bar{L}^{-1} - \gamma^{-2}\bar{T} \geq 0, \qquad \gamma^2 I - \bar{E}\left(\bar{L}^{-1} - \gamma^{-2}\bar{T}\right)^{-1}\bar{E}' \geq 0,$$

which can be equivalently expressed as

$$\begin{pmatrix} \bar{L}^{-1} - \gamma^{-2}\bar{T} & \bar{E}' \\ \bar{E} & \gamma^2 I \end{pmatrix} \geq 0$$

or

$$\bar{L}^{-1} - \gamma^{-2}\bar{T} - \gamma^{-2}\bar{E}\bar{E}' \geq 0.$$

We also need a discrete-time version of Lemma 11.11 as follows.

Lemma 11.30 *Let s_i be an invariant zero on the unit circle of the subsystem characterized by (A, B, C, D). Define*

$$\gamma_{0,i} = \inf_K \left\{ \| E(s_i I - A + KC)^{-1}(B - KD)\| \mid s_i I - A + KC \text{ is invertible} \right\}. \tag{11.34}$$

Then, we have

$$\gamma_{0,i} = \|L_{i,1} L_{i,2}^+\|, \tag{11.35}$$

where

$$L_{i,1} = E(s_i - A + K_1 C)^{-1}(B - K_1 D),$$
$$L_{i,2} = C(s_i - A + K_1 C)^{-1}(B - K_1 D) + D,$$

with K_1 an arbitrary matrix such that $A - K_1 C$ is Schur-stable, whereas $L_{i,2}^+$ is such that

$$(L_{i,2}^+)' L_{i,2}^+ = I, \quad \operatorname{im} L_{i,2}^+ = \ker L_{i,2}.$$

Proof : The proof follows along the same lines as in the continuous-time version given in Lemma 11.11. ∎

The above development, in view of Theorem 11.39, leads to the following result that pertains to strictly proper case.

Theorem 11.31 *Consider the discrete-time system (11.1) and the associated OID filtering performance via strictly proper filters, i.e., γ_{sp}^*. We assume that (11.24) is satisfied while $F = 0$. Let $\bar{L}$ and $\bar{T}$ be the solution of the H_2 DARE (11.30) and the Lyapunov equation (11.31), respectively. In this case, we define*

$$\gamma_{sp,1} = \sqrt{\lambda_{\max}\left((\bar{T} + \bar{E}\bar{E}')\bar{L})\right)}.$$

Then, the following hold:

(i) For the case when no invariant zeros exist on the unit circle for the system characterized by the quadruple (A, B, C, D), we have $\gamma_{sp}^* = \gamma_{sp,1}$.

(ii) Otherwise, let $s_1, \ldots, s_v$ be the invariant zeros on the unit circle of the subsystem characterized by (A, B, C, D) and we define $\gamma_{0,i}$ by (11.34). Then, we have

$$\gamma_{sp}^* = \sup\{\gamma_{sp,1}, \gamma_{0,1}, \ldots, \gamma_{0,v}\}.$$

Proof : The proof follows from the development given before the theorem. ∎

Next, we consider the explicit computation of γ_p^*. We note that in view of our assumption (11.25), a matrix K exists such that $\hat{F} = F - KD = 0$ and

$$(E - KC)\Gamma_s^{-1} = \begin{pmatrix} E_a^- & E_a^0 & E_a^+ & E_b & 0 & E_d \end{pmatrix}. \tag{11.36}$$

For the computation of γ_p^*, instead of estimating z, we estimate $z - Ky$. Also, in this case, all data of the given system Σ as in (11.1) remain as is except that we use $\hat{E} = E - KC$ instead of E while F is replaced by $\hat{F}$, which is zero. With such data, all development given earlier for the computation of γ_{sp}^* carries over except for the crucial difference that $\hat{E}$ has to be used instead of E. Thus, we again use the reduced-order H_∞^1 DARE (11.29). Therefore, we find that, for all but finitely many γ, Q is of the form (11.33) where

$$\bar{Q} = (\bar{L}^{-1} - \gamma^{-2}\bar{T})^{-1}$$

and $\bar{L}$ and $\bar{T}$ are the solutions of the H_2 DARE (11.30) and the Lyapunov equation (11.31), respectively.

In view of $\hat{F} = F - KD = 0$, the additional condition (11.41) of Theorem 11.39 simplifies to

$$\hat{E}Q\hat{E}' - CQ\hat{E}'(DD' + CQC')^\dagger\hat{E}QC' \leqslant \gamma^2 I.$$

The above requirement leads to the requirement that

$$\bar{E}\bar{Q}\bar{E}' - \bar{C}\bar{Q}\bar{E}'(I + \bar{C}\bar{Q}\bar{C}')^{-1}\bar{E}\bar{Q}\bar{C}' \leqslant \gamma^2 I.$$

In turn, the above requirement together with $\bar{Q} \geqslant 0$ results in the condition that

$$\bar{L}^{-1} + \bar{C}'\bar{C} \geqslant \gamma^{-2}(\bar{T} + \bar{E}'\bar{E})$$

or

$$\lambda_{\max}\left[(\bar{L}^{-1} + \bar{C}'\bar{C})^{-1}(\bar{T} + \bar{E}'\bar{E})\right] \leqslant \gamma^2.$$

This leads to the following theorem.

Theorem 11.32 *Consider the discrete-time system* (11.1) *and the associated OID filtering performance via proper filters, i.e.,* γ_p^*. *Assume that a matrix* K *exists such that* $F - KD = 0$ *while* (11.25) *is satisfied. Choose* K *such that* $F - KD = 0$ *while* $E - KC$ *satisfies* (11.36). *Let* $\bar{L}$ *and* $\bar{T}$ *be the solution of the* H_2 *DARE* (11.30) *and the Lyapunov equation* (11.31), *respectively. In this case, we define*

$$\gamma_{p,1} = \sqrt{\lambda_{\max}(\bar{L}^{-1} + \bar{C}'\bar{C})^{-1}(\bar{T} + \bar{E}'\bar{E})}.$$

Then, the following hold:

(i) *For the case when no invariant zeros exist on the unit circle for the system characterized by the quadruple* (A, B, C, D), *we have* $\gamma_p^* = \gamma_{p,1}$.

(ii) *Otherwise define* $\gamma_{0,1}, \ldots, \gamma_{0,v}$ *as in Theorem 11.31 with* E *replaced by* $E - KC$. *In this case, we obtain*

$$\gamma_p^* = \sup\{\gamma_{p,1}, \gamma_{0,1}, \ldots, \gamma_{0,v}\}.$$

Proof : The proof follows from the development given before the theorem. ∎

11.4.4 *Numerical computation of* γ_{sp}^* *and* γ_p^*—*discrete-time systems*

Previous subsection deals with explicit computation of γ_{sp}^* and γ_p^* for a subclass of systems satisfying certain conditions. As in continuous-time systems, whenever such conditions are not satisfied, the existing methods to compute γ_{sp}^* and γ_p^* are numerical in nature. In what follows, we present below such numerical methods. As in continuous-time systems, it turns out that such numerical methods are much more complex for the general singular case of γ-level H_∞ SOID problems than for the regular case. In view of this, once again for clarity as well as transparency of the results, we first present the regular case and then the singular case. As before, such a presentation also highlights the differences between the regular and singular cases while pointing out the complexity involved in the singular case.

Numerical computation of γ_{sp}^* and γ_p^*—regular case of discrete-time systems:

We consider here the regular γ-level H_∞ SOID problems. The following theorem deals with strictly proper case.

Theorem 11.33 *Consider the discrete-time system* (11.1) *and the associated OID filtering performance via strictly proper filters, i.e.,* γ_{sp}^*. *Assume that we have a regular filtering problem where* (A, B, C, D) *is right-invertible and has no zeros on the unit circle. In this case,* $\gamma > \gamma_{sp}^*$ *if and only if a stabilizing solution* $Q \geq 0$ *exists of the* H_∞^1 *DARE:*

$$Q = AQA' + BB' - \begin{pmatrix} CQA' + DB' \\ EQA' + FB' \end{pmatrix}' G(Q)^{-1} \begin{pmatrix} CQA' + DB' \\ EQA' + FB' \end{pmatrix} \quad (11.37)$$

such that

$$FF' + EQE' < \gamma^2 I, \quad (11.38)$$

where

$$G(Q) = \begin{pmatrix} DD' + CQC' & DF' + CQE' \\ FD' + EQC' & FF' + EQE' - \gamma^2 I \end{pmatrix}.$$

Proof : The proof will be presented in Subsection 11.6.1.   ■

Remark 11.34 *The Riccati equation* (11.37), *which we claimed as* H_∞^1 *DARE, looks different from the form of* H_∞^1 *DARE studied in Chapter 4. However, it can be rewritten in the form of an* H_∞^1 *DARE when we exploit* (11.38).

Remark 11.35 *The above theorem can be viewed as an existing condition for a* γ-*level* H_∞ *SOID filter. This is because it yields a test for any given* γ *whether* $\gamma < \gamma_{sp}^*$ *or* $\gamma \geq \gamma_{sp}^*$. *As in the continuous-time case, the above theorem also provides a simple bisection algorithm that enables us to compute* γ_{sp}^* *numerically to any degree of numerical accuracy:*

(i) *Set* $i = 0$, *and set* $\gamma_\ell = \|F\|$.

(ii) *Choose any* $\gamma_0 > \gamma_\ell$.

(iii) *Use Theorem 11.33 to test whether* $\gamma_i < \gamma_{sp}^*$. *If so, set* $\gamma_{i+1} = 2\gamma_i$ *and* $i = i + 1$ *and continue with step* (iii). *Otherwise set* $\gamma_u = \gamma_i$ *and continue with step* (iv).

(iv) *Set* $\gamma = (\gamma_u + \gamma_\ell)/2$. *Use Theorem 11.33 to test whether* $\gamma < \gamma_{sp}^*$. *If so, set* $\gamma_\ell = \gamma$; *otherwise set* $\gamma_u = \gamma$, *and repeat step* (iv).

In step (iv) *we always know that* $\gamma_{sp}^* \in [\gamma_\ell, \gamma_u]$ *and after each repetition of that step, the size of the interval is divided in half. For continuous time, the above algorithm has been refined in* [84] *and quadratic convergence has been established. A similar refinement can also be achieved in discrete time.*

The following theorem deals with proper case.

Theorem 11.36 *Consider the discrete-time system* (11.1) *and the associated OID filtering performance via proper filters, i.e.,* γ_p^*. *Assume that we have a regular filtering problem where the system characterized by* (A, B, C, D) *is right-invertible and has no zeros on the unit circle. In this case,* $\gamma > \gamma_p^*$ *if and only if a stabilizing solution* $Q \geq 0$ *exists of the* H_∞^1 *DARE* (11.37) *such that*

$$FF' + EQE' - (DF' + CQE')(DD' + CQC')^{-1}(FD' + EQC') < \gamma^2 I.$$
$$(11.39)$$

Proof : The proof will be presented in Subsection 11.6.1.   ■

Remark 11.37 *Once again, the Riccati equation (11.37), which we claimed as H_∞^1 DARE, looks different from the form of H_∞^1 DARE studied in Chapter 4. However, it can be rewritten in the form of an H_∞^1 DARE when we exploit (11.39).*

Remark 11.38 *As before, the above theorem can be viewed as an existing condition for a γ-level H_∞ SOID filter. This is because it yields a test for any given γ whether $\gamma < \gamma_p^*$ or $\gamma \geq \gamma_p^*$. As noted, the above theorem also provides a simple bisection algorithm that enables us to compute γ_p^* numerically to any degree of numerical accuracy.*

Numerical computation of γ_{sp}^* and γ_p^* —singular case of discrete-time systems:

In the previous subsubsection, we presented a numerical algorithm of computing γ_{sp}^* and γ_p^* for the regular case of γ-level H_∞ SOID problems. We address here the same issue, however, for the singular case.

The following theorem deals with the strictly proper case.

Theorem 11.39 *Consider the discrete-time system (11.1) and the associated OID filtering performance via strictly proper filters, i.e., γ_{sp}^*. In this case, $\gamma > \gamma_{sp}^*$ if and only if*

- *A semi-stabilizing solution $Q \geq 0$ exists of the H_∞^1 GDARE:*

$$Q = AQA' + BB' - \begin{pmatrix} CQA' + DB' \\ EQA' + FB' \end{pmatrix}' G(Q)^\dagger \begin{pmatrix} CQA' + DB' \\ EQA' + FB' \end{pmatrix}, \quad (11.40)$$

where

$$G(Q) = \begin{pmatrix} DD' + CQC' & DF' + CQE' \\ FD' + EQC' & FF' + EQE' - \gamma^2 I \end{pmatrix}$$

such that

$$FF' + EQE' < \gamma^2 I. \quad (11.41)$$

- *For each invariant zero s_0 on the unit circle, a matrix K exists such that*

$$\| E(s_0 I - A + KC)^{-1}(B - KD) + F \| < \gamma,$$

and such that $s_0 I - A + KC$ is invertible.

Proof : The proof will be presented in Subsections 11.6.2 and 11.6.3. ∎

Remark 11.40 *Using the same transformation as in Theorem 11.12, we can express the final condition of the above theorem in terms of $\gamma_{0,i} < \gamma$ for $i = 1, \ldots v$.*

Remark 11.41 *The Riccati equation (11.40), which we claimed as H_∞^1 GDARE, looks different from the form of H_∞^1 GDARE studied in Chapter 4. However, it is straightforward to rewrite it in the form of H_∞^1 GDARE when we exploit (11.41).*

Remark 11.42 *As before, the above theorem can be viewed as an existing condition for a γ-level H_∞ SOID filter. This is because it yields a test for any given γ whether $\gamma < \gamma_{sp}^*$ or $\gamma \geq \gamma_{sp}^*$. As before, the above theorem also provides a simple bisection algorithm that enables us to compute γ_{sp}^* numerically to any degree of numerical accuracy as described earlier in Remark 11.18.*

The following theorem deals with the proper case.

Theorem 11.43 *Consider the discrete-time system (11.1) and the associated OID filtering performance via proper filters, i.e., γ_p^*. In this case, $\gamma > \gamma_p^*$ if and only if*

- *A semi-stabilizing solution $Q \geq 0$ exists of the H_∞^1 GDARE (11.40) such that*

$$FF' + EQE' - (DF' + CQE')(DD' + CQC')^\dagger(FD' + EQC') < \gamma^2 I$$
$$(11.42)$$

  is satisfied.

- *For each invariant zero s_0 on the unit circle, matrices K and P exist such that*

$$\|(E - PC)(s_0 I - A + KC)^{-1}(B - KD) + (F - PD)\| < \gamma,$$

  where K is such that $s_0 I - A + KC$ is invertible.

Proof : The proof will be presented in Subsections 11.6.2 and 11.6.3. ∎

Remark 11.44 *The above theorem yields a test for any given γ whether $\gamma < \gamma_p^*$ or $\gamma \geq \gamma_p^*$. As before, the above theorem also provides a simple bisection algorithm that enables us to compute γ_p^* numerically to any degree of numerical accuracy.*

11.5   Design of γ-level H_∞ SOID filters—continuous-time systems

In this section, for continuous-time systems, we present methods of designing γ-level H_∞ SOID filters for any given $\gamma > \gamma_{sp}^*$ or $\gamma > \gamma_p^*$ depending on whether strictly proper or proper filters are sought. As in the last two sections, for clarity as well as transparency, our development here is divided into three layers that consider progressively a more complex class of problems, namely regular H_∞ SOID problems, singular H_∞ SOID problems with the restriction that the given system does not have any invariant zeros on the imaginary axis, and finally a general class of singular H_∞ SOID problems without any restrictions on the invariant zeros. Such a division also highlights the differences between the regular and the singular cases while pointing out the complexity involved in the singular case. We would like to emphasize here one fundamental aspect of our design philosophy; namely we transform the design of an H_∞ SOID filters for a given system to the design of exact input-decoupling (EID) filters or H_∞ almost input-decoupling (H_∞ AID) filters for an auxiliary system constructed from the data of the given system. Indeed such a transformation emerges as a by-product while proving sufficiency of certain conditions for the existence of H_∞ SOID filters (see Theorems 11.16, 11.19, 11.22, and 11.26). Irrespective of the origin of such a transformation, it offers profound advantages. This is because it lays a clear roadmap for design while enabling us to use directly all earlier detailed development regarding the design of EID and H_∞ AID filters as given in Chapters 7 and 9. In so doing, it also offers us structural insight while displaying all flexibility available in the design process.

11.5.1   Regular γ-level H_∞ SOID filters

In this subsection, we consider the design of regular H_∞ SOID filters. To start with, we need to recall the following bounded-real lemma, which is also known as the Kalman–Yakubovich–Popov lemma. Our version is not the most general result. We have a slight modification, which will be more useful for our purposes. For a very elementary proof of the bounded-real lemma, we refer to the paper [65].

Theorem 11.45 *Consider a continuous-time system Σ parameterized by the quadruple (A, B, C, D) with associated transfer matrix G. Then we have:*

- *Assume that the observable eigenvalues of (C, A) are in the open left-half plane. In this case, the system is input–output stable with associated H_∞ norm $\|G\|_\infty < \gamma$ if and only if a matrix $P \geqslant 0$ exists satisfying the following CLMI:*

$$F(P) = \begin{pmatrix} A'P + PA + C'C & PB + C'D \\ B'P + D'C & D'D - \gamma^2 I \end{pmatrix} \leqslant 0 \qquad (11.43)$$

such that $\ker F(P) = \ker P \oplus \{0\}$.

If A is Hurwitz-stable, then a matrix $P > 0$ *exists such that* $F(P) < 0$.

- *Assume that the controllable eigenvalues of* (A, B) *are in the open left-half plane. In this case, the system is input–output stable with associated* H_∞ *norm* $\|G\|_\infty < \gamma$ *if and only if a matrix* $Q \geq 0$ *exists satisfying the following dual CLMI:*

$$G(Q) = \begin{pmatrix} QA' + AQ + BB' & QC' + BD' \\ CQ + DB' & DD' - \gamma^2 I \end{pmatrix} \leq 0 \qquad (11.44)$$

such that $\ker G(Q) = \ker Q \oplus \{0\}$.

If A is Hurwitz-stable, then a matrix $Q > 0$ *exists such that* $G(Q) < 0$.

We begin our development by first providing the proofs of necessity parts of Theorems 11.16 and 11.19, which present existence conditions for regular γ-level H_∞ SOID filters. As we said, the proofs of sufficiency parts of these theorems given subsequently enable us to transform the design of appropriate H_∞ SOID filters for the given system to the design of EID filters for an auxiliary system constructed from the data of the given system.

Proof of necessity for Theorem 11.16 : Given the system (11.1) and a filter of the form (11.2) with $P = 0$, we obtain the following interconnection:

$$\begin{pmatrix} \dot{x} \\ \dot{\xi} \end{pmatrix} = A_e \begin{pmatrix} x \\ \xi \end{pmatrix} + B_e u,$$

$$e_z = C_e \begin{pmatrix} e \\ \xi \end{pmatrix} + F u,$$

with $e_z = z - \hat{z}$, where

$$A_e = \begin{pmatrix} A & 0 \\ MC & L \end{pmatrix}, \quad B_e = \begin{pmatrix} B \\ MD \end{pmatrix}, \quad C_e = \begin{pmatrix} E & -N \end{pmatrix}, \qquad (11.45)$$

where the fact that we have an unbiased filter implies that the observable eigenvalues of (C_e, A_e) are in the open left-half plane. We know from Theorem 11.45 that this guarantees that a matrix $P \geq 0$ exists satisfying

$$F(P) = \begin{pmatrix} A_e'P + PA_e + C_e'C_e & PB_e + C_e'F \\ B_e'P + F'C_e & F'F - \gamma^2 I \end{pmatrix} \leq 0$$

and $\ker F(P) = \ker P \oplus \{0\}$. We decompose P compatible with (11.45):

$$P = \begin{pmatrix} P_{11} & P_{12} \\ P_{21} & P_{22} \end{pmatrix}.$$

Given that $P \geqslant 0$, a matrix X exists such that $P_{21} = XP_{11}$. We find that $P_{11} \geqslant 0$ satisfies $F_{11}(P_{11}) \leqslant 0$, where

$$F_{11}(P_{11}) =$$
$$\begin{pmatrix} P_{11}(A - KC) + (A - KC)'P_{11} + E'E & P_{11}(B - KD) + E'F \\ (B - KD)'P_{11} + F'E & F'F - \gamma^2 I \end{pmatrix}$$

with $K = -XM$ and $\ker F_{11}(P_{11}) = \ker P_{11} \oplus \{0\}$. This implies that the system characterized by the parameters

$$(A - KC, B - KD, E, F) \tag{11.46}$$

has H_∞ norm less than γ according to Theorem 11.45. Moreover, $(A - KC)x = \lambda x$ with $\operatorname{Re} \lambda \geqslant 0$ yields $Ex = 0$ when we use that $F_{11}(P_{11}) \leqslant 0$, and hence, the unstable dynamics are unobservable for the system (11.46). As (C, A) is $\mathbb{C}^-$-detectable, it is easily checked that a matrix K_1 exists such that

$$(A - KC - K_1C, B - KD - K_1D, E, F) \tag{11.47}$$

is asymptotically stable and the transfer matrix is equal to the transfer matrix of (11.46) and hence, we have an H_∞ norm less than γ. Define $\tilde{K} = K + K_1$. Using Theorem 11.45, we find that a matrix $\tilde{Q} > 0$ exists such that

$$\begin{pmatrix} (A - \tilde{K}C)\tilde{Q} + \tilde{Q}(A - \tilde{K}C)' + (B - \tilde{K}D)(B - \tilde{K}D)' & \tilde{Q}E' + (B - \tilde{K}D)F' \\ E\tilde{Q} + F(B - \tilde{K}D)' & FF' - \gamma^2 I \end{pmatrix} < 0.$$

This implies that

$$(A - \tilde{K}C)\tilde{Q} + \tilde{Q}(A - \tilde{K}C)' + (B - \tilde{K}D)(B - \tilde{K}D)'$$
$$- (\tilde{Q}E' + (B - \tilde{K}D)F')(FF' - \gamma^2 I)^{-1}(E\tilde{Q} + F(B - \tilde{K}D)') < 0.$$

This is a quadratic matrix inequality in $\tilde{K}$. We already have a specific choice of $\tilde{K}$ for which this matrix is negative definite. If we release this specific choice of $\tilde{K}$ and instead minimize over $\tilde{K}$, then this matrix should obviously remain negative definite. This results in

$$\tilde{K} = -\left[(\tilde{Q}C' + BD') - (\tilde{Q}E' + BF')(FF' - \gamma^2 I)^{-1}FD'\right]$$
$$(DD' - DF'(FF' - \gamma^2 I)^{-1}FD')^{-1}$$

and

$$0 > A\tilde{Q} + \tilde{Q}A' + BB'$$
$$- \begin{pmatrix} C\tilde{Q} + DB' \\ E\tilde{Q} + FB' \end{pmatrix}' \begin{pmatrix} DD' & DF' \\ FD' & FF' - \gamma^2 I \end{pmatrix}^{-1} \begin{pmatrix} C\tilde{Q} + DB' \\ E\tilde{Q} + FB' \end{pmatrix}.$$

Using Proposition 4.30, we find that a positive semi-definite strongly semi-stabilizing solution exists of the H_∞^1 CARE (11.17). As (A, B, C, D) has no invariant zeros on the imaginary axis, we conclude that we actually have a positive semi-definite stabilizing solution.   ∎

The above proof of necessity for strictly proper filters quickly yields the result for proper filters.

Proof of necessity for Theorem 11.19 : We note that applying a proper filter of the form (11.2) to the system (11.1) is equivalent to applying the strictly proper filter:

$$\begin{aligned} \dot{\xi} &= L\xi + My \\ \hat{z} &= N\xi \end{aligned} \qquad (11.48)$$

to the system

$$\Sigma : \begin{cases} \dot{x} = Ax + Bu \\ y = Cx + Du \\ z = E^*x + F^*u, \end{cases} \qquad (11.49)$$

where $E^* = E - PC$ and $F^* = F - PD$. Therefore, if this filter is a γ-level H_∞ SOID filter, then using Theorem 11.16, we note that a stabilizing solution Q exists of the following H_∞^1 CARE:

$$0 = AQ + QA' + BB'$$
$$- \begin{pmatrix} CQ + DB' \\ E^*Q + F^*B' \end{pmatrix}' \begin{pmatrix} DD' & D(F^*)' \\ F^*D' & F^*(F^*)' - \gamma^2 I \end{pmatrix}^{-1} \begin{pmatrix} CQ + DB' \\ E^*Q + F^*B' \end{pmatrix}.$$

But it then only takes some algebraic manipulations to establish that Q is also a stabilizing solution of the original H_∞^1 CARE given in (11.17). In other words, the matrix P does not affect the H_∞^1 CARE.

The necessity proof is completed by noting that when a filter is a γ-level H_∞ SOID filter, then the direct feedthrough matrix from u to e_z must have a norm less than γ. Hence, a matrix P exists such that

$$\|F - PD\| < \gamma,$$

which immediately implies that (11.18) is satisfied. This completes the necessity part of the proof.   ∎

Proofs of sufficiency parts of Theorems 11.16 and 11.19:

So far we have proved the necessity parts of Theorems 11.16 and 11.19. We proceed now to prove the sufficiency parts. In this regard, as we said, our design methodology arises as a by-product while proving the sufficiency parts. As in the previous chapters, we explore here the design of strictly proper, proper, as well as reduced-order γ-level H_∞ SOID filters of CSS architecture.

We first start with two preliminary lemmas useful in proving the sufficiency parts of Theorems 11.16 and 11.19. Consider a system Σ characterized by (A, B, C, D) with input–output operator $\mathscr{G}$ (i.e., the map that associates to every input d an output z, given zero initial state). Σ is called *inner* if the system is internally stable, and the input–output operator $\mathscr{G}$ maps L_2^m into itself and has the property that for all $f \in L_2^m$, we have

$$\|\mathscr{G} f\|_2 = \|f\|_2.$$

Often, inner is defined as a property of the transfer matrix, but in our setting, the above is a more natural definition. It can be shown that $\mathscr{G}$ being inner implies that the transfer matrix of the system, denoted by G, satisfies

$$G'(-s)G(s) = I. \tag{11.50}$$

A transfer matrix G satisfying (11.50) is called *unitary*. Note that if G is unitary, then G need not be $\mathbb{C}^-$ stable. The transfer matrix G is unitary and the system is internally stable if and only if the system is inner. In general, for an operator from L_2^m to L_2^p, in the literature two concepts, inner and co-inner are defined. A system Σ characterized by (A, B, C, D) is called *co-inner* if the dual system Σ' characterized by (A', C', B', D') is inner. In other words, the system is co-inner if it is internally stable and its transfer matrix satisfies

$$G(s)G'(-s) = I.$$

Note that for square systems, the concepts of inner and co-inner coincide. We now formulate a lemma that yields a test to check whether a system is co-inner.

Lemma 11.46 *Consider the system Σ described by*

$$\Sigma \; : \; \begin{cases} \dot{x} = Ax + Bu \\ z = Cx + Du, \end{cases} \tag{11.51}$$

with A Hurwitz-stable. The system Σ is co-inner if a matrix X exists satisfying,

(i) $AY + YA' + BB' = 0$

(ii) $DB' + CY = 0$

(iii) $DD' = I$.

Inner and co-inner systems are very important in H_∞ control. Before we present a lemma that is a main ingredient in the proof of sufficiency conditions for the existence of regular γ-level H_∞ SOID filters, we first need the following preliminary lemma often referred to as "Redheffer's lemma." Note that we present a dual version of the lemma, the original version of which for instance can be found in [20].

Lemma 11.47 *Consider the continuous-time, linear time-invariant systems Σ_U and Ψ. Suppose Σ_U has inputs u and e_s and outputs e and u_s, whereas Ψ has input u_s and output e_s. Consider the interconnection depicted in the diagram in Figure 11.2.*

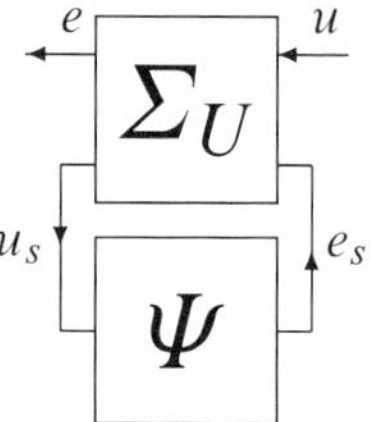

Figure 11.2: Interconnection of Σ_U and Ψ

Assume that Σ_U is co-inner, and its input–output operator $\mathcal{G}$ has the following decomposition:

$$\mathcal{G}\begin{pmatrix} u \\ e_s \end{pmatrix} =: \begin{pmatrix} \mathcal{G}_{11} & \mathcal{G}_{12} \\ \mathcal{G}_{21} & \mathcal{G}_{22} \end{pmatrix}\begin{pmatrix} u \\ e_s \end{pmatrix} = \begin{pmatrix} e \\ u_s \end{pmatrix}, \tag{11.52}$$

which is compatible with the sizes of u, e, u_s, and e_s, such that the $\mathcal{G}_{12}$ has a stable inverse.

Under the above assumptions, the following two statements are equivalent:

(i) *The interconnection in Figure 11.2 is internally stable and its closed-loop transfer matrix has H_∞ norm less than 1.*

(ii) *The system Ψ is internally stable, and its transfer matrix has H_∞ norm less than 1.*

Proof : Suppose part (*ii*) is satisfied. As $\|\Psi\|_\infty < 1$ and $\|\mathcal{G}_{22}\|_\infty \leqslant \|\mathcal{G}\|_\infty = 1$ an application of the small gain theorem implies that $(I - \mathcal{G}_{22}\Psi)^{-1}$ exists and has a finite L_2 induced operator norm. This implies the internal stability of the interconnection (11.2) because both $\mathcal{G}$ and Ψ are stable. Next, note that, because $\mathcal{G}$ is co-inner, we have

$$\mathcal{G}_{11}\mathcal{G}_{11}^* + \mathcal{G}_{12}\mathcal{G}_{12}^* = 1.$$

As $\mathcal{G}_{12}$ has a stable inverse, this implies that $\|\mathcal{G}_{11}\| < 1$. But the fact that $\mathcal{G}$ is co-inner, also implies that

$$\|e\|_2^2 + \|u_s\|_2^2 \leqslant \|u\|_2^2 + \|e_s\|_2^2.$$

This results in

$$\|e\|_2^2 - \|u\|_2^2 \leqslant -\varepsilon\|e_s\|_2^2, \tag{11.53}$$

where we use that $\|\Psi\|_\infty < 1$, and hence an $\varepsilon > 0$ exists such that

$$(1 + \varepsilon)\|e_s\|_2^2 \leqslant \|u_s\|_2^2.$$

Let $\beta = 1 - \|\mathcal{G}_{11}\|$. Clearly $\beta > 0$. Assume that

$$\|e\|_2 \geqslant (1 - \tfrac{\beta}{2})\|u\|_2. \tag{11.54}$$

We get

$$\|e\|_2 = \|\mathcal{G}_{11}u + \mathcal{G}_{12}e_s\|_2 \leqslant \|\mathcal{G}_{11}\|\|u\|_2 + \|\mathcal{G}_{12}e_s\|_2.$$

This requires that

$$\|\mathcal{G}_{12}e_s\|_2 \geqslant \tfrac{\beta}{2}\|u\|_2,$$

and hence,

$$\|e_s\|_2 \geqslant \tfrac{\beta}{2}\|\mathcal{G}_{12}\|^{-1}\|u\|_2.$$

Combined with (11.53) this yields

$$\|e\|_2^2 \leqslant \left(1 - \tfrac{\varepsilon\beta^2}{4}\|\mathcal{G}_{12}\|^{-2}\right)\|u\|_2^2$$

provided (11.54) holds. Hence,

$$\|e\|_2^2\| \leqslant \max\left\{\left(1 - \tfrac{\varepsilon\beta^2}{4}\|\mathcal{G}_{12}\|^{-2}\right), \left(1 - \tfrac{\beta}{2}\right)^2\right\}\|u\|_2^2,$$

and the closed-loop transfer matrix of the interconnection in Figure 11.2 has an H_∞ norm less than 1.

Conversely, assume that part (i) is satisfied. Note that the fact that $\mathcal{G}_{12}$ has a stable inverse guarantees that for any $e_s \in L_2$, a u exists such that

$$\begin{pmatrix} u \\ e_s \end{pmatrix} \in [\ker \mathcal{G}]^\perp.$$

In this case, we have

$$\|e\|_2^2 + \|u_s\|_2^2 = \|u\|_2^2 + \|e_s\|_2^2.$$

This yields

$$\|e_s\|_2^2 - \|u_s\|_2^2 = \|e\|_2^2 - \|u\|_2^2 \leqslant -\varepsilon\|u\|_2^2. \tag{11.55}$$

Choose $\beta = \tfrac{1}{4}\|\mathcal{G}_{12}^{-1}\|^{-1}$. Assume that $\|u\|_2 \leqslant \beta\|e_s\|_2$. In this case, we also get $\|e\|_2 \leqslant \beta\|e_s\|_2$ and

$$\|e\|_2 = \|\mathcal{G}_{11}u + \mathcal{G}_{12}e_s\|_2 \geqslant \|G_{12}e_s\|_2 - \|G_{11}u\|_2 \geqslant \|G_{12}e_s\|_2 - \beta\|e_s\|_2.$$

This yields

$$2\beta\|e_s\|_2 \geqslant \|G_{12}e_s\|,$$

which contradicts with the definition of β. Hence, our assumption was wrong and we have

$$\|u\|_2 \geqslant \beta \|e_s\|_2,$$

which together with (11.55) shows that the H_∞ norm of Ψ is strictly less than 1.

$\blacksquare$

We established the above two preliminary lemmas in our quest to prove the sufficiency parts of Theorems 11.16 and 11.19. We will now proceed to show that the problem of finding a suitable H_∞ SOID filter for the system Σ given in (11.1) is equivalent to finding a suitable H_∞ SOID filter for a *new system* Σ_Q, which has some very nice structural properties. It turns out that we can construct a slightly modified version of Σ_Q (namely, $\check{\Sigma}_Q$) for which the EID filtering problem is solvable by a strictly proper filter. This implies that we can directly use the design methodology developed in Chapter 7 to first find an EID filter for $\check{\Sigma}_Q$. Also, it turns out that such an EID filter for $\check{\Sigma}_Q$ can be translated as a γ-level H_∞ SOID filter for Σ_Q, and that in turn can be translated as a γ-level H_∞ SOID filter for the given system Σ. Thus, the construction of new systems Σ_Q and $\check{\Sigma}_Q$ have definite advantages.

We proceed now to construct the new system Σ_Q. Assume that the H_∞ SOID filtering problem is solvable by a strictly proper or proper filter. Then we know from our necessity proof of Theorem 11.16 that a stabilizing solution Q exists of the H_∞^1 CARE (11.17) such that (11.16) or (11.18), respectively, are satisfied. Using this matrix Q, we can define the following system:

$$\Sigma_Q : \begin{cases} \dot{x}_q = Ax_q + B_Q u_q + G_Q z_q \\ \quad y = Cx_q + Du_q \\ \quad z_q = Ex_q + Fu_q, \end{cases} \tag{11.56}$$

where B_Q and G_Q are given by

$$B_Q = -\begin{pmatrix} QC' + BD' & QE' + BF' \end{pmatrix} \begin{pmatrix} DD' & DF' \\ FD' & FF' - \gamma^2 I \end{pmatrix}^{-1} \begin{pmatrix} D \\ F \end{pmatrix}, \tag{11.57}$$

$$G_Q = -\begin{pmatrix} QC' + BD' & QE' + BF' \end{pmatrix} \begin{pmatrix} DD' & DF' \\ FD' & FF' - \gamma^2 I \end{pmatrix}^{-1} \begin{pmatrix} 0 \\ I \end{pmatrix}. \tag{11.58}$$

We will establish shortly that there is a direct connection between both the strictly proper as well as the proper γ-level H_∞ SOID filters designed for Σ_Q and the corresponding filters designed for Σ.

We will first show that the strictly proper (or proper) γ-level H_∞ SOID filtering problem for Σ_Q is solvable whenever the strictly proper (or proper) γ-level H_∞ SOID filtering problem for Σ is solvable. To do so, we first introduce a slightly modified version of Σ_Q, namely a system $\check{\Sigma}_Q$ whose desired output signal $\check{z}_q$ that needs to be estimated is slightly different from z_q:

$$\check{\Sigma}_Q : \begin{cases} \dot{x}_q = (A + G_Q E)x_q + (B_Q + G_Q F)u_q \\ y = C x_q + D u_q \\ \check{z}_q = (E - PC)x_q, \end{cases} \qquad (11.59)$$

where the matrix P is different for the two different cases: strictly proper and proper:

- For the case when Σ satisfies the conditions for the existence of strictly proper γ-level H_∞ SOID filters (i.e., besides the existence of a stabilizing solution Q of the H_∞^1 CARE (11.17), we have $\|F\| < \gamma$), a suitable choice for the matrix P is $P = 0$.

- When Σ satisfies the conditions for the existence of proper γ-level H_∞ SOID filters (i.e., besides the existence of a stabilizing solution Q of the H_∞^1 CARE (11.17), a matrix P exists such that $\|F - PD\| < \gamma$), obviously a suitable choice for the matrix P in (11.59) is such that $\|F - PD\| < \gamma$.

Clearly, in what follows, we can distinguish these two cases of defining $\check{\Sigma}_Q$ by selecting appropriate P in (11.59) as strictly proper case or proper case.

We make next two claims. First, for the system $\check{\Sigma}_Q$ the EID filtering problem is solvable by a strictly proper filter irrespective of whether $P = 0$ or P is such that $\|F - PD\| < \gamma$; that is, irrespective of whether the definition of $\check{\Sigma}_Q$ is for the class of strictly proper filters or for the class of proper filters. Second, any EID filter designed for $\check{\Sigma}_Q$ using the methodology of Chapter 7 results in a γ-level H_∞ SOID filter for Σ_Q. These claims are, respectively, formalized in the following two lemmas.

Lemma 11.48 *Consider the regular γ-level H_∞ SOID filtering problem for the continuous-time system Σ described by (11.1). That is, assume that D is surjective and that the system characterized by (A, B, C, D) has no invariant zeros on the imaginary axis. Also, assume that $\gamma > \gamma_{sp}^*$ (or $\gamma > \gamma_p^*$); that is, assume that a stabilizing solution Q of the H_∞^1 CARE (11.17) exists, and $\|F\| < \gamma$ (or a matrix P exists such that $\|F - PD\| < \gamma$). Define $\check{\Sigma}_Q$ by (11.59) with $P = 0$ (or with P such that $\|F - PD\| < \gamma$).*

Then, the EID filtering problem is solvable for $\check{\Sigma}_Q$ by a strictly proper filter irrespective of $P = 0$ (strictly proper case) or P is such that $\|F - PD\| < \gamma$ (proper case).

Proof : It can be easily verified that

$$K = \begin{pmatrix} QC' + BD' & QE' + BF' \end{pmatrix} \begin{pmatrix} DD' & DF' \\ FD' & FF' - \gamma^2 I \end{pmatrix}^{-1} \begin{pmatrix} I \\ 0 \end{pmatrix}$$

is such that $B_Q + G_Q F - KD = 0$, whereas $A + G_Q E - KC$ equals

$$A - \begin{pmatrix} CQ + DB' \\ EQ + FB' \end{pmatrix}' \begin{pmatrix} DD' & DF' \\ FD' & FF' - \gamma^2 I \end{pmatrix}^{-1} \begin{pmatrix} C \\ E \end{pmatrix},$$

which is Hurwitz-stable because Q is a stabilizing solution of the H_∞^1 CARE given in (11.17). This clearly implies that the EID filtering problem is solvable for $\check{\Sigma}_Q$ by the strictly proper filter:

$$\begin{cases} \dot{\xi} = (A + G_Q E)\xi + K(y - C\xi) \\ \hat{\check{z}}_q = (E - PC)\xi. \end{cases} \qquad \blacksquare$$

Lemma 11.49 *Consider the regular γ-level H_∞ SOID filtering problem for the continuous-time system Σ described by (11.1). That is, assume that D is surjective and that the system characterized by (A, B, C, D) has no invariant zeros on the imaginary axis. Also, assume that $\gamma > \gamma_{sp}^*$ (or $\gamma > \gamma_p^*$); that is, assume that a stabilizing solution Q exists of the H_∞^1 CARE (11.17) and $\|F\| < \gamma$ (or a matrix P exists such that $\|F - PD\| < \gamma$). Define Σ_Q by (11.56). Also, define $\check{\Sigma}_Q$ by (11.59) with $P = 0$ (or with P such that $\|F - PD\| < \gamma$).*
Then the following hold:

(i) *Let a strictly proper filter of the form (11.2) characterized by a matrix triple (L, M, N) solve the EID filtering problem for the system $\check{\Sigma}_Q$ with $P = 0$ (strictly proper case). Then such a filter is a strictly proper γ-level H_∞ SOID filter for Σ_Q.*

(ii) *Let a strictly proper filter of the form (11.2) characterized by a matrix triple (L, M, N) solve the EID filtering problem for the system $\check{\Sigma}_Q$ with P such that $\|F - PD\| < \gamma$ (proper case). Consider a proper filter of the form (11.2) characterized by the matrix quadruple (L, M, N, P), where the matrix triple (L, M, N) corresponds to the EID filter for $\check{\Sigma}_Q$ and the matrix P corresponds to the one that is used in defining $\check{\Sigma}_Q$. Then such a proper filter is a proper γ-level H_∞ SOID filter for Σ_Q.*

Proof : For the strictly proper case, it is clear that a filter that solves the EID filtering problem for the system $\check{\Sigma}_Q$ is obviously unbiased when applied to $\check{\Sigma}_Q$. Then the same filter is also unbiased when applied to Σ_Q. Moreover because such a filter solves the EID filtering problem for the system $\check{\Sigma}_Q$ for the strictly proper case, it is easily verified that the same filter when applied to Σ_Q results in the estimation error

$$e_q = z_q - \hat{z}_q = Fu_q.$$

As $\|F\| < \gamma$, this implies that such a filter solves the γ-level H_∞ SOID filtering problem for Σ_Q.

Similarly, in the proper case, a strictly proper filter characterized by a matrix triple (L, M, N) that solves the EID filtering problem for the system $\breve{\Sigma}_Q$ is obviously unbiased when applied to $\breve{\Sigma}_Q$. Consider a new proper filter of the form (11.2) with this matrix triple (L, M, N) and with matrix P used in defining $\breve{\Sigma}_Q$. Such a new filter is an unbiased filter when applied to the system Σ_Q. Also, because the strictly proper filter characterized by the matrix triple (L, M, N) solves the EID filtering problem for the system $\breve{\Sigma}_Q$ for proper case, it is easily verified that the new proper filter when applied to Σ_Q results in the estimation error

$$e_q = z_q - \hat{z}_q = (F - PD)u_q.$$

As such, because in this case $\|F - PD\| < \gamma$, it implies that the new proper filter is a proper γ-level H_∞ SOID filter for Σ_Q.   ∎

The above lemma guarantees that the strictly proper (or proper) γ-level H_∞ SOID filtering problem is solvable for Σ_Q whenever the strictly proper (or proper) γ-level H_∞ SOID filtering problem is solvable for Σ. We will shortly and formally establish a relationship between the γ-level H_∞ SOID filters for Σ and the γ-level H_∞ SOID filters for Σ_Q. This result will be a consequence of "Redheffer's lemma" (Lemma 11.47). The following lemma will be a crucial component in establishing such a result.

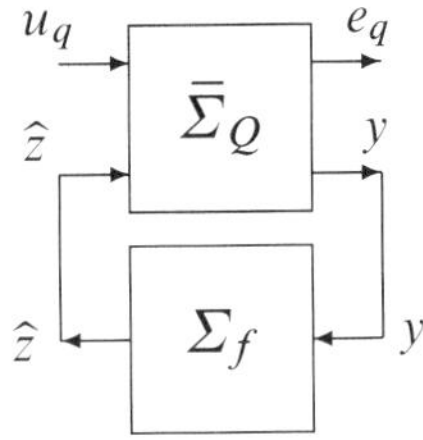

Figure 11.3: Interconnection of $\overline{\Sigma}_Q$ and Σ_f

Lemma 11.50 *Consider the regular γ-level H_∞ SOID filtering problem for the continuous-time system Σ described by (11.1). That is, assume that D is surjective and that the system characterized by (A, B, C, D) has no invariant zeros on the imaginary axis. Also, assume that $\gamma > \gamma_{sp}^*$ (or $\gamma > \gamma_p^*$); that is, assume that a stabilizing solution Q of the H_∞^1 CARE (11.17) exists and $\|F\| < \gamma$ (or a matrix P exists such that $\|F - PD\| < \gamma$). Define the following system $\overline{\Sigma}_Q$ with two* *inputs u_q and $\hat{z}$ and two outputs y and e_q as shown in Figure 11.3:*

$$\bar{\Sigma}_Q : \begin{cases} \dot{x}_q = Ax_q + B_Q u_q + G_Q e_q \\ y = Cx_q + Du_q \\ e_q = Ex_q + Fu_q - \hat{z}, \end{cases}$$

where B_Q and G_Q are as defined in (11.57) and (11.58). Then the following hold:

(i) *Suppose a filter Σ_f of the form (11.2) and characterized by the quadruple (L, M, N, P) (with $P = 0$ for strictly proper case) is a γ-level H_∞ SOID filter for the given system Σ. Then the interconnection of Σ_f and $\bar{\Sigma}_Q$ (that is, $\bar{\Sigma}_Q \times \Sigma_f$ as shown in Figure 11.3) is such that the H_∞ norm of u_q to e_q is strictly less than γ, and for all initial conditions for (11.2) and $\bar{\Sigma}_Q$ whenever $u_q = 0$, we have that $e_q(t) \to 0$ as $t \to \infty$.*

(ii) *Conversely, suppose a system Σ_f of the form (11.2) and characterized by the quadruple (L, M, N, P) (with $P = 0$ for strictly proper case) interconnected with the system $\bar{\Sigma}_Q$ (that is, $\bar{\Sigma}_Q \times \Sigma_f$ as shown in Figure 11.3) is such that the H_∞ norm of u_q to e_q is strictly less than γ, and for all initial conditions for (11.2) and $\bar{\Sigma}_Q$, whenever $u_q = 0$, we have that $e_q(t) \to 0$ as $t \to \infty$. Then (11.2) describes a filter that is a γ-level H_∞ SOID filter for the system Σ.*

Proof : We derive a proof only for the case $\gamma = 1$. The proof for the general case is then obtained by simply scaling.

First we apply a preliminary transformation $z = \bar{z} + P^*y$, where P^* is such that $D\bar{F}' = 0$ and

$$\bar{F} = F - P^*D \quad \text{and} \quad \bar{E} = E - P^*C.$$

Note that the existence of P such that $\|F - PD\| < 1$ guarantees that $\|\bar{F}\| < 1$. We define the following system $\bar{\Sigma}$:

$$\bar{\Sigma} : \begin{cases} \dot{x} = Ax + Bu \\ y = Cx + Du \\ \tilde{z} = \bar{E}x + \bar{F}u, \end{cases}$$

and then obtain the filter $\bar{\Sigma}_f$ for $\bar{\Sigma}$ as

$$\bar{\Sigma}_f : \begin{cases} \dot{\xi} = L\xi + My \\ \bar{z} = N\xi + (P - P^*)y. \end{cases}$$

Clearly $\bar{\Sigma}_f$ applied to $\bar{\Sigma}$ with error $e = \tilde{z} - \bar{z}$ yields the same system as Σ_f applied to Σ with error $e = z - \bar{z}$. We define the next two systems Σ_ℓ and Σ_U.

We have

$$\Sigma_\ell : \begin{cases} \dot{x}_\ell = A_\ell x_\ell + B_\ell u_\ell + G_\ell \bar{z} \\ \quad y = C x_\ell + (DD')^{1/2} u_\ell \\ \quad e_\ell = E_\ell x_\ell - H_\ell \bar{z}, \end{cases}$$

where

$$A_\ell = A + (Q\bar{E}' + B\bar{F}')(I - \bar{F}\bar{F}')^{-1}\bar{E},$$
$$B_\ell = (QC' + BD')(DD')^{-1/2},$$
$$G_\ell = -(Q\bar{E}' + B\bar{F}')(I - \bar{F}\bar{F}')^{-1},$$
$$E_\ell = (I - \bar{F}\bar{F}')^{-1/2}\bar{E},$$
$$H_\ell = (I - \bar{F}\bar{F}')^{-1/2},$$

and

$$\Sigma_U : \begin{cases} \dot{x}_u = (A - KC)x_u + (B - KD)u \\ \qquad\qquad - (Q\bar{E}' + \bar{B}\bar{F}')(I - \bar{F}\bar{F}')^{-1/2}e_\ell \\ \quad e = \bar{E}x_u + \bar{F}u + (I - \bar{F}\bar{F}')^{1/2}e_\ell \\ \quad u_\ell = (DD')^{-1/2}Cx_u + (DD')^{-1/2}Du, \end{cases} \qquad (11.60)$$

where

$$K = (QC' + BD')(DD')^{-1}.$$

The system Σ_U is co-inner. This is seen by noting that Q satisfies the conditions of Lemma 11.46 for the system Σ_U. Moreover, the subsystem from e_ℓ to e has a stable inverse (using that Q is a stabilizing solution of the H_∞^1 CARE). Then we can apply Lemma 11.47 where Ψ is the interconnection of $\bar{\Sigma}_f$ with Σ_ℓ.

Note that it can easily be checked that the interconnection of Σ_U and Σ_ℓ results in the same system as $\bar{\Sigma}$ when seen as a system with inputs u and $\bar{z}$ and outputs y and $e = \tilde{z} - \bar{z}$ but with some added unobservable and stable dynamics. Lemma 11.47 then yields that a filter of the form (11.2) is unbiased and results in an H_∞ norm less than 1 when applied to Σ if and only if the system $\bar{\Sigma}_f$ when applied to Σ_ℓ is such that the H_∞ norm of u_ℓ to e_ℓ is strictly less than 1, and for all initial conditions of $\bar{\Sigma}_f$ and Σ_ℓ, whenever $u_\ell = 0$, we have that $e_\ell(t) \to 0$ as $t \to \infty$.

If we replace u_ℓ by $(DD')^{-1/2}D\tilde{u}_\ell$, then clearly the induced norm from u_ℓ to e_ℓ is the same as the induced norm from $\tilde{u}_\ell$ to e_ℓ. We obtain the system:

$$\tilde{\Sigma}_\ell : \begin{cases} \dot{x}_\ell = A_\ell x_\ell + \tilde{B}_\ell \tilde{u}_\ell + G_\ell \bar{z} \\ \quad y = C x_\ell + D\tilde{u}_\ell \\ \quad e_\ell = \bar{E}_\ell x_\ell - H_\ell \bar{z}, \end{cases}$$

where

$$\tilde{B}_\ell = (QC' + BD')(DD')^{-1}D.$$

Hence, a filter of the form (11.2) is unbiased and results in an H_∞ norm less than 1, when applied to Σ if and only if the system $\bar{\Sigma}_f$ when applied to $\tilde{\Sigma}_\ell$ is such that the H_∞ norm of u_ℓ to e_ℓ is strictly less than 1, and for all initial conditions of $\bar{\Sigma}_f$ and $\tilde{\Sigma}_\ell$, whenever $u_\ell = 0$, we have that $e_\ell(t) \to 0$ as $t \to \infty$. Next, we define the static system:

$$\Sigma_F \: : \: \begin{pmatrix} e_q \\ \tilde{u}_\ell \end{pmatrix} = \begin{pmatrix} F & (I - FF')^{1/2} \\ (I - F'F)^{1/2} & -F' \end{pmatrix} \begin{pmatrix} u_q \\ e_\ell \end{pmatrix}.$$

Clearly Σ_F is co-inner with the subsystem from e_ℓ to e_q having a (stable) inverse. Note that the interconnection of Σ_F and $\tilde{\Sigma}_\ell$ equals the system $\tilde{\Sigma}_Q$, where

$$\tilde{\Sigma}_Q \: : \: \begin{cases} \dot{x}_q = A x_q + B_Q u_q + G_Q e_q \\ y = C x_q + D u_q \\ e_q = \bar{E} x_q + \bar{F} u_q - \bar{z}, \end{cases}$$

with

$$B_Q = \begin{pmatrix} CQ + DB' \\ \bar{E}Q + \bar{F}B' \end{pmatrix}' \begin{pmatrix} DD' & 0 \\ 0 & \bar{F}\bar{F}' - I \end{pmatrix}^{-1} \begin{pmatrix} D \\ \bar{F} \end{pmatrix},$$

$$G_Q = -(Q\bar{E}' + B\bar{F}')(\bar{F}\bar{F}' - I)^{-1} = -G_\ell.$$

Note that this definition can be shown to be consistent with the definition given in (11.57) and (11.58). Therefore, we can apply Lemma 11.47 once again with Σ_U replaced by Σ_F and Ψ the interconnection of $\tilde{\Sigma}_\ell$ and $\bar{\Sigma}_f$.

This yields that a filter of the form (11.2) is unbiased and results in an H_∞ norm less than 1 when applied to Σ if and only if the system $\bar{\Sigma}_f$ when applied to $\tilde{\Sigma}_Q$ is such that the H_∞ norm of u_q to e_q is strictly less than 1, and for all initial conditions of $\bar{\Sigma}_f$ and $\tilde{\Sigma}_Q$, whenever $u_q = 0$, we have that $e_q(t) \to 0$ as $t \to \infty$.

Finally, we note that the filter (11.2) applied to $\bar{\Sigma}_Q$ achieves the same system as the filter $\bar{\Sigma}_f$ applied to $\tilde{\Sigma}_Q$. The result then follows immediately. ■

The following lemma that connects Σ_Q and Σ is an important consequence of Lemma 11.50.

Lemma 11.51 *Consider the regular γ-level H_∞ SOID filtering problem for the continuous-time system Σ described by (11.1). That is, assume that D is surjective and that the system characterized by (A, B, C, D) has no invariant zeros on the imaginary axis. Also, assume that $\gamma > \gamma_{sp}^*$ (or $\gamma > \gamma_p^*$), which implies that a stabilizing solution Q of the H_∞^1 CARE (11.17) exists. Finally, assume that A is Hurwitz-stable. Let Σ_Q be as in (11.56).*

Then the following hold:

(*i*) *Suppose a filter of the form* (11.2) *and characterized by the quadruple* (L, M, N, P) *(with* $P = 0$ *for the strictly proper case) is a* γ*-level* H_∞ *SOID filter for the given system* Σ. *Then the following filter is a* γ*-level* H_∞ *SOID filter for the system* Σ_Q:

$$\begin{aligned}
\dot{x}_1 &= (A + G_Q E)x_1 - G_Q \hat{z}_q, \\
\dot{\xi} &= L\xi + M(y + Cx_1), \\
\hat{z}_q &= N\xi + P(y + Cx_1).
\end{aligned} \tag{11.61}$$

(*ii*) *Conversely, suppose a filter of the form* (11.2) *and characterized by the quadruple* (L, M, N, P) *(with* $P = 0$ *for the strictly proper case) is a* γ*-level* H_∞ *SOID filter for the system* Σ_Q. *Then the following filter is a* γ*-level* H_∞ *SOID filter for the given system* Σ:

$$\begin{aligned}
\dot{x}_1 &= (A + G_Q E)x_1 - G_Q \hat{z}, \\
\dot{\xi} &= L\xi + M(y - Cx_1), \\
\hat{z} &= N\xi + P(y - Cx_1).
\end{aligned} \tag{11.62}$$

Proof : We again prove this for the case of $\gamma = 1$. The general result follows from a scaling argument. By Lemma 11.50, a filter (11.2) applied to Σ is unbiased and achieves an H_∞ norm less than 1 if and only if the system (11.2) applied to $\bar{\Sigma}_Q$ achieves an H_∞ norm less than 1 from u_q to e_q, and for all initial conditions for (11.2) and $\bar{\Sigma}_Q$, whenever $u_q = 0$, we have that $e_q(t) \to 0$ as $t \to \infty$.

It remains to connect the interconnection of (11.2) and $\bar{\Sigma}_Q$ with the filter (11.61) for Σ_Q. But it is easily checked that a filter (11.61) applied to Σ_Q yields the same error dynamics (except for some additional stable but unobservable dynamics) as (11.2) applied to $\bar{\Sigma}_Q$. Conversely, a filter (11.62) applied to $\bar{\Sigma}_Q$ yields the same error dynamics (except for some additional stable but unobservable dynamics) as (11.2) applied to Σ_Q. The lemma then follows immediately.   ∎

By now we have completed the proofs of sufficiency parts of Theorems 11.16 and 11.19. The above development of Lemmas 11.48 to 11.51 clearly lays a roadmap for the design of γ-level H_∞ SOID filters for the given system Σ. That is, we construct new systems Σ_Q and $\check{\Sigma}_Q$. We design an EID filter for $\check{\Sigma}_Q$, which can then be translated as an H_∞ AID filter for Σ_Q. Then in turn such an H_∞ AID filter for Σ_Q can be translated as a γ-level H_∞ SOID filter for the given system Σ. We observe that a filter of any arbitrary architecture can be used to design an EID filter for $\check{\Sigma}_Q$. However, by using the design methodology developed in Chapter 7, one can easily design an EID filter of CSS architecture for $\check{\Sigma}_Q$.

We note that Lemma 11.51 assumes that the matrix A is Hurwitz-stable. For a filtering problem, the assumption of matrix A being Hurwitz-stable is natural but may not be very elegant. However, we can remove such an assumption when we focus below on the design of γ-level H_∞ SOID filters of CSS architecture for the given system Σ.

We proceed now to design γ-level H_∞ SOID filters of CSS architecture for Σ. As in the case of previous chapters, we consider strictly proper, proper, and reduced-order filters of CSS architecture, one at a time.

Design of γ-level strictly proper H_∞ SOID filters of CSS architecture.

As we said above, Lemma 11.51 plays a fundamental role in proving the sufficiency parts of Theorems 11.16 and 11.19, while suggesting a methodology of designing γ-level H_∞ SOID filters. The following lemma explores this further by using strictly proper filters of CSS architecture.

Lemma 11.52 *Consider the regular γ-level H_∞ SOID filtering problem for the continuous-time system Σ described by (11.1). That is, assume that D is surjective and that the system characterized by (A, B, C, D) has no invariant zeros on the imaginary axis. Also, assume that $\gamma > \gamma_{sp}^*$, which implies that a stabilizing solution Q of the H_∞^1 CARE (11.17) exists. Also, define Σ_Q by (11.56).*
Then the following hold:

(i) *Let the following strictly proper filter, which is of CSS architecture, be a γ-level H_∞ SOID filter for Σ:*

$$\begin{aligned}
\dot{\xi} &= A\xi + K(y - C\xi), \\
\hat{z} &= E\xi.
\end{aligned} \qquad (11.63)$$

Then, the following strictly proper filter of CSS architecture is a γ-level H_∞ SOID filter for the system Σ_Q:

$$\begin{aligned}
\dot{\xi} &= A\xi + K(y - C\xi) + G_Q\hat{z}_q, \\
\hat{z}_q &= E\xi.
\end{aligned} \qquad (11.64)$$

(ii) *Conversely, let a strictly proper filter of the form (11.64), which is of CSS architecture, be a γ-level H_∞ SOID filter for Σ_Q. Then, the strictly proper filter (11.63) is a γ-level H_∞ SOID filter for Σ.*

Proof : The proof relies on Lemma 11.50. It is easily checked that the error dynamics that results when the filter (11.64) is applied to Σ_Q is the same as the dynamics of e_q when (11.63) is applied to $\bar{\Sigma}_Q$ as shown in Figure 11.3. Lemma 11.50 then immediately yields that (11.64) is a γ-level H_∞ SOID filter for Σ_Q if and only if (11.63) is a γ-level H_∞ SOID filter for Σ.

The fact that for Σ_Q we can obtain a γ-level H_∞ SOID filter is a direct consequence of Lemmas 11.48 and 11.49. ∎

We can now summarize the methodology of designing a strictly proper γ-level H_∞ SOID filter for Σ:

- Consider $\check{\Sigma}_Q$ of (11.59) with $P = 0$. That is, consider

$$\check{\Sigma}_Q : \begin{cases} \dot{x}_q = A x_q + (B_Q + G_Q F)u_q + G_Q z_q \\ \;\; y = C x_q + D u_q \\ \;\; \check{z}_q = E x_q. \end{cases} \tag{11.65}$$

By Lemma 11.48, a strictly proper EID filter of CSS architecture can be designed for $\check{\Sigma}_Q$. This can be done by using the design procedure developed in Chapter 7.

- Lemma 11.49 shows that the filter designed above is a γ-level H_∞ SOID filter for Σ_Q.

- We can then use Lemma 11.52 to obtain a strictly proper γ-level H_∞ SOID filter of CSS architecture (which can be denoted by Σ_{sp-CSS}) for Σ.

Remark 11.53 *We would like to point out one aspect of the above design procedure. Even if we can characterize all possible EID filters for $\check{\Sigma}_Q$, the above procedure does not enable us to characterize all possible γ-level H_∞ SOID filters for Σ. Also it should be stressed that minimizing the H_∞ norm of the error dynamics for Σ_Q does not mean that we minimize the H_∞ norm of the error dynamics when considering the original system Σ. The above design yields so-called error dynamics with minimum entropy as studied in [52]. In can also be interpreted in the sense of minimizing a linear exponential quadratic Gaussian (LEQG) criterion as studied in, for instance, [101]. However, it should be noted that both of these references look at this problem from a control perspective.*

Design of γ-level proper H_∞ SOID filters of CSS architecture.
We now proceed to the case of designing γ-level proper H_∞ SOID filters of CSS architecture. We can pursue two methods of designing such filters. One method that can be called a direct method arises from the proofs of Lemmas 11.48 to 11.51, that is, by computing the systems $\check{\Sigma}_Q$ and Σ_Q from the data of the system Σ, designing a strictly proper EID filter for $\check{\Sigma}_Q$ by using the methodology developed in Chapter 7 (which results as a proper γ-level H_∞ SOID filter for Σ_Q by Lemma 11.49), and then converting such a filter to a proper γ-level H_∞ SOID filter for the given system Σ by Lemma 11.51. The second method follows the footsteps of our design philosophy in Chapters 7–10. In this method, by using the methodology we just developed above to design a *strictly* proper γ-level H_∞ SOID filter, we first design a strictly proper γ-level H_∞ SOID filter for an auxiliary system $\widetilde{\Sigma}^*$, which is essentially the same as the given system Σ except that it has some additional measurement output. It is shown that such a strictly proper filter for $\widetilde{\Sigma}^*$ exists whenever a proper γ-level H_∞ SOID filter exists for Σ. We then transform such a strictly proper filter for $\widetilde{\Sigma}^*$ to a proper γ-level H_∞ SOID

filter for the given system Σ. It is shown that these two methods of design lead to the same proper filter that can be denoted by Σ_{p-CSS}.

To explain explicitly the above two methods and to show that they lead to the same filter Σ_{p-CSS}, we first need to construct the auxiliary system $\tilde{\Sigma}^*$. To do so, as in the previous chapters, for ease of presentation and without loss of generality, we decompose the measured output y into two parts y_0 and y_1 in such a way that y_0 contains explicitly the unknown input u in it, whereas y_1 does not contain any input u in it. That is, we write

$$y = \begin{pmatrix} y_0 \\ y_1 \end{pmatrix} = Cx + Du \quad \text{and} \quad C = \begin{pmatrix} C_0 \\ C_1 \end{pmatrix}, \quad D = \begin{pmatrix} D_0 \\ 0 \end{pmatrix}, \tag{11.66}$$

where matrix D_0 has rank m_0. We can then rewrite the given system (11.1) as

$$\Sigma : \begin{cases} \dot{x} \;\;\;\; = Ax + Bu \\ \begin{pmatrix} y_0 \\ y_1 \end{pmatrix} = \begin{pmatrix} C_0 \\ C_1 \end{pmatrix} x + \begin{pmatrix} D_0 \\ 0 \end{pmatrix} u \\ z \;\;\;\; = Ex + Fu, \end{cases} \tag{11.67}$$

where matrix D_0 has rank m_0. Assume (11.18) is satisfied, which implies that we can choose P^* such that $\|F^*\| < \gamma$, where

$$E^* = E - P^*C \quad \text{and} \quad F^* = F - P^*D.$$

Finally, we define the auxiliary system $\tilde{\Sigma}^*$ as

$$\tilde{\Sigma}^* : \begin{cases} \dot{x} \;\; = Ax + Bu \\ \tilde{y} \;\; = \begin{pmatrix} y_0 \\ y_1 \\ \dot{y}_1 \end{pmatrix} = \tilde{C}x + \tilde{D}u \\ z^* = E^*x + F^*u, \end{cases} \tag{11.68}$$

where

$$\tilde{C} = \begin{pmatrix} C_0 \\ C_1 \\ C_1 A \end{pmatrix} = \begin{pmatrix} C \\ C_1 A \end{pmatrix}, \tag{11.69}$$

$$\tilde{D} = \begin{pmatrix} D_0 \\ 0 \\ C_1 B \end{pmatrix} = \begin{pmatrix} D \\ C_1 B \end{pmatrix}. \tag{11.70}$$

We observe that the desired output z^* that needs to be estimated in system $\tilde{\Sigma}^*$ is obtained from the desired output z that needs to be estimated in the given system

Σ; that is, $z^* = z - P^* y$. Thus, whenever an estimate of z^* is available, one can form easily an estimate of z. Also, as will be shown, a strictly proper γ-level H_∞ SOID filter exists for the auxiliary system $\widetilde{\Sigma}^*$ whenever a proper γ-level H_∞ SOID filter exists for the given system Σ.

We are now ready to restate the following two design methodologies that we alluded to earlier:

- **First method:** Compute the systems $\widecheck{\Sigma}_Q$ and Σ_Q from the data of the given system Σ. Then design a strictly proper EID filter of CSS architecture for $\widecheck{\Sigma}_Q$ by using the methodology developed in Chapter 7. Such a filter is a proper γ-level H_∞ SOID filter of CSS architecture for Σ_Q (by Lemma 11.49). Then convert this to a proper γ-level H_∞ SOID filter of CSS architecture for Σ using the previously described technique given in Lemma 11.51.

- **Second method:** Construct the systems $\widecheck{\widetilde{\Sigma}}_{\widetilde{Q}}$ and $\widetilde{\Sigma}_{\widetilde{Q}}$ from the data of the system $\widetilde{\Sigma}^*$ by first obtaining a stabilizing solution $\widetilde{Q}$ of the H_∞^1 CARE associated with the system $\widetilde{\Sigma}^*$ by following the same procedure as that of constructing the systems $\widecheck{\Sigma}_Q$ and Σ_Q from the data of Σ. Design a strictly proper EID filter of CSS architecture for $\widecheck{\widetilde{\Sigma}}_{\widetilde{Q}}$ by using the methodology developed in Chapter 7 and then convert this to a strictly proper γ-level H_∞ SOID filter of CSS architecture for $\widetilde{\Sigma}^*$ (This is the same as applying Lemma 11.52 to the system $\widetilde{\Sigma}^*$ instead of to the system Σ, and following the design method for strictly proper filters of CSS architecture developed right after Lemma 11.52). Finally convert this to a proper γ-level H_∞ SOID filter of CSS architecture for Σ.

We emphasize that both of these two design methods actually result in the same filters. This follows quickly when we note that $\widetilde{Q} = Q$. For details of these two different design philosophies, we refer to the previous chapter where these two routes have been explained in more detail. The arguments for H_∞ SOID are basically the same as in the H_2 OID case.

We now proceed with the development of the second method. Applying Lemma 11.52 to the system $\widetilde{\Sigma}^*$ instead of to the system Σ, and following the design method developed right after Lemma 11.52, we can design a γ-level strictly proper H_∞ SOID filter of CSS architecture for $\widetilde{\Sigma}^*$ as

$$\widetilde{\Sigma}^*_{sp-CSS} : \begin{cases} \dot{\widetilde{\xi}} = A\widetilde{\xi} + K(\widetilde{y} - \widetilde{C}\widetilde{\xi}) \\ \widehat{z}^* = E^*\widetilde{\xi}, \end{cases} \tag{11.71}$$

where the matrix K is a filter gain. Existence of gain K and thus the existence of the above filter follows from Lemma 11.54 that is to be developed soon shortly. The above filter requires $\dot{y}_1$ as one of its inputs. To avoid this, we modify the above filter. To do so, we can partition K in conformity with the partitioning of $\widetilde{y}$. That is,

$$K = \begin{pmatrix} K_0 & K_1 & K_2 \end{pmatrix}.$$

We can then rewrite the filter equation (11.71) as

$$
\begin{cases}
\dot{\xi} = (A - K\widetilde{C})\xi + \begin{pmatrix} K_0 & K_1 + (A - K\widetilde{C})K_2 \end{pmatrix} y \\
\widetilde{\xi} = \xi + K_2 y_1 \\
\widehat{z}^* = E^*\widetilde{\xi} = E^*(\xi + K_2 y_1).
\end{cases}
\tag{11.72}
$$

As this filter does not use $\dot{y}_1$, it is easy to convert this filter to a filter for Σ by noting the relationship between z^* and z, namely $z^* = z - P^* y$. We thus obtain

$$
\Sigma_{p-css} : \begin{cases}
\dot{\xi} = (A - K\widetilde{C})\xi + \begin{pmatrix} K_0 & K_1 + (A - K\widetilde{C})K_2 \end{pmatrix} y \\
\widehat{z} = E^*(\xi + K_2 y_1) + P^* y.
\end{cases}
\tag{11.73}
$$

We have the following lemma.

Lemma 11.54 *Consider the regular γ-level H_∞ SOID filtering problem for the continuous-time system Σ described by (11.1). That is, assume that D is surjective and that the system characterized by (A, B, C, D) has no invariant zeros on the imaginary axis. Also, assume that $\gamma > \gamma_p^*$. Choose P^* such that $\|F - P^*C\| < \gamma$, and define $\widetilde{\Sigma}^*$ by (11.68). Then the γ-level H_∞ SOID filtering problem for $\widetilde{\Sigma}^*$ is solvable by a strictly proper filter of the form (11.71) having the CSS architecture. Moreover, any strictly proper γ-level H_∞ SOID filter of the form (11.71) for $\widetilde{\Sigma}^*$ results in a proper γ-level H_∞ SOID filter (11.73) with CSS architecture for the system Σ.*

Proof : The fact that F^* satisfies $\|F^*\| < \gamma$ and that Q satisfies the H_∞^1 CARE associated with the system $\widetilde{\Sigma}^*$ shows that the strictly proper γ-level H_∞ SOID filtering problem for $\widetilde{\Sigma}^*$ is solvable.

The error dynamics of the filter (11.71) applied to $\widetilde{\Sigma}^*$ is equal to the error dynamics of (11.73) applied to Σ. Therefore, it is clear that (11.73) is a proper γ-level H_∞ SOID filter for Σ whenever (11.71) is a strictly proper γ-level H_∞ SOID filter for $\widetilde{\Sigma}^*$. ∎

Just like in the strictly proper case, the above design of course does not result in a complete characterization of all γ-level H_∞ SOID filters.

Design of reduced-order γ-level H_∞ SOID filters of CSS architecture.

We now proceed with the design of reduced-order γ-level H_∞ SOID filters of CSS architecture. To do this, as in Chapters 7–10, we first extract a reduced-order system Σ_r^* from the given system Σ and then design a strictly proper γ-level H_∞ SOID filter for it. It is shown that such a strictly proper γ-level H_∞ SOID filter exists for Σ_r^* whenever a proper γ-level H_∞ SOID filter exists for the given system Σ. The strictly proper filter designed for Σ_r^* is then translated to form a reduced-order γ-level H_∞ SOID filter for the given system Σ.

We now proceed to extract the reduced-order system Σ_r^* from the given system Σ. To do so, to start with, let us rewrite the matrices C and D of (11.1) as

$$C = \begin{pmatrix} 0 & C_{02} \\ I_{p-m_0} & 0 \end{pmatrix}, \quad D = \begin{pmatrix} D_0 \\ 0 \end{pmatrix},$$

where again rank D = rank D_0 = m_0. This can always be done without any loss of generality by appropriate coordinate transformations. In view of the above partitioning of C and D, we can partition the given system Σ as

$$\Sigma : \begin{cases} \begin{pmatrix} \dot{x}_1 \\ \dot{x}_2 \end{pmatrix} = \begin{pmatrix} A_{11} & A_{12} \\ A_{21} & A_{22} \end{pmatrix} \begin{pmatrix} x_1 \\ x_2 \end{pmatrix} + \begin{pmatrix} B_{11} \\ B_{22} \end{pmatrix} u \\[2mm] y = \begin{pmatrix} y_0 \\ y_1 \end{pmatrix} = \begin{pmatrix} 0 & C_{02} \\ I & 0 \end{pmatrix} \begin{pmatrix} x_1 \\ x_2 \end{pmatrix} + \begin{pmatrix} D_0 \\ 0 \end{pmatrix} u \\[2mm] z = Ex + Fu, \end{cases} \tag{11.74}$$

where different variables have obvious meanings.

Let us next assume that (11.18) is satisfied, which implies that we can choose P^* such that $\|F^*\| < \gamma$, where

$$E^* = E - P^*C \quad \text{and} \quad F^* = F - P^*D.$$

We define a new desired output z^* that needs to be estimated as

$$z^* = z - P^*y = (E - P^*C)x + (F - P^*D)u = E^*x + F^*u. \tag{11.75}$$

Note that whenever an estimate of z^* is available, one can form easily an estimate of z. Thus, in view of (11.74) and (11.75), we can define a new system Σ^* as

$$\Sigma^* : \begin{cases} \begin{pmatrix} \dot{x}_1 \\ \dot{x}_2 \end{pmatrix} = \begin{pmatrix} A_{11} & A_{12} \\ A_{21} & A_{22} \end{pmatrix} \begin{pmatrix} x_1 \\ x_2 \end{pmatrix} + \begin{pmatrix} B_{11} \\ B_{22} \end{pmatrix} u \\[2mm] y = \begin{pmatrix} y_0 \\ y_1 \end{pmatrix} = \begin{pmatrix} 0 & C_{02} \\ I & 0 \end{pmatrix} \begin{pmatrix} x_1 \\ x_2 \end{pmatrix} + \begin{pmatrix} D_0 \\ 0 \end{pmatrix} u \\[2mm] z^* = E^*x = E_1^*x_1 + E_2^*x_2 + F^*u, \end{cases} \tag{11.76}$$

with $E^* = \begin{pmatrix} E_1^* & E_2^* \end{pmatrix}$.

We note that the y_1 is not contaminated by the input u, and hence, $x_1 = y_1$ is known exactly from the measurement y. Thus, all we need to do next is to estimate the state x_2. To proceed further, let us rewrite the state equation for x_1 in terms of the output y_1 and the state x_2 as

$$\dot{y}_1 = A_{11}y_1 + A_{12}x_2 + B_{11}u. \tag{11.77}$$

The above equation can be rewritten as

$$\dot{y}_1 - A_{11}y_1 = A_{12}x_2 + B_{11}u.$$

Treating $\dot{y}_1$ as known, we can define a new measurement variable y_r:

$$y_r = \begin{pmatrix} y_0 \\ \dot{y}_1 - A_{11}y_1 \end{pmatrix}.$$

Although $\dot{y}_1$ is not directly available, as we did in previous chapters, we can eliminate it from any filter that is constructed using y_r as a measured output. With this in mind, we form the following auxiliary system:

$$\Sigma_r^* : \begin{cases} \dot{x}_r = A_r x_r + B_r u + A_{21}y_1 \\ y_r = C_r x_r + D_r u \\ z_r^* = E_r^* x_r + F_r^* u, \end{cases} \tag{11.78}$$

where $x_r = x_2$ and

$$A_r = A_{22}, \quad B_r = B_{22}, \quad C_r = \begin{pmatrix} C_{02} \\ A_{12} \end{pmatrix}, \quad D_r = \begin{pmatrix} D_0 \\ B_{11} \end{pmatrix},$$

$$E_r^* = E_2^*, \quad F_r^* = F^*. \tag{11.79}$$

We note that the dynamic order n_r of the above Σ_r^* is less than the dynamic order n of the given system Σ by a number equal to the dimension of $x_1 = y_1$.

Next, to construct the required filter for Σ, in the spirit of the above development, we first construct a strictly proper γ-level H_∞ SOID filter of CSS architecture for the reduced-order system Σ_r^* as

$$\begin{cases} \dot{\tilde{\xi}}_r = A_r \tilde{\xi}_r + A_{21}y_1 + K_r(y_r - C_r\tilde{\xi}_r) \\ \hat{z}_r^* = E_r^* \tilde{\xi}_r, \end{cases} \tag{11.80}$$

where K_r is termed as the reduced-order filter gain. As shown shortly in Lemma 11.55, the gain K_r and thus the filter (11.80) always exist whenever a proper γ-level H_∞ SOID filter exists for Σ.

As $\dot{y}_1$ is not available, we need to modify the above filter. To this end, let us partition $K_r = (K_{r0} \quad K_{r1})$ so as to be compatible with the partitioning of y_r. Also, let

$$\xi_r = \tilde{\xi}_r - K_{r1}y_1. \tag{11.81}$$

We can then easily rewrite the filter (11.80) as a proper filter for Σ as

$$\Sigma_{r-css} : \begin{cases} \dot{\xi}_r = (A_r - K_r C_r)\xi_r + \tilde{K}_r y \\ \tilde{\xi}_r = \xi_r + K_{r1}y_1 \\ \hat{z} = E_1^* x_1 + E_r^* \tilde{\xi}_r + P^* y = E_r^* \xi_r + \tilde{P}_r y, \end{cases} \tag{11.82}$$

where

$$\tilde{K}_r = \begin{pmatrix} K_{r0} & A_{21} - K_{r1}A_{11} + (A_r - K_rC_r)K_{r1} \end{pmatrix}$$

and

$$\tilde{P}_r = \begin{pmatrix} 0 & E_1^* + E_r^*K_{r1} \end{pmatrix} + P^*.$$

The above development leads to the following lemma.

Lemma 11.55 *Consider the regular γ-level H_∞ SOID filtering problem for the continuous-time system Σ described by (11.1). That is, assume that D is surjective and that the system characterized by (A, B, C, D) has no invariant zeros on the imaginary axis. Also, assume that $\gamma > \gamma_p^*$.*

Then, the strictly proper γ-level H_∞ SOID filtering problem is solvable for the system Σ_r^*. Moreover, any strictly proper γ-level H_∞ SOID filter of the form (11.80) with CSS architecture for Σ_r^* results in a reduced-order γ-level H_∞ SOID filter (11.82) with CSS architecture for the system Σ.

Proof : Let Q be the stabilizing solution of the H_∞^1 CARE associated with Σ. Then it is easy to verify that $C_1 Q = 0$, and using this, it becomes clear that Q has the form:

$$Q = \begin{pmatrix} Q_r & 0 \\ 0 & 0 \end{pmatrix},$$

where Q_r is a stabilizing solution of the H_∞^1 CARE associated with Σ_r^*. This shows that the the strictly proper γ-level H_∞ SOID filtering problem for Σ_r^* is solvable. The error dynamics of a strictly proper H_∞ SOID filter of the form (11.80) for Σ_r^* is the same as the error dynamics resulting from applying the filter (11.82) to the system Σ. It is then immediately obvious that (11.82) is a reduced-order γ-level H_∞ SOID for the system Σ. ∎

Again, we note that the above design does not result in a complete characterization of all reduced-order γ-level H_∞ SOID filters.

We should note that the alternative route of first transforming Σ to Σ_Q, designing a reduced-order γ-level H_∞ SOID filter of CSS architecture for Σ_Q, and then converting it back to a filter for Σ results in the same family of reduced-order γ-level H_∞ SOID filters.

11.5.2 *Singular γ-level H_∞ SOID filters—the system characterized by (A, B, C, D) has no invariant zeros on the imaginary axis*

The previous subsection deals with the regular γ-level H_∞ SOID filtering. In this subsection, we consider the singular γ-level H_∞ SOID filtering for the special case when the system characterized by (A, B, C, D) has no invariant zeros on the imaginary axis. Following the spirit of the regular case, we first prove the necessity parts of proofs of Theorems 11.22 and 11.26. Then, we prove the sufficiency parts of the same theorems. Once again, as in the previous subsection, the proofs of sufficiency parts lay a roadmap for the design of appropriate γ-level H_∞ SOID filters for the given system Σ. Then, as usual, we explore the design of strictly proper, proper, as well as reduced-order γ-level H_∞ SOID filters of CSS architecture.

We start with a necessity proof for Theorem 11.22.

Proof of necessity for Theorem 11.22 (no invariant zeros of the system (A, B, C, D) on the imaginary axis) : We use the dual decomposition from the one presented in (4.222). Hence, a matrix K exists such that in a suitable basis (the compact form of the SCB), we have

$$
A - KC = \begin{pmatrix} A_{11} & B_1 F \\ A_{21} & A_{22} \end{pmatrix}, \quad B - KD = \begin{pmatrix} B_1 \\ B_2 \end{pmatrix},
$$

$$
C = \begin{pmatrix} C_{11} & 0 \\ C_{21} & C_{22} \end{pmatrix}, \quad D = \begin{pmatrix} 0 \\ D_2 \end{pmatrix}, \quad E = \begin{pmatrix} E_1 & E_2 \end{pmatrix},
$$

where K is such that $(B - KD)'D = 0$. Moreover, D_2 is surjective, (C_1, A_{11}) is $\mathbb{C}^-$-detectable, $D_2 B_2' = 0$, and $D_2 B_1' = 0$. An output injection does not change the achievable H_∞ norm of the error dynamics. Moreover, if we assume that x_1 is additionally measured without noise, then the achievable H_∞ norm of the error dynamics by an unbiased filter stays the same or can be made even smaller. Therefore, for the following system, we can achieve an H_∞ norm of the error dynamics less than γ by an unbiased filter:

$$
\begin{cases}
\dot{x}_2 = A_{22}x_2 + B_2 u \\
y = \begin{pmatrix} L B_1 F \\ C_{22} \end{pmatrix} x_2 + \begin{pmatrix} L B_1 \\ D_2 \end{pmatrix} u \\
z = E_2 x_2 + F u,
\end{cases}
$$

where L is such that $\operatorname{rank} L B_1 = \operatorname{rank} B_1$ while $L B_1$ is surjective. As this is a regular H_∞ SOID filtering problem, we can apply the results from Theorem 11.16, which guarantees the existence of a stabilizing solution of the H_∞^1 CARE:

$$
\bar{Y} \bar{A}_r' + \bar{A}_r \bar{Y} - \begin{pmatrix} C_r \bar{Y} + D_r B_r' \\ E_r \bar{Y} + F_r B_r' \end{pmatrix}' \begin{pmatrix} D_r D_r' & D_r F_r' \\ F_r D_r' & F_r F_r' - \gamma^2 I \end{pmatrix}^{-1} \begin{pmatrix} C_r \bar{Y} + D_r B_r' \\ E_r \bar{Y} + F_r B_r' \end{pmatrix}
$$
$$
+ B_r B_r' = 0,
$$

where

$$
\bar{A}_r = A_{22}, B_r = B_2, C_r = \begin{pmatrix} L B_1 F \\ C_{22} \end{pmatrix}, D_r = \begin{pmatrix} L B_1 \\ D_2 \end{pmatrix}, E_r = E_2. \quad (11.83)
$$

Using a dual version of Theorem 4.158, we then obtain the existence of a stabilizing solution of the CQMI (11.20).

The fact that (11.19) must be satisfied when using strictly proper filters should by now be obvious. The conditions regarding the invariant zeros on the imaginary axis are clearly empty because we are looking at the case where there are no invariant zeros on the imaginary axis. ∎

We continue next with a necessity proof for the equivalent theorem when proper filters are used.

Proof of necessity for Theorem 11.26 (no invariant zeros of the system (A, B, C, D) on the imaginary axis) : The proof follows along the same lines as in the necessity proof of Theorem 11.22. The fact that (11.22) is a necessary condition should be obvious.

We again assume that x_1 is available to us and construct a new system. By applying Theorem 11.19, we obtain the existence of a stabilizing solution of the associated H_∞^1 CARE. The dual version of Theorem 4.158 then results in the existence of a stabilizing solution of the CQMI (11.23). ∎

Proofs of sufficiency parts of Theorems 11.22 and 11.26 (no invariant zeros of the system (A, B, C, D) on the imaginary axis).

So far we have proved the necessity parts of Theorems 11.22 and 11.26. We proceed now to prove the sufficiency parts. In this regard, as we said, our design methodology arises as a by-product while proving the sufficiency parts. As in the previous subsection, we can then explore the design of strictly proper, proper, as well as reduced-order γ-level H_∞ SOID filters of CSS architecture.

As in the previous subsection, we define two auxiliary systems Σ_Q and $\check{\Sigma}_Q$. We define Σ_Q by

$$\Sigma_Q : \begin{cases} \dot{x}_q = A x_q + B_Q u_q + G_Q z_q \\ y = C x_q + D u_q \\ z_q = E x_q + F u_q, \end{cases} \tag{11.84}$$

where B_Q and G_Q are defined by

$$B_Q = -\begin{pmatrix} QC' + BD' & QE' + BF' \end{pmatrix} \begin{pmatrix} DD' & DF' \\ FD' & FF' - \gamma^2 I \end{pmatrix}^{\dagger} \begin{pmatrix} D \\ F \end{pmatrix}, \tag{11.85a}$$

$$G_Q = -\begin{pmatrix} QC' + BD' & QE' + BF' \end{pmatrix} \begin{pmatrix} DD' & DF' \\ FD' & FF' - \gamma^2 I \end{pmatrix}^{\dagger} \begin{pmatrix} 0 \\ I \end{pmatrix}. \tag{11.85b}$$

Here $M^\dagger$ denotes the generalized inverse of M. This is the same construction as for the regular case in (11.56), but this time Q is a solution of the CQMI (11.23). Note that in the strictly proper case, we have (11.19), and hence, as noted in Remark 11.29, the CQMIs (11.20) and (11.23) are equal.

We can show that the strictly proper (or proper) γ-level H_∞ SOID filtering problem for Σ_Q is solvable whenever the strictly proper (or proper) γ-level H_∞ SOID filtering problem for Σ is solvable. To do so, as in the regular case, we introduce next a slightly modified version of Σ_Q, namely a system $\check{\Sigma}_Q$ whose desired output signal $\check{z}_q$ that needs to be estimated is slightly different from z_q:

$$\check{\Sigma}_Q : \begin{cases} \dot{x}_q = (A + G_Q E)x_q + (B_Q + G_Q F)u_q \\ y = C x_q + D u_q, \\ \check{z}_q = (E - PC)x_q, \end{cases} \tag{11.86}$$

where the matrix P, as in the previous section, is different for the two different cases, strictly proper and proper. For the case when Σ satisfies the conditions for the existence of strictly proper γ-level H_∞ SOID filters, we have $\|F\| < \gamma$. Then a suitable choice for the matrix P is $P = 0$. On the other hand, when Σ satisfies the conditions for the existence of proper γ-level H_∞ SOID filters, a matrix P exists such that $\|F - PD\| < \gamma$. Then, obviously a suitable choice for the matrix P in (11.86) is such that $\|F - PD\| < \gamma$.

As before, we make two claims. First, for the system $\check{\Sigma}_Q$ the H_∞ AID filtering problem is solvable by a sequence of strictly proper filters irrespective of the choice of P. Second, any sequence of H_∞ AID filters designed for $\check{\Sigma}_Q$ using the methodology of Chapter 9 can be translated as γ-level H_∞ SOID filters for Σ_Q. These claims are formalized in the following two lemmas.

Lemma 11.56 *Consider the singular γ-level H_∞ SOID filtering problem for the continuous-time system Σ described by (11.1) such that the system characterized by (A, B, C, D) has no invariant zeros on the imaginary axis. Assume that $\gamma > \gamma_{sp}^*$ (or $\gamma > \gamma_p^*$); that is, assume that (11.19) is satisfied, and that a stabilizing solution Q of (11.20)) exists [or assume that (11.22) is satisfied and that a stabilizing solution Q of (11.23) exists]. Define $\check{\Sigma}_Q$ by (11.86) with $P = 0$ (or with P such that $\|F - PD\| < \gamma$).*

Then, the H_∞ AID filtering problem is solvable for $\check{\Sigma}_Q$ by a sequence of strictly proper filters irrespective of $P = 0$ (strictly proper case) or P is such that $\|F - PD\| < \gamma$ (proper case).

Proof : The fact that Q is a stablizating solution of the CQMI (11.23) implies that

$$\text{rank} \begin{pmatrix} A + G_Q E & B_Q + G_Q F \\ C & D \end{pmatrix} = n + \text{rank} \begin{pmatrix} B_Q + G_Q F \\ D \end{pmatrix}.$$

This implies that

$$\mathcal{V}^*(\Sigma_*) \cap \mathcal{S}^{-0}(\Sigma_*) = \{0\},$$

where Σ^* is the subsystem described by the quadruple:

$$(A + G_Q E, B_Q + G_Q F, C, D).$$

This in turn shows that the H_∞ AID filtering problem is solvable for $\breve{\Sigma}_Q$ according to Theorem 9.5. ∎

Lemma 11.57 *Consider the singular γ-level H_∞ SOID filtering problem for the continuous-time system Σ described by (11.1) such that the system characterized by (A, B, C, D) has no invariant zeros on the imaginary axis. Assume that $\gamma > \gamma_{sp}^*$ (or $\gamma > \gamma_p^*$); that is, assume that (11.19) is satisfied and that a stabilizing solution Q of (11.20) exists [or assume that (11.22) is satisfied and that a stabilizing solution Q of (11.23) exists]. Define Σ_Q by (11.84). Also, define $\breve{\Sigma}_Q$ by (11.86) with $P = 0$ for the strictly proper case (or with P such that $\|F - PD\| < \gamma$ for the proper case).*
Then the following hold:

- (i) *Let a sequence of strictly proper filters of the form (11.2) characterized by a parameterized matrix triple $(L_\varepsilon, M_\varepsilon, N_\varepsilon)$ solve the H_∞ AID filtering problem for the system $\breve{\Sigma}_Q$ with $P = 0$ (strictly proper case). Then, for ε sufficiently small, any member of such a sequence is a strictly proper γ-level H_∞ SOID filter for Σ_Q.*

- (ii) *Let a sequence of strictly proper filters of the form (11.2) characterized by a parameterized matrix triple $(L_\varepsilon, M_\varepsilon, N_\varepsilon)$ solve the H_∞ AID filtering problem for the system $\breve{\Sigma}_Q$ with P such that $\|F - PD\| < \gamma$ (proper case). For ε sufficiently small, consider a proper filter of the form (11.2) characterized by the matrix quadruple $(L_\varepsilon, M_\varepsilon, N_\varepsilon, P)$, where the matrix triple $(L_\varepsilon, M_\varepsilon, N_\varepsilon)$ corresponds to the sequence of strictly proper H_∞ AID filters for $\breve{\Sigma}_Q$ and the matrix P corresponds to the one that is used in defining $\breve{\Sigma}_Q$. Then such a proper filter is a proper γ-level H_∞ SOID filter for Σ_Q.*

Proof : This lemma is analogous to Lemma 11.49, and the proof follows along the same lines as the proof of Lemma 11.49. ∎

The above lemmas guarantee that the strictly proper (or proper) γ-level H_∞ SOID filtering problem is solvable for Σ_Q whenever the strictly proper (or proper) γ-level H_∞ SOID filtering problem is solvable for Σ. We will shortly and formally establish a relationship between the γ-level H_∞ SOID filters for Σ and the γ-level H_∞ SOID filters for Σ_Q. This result will be obtained from the result for the regular case (Lemmas 11.50 and 11.51). The following lemma will be a crucial component in establishing such a result.

Lemma 11.58 *Consider the singular γ-level H_∞ SOID filtering problem for the continuous-time system Σ described by (11.1) such that the system characterized by (A, B, C, D) has no invariant zeros on the imaginary axis. Assume that $\gamma > \gamma_{sp}^*$ (or $\gamma > \gamma_p^*$); that is, assume that (11.19) is satisfied, and that a stabilizing solution Q of (11.20)) exists [or assume that (11.22) is satisfied and that a stabilizing solution Q of (11.23)) exists]. Define the following system $\bar{\Sigma}_Q$ with two inputs u_q and $\hat{z}$ and two outputs y and e_q as shown in Figure 11.3:*

$$\bar{\Sigma}_Q : \begin{cases} \dot{x}_q = Ax_q + B_Q u_q + G_Q e_q \\ \quad y = Cx_q + Du_q \\ \quad e_q = Ex_q + Fu_q - \hat{z}. \end{cases}$$

Then the following hold:

(i) *Suppose a filter Σ_f of the form (11.2) and characterized by the quadruple (L, M, N, P) (with $P = 0$ for strictly proper case) is a γ-level H_∞ SOID filter for the given system Σ. Then the interconnection of Σ_f and $\bar{\Sigma}_Q$ (that is, $\bar{\Sigma}_Q \times \Sigma_f$ as shown in Figure 11.3) is such that the H_∞ norm of u_q to e_q is strictly less than γ, and for all initial conditions for (11.2) and $\bar{\Sigma}_Q$, whenever $u_q = 0$, we have that $e_q(t) \to 0$ as $t \to \infty$.*

(ii) *Conversely, suppose a system Σ_f of the form (11.2) and characterized by the quadruple (L, M, N, P) (with $P = 0$ for strictly proper case) interconnected with the system $\bar{\Sigma}_Q$ (that is, $\bar{\Sigma}_Q \times \Sigma_f$ as shown in Figure 11.3) is such that the H_∞ norm of u_q to e_q is strictly less than γ, and for all initial conditions for (11.2) and $\bar{\Sigma}_Q$, whenever $u_q = 0$, we have that $e_q(t) \to 0$ as $t \to \infty$. Then (11.2) describes a filter that is a γ-level H_∞ SOID filter for the system Σ.*

Proof : First of all, it is easy to verify that a filter Σ_f is unbiased for Σ if and only if the interconnection of Σ_f and $\bar{\Sigma}_Q$ is such that for all initial conditions for (11.2) and $\bar{\Sigma}_Q$, whenever $u_q = 0$, we have that $e_q(t) \to 0$ as $t \to \infty$.

Next, we perturb B, D and F as

$$B_\varepsilon = \begin{pmatrix} B & \varepsilon I & 0 \end{pmatrix}, \quad D_\varepsilon = \begin{pmatrix} D & 0 & \varepsilon I \end{pmatrix}, \quad F_\varepsilon = \begin{pmatrix} F & 0 & 0 \end{pmatrix},$$

and we obtain a perturbed system Σ_ε and look at the associated Q_ε, which is the solution of the H_∞^1 CARE associated with this regular problem and define the associated system $\bar{\Sigma}_{\varepsilon, Q_\varepsilon}$ according to Lemma 11.50. Then a γ-level H_∞ SOID filter for the given system Σ is also a γ-level H_∞ SOID filter for the perturbed system Σ_ε for ε small enough. Then we can apply Lemma 11.50. We obtain that the interconnection of Σ_f and $\bar{\Sigma}_{\varepsilon, Q_\varepsilon}$ has an H_∞ norm less than γ for all ε small enough. Letting $\varepsilon \to 0$, we note that $Q_\varepsilon \to Q$. Hence, as ε converges to zero, the system parameters of $\bar{\Sigma}_{\varepsilon, Q}$ converge to the system parameters of $\bar{\Sigma}_Q$ except for some additional zero columns added to B_Q, D, and F. We also know that for all initial conditions for Σ_f and $\bar{\Sigma}_Q$, whenever $u_q = 0$, we have that $e_q(t) \to 0$ as $t \to \infty$. Using these facts, we obtain, by letting $\varepsilon \to 0$, that the interconnection of Σ_f and $\bar{\Sigma}_Q$ is such that the H_∞ norm from u_q to e_q is less than γ.   ∎

The following lemma that connects Σ_Q and Σ is an important consequence of Lemma 11.58.

Lemma 11.59 *Consider the singular γ-level H_∞ SOID filtering problem for the continuous-time system Σ described by (11.1) such that the system characterized by (A, B, C, D) has no invariant zeros on the imaginary axis. Assume that $\gamma > \gamma_{sp}^*$ (or $\gamma > \gamma_p^*$), which implies that a semi-stabilizing solution Q of the CQMI (11.20) exists [or which implies that a semi-stabilizing solution Q of the CQMI (11.23) exists]. Finally, assume that A is Hurwitz-stable. Let Σ_Q be as in (11.84). Then the following hold:*

(i) Suppose a filter of the form (11.2) and characterized by the quadruple (L, M, N, P) (with $P = 0$ for the strictly proper case) is a γ-level H_∞ SOID filter for the given system Σ. Then the following filter is a γ-level H_∞ SOID filter for the system Σ_Q;

$$
\begin{aligned}
\dot{x}_1 &= (A + G_Q E)x_1 - G_Q \hat{z}_q, \\
\dot{\xi} &= L\xi + M(y + Cx_1), \\
\hat{z}_q &= N\xi + P(y + Cx_1).
\end{aligned}
\tag{11.87}
$$

(ii) Conversely, suppose a filter of the form (11.2) and characterized by the quadruple (L, M, N, P) (with $P = 0$ for the strictly proper case) is a γ-level H_∞ SOID filter for the system Σ_Q. Then the following filter is a γ-level H_∞ SOID filter for the given system Σ:

$$
\begin{aligned}
\dot{x}_1 &= (A + G_Q E)x_1 - G_Q \hat{z}, \\
\dot{\xi} &= L\xi + M(y - Cx_1), \\
\hat{z} &= N\xi + P(y - Cx_1).
\end{aligned}
\tag{11.88}
$$

Proof : For the strictly proper case, it is clear that a sequence of filters that solves the H_∞ AID filtering problem for the system $\breve{\Sigma}_Q$ is obviously unbiased when applied to $\breve{\Sigma}_Q$. Then the same filters are also unbiased when applied to Σ_Q. Moreover, because such a sequence of filters solves the H_∞ AID filtering problem for the system $\breve{\Sigma}_Q$ for strictly proper case, it is easily verified that the same sequence of filters when applied to Σ_Q results asymptotically (as $\varepsilon \to 0$) in an estimation error

$$
e_q = z_q - \hat{z}_q = F u_q.
$$

As $\|F\| < \gamma$, this implies that the filters from this sequence solve the γ-level H_∞ SOID filtering problem for Σ_Q when ε is small enough.

Similarly, in the proper case, a sequence of strictly proper filters characterized by a matrix triple $(L_\varepsilon, M_\varepsilon, N_\varepsilon)$ that solves the H_∞ AID filtering problem for the system $\check{\Sigma}_Q$ is obviously unbiased when applied to $\check{\Sigma}_Q$. Consider a new sequence of proper filters of the form (11.2) with the same matrix triple $(L_\varepsilon, M_\varepsilon, N_\varepsilon)$ and with the matrix P used in defining $\check{\Sigma}_Q$. Such a new filter is an unbiased filter when applied to the system Σ_Q. Also, because the sequence of strictly proper filters characterized by the matrix triple $(L_\varepsilon, M_\varepsilon, N_\varepsilon)$ solves the H_∞ AID filtering problem for the system $\check{\Sigma}_Q$ for proper case, it is easily verified that the new sequence of proper filters, when applied to Σ_Q, results asymptotically (as $\varepsilon \to 0$) in the estimation error

$$e_q = z_q - \hat{z}_q = (F - PD)u_q.$$

As such, because in this case $\|F - PD\| < \gamma$, it implies that the new sequence of proper filters result in proper γ-level H_∞ SOID filters for Σ_Q when ε small enough.   ∎

By now we have completed the proofs of sufficiency parts of Theorems 11.22 and 11.26. The above development of Lemmas 11.56 to 11.59 clearly lays a roadmap for the design of γ-level H_∞ SOID filters for the given system Σ. That is, we construct new systems Σ_Q and $\check{\Sigma}_Q$. We design a sequence of H_∞ AID filters for $\check{\Sigma}_Q$, which can then be translated as an H_∞ AID filter sequence for Σ_Q. Then in turn such an H_∞ AID filter sequence for Σ_Q can be translated as a H_∞ SOID filter sequence for the given system Σ. We observe that filters of any arbitrary architecture can be used to design the sequence of H_∞ AID filters for $\check{\Sigma}_Q$. However, by using the design methodology developed in Chapter 9, one can easily design a sequence of AID filters of CSS architecture for $\check{\Sigma}_Q$.

Design of γ-level H_∞ SOID filters of CSS architecture.

In what follows we consider the design γ-level H_∞ SOID filters of CSS architecture for Σ. As in the case of previous chapters, we consider strictly proper, proper, and reduced-order filters of CSS architecture. Our design methodology follows exactly as in the previous subsection that corresponds to regular γ-level H_∞ SOID filtering. The first essential difference is that instead of using the stabilizing solution Q of the H_∞^1 CARE (11.17), we need to use the stabilizing solution Q of the CQMI (11.20) or (11.23), and then we define Σ_Q and $\check{\Sigma}_Q$ accordingly by (11.84) and (11.86). Also, in the regular case, we need to design EID filters for $\check{\Sigma}_Q$ or for Σ_Q, which can then be translated as γ-level H_∞ SOID filters for the given system Σ. However, in the singular case being considered now, we need to design H_∞ AID filters for $\check{\Sigma}_Q$ or for Σ_Q, which can then be translated as γ-level H_∞ SOID filters for the given system Σ. Moreover, in the regular case, the technical lemmas, which were essential components in designing γ-level H_∞ SOID filters are Lemmas 11.52, 11.54, and 11.55. In the singular case being considered now, the corresponding lemmas are to be redeveloped as given below.

The following lemma is analogous to Lemma 11.52.

Lemma 11.60 *Consider the singular γ-level H_∞ SOID filtering problem for the continuous-time system Σ described by (11.1) such that the system characterized by (A, B, C, D) has no invariant zeros on the imaginary axis. Assume that $\|F\| < \gamma$ and that a stabilizing solution Q of the CQMI (11.20) exists. Let Σ_Q be as in (11.84).*

Then the following hold:

(i) *Let the following strictly proper filter, which is of CSS architecture, be a γ-level H_∞ SOID filter for Σ:*

$$\dot{\xi} = A\xi + K(y - C\xi),$$
$$\hat{z} = E\xi. \tag{11.89}$$

Then, the following strictly proper filter of CSS architecture is a γ-level H_∞ SOID filter for the system Σ_Q:

$$\dot{\xi} = A\xi + K(y - C\xi) + G_Q\hat{z},$$
$$\hat{z}_q = E\xi. \tag{11.90}$$

(ii) *Conversely, let a strictly proper filter of the form (11.90) be a γ-level H_∞ SOID filter for the system Σ_Q. Then, the strictly proper filter (11.89) is a γ-level H_∞ SOID filter for the system Σ.*

Proof : The proof follows along the lines of the proof of Lemma 11.52. ∎

We find that the results presented earlier in Lemmas 11.54 and 11.55 still apply with hardly any modifications, and we find the following lemmas.

Lemma 11.61 *Consider the singular γ-level H_∞ SOID filtering problem for the continuous-time system Σ described by (11.1) such that the system characterized by (A, B, C, D) has no invariant zeros on the imaginary axis. Assume that $\gamma > \gamma_p^*$; that is, assume that (11.22) is satisfied and that a stabilizing solution Q of the CQMI (11.23) exists. Choose P^* such that $\|E - P^*F\| < \gamma$, and define $\tilde{\Sigma}^*$ by (11.68).*

Then the H_∞ SOID filtering problem for $\tilde{\Sigma}^*$ is solvable by a strictly proper filter of the form (11.71) having the CSS architecture. Moreover, any strictly proper H_∞ SOID filter of the form (11.71) for $\tilde{\Sigma}^*$ results in a proper H_∞ SOID filter (11.73) with CSS architecture for the system Σ.

Lemma 11.62 *Consider the system* (11.1) *where the system characterized by the quadruple* (A, B, C, D) *has no zeros on the imaginary axis. Assume that* $\gamma > \gamma_p^*$; *that is, assume that* (11.22) *is satisfied and that a stabilizing solution* Q *of the CQMI* (11.23) *exists.*

Then, the strictly proper γ*-level* H_∞ *SOID filtering problem for the system* Σ_r^* *of* (11.78) *is solvable by a strictly proper filter of the form* (11.71) *having the CSS architecture. Moreover, any strictly proper* H_∞ *SOID filter of the form* (11.80) *for* Σ_r^* *results in a reduced-order* γ*-level* H_∞ *SOID filter* (11.82) *with CSS architecture for the system* Σ.

Again, we note that the above design of course does not result in a complete characterization of all reduced-order H_∞ SOID filters. Also, we stress once more that minimizing the H_∞ norm of the error dynamics for Σ_Q does not mean that we minimize the H_∞ norm of the error dynamics when considering the original system Σ.

11.5.3 Singular γ-level H_∞ SOID filters—the system characterized by (A, B, C, D) has invariant zeros on the imaginary axis

In the previous subsection, we already proved Theorem 11.22 for the case when there are no invariant zeros on the imaginary axis. We start this subsection with the necessity proof for this theorem for the case that the system characterized by (A, B, C, D) has invariant zeros on the imaginary axis.

Proof of necessity for Theorem 11.22 (invariant zeros of the system (A, B, C, D) on the imaginary axis) : We use the same arguments as in the necessity proof for Theorem 11.22 in the previous subsection. We first use a reduction to an H_∞ SOID filtering problem with a surjective direct feedthrough matrix from u to y. From the proof of Theorem 11.16, we can conclude the existence of a strongly semi-stabilizing solution to the corresponding H_∞^1 CARE. Then using the dual version of Theorem 4.158, we then obtain the existence of a positive semi-definite and semi-stabilizing solution of the CQMI (11.20).

The fact that (11.19) must be satisfied when using strictly proper filters should by now be obvious. The conditions regarding invariant zeros on the imaginary axis still need to be established. From the necessity proof of Theorem 11.16, we know that matrices K and K_1 exist such that (11.47) is asymptotically stable and the transfer matrix has an H_∞ norm less than γ. This immediately implies that for $\bar{K} = K + K_1$, we have for any s_0 on the imaginary axis,

$$\|E(s_0 I - A + \bar{K} C)^{-1}(B - \bar{K} D) + F\| < \gamma,$$

whereas $\bar{K}$ is such that $s_0 I - A + \bar{K} C$ is invertible. This in turn immediately implies that the conditions regarding the invariant zeros on the imaginary axis are satisfied.   ■

We continue with a necessity proof for the equivalent theorem when using proper filters.

Proof of necessity for Theorem 11.26 (invariant zeros of the system (A, B, C, D) on the imaginary axis) : The proof follows along the same line as the necessity proof of Theorem 11.22 and the proof of Theorem 11.26 as given in the previous subsection for the case without invariant zeros on the imaginary axis. The fact that (11.22) is necessary should be obvious.

Also, the same argument as before yields a solution of the CQMI (11.23). From the existence of a proper H_∞ SOID filter, it is easily seen that matrices K and P exist such that for any s_0 on the imaginary axis, we have

$$\|(E - PC)(s_0 I - A + KC)^{-1}(B - KD) + (F - PD)\| < \gamma,$$

where K is such that $s_0 I - A + KC$ is invertible.   ∎

We need to prove the sufficiency parts of Theorems 11.22 and 11.26, and then we present methods of designing appropriate γ-level H_∞ SOID filters. In this regard, at the beginning of this section, we noted that our design philosophy is to transform the design of γ-level H_∞ SOID filters for a given system to the design of EID filters or H_∞ AID filters for an auxiliary system constructed from the data of the given system. We also emphasized that such a transformation emerges as a by-product while proving sufficiency parts of Theorems 11.22 and 11.26. We had been successful in doing so for regular H_∞ SOID filtering as well as for singular H_∞ SOID filtering, with the restriction that the system characterized by (A, B, C, D) has no invariant zeros on the imaginary axis. However, we have not succeeded yet in doing so for the case of singular H_∞ SOID filtering when the system characterized by (A, B, C, D) has indeed invariant zeros on the imaginary axis. Nevertheless we still conjuncture that it can be done so.

We proceed now to indicate the proofs of the sufficiency parts of Theorems 11.22 and 11.26. To do so, we need to indicate the existence of suitable H_∞ SOID filters for the given system system Σ under the given conditions. In this regard, two methods are available in the literature. One is based on a perturbation argument where a small perturbation results in a regular problem for which we have a design methodology available. These perturbation arguments are worked out in detail in the books [75,92]. The alternative, and numerically more attractive, method is worked out in [86]. These references are for the design of state feedback controllers, but by duality arguments, these can be used to obtain strictly proper γ-level H_∞ SOID filters of CSS architecture. This duality connection has been explored before to connect the EDD problem and the EID filtering problem and the ADD problem and the AID filtering problem. Finally, to design proper filters or reduced-order filters, we can follow the methodology of the previous two subsections. We can construct two auxiliary systems such that strictly proper filters for the auxiliary systems can be converted to proper or reduced-order filters, respectively, for the original system.

11.6   Design of γ-level H_∞ SOID filters—discrete-time systems

This section has the same outline as Section 11.5 while considering discrete-time systems. That is, for discrete-time systems, we present here methods of designing γ-level H_∞ SOID filters for any given $\gamma > \gamma_{sp}^*$ or $\gamma > \gamma_p^*$ depending on whether strictly proper or proper filters are sought. Once again, for clarity as well as transparency, our development here is divided into three layers, which consider progressively more complex class of problems, namely regular H_∞ SOID problems, singular H_∞ SOID problems with the restriction that the given system does not have any invariant zeros on the unit circle, and finally a general class of singular H_∞ SOID problems without any restrictions on the invariant zeros. As expected, such a division also highlights the differences between the regular and the singular cases while pointing out the complexity involved in the singular case. As in the previous section, our design philosophy transforms the design of an H_∞ SOID filters for a given system to the design of EID filters or H_∞ AID filters for an auxiliary system constructed from the data of the given system. Once again, such a transformation emerges as a by-product while proving sufficiency of certain conditions for the existence of H_∞ SOID filters (see Theorems 11.33, 11.36, 11.39, and 11.43), and it lays a clear roadmap for design while enabling us to use directly all earlier detailed development regarding the design of EID and H_∞ AID filters as given in Chapters 7 and 9. In so doing, it also offers us structural insight while displaying all the flexibility available in the design process.

11.6.1   Regular γ-level H_∞ SOID filters

In this subsection, we consider the design of regular H_∞ SOID filters. To start with we need to recall the following bounded-real lemma, which is also known as the Kalman–Yakubovich–Popov lemma. As in the previous section, our version is not the most general result. We have a slight modification, which will be more useful for our purposes. For a very elementary proof of the bounded-real lemma, we refer to [65].

Theorem 11.63 *Consider a discrete-time system Σ parameterized by the quadruple (A, B, C, D) with associated transfer matrix G. Then we have:*

- *Assume that the observable eigenvalues of (C, A) are inside the unit circle. In this case, the system is input–output stable with associated H_∞ norm $\|G\|_\infty < \gamma$ if and only if a matrix $P \geqslant 0$ exists satisfying the DLMI:*

$$F(P) = \begin{pmatrix} A'PA - P + C'C & A'PB + C'D \\ B'PA + D'C & B'PB + D'D - \gamma^2 I \end{pmatrix} \leqslant 0, \quad (11.91)$$

such that $\ker F(P) = \ker P \oplus \{0\}.$

If A is Schur-stable, then there exists a $P > 0$ such that $F(P) < 0$.

- *Assume that the controllable eigenvalues of (A, B) are inside the unit circle. In this case, the system is input–output stable with associated H_∞ norm $\|G\|_\infty < \gamma$ if and only if a matrix $Q \geqslant 0$ exists satisfying the dual DLMI:*

$$G(Q) = \begin{pmatrix} AQA' - Q + BB' & AQC' + BD' \\ CQA' + DB' & CQC' + DD' - \gamma^2 I \end{pmatrix} \leqslant 0, \quad (11.92)$$

such that $\ker G(Q) = \ker Q \oplus \{0\}$.

If A is Schur-stable, then a $Q > 0$ exists such that $G(Q) < 0$.

We begin our development by first providing the proofs of necessity parts of Theorems 11.33 and 11.36, which present existence conditions for regular γ-level H_∞ SOID filters. As we said, the proofs of sufficiency parts of these theorems given subsequently enable us to transform the design of appropriate H_∞ SOID filters for the given system to the design of EID filters for an auxiliary system constructed from the data of the given system.

Proof of necessity for Theorem 11.33 : Given the system (11.1) and a filter of the form (11.2) with $P = 0$, we obtain the following interconnection:

$$\begin{pmatrix} \sigma x \\ \sigma \xi \end{pmatrix} = A_e \begin{pmatrix} x \\ \xi \end{pmatrix} + B_e u,$$

$$e_z = C_e \begin{pmatrix} e \\ \xi \end{pmatrix} + F u,$$

with $e_z = z - \hat{z}$, where

$$A_e = \begin{pmatrix} A & 0 \\ MC & L \end{pmatrix}, \quad B_e = \begin{pmatrix} B \\ MD \end{pmatrix}, \quad C_e = \begin{pmatrix} E & -N \end{pmatrix}, \quad (11.93)$$

and the fact that we have an unbiased filter implies that the observable eigenvalues of (C_e, A_e) are asymptotically stable. We know from Theorem 11.63 that this guarantees that a matrix $P \geqslant 0$ exists satisfying,

$$F(P) = \begin{pmatrix} A_e' P A_e - P + C_e' C_e & A_e' P B_e + C_e' F \\ B_e' P A_e + F' C_e & B_e' P B_e + F' F - \gamma^2 I \end{pmatrix} \leqslant 0,$$

and $\ker F(P) = \ker P \oplus \{0\}$. We decompose P compatible with (11.93):

$$P = \begin{pmatrix} P_{11} & P_{12} \\ P_{21} & P_{22} \end{pmatrix}.$$

Given that $P \geq 0$, a matrix X exists such that $P_{21} = XP_{11}$. We find that $P_{11} \geq 0$ satisfies $F_{11}(P_{11}) \leq 0$, where

$$F_{11}(P_{11}) = \begin{pmatrix} A_K' P_{11} A_K - P_{11} + E'E & A_K' P_{11} B_K + E'F \\ B_K' P_{11} A_K + F'E & B_K' P_{11} B_K + F'F - \gamma^2 I \end{pmatrix},$$

with $A_K = A - KC$, $B_K = B - KD$ and $\ker F_{11}(P_{11}) = \ker P_{11} \oplus \{0\}$. This implies that the system parameterized by

$$(A - KC, B - KD, E, F) \tag{11.94}$$

has H_∞ norm less than γ according to Theorem 11.63. Moreover, $(A - KC)x = \lambda x$ with $|\lambda| \geq 1$ yields $Ex = 0$ when we use that $F_{11}(P_{11}) \leq 0$, and hence, the unstable dynamics are unobservable for the system (11.94). As (C, A) is $\mathbb{C}^\ominus$-detectable, it is easily checked that a matrix K_1 exists such that

$$(A - KC - K_1 C, B - KD - K_1 D, E, F) \tag{11.95}$$

is asymptotically stable and the transfer matrix is equal to the transfer matrix of (11.94), and hence, we have an H_∞ norm less than γ. Define $\tilde{K} = K + K_1$. Using Theorem 11.63, we find that a matrix $\tilde{Q} > 0$ exists such that

$$\begin{pmatrix} A_{\tilde{K}} \tilde{Q} A_{\tilde{K}}' - \tilde{Q} + B_{\tilde{K}} B_{\tilde{K}}' & A_{\tilde{K}} \tilde{Q} E' + B_{\tilde{K}} F' \\ E \tilde{Q} A_{\tilde{K}}' + F B_{\tilde{K}}' & E \tilde{Q} E' + FF' - \gamma^2 I \end{pmatrix} < 0$$

with $A_{\tilde{K}} = A - \tilde{K}C$ and $B_{\tilde{K}} = B - \tilde{K}D$. This implies that

$$A_{\tilde{K}} \tilde{Q} A_{\tilde{K}}' - \tilde{Q} + B_{\tilde{K}} B_{\tilde{K}}'$$
$$- (A_{\tilde{K}} \tilde{Q} E' + B_{\tilde{K}} F')(E \tilde{Q} E' + FF' - \gamma^2 I)^{-1}(E \tilde{Q} A_{\tilde{K}}' + F B_{\tilde{K}}') < 0.$$

This is a quadratic matrix inequality in $\tilde{K}$. We already have a specific choice of $\tilde{K}$ for which this matrix is negative definite. If we release this specific choice of $\tilde{K}$ and instead minimize over $\tilde{K}$, then this matrix should obviously remain negative definite. This results in

$$\tilde{K} = LS^{-1},$$

where

$$L = A\tilde{Q}C' + BD'$$
$$- (A\tilde{Q}E' + BF')(E\tilde{Q}E' + FF' - \gamma^2 I)^{-1}(E\tilde{Q}C' + FD'),$$
$$S = C\tilde{Q}C' + DD'$$
$$- (C\tilde{Q}E' + DF')(E\tilde{Q}E' + FF' - \gamma^2 I)^{-1}(E\tilde{Q}C' + FD'),$$

and

$$0 > A\tilde{Q}A' - \tilde{Q} + BB' - \begin{pmatrix} C\tilde{Q}A' + DB' \\ E\tilde{Q}A' + FB' \end{pmatrix}'$$

$$\times \begin{pmatrix} C\tilde{Q}C' + DD' & C\tilde{Q}E' + DF' \\ E\tilde{Q}C' + FD' & E\tilde{Q}E' + FF' - \gamma^2 I \end{pmatrix}^{-1} \begin{pmatrix} C\tilde{Q}A' + DB' \\ E\tilde{Q}A' + FB' \end{pmatrix}.$$

Using Proposition 4.88, we find that a strongly semi-stabilizing solution of the H_∞^1 DARE (11.37) exists. Since the system characterized by (A, B, C, D) has no invariant zeros on the unit circle, we conclude that we actually have a stabilizing solution. $\blacksquare$

The above proof of necessity for strictly proper filters quickly yields the result for proper filters.

Proof of necessity for Theorem 11.36 : We note that applying a proper filter of the form (11.2) to the system (11.1) is equivalent to applying the strictly proper filter:

$$\begin{aligned} \sigma\xi &= L\xi + My, \\ \hat{z} &= N\xi, \end{aligned} \tag{11.96}$$

to the system

$$\Sigma : \begin{cases} \sigma x = Ax + Bu \\ y\ \ = Cx + Du \\ z\ \ = E^*x + F^*u, \end{cases} \tag{11.97}$$

where $E^* = E - PC$ and $F^* = F - PD$. Therefore, if this filter is a γ-level H_∞ SOID filter, then using Theorem 11.33, we note that a stabilizing solution Q exists of the following H_∞^1 DARE:

$$Q = AQA' + BB' - \begin{pmatrix} CQA' + DB' \\ E^*QA' + F^*B' \end{pmatrix}' \times$$

$$\begin{pmatrix} CQC' + DD' & CQ(E^*)' + D(F^*)' \\ E^*QC' + F^*D' & E^*Q(E^*)' + F^*(F^*)' - \gamma^2 I \end{pmatrix}^{-1} \begin{pmatrix} CQA' + DB' \\ E^*QA' + F^*B' \end{pmatrix}.$$

But it then only takes some algebraic manipulations to establish that Q is also a stabilizing solution of the original H_∞^1 DARE given in (11.37). In other words, the matrix P does not affect the H_∞^1 DARE.

Also note that

$$E^*Q(E^*)' + F^*(F^*)' < \gamma^2 I$$

implies that (11.39) is satisfied. This completes the necessity part of the proof. $\blacksquare$

Proofs of sufficiency parts of Theorems 11.33 and 11.36.

So far we have proved the necessity parts of Theorems 11.33 and 11.36. We proceed now to prove the sufficiency parts. In this regard, as we said earlier and as in the previous section, our design methodology arises as a by-product while proving the sufficiency parts. As in the previous chapters, we explore here the design of strictly proper, proper, as well as reduced-order γ-level H_∞ SOID filters of CSS architecture.

As in the continuous-time case, we first start with two preliminary lemmas useful in proving the sufficiency parts of Theorems 11.33 and 11.36. Consider a discrete-time system Σ characterized by (A, B, C, D) with input–output operator $\mathscr{G}$ (i.e., the map that associates with every input d an output z, given zero initial state). Σ is called *inner* if the system is internally stable, and the input–output operator $\mathscr{G}$ maps ℓ_2^m into itself and has the property that for all $f \in \ell_2^m$, we have

$$\|\mathscr{G} f\|_2 = \|f\|_2.$$

Often, inner is defined as a property of the transfer matrix, but in our setting, the above is a more natural definition. It can be shown that $\mathscr{G}$ being inner implies that the transfer matrix of the system, denoted by G, satisfies

$$G'(z^{-1})G(z) = I. \tag{11.98}$$

A transfer matrix G satisfying (11.50) is called *unitary*. Note that if G is unitary, then G need not be $\mathbb{C}^\ominus$ stable. The transfer matrix G is unitary, and the system is internally stable if and only if the system is inner. In general, for an operator from ℓ_2^m to ℓ_2^p, in the literature, two concepts, inner and co-inner, are defined. A system Σ characterized by (A, B, C, D) is called *co-inner* if the dual system $\Sigma' = (A', C', B', D')$ is inner. In other words, the system is co-inner if it is internally stable and its transfer matrix satisfies

$$G(z)G'(z^{-1}) = I.$$

Note that for square systems, the concepts of inner and co-inner coincide. We now formulate a lemma that yields a test to check whether a system is co-inner.

Lemma 11.64 *Consider the discrete-time system Σ described by*

$$\Sigma : \begin{cases} \sigma x = Ax + Bu \\ z = Cx + Du, \end{cases} \tag{11.99}$$

with A Schur-stable. The system Σ is co-inner if a matrix X exists satisfying

(i) $AYA' + BB' = Y$,

(ii) $CYA' + DB' = 0$,

(iii) $CYC' + DD' = I$.

Inner and co-inner systems are very important in H_∞ control. We will present a lemma that is a main ingredient in the proof of our design method for H_∞ SOID filters. However, we first need the following preliminary lemma.

The following result is often referred to as "Redheffer's lemma." Note that we present a dual and discrete-time version of this lemma. The continuous-time version was given earlier in Lemma 11.47.

Lemma 11.65 *Consider the discrete-time, linear time-invariant systems Σ_U and Ψ. Suppose Σ_U has inputs u and e_s and outputs e and u_s, whereas Ψ has input u_s and output e_s. Consider the interconnection depicted in the diagram in Figure 11.4. Assume that Σ_U is co-inner and its input–output operator $\mathcal{G}$ has the decomposition:*

$$\mathcal{G}\begin{pmatrix} u \\ e_s \end{pmatrix} =: \begin{pmatrix} \mathcal{G}_{11} & \mathcal{G}_{12} \\ \mathcal{G}_{21} & \mathcal{G}_{22} \end{pmatrix}\begin{pmatrix} u \\ e_s \end{pmatrix} = \begin{pmatrix} e \\ u_s \end{pmatrix}, \tag{11.100}$$

which is compatible with the sizes of u, e, u_s and e_s, such that the $\mathcal{G}_{12}$ has a stable inverse.

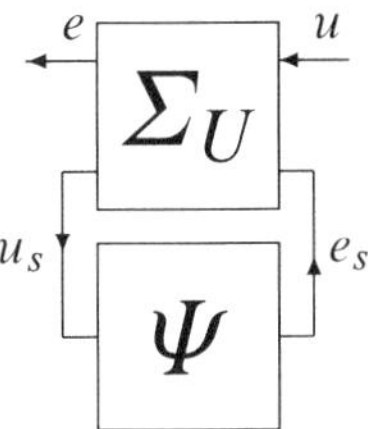

Figure 11.4: Interconnection of Σ_U and Ψ

Under the above assumptions, the following two statements are equivalent:

(i) The interconnection in Figure 11.4 is internally stable and its closed-loop transfer matrix has H_∞ norm less than 1.

(ii) The system Ψ is internally stable and its transfer matrix has H_∞ norm less than 1.

Proof : The proof is basically identical to the proof of the continuous-time version given in Lemma 11.47. $\blacksquare$

We established the above two preliminary lemmas in our quest to prove the sufficiency parts of Theorems 11.33 and 11.36. As in the continuous-time case, we will now proceed to show that the problem of finding a suitable H_∞ SOID filter for the system Σ given in (11.1) is equivalent to finding a suitable H_∞ SOID filter for a *new system* Σ_Q, which as before has some very nice structural properties. Again, it turns out that we can construct a slightly modified version of Σ_Q (namely, $\check{\Sigma}_Q$) for which the EID filtering problem is solvable by a strictly proper filter. This implies that we can directly use the design methodology developed in Chapter 7 to first find an EID filter for $\check{\Sigma}_Q$. Also, it turns out that such an EID filter for $\check{\Sigma}_Q$ can be translated as a γ-level H_∞ SOID filter for Σ_Q, and that in turn can be translated as a γ-level H_∞ SOID filter for the given system Σ. Thus, as before, the construction of new systems Σ_Q and $\check{\Sigma}_Q$ have definite advantages.

We proceed now to construct the new system Σ_Q. Assume that the H_∞ SOID filtering problem is solvable by a strictly proper filter. Then we know from our necessity proof of Theorem 11.33 that a stabilizing solution Q of the H_∞^1 DARE (11.37) exists. Using this matrix Q, we can define the following system:

$$\Sigma_Q : \begin{cases} \sigma x_q = A x_q + B_Q u_q + G_Q z_q \\ \quad y \;\; = C x_q + D_Q u_q \\ \quad z_q \;\; = E x_q + F_Q u_q, \end{cases} \tag{11.101}$$

where B_Q and G_Q are given by

$$B_Q = \begin{pmatrix} CQA' + DB' \\ EQA' + FB' \end{pmatrix}' \begin{pmatrix} CQC' + DD' & CQE' + DF' \\ EQC' + FD' & EQE' + FF' - \gamma^2 I \end{pmatrix}^{-1} \begin{pmatrix} D_Q \\ F_Q \end{pmatrix}, \tag{11.102a}$$

$$G_Q = -\begin{pmatrix} CQA' + DB' \\ EQA' + FB' \end{pmatrix}' \begin{pmatrix} CQC' + DD' & CQE' + DF' \\ EQC' + FD' & EQE' + FF' - \gamma^2 I \end{pmatrix}^{-1} \begin{pmatrix} 0 \\ I \end{pmatrix}, \tag{11.102b}$$

whereas D_Q and F_Q are such that

$$\begin{pmatrix} D_Q \\ F_Q \end{pmatrix} \begin{pmatrix} D_Q \\ F_Q \end{pmatrix}' = \begin{pmatrix} CQC' + DD' & CQE' + DF' \\ EQC' + FD' & EQE' + FF' \end{pmatrix}. \tag{11.102c}$$

We will first show that the strictly proper (or proper) γ-level H_∞ SOID filtering problem for Σ_Q is solvable whenever the strictly proper (or proper) γ-level H_∞ SOID filtering problem for Σ is solvable. To do so, we first introduce a slightly modified version of Σ_Q, namely a system $\check{\Sigma}_Q$ whose desired output signal $\check{z}_q$ that needs to be estimated is slightly different from z_q:

$$\check{\Sigma}_Q : \begin{cases} \sigma x_q = (A + G_Q E) x_q + (B_Q + G_Q F_Q) u_q \\ \quad y \;\; = C x_q + D_Q u_q \\ \quad \check{z}_q \;\; = (E - PC) x_q, \end{cases} \tag{11.103}$$

where the matrix P is different for the two different cases, strictly proper and proper:

- For the case when Σ satisfies the conditions for the existence of strictly proper γ-level H_∞ SOID filters [i.e., besides the existence of a stabilizing solution Q of the H_∞^1 DARE (11.37), we have $\|F_Q\| < \gamma$]. A suitable choice for the matrix P is $P = 0$.

- When Σ satisfies the conditions for the existence of proper γ-level H_∞ SOID filters [i.e., besides the existence of a stabilizing solution Q of the H_∞^1 DARE (11.37), we have (11.39)]. The condition (11.39) yields that a matrix P exists such that $\|F_Q - PD_Q\| < \gamma$, which is then a suitable choice for the matrix P in (11.103).

Clearly, in what follows, we can distinguish these two cases of defining $\check{\Sigma}_Q$ by selecting appropriate P in (11.103) as strictly proper case or proper case.

We make two claims. First, for the system $\check{\Sigma}_Q$, the EID filtering problem is solvable by a strictly proper filter irrespective of the choice of P. Second, any EID filter designed for $\check{\Sigma}_Q$ using the methodology of Chapter 7 results in a γ-level H_∞ SOID filter for Σ_Q. These claims are formalized in the following two lemmas.

Lemma 11.66 *Consider the regular γ-level H_∞ SOID filtering problem for the discrete-time system Σ described by (11.1). That is, assume that the system characterized by (A, B, C, D) is right-invertible and has no invariant zeros on the unit circle. Also, assume that $\gamma > \gamma_{sp}^*$ (or $\gamma > \gamma_p^*$); that is, assume that a stabilizing solution Q of the H_∞^1 DARE (11.37) exists such that (11.38) is satisfied [or a matrix P exists such that (11.39) is satisfied]. Define $\check{\Sigma}_Q$ by (11.103) with $P = 0$ (or with P such that $\|F_Q - PD_Q\| < \gamma$).*

Then, the EID filtering problem is solvable for $\check{\Sigma}_Q$ by a strictly proper filter irrespective of $P = 0$ (strictly proper case) or P is such that $\|F_Q - PD_Q\| < \gamma$ (proper case).

Proof : It can be easily verified that

$$K = \begin{pmatrix} CQA' + DB' \\ EQA' + FB' \end{pmatrix}' \begin{pmatrix} CQC' + DD' & CQE' + DF' \\ EQC' + FD' & EQE' + FF' - \gamma^2 I \end{pmatrix}^{-1} \begin{pmatrix} I \\ 0 \end{pmatrix}$$

is such that $B_Q + G_Q F_Q - KD_Q = 0$, whereas $A + G_Q E - KC$ equals

$$A - \begin{pmatrix} CQA' + DB' \\ EQA' + FB' \end{pmatrix}' \begin{pmatrix} CQC' + DD' & CQE' + DF' \\ EQC' + FD' & EQE' + FF' - \gamma^2 I \end{pmatrix}^{-1} \begin{pmatrix} C \\ E \end{pmatrix},$$

which is Schur-stable because Q is defined as the stabilizing solution of the H_∞^1 DARE (11.37). This clearly implies that the EID filtering problem is solvable for

$\check{\Sigma}_Q$ by the strictly proper filter:

$$\begin{cases} \sigma\xi = (A + G_Q E)\xi + K(y - C\xi) \\ \hat{\check{z}}_q = (E - PC)\xi. \end{cases} \qquad \blacksquare$$

The following lemma connects H_∞ SOID filtering for a system to EID filtering for a related system.

Lemma 11.67 *Consider the regular γ-level H_∞ SOID filtering problem for the discrete-time system Σ described by (11.1). That is, assume that the system characterized by (A, B, C, D) is right-invertible and has no invariant zeros on the unit circle. Also, assume that $\gamma > \gamma_{sp}^*$ (or $\gamma > \gamma_p^*$); that is, assume that a stabilizing solution Q of the H_∞^1 DARE (11.37) exists such that (11.38) is satisfied [or a matrix P exists such that (11.39) is satisfied]. Define Σ_Q by (11.101). Also, define $\check{\Sigma}_Q$ by (11.103) with $P = 0$ for the strictly proper case (or with P such that $\|F_Q - PD_Q\| < \gamma$ for the proper case).*
Then the following hold:

   (i) Let a strictly proper filter of the form (11.2) characterized by a matrix triple (L, M, N) solve the EID filtering problem for the system $\check{\Sigma}_Q$ with $P = 0$ (strictly proper case). Then such a filter is a strictly proper γ-level H_∞ SOID filter for Σ_Q.

   (ii) Let a strictly proper filter of the form (11.2) characterized by a matrix triple (L, M, N) solve the EID filtering problem for the system $\check{\Sigma}_Q$ with P such that $\|F_Q - PD_Q\| < \gamma$ (proper case). Consider a proper filter of the form (11.2) characterized by the matrix quadruple (L, M, N, P) where the matrix triple (L, M, N) corresponds to the EID filter for $\check{\Sigma}_Q$ and the matrix P corresponds to the one that is used in defining $\check{\Sigma}_Q$. Then such a proper filter is a proper γ-level H_∞ SOID filter for Σ_Q.

Proof : This proof follows the same arguments as the proof of Lemma 11.49 for continuous time. $\qquad \blacksquare$

The above lemma guarantees that the strictly proper (or proper) γ-level H_∞ SOID filtering problem is solvable for Σ_Q whenever the strictly proper (or proper) γ-level H_∞ SOID filtering problem is solvable for Σ. We will shortly and formally establish a relationship between the γ-level H_∞ SOID filters for Σ and the γ-level H_∞ SOID filters for Σ_Q. This result will be a consequence of "Redheffer's lemma" (Lemma 11.65). The following lemma will be the crucial component in establishing such a result.

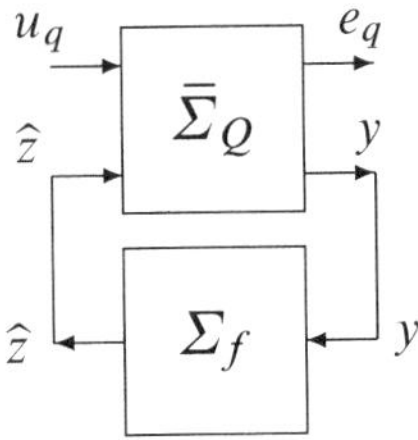

Figure 11.5: Interconnection of $\bar{\Sigma}_Q$ and Σ_f

Lemma 11.68 *Consider the regular γ-level H_∞ SOID filtering problem for the discrete-time system Σ described by* (11.1). *That is, assume that the system characterized by* (A, B, C, D) *is right-invertible and has no invariant zeros on the unit circle. Also, assume that $\gamma > \gamma_{sp}^*$ (or $\gamma > \gamma_p^*$); that is, assume that a stabilizing solution Q of the H_∞^1 DARE* (11.37) *exists such that* (11.38) *is satisfied (or a matrix P exists such that* (11.39) *is satisfied). We define the following system $\bar{\Sigma}_Q$ with two inputs u_q and $\hat{z}$ and two outputs y and e_q as shown in Figure 11.5:*

$$
\bar{\Sigma}_Q : \begin{cases}
\sigma x_q = A x_q + B_Q u_q + G_Q e_q \\
\quad y = C x_q + D u_q \\
\quad e_q = E x_q + F u_q - \hat{z},
\end{cases}
$$

where B_Q, D_Q, F_Q, and G_Q are defined by (11.102). *Then the following hold:*

(i) *Suppose a filter Σ_f of the form* (11.2) *and characterized by the quadruple (L, M, N, P) (with $P = 0$ for strictly proper case) is a γ-level H_∞ SOID filter for the given system Σ. Then the interconnection of Σ_f and $\bar{\Sigma}_Q$ (that is, $\bar{\Sigma}_Q \times \Sigma_f$ as shown in Figure 11.5) is such that the H_∞ norm of u_q to e_q is strictly less than γ, and for all initial conditions for* (11.2) *and $\bar{\Sigma}_Q$, whenever $u_q = 0$, we have that $e_q(k) \to 0$ as $k \to \infty$.*

(ii) *Conversely, suppose a system Σ_f of the form* (11.2) *and characterized by the quadruple (L, M, N, P) (with $P = 0$ for strictly proper case) interconnected with the system $\bar{\Sigma}_Q$ (that is, $\bar{\Sigma}_Q \times \Sigma_f$ as shown in Figure 11.5) is such that the H_∞ norm of u_q to e_q is strictly less than γ, and for all initial conditions for* (11.2) *and $\bar{\Sigma}_Q$, whenever $u_q = 0$, we have that $e_q(k) \to 0$ as $k \to \infty$. Then* (11.2) *describes a filter that is a γ-level H_∞ SOID filter for the system Σ.*

Proof : We derive a proof only for the case $\gamma = 1$. The proof for the general case is then obtained by simply scaling.

First we apply a preliminary transformation where $\hat{z} = \bar{z} + P^* y$ with

$$
P^* = (E Q C' + F D')(C Q C' + F D')^{-1}.
$$

This guarantees that $\bar{E}QC' + \bar{F}D' = 0$, where

$$\bar{F} = F - P^*D \ \text{ and } \ \bar{E} = E - P^*C.$$

Note that for the H_∞^1 DARE, we have the additional condition:

$$EQE' + FF' - (EQC' + FD')(CQC' + DD')^{-1}(CQE' + DF') < I,$$

which guarantees that $\bar{E}Q\bar{E}' + \bar{F}\bar{F}' < I$. We define the following system:

$$\bar{\Sigma} : \begin{cases} \sigma x = Ax + Bu \\ y \ = Cx + Du \\ \tilde{z} \ = \bar{E}x + \bar{F}u, \end{cases}$$

and then we obtain the filter $\bar{\Sigma}_f$ for $\bar{\Sigma}$ as

$$\bar{\Sigma}_f : \begin{cases} \sigma\xi = L\xi + My \\ \bar{z} \ = N\xi + (P - P^*)y. \end{cases}$$

Clearly $\bar{\Sigma}_f$ applied to $\bar{\Sigma}$ yields the same system as Σ_f applied to Σ. We define the next two systems Σ_ℓ and Σ_U. We have

$$\Sigma_\ell : \begin{cases} \sigma x_\ell = A_\ell x_\ell + B_\ell u_\ell + G_\ell \bar{z} \\ \quad y = C x_\ell + (CQC' + DD')^{1/2} u_\ell \\ \quad e_\ell = E_\ell x_\ell - H_\ell \bar{z}, \end{cases}$$

where

$$\begin{aligned}
A_\ell &= A + (AQ\bar{E}' + B\bar{F}')(I - \bar{E}Q\bar{E}' - \bar{F}\bar{F}')^{-1}\bar{E} \\
B_\ell &= (AQC' + BD')(CQC' + DD')^{-1/2} \\
G_\ell &= -(AQ\bar{E}' + B\bar{F}')(I - \bar{E}Q\bar{E}' - \bar{F}\bar{F}')^{-1} \\
E_\ell &= (I - \bar{E}Q\bar{E}' - \bar{F}\bar{F}')^{-1/2}\bar{E} \\
H_\ell &= (I - \bar{E}Q\bar{E}' - \bar{F}\bar{F}')^{-1/2},
\end{aligned}$$

and

$$\Sigma_U : \begin{cases} \sigma x_u = (A - KC)x_u + (B - KD)u \\ \qquad\qquad -(AQ\bar{E}' + \bar{B}\bar{F}')(I - \bar{E}Q\bar{E}' - \bar{F}\bar{F}')^{-1/2}e_\ell \\ \quad e \ = \bar{E}x_u + \bar{F}u + (I - \bar{E}Q\bar{E}' - \bar{F}\bar{F}')^{1/2}e_\ell \\ \quad u_\ell \ = (CQC' + DD')^{-1/2}Cx_u + (CQC' + DD')^{-1/2}Du, \end{cases}$$

$$(11.104)$$

where

$$K = (AQC' + BD')(CQC' + DD')^{-1}.$$

The system Σ_U is co-inner. This is seen by noting that Q satisfies the conditions of Lemma 11.64 for the system Σ_U. Moreover, the subsystem from e_l to e has a stable inverse (using that Q is a stabilizing solution of the H_∞^1 DARE). Then we can apply Lemma 11.65, where Ψ is the interconnection of $\tilde{\Sigma}_f$ with Σ_ℓ.

Note that it can easily be checked that the interconnection of Σ_U and Σ_ℓ results in the same system as $\tilde{\Sigma}$ but with some added unobservable and stable dynamics.

Lemma 11.65 then yields that a filter of the form (11.2) is unbiased and results in an H_∞ norm less than 1 when applied to Σ if and only if the filter $\bar{\Sigma}_f$ when applied to Σ_ℓ is such that the H_∞ norm of u_q to e_q is strictly less than γ, and for all initial conditions of $\bar{\Sigma}_f$ and Σ_ℓ, whenever $u_q = 0$, we have that $e_q(k) \to 0$ as $k \to \infty$.

If we replace u_ℓ by $(I \quad 0)\tilde{u}_\ell$, then clearly the induced norm from u_ℓ to e_ℓ is the same as the induced norm from $\tilde{u}_\ell$ to e_ℓ. We obtain the system:

$$\tilde{\Sigma}_\ell : \begin{cases} \sigma x_\ell = A_\ell x_\ell + \tilde{B}_\ell \tilde{u}_\ell + G_\ell \bar{z} \\ \quad y = C x_\ell + D_\ell \tilde{u}_\ell \\ \quad e_\ell = \bar{E}_\ell x_\ell - H_\ell \bar{z}, \end{cases}$$

where

$$\tilde{B}_\ell = \left((AQC' + BD')(CQC' + DD')^{-1/2} \quad 0 \right),$$

$$D_\ell = \left((CQC' + DD')^{1/2} \quad 0 \right).$$

Hence, a filter of the form (11.2) is unbiased and results in an H_∞ norm less than 1 when applied to Σ if and only if the system $\bar{\Sigma}_f$ when applied to $\tilde{\Sigma}_\ell$ is such that the H_∞ norm of u_q to e_q is strictly less than 1, and for all initial conditions of $\bar{\Sigma}_f$ and $\tilde{\Sigma}_\ell$, whenever $u_q = 0$, we have that $e_q(k) \to 0$ as $k \to \infty$. Next, we define the static system:

$$\Sigma_F : \begin{pmatrix} e_q \\ \tilde{u}_\ell \end{pmatrix} = \left(\begin{pmatrix} 0 & (I-L)^{1/2} \\ \begin{pmatrix} I & 0 \\ 0 & L^{1/2} \end{pmatrix} \end{pmatrix} \quad \begin{pmatrix} L^{1/2} \\ \begin{pmatrix} 0 \\ (I-L)^{1/2} \end{pmatrix} \end{pmatrix} \right) \begin{pmatrix} \tilde{u}_q \\ e_\ell \end{pmatrix},$$

where

$$L = I - \bar{E} Q \bar{E}' - \bar{F} \bar{F}'.$$

Clearly Σ_F is co-inner with the subsystem from e_ℓ to e_q having a (stable) inverse. Note that the interconnection of Σ_F and $\tilde{\Sigma}_\ell$ equals

$$\tilde{\Sigma}_Q : \begin{cases} \sigma x_q = A x_q + \tilde{B}_Q u_q + G_Q e_q \\ \quad y = C x_q + \tilde{D}_Q u_q \\ \quad e_q = \bar{E} x_q + \tilde{F}_Q u_q - \bar{z}, \end{cases}$$

where

$$\tilde{B}_Q = \begin{pmatrix} CQA' + DB' \\ EQA' + FB' \end{pmatrix}' \begin{pmatrix} CQC' + DD' & 0 \\ 0 & \bar{E}Q\bar{E}' + \bar{F}\bar{F}' - I \end{pmatrix}^{-1/2},$$

$$\tilde{D}_Q = \begin{pmatrix} (CQC' + DD')^{1/2} & 0 \end{pmatrix},$$

$$\tilde{F}_Q = \begin{pmatrix} 0 & (\bar{E}Q\bar{E}' + \bar{F}\bar{F}')^{1/2} \end{pmatrix}.$$

Therefore we can apply Lemma 11.65 with Σ_U replaced by Σ_F and Ψ the interconnection of $\tilde{\Sigma}_\ell$ and $\bar{\Sigma}_f$.

This yields that a filter of the form (11.2) is unbiased and results in an H_∞ norm less than 1 when applied to Σ if and only if the system $\bar{\Sigma}_f$ when applied to $\tilde{\Sigma}_Q$ is such that the H_∞ norm of u_q to e_q is strictly less than 1, and for all initial conditions of $\bar{\Sigma}_f$ and $\tilde{\Sigma}_Q$, whenever $u_q = 0$, we have that $e_q(k) \to 0$ as $k \to \infty$.

It can be verified that the filter (11.2) applied to Σ achieves the same system as the filter $\bar{\Sigma}_f$ applied to $\tilde{\Sigma}_Q$. This proves the result.   ∎

The following lemma that connects Σ_Q and Σ is an important consequence of Lemma 11.68.

Lemma 11.69 *Consider the regular γ-level H_∞ SOID filtering problem for the discrete-time system Σ described by (11.1). That is, assume that the system characterized by (A, B, C, D) is right-invertible and has no invariant zeros on the unit circle. Also, assume that $\gamma > \gamma_{sp}^*$ (or $\gamma > \gamma_p^*$), which implies that a stabilizing solution Q of the H_∞^1 DARE (11.37) exists. Finally, assume that A is Schur-stable. Let Σ_Q be as in (11.101).*
Then the following hold:

(i) *Suppose a filter of the form (11.2) and characterized by the quadruple (L, M, N, P) (with $P = 0$ for the strictly proper case) is a γ-level H_∞ SOID filter for the given system Σ. Then the following filter is a γ-level H_∞ SOID filter for the system Σ_Q:*

$$\begin{aligned} \sigma x_1 &= (A + G_Q E)x_1 - G_Q \hat{z}, \\ \sigma \xi &= L\xi + M(y + Cx_1), \\ \hat{z} &= N\xi + P(y + Cx_1). \end{aligned} \tag{11.105}$$

(ii) *Conversely, suppose a filter of the form (11.2) and characterized by the quadruple (L, M, N, P) (with $P = 0$ for the strictly proper case) is a γ-level H_∞ SOID filter for the system Σ_Q. Then the following filter is a γ-level H_∞ SOID filter for the given system Σ:*

$$\begin{aligned} \sigma x_1 &= (A + G_Q E)x_1 - G_Q \hat{z}, \\ \sigma \xi &= L\xi + M(y - Cx_1), \\ \hat{z} &= N\xi + P(y - Cx_1). \end{aligned} \tag{11.106}$$

Proof : We again only prove this for $\gamma = 1$. The general result follows from a scaling argument.

A filter (11.2) applied to Σ is unbiased and achieves an H_∞ norm less than 1 if and only if the system (11.2) applied to $\bar{\Sigma}_Q$ achieves an H_∞ norm less than 1 from u_q to e_q, and for all initial conditions for (11.2) and $\bar{\Sigma}_Q$, whenever $u_q = 0$, we have that $e_q(k) \to 0$ as $k \to \infty$.

It remains to connect the interconnection of (11.2) and $\bar{\Sigma}_Q$ with the filter (11.105) for Σ_Q. But it is easily checked that a filter (11.105) applied to Σ_Q yields the same error dynamics (except for some additional stable but unobservable dynamics) as (11.2) applied to $\bar{\Sigma}_Q$. Conversely, a filter (11.106) applied to $\bar{\Sigma}_Q$ yields the same error dynamics (except for some additional stable but unobservable dynamics) as (11.2) applied to Σ_Q. The lemma then follows immediately. ∎

By now we have completed the proofs of sufficiency parts of Theorems 11.33 and 11.36. The above development of Lemmas 11.66 to 11.69 clearly lays a roadmap for the design of γ-level H_∞ SOID filters for the given system Σ. That is, we construct new systems Σ_Q and $\check{\Sigma}_Q$. We design an EID filter for $\check{\Sigma}_Q$, which can then be translated as an H_∞ AID filter for Σ_Q. Then in turn such an H_∞ AID filter for Σ_Q can be translated as a γ-level H_∞ SOID filter for the given system Σ. We observe that a filter of any arbitrary architecture can be used to design an EID filter for $\check{\Sigma}_Q$. However, by using the design methodology developed in Chapter 7, one can easily design an EID filter of CSS architecture for $\check{\Sigma}_Q$.

We note that Lemma 11.69 assumes that the matrix A is Schur-stable. For a filtering problem, the assumption of matrix A being Schur-stable is natural but may not be very elegant. However, we can remove such an assumption when we focus below on the design of γ-level H_∞ SOID filters of CSS architecture for the given system Σ.

We proceed now to design γ-level H_∞ SOID filters of CSS architecture for Σ. As in the case of previous chapters, we consider strictly proper, proper, and reduced-order filters of CSS architecture, one at a time.

Design of γ-level strictly proper H_∞ SOID filters of CSS architecture.

As we said above, Lemma 11.69 plays a fundamental role in proving the sufficiency parts of Theorems 11.33 and 11.36, while suggesting a methodology of designing γ-level H_∞ SOID filters. The following lemma explores this further by using strictly proper filters of CSS architecture.

Lemma 11.70 *Consider the regular γ-level H_∞ SOID filtering problem for the discrete-time system Σ described by (11.1). That is, assume that the system characterized by (A, B, C, D) is right-invertible and has no invariant zeros on the unit circle. Also, assume that $\gamma > \gamma_{sp}^*$ (or $\gamma > \gamma_p^*$), which implies that a stabilizing solution Q of the H_∞^1 DARE (11.37) exists. Also, define Σ_Q by (11.101).*

Then the following hold:

(*i*) *Let the following strictly proper filter, which is of CSS architecture, be a*
γ-*level* H_∞ *SOID filter for* Σ:

$$\begin{aligned}
\sigma\xi &= A\xi + K(y - C\xi), \\
\hat{z} &= E\xi.
\end{aligned}$$

(11.107)

Then, the following strictly proper filter of CSS architecture is a γ-*level*
H_∞ *SOID filter for the system* Σ_Q:

$$\begin{aligned}
\sigma\xi &= A\xi + K(y - C\xi) + G_Q\hat{z}_q, \\
\hat{z}_q &= E\xi.
\end{aligned}$$

(11.108)

(*ii*) *Conversely, let a strictly proper filter of the form* (11.108), *which is of CSS*
architecture, be a γ-*level* H_∞ *SOID filter for* Σ_Q. *Then, the strictly proper*
filter (11.107) *is a* γ-*level* H_∞ *SOID filter for* Σ.

Proof : The proof relies on Lemma 11.68. It is easily checked that the error dynamics that results when the filter (11.108) applied to Σ_Q results in the same
dynamics of e_q as when (11.107) is applied to $\bar{\Sigma}_Q$. The proof of Lemma 11.69
then immediately yields that (11.108) is an H_∞ SOID filter for Σ_Q if and only if
(11.107) is an H_∞ SOID filter for Σ.

The fact that for Σ_Q we can obtain a γ-level H_∞ SOID filter is a direct consequence of Lemmas 11.66 and 11.67. ∎

We can now summarize the methodology of designing a strictly proper γ-level
H_∞ SOID filter for Σ:

- Consider $\check{\Sigma}_Q$ of (11.103) with $P = 0$. That is, consider

$$\check{\Sigma}_Q : \begin{cases} \sigma x_q = Ax_q + (B_Q + G_Q F)u_q + G_Q z_q \\ y = Cx_q + Du_q \\ \check{z}_q = Ex_q. \end{cases}$$

(11.109)

By Lemma 11.66, a strictly proper EID filter of CSS architecture can be
designed for $\check{\Sigma}_Q$. This can be done by using the design procedure developed
in Chapter 7.

- Lemma 11.67 shows that the filter designed above is a γ-level H_∞ SOID
filter for Σ_Q.

- We can then use Lemma 11.70 to obtain a strictly proper γ-level H_∞ SOID
filter (which can be denoted by Σ_{sp-CSS}) for Σ.

Remark 11.71 *As in the continuous-time case, we would like to point out one aspect of the above design procedure. Even if we can characterize all possible EID filters for $\breve{\Sigma}_Q$, the above procedure does not enable us to characterize all possible γ-level H_∞ SOID filters for Σ. Also it should be stressed that minimizing the H_∞ norm of the error dynamics for Σ_Q does not mean that we minimize the H_∞ norm of the error dynamics when considering the original system Σ. The above design yields so-called error dynamics with minimum entropy as studied in [52]. In can also be interpreted in the sense of minimizing a linear exponential quadratic Gaussian (LEQG) criterion as studied in, for instance, [101]. However, it should be noted that both of these references look at this problem from a control perspective.*

Design of γ-level proper H_∞ SOID filters of CSS architecture.

We now proceed to the case of designing γ-level proper H_∞ SOID filters of CSS architecture. As in the previous chapter, we can pursue two methods of designing such filters. One method that can be called a direct method arises from the proofs of Lemmas 11.66 to 11.69, that is, by computing the systems $\breve{\Sigma}_Q$ and Σ_Q from the data of the system Σ, designing a strictly proper EID filter for $\breve{\Sigma}_Q$ by using the methodology developed in Chapter 7 (which results as a proper γ-level H_∞ SOID filter for Σ_Q by Lemma 11.67), and then converting such a filter to a proper γ-level H_∞ SOID filter for the given system Σ by Lemma 11.69. The second method follows the footsteps of our design philosophy in Chapters 7–10. In this method, by using the methodology we just developed above to design a *strictly* proper γ-level H_∞ SOID filter, we first design a strictly proper γ-level H_∞ SOID filter for an auxiliary system $\widetilde{\Sigma}^*$, which is essentially the same as the given system Σ except that it has some additional measurement output. It is shown that such a strictly proper filter for $\widetilde{\Sigma}^*$ exists whenever a proper γ-level H_∞ SOID filter exists for Σ. We then transform such a strictly proper filter for $\widetilde{\Sigma}^*$ to a proper γ-level H_∞ SOID filter for the given system Σ.

To explain explicitly the above two methods, we first need to construct the auxiliary system $\widetilde{\Sigma}^*$. To do so, as in the previous chapters, for ease of presentation and without loss of generality, we decompose the measured output y into two parts y_0 and y_1 in such a way that y_0 contains explicitly the unknown input u in it, whereas y_1 does not contain any input u in it. That is, we write

$$y = \begin{pmatrix} y_0 \\ y_1 \end{pmatrix} = Cx + Du \quad \text{and} \quad C = \begin{pmatrix} C_0 \\ C_1 \end{pmatrix}, \quad D = \begin{pmatrix} D_0 \\ 0 \end{pmatrix}, \quad (11.110)$$

where matrix D_0 has rank m_0. We can then rewrite the given system (11.1) as

$$\Sigma : \begin{cases} \sigma x \;\;\; = Ax + Bu \\ \begin{pmatrix} y_0 \\ y_1 \end{pmatrix} = \begin{pmatrix} C_0 \\ C_1 \end{pmatrix} x + \begin{pmatrix} D_0 \\ 0 \end{pmatrix} u \\ z \;\;\;\;\; = Ex + Fu, \end{cases} \quad (11.111)$$

where matrix D_0 has rank m_0. If the H_∞ SOID filtering problem is solvable by a proper filter, then we know from the necessity part of the proof of Theorem 11.36 that a stabilizing solution $Q \geqslant 0$ of the DARE (11.37) exists such that (11.39) is satisfied. Some algebraic manipulations then yield that a matrix P^* exists such that

$$E^* Q (E^*)' + F^* (F^*)' < \gamma^2 I,$$

where

$$E^* = E - P^* C \quad \text{and} \quad F^* = F - P^* D.$$

Finally, we define the auxiliary system $\widetilde{\Sigma}^*$ as

$$\widetilde{\Sigma}^* : \begin{cases} \sigma x = A x + B u \\ \widetilde{y} = \begin{pmatrix} y_0 \\ \sigma y_1 \end{pmatrix} = \widetilde{C} x + \widetilde{D} u \\ z^* = E^* x + F^* u, \end{cases} \tag{11.112}$$

where

$$\widetilde{C} = \begin{pmatrix} C_0 \\ C_1 A \end{pmatrix}, \tag{11.113}$$

$$\widetilde{D} = \begin{pmatrix} D_0 \\ C_1 B \end{pmatrix}. \tag{11.114}$$

We observe that the desired output z^* that needs to be estimated in system $\widetilde{\Sigma}^*$ is obtained from the desired output z that needs to be estimated in the given system Σ; that is, $z^* = z - P^* y$. Thus, whenever an estimate of z^* is available, one can form easily an estimate of z. Also, as will be shown shortly, a strictly proper γ-level H_∞ SOID filter exists for the auxiliary system $\widetilde{\Sigma}^*$ whenever a proper γ-level H_∞ SOID filter exists for the given system Σ.

We are now ready to restate the following two design methodologies that we alluded to earlier:

- **First method:** Compute the systems $\widecheck{\Sigma}_Q$ and Σ_Q from the data of the given system Σ. Then design a strictly proper EID filter of CSS architecture for $\widecheck{\Sigma}_Q$ by using the methodology developed in Chapter 7. Such a filter is a proper γ-level H_∞ SOID filter of CSS architecture for Σ_Q (by Lemma 11.67). Then convert this to a proper γ-level H_∞ SOID filter of CSS architecture for Σ using the previously described technique given in Lemma 11.69.

- **Second method:** Construct the systems $\widecheck{\widetilde{\Sigma}}_{\widetilde{Q}}$ and $\widetilde{\Sigma}_{\widetilde{Q}}$ from the data of the system $\widetilde{\Sigma}^*$ by first obtaining a stabilizing solution $\widetilde{Q}$ of the H_∞^1 DARE associated with the system $\widetilde{\Sigma}^*$ by following the same procedure as that of constructing the systems $\widecheck{\Sigma}_Q$ and Σ_Q from the data of Σ. Design a strictly proper EID filter of CSS architecture for $\widetilde{\widecheck{\Sigma}}_{\widetilde{Q}}$ by using the methodology developed in Chapter 7 and then convert this to a strictly proper γ-level H_∞ SOID filter of CSS architecture for $\widetilde{\Sigma}^*$ (This is the same as applying Lemma 11.70 to the system $\widetilde{\Sigma}^*$ instead of to the system Σ, and following the design method for strictly proper filters of CSS architecture developed right after Lemma 11.70). Finally convert this to a proper γ-level H_∞ SOID filter of CSS architecture for Σ.

We emphasize that, unlike in the continuous-time case, the above two design methods do not result in the same family of filters. This is because D_Q is of full rank (in the regular case that is being studied) and hence, a decomposition of the form (11.110) for Σ_Q would not yield a C_1 component. Therefore, the second design methodology is preferable because it gives more flexibility.

We now proceed with the development of the second method. Applying Lemma 11.70 to the system $\widetilde{\Sigma}^*$ instead of to the system Σ, and following the design method developed right after Lemma 11.70, we can design a γ-level strictly proper H_∞ SOID filter for $\widetilde{\Sigma}^*$ as

$$\begin{cases} \sigma\widetilde{\xi} = A\widetilde{\xi} + K(\widetilde{y} - \widetilde{C}\widetilde{\xi}) \\ \widehat{z}^* = E^*\widetilde{\xi}, \end{cases} \qquad (11.115)$$

where the matrix K is a filter gain. We partition K in conformity with the partitioning of $\widetilde{y}$. That is,

$$K = \begin{pmatrix} K_0 & K_2 \end{pmatrix}.$$

We can rewrite the filter equation (11.115) as

$$\begin{cases} \sigma\xi = (A - K\widetilde{C})\xi + \begin{pmatrix} K_0 & (A - K\widetilde{C})K_2 \end{pmatrix} y \\ \widetilde{\xi} \ = \xi + K_2 y_1 \\ \widehat{z}^* = E^*\widetilde{\xi} = E^*(\xi + K_2 y_1). \end{cases} \qquad (11.116)$$

As this filter does not use σy_1, it is easy to convert this from a filter for $\widetilde{\Sigma}^*$ to a filter for Σ by noting the relationship between z^* and z, namely $z^* = z - P^* y$. We thus obtain:

$$\begin{cases} \sigma\xi = (A - K\widetilde{C})\xi + \begin{pmatrix} K_0 & (A - K\widetilde{C})K_2 \end{pmatrix} y \\ \widehat{z}^* = E^*(\xi + K_2 y_1) + P^* y. \end{cases} \qquad (11.117)$$

We have the following lemma.

Lemma 11.72 *Consider the regular γ-level H_∞ SOID filtering problem for the discrete-time system Σ described by (11.1). That is, assume that the system characterized by (A, B, C, D) is right-invertible and has no invariant zeros on the unit* *circle. Also, assume that $\gamma > \gamma_p^*$; that is, assume that a stabilizing solution Q of the H_∞^1 DARE (11.37) exists such that (11.39) is satisfied. Choose P^* such that*

$$(E - P^*C)Q(E - P^*C)' + (F - P^*D)(F - P^*D)' < \gamma^2 I,$$

and define $\widetilde{\Sigma}^*$ by (11.112).

Then the H_∞ SOID filtering problem for $\widetilde{\Sigma}^*$ is solvable by a strictly proper filter of the form (11.115) having the CSS architecture. Moreover, any strictly proper H_∞ SOID filter of the form (11.115) for $\widetilde{\Sigma}^*$ results in a proper H_∞ SOID filter (11.117) with CSS architecture for the system Σ.

Proof : The fact that $E^*Q(E^*)' + F^*(F^*)' < \gamma^2 I$ and that Q also satisfies the the H_∞^1 DARE associated with the system $\widetilde{\Sigma}^*$ shows that the strictly proper H_∞ SOID filtering problem for $\widetilde{\Sigma}^*$ is solvable.

The error dynamics of the filter (11.115) applied to $\widetilde{\Sigma}^*$ is equal to the error dynamics of (11.117) applied to Σ. Therefore, it is clear that (11.117) is a proper H_∞ SOID filter for Σ whenever (11.115) is a strictly proper H_∞ SOID filter for $\widetilde{\Sigma}^*$. ∎

Just like in the strictly proper case, the above design of course does not result in a complete characterization of all γ-level H_∞ SOID filters.

Design of reduced-order γ-level H_∞ SOID filters of CSS architecture.

We now proceed with the design of reduced-order γ-level H_∞ SOID filters of CSS architecture. To do this, as in Chapters 7 to 10, we first extract a reduced-order system Σ_r^* from the given system Σ and then design a strictly proper γ-level H_∞ SOID filter for it. It is shown that such a strictly proper γ-level H_∞ SOID filter exists for Σ_r^* whenever a proper γ-level H_∞ SOID filter exists for the given system Σ. The strictly proper filter designed for Σ_r^* is then translated to form a reduced-order γ-level H_∞ SOID filter for the given system Σ.

We now proceed to extract the reduced-order system Σ_r^* from the given system Σ. To do so, to start with, let us rewrite the matrices C and D of (11.1) as

$$C = \begin{pmatrix} 0 & C_{02} \\ I_{p-m_0} & 0 \end{pmatrix}, \quad D = \begin{pmatrix} D_0 \\ 0 \end{pmatrix},$$

where again rank D = rank D_0 = m_0. This can always be done without any loss of generality by appropriate coordinate transformations. In view of the above partitioning of C and D, we can partition the given system Σ as

$$\Sigma : \begin{cases} \begin{pmatrix} \sigma x_1 \\ \sigma x_2 \end{pmatrix} = \begin{pmatrix} A_{11} & A_{12} \\ A_{21} & A_{22} \end{pmatrix} \begin{pmatrix} x_1 \\ x_2 \end{pmatrix} + \begin{pmatrix} B_{11} \\ B_{22} \end{pmatrix} u \\[2mm] y = \begin{pmatrix} y_0 \\ y_1 \end{pmatrix} = \begin{pmatrix} 0 & C_{02} \\ I & 0 \end{pmatrix} \begin{pmatrix} x_1 \\ x_2 \end{pmatrix} + \begin{pmatrix} D_0 \\ 0 \end{pmatrix} u \\[2mm] z = Ex + Fu, \end{cases} \qquad (11.118)$$

where different variables have obvious meanings.

If the H_∞ SOID filtering problem is solvable by reduced-order filters, then we know from the necessity part of the proof of Theorem 11.36 that a stabilizing solution $Q \geq 0$ of the H_∞^1 DARE (11.37) exists such that (11.39) is satisfied. Some algebraic manipulations then yield that a matrix P^* exists such that

$$E^* Q (E^*)' + F^* (F^*)' < \gamma^2 I,$$

where

$$E^* = E - P^* C \quad \text{and} \quad F^* = F - P^* D.$$

We define a new desired output z^* as

$$z^* = z - P^* y = (E - P^* C)x + (F - P^* D)u = E^* x + F^* u. \quad (11.119)$$

In view of (11.118) and (11.119), we can define a new system Σ^* as

$$\Sigma^* : \begin{cases} \begin{pmatrix} \sigma x_1 \\ \sigma x_2 \end{pmatrix} = \begin{pmatrix} A_{11} & A_{12} \\ A_{21} & A_{22} \end{pmatrix} \begin{pmatrix} x_1 \\ x_2 \end{pmatrix} + \begin{pmatrix} B_{11} \\ B_{22} \end{pmatrix} u \\[12pt] y = \begin{pmatrix} y_0 \\ y_1 \end{pmatrix} = \begin{pmatrix} 0 & C_{02} \\ I & 0 \end{pmatrix} \begin{pmatrix} x_1 \\ x_2 \end{pmatrix} + \begin{pmatrix} D_0 \\ 0 \end{pmatrix} u \\[12pt] z^* = E^* x = E_1^* x_1 + E_2^* x_2 + F^* u, \end{cases} \quad (11.120)$$

with $E^* = \begin{pmatrix} E_1^* & E_2^* \end{pmatrix}$.

We note that the y_1 is not contaminated by the input u, and hence, $x_1 = y_1$ is known exactly from the measurement y. Thus, all we need to do next is to estimate the state x_2. To proceed further, let us rewrite the state equation for x_1 in terms of the output y_1 and the state x_2 as

$$\sigma y_1 = A_{11} y_1 + A_{12} x_2 + B_{11} u. \quad (11.121)$$

The above equation can be rewritten as

$$\sigma y_1 - A_{11} y_1 = A_{12} x_2 + B_{11} u.$$

Treating σy_1 as known, we can define a new measurement variable y_r:

$$y_r = \begin{pmatrix} y_0 \\ \sigma y_1 - A_{11} y_1 \end{pmatrix}.$$

Although σy_1 is not directly available, as we did earlier, we can eliminate it from any filter that is constructed using y_r as a measured output. With this in mind, we form the following auxiliary system:

$$\Sigma_r^* : \begin{cases} \sigma x_r = A_r x_r + B_r u + A_{21} y_1 \\ y_r = C_r x_r + D_r u \\ z_r^* = E_r^* x_r + F_r^* u, \end{cases} \quad (11.122)$$

where $x_r = x_2$ and

$$A_r = A_{22}, \quad B_r = B_{22}, \quad C_r = \begin{pmatrix} C_{02} \\ A_{12} \end{pmatrix}, \quad D_r = \begin{pmatrix} D_0 \\ B_{11} \end{pmatrix},$$

$$E_r^* = E_2^*, \quad F_r^* = F^*. \quad (11.123)$$

We note that the dynamic order n_r of the above Σ_r^* is less than the dynamic order n of the given system Σ by a number equal to the dimension of $x_1 = y_1$.

Next, to construct the required filter for Σ, in the spirit of the above development, we first construct a strictly proper filter of CSS architecture for the reduced-order system Σ_r^* as

$$\begin{cases} \sigma \tilde{\xi}_r = A_r \tilde{\xi}_r + A_{21} y_1 + K_r (y_r - C_r \tilde{\xi}_r) \\ \hat{z}_r^* = E_r^* \tilde{\xi}_r, \end{cases} \quad (11.124)$$

where K_r is termed as the reduced-order filter gain. As shown shortly in Lemma 11.73, the gain K_r and thus the filter (11.124) always exist whenever a proper γ-level H_∞ SOID filter exists for Σ.

As σy_1 is not available, we need to modify the above filter. To this end, let us partition $K_r = (K_{r0} \quad K_{r1})$ so as to be compatible with the partitioning of y_r. Also, let

$$\xi_r = \tilde{\xi}_r - K_{r1} y_1. \quad (11.125)$$

We can then easily rewrite the filter (11.124) as a proper filter for Σ as shown below.

The reduced-order proper filter of CSS architecture for Σ is given by

$$\Sigma_{r-CSS} : \begin{cases} \sigma \xi_r = (A_r - K_r C_r) \xi_r + \tilde{K}_r y \\ \tilde{\xi}_r = \xi_r + K_{r1} y_1 \\ \hat{z} = E_1^* x_1 + E_r^* \tilde{\xi}_r + P^* y = E_r^* \xi_r + \tilde{P}_r y, \end{cases} \quad (11.126)$$

where

$$\tilde{K}_r = \begin{pmatrix} K_{r0} & A_{21} - K_{r1} A_{11} + (A_r - K_r C_r) K_{r1} \end{pmatrix}$$

and

$$\tilde{P}_r = \begin{pmatrix} 0 & E_1^* + E_r^* K_{r1} \end{pmatrix} + P^*.$$

The above development leads to the following lemma.

Lemma 11.73 *Consider the regular γ-level H_∞ SOID filtering problem for the discrete-time system Σ described by (11.1). That is, assume that the system characterized by (A, B, C, D) is right-invertible and has no invariant zeros on the unit circle. Also, assume that $\gamma > \gamma_p^*$. Then, the strictly proper γ-level H_∞ SOID filtering problem is solvable for the system Σ_r^*. Moreover, any strictly proper γ-level H_∞ SOID filter of the form (11.124) with CSS architecture for Σ_r^* results in a reduced-order γ-level H_∞ SOID filter (11.126) with CSS architecture for the system Σ.*

Proof : Let Q be the stabilizing solution of the H_∞^1 DARE associated with the system Σ. Then it is easy to verify that we have $C_1 Q = 0$, and using this, it becomes clear that Q has the form:

$$Q = \begin{pmatrix} Q_r & 0 \\ 0 & 0 \end{pmatrix},$$

where Q_r is a stabilizing solution of the H_∞^1 DARE associated with Σ_r^*. This shows that the the strictly proper H_∞ SOID filtering problem is solvable. The error dynamics of a strictly proper H_∞ SOID filter of the form (11.124) for Σ_r^* is the same as the error dynamics resulting from applying the filter (11.126) to the system Σ. It is then immediately obvious that (11.126) is a reduced-order H_∞ SOID for the system Σ. ∎

Again, we note that the above design does not result in a complete characterization of all reduced-order H_∞ SOID filters.

We should note that, unlike in continuous-time systems, the alternative route of first transforming Σ to Σ_Q and then designing a reduced-order filter of CSS architecture for Σ_Q, which is then converted back to a filter for Σ, does not result in the same family of reduced-order filters. In fact, as D_Q has full rank, there does not exist any order reduction in any attempt to design reduced-order filters for Σ_Q.

11.6.2 *Singular γ-level H_∞ SOID filters—the system characterized by (A, B, C, D) has no invariant zeros on the unit circle*

The previous subsection deals with the regular γ-level H_∞ SOID filtering. In this subsection, we consider the singular γ-level H_∞ SOID filtering for the special case when the system characterized by (A, B, C, D) has no invariant zeros on the unit circle. Following the spirit of the regular case, we first prove the necessity parts of proofs of Theorems 11.39 and 11.43. Then, we prove the sufficiency parts of the same theorems. Once again, as in the previous subsection, the proofs of sufficiency parts lay a roadmap for the design of appropriate γ-level H_∞ SOID filters for the given system Σ. Then, as usual, we explore the design of strictly proper, proper, as well as reduced-order γ-level H_∞ SOID filters of CSS architecture.

We start with a necessity proof for Theorem 11.39.

Proof of necessity for Theorem 11.39 (no invariant zeros of the system (A, B, C, D) on the unit circle) : We use the dual decomposition of the one presented in (4.153). Hence, a matrix K exists such that in a suitable basis (the compact form of the SCB), we have

$$A - KC = \begin{pmatrix} A_{11} & 0 \\ A_{21} & A_{22} \end{pmatrix}, \quad B - KD = \begin{pmatrix} 0 \\ B_2 \end{pmatrix},$$

$$C = \begin{pmatrix} C_{11} & 0 \\ C_{21} & C_{22} \end{pmatrix}, \quad D = \begin{pmatrix} 0 \\ D_2 \end{pmatrix}, E = \begin{pmatrix} E_1 & E_2 \end{pmatrix},$$

where K is such that $(B - KD)'D = 0$. Moreover, $(A_{22}, B_2, C_{22}, D_2)$ is left-invertible, (C_1, A_{11}) is $\mathbb{C}^\ominus$-detectable, and $D_2 B_2' = 0$. An output injection does not change the achievable H_∞ norm of the error dynamics. Moreover, if we assume that x_1 is additionally measured without noise, then the achievable H_∞ norm of the error dynamics by an unbiased filter can be made even smaller. Therefore, for the following system, we can achieve an H_∞ norm of the error dynamics less than γ by an unbiased filter:

$$\begin{cases} \sigma x_2 = A_{22}x_2 + B_2 u \\ \quad y = C_{22}x_2 + D_2 u \\ \quad z = E_2 x_2 + Fu. \end{cases}$$

As this is a regular H_∞ SOID filtering problem, we can apply the results from Theorem 11.33, which guarantees the existence of a stabilizing solution of the H_∞^1 DARE:

$$\bar{Y} = \bar{A}_r \bar{Y} \bar{A}_r' + B_r B_r'$$

$$- \begin{pmatrix} C_r \bar{Y} \bar{A}_r' + D_r B_r \bar{A}_r' \\ E_r \bar{Y} \bar{A}_r' + F_r B_r' \end{pmatrix} G_r(\bar{Y}) \begin{pmatrix} C_r \bar{Y} \bar{A}_r' + D_r B_r' \\ E_r \bar{Y} \bar{A}_r' + F_r B_r' \end{pmatrix},$$

where

$$G_r(\bar{Y}) = \begin{pmatrix} B_r \bar{Y} B_r' + D_r D_r' & B_r \bar{Y} E_r' + D_r F_r' \\ E_r \bar{Y} B_r' + F_r D_r' & E_r \bar{Y} E_r' + F_r F_r' - \gamma^2 I \end{pmatrix}^{-1}$$

and

$$\bar{A}_r = A_{22}, B_r = B_2, C_r = C_{22}, D_r = D_2, E_r = E_2. \tag{11.127}$$

Using a dual version of Proposition 4.93 and the discussion after this proposition, we then obtain the existence of a stabilizing solution of the GDARE (11.40).

The fact that (11.41) must be satisfied when using strictly proper filters should by now be obvious. The conditions regarding the invariant zeros on the unit circle are clearly empty because we are looking at the case where there are no invariant zeros on the unit circle.   ■

We continue with a necessity proof for the equivalent theorem when proper filters are used.

Proof of necessity for Theorem 11.43 (no invariant zeros of the system (A, B, C, D) on the unit circle) : The proof follows along the same lines as in the necessity proof of Theorem 11.22. The fact that (11.42) is a necessary condition should be obvious.

We again assume that x_1 is available to us and construct a new system. By applying Theorem 11.36, we obtain the existence of a stabilizing solution of the associated H_∞^1 DARE. Using a dual version of Proposition 4.93 and the discussion after this proposition, we then obtain the existence of a stabilizing solution of the GDARE (11.40).   ∎

Proofs of sufficiency parts of Theorems 11.39 and 11.43 (no invariant zeros of the system (A, B, C, D) on the unit circle).

So far we have proved the necessity parts of Theorems 11.39 and 11.43. We proceed now to prove the sufficiency parts. In this regard, as we said earlier, our design methodology arises as a by-product while proving the sufficiency parts. As in the previous subsection, we can then explore the design of strictly proper, proper, as well as reduced-order γ-level H_∞ SOID filters of CSS architecture.

As in the previous subsection, we define two auxiliary systems Σ_Q and $\breve{\Sigma}_Q$. We define Σ_Q by

$$
\Sigma_Q : \begin{cases} \sigma x_q = A x_q + B_Q u_q + G_Q z_q \\ \quad y \ = C x_q + D_Q u_q \\ \quad z_q \ = E x_q + F_Q u_q, \end{cases} \tag{11.128}
$$

where D_Q, F_Q are defined by (11.102c), whereas B_Q and G_Q are defined by

$$
B_Q = \begin{pmatrix} CQA' + DB' \\ EQA' + FB' \end{pmatrix}' \begin{pmatrix} CQC' + DD' & CQE' + DF' \\ EQC' + FD' & EQE' + FF' - \gamma^2 I \end{pmatrix}^{\dagger} \begin{pmatrix} D_Q \\ F_Q \end{pmatrix}, \tag{11.129a}
$$

$$
G_Q = -\begin{pmatrix} CQA' + DB' \\ EQA' + FB' \end{pmatrix}' \begin{pmatrix} CQC' + DD' & CQE' + DF' \\ EQC' + FD' & EQE' + FF' - \gamma^2 I \end{pmatrix}^{\dagger} \begin{pmatrix} 0 \\ I \end{pmatrix}. \tag{11.129b}
$$

This is the same construction as for the regular case in (11.101).

We can show that the strictly proper (or proper) γ-level H_∞ SOID filtering problem for Σ_Q is solvable whenever the strictly proper (or proper) γ-level H_∞ SOID filtering problem for Σ is solvable. To do so, as in the regular case, we introduce next a slightly modified version of Σ_Q, namely a system $\breve{\Sigma}_Q$ whose desired output signal $\breve{z}_q$ that needs to be estimated is slightly different from z_q:

$$
\breve{\Sigma}_Q : \begin{cases} \sigma x_q = (A + G_Q E) x_q + (B_Q + G_Q F_Q) u_q, \\ \quad y \ = C x_q + D_Q u_q, \\ \quad \breve{z}_q \ = (E - PC) x_q, \end{cases} \tag{11.130}
$$

where the matrix P, as in the previous section, is different for the two different cases, strictly proper and proper. For the case when Σ satisfies the conditions for the existence of strictly proper γ-level H_∞ SOID filters, a suitable choice for the matrix P is $P = 0$, which implies that $\|F_Q\| < \gamma$. On the other hand, when Σ satisfies the conditions for the existence of proper γ-level H_∞ SOID filters, a matrix P exists such that $\|F_Q - PD_Q\| < \gamma$. Then, obviously a suitable choice for the matrix P in (11.130) is such that $\|F_Q - PD_Q\| < \gamma$.

As before, we make two claims. First, for the system $\check{\Sigma}_Q$ the H_∞ AID filtering problem is solvable by a sequence of strictly proper filters irrespective of the choice of P. Second, any sequence of H_∞ AID filters designed for $\check{\Sigma}_Q$ using the methodology of Chapter 9 can be translated as γ-level H_∞ SOID filters for Σ_Q. These claims are formalized in the following two lemmas.

Lemma 11.74 *Consider the singular γ-level H_∞ SOID filtering problem for the discrete-time system Σ given in (11.1) such that the system characterized by (A, B, C, D) has no invariant zeros on the unit circle. Assume that a stabilizing solution Q of the H_∞^1 GDARE (11.40) exists such that (11.41) is satisfied [or such that (11.42) is satisfied]. In this case, we can define $\check{\Sigma}_Q$ by (11.130) with $P = 0$ (or with P such that $\|F_Q - PD_Q\| < \gamma$).*

Then, the H_∞ AID filtering problem is solvable for $\check{\Sigma}_Q$ by a sequence of strictly proper filters irrespective of $P = 0$ (strictly proper case) or P is such that $\|F_Q - PD_Q\| < \gamma$ (proper case).

Proof : The fact that Q is a stabilizing solution of the H_∞^1 GDARE (11.40) implies that

$$\text{rank} \begin{pmatrix} A + G_Q E & B_Q + G_Q F_Q \\ C & D_Q \end{pmatrix} = n + \text{rank} \begin{pmatrix} B_Q + G_Q F_Q \\ D_Q \end{pmatrix}.$$

This in turn implies that

$$\mathcal{V}^*(\Sigma_*) \cap \mathcal{S}^{-0}(\Sigma_*) = \{0\},$$

where Σ^* is the subsystem described by the quadruple,

$$(A + G_Q E, B_Q + G_Q F_Q, C, D_Q).$$

This in turn shows that the H_∞ AID filtering problem is solvable for $\check{\Sigma}_Q$ according to Theorem 9.14. $\blacksquare$

Lemma 11.75 *Consider the singular γ-level H_∞ SOID filtering problem for the discrete-time system Σ given in (11.1) such that the system characterized by (A, B, C, D) has no invariant zeros on the unit circle. Assume that a stabilizing*

solution Q of the H_∞^1 GDARE (11.40) exists such that (11.41) is satisfied [or such that (11.42) is satisfied]. Define Σ_Q by (11.128). Also, define $\check{\Sigma}_Q$ by (11.130) with $P = 0$ for the strictly proper case (or with P such that $\|F_Q - PD_Q\| < \gamma$ for the proper case).
Then the following hold:

(*i*) *Let a sequence of strictly proper filters of the form (11.2) characterized by a parameterized matrix triple $(L_\varepsilon, M_\varepsilon, N_\varepsilon)$ solve the H_∞ AID filtering problem for the system $\check{\Sigma}_Q$ with $P = 0$ (strictly proper case). Then, for ε sufficiently small, any member of such a sequence is a strictly proper γ-level H_∞ SOID filter for Σ_Q.*

(*ii*) *Let a sequence of strictly proper filters of the form (11.2) characterized by a parameterized matrix triple $(L_\varepsilon, M_\varepsilon, N_\varepsilon)$ solve the H_∞ AID filtering problem for the system $\check{\Sigma}_Q$ with P such that $\|F_Q - PD_Q\| < \gamma$ (proper case). For ε sufficiently small, consider a proper filter of the form (11.2) characterized by the matrix quadruple $(L_\varepsilon, M_\varepsilon, N_\varepsilon, P)$, where the matrix triple $(L_\varepsilon, M_\varepsilon, N_\varepsilon)$ corresponds to the sequence of strictly proper H_∞ AID filters for $\check{\Sigma}_Q$ and the matrix P corresponds to the one that is used in defining $\check{\Sigma}_Q$. Then such a proper filter is a proper γ-level H_∞ SOID filter for Σ_Q.*

Proof : This lemma is analogous to Lemma 11.57 for the continuous-time case, and the proof follows along the same lines as the proof of Lemma 11.49.   ∎

The above lemmas guarantee that the strictly proper (or proper) γ-level H_∞ SOID filtering problem is solvable for Σ_Q whenever the strictly proper (or proper) γ-level H_∞ SOID filtering problem is solvable for Σ. We will shortly and formally establish a relationship between the γ-level H_∞ SOID filters for Σ and γ-level H_∞ SOID filters for Σ_Q. This result will be obtained from the result for the regular case (Lemmas 11.68 and 11.69). The following lemma will be a crucial component in establishing such a result.

Lemma 11.76 *Consider the singular γ-level H_∞ SOID filtering problem for the discrete-time system Σ given in (11.1) such that the system characterized by (A, B, C, D) has no invariant zeros on the unit circle. Also, assume that $\gamma > \gamma_{sp}^*$ (or $\gamma > \gamma_p^*$); that is, assume that a stabilizing solution Q of the H_∞^1 GDARE (11.40) exists such that (11.41) is satisfied [or such that (11.42) is satisfied]. We define the following system $\bar{\Sigma}_Q$ with two inputs u_q and $\hat{z}$ and two outputs y and e_q as shown in Figure 11.5:*

$$\bar{\Sigma}_Q : \begin{cases} \sigma x_q = A x_q + B_Q u_q + G_Q e_q \\ \quad y \ = C x_q + D u_q \\ \quad e_q = E x_q + F u_q - \hat{z}. \end{cases}$$

Then the following hold:

(i) *Suppose a filter Σ_f of the form (11.2) and characterized by the quadruple (L, M, N, P) (with $P = 0$ for strictly proper case) is a γ-level H_∞ SOID filter for the given system Σ. Then the interconnection of Σ_f and $\bar{\Sigma}_Q$ (that is, $\bar{\Sigma}_Q \times \Sigma_f$ as shown in Figure 11.5) is such that the H_∞ norm of u_q to e_q is strictly less than γ, and for all initial conditions for (11.2) and $\bar{\Sigma}_Q$, whenever $u_q = 0$, we have that $e_q(k) \to 0$ as $k \to \infty$.*

(ii) *Conversely, suppose a system Σ_f of the form (11.2) and characterized by the quadruple (L, M, N, P) (with $P = 0$ for strictly proper case) interconnected with the system $\bar{\Sigma}_Q$ (that is, $\bar{\Sigma}_Q \times \Sigma_f$ as shown in Figure 11.5) is such that the H_∞ norm of u_q to e_q is strictly less than γ, and for all initial conditions for (11.2) and $\bar{\Sigma}_Q$, whenever $u_q = 0$, we have that $e_q(k) \to 0$ as $k \to \infty$. Then (11.2) describes a filter that is a γ-level H_∞ SOID filter for the system Σ.*

Proof : The proof follows analogous to the continuous-time result in Lemma 11.58. ∎

The following lemma that connects Σ_Q and Σ is an important consequence of Lemma 11.76.

Lemma 11.77 *Consider the singular γ-level H_∞ SOID filtering problem for the discrete-time system Σ described by (11.1) such that the system characterized by (A, B, C, D) has no invariant zeros on the unit circle. Assume that $\gamma > \gamma_{sp}^*$ (or $\gamma > \gamma_p^*$) which implies that a semi-stabilizing solution Q of the H_∞^1 GDARE (11.37) exists. Finally, assume that A is Schur-stable. Let Σ_Q be as in (11.128). Then the following hold:*

(i) *Suppose a filter of the form (11.2) and characterized by the quadruple (L, M, N, P) (with $P = 0$ for the strictly proper case) is a γ-level H_∞ SOID filter for the given system Σ. Then the following filter is a γ-level H_∞ SOID filter for the system Σ_Q:*

$$\sigma x_1 = (A + G_Q E)x_1 - G_Q \hat{z}_q$$
$$\sigma \xi = L\xi + M(y + Cx_1) \tag{11.131}$$
$$\hat{z}_q = N\xi + P(y + Cx_1).$$

(ii) *Conversely, suppose a filter of the form (11.2) and characterized by the quadruple (L, M, N, P) (with $P = 0$ for the strictly proper case) is a*

γ-level H_∞ SOID filter for the system Σ_Q. Then the following filter is a γ-level H_∞ SOID filter for the given system Σ:

$$\begin{aligned}
\sigma x_1 &= (A + G_Q E)x_1 - G_Q \hat{z} \\
\sigma \xi &= L\xi + M(y - Cx_1) \\
\hat{z} &= N\xi + P(y - Cx_1).
\end{aligned} \qquad (11.132)$$

Proof : The proof follows analogous to the continuous-time case as in Lemma 11.59. ∎

By now we have completed the proofs of sufficiency parts of Theorems 11.39 and 11.43. The above development of Lemmas 11.74 to 11.77 clearly lays a roadmap for the design of γ-level H_∞ SOID filters for the given system Σ. That is, we construct new systems Σ_Q and $\breve{\Sigma}_Q$. We design a sequence of H_∞ AID filters for $\breve{\Sigma}_Q$ that can then be translated as an H_∞ AID filter for Σ_Q. Then in turn such an H_∞ AID filter for Σ_Q can be translated as a γ-level H_∞ SOID filter for the given system Σ. We observe that filters of any arbitrary architecture can be used to design the sequence of H_∞ AID filters for $\breve{\Sigma}_Q$. However, by using the design methodology developed in Chapter 9, one can easily design a sequence of AID filters of CSS architecture for $\breve{\Sigma}_Q$.

Design of γ-level H_∞ SOID filters of CSS architecture.

In what follows, we consider the design of γ-level H_∞ SOID filters of CSS architecture for Σ while providing the proofs of sufficiency parts of Theorems 11.39 and 11.43. As in the case of previous chapters, we consider strictly proper, proper, and reduced-order filters of CSS architecture, one at a time. Our design methodology follows exactly as in the previous subsection that corresponds to regular γ-level H_∞ SOID filtering. The first essential difference is that instead of using the stabilizing solution Q of the H_∞^1 DARE (11.37), we need to use the stabilizing solution Q of the H_∞^1 GDARE (11.40), and then define Σ_Q and $\breve{\Sigma}_Q$ accordingly by (11.128) and (11.130). Also, in the regular case, we need to design EID filters for $\breve{\Sigma}_Q$ or for Σ_Q that can then be translated as γ-level H_∞ SOID filters for the given system Σ. However, in the singular case being considered now, we need to design H_∞ AID filters for $\breve{\Sigma}_Q$ or for Σ_Q that can then be translated as γ-level H_∞ SOID filters for the given system Σ. Moreover, in the regular case, the technical lemmas that were essential components in designing γ-level H_∞ SOID filters are Lemmas 11.70, 11.72, and 11.73. In the singular case being considered now, the corresponding lemmas are to be redeveloped as given below.

The following lemma is analogous to Lemma 11.70.

Lemma 11.78 *Consider the singular γ-level H_∞ SOID filtering problem for the discrete-time system Σ given in (11.1) such that the system characterized by*

(A, B, C, D) *has no invariant zeros on the unit circle. Assume that a stabilizing solution* Q *of the* H_∞^1 *GDARE* (11.40) *exists such that* (11.38) *is satisfied. Define* $B_Q, D_Q, F_Q,$ *and* G_Q *by* (11.102).

Then the following hold:

(i) *Let the following strictly proper filter, which is of CSS architecture, be a* γ-*level* H_∞ *SOID filter for* Σ:

$$
\begin{aligned}
\sigma\xi &= A\xi + K(y - C\xi), \\
\hat{z} &= E\xi.
\end{aligned}
\tag{11.133}
$$

Then, the following strictly proper filter of CSS architecture is a γ-*level* H_∞ *SOID filter for the system* Σ_Q:

$$
\begin{aligned}
\sigma\xi &= A\xi + K(y - C\xi) + G_Q\hat{z}_q, \\
\hat{z}_q &= E\xi.
\end{aligned}
\tag{11.134}
$$

(ii) *Conversely, let a strictly proper filter of the form* (11.134)*, which is of CSS architecture, be a* γ-*level* H_∞ *SOID filter for* Σ_Q. *Then, the strictly proper filter* (11.133) *is a* γ-*level* H_∞ *SOID filter for the system* Σ.

Proof : The proof follows along the lines of the proof of Lemma 11.70. ∎

We find that the results presented earlier in Lemmas 11.72 and 11.73 still apply with hardly any modifications, and we find the following lemmas.

Lemma 11.79 *Consider the singular* γ-*level* H_∞ *SOID filtering problem for the discrete-time system* Σ *given in* (11.1) *such that the system characterized by* (A, B, C, D) *has no invariant zeros on the unit circle. Assume that a stabilizing solution* Q *of the* H_∞^1 *GDARE* (11.40) *exists such that* (11.42) *is satisfied. Choose* P^* *such that*

$$
(E - P^*C)Q(E - P^*C)' + (F - P^*D)(F - P^*D)' < \gamma^2 I,
$$

and define $\widetilde{\Sigma}^*$ *by* (11.112).

Then the H_∞ *SOID filtering problem for* $\widetilde{\Sigma}^*$ *of* (11.112) *is solvable by a strictly proper filter of the form* (11.115) *having the CSS architecture. Moreover, any strictly proper* H_∞ *SOID filter of the form* (11.115) *for* $\widetilde{\Sigma}^*$ *results in a proper* H_∞ *SOID filter* (11.117) *for the system* Σ.

Lemma 11.80 *Consider the singular* γ-*level* H_∞ *SOID filtering problem for the discrete-time system* Σ *given in* (11.1) *such that the system characterized by*

(A, B, C, D) *has no invariant zeros on the unit circle. Assume that a stabilizing solution Q of the H_∞^1 GDARE (11.40) exists such that (11.42) is satisfied. Let Σ_r^* be as in (11.122).*

In this case, for the system Σ_r^* of (11.122), the strictly proper H_∞ SOID filtering problem is solvable. Moreover, any strictly proper H_∞ SOID filter of the form (11.124) with the CSS architecture for Σ_r^* results in a reduced-order H_∞ SOID filter (11.126) with the CSS architecture for the system Σ.

Again, we note that the above design of course does not result in a complete characterization of all reduced-order H_∞ SOID filters. Also, we stress once more that minimizing the H_∞ norm of the error dynamics for Σ_Q does not mean that we minimize the H_∞ norm of the error dynamics when considering the original system Σ.

11.6.3 Singular γ-level H_∞ SOID filters—the system characterized by (A, B, C, D) has invariant zeros on the unit circle

In the previous subsection, we already proved Theorem 11.39 in case there are no invariant zeros on the unit circle. We start this subsection with a necessity proof for this theorem for the case in which the system characterized by (A, B, C, D) has invariant zeros on the unit circle.

Proof of necessity for Theorem 11.39 (invariant zeros of the system (A, B, C, D) on the unit circle) : We use the same arguments as in the necessity proof for Theorem 11.39 in the previous subsection. We first use a reduction to an H_∞ SOID filtering problem with a right-invertible subsystem from u to y. From the proof of Theorem 11.33, we can conclude the existence of a strongly semi-stabilizing solution to the corresponding H_∞^1 DARE. Then using the dual version of Proposition 4.93 and the discussion after this proposition, we then obtain the existence of a positive semi-definite and semi-stabilizing solution of the H_∞^1 GDARE (11.40).

The fact that (11.41) must be satisfied when using strictly proper filters should by now be obvious. The conditions regarding invariant zeros on the unit circle still need to be established. From the necessity proof of Theorem 11.33, we know that matrices K and K_1 exist such that (11.95) is Schur-stable and the transfer matrix has an H_∞ norm less than γ. This immediately implies that for $\bar{K} = K + K_1$, we have for any s_0 on the unit circle,

$$\| E(s_0 I - A + \bar{K}C)^{-1}(B - \bar{K}D) + F \| < \gamma,$$

whereas $\bar{K}$ is such that $s_0 I - A + \bar{K}C$ is invertible. This immediately implies that the conditions regarding the invariant zeros on the unit circle are satisfied. ∎

We continue with a necessity proof for the equivalent theorem when using proper filters.

Proof of necessity for Theorem 11.43 (invariant zeros of the system $(A$, B, C, $D)$ on the unit circle) : The proof follows the same line as the necessity proof of Theorem 11.39 and the proof of Theorem 11.43 as given in the previous subsection for the case without invariant zeros on the unit circle. The fact that (11.42) is a necessary condition should be obvious.

Also the same argument as before yields a solution of the H_∞^1 GDARE (11.40). From the existence of a proper H_∞ SOID filter, it is easily seen that matrices K and P exist such that for any s_0 on the unit circle,

$$\|(E - PC)(s_0 I - A + KC)^{-1}(B - KD) + (F - PD)\| < \gamma,$$

where K is such that $s_0 I - A + KC$ is invertible.   ∎

We need to prove the sufficiency parts of Theorems 11.39 and 11.43, and then we present methods of designing appropriate γ-level H_∞ SOID filters. In this regard, at the beginning of this section, we noted that our design philosophy is to transform the design of γ-level H_∞ SOID filters for a given system to the design of EID filters or H_∞ AID filters for an auxiliary system constructed from the data of the given system. We also emphasized that such a transformation emerges as a by-product while proving sufficiency parts of Theorems 11.39 and 11.43. We had been successful in doing so for regular H_∞ SOID filtering as well as for singular H_∞ SOID filtering with the restriction that the system characterized by (A, B, C, D) has no invariant zeros on the unit circle. However, as in the case of continuous-time systems, we have not succeeded yet in doing so for the case of singular H_∞ SOID filtering when the system characterized by (A, B, C, D) has indeed invariant zeros on the unit circle. Nevertheless, we still conjecture that it can be done.

We proceed now to indicate the proofs of the sufficiency parts of Theorems 11.39 and 11.43. To do so, we need to indicate the existence of suitable H_∞ SOID filters for the given system system Σ under the given conditions. In this regard, like in the continuous-time case, two methods are available in the literature. One is based on a perturbation argument where a small perturbation results in a regular problem for which we have a design methodology available. These perturbation arguments are worked out in detail in the books [75, 92] for the continuous-time case but apply as well for the discrete-time case The alternative, and numerically more attractive, method is worked out in [86]. However, this is a continuous-time result. To apply this in our case, we need to apply a bilinear transform (which preserves the H_∞ norm). These references are for the design of state feedback controllers, but by duality arguments, these can be used to obtain strictly proper γ-level H_∞ SOID filters of CSS architecture. This duality connection has been explored before to connect the EDD problem and the EID filtering problem and the ADD problem and the AID filtering problem. Finally, to design proper filters or reduced-order filters, we can follow the methodology of the previous two subsections. We can construct two auxiliary systems such that strictly proper filters for the auxiliary systems can be converted to proper or reduced-order filters, respectively, for the original system.

12

Generalized H_2 suboptimally input-decoupled filtering

12.1 Introduction

In the previous chapters, we have studied EID, H_2 AID, H_∞ AID, H_2 OID, H_2 SOID, as well as γ-level H_∞ SOID filtering problems. These filtering problems are reconsidered in this chapter and the next with additional unknown input signals containing linear combination of sinusoidal signals, each of which has an unknown amplitude and phase but known frequency. We will use the qualifier "generalized" to refer to each of the above problems when such additional inputs are present. For instance, the H_2 OID, H_2 SOID, and γ-level H_∞ SOID filtering problems described above when additional sinusoidal inputs (of unknown amplitude and phase but known frequency) are present are, respectively, referred to as generalized H_2 OID, generalized H_2 SOID, and generalized γ-level H_∞ SOID filtering problems.

The additional sinusoidal inputs of unknown amplitude and phase but known frequency are modeled by an exosystem. Such an exosystem can then be appended to the given system to form an expanded system. This lets us analyze the effects of such additional sinusoidal inputs by using the formalism developed in the previous chapters. Also, in the presence of such additional sinusoidal inputs, the filters used must be unbiased in a generalized sense. A filter is unbiased in the generalized sense if the effect of the sinusoidal signal of known frequency but unknown amplitude and phase reduces to zero asymptotically.

Our primary focus in this chapter at first is on the generalized H_2 OID and the generalized H_2 SOID filtering. Once we do so, we consider the generalized EID and the generalized H_2 AID filtering problems. The reason to do so is simple. It turns out that, as discussed in detail in the body of the chapter, the results on the generalized EID and the generalized H_2 AID filtering are a consequence of some of the results on generalized H_2 OID and generalized H_2 SOID filtering. The generalized γ-level H_∞ SOID filtering as well as the generalized H_∞ AID filtering are considered in Chapter 13.

After formulating formally the generalized H_2 OID and the generalized H_2 SOID filtering problems, we first study and resolve several issues associated with them. These issues include performance, existence, and uniqueness of generalized H_2 OID filters; design of generalized H_2 OID and generalized H_2 SOID filters; and the fixed modes associated with them. We then study carefully the cost incurred by the additional requirement of rejecting sinusoidal signals of known frequency but unknown amplitude and phase. We will show that the infimum of the root-mean-square (RMS) norm is not affected. In general, the solvability conditions of the generalized H_2 OID filtering problem might be stronger than the solvability conditions of the H_2 OID filtering problem. Next, we look at the effect on the fixed modes because of the additional requirement of rejecting sinusoidal signals of known frequency but unknown amplitude and phase. Provided that the generalized H_2 OID filtering problem is solvable, we show that the fixed modes remain the same. However, if the generalized H_2 OID filtering problem is not solvable and when we look at the generalized H_2 SOID filtering problem, then in general the set of asymptotic fixed modes is expanded with some of the modes being those of the external sinusoidal signals.

For the RMS norm of the estimation error signal to be finite, we must use filters that are unbiased in the generalized sense. The cost of unbiasedness (in the generalized sense) can be expressed in terms of the energy of the estimation error signal. Similarly, the cost of rejecting a white noise input is expressed in terms of the RMS norm of the estimation error signal. We will see that both the cost of rejecting a white noise input and the cost of unbiasedness (in the generalized sense) are related to the locations of the non-minimum-phase zeros of the system. Moreover, we will uncover a peculiar property that the cost of rejecting a white noise input *reduces* when the non-minimum-phase zeros are moved closer to the boundary of the stability domain (imaginary axis in continuous-time and unit circle in discrete time), whereas the cost of unbiasedness (in the generalized sense) *increases to infinity* when the non-minimum-phase zeros are moved closer to the boundary of the stability domain.

After studying several issues outlined above that are associated with the generalized H_2 OID and the generalized H_2 SOID filtering problems, we turn our attention here to the generalized EID and the generalized H_2 AID filtering problems.

12.2   Preliminaries

Let us consider the plant or system model:

$$\Sigma : \begin{cases} \sigma x = Ax + Bu \\ y \ \ = Cx + Du \\ z \ \ = Ex + Fu. \end{cases} \tag{12.1}$$

As before, $u \in \mathbb{R}^m$ is the input, $x \in \mathbb{R}^n$ is the state, $y \in \mathbb{R}^p$ is the measured output, and $z \in \mathbb{R}^q$ is the desired output signal to be estimated. The input u is decomposed into two parts:

$$u = \begin{pmatrix} u_1 \\ u_2 \end{pmatrix}, \tag{12.2}$$

where $u_1 \in \mathbb{R}^{m_1}$ and $u_2 \in \mathbb{R}^{m_2}$. The first type of input, denoted by u_1, is assumed as in Chapter 10 to be a zero-mean wide-sense stationary white noise of unit intensity. On the other hand, the second type of input, denoted by u_2, is assumed to be a linear combination of sinusoidal signals, each of which has an unknown amplitude and phase but known frequency. Clearly, such a signal u_2 can be modeled as the output of a known linear autonomous system with unknown initial conditions. We call such a system an exogenous system or, for short, exosystem. Thus, consider

$$\Sigma_a : \quad \sigma x_a = S x_a, \quad u_2 = C_a x_a, \tag{12.3}$$

where $x_a \in \mathbb{R}^{n_a}$ for some n_a. An important special case where this type of problem arises is the case of having a system driven by a wide-sense stationary white noise input with unknown intensity (variance) and mean. It is easily verified that the filters designed in the previous chapters are independent of the intensity level of the noise. After all, if we change B into BV with V invertible, then the class of H_2 OID filters remains the same even though V does effect the RMS gain. However, having a nonzero mean of the external input requires a modification of our filter as we will see from the results of this chapter. The first person who looked into this modification with nonzero mean disturbances for the H_2 OID filter was James Blight in [6].

Let us next partition the matrices, B, D, and F in conformity with the partitioning of u:

$$B = \begin{pmatrix} B_1 & B_2 \end{pmatrix}, \quad D = \begin{pmatrix} D_1 & D_2 \end{pmatrix}, \quad F = \begin{pmatrix} F_1 & F_2 \end{pmatrix}. \tag{12.4}$$

For future reference, we note that the system then has the structure:

$$\Sigma : \begin{cases} \sigma x = Ax + B_1 u_1 + B_2 u_2 \\ y \;\;= Cx + D_1 u_1 + D_2 u_2 \\ z \;\;= Ex + F_1 u_1 + F_2 u_2. \end{cases} \tag{12.5}$$

As before, our interest lies in estimating the desired output signal z while using only the measured output y but not the input u. As usual, let $\hat{z}$ be the estimate of z as given by a filter, and let e_z be the estimation error, $e_z = z - \hat{z}$ as depicted in Figure 12.1.

As before, it is natural to use the following assumption throughout this chapter as well.

Assumption 12.1 *The matrix pair (C, A) is $\mathbb{C}^-$-detectable for continuous-time systems and $\mathbb{C}^\ominus$-detectable for discrete-time systems.*

Again, as before, we consider a general proper filter of the form (10.2), which is repeated below:

$$\Sigma_f : \begin{cases} \sigma \xi = L\xi + My \\ \hat{z} \;= N\xi + Py. \end{cases} \tag{12.6}$$

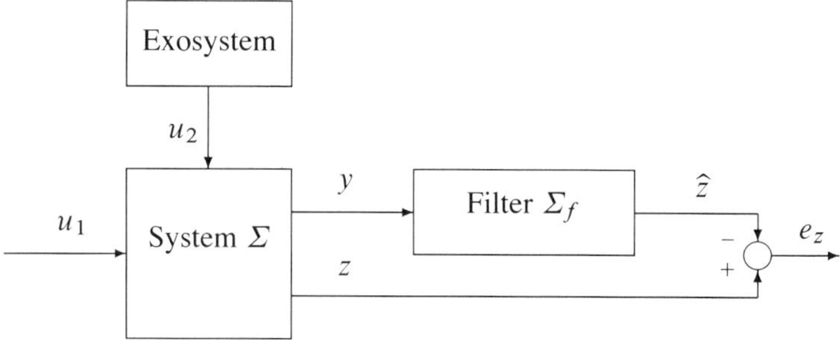

Figure 12.1: General block diagram of filtering for Σ

Whenever $P = 0$, the above filter is said to be a strictly proper filter. As in the previous chapters, we require that the filter (12.6) be internally stable. When the above filter is used as shown in Figure 12.1, the dynamic equations of the error e_z are described by

$$\Sigma_{ue} : \begin{cases} \sigma x = Ax + B_1 u_1 + B_2 u_2 \\ \sigma \xi = MCx + L\xi + MD_1 u_1 + MD_2 u_2 \\ \sigma x_a = S x_a, \qquad u_2 = C_a x_a \\ e_z = (E - PC)x - N\xi + (F_1 - PD_1)u_1 + (F_2 - PD_2)u_2. \end{cases}$$
(12.7)

The transfer matrices from u_i to e_z, $i = 1, 2$, can be computed easily as

$$G_{u_i e_z}(\sigma) = \begin{pmatrix} E - PC & -N \end{pmatrix} \begin{pmatrix} \sigma I - A & 0 \\ -MC & \sigma I - L \end{pmatrix}^{-1} \begin{pmatrix} B_i \\ MD_i \end{pmatrix} + (F_i - PD_i).$$
(12.8)

As we did in other chapters, sometimes we seek a family of filters parameterized in a positive parameter ε. In this case, a family of filters is described by

$$\Sigma_f^\varepsilon : \begin{cases} \sigma \xi = L^\varepsilon \xi + M^\varepsilon y \\ \widehat{z} = N^\varepsilon \xi + P^\varepsilon y, \end{cases}$$
(12.9)

where L^ε, M^ε, N^ε, and P^ε are matrices parameterized in a positive parameter ε.

12.3   Problem statements

We focus first on generalized H_2 OID and generalized H_2 SOID filtering. Before we formulate formally the generalized H_2 OID and the generalized H_2 SOID filtering problems, we first define what we mean by unbiased filters in the generalized setting. The following is an extension of the earlier definition, for instance, given in Definition 10.2.

Definition 12.2 *(Generalized unbiasedness) Consider a continuous- or discrete-time system Σ as given in (12.5) where the input u_1 is a zero-mean wide-sense stationary white noise, whereas u_2 is generated by a linear exosystem given in (12.3). We say a linear stable strictly proper (or proper) filter (12.6) is unbiased in the generalized sense if, in the absence of the input u_1, the estimation error e_z decays asymptotically to zero for all possible initial conditions of the given system (12.5) and the filter (12.6), and for all input signals u_2 generated by (12.3). Equivalently, a filter is unbiased if, in the absence of the input u_1, the estimation error e_z decays asymptotically to zero for all possible initial conditions of the exosystem (12.3), the given system (12.5), and the filter (12.6).*

We are now ready to define formally the following generalized OID filtering problem under white noise input.

Problem 12.3 (Generalized OID filtering problem under white noise input) Consider a continuous- or discrete-time system Σ as given in (12.5) where the input u_1 is a zero-mean wide-sense stationary white noise of unit intensity, and u_2 is the output of an exosystem as given in (12.3). Then, the generalized OID filtering problem under white noise input is defined as follows: Find, whenever it exists, a linear stable strictly proper (or proper) filter such that

(i) **(Generalized unbiasedness)** the estimation error e_z, in the absence of the input u_1, decays asymptotically to zero for all possible finite initial values of the exosystem (12.3), the given system (12.5), and the filter (12.6), and

(ii) **(Performance)** the RMS norm of the error signal, namely $\|e_z\|_{RMS}$, is as small as possible.

Remark 12.4 *The name "generalized OID filtering problem under white noise input" is appropriate to the above problem because whenever the input u_2 is set to zero, it reduces to the OID filtering problem under white noise input (see Problem 10.4).*

We can now define the generalized OID filtering performance under white noise input associated with the above generalized OID filtering problem.

Definition 12.5 *(Generalized OID filtering performance under white noise input) For the continuous- or discrete-time system Σ given in (12.5), where the input u_1 is a zero-mean wide-sense stationary white noise of unit intensity, and u_2 is the output of an exosystem as given in (12.3), the infimum of the RMS norm of the error signal e_z over the set of all linear stable strictly proper (or proper)* *unbiased (in the generalized sense) filters is called the generalized OID filtering performance under white noise input via linear stable strictly proper (or proper) filters and is denoted by $\gamma^*_{g,sp}$ (or $\gamma^*_{g,p}$).*

As noted, whenever the input u_2 is set to zero, the above generalized OID filtering problem under white noise input for the given system Σ reduces to the OID filtering problem under white noise input for a system Σ_0 given by

$$\Sigma_0 : \begin{cases} \sigma x = Ax + B_1 u_1 \\ y \;\;= Cx + D_1 u_1 \\ z \;\;= Ex + F_1 u_1. \end{cases} \tag{12.10}$$

The block diagram of filtering for the system Σ_0 is depicted in Figure 12.2. Also, in accordance with Definition 10.3, we denote the infimum of the RMS norm of the error signal over all linear unbiased stable filters for the system Σ_0 by γ^*_{sp} or γ^*_p depending on whether we use strictly proper or proper filters. We emphasize that

$$\gamma^*_{g,p} \geq \gamma^*_p \quad \text{and} \quad \gamma^*_{g,sp} \geq \gamma^*_{sp}.$$

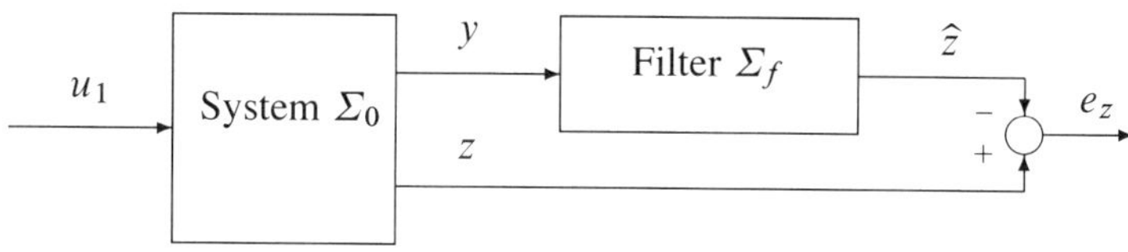

Figure 12.2: General block diagram of filtering for Σ_0

As argued in previous chapters, the generalized OID filtering problem under white noise input can be given a deterministic interpretation because the RMS norm of the error signal e_z is equal to the H_2 norm of the transfer matrix from the input u_1 to the error e_z. Thus, we can interpret the generalized OID filtering problem under white noise input as the **generalized H_2 OID filtering** problem, and similarly $\gamma^*_{g,sp}$ (or $\gamma^*_{g,p}$) as the **generalized H_2 OID filtering performance** via linear stable strictly proper (or proper) filters.

We classify below the generalized H_2 OID filtering problem into two major classes, *regular* and *singular*. We have the following definitions.

Definition 12.6 *Consider a continuous- or discrete-time system Σ as in (12.5) where the input u_1 is a zero-mean wide-sense stationary white noise of unit intensity and u_2 is the output of an exosystem as given in (12.3).*

(*i*) *For a continuous-time system Σ, a regular generalized H_2 OID filtering problem refers to a generalized H_2 OID filtering problem in which the matrix D_1 is surjective, and the subsystem characterized by the quadruple (A, B_1, C, D_1) has no invariant zeros on the imaginary axis.*

(*ii*) *For a discrete-time system Σ, a regular generalized H_2 OID filtering problem refers to a generalized H_2 OID filtering problem in which the subsystem characterized by the quadruple (A, B_1, C, D_1) is right invertible and has no invariant zeros on the unit circle.*

A generalized H_2 OID filtering problem is said to be a singular generalized H_2 OID filtering problem if it is not a regular generalized H_2 OID filtering problem.

Clearly, if the generalized H_2 OID filtering problem is not solvable, the next question is to try to find a family of linear stable unbiased (in the generalized sense) filters that make the RMS norm of the error signal close to $\gamma_{g,sp}^*$ or $\gamma_{g,p}^*$ depending on the class of filters one uses. This is formally defined in the following problem.

Problem 12.7 Consider a continuous- or discrete-time system Σ as in (12.5) where the input u_1 is a zero-mean wide-sense stationary white noise of unit intensity and u_2 is the output of an exosystem as given in (12.3). Then, the **generalized SOID filtering problem under white noise input** is defined as follows: Find, whenever it exists, a family of linear stable strictly proper (or proper) filters parameterized in positive ε such that

(*i*) (**Generalized unbiasedness**) for any given filter in the family, the estimation error e_z, in the absence of the input u_1, decays asymptotically to zero for all possible finite initial values of the exosystem (12.3), the given system (12.5), and the filter (12.6), and

(*ii*) (**Performance**) the RMS norm of the error signal, namely $\|e_z\|_{\mathrm{RMS}}$, approaches $\gamma_{g,sp}^*$ (or $\gamma_{g,p}^*$) as ε tends to zero.

Clearly, as the RMS norm of the error signal e_z is equal to the H_2 norm of the transfer matrix from the input u_1 to the error e_z, we can interpret the above problem as the **generalized H_2 SOID filtering problem**.

12.4   Performance, existence, and uniqueness conditions, design, and fixed modes

As in Chapter 10, we need to investigate several issues pertaining to generalized H_2 OID (H_2 SOID) filtering. These issues include computing the generalized H_2

OID filtering performance $\gamma^*_{g,sp}$ (or $\gamma^*_{g,p}$), developing the existence and uniqueness conditions for the generalized H_2 OID (SOID) filters, designing the generalized H_2 OID filters, and determining the fixed modes of the generalized H_2 OID filters. In this section, we relate the above issues pertaining to the generalized H_2 OID (H_2 SOID) filtering for the given system to the corresponding issues of H_2 OID (H_2 SOID) filtering, however, for an expanded system constructed from the data of the given system. In this way, the results of Chapter 10 can be of direct help in studying such issues. The required expanded system $\widetilde{\Sigma}$ is constructed by viewing together the given system Σ and the exosystem Σ_a as one system, and it is given by

$$\widetilde{\Sigma} : \begin{cases} \sigma \bar{x} = A_e \bar{x} + B_e u_1 \\ y \;\;= C_e \bar{x} + D_1 u_1 \\ z \;\;= E_e \bar{x} + F_1 u_1, \end{cases} \tag{12.11}$$

where

$$A_e = \begin{pmatrix} A & B_2 C_a \\ 0 & S \end{pmatrix}, \quad B_e = \begin{pmatrix} B_1 \\ 0 \end{pmatrix},$$

$$C_e = \begin{pmatrix} C & D_2 C_a \end{pmatrix}, \quad E_e = \begin{pmatrix} E & F_2 C_a \end{pmatrix}. \tag{12.12}$$

We need the following assumption for the expanded system $\widetilde{\Sigma}$, which is similar to the Assumption 12.1 for Σ.

Assumption 12.8 *The matrix pair (C_e, A_e) is $\mathbb{C}^-$-detectable for continuous-time systems and $\mathbb{C}^\ominus$-detectable for discrete-time systems.*

Before we state our main result of this section, let us draw the general filtering block diagram for the expanded system $\widetilde{\Sigma}$. Such a block diagram is given in Figure 12.3.

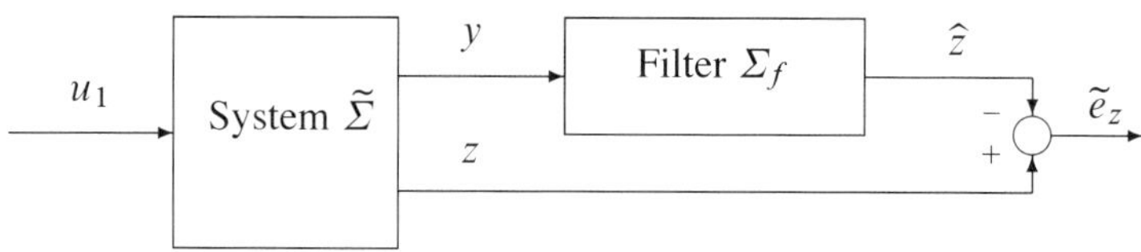

Figure 12.3: General block diagram of filtering for $\widetilde{\Sigma}$

The following theorem addresses the issue of how to compute $\gamma^*_{g,sp}$ or $\gamma^*_{g,p}$.

Theorem 12.9 *Consider the system Σ given in (12.5) together with the exosystem Σ_a given in (12.3). Let Assumption 12.1 be satisfied by Σ. Also, consider the generalized H_2 OID filtering performance associated with Σ, namely $\gamma^*_{g,sp}$ or $\gamma^*_{g,sp}$, depending on whether strictly proper or proper filters are used. Consider next the expanded system $\widetilde{\Sigma}$ given in (12.11). Let Assumption 12.8 be satisfied by $\widetilde{\Sigma}$. Let the H_2 OID filtering performance associated with the system $\widetilde{\Sigma}$ be denoted by $\gamma^*_{sp}(\widetilde{\Sigma})$ or $\gamma^*_p(\widetilde{\Sigma})$ depending on whether we use strictly proper or proper filters. Then, we have*

$$\gamma^*_{g,p} = \gamma^*_p(\widetilde{\Sigma}) \ \ and \ \ \gamma^*_{g,sp} = \gamma^*_{sp}(\widetilde{\Sigma}).$$

Proof : The proof follows directly from the proof of Theorem 12.10 given below.
∎

The above theorem lays a roadmap to compute $\gamma^*_{g,sp}$ or $\gamma^*_{g,p}$. Indeed, by using the procedure developed in Chapter 10, $\gamma^*_{g,sp}$ or $\gamma^*_{g,p}$ can be computed as $\gamma^*_{sp}(\widetilde{\Sigma})$ or $\gamma^*_p(\widetilde{\Sigma})$, which is associated with the expanded system $\widetilde{\Sigma}$. We present a theorem that provides a roadmap to study (a) the existence and uniqueness, (b) the design, and (c) the fixed modes of the generalized H_2 OID filters.

Theorem 12.10 *Consider the system Σ given in (12.5) together with the exosystem Σ_a given in (12.3) and the associated generalized H_2 OID filtering problem. Let Assumption 12.1 be satisfied by Σ. Consider next the expanded system $\widetilde{\Sigma}$ given in (12.11) and the associated H_2 OID filtering problem. Let Assumption 12.8 be satisfied by $\widetilde{\Sigma}$. Then, the following two statements are equivalent:*

    (i) The filter Σ_f given in (12.6) is a proper (or strictly proper) generalized H_2 OID filter for Σ.

    (ii) The filter Σ_f given in (12.6) is a proper (or strictly proper) H_2 OID filter for $\widetilde{\Sigma}$.

Proof : Assume a filter Σ_f of the form (12.6) is unbiased in the sense of Definition 12.2 for the system Σ along with the associated exosystem Σ_a and yields a stable transfer matrix $G_{u_1 e_z}$ from u_1 to $e_z = z - \hat{z}$. Then it can be trivially verified that such a filter when applied to the expanded system is unbiased in the sense of Definition 10.2 and results in the same stable transfer matrix $G_{u_1 e_z}$ from u_1 to $e_z = z - \hat{z}$. The converse of the above implication is also trivially satisfied. But then it is immediate that a filter is a generalized H_2 OID filter for the system (12.5) and the associated exosystem (12.3) if and only if it is a H_2 OID filter for the system (12.11). ∎

Theorem 12.10 is indeed a powerful tool to study various aspects of the generalized H_2 OID filtering problem. Let us expand on this. To start with, let us first discuss the existence and uniqueness conditions of generalized H_2 OID filters. Clearly, in view of Theorem 12.10, the study of existence and uniqueness conditions of generalized H_2 OID filters for the given system simply reduces to the study of H_2 OID filters for the expanded system. Let us next make a few comments on the design of generalized H_2 OID filters. As we said in previous chapters, any filter design starts with a fixed architecture for a filter. For H_2 OID filters, we have considered earlier in Chapter 10 three types of architectures, full-order strictly proper CSS architecture [see (10.66)], full-order proper CSS architecture [see (10.85) or (10.152)], and reduced-order proper CSS architecture [see (10.101) or (10.168)]. Once again in view of Theorem 12.10, for the generalized H_2 OID filtering for Σ, to obtain filters of full-order strictly proper CSS architecture or full-order proper CSS architecture or reduced-order proper CSS architecture, we can simply form the corresponding architectures of the H_2 OID filters, however, for the expanded system $\tilde{\Sigma}$. For instance, following (10.66), a filter of full-order strictly proper CSS architecture for generalized H_2 OID filtering can be formed as

$$
\Sigma_{\text{sp-CSS}}^{g} : \begin{cases} \sigma\xi = (A_e - KC_e)\xi + Ky, & \xi(0) = \xi_0 \in \mathbb{R}^{n+n_a}, \\ \hat{z} \;= E_e\xi, \end{cases} \tag{12.13}
$$

where K is the filter gain. In a similar way, following (10.85) or (10.152), a filter of full-order proper CSS architecture for generalized H_2 OID filtering can be formed. Moreover, a filter of reduced-order proper CSS architecture for generalized H_2 OID filtering can also be formed by following (10.101) or (10.168). Each one of these filters can be designed by simply following the design procedure given in Chapter 10, however, using the expanded system $\tilde{\Sigma}$ in place of the given system Σ.

As we discussed in Chapter 10, all H_2 OID filters have some common fixed modes. As generalized H_2 OID filters for the given system Σ are the H_2 OID filters for the expanded system $\tilde{\Sigma}$, we note that the generalized H_2 OID filters also have certain common fixed modes. Depending on the architecture of a filter, we define below formally the notion of fixed modes of generalized H_2 OID filters.

Definition 12.11 *(Fixed modes of generalized H_2 OID filters) Consider the generalized H_2 OID filtering problem for the system Σ of (12.5) along with the associated exosystem Σ_a of (12.3). Also, consider the H_2 OID filtering problem 10.6 for the expanded system $\tilde{\Sigma}$ characterized by the matrix octuple $(A, B, C, D, E, F, S, C_a)$. Assume that the solvability conditions of the generalized H_2 OID filtering problem as defined in Problem 12.3 are satisfied by a strictly proper filter. Then, a scalar $\lambda \in \mathbb{C}^-$ for continuous-time systems or $\lambda \in \mathbb{C}^\ominus$ for discrete-time systems is said to be the fixed mode of a generalized H_2 OID filter with the strictly proper full-order CSS architecture if λ is a pole of every H_2 OID filter of such an* *architecture. The set of all such generalized H_2 OID filter fixed modes is denoted here by*

$$\Omega_{\text{sp-CSS}}^{h2-\text{goid}}(A, B, C, D, E, F, S, C_a).$$

Similarly, if the solvability conditions of the generalized H_2 OID filtering problem as defined in Problem 12.3 are satisfied by a proper filter, then a scalar $\lambda \in \mathbb{C}^-$ for continuous-time systems or $\lambda \in \mathbb{C}^\ominus$ for discrete-time systems is said to be the fixed mode of a generalized H_2 OID filter with a proper full-order CSS architecture if λ is a pole of every H_2 OID filter of such an architecture. The set of all such generalized H_2 OID filter fixed modes is denoted here by

$$\Omega_{\text{p-CSS}}^{h2-\text{goid}}(A, B, C, D, E, F, S, C_a).$$

Moreover, if the solvability conditions of the generalized H_2 OID filtering problem as defined in Problem 12.3 are satisfied by a proper filter, then a scalar $\lambda \in \mathbb{C}^-$ for continuous-time systems or $\lambda \in \mathbb{C}^\ominus$ for discrete-time systems is said to be the fixed mode of a generalized H_2 OID filter with a proper reduced-order CSS architecture if λ is a pole of every H_2 OID filter of such an architecture. The set of all such generalized H_2 OID filter fixed modes is denoted here by

$$\Omega_{r-CSS}^{h2-\text{goid}}(A, B, C, D, E, F, S, C_a).$$

Again, in view of Theorem 12.10, the generalized H_2 OID filter fixed modes for the given system Σ can be determined by determining the H_2 OID filter fixed modes for the expanded system $\widetilde{\Sigma}$. That is, we have

$$\Omega_{\text{sp-CSS}}^{h2-\text{goid}}(A, B, C, D, E, F, S, C_a) = \Omega_{\text{sp-CSS}}^{h2-\text{oid}}(A_e, B_e, C_e, D_1, E_e, F_1),$$

$$\Omega_{\text{p-CSS}}^{h2-\text{goid}}(A, B, C, D, E, F, S, C_a) = \Omega_{\text{p-CSS}}^{h2-\text{oid}}(A_e, B_e, C_e, D_1, E_e, F_1),$$

$$\Omega_{r-CSS}^{h2-\text{goid}}(A, B, C, D, E, F, S, C_a) = \Omega_{r-CSS}^{h2-\text{oid}}(A_e, B_e, C_e, D_1, E_e, F_1).$$

The above development pertains to the generalized H_2 OID filtering. To establish similar results for the generalized H_2 SOID filtering, let us consider a family of parameterized filters of the form given in (12.9). Then, we have the following theorem pertaining to H_2 SOID filtering.

Theorem 12.12 *Consider the system Σ given in (12.5) together with the exosystem Σ_a given in (12.3) and the associated generalized H_2 SOID filtering problem. Let Assumption 12.1 be satisfied by Σ. Consider next the expanded system $\widetilde{\Sigma}$ given in (12.11) and the associated H_2 OID filtering problem. Let Assumption 12.8 be satisfied by $\widetilde{\Sigma}$. Then, the following two statements are equivalent:*

    (i) The family of filters Σ_f^ε given in (12.9) is a family of proper (or strictly proper) generalized H_2 SOID filters for Σ.

> *(ii) The family of filters Σ_f^ε given in (12.9) is a family of proper (or strictly proper) H_2 SOID filters for $\widetilde{\Sigma}$.*

Proof : The proof follows along the same lines as that of Theorem 12.10.   ■

As Theorem 12.10 did for the case of generalized H_2 OID filtering, Theorem 12.12 does the same for the generalized H_2 SOID filtering. To be explicit, consider families of strictly proper, proper, and reduced-order CSS architecture filters that are defined in the same way as H_2 OID filters with the corresponding architecture, except each one of them is parameterized in a parameter ε. Then, one can design such families of filters to be generalized H_2 SOID filters for the given system Σ by designing them as H_2 SOID filters for the expanded system $\widetilde{\Sigma}$. Also, one can study the *finite* asymptotic fixed modes of such generalized H_2 SOID filters. That is, following the concepts of defining fixed modes of a filter of a given architecture as done in Chapter 10 and as done in Definition 12.11, one can easily introduce the notion of the finite asymptotic fixed modes of generalized H_2 SOID filters for each one of the filters of generalized CSS architecture. Then, it follows from Theorem 12.12 that such finite asymptotic fixed modes coincide with the finite asymptotic fixed modes of H_2 SOID filters of respective architecture corresponding to the expanded system $\widetilde{\Sigma}$.

The above development is based on the assumption that the pair (C_e, A_e) is $\mathbb{C}^-$-detectable for continuous-time systems or $\mathbb{C}^\ominus$-detectable for discrete-time systems. Then, to complete our study, we need to examine the implications when it is not so. As argued, it is natural to assume that (C, A) is $\mathbb{C}^-$-detectable for continuous-time systems or $\mathbb{C}^\ominus$-detectable for discrete-time systems. Moreover, if there are unstable dynamics that are not observable from y but that are observable from z, then clearly we can never be able to obtain an unbiased (in the generalized sense) filter. Using the Hautus test for detectability, this can be formally expressed by the following necessary condition:

Assumption 12.13 *For all $\lambda \in \mathbb{C}$ with $\operatorname{Re} \lambda \geq 0$ (for continuous-time systems) or $|\lambda| \geq 1$ (for discrete-time systems), we have*

$$\operatorname{rank} \begin{pmatrix} \lambda I - A & -B_2 C_a \\ 0 & \lambda I - S \\ C & D_2 C_a \end{pmatrix} = \operatorname{rank} \begin{pmatrix} \lambda I - A & -B_2 C_a \\ 0 & \lambda I - S \\ C & D_2 C_a \\ E & F_2 C_a \end{pmatrix}.$$

If (C, A) is $\mathbb{C}^-$-detectable for continuous-time systems and $\mathbb{C}^\ominus$-detectable for discrete-time systems, and the above assumption is satisfied, then we can use a reduction technique to get into a situation where (C_e, A_e) is $\mathbb{C}^-$-detectable for continuous-time systems or $\mathbb{C}^{\ominus}$-detectable for discrete-time systems. We first find V_1 and V_2 such that

$$\mathrm{im}\begin{pmatrix} V_1 \\ V_2 \end{pmatrix}$$

represents the unstable, unobservable dynamics of the pair (C_e, A_e). The fact that (C, A) is $\mathbb{C}^{-}$-detectable for continuous-time systems or $\mathbb{C}^{\ominus}$-detectable for discrete-time systems implies that V_2 must be injective. Moreover, Assumption 12.13 implies that we must have

$$\begin{pmatrix} E & F_2 C_a \end{pmatrix}\begin{pmatrix} V_1 \\ V_2 \end{pmatrix} = 0.$$

Then a suitable basis transformation for the exosystem exists such that

$$V_2 = \begin{pmatrix} 0 \\ I \end{pmatrix}, \quad S = \begin{pmatrix} S_{11} & 0 \\ S_{21} & S_{22} \end{pmatrix}, \quad C_a = \begin{pmatrix} C_{a1} & C_{a2} \end{pmatrix}.$$

Consider the following exosystem:

$$\Sigma_{a1} : \sigma x_{a1} = S_{11} x_{a1} \quad u_2 = C_{a1} x_{a1}. \tag{12.14}$$

Then the above implies that the system (12.5) with the original exosystem (12.3) and the same system (12.5) with the new exosystem (12.14) result in the same outputs y and z provided we modify the initial conditions of the system $x(0)$ and the exosystem $x_a(0)$ in the first case to the initial conditions

$$x(0) - V_1\begin{pmatrix} 0 & I \end{pmatrix} x_a(0), \text{ and } \begin{pmatrix} I & 0 \end{pmatrix} x_a(0)$$

for the system and exosystem, respectively. From this, it is clear that a filter design for the original system and exosystem can be reduced to a filter design for the same system but with a modified (reduced) exosystem. After this reduction, we obtain a system and exosystem that when viewed together are $\mathbb{C}^{-}$-detectable for continuous-time systems or $\mathbb{C}^{\ominus}$-detectable for discrete-time systems from y.

The above development indeed weakens Assumption 12.8 to Assumptions 12.1 and 12.13.

12.5 Dependence of performance, existence, and uniqueness conditions and fixed modes on the input u_2

As discussed in the previous section, various aspects of the generalized H_2 OID filtering problem for the given system Σ are tantamount to the corresponding aspects of the H_2 OID filtering for the expanded system $\tilde{\Sigma}$ that incorporates the given system Σ and the exosystem Σ_a, which models the input signal u_2. As the generalized H_2 OID filtering has an additional requirement of rejecting the input u_2 (generalized unbiased requirement) over and above the requirement of H_2 OID filtering (namely, minimizing the RMS value of the estimation error signal), in this section, we ask ourselves several fundamental questions regarding the dependence on the input signal u_2 of various aspects of the generalized H_2 OID filtering problem such as the existence and uniqueness conditions, performance, and fixed modes. We remark that when the signal u_2 is set to zero, the generalized H_2 OID filtering problem for the given system Σ simply reduces to the H_2 OID filtering problem for the system Σ_0 given in (12.10). The questions we pose are as follows:

(*i*) How does the performance of generalized H_2 OID filtering for Σ differ from the performance of H_2 OID filtering for Σ_0?

(*ii*) How do the solvability conditions of generalized H_2 OID filtering problem for Σ differ from the solvability conditions of H_2 OID filtering problem for Σ_0?

(*iii*) How do the generalized H_2 OID filter fixed modes for Σ differ from the H_2 OID filter fixed modes for Σ_0?

We answer the above questions one at a time in the following subsections.

12.5.1   *Dependency of performance on the input u_2*

In this subsection, we investigate the effect of u_2 on the generalized H_2 OID filtering performance. In fact, we show here that the generalized H_2 OID filtering performance is not affected at all by the input signal u_2 provided Assumption 12.8 is satisfied. To be precise, we have the following theorem.

Theorem 12.14 *Consider the generalized H_2 OID filtering problem as defined in Problem 12.3 for the system Σ of (12.5) along with the associated exosystem Σ_a of (12.3) whose performance is indicated by $\gamma^*_{g,sp}$ or $\gamma^*_{g,p}$ depending on whether the class of strictly proper or proper filters are used. Also, consider the H_2 OID filtering problem for the system Σ_0 of (12.10) whose performance is indicated by γ^*_{sp} or γ^*_p depending on whether the class of strictly proper or proper filters are used. Then, under Assumptions 12.1 and 12.8, we have*

$$\gamma^*_{g,sp} = \gamma^*_{sp} \quad and \quad \gamma^*_{g,p} = \gamma^*_p.$$

Proof : Using the reduction technique presented, we can assume without loss of generality that Assumption 12.8 is satisfied. In view of Theorem 12.9, $\gamma^*_{g,sp}$ (or $\gamma^*_{g,p}$) is the infimum of the RMS norm of the estimation error signal $\tilde{e}_z$ (see

Fig. 12.3) over all linear unbiased stable strictly proper (or proper) filters for the expanded system $\tilde{\Sigma}$ given in (12.11). Also, γ_{sp}^* (or γ_p^*) is the infimum of the RMS norm of the estimation error signal e_z (see Fig. 12.2) over all linear unbiased stable strictly proper (or proper) filters for the system Σ_0 given in (12.10). Let us consider continuous-time systems, and to facilitate the comparison of $\gamma_{g,sp}^*$ (or $\gamma_{g,p}^*$) with γ_{sp}^* (or γ_p^*), let us consider the semi-stabilizing solution Q of the CLMI:

$$\begin{pmatrix} AQ + QA' + B_1 B_1' & QC' + B_1 D_1' \\ CQ + D_1 B_1' & D_1 D_1' \end{pmatrix} \geq 0, \qquad (12.15)$$

and the semi-stabilizing solution Q_e of the CLMI:

$$\begin{pmatrix} A_e Q_e + Q_e A_e' + B_e B_e' & Q_e C_e' + B_e D' \\ C_e Q_e + DB_e' & DD' \end{pmatrix} \geq 0.$$

Then it is easily verified that

$$Q_e = \begin{pmatrix} Q & 0 \\ 0 & 0 \end{pmatrix}.$$

But then according to Corollary 10.16, we have

$$\gamma_{sp}^* = \left(\text{trace } EQE'\right)^{1/2} = \left(\text{trace } \begin{pmatrix} E & F_2 \end{pmatrix} Q_e \begin{pmatrix} E & F_2 \end{pmatrix}\right)^{1/2} = \gamma_{g,sp}^*.$$

On the other hand, for proper filters, we need to find P^* such that $F_1 - P^* D_1 = 0$. But then, according to Corollary 10.18,

$$\gamma_p^* = \left(\text{trace}(E - P^*C)Q(E - P^*C)'\right)^{1/2}$$

$$= \left(\text{trace } \begin{pmatrix} E - P^*C & F_2 - P^* D_2 \end{pmatrix} Q_e \begin{pmatrix} E - P^*C & F_2 - P^* D_2 \end{pmatrix}'\right)^{1/2}$$

$$= \gamma_{g,p}^*.$$

The discrete-time result can be obtained similarly but this time relying on Corollaries 10.39 and 10.41. ∎

12.5.2   Dependency of the solvability conditions on the input u_2

In this subsection, we investigate the dependency of the solvability conditions of the generalized H_2 OID (H_2 SOID) filtering problem on the input signal u_2. That is, we enquire whether any relationship exists between the solvability of the H_2 OID filtering problem studied in Chapter 10 and the solvability of the generalized H_2 OID filtering problem currently being studied. The following example answers this question.

Example 12.15 Consider a continuous-time system:

$$\Sigma : \begin{cases} \dot{x} = u_1 + b_2 u_2 \\ y = x + u_1 + d_2 u_2 \\ z = x + f_2 u_2, \end{cases} \qquad (12.16)$$

where b_2, d_2, and f_2 are some constants. Also, let the exosystem be given by

$$\Sigma_a : \dot{x}_a = 0, \quad \text{and} \quad u_2 = x_a. \qquad (12.17)$$

Then, from the above give data, the expanded system $\tilde{\Sigma}$ can be constructed as

$$\tilde{\Sigma} : \begin{cases} \dot{\bar{x}} = \begin{pmatrix} 0 & b_2 \\ 0 & 0 \end{pmatrix} \bar{x} + \begin{pmatrix} 1 \\ 0 \end{pmatrix} u_1 \\ y = \begin{pmatrix} 1 & d_2 \end{pmatrix} \bar{x} + u_1 \\ z = \begin{pmatrix} 1 & f_2 \end{pmatrix} \bar{x}. \end{cases} \qquad (12.18)$$

The above systems satisfy the detectability Assumptions 12.1 and 12.8 provided b_2 is not zero. It is easy to show that the H_2 OID filtering problem for the expanded system $\tilde{\Sigma}$ is solvable by a strictly proper filter if and only if

$$f_2 = d_2 - b_2.$$

If $f_2 \neq d_2 - b_2$, then the H_2 OID filtering problem for the expanded system $\tilde{\Sigma}$ is solvable by neither a strictly proper filter nor a proper filter. Next, to see whether the H_2 OID filtering problem for the given system Σ is solvable if the input signal $u_2 = 0$, let us construct Σ_0 that corresponds to this example. It is given by

$$\Sigma_0 : \begin{cases} \dot{x} = u_1 \\ y = x + u_1 \\ z = x. \end{cases} \qquad (12.19)$$

One can verify easily that the EID filtering problem is solvable for the system Σ_0 by a strictly proper filter, and hence, in particular, the H_2 OID filtering problem is solvable. To conclude, for the system Σ given in (12.16) and the exosystem given in (12.17), the generalized H_2 OID filtering problem is not solvable via either a proper or strictly proper filter if $f_2 \neq d_2 - b_2$, whereas the H_2 OID filtering problem is always solvable by a strictly proper filter when u_2 is not present. This demonstrates that, in general, the solvability of the H_2 OID filtering problem does depend on the input signal u_2 or, equivalently, on the exosystem Σ_a.

We can also create a similar example for the discrete time case.

Example 12.16 We consider the discrete-time system:

$$\Sigma : \begin{cases} \sigma x = x + u_1 + b_2 u_2 \\ y \ = x + u_1 + d_2 u_2 \\ z \ = x + f_2 u_2, \end{cases} \qquad (12.20)$$

where b_2, d_2, and f_2 are some constants. Also, let the exosystem be given by

$$\Sigma_a : \ \sigma x_a = x_a, \quad \text{and} \quad u_2 = x_a. \qquad (12.21)$$

In this case, using the same arguments as in Example 12.15, we find that the generalized H_2 OID filtering is solvable by a strictly proper filter if and only if $f_2 = d_2 - b_2$, whereas H_2 OID filtering is always solvable by a strictly proper filter for the above system Σ with $u_2 = 0$. Moreover, to emphasize, we note that the generalized H_2 OID filtering problem is not solvable via either a proper or strictly proper filter if $f_2 \neq d_2 - b_2$.

Let us next investigate the effect of the input signal u_2 on the generalized H_2 SOID filtering. We note that comparing the solvability conditions for the generalized H_2 SOID filtering problem with those of the H_2 SOID filtering problem is easy. In both cases, we need a detectability assumption, Assumption 12.8 for the generalized H_2 SOID filtering problem and Assumption 12.1 for H_2 SOID filtering problem. We know that we can weaken Assumption 12.8 to Assumptions 12.1 and 12.13 through a reduction technique. So basically we need one additional assumption for the solvability of generalized H_2 SOID filtering problem, namely Assumption 12.13. Thus, under Assumptions 12.1 and 12.13, the existence conditions for H_2 SOID filters are not dependent on the input signal u_2.

12.5.3   Dependency of the fixed modes on the input u_2

In this subsection, we investigate the dependency of the generalized H_2 OID filter fixed modes on the input signal u_2. We have the following result, which shows that the fixed modes of generalized H_2 OID filters, whenever they exist, are unaffected by the presence of the input signal u_2 provided Assumptions 12.1 and 12.8 are satisfied.

Theorem 12.17 *Consider the generalized H_2 OID filtering problem as defined in Problem 12.3 for the system Σ of (12.5) along with the associated exosystem Σ_a of (12.3). Let Assumptions 12.1 and 12.8 be satisfied. Then, assuming that the solvability conditions of the generalized H_2 OID filtering problem are satisfied by a strictly proper filter, we have*

$$\Omega_{\text{sp-CSS}}^{h2-\text{goid}}(A, B, C, D, E, F, S, C_a) = \Omega_{\text{sp-CSS}}^{h2-\text{oid}}(A, B, C, D, E, 0).$$

Similarly, if the solvability conditions of the generalized H_2 OID filtering problem as defined in Problem 12.3 are satisfied by a proper filter, then we have

$$\Omega_{\text{p-CSS}}^{h2-\text{goid}}(A, B, C, D, E, F, S, C_a) = \Omega_{\text{p-CSS}}^{h2-\text{oid}}(A, B, C, D, E, F).$$

Proof : Let us first consider the case of continuous-time systems while using strictly proper filters. Then the fixed modes for the standard and generalized H_2 OID filtering problem are equal to the fixed modes for the EID filtering problem for the systems:

$$\begin{cases} \sigma \tilde{x} = A\tilde{x} + B_\varrho \tilde{u} \\ \tilde{y} = C\tilde{x} + D_\varrho \tilde{u} \\ \tilde{z} = E\tilde{x} \end{cases} \tag{12.22}$$

and

$$\begin{cases} \sigma \bar{x} = \begin{pmatrix} A & B_2 C_a \\ 0 & S \end{pmatrix} \bar{x} + \begin{pmatrix} B_\varrho \\ 0 \end{pmatrix} u_1 \\ y = \begin{pmatrix} C & D_2 C_a \end{pmatrix} \bar{x} + D_\varrho u_1 \\ z = \begin{pmatrix} E & F_2 C_a \end{pmatrix} \bar{x}, \end{cases} \tag{12.23}$$

respectively. EID filtering is dual to exact disturbance decoupling, and in this case, it is convenient to use the characterization of the fixed modes given in Remark 5.18. We find that the fixed modes for (12.22) are given by $\Omega_s^1 \cup \Omega_s^2$, where

$$\Omega_s^1 := \{ \text{ invariant zeros of } (A, B_\varrho, C, D_\varrho) \}$$
$$- \left\{ \text{ invariant zeros of } \left(A, B_\varrho, \begin{pmatrix} C \\ E \end{pmatrix}, \begin{pmatrix} D_\varrho \\ 0 \end{pmatrix}\right) \right\}$$

and

$$\Omega_s^2 := \left\{ \text{ output decoupling zeros of } \left(\begin{pmatrix} C \\ E \end{pmatrix}, A\right) \right\}.$$

Similarly, the fixed modes for (12.23) are given by $\tilde{\Omega}_s^1 \cup \tilde{\Omega}_s^2$, where

$$\tilde{\Omega}_s^1 := \left\{ \text{ invariant zeros of } \left(\begin{pmatrix} A & B_2 C_a \\ 0 & S \end{pmatrix}, \begin{pmatrix} B_\varrho \\ 0 \end{pmatrix}, \begin{pmatrix} C & D_2 C_a \end{pmatrix}, D_\varrho\right) \right\}$$
$$- \left\{ \text{ invariant zeros of } \left(\begin{pmatrix} A & B_2 C_a \\ 0 & S \end{pmatrix}, \begin{pmatrix} B_\varrho \\ 0 \end{pmatrix}, \begin{pmatrix} C & D_2 C_a \\ E & F_2 C_a \end{pmatrix}, \begin{pmatrix} D_\varrho \\ 0 \end{pmatrix}\right) \right\},$$

and

$$\tilde{\Omega}_s^2 := \left\{ \text{ output decoupling zeros of } \left(\begin{pmatrix} C & D_2 C_a \\ E & F_2 C_a \end{pmatrix}, \begin{pmatrix} A & B_2 C_a \\ 0 & S \end{pmatrix}\right) \right\}.$$

Clearly as the generalized H_2 OID filtering problem is solvable, the fixed modes are stable and hence are in the open left-half plane. But S has only eigenvalues on the imaginary axis. Stable eigenvectors of

$$\begin{pmatrix} A & B_2 C_a \\ 0 & S \end{pmatrix}$$

are therefore of the form $(x' \quad 0)'$, and hence, it is then immediate that $\boldsymbol{\Omega}_s^1 = \tilde{\boldsymbol{\Omega}}_s^1$.
To establish that $\boldsymbol{\Omega}_s^2 = \tilde{\boldsymbol{\Omega}}_s^2$, we look at the Rosenbrock system matrix and we note that for stable λ, the matrix $\lambda I - S$ is invertible and hence

$$\operatorname{rank} \begin{pmatrix} \lambda I - A & B_2 C_a & -B_\varrho \\ 0 & \lambda I - S & 0 \\ C & D_2 C_a & D_\varrho \end{pmatrix} = \operatorname{rank} \begin{pmatrix} \lambda I - A & -B_\varrho \\ C & D_\varrho \end{pmatrix},$$

and

$$\operatorname{rank} \begin{pmatrix} \lambda I - A & B_2 C_a & -B_\varrho \\ 0 & \lambda I - S & 0 \\ C & D_2 C_a & D_\varrho \\ E & F_2 C_a & 0 \end{pmatrix} = \operatorname{rank} \begin{pmatrix} \lambda I - A & -B_\varrho \\ C & D_\varrho \\ E & 0 \end{pmatrix}.$$

The result then follows immediately. Also, the results for the discrete-time case or the case of proper filters can be established using similar techniques.   ■

The above theorem shows that the fixed modes of generalized H_2 OID filters, whenever they exist, are unaffected by the presence of the input signal u_2 provided Assumptions 12.1 and 12.8 are satisfied. Next, we turn our attention to the *finite* asymptotic fixed modes of generalized H_2 SOID filters. The impact of the input signal u_2 on such finite asymptotic fixed modes can be studied by studying the finite asymptotic fixed modes of H_2 SOID filters of the expanded system. It turns out that the set of finite asymptotic fixed modes of the generalized H_2 SOID filters can be larger than the set of finite asymptotic fixed modes of the H_2 SOID filters. It can be shown that all additional finite asymptotic fixed modes that are introduced due to the requirements of rejecting the signal u_2 must be the eigenvalues of the matrix S. These issues are illustrated in the next example.

Example 12.18 We will take a second look at the example already considered before in Example 12.15. Consider the system (12.16) with b_2, d_2, and f_2 as arbitrary constants. In Example 12.15, we already noted that for this system H_2 OID filtering is solvable. Consider the exosystem (12.17). We have also seen that the generalized H_2 OID filtering is solvable only if we have $f_2 = 0$. Let us now consider the case of $f_2 \neq 0$. In this case, the generalized H_2 SOID filtering problem is solvable. Let us next look at the finite asymptotic fixed modes. We find, using the techniques from Chapter 10, that we have two finite asymptotic fixed modes at -1 and 0 except in the special case when $d_2 - b_2 = f_2$, where we have only one asymptotic fixed mode at -1. Note that the condition $d_2 - b_2 = f_2$ is exactly the condition under which the generalized H_2 OID filtering problem is solvable.

We note, based on this example, that in many cases, the set of asymptotic fixed modes for the generalized H_2 SOID filtering problem is larger than the set of asymptotic fixed modes for the H_2 SOID filtering problem. However, this is not always the case (as illustrated for the case when $d_2 - b_2 = f_2$).

12.6   Performance limitations due to structural properties of a system

We study in this section the performance limitations of generalized H_2 OID filtering due to the structural properties of a system. As we said earlier, the generalized H_2 OID filtering has to satisfy simultaneously two requirements. One is the generalized unbiased requirement where, when the input $u_1 = 0$, the error signal needs to converge to zero asymptotically for all possible initial conditions of the given system, exosystem, and the filter. The second one is to minimize the RMS norm of the error signal due to the white noise input u_1. Both of these requirements are on the asymptotic behavior of the system. Regarding the second requirement, we note that the generalized H_2 OID filtering performance measure, namely $\gamma^*_{g,sp}$ or $\gamma^*_{g,p}$, measures simply asymptotic performance. Such a performance measure is blind to the transient behavior of the filter. As in Chapter 10, a good way of bringing the transient performance into picture is to consider the performance associated with the generalized unbiased requirement. Note that for an unbiased (in the generalized sense) filter, the asymptotic effect of the input u_2 must be zero. Hence, the effect of this input is an energy signal, and as such, one can formulate the performance measure due to the generalized unbiasedness as the energy of the estimation error signal under the condition that u_1 is zero. Such a measure is independent of the second requirement of the generalized H_2 OID filtering performance.

In this section, we examine the dependence of $\gamma^*_{g,sp}$ and $\gamma^*_{g,p}$ on the structural properties of the given system. Also, we will define the performance measure associated with the generalized unbiased requirement and examine its dependence on the structural properties of the given system.

12.6.1   Dependence of performance on structural properties of the given system

Here we examine the dependence of $\gamma^*_{g,sp}$ and $\gamma^*_{g,p}$ on the structural properties of the given system. In this regard, Theorem 12.14 developed earlier states that $\gamma^*_{g,sp} = \gamma^*_{sp}$ and $\gamma^*_{g,p} = \gamma^*_p$, where γ^*_{sp} and γ^*_p are the H_2 OID performance measures for the system Σ_0 of (12.10). This implies that studying $\gamma^*_{g,sp}$ and $\gamma^*_{g,p}$ is tantamount to studying γ^*_{sp} and γ^*_p. Such a study of γ^*_{sp} and γ^*_p has already been discussed in Subsections 10.5.1 and 10.8.1. Let us summarize that study here. For continuous-time systems, both γ^*_{sp} and γ^*_p can be decomposed into two parts, one arising due to non-left-invertible dynamics and the other due to unstable zero dynamics (i.e., the dynamics dictated by the invariant zeros in the open right-half plane) of the subsystem characterized by (A, B_1, C, D_1). The part that is contributed by the unstable zero dynamics to γ^*_{sp} and γ^*_p tends to zero as the open right-half plane invariant zeros move toward the imaginary axis, and in the same way, it increases as the open right-half plane invariant zeros move away from the imaginary axis. If the subsystem characterized by (A, B_1, C, D_1) is left-invertible and has no unstable zero dynamics, both γ^*_{sp} and γ^*_p equal zero. Similarly, for discrete-time systems, both γ^*_{sp} and γ^*_p can be decomposed into two parts, one arising due to unstable zero dynamics (i.e., the dynamics dictated by the invariant zeros outside the unit circle) and the other due to the non-left-invertible dynamics as well as the dynamics dictated by the infinite zero structure of order greater than or equal to one of the subsystem characterized by (A, B_1, C, D_1). The part that is contributed by the unstable zero dynamics to γ^*_{sp} and γ^*_p tends to zero as the invariant zeros outside the unit circle move toward the unit circle, and in the same way, it increases as the invariant zeros outside the unit circle move away from the unit circle.

12.6.2   *Performance issues of generalized unbiased filtering*

In this subsection, we will define the performance measure associated with the generalized unbiased requirement and then compute it.

As we said earlier, the performance measure due to the generalized unbiasedness can be considered as the energy of the estimation error signal under the condition that u_1 is zero. Thus, consider the filtering block diagram of Figure 12.1 for the system Σ given in (12.5), the filter Σ_f given in (12.6), and the exosystem Σ_a given in (12.3). By combining the given system Σ and the exosystem Σ_a together, we form the expanded system $\widetilde{\Sigma}$ as in (12.11), except we set there $u_1 = 0$; i.e., let

$$\begin{cases} \sigma \bar{x} = A_e \bar{x} \\ y \;\; = C_e \bar{x} \\ z \;\; = E_e \bar{x}, \end{cases} \tag{12.24}$$

where the matrix triple (A_e, C_e, E_e) is as in (12.12). Whenever the generalized unbiased requirement is satisfied by the filter Σ_f, the error e_z then is an energy signal, and thus, we can define the generalized unbiasedness performance measure J^g as follows: For continuous-time systems,

$$J^g(\bar{x}_0, \xi_0, \Sigma_f) = \int_0^\infty e_z(t)' e_z(t)\, dt,$$

and for discrete-time systems,

$$J^g(\bar{x}_0, \xi_0, \Sigma_f) = \sum_{i=0}^{\infty} e_z(i)' e_z(i).$$

In the above equations,

$$\bar{x} = \begin{pmatrix} x \\ x_a \end{pmatrix}, \quad \bar{x}_0 = \bar{x}(0), \quad \xi_0 = \xi(0).$$

Clearly, the performance measure due to the generalized unbiasedness depends on the filter used. Our aim is to study $J^g(\bar{x}_0, \xi_0, \Sigma_f)$ when Σ_f has the CSS architecture, either strictly proper, proper, or reduced-order type.

Strictly proper filter of CSS architecture.

We will first study in detail $J^g(\bar{x}_0, \xi_0, \Sigma_f)$ when the filter Σ_f considered is the filter with strictly proper CSS architecture, namely $\Sigma^g_{\text{sp-CSS}}$ as given in (12.13). The dynamics of the error $e_z = z - \hat{z}$, in the absence of input signal u_1 (as we assumed), is then given by

$$\begin{cases} \sigma e = (A_e - KC_e)e, \quad e(0) = e_0 = (\bar{x}_0 - \xi_0) \in \mathbb{R}^{n+n_a}, \\ e_z = E_e e. \end{cases} \tag{12.25}$$

We would like to study the limitations imposed by the given system on the performance measure $J^g(\bar{x}_0, \xi_0, \Sigma^g_{\text{sp-CSS}})$. To do so, we can define its infimum over all possible filter gains subject to the constraint (12.25) and denote such an infimum by $J^{*g}(\bar{x}_0, \xi_0, \Sigma^g_{\text{sp-CSS}})$. We can compute $J^{*g}(\bar{x}_0, \xi_0, \Sigma^g_{\text{sp-CSS}})$ as follows. Consider an auxiliary system:

$$\Sigma_{\text{aux}} : \begin{cases} \sigma x_{\text{aux}} = A_e x_{\text{aux}} + B_{\text{aux}} v \\ y_{\text{aux}} = C_e x_{\text{aux}} \\ z_{\text{aux}} = E_e x_{\text{aux}}, \end{cases} \tag{12.26}$$

where $B_{\text{aux}} = \bar{x}_0 - \xi_0$ and v is an unknown white noise input. Then, in view of Theorem 12.10, we have the following result whose proof can be written easily. Notationally, whenever we refer to $\gamma^*_{sp}(A, B, C, D, E, F)$, we mean by it γ^*_{sp} associated with the H_2 OID filtering problem characterized by the sextuple (A, B, C, D, E, F).

Lemma 12.19 *Consider the generalized H_2 OID filtering problem as defined in Problem 12.3 for the system Σ of (12.5) along with the associated exosystem Σ_a of (12.3). Let Assumptions 12.1 and 12.8 be satisfied. Also, consider the strictly proper filter $\Sigma^g_{\text{sp-CSS}}$ given in (12.13). Let Σ_{aux} be as in (12.26). Then the infimum of the performance measure due to the generalized unbiasedness is given by*

$$J^{*g}(\bar{x}_0, \xi_0, \Sigma^g_{\text{sp-CSS}}) = (\gamma^*_{sp}(A_e, B_{\text{aux}}, C_e, 0, E_e, 0))^2.$$

We note that the initial condition $\bar{x}_0$ of the given system is usually unknown, whereas one can set the initial condition ξ_0 of the filter as one likes. As in Chapter 10, this suggests that one can generate an average performance measure for the generalized unbiasedness requirement. Let e_i, $i = 1, \cdots, n$, form a basis for the state space of Σ. Also, f_i, $i = 1, \cdots, n_a$, form a basis for the state space of Σ_a. Moreover, assume that one always sets the initial condition ξ_0 of the filter to zero. Then, as in Chapter 10, we can define a new average performance measure for the generalized unbiasedness requirement while using a strictly proper filter of CSS architecture as

$$\tilde{J}^g(\Sigma, \Sigma_a, \Sigma^g_{\text{sp-CSS}}) = \tilde{J}^g_1(\Sigma, \Sigma_a, \Sigma^g_{\text{sp-CSS}}) + \tilde{J}^g_2(\Sigma, \Sigma_a, \Sigma^g_{\text{sp-CSS}}), \quad (12.27)$$

where

$$\tilde{J}^g_1(\Sigma, \Sigma_a, \Sigma^g_{\text{sp-CSS}}) = \sum_{i=1}^{n} J^g\left(\begin{pmatrix} e_i \\ 0 \end{pmatrix}, 0, \Sigma^g_{\text{sp-CSS}}\right)$$

and

$$\tilde{J}^g_2(\Sigma, \Sigma_a, \Sigma^g_{\text{sp-CSS}}) = \sum_{i=1}^{n_a} J^g\left(\begin{pmatrix} 0 \\ f_i \end{pmatrix}, 0, \Sigma^g_{\text{sp-CSS}}\right).$$

To study the limitations imposed by the given system on $\tilde{J}^g(\Sigma, \Sigma_a, \Sigma^g_{\text{sp-CSS}})$, we can define $\tilde{J}^{*g}(\Sigma, \Sigma_a, \Sigma^g_{\text{sp-CSS}})$ as its infimum over all possible filter gains K subject to the constraint (12.25). Once again it is straightforward to show that $\tilde{J}^{*g}(\Sigma, \Sigma_a, \Sigma^g_{\text{sp-CSS}})$ is also related to the H_2 OID performance of an appropriately defined auxiliary system. Let

$$\tilde{\Sigma}_{\text{aux}} : \begin{cases} \sigma \tilde{x}_{\text{aux}} = A_e \tilde{x}_{\text{aux}} + I v \\ \tilde{y}_{\text{aux}} = C_e \tilde{x}_{\text{aux}} \\ \tilde{z}_{\text{aux}} = E_e \tilde{x}_{\text{aux}}, \end{cases} \quad (12.28)$$

where v is an unknown white noise input. We have the following result whose proof can be written easily.

Lemma 12.20 *Consider the generalized H_2 OID filtering problem as defined in Problem 12.3 for the system Σ of (12.5) along with the associated exosystem Σ_a of (12.3). Let Assumptions 12.1 and 12.8 be satisfied. Also, consider the strictly proper filter $\Sigma^g_{\text{sp-CSS}}$ given in (12.13). Let Σ_{aux} be as in (12.28). Then the infimum of the average performance measure due to the generalized unbiasedness is given by*

$$\tilde{J}^{*g}(\Sigma, \Sigma_a, \Sigma^g_{\text{sp-CSS}}) = (\gamma^*_{sp}(A_e, I, C_e, 0, E_e, 0))^2.$$

Remark 12.21 *Following the results of Subsection 10.5.1 for continuous time systems, we note that only the unstable zero dynamics and the non-left-invertible* *dynamics of the subsystem characterized by $(A_e, I, C_e, 0)$ contribute to the value of γ_{sp}^*. However, it is easy to see that the said subsystem does not have any zero dynamics, and moreover, it is left-invertible only if* $\operatorname{rank} C_e = (n + n_a)$. *As* $\operatorname{rank} C_e \neq (n + n_a)$, *the said subsystem is always non-left-invertible. This implies that* $\gamma_{sp}^*(A_e, I, C_e, 0, E_e, 0)$ *is always nonzero. In other words, the average performance measure* $\widetilde{J}^g(\Sigma, \Sigma_a, \Sigma_{sp\text{-}CSS}^g)$ *due to the generalized unbiasedness of a strictly proper filter of CSS architecture is always nonzero. Following similar analysis, one can deduce the same result for discrete-time systems as well.*

Proper filter of CSS architecture: The above development studies in detail $J^g(\bar{x}_0, \xi_0, \Sigma_f)$ when the filter Σ_f considered is the filter with strictly proper CSS architecture. A similar study can be undertaken easily for a filter with proper CSS architecture. As we said earlier below (12.13), following (10.85) or (10.152), a filter of full-order proper CSS architecture for generalized H_2 OID filtering, namely $\Sigma_{p\text{-}CSS}^g$, can be formed. Also, Lemma 10.72 for continuous-time systems and Lemma 10.133 for discrete-time systems show how a full-order proper filter of CSS architecture can be designed for one system via the design of a full order strictly proper filter of CSS architecture for a related system. This crucial result enables us to study in detail $J^g(\bar{x}_0, \xi_0, \Sigma_{p\text{-}CSS}^g)$ by studying a similar measure while using a strictly proper filter (rather than a proper filter), however, for a system related to the expanded system $\widetilde{\Sigma}$. Such a study is straightforward, and hence, it is omitted here. One obvious result of such a study is that the average of performance measure $J^{*g}(\bar{x}_0, \xi_0, \Sigma_{p\text{-}CSS}^g)$, namely $\widetilde{J}^g(\Sigma, \Sigma_a, \Sigma_{p\text{-}CSS}^g)$, due to the generalized unbiasedness of a proper filter of CSS architecture is always nonzero.
Reduced-order filter of CSS architecture.

Next, as we said earlier below (12.13), a filter of reduced-order proper CSS architecture for generalized H_2 OID filtering, namely Σ_{r-CSS}^g, can also be formed by following (10.101) or (10.168). Then using Lemma 10.89 for continuous-time systems and Lemma 10.150 for discrete-time systems, one can study in detail the performance measure due to generalized unbiasedness of the reduced-order filter by studying a similar measure while using a strictly proper filter, however, for a system related to the expanded system $\widetilde{\Sigma}$. Such a study is once again straightforward, and hence, it is omitted here. One obvious result of such a study is that the average of performance measure due to generalized unbiasedness of the reduced-order filter of CSS architecture is always nonzero.

12.6.3 Impact of the structural properties of Σ on $\widetilde{J}^{*g}(\Sigma, \Sigma_a, \Sigma_{sp\text{-}CSS}^g)$ and $\widetilde{J}^{*g}(\Sigma, \Sigma_a, \Sigma_{p\text{-}CSS}^g)$

In this subsection, we consider the impact of the structural properties of Σ on $\widetilde{J}^{*g}(\Sigma, \Sigma_a, \Sigma_{sp\text{-}CSS}^g)$ and $\widetilde{J}^{*g}(\Sigma, \Sigma_a, \Sigma_{p\text{-}CSS}^g)$. In this regard, a relevant question we pose is under what circumstances, $\widetilde{J}^{*g}(\Sigma, \Sigma_a, \Sigma_{sp\text{-}CSS}^g)$ and $\widetilde{J}^{*g}(\Sigma, \Sigma_a, \Sigma_{p\text{-}CSS}^g)$ are unbounded when we move invariant zeros closer to the boundary of the stability domain. Apparently under such circumstances, estimation becomes near impossible when the invariant zeros are too close to the boundary of the stability domain. We focus here on developing a relationship that shows how $\widetilde{J}^{*g}(\Sigma, \Sigma_a, \Sigma^g_{\text{sp-CSS}})$ and $\widetilde{J}^{*g}(\Sigma, \Sigma_a, \Sigma^g_{\text{p-CSS}})$ depend on the locations/direction of the invariant zeros of the subsystem characterized by $(A_e, I, C_e, 0)$. It turns out that the non-minimum-phase dynamics and the exosystem dynamics play significant roles in dictating the behavior of $\widetilde{J}^{*g}$. Basically, we find out that $\widetilde{J}^{*g}$ is inversely related to the distance between the invariant zeros and the modes of the exosystem, and indeed it could go to infinity when the minimal distance of poles of the exosystem and the invariant zeros of the system goes to zero. There are two possible exceptions to this behavior. First, when the effect of the invariant zeros of the system are asymptotically invisible from the output z (i.e., the non-minimum-phase dynamics is asymptotically unobservable from the desired output to be estimated). Second, if u_2 is a vector, then the input direction of an invariant zero and the direction of a pole of the exosystem need to be misaligned (to be made precise soon) in order to have the cost bounded when the pole and the invariant zero get close to each other. Due to lack of space, we illustrate our findings by considering two special but important cases and an example.

We proceed now to illustrate the above-discussed results. Let λ be any unstable invariant zero of (A, B_2, C, D_2). Hence, vectors $\widetilde{p}$ and $\widetilde{q}$ exist such that

$$\text{rank} \begin{pmatrix} \lambda I - A & -B_2 \\ C & D_2 \end{pmatrix} \begin{pmatrix} \widetilde{p} \\ \widetilde{q} \end{pmatrix} = 0.$$

Detectability of (C, A) guarantees that $\widetilde{q} \neq 0$. As seen from (12.3), the exosystem is characterized by the matrices S and C_a. If for an eigenvalue μ of S, we can choose an eigenvector s with $Ss = \mu s$ such that $C_a s = \widetilde{q}$, then we call the pole of the exosystem μ and the invariant zero λ of the system *aligned*; otherwise, they are misaligned. Note that by scaling $\widetilde{p}$ and $\widetilde{q}$, we can guarantee, without loss of generality, that $\|s\| = 1$. We will show that if an invariant zero of the system moves toward an aligned pole of the exosystem, then the average transient performance measure will go to infinity. For two special cases,

Case 1: $S = 0$ and $C_a = I$ (input u_2 is a vector DC signal),

Case 2: $m_2 = 1$ (input u_2 is a scalar signal),

all poles of the exosystem are aligned to all invariant zeros of the system characterized by (A, B_2, C, D_2). But in general this might clearly be not the case. Note that, in the above two cases, it can be shown that Assumption 12.8 implies that (A, B_2, C, D_2) is left-invertible.

As we discussed, the minimal achievable H_2 norm of Σ_{au} is indeed the minimal average transient performance measure. To simplify our study of the H_2 norm of Σ_{au}, we restrict v by setting $v = (0 \quad \widetilde{q}')'\omega$, and we add an additional measurement $y_1 = (\widetilde{m} \quad 0)x_{\text{aux}} + \widetilde{n}\omega$, where $\widetilde{m}$ is such that $\widetilde{m}\widetilde{p} = 1$ and $\widetilde{n} = (\mu - \lambda)^{-1}$. Obviously both of these actions reduce the achievable H_2 norm, and hence, we are investigating a lower bound for the achievable H_2 norm of Σ_{au}. The above restrictions imply that we will study the design of an observer for the system:

$$
\bar{\Sigma}_{au} : \begin{cases}
\dot{x}_{au} = \begin{pmatrix} A & BC_a \\ 0 & S \end{pmatrix} x_{au} + \begin{pmatrix} 0 \\ \tilde{q} \end{pmatrix} \omega \\[2mm]
\bar{y}_{au} = \begin{pmatrix} \tilde{m} & 0 \\ C & D_2 C_a \end{pmatrix} x_{au} + \begin{pmatrix} \tilde{n} \\ 0 \end{pmatrix} \omega \\[2mm]
z_{au} = \begin{pmatrix} E & F_2 C_a \end{pmatrix} x_{au}.
\end{cases}
$$

Note that an output injection does not change the achievable H_2 norm for the error dynamics, and hence, we can equally well study the system:

$$
\begin{aligned}
\dot{\tilde{x}}_{au} &= \begin{pmatrix} A + (\mu - \lambda)\tilde{p}\tilde{m} & BC_a \\ 0 & S \end{pmatrix} \tilde{x}_{au} + \begin{pmatrix} \tilde{p} \\ \tilde{q} \end{pmatrix} \omega \\[2mm]
\tilde{y}_{au} &= \begin{pmatrix} \tilde{m} & 0 \\ C & D_2 C_a \end{pmatrix} \tilde{x}_{au} + \begin{pmatrix} \tilde{n} \\ 0 \end{pmatrix} \omega \\[2mm]
\tilde{z}_{au} &= \begin{pmatrix} E & F_2 C_a \end{pmatrix} \tilde{x}_{au}.
\end{aligned}
$$

It is easy to see that the state of this system (given zero initial conditions) will satisfy $\tilde{x}_{au}(t) = (\tilde{p}' \quad \tilde{q}')'r(t)$ for some scalar valued function r. Next we derive a differential equation for r and express the whole system in terms of the function r:

$$
\begin{aligned}
\dot{r} &= \mu r + \omega, \\
\tilde{y}_{au} &= \begin{pmatrix} 1 & 0 \end{pmatrix}' r + \begin{pmatrix} (\mu - \lambda)^{-1} & 0 \end{pmatrix}' \omega, \\
\tilde{z}_{au} &= \begin{pmatrix} E\tilde{p} & F_2 C_a \tilde{q} \end{pmatrix} r.
\end{aligned}
$$

However, for this scalar system, the achievable performance measure can very easily be computed. For both strictly proper and proper filters, we obtain as the optimal performance measure,

$$
2 \left\| \begin{pmatrix} E\tilde{p} & F_2 C_a \tilde{q} \end{pmatrix} \right\|^2 \frac{\operatorname{Re}\lambda}{|\mu - \lambda|^2}. \tag{12.29}
$$

The expression given in (12.29) is a lower bound for the average transient performance measure of $\tilde{J}^{*g}(\Sigma, \Sigma_a, \Sigma_{\text{sp-CSS}}^g)$ as well as $\tilde{J}^{*g}(\Sigma, \Sigma_a, \Sigma_{\text{p-CSS}}^g)$. We clearly see that $\tilde{J}^{*g}(\Sigma, \Sigma_a, \Sigma_{\text{sp-CSS}}^g)$ and $\tilde{J}^{*g}(\Sigma, \Sigma_a, \Sigma_{\text{p-CSS}}^g)$ are inversely related to the distance between the poles of an exosystem and the non-minimum-phase invariant zeros. That is, when λ gets close to an aligned eigenvalue of S, then the achievable performance measure goes to infinity. However, there is one exception to this unbounded behavior. That is, when $E\tilde{p}$ and $F_2 C_a \tilde{q}$ converge asymptotically to zero, $\tilde{J}_{sp}^{*g}$ as well as $\tilde{J}_p^{*g}$ can be bounded. Under this circumstance, the effect of the invariant zero is asymptotically invisible in the to-be-estimated output. Note that, in the special cases we considered above, poles of exosystem and non-minimum-phase invariant zeros are always aligned.

We consider next an example in which the poles of the exosystem and the non-minimum-phase invariant zeros are misaligned. As seen in this example, $\tilde{J}^{*g}(\Sigma, \Sigma_a, \Sigma^g_{\text{sp-CSS}})$ as well as $\tilde{J}^{*g}(\Sigma, \Sigma_a, \Sigma^g_{\text{p-CSS}})$ need not be unbounded as the distance between the poles of an exosystem and the non-minimum-phase invariant zeros goes to zero. Thus, the alignment of poles of the exosystem and the non-minimum-phase invariant zeros as mentioned in the beginning of this section plays a crucial role.

Example 12.22 Consider the system (12.5), where

$$A = \begin{pmatrix} -1 & 0 \\ 0 & -1 \end{pmatrix}, \quad B_1 = \begin{pmatrix} 1 \\ 1 \end{pmatrix}, \quad B_2 = \begin{pmatrix} -1-\varepsilon & 0 \\ 0 & 1 \end{pmatrix},$$

$$C = \begin{pmatrix} 1 & 0 \\ 0 & 1 \end{pmatrix}, \quad D_1 = \begin{pmatrix} 0 \\ 0 \end{pmatrix}, \quad D_2 = \begin{pmatrix} -1 & 0 \\ 0 & -1 \end{pmatrix},$$

whereas the exosystem is given by

$$S = \begin{pmatrix} 0 & 0 & 0 \\ 0 & 0 & 1 \\ 0 & -1 & 0 \end{pmatrix}, \quad C_a = \begin{pmatrix} 1 & 1 & 1 \\ 1 & 1 & -1 \end{pmatrix}.$$

The system has an invariant zero in ε which, when $\varepsilon \to 0$, converges to a pole 0 of the exosystem without exhibiting the alignment property. In contrast to the aligned case, we see here that the average transient performance measure $\tilde{J}^{*g}$ does not go to ∞. This can be seen by realizing that removing the first part of the measurement removes the invariant zero, whereas relying simply only on the second measurement, the system is still detectable. Therefore, an observer design based on the second measurement only will not have a transient performance measure that converges to ∞. Although due to space limitations we are not showing here all the details, the reader can easily work them out.

12.7   Generalized EID filtering problem

We first formulate below the generalized EID filtering problem.

Problem 12.23 Consider a continuous- or discrete-time system Σ as given in (12.5) where the input u_1 is an unknown input, and u_2 is the output of an exosystem as given in (12.3). Then, the **generalized EID filtering problem** is defined as follows: Find, whenever it exists, a linear stable strictly proper (or proper) filter such that (*i*) (**Generalized unbiasedness**) the estimation error e_z, in the absence of the input u_1, decays asymptotically to zero for all possible finite initial values of the exosystem (12.3), the given system (12.5), and the filter (12.6), and

(*ii*) (**Performance**) the transfer matrix $G_{u_1 e}$ from u_1 to e_z is zero.

We have the following theorem stating the solvability conditions of the above problem.

Theorem 12.24 *Consider a continuous- or discrete-time system Σ as given in (12.5) together with the exosystem Σ_a given in (12.3), and the associated generalized EID filtering problem 12.23. Let Assumption 12.1 be satisfied by Σ. Consider next the expanded system $\widetilde{\Sigma}$ given in (12.11) and the associated EID filtering problem 7.3 for $\widetilde{\Sigma}$. Let Assumption 12.8 be satisfied by it. Then, the generalized EID filtering problem 12.23 for Σ is solvable via strictly proper (proper) filters if and only if the EID filtering problem for $\widetilde{\Sigma}$ is solvable via strictly proper (proper) filters.*

Proof : The proof follows from Theorem 12.14 and Remarks 10.28 and 10.50. ∎

Whenever the generalized EID filtering problem is solvable, the proofs of Theorem 12.24 and Theorem 12.14 immediately point out that a generalized EID filter (i.e., the filter that solves the generalized EID filtering problem) for a given system is the same as a generalized H_2 OID filter for the same system. This implies that both the design of generalized EID filters and the fixed modes associated with them for a given system coincide, respectively, with the design of generalized H_2 OID filters and the fixed modes associated with them for the same system.

12.8   Generalized H_2 AID filtering problem

We first formulate below the generalized H_2 AID filtering problem.

Problem 12.25 Consider a continuous- or discrete-time system Σ as given in (12.5) where the input u_1 is a zero-mean wide-sense stationary white noise of unit intensity and u_2 is the output of an exosystem as given in (12.3). Then, the **generalized H_2 AID filtering problem** is defined as follows: Find, whenever it exists, a family of linear stable strictly proper (or proper) filters of the type Σ_f^ε given in (12.9) and parameterized in positive $\varepsilon \in (0, \varepsilon^*]$ such that

(i) (**Generalized unbiasedness**) for any given filter in the family, the estimation error e_z, in the absence of the input u_1, decays asymptotically to zero for all possible finite initial values of the exosystem (12.3), the given system (12.5), and the filter (12.9), and

(ii) (**Performance**) the H_2 norm of the transfer matrix $G_{u_1 e}$ from u_1 to e_z tends to zero as ε tends to zero.

We have the following theorem stating the solvability conditions of the above problem.

Theorem 12.26 *Consider a continuous- or discrete-time system Σ as given in (12.5) together with the exosystem Σ_a given in (12.3) and the associated generalized H_2 AID filtering problem 12.25. Let Assumption 12.1 be satisfied by Σ. Also, consider the system Σ_0 given in (12.10) and the associated H_2 AID filtering problem 8.3. Consider next the expanded system $\widetilde{\Sigma}$ given in (12.11), and let Assumption 12.8 be satisfied by it. Then, the generalized H_2 AID filtering problem 12.25 for Σ is solvable via a family of strictly proper (proper) filters if and only if the H_2 AID filtering problem for Σ_0 is solvable via a family of strictly proper (proper) filters.*

Proof : The proof follows from Theorem 12.14 and Remarks 10.17 and 10.20 for the continuous-time case, and from Remarks 10.40 and 10.42 for the discrete-time case. ∎

Whenever the generalized H_2 AID filtering problem is solvable, the proofs of Theorem 12.26 and Theorem 12.14 immediately point out that a family of filters that solves the generalized H_2 AID filtering problem for a given system is the same as a family of filters that solves the generalized H_2 SOID filtering problem for the same system. This implies that both the design of families of filters that solve the generalized H_2 AID filtering problem and the asymptotic fixed modes associated with them for a given system coincide, respectively, with the design of families of generalized H_2 SOID filters and the asymptotic fixed modes associated with them for the same system.

13

Generalized H_∞ suboptimally input-decoupled filtering

13.1 Introduction

The generalized H_2 OID, generalized H_2 SOID, generalized EID, and the generalized H_2 AID filtering problems are considered in depth in Chapter 12. We consider in this chapter the generalized γ-level H_∞ SOID as well as the generalized H_∞ AID filtering problems. Let us recall that the generalization is due to the presence of additional unknown input signals containing linear combinations of sinusoidal signals, each of which has an unknown amplitude and phase but known frequency.

As in Chapter 11, we do not assume any statistical information on the first type of input except that it has a finite RMS value. As said above, the second type or additional unknown input signals are linear combinations of sinusoidal signals, each of which has an unknown amplitude and phase but known frequency. As in Chapter 12, the additional unknown sinusoidal inputs are modeled by an exosystem. Such an exosystem can then be appended to the given system to form an expanded system. Once again, this lets us analyze the effects of such additional sinusoidal inputs by using the formalism developed in the previous chapters. Also, as in Chapter 12, in the presence of such additional sinusoidal inputs, the filters used must be unbiased in a generalized sense. Let us recall that a filter is unbiased in the generalized sense if the effect of the sinusoidal signal of known frequency but unknown amplitude and phase reduces to zero asymptotically.

At first, our primary focus in this chapter is on the generalized γ-level H_∞ SOID filtering. Once we study in depth the generalized γ-level H_∞ SOID filtering, we consider the generalized H_∞ AID filtering. The reason to do so is simple. It turns out that, as discussed in detail in the body of the chapter, the results on the generalized H_∞ AID filtering are a consequence of some of the results on the generalized γ-level H_∞ SOID filtering.

13.2 Preliminaries

Let us consider the plant or system model:

$$\Sigma : \begin{cases} \sigma x = Ax + Bu \\ y \;\; = Cx + Du \\ z \;\; = Ex + Fu. \end{cases} \qquad (13.1)$$

As before, $u \in \mathbb{R}^m$ is the input, $x \in \mathbb{R}^n$ is the state, $y \in \mathbb{R}^p$ is the measured output, and $z \in \mathbb{R}^q$ is the desired output signal to be estimated. The input u is decomposed into two parts:

$$u = \begin{pmatrix} u_1 \\ u_2 \end{pmatrix}, \qquad (13.2)$$

where $u_1 \in \mathbb{R}^{m_1}$ and $u_2 \in \mathbb{R}^{m_2}$. As mentioned in Section 13.1, regarding the first input u_1, we do not assume any statistical information except that it has a finite but unknown RMS value. On the other hand, we assume that the second input is a linear combination of sinusoidal signals, each of which has an unknown amplitude and phase but known frequency. As in Chapter 12, such a signal u_2 can be modeled as the output of a known linear autonomous system with unknown initial conditions. We call once again such a system an exogenous system or, for short, an exosystem. Thus, consider

$$\Sigma_a : \;\; \sigma x_a = S x_a, \quad u_2 = C_a x_a. \qquad (13.3)$$

Let us next partition the matrices, B, D, and F in conformity with the partitioning of u:

$$B = \begin{pmatrix} B_1 & B_2 \end{pmatrix}, \quad D = \begin{pmatrix} D_1 & D_2 \end{pmatrix}, \quad F = \begin{pmatrix} F_1 & F_2 \end{pmatrix}. \qquad (13.4)$$

As before, our interest lies in estimating the desired output signal z while using only the measured output y but not the input u. As usual, let $\hat{z}$ be the estimate of z as given by a filter, and let e_z be the estimation error, $e_z = z - \hat{z}$, as depicted in Figure 13.1.

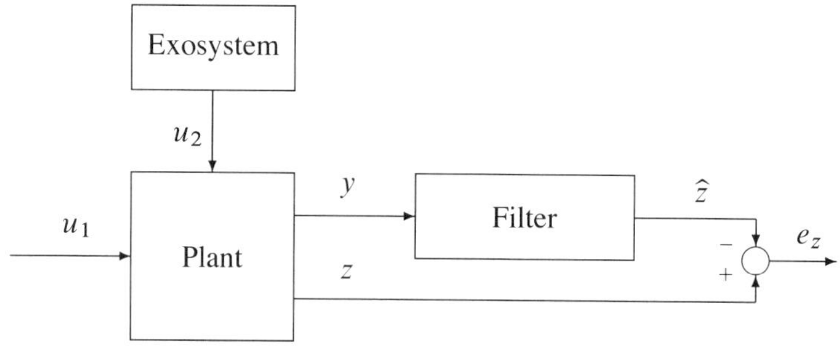

Figure 13.1: General block diagram

As before, it is necessary to use the following assumption throughout this chapter.

Assumption 13.1 *The matrix pair (C, A) is $\mathbb{C}^-$-detectable for continuous-time systems and $\mathbb{C}^\ominus$-detectable for discrete-time systems.*

Again, we consider a general proper filter of the form (7.2), which is repeated as follows:

$$\begin{aligned}
\sigma\xi &= L\xi + My, \\
\hat{z} &= N\xi + Py.
\end{aligned} \tag{13.5}$$

Whenever $P = 0$, the above filter is said to be a strictly proper filter. As in the previous chapters, we require that the filter (13.5) be internally stable. As we will see later this is a consequence of the stronger concept that the filter needs to be unbiased. When the above filter is used as shown in Figure 13.1, the dynamic equations of the error e_z are described by

$$\Sigma_{ue} : \begin{cases}
\sigma x &= Ax + B_1 u_1 + B_2 u_2 \\
\sigma\xi &= MCx + L\xi + MD_1 u_1 + MD_2 u_2 \\
\sigma x_a &= S x_a, \quad u_2 = C_a x_a \\
e_z &= (E - PC)x - N\xi + (F_1 - PD_1)u_1 + (F_2 - PD_2)u_2.
\end{cases} \tag{13.6}$$

The transfer matrices from u_i to e_z, $i = 1, 2$, can be computed easily as

$$G_{u_i e_z}(\sigma) = \begin{pmatrix} E - PC & -N \end{pmatrix} \begin{pmatrix} \sigma I - A & 0 \\ -MC & \sigma I - L \end{pmatrix}^{-1} \begin{pmatrix} B_i \\ MD_i \end{pmatrix} + F_i - PD_i. \tag{13.7}$$

As we did in other chapters, sometimes we seek a family of filters parameterized in a positive parameter ε. In this case, a family of filters is described by

$$\Sigma_f^\varepsilon : \begin{cases}
\sigma\xi = L^\varepsilon \xi + M^\varepsilon y \\
\hat{z} = N^\varepsilon \xi + P^\varepsilon y,
\end{cases} \tag{13.8}$$

where L^ε, M^ε, N^ε, and P^ε are matrices parameterized in a positive parameter ε.

13.3 γ-level generalized H_∞ SOID filtering problem statement

Our first goal in this chapter is to seek and design filters that render the RMS value of the estimation error signal less than γ times the RMS norm of the first input u_1 where γ is any prescribed number. This is formalized in the following problem statement.

Problem 13.2 Consider a continuous- or discrete-time system Σ of the form (13.1), where the first input u_1 is an unknown signal with finite but unknown

RMS value, and the second input u_2 is the output of a linear exosystem given in (13.3). Then, the **γ-level generalized SOID filtering problem without any statistical information about the input u_1** is defined as follows: For a given positive number γ, find, whenever it exists, a linear stable strictly proper (or proper) filter such that the following two conditions are satisfied:

(i) (**Generalized unbiasedness**) the estimation error e_z, in the absence of the input u_1, decays asymptotically to zero for all possible finite initial values of the linear exosystem (13.3), the given system (13.1), and the filter (13.5), and

(ii) (**Performance**) in the presence of both the inputs u_1 and u_2, we have $\|e_z\|_{\mathrm{RMS}} \leq \gamma \|u_1\|_{\mathrm{RMS}}$.

As argued in previous chapters, the above γ-level generalized OID filtering problem can be given a deterministic interpretation. This is because the gain γ from $\|u_1\|_{\mathrm{RMS}}$ to $\|e_z\|_{\mathrm{RMS}}$ is indeed the H_∞ norm of the transfer function from u_1 to e_z. We therefore can replace the condition (ii) in the above problem by the following equivalent condition:

(ii) (**Performance**) we have $\|G_{u_1 e_z}\|_\infty \leqslant \gamma$,

where $G_{u_1 e_z}$ is the resulting transfer matrix from the input u_1 to the error signal e_z. Because of this equivalence, we will refer to Problem 13.2 as the **γ-level generalized H_∞ SOID filtering** problem. Moreover, a filter that solves such a problem is obviously called a γ-level generalized H_∞ SOID filter.

We have the following definition.

Definition 13.3 *Consider a continuous- or discrete-time system Σ as given in (13.1), where the first input u_1 is an unknown signal with finite but unknown RMS value, and the second input u_2 is the output of a linear exosystem given in (13.3). For such a system, the infimum over all γ for which a γ-level generalized H_∞ SOID filter exists is called the **generalized H_∞ OID filtering performance**. For the class of strictly proper filters, such an infimum performance is denoted by $\gamma^*_{g,sp}$, whereas for the class of proper filters, it is denoted by $\gamma^*_{g,p}$.*

As can be seen easily, whenever the input u_2 is set to zero, the above γ-level generalized H_∞ SOID filtering problem for the given system Σ reduces to the γ-level H_∞ SOID filtering problem for a system Σ_0 given by

$$\Sigma_0 : \begin{cases} \sigma x = Ax + B_1 u_1 \\ y = Cx + D_1 u_1 \\ z = Ex + F_1 u_1. \end{cases} \tag{13.9}$$

The block diagram of filtering for the system Σ_0 is depicted in Figure 13.2. We obviously have

$$\gamma^*_{g,p} \geq \gamma^*_p \quad \text{and} \quad \gamma^*_{g,sp} \geq \gamma^*_{sp},$$

where, in accordance with the previous chapters, the H_∞ OID filtering performance associated with the system Σ_0 is denoted by γ^*_{sp} or γ^*_p depending on whether we use strictly proper or proper filters.

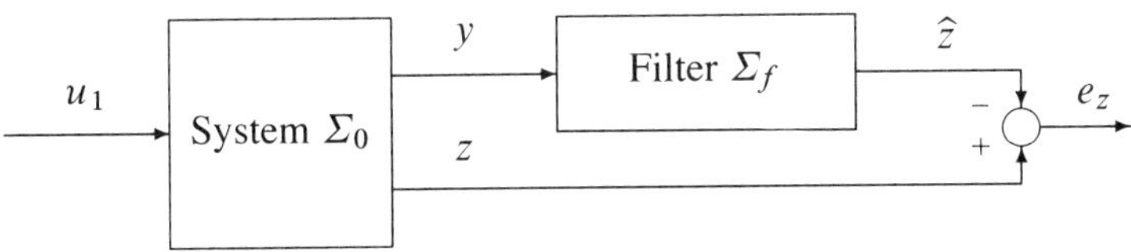

Figure 13.2: General Block diagram of filtering for Σ_0

13.4 Computation of $\gamma^*_{g,sp}$ and $\gamma^*_{g,p}$ and the design of γ-level generalized H_∞ SOID filters

Clearly a γ-level generalized H_∞ SOID filter exists for a given γ if and only if $\gamma > \gamma^*_{g,sp}$ (or $\gamma > \gamma^*_{g,p}$). Thus, the determination of $\gamma^*_{g,sp}$ and $\gamma^*_{g,p}$ is essential. One goal in this section is to compute them. Another goal is to present a method of designing γ-level generalized H_∞ SOID filters. We relate both of these issues pertaining to the generalized H_∞ SOID filtering for the given system to the corresponding issues of H_∞ SOID filtering, however, for an expanded system constructed from the data of the given system. In this way, the results of Chapter 11 can be of direct help in studying such issues. As in Chapter 12, the required expanded system $\tilde{\Sigma}$ is constructed by viewing together the given system Σ and the exosystem Σ_a as one system, and it is given by

$$\tilde{\Sigma} : \begin{cases} \sigma\bar{x} = A_e\bar{x} + B_e u_1 \\ y \;\;= C_e\bar{x} + D_1 u_1 \\ z \;\;= E_e\bar{x} + F_1 u_1, \end{cases} \tag{13.10}$$

where

$$A_e = \begin{pmatrix} A & B_2 C_a \\ 0 & S \end{pmatrix}, \quad B_e = \begin{pmatrix} B_1 \\ 0 \end{pmatrix},$$

$$C_e = \begin{pmatrix} C & D_2 C_a \end{pmatrix}, \quad E_e = \begin{pmatrix} E & F_2 C_a \end{pmatrix}. \tag{13.11}$$

We need the following assumption for the expanded system $\tilde{\Sigma}$, which is similar to the Assumption 13.1 for Σ.

Assumption 13.4 *The matrix pair (C_e, A_e) is $\mathbb{C}^-$-detectable for continuous-time systems and $\mathbb{C}^\ominus$-detectable for discrete-time systems.*

Before we state our main result of this section, let us draw the general filtering block diagram for the expanded system $\widetilde{\Sigma}$. Such a block diagram is given in Figure 13.3.

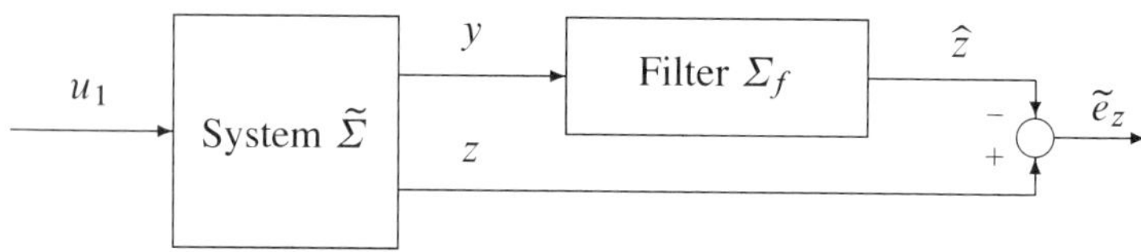

Figure 13.3: General block diagram of filtering for $\widetilde{\Sigma}$

The following theorem addresses the issue of how to compute $\gamma^*_{g,sp}$ or $\gamma^*_{g,p}$.

Theorem 13.5 *Consider the system Σ given in (13.1) together with the exosystem Σ_a given in (13.3) and the associated γ-level generalized H_∞ SOID filtering problem as stated in Problem 13.2. Let Assumption 13.1 be satisfied by Σ. Also, as defined in Definition 13.3, let the generalized H_∞ OID filtering performance associated with Σ be denoted by $\gamma^*_{g,sp}$ or by $\gamma^*_{g,sp}$ depending on whether strictly proper or proper filters are used. Consider next the expanded system $\widetilde{\Sigma}$ given in (13.10), and let Assumption 13.4 be satisfied by $\widetilde{\Sigma}$. Let the H_∞ OID filtering performance associated with the system $\widetilde{\Sigma}$ be denoted by $\gamma^*_{sp}(\widetilde{\Sigma})$ or $\gamma^*_p(\widetilde{\Sigma})$ depending on whether we use strictly proper or proper filters. Then, we have*

$$\gamma^*_{g,p} = \gamma^*_p(\widetilde{\Sigma}) \quad \text{and} \quad \gamma^*_{g,sp} = \gamma^*_{sp}(\widetilde{\Sigma}).$$

Proof : The proof follows directly from the proof of Theorem 13.6 given below.
∎

The above theorem lays a roadmap to compute $\gamma^*_{g,sp}$ or $\gamma^*_{g,p}$. Indeed, by using the procedure developed in Chapter 11, $\gamma^*_{g,sp}$ or $\gamma^*_{g,p}$ can be computed as $\gamma^*_{sp}(\widetilde{\Sigma})$ or $\gamma^*_p(\widetilde{\Sigma})$, which is associated with the expanded system $\widetilde{\Sigma}$.

We present next a theorem that provides a roadmap to design γ-level generalized H_∞ SOID filters.

Theorem 13.6 *Consider the system Σ given in (13.1) together with the exosystem Σ_a given in (13.3) and the associated γ-level generalized H_∞ SOID filtering* *problem as stated in Problem 13.2. Let Assumption 13.1 be satisfied by Σ. Also, as defined in Definition 13.3, let the generalized H_∞ OID filtering performance associated with Σ be denoted by $\gamma^*_{g,sp}$ or by $\gamma^*_{g,sp}$ depending on whether strictly proper or proper filters are used. Consider next the expanded system $\widetilde{\Sigma}$ given in (13.10), and let Assumption 13.4 be satisfied by $\widetilde{\Sigma}$. Consider a number $\gamma > \gamma^*_{g,p}$ ($\gamma > \gamma^*_{g,sp}$) when the class of proper (strictly proper) filters are considered. Then, the following two statements are equivalent:*

(i) *The filter Σ_f given in (13.5) is a proper (or strictly proper) γ-level generalized H_∞ SOID filter for Σ.*

(ii) *The filter Σ_f given in (13.5) is a proper (or strictly proper) γ-level H_∞ SOID filter for $\widetilde{\Sigma}$.*

Proof : Assume a filter Σ_f of the form (13.5) is unbiased (in a generalized sense) for the system Σ along with the associated exosystem Σ_a and yields a stable transfer matrix $G_{u_1 e_z}$ from u_1 to $e_z = z - \hat{z}$. Then it can be trivially verified that such a filter when applied to the expanded system is unbiased and results in the same stable transfer matrix $G_{u_1 e_z}$ from u_1 to $\tilde{e}_z$ in the block diagram of Fig. 13.3. The converse of the above implication is also trivially satisfied. But then it is immediate that a filter is a γ-level generalized H_∞ SOID filter for the system (13.1) and the associated exosystem (13.3) if and only if it is a γ-level H_∞ SOID filter for the system (13.10). ∎

Theorem 13.6 implies that the design procedures developed in Chapter 11 can be directly used to design γ-level generalized H_∞ SOID filters for the given system Σ by designing γ-level H_∞ SOID filters for the expanded system $\widetilde{\Sigma}$.

The above development is based on the assumption that the pair (C_e, A_e) is $\mathbb{C}^-$-detectable for continuous-time systems or $\mathbb{C}^\ominus$-detectable for discrete-time systems. Then, to complete our study, we need to examine the implications when it is not so. As argued earlier, it is natural to assume that (C, A) is $\mathbb{C}^-$-detectable for continuous-time systems or $\mathbb{C}^\ominus$-detectable for discrete-time systems. Moreover, if unstable dynamics are not observable from y but are observable from z, then clearly we can never be able to obtain an unbiased filter. Using the Hautus test for detectability, this can be formally expressed by the following necessary condition:

Assumption 13.7 *For all $\lambda \in \mathbb{C}$ with* $\mathrm{Re}\,\lambda \geq 0$ *(for continuous-time systems) or* $|\lambda| \geq 1$ *(for discrete-time systems), we have*

$$\mathrm{rank} \begin{pmatrix} \lambda I - A & -B_2 C_a \\ 0 & \lambda I - S \\ C & D_2 C_a \end{pmatrix} = \mathrm{rank} \begin{pmatrix} \lambda I - A & -B_2 C_a \\ 0 & \lambda I - S \\ C & D_2 C_a \\ E & F_2 C_a \end{pmatrix}.$$

If (C, A) is $\mathbb{C}^-$-detectable (or $\mathbb{C}^\ominus$-detectable) and the above assumption is satisfied, then we can use the same reduction technique given in Chapter 12 to get into a situation where (C_e, A_e) is $\mathbb{C}^-$-detectable (or $\mathbb{C}^\ominus$-detectable).

13.5   Dependence of performance on the input u_2

As we alluded to earlier, $\gamma_{g,p}^* \geq \gamma_p^*$ and $\gamma_{g,sp}^* \geq \gamma_{sp}^*$ where γ_{sp}^* or γ_p^* is associated with the system Σ_0, which is obtained from the given system Σ by setting the input $u_2 = 0$. Obviously, if u_2 were to be always zero, $\gamma_{g,p}^* = \gamma_p^*$ and $\gamma_{g,sp}^* = \gamma_{sp}^*$. However, it is not so if u_2 were to be nonzero. The goal of this section is to get a clear insight as to how $\gamma_{g,sp}^*$ (or $\gamma_{g,p}^*$) depends on the input signal u_2. The following theorem provides the needed insight.

Theorem 13.8 *Consider the system Σ given in (13.1) together with the exosystem Σ_a given in (13.3) and the associated γ-level generalized H_∞ SOID filtering problem as stated in Problem 13.2. As defined in Definition 13.3, let the associated generalized H_∞ OID filtering performance be $\gamma_{g,sp}^*$ or $\gamma_{g,p}^*$ depending on whether the class of strictly proper or proper filters are used. Also, let the H_∞ OID filtering performance associated with the system Σ_0 be γ_{sp}^* or γ_p^* depending on whether the class of strictly proper or proper filters are used. Then, under Assumptions 13.1 and 13.4, we have*

$$\gamma_{g,sp}^* = \max\{\gamma_{sp}^*, \gamma_{sp}^0\}, \qquad \gamma_{g,p}^* = \max\{\gamma_p^*, \gamma_p^0\},$$

where γ_{sp}^0 and γ_p^0 are defined as

$$\gamma_{sp}^0 = \max_{s_0 \in \lambda(S)} \ \inf_{K_e} \ \big\{ \, \| E_e(s_0 I - A_e + K_e C_e)^{-1}(B_e - K_e D_1) + F_1 \|$$

$$| \, s_0 I - A_e + K_e C_e \text{ is invertible} \, \big\}, \quad (13.12)$$

and

$$\gamma_p^0 = \max_{s_0 \in \lambda(S)} \ \inf_{K_e} \ \big\{ \, \| \bar{E}_e(s_0 I - A_e + K_e C_e)^{-1}(B_e - K_e D_1) + \bar{F}_1 \|$$

$$| \, s_0 I - A_e + K_e C_e \text{ is invertible} \, \big\}, \quad (13.13)$$

where

$$\bar{E}_e := E_e - F_1 D_1'(D_1 D_1')^\dagger C_e \quad and \quad \bar{F}_1 := F_1 - F_1 D_1'(D_1 D_1')^\dagger D_1.$$

Proof : Let us consider continuous-time systems to facilitate the comparison of $\gamma_{g,sp}^*$ and $\gamma_{g,p}^*$ with γ_{sp}^* and γ_p^* respectively. We will use the characterization presented in Theorem 11.22 to see whether an arbitrary γ satisfies $\gamma > \gamma^*_{g,sp}$. We need the existence of a strongly stabilizing semi-solution of the dual CQMI:

$$G^e_\infty(Q_e) \geq 0,$$

where

$$G^e_\infty(Q_e) := \begin{pmatrix} A_e Q_e + Q_e A'_e + B_e B'_e & Q_e C'_e + B_e D'_1 \\ C_e Q_e + D_1 B'_e & D_1 D'_1 \end{pmatrix}$$
$$+ \begin{pmatrix} Q_e E'_e + B_e F'_1 \\ D_1 F'_1 \end{pmatrix} (\gamma^2 I - F_1 F'_1)^{-1} \begin{pmatrix} E_e Q_e + F_1 B'_e & F_1 D'_1 \end{pmatrix}.$$

The requirement that $\gamma > \gamma^*_{sp}$ requires the existence of a strongly semi-stabilizing solution of the dual CQMI $G_\infty(Q) \geq 0$, where

$$G_\infty(Q) := \begin{pmatrix} AQ + QA' + B_1 B'_1 & QC' + B_1 D'_1 \\ CQ + D_1 B'_1 & D_1 D'_1 \end{pmatrix}$$
$$+ \begin{pmatrix} QE' + B_1 F'_1 \\ D_1 F'_1 \end{pmatrix} (\gamma^2 I - F_1 F'_1)^{-1} \begin{pmatrix} EQ + F_1 B'_1 & F_1 D'_1 \end{pmatrix}. \quad (13.14)$$

Using the definition of strongly semi-stabilizing solutions as presented in Definition 4.155, it is easily seen that any strongly semi-stabilizing solution Q_e of $G^e_\infty(Q_e) \geq 0$ has the form:

$$Q_e = \begin{pmatrix} Q & 0 \\ 0 & 0 \end{pmatrix}.$$

It then follows that a semi-stabilizing solution of $G^e_\infty(Q_e) \geq 0$ exists if and only if a semi-stabilizing solution of $G_\infty(Q) \geq 0$ exists. This immediately shows that a possible difference between $\gamma^*_{g,sp}$ and γ^*_{sp} must be due to the condition (11.21) of Theorem 11.22. It is obvious that $\gamma^*_{g,sp} \geq \gamma^*_{sp}$. Consider $\gamma > \gamma^*_{sp}$ while we want to check whether $\gamma > \gamma^*_{g,sp}$. We know that the invariant zeros of (A_e, B_e, C_e, D_1) are either the invariant zeros of (A, B_1, C, D_1) or the eigenvalues of S. Assume that s_0 is an invariant zero of (A, B_1, C, D_1) on the imaginary axis, which is not an eigenvalue of S. As $\gamma > \gamma^*_{sp}$, a matrix K exists such that $s_0 I - A + KC$ is invertible and

$$\| E(s_0 I - A + KC)^{-1}(B_1 - KD_1) + F \| < \gamma.$$

In this case, for

$$K_e = \begin{pmatrix} K \\ 0 \end{pmatrix},$$

we have $s_0 I - A_e + K_e C_e$ invertible and

$$\| E_e(s_0 I - A_e + K_e C_e)^{-1}(B_e - K_e D_1) + F_1 \| < \gamma.$$

Therefore, we know that for s_0, we need to consider only the eigenvalues of S and then the result follows immediately.

The arguments when proper filters are used and for discrete-time systems follow along similar lines. ∎

Finally we make a remark that in general $\gamma^*_{g,sp} > \gamma^*_{sp}$ and $\gamma^*_{g,p} > \gamma^*_p$ as the following example shows.

Example 13.9 Consider the following system:

$$\Sigma : \begin{cases} \dot{x} = x + 2u_1 + u_2 \\ y = x + u_1 \\ z = x + u_2, \end{cases}$$

along with the associated exosystem

$$\Sigma_a : \dot{x}_a = 0, \quad u_2 = x_a.$$

We first consider the associated system Σ_0:

$$\Sigma_0 : \begin{cases} \dot{x} = x + 2u_1 \\ y = x + u_1 \\ z = x. \end{cases}$$

For this system, EID filtering is possible with a strictly proper filter, and hence, $\gamma^*_{sp} = 0$ and $\gamma^*_p = 0$. If we use Theorem 13.8, then it is not difficult to establish that $\gamma^*_{g,sp} = 2$ and $\gamma^*_{g,p} = 2$. Hence, in contrast with the generalized H_2 SOID filtering (see Chapter 12), the requirement of asymptotically rejecting the effects of the exosystem results in a worse performance for the generalized H_∞ SOID filtering compared with just H_∞ SOID filtering.

13.6   Performance limitations due to structural properties of a system

The intent of this section is to study performance limitations of generalized H_∞ OID filtering due to the structural properties of a given system. In this regard, one can study transient performance limitations as well as steady-state performance limitations. The study of transient performance limitations is exact as discussed in Section 12.6. Thus, we focus here on steady-state performance limitations, namely limitations on $\gamma^*_{g,p}$ or $\gamma^*_{g,sp}$ due to the structural properties of a given system. We limit our analysis to continuous-time systems, although similar analysis can be done in discrete time.

To be explicit, we examine here the impact of unstable zero dynamics of Σ_0 on $\gamma_{g,p}^*$ or $\gamma_{g,sp}^*$. In fact, in view of Theorem 13.8, studying the impact of unstable zero dynamics on $\gamma_{g,p}^*$ or $\gamma_{g,sp}^*$ is tantamount to studying the impact of unstable zero dynamics on γ_p^* or γ_{sp}^* associated with Σ_0.

To proceed with our development, we note that the system Σ_0 given in (13.9) is characterized by the sextuple (A, B_1, C, D_1, E, F_1). As in Chapter 7, the study of a filtering problem for a given system Σ_0 can be transformed to the study of a control problem for a dual system, represented here by $\bar{\Sigma}$ and characterized by the sextuple $(\bar{A}, \bar{B}, \bar{C}, \bar{D}, \bar{E}, \bar{F})$, where

$$\bar{A} = A', \quad \bar{B} = C', \quad \bar{C} = B_1', \quad \bar{D} = D', \quad \bar{E} = E_1', \text{ and } \bar{F} = F_1'. \tag{13.15}$$

To view the structural details of the dual system $\bar{\Sigma}$, we can transform its subsystem characterized by the quadruple $(\bar{A}, \bar{B}, \bar{C}, \bar{D})$ to SCB. Let $(\bar{\Gamma}_s, \bar{\Gamma}_i, \bar{\Gamma}_o)$ be the related state, input, and output transformation matrices. Let

$$\bar{\Gamma}_s^{-1}\bar{E} = \left((\bar{E}_a^-)' \quad (\bar{E}_a^0)' \quad (\bar{E}_a^+)' \quad (\bar{E}_b)' \quad (\bar{E}_c)' \quad (\bar{E}_d)' \right)'. \tag{13.16}$$

We start with the case when we are using strictly proper filters and hence we want to characterize γ_{sp}^*. We clearly will rely on Theorem 11.22. Using similar techniques as in Section 10.5, we intend to write the unique semi-stabilizing solution Q of the CQMI $G_\infty(Q) \geq 0$, where G_∞ is given by (13.14) in terms of a certain CARE. To do so, we extract the following matrices from the SCB constructed above:

$$\bar{A}_s := \begin{pmatrix} \bar{A}_{aa}^+ & \bar{L}_{ab}^+ \bar{C}_b \\ 0 & \bar{A}_{bb} \end{pmatrix}, \qquad \bar{B}_s := \begin{pmatrix} \bar{B}_{a0}^+ & \bar{L}_{ad}^+ \\ \bar{B}_{b0} & \bar{L}_{bd} \end{pmatrix}, \tag{13.17}$$

$$\bar{C}_s := \bar{\Gamma}_o \begin{pmatrix} 0 & 0 \\ 0 & 0 \\ 0 & \bar{C}_b \end{pmatrix}, \qquad \bar{D}_s := \bar{\Gamma}_o \begin{pmatrix} I_{\bar{m}_0} & 0 \\ 0 & \bar{C}_d \bar{C}_d' \\ 0 & 0 \end{pmatrix}, \tag{13.18}$$

and

$$\bar{E}_s = \begin{pmatrix} \bar{E}_a^+ \\ \bar{E}_b \end{pmatrix}. \tag{13.19}$$

We remark that various submatrices in the above definitions come from the SCB as applied to the subsystem characterized by the quadruple $(\bar{A}, \bar{B}, \bar{C}, \bar{D})$ (for details, see Chapter 3). Then, in view of Chapter 4, any semi-stabilizing solution Q of the CQMI $G_\infty(Q) \geq 0$ where G_∞ is given by (13.14) must be of the form

$$Q = (\bar{\Gamma}_s^{-1})' \begin{pmatrix} 0 & 0 & 0 & 0 & 0 \\ 0 & 0 & 0 & 0 & 0 \\ 0 & 0 & Q_s & 0 & 0 \\ 0 & 0 & 0 & 0 & 0 \\ 0 & 0 & 0 & 0 & 0 \end{pmatrix} \bar{\Gamma}_s^{-1}, \tag{13.20}$$

where $Q_s \geq 0$ is the stabilizing solution of the H_∞ CARE:

$$Q_s \bar{A}_s + \bar{A}_s' Q_s + \bar{C}_s' \bar{C}_s + \gamma^{-2} Q_s \bar{E}_s \bar{E}_s' Q_s$$
$$- (Q_s \bar{B}_s + \bar{C}_s' \bar{D}_s)(\bar{D}_s' \bar{D}_s)^{-1}(\bar{B}_s' Q_s + \bar{D}_s' \bar{C}_s) = 0. \quad (13.21)$$

Moreover, a positive semi-definite semi-stabilizing solution exists of the CQMI $G_\infty(Q) \geq 0$ if and only if a positive semi-definite semi-stabilizing solution exists of (13.21).

We will consider a special case where (A, B_1, C, D_1) is left-invertible and then investigate the general case.

Case A: Left-invertible systems

We first look at left-invertible systems for which the dual system $(\bar{A}, \bar{B}, \bar{C}, \bar{D})$ is obviously right-invertible, and hence, in the SCB format, the state x_b is not present in the dual system. This implies that the H_∞ CARE (13.21) can be simplified to

$$Q_s \bar{A}_{aa}^+ + (\bar{A}_{aa}^+)' Q_s + \gamma^{-2} Q_s \bar{E}_a^+ (\bar{E}_a^+)' Q_s$$
$$- Q_s \bar{B}_{s,1}(\bar{D}_s' \bar{D}_s)^{-1} \bar{B}_{s,1}' Q_s = 0, \quad (13.22)$$

where

$$\bar{B}_{s,1} = \begin{pmatrix} \bar{B}_{a0}^+ & \bar{L}_{ad}^+ \end{pmatrix}.$$

Next, we note the fact that $\bar{A}_{aa}^+$ is anti-stable. This together with the fact that Q_s must be such that

$$\bar{A}_{aa}^+ + \gamma^{-2} \bar{E}_a^+ (\bar{E}_a^+)' Q_s - \bar{B}_{s,1}(\bar{D}_s' \bar{D}_s)^{-1} \bar{B}_{s,1}' Q_s$$

is asymptotically stable yields that Q_s must be invertible. It is also easily verified that the solutions V_1 and V_2 of the following two Lyapunov equations exist and are moreover unique:

$$\bar{A}_{aa}^+ V_1 + V_1 (\bar{A}_{aa}^+)' = \bar{E}_a^+ (\bar{E}_a^+)' \quad (13.23)$$
$$\bar{A}_{aa}^+ V_2 + V_2 (\bar{A}_{aa}^+)' = \bar{B}_{s,1}(\bar{D}_s' \bar{D}_s)^{-1} \bar{B}_{s,1}'. \quad (13.24)$$

We then find that

$$Q_s^{-1} = V_2 - \gamma^{-2} V_1.$$

Stabilizability of $(\bar{A}, \bar{B})$ [or equivalently the detectability of (C, A)] guarantees that $V_2 > 0$. It is an almost immediate consequence that a positive semi-definite and semi-stabilizing solution of the H_∞ CARE exists if and only if γ is such that

$$V_2 - \gamma^{-2} V_1 > 0.$$

We recall that a positive semi-definite and semi-stabilizing solution of this H_∞ CARE immediately yields a positive semi-definite and semi-stabilizing solution of the CQMI $G_\infty(Q) \geq 0$. We obtain

$$\gamma_{sp}^* = \max\{ \gamma_{sp}^{*+}, \gamma_{sp}^{*0}, \|F_1\| \}, \quad (13.25)$$

where

$$\gamma_{sp}^{*+} = \|V_1 V_2^{-1}\|^{1/2}$$

and γ_{sp}^{*0} is the infimum over all γ for which for any invariant zero s_0 on the imaginary axis of the system characterized by (A, B_1, C, D_1), a matrix K exists such that $s_0 I - A + KC$ is invertible and

$$\|E(s_0 I - A + KC)^{-1}(B_1 - KD_1) + F_1\| < \gamma.$$

Note that the characterization in (13.25) can hence be characterized as consisting of three components. The first one is associated with the invariant zeros in the right-half plane. The second one is associated with the invariant zeros on the imaginary axis, whereas the last one looks at infinity.

For the case of proper filters, we need to characterize γ_p^*. In this regard, we note that the only thing, compared with our analysis for the strictly proper case, that we need to change is to replace E and F_1 by

$$\hat{E} := E - F_1 D_1'(D_1 D_1')^\dagger C \quad \text{and} \quad \hat{F}_1 := F_1 - F_1 D_1'(D_1 D_1')^\dagger D_1,$$

respectively. We then define

$$\bar{\Gamma}_s^{-1} \hat{E}' = \left((\hat{E}_a^-)' \quad (\hat{E}_a^0)' \quad (\hat{E}_a^+)' \quad (\hat{E}_b)' \quad (\hat{E}_c)' \quad (\hat{E}_d)' \right)'. \qquad (13.26)$$

The equation we earlier obtained for V_1 is then different, and we define $\hat{V}_1$ as the unique solution of

$$\bar{A}_{aa}^+ \hat{V}_1 + \hat{V}_1 (\bar{A}_{aa}^+)' = \hat{E}_a^+ (\hat{E}_a^+)'. \qquad (13.27)$$

We then obtain

$$\gamma_p^* = \max\{ \gamma_p^{*+}, \gamma_p^{*0}, \|\hat{F}_1\| \}, \qquad (13.28)$$

where

$$\gamma_p^{*+} = \|\hat{V}_1 V_2^{-1}\|^{1/2}$$

and γ_p^{*0} is the infimum over all γ for which for any invariant zero s_0 on the imaginary axis of the system characterized by (A, B_1, C, D_1), a matrix K exists such that $s_0 I - A + KC$ is invertible and

$$\|\hat{E}(s_0 I - A + KC)^{-1}(B_1 - KD_1) + \hat{F}_1\| < \gamma.$$

Equations (13.25) and (13.28) clearly clarify the contribution of non-minimum-phase unstable zeros (the eigenvalues of A_{aa}^+) on γ_{sp}^* and γ_p^*, respectively. In particular, these equations show that if we move all non-minimum-phase invariant zeros toward the imaginary axis (i.e., if we replace A_{aa}^+ by $A_{aa}^+ - \alpha I$ with $\alpha > 0$), then both γ_{sp}^* and γ_p^* decrease and thus the performance improves.

Case B: Non-left-invertible systems

We would like to investigate next how the location of the open right-half plane invariant zeros (non-minimum-phase zeros) of the subsystem characterized by (A, B_1, C, D_1) affect γ_{sp}^* and γ_p^* in the case when (A, B_1, C, D_1) is no longer left-invertible. We focus on strictly proper generalized H_∞ OID filtering for conti-nuous-time systems, and our starting point is the CARE (13.21). We decompose Q_s below into two parts, one arising due to unstable zero dynamics (i.e., the dynamics dictated by the open right-half plane invariant zeros) and the other due to non-left-invertible dynamics. For ease of presentation, we define

$$\bar{C}_s = \begin{pmatrix} 0 & \bar{C}_{s,2} \end{pmatrix}, \quad \bar{B}_s = \begin{pmatrix} \bar{B}_{s,1} \\ \bar{B}_{s,2} \end{pmatrix}.$$

Let

$$Q_s = \begin{pmatrix} 0 & 0 \\ 0 & Q_b \end{pmatrix} + Q_r. \tag{13.29}$$

Here Q_b is the stabilizing solution of the following H_∞^1 CARE:

$$\bar{A}_{bb}' Q_b + Q_b \bar{A}_{bb} + \bar{C}_{s,2}' \bar{C}_{s,2} + \gamma^{-2} Q_b \bar{E}_b \bar{E}_b' Q_b$$
$$- (Q_b \bar{B}_{s,2} + \bar{C}_{s,2}' \bar{D}_s)(\bar{D}_s' \bar{D}_s)^{-1}(\bar{B}_{s,2}' Q_b + \bar{D}_s' \bar{C}_{s,2}) = 0. \tag{13.30}$$

It is easy to verify that the existence of the stabilizing solution to the H_∞^1 CARE (13.30) is necessary for the existence of the semi-stabilizing solution Q to the original CQMI given in (13.14). Define γ_b^* as the infimum over all γ for which a positive semi-definite semi-stabilizing solution to the H_∞^1 CARE (13.30) exists.

In view of (13.29) and (13.30), it is easy to show that Q_r satisfies the CARE:

$$Q_r V + V' Q_r - Q_r \bar{B}_s (\bar{D}_s' \bar{D}_s)^{-1} \bar{B}_s' Q_r = 0 \tag{13.31}$$

with V given by

$$V = \begin{pmatrix} \bar{A}_{aa}^+ & \bar{L}_{ab}^+ \bar{C}_b - \bar{B}_{s,1}(\bar{D}_s' \bar{D}_s)^{-1}(\bar{B}_{s,2}' Q_b + \bar{D}_s' \bar{C}_{s,2}) + \gamma^{-2} \bar{E}_a^+ \bar{E}_b' Q_b \\ 0 & \bar{A}_{bb} - \bar{B}_{s,2}(\bar{D}_s' \bar{D}_s)^{-1}(\bar{B}_{s,2}' Q_b + \bar{D}_s' \bar{C}_{s,2}) + \gamma^{-2} \bar{E}_b \bar{E}_b' Q_b \end{pmatrix}$$
$$= \begin{pmatrix} V_{11} & V_{12} \\ 0 & V_{22} \end{pmatrix}.$$

As Q_b is a stabilizing solution of (13.30), we know that V_{22} is asymptotically stable. Next, we would like to rewrite Q_r in terms of another matrix $\bar{Q}_r$. To do so, let R be the solution of

$$\bar{A}_{aa}^+ R - R V_{22} + V_{12} = 0.$$

Note that R is bounded because V_{12} is bounded and the eigenvalues of $\bar{A}_{aa}^+$ and V_{22} are in the open right- and left-half plane, respectively, and hence bounded away from each other. We then find that

$$Q_r = \begin{pmatrix} I \\ R \end{pmatrix} \bar{Q}_r \begin{pmatrix} I & R' \end{pmatrix}$$

with $\bar{Q}_r$ being a stabilizing solution of the CARE:

$$\bar{Q}_r \bar{A}_{aa}^+ + (\bar{A}_{aa}^+)' \bar{Q}_r + \gamma^{-2} \bar{Q}_r \bar{E}_{s,1} \bar{E}_{s,1}' \bar{Q}_r$$
$$- \bar{Q}_r \hat{B}_{s,1} (\bar{D}_s' \bar{D}_s)^{-1} \hat{B}_{s,1}' \bar{Q}_r = 0, \quad (13.32)$$

where

$$\begin{pmatrix} I & -R \\ 0 & I \end{pmatrix} \bar{B}_s = \begin{pmatrix} \hat{B}_{s,1} \\ \hat{B}_{s,2} \end{pmatrix}, \quad \begin{pmatrix} I & -R \\ 0 & I \end{pmatrix} \bar{E}_s = \begin{pmatrix} \bar{E}_{s,1} \\ \bar{E}_{s,2} \end{pmatrix}.$$

The above analysis implies that

$$Q_s = \begin{pmatrix} 0 & 0 \\ 0 & Q_b \end{pmatrix} + \begin{pmatrix} I \\ R \end{pmatrix} \bar{Q}_r \begin{pmatrix} I & R' \end{pmatrix}. \tag{13.33}$$

Next, we note the fact that $\bar{A}_{aa}^+$ is anti-stable. This together with the fact that $\bar{Q}_r$ must be such that

$$\bar{A}_{aa}^+ + \gamma^{-2} \bar{E}_{s,1} \bar{E}_{s,1}' \bar{Q}_r - \hat{B}_{s,1} (\bar{D}_s' \bar{D}_s)^{-1} \hat{B}_{s,1}' \bar{Q}_r$$

is asymptotically stable yield that $\bar{Q}_r$ must be invertible. It is also easily verified that the solutions W_1 and W_2 of the following two Lyapunov equations exist and are moreover unique:

$$\bar{A}_{aa}^+ W_1 + W_1 (\bar{A}_{aa}^+)' = \bar{E}_{s,1} \bar{E}_{s,1}', \tag{13.34}$$
$$\bar{A}_{aa}^+ W_2 + W_2 (\bar{A}_{aa}^+)' = \hat{B}_{s,1} (\bar{D}_s' \bar{D}_s)^{-1} \hat{B}_{s,1}'. \tag{13.35}$$

We then find that

$$\bar{Q}_r^{-1} = W_2 - \gamma^{-2} W_1.$$

Stabilizability of $(\bar{A}, \bar{B})$ [or equivalently the detectability of (C, A)] again guarantees that $W_2 > 0$, and therefore, our existence condition becomes

$$W_2 - \gamma^{-2} W_1 > 0$$

or

$$\gamma \geqslant \| W_1 W_2^{-1} \|.$$

The difficulty with this characterization is that W_1 and W_2 themselves depend on γ (because R depends on γ), and hence, this is a highly implicit characterization. It definitely does not yield the clear characterization we obtained before for the left-invertible case. However, we can conclude as we did in the left-invertible case that moving all non-minimum-phase invariant zeros toward the imaginary axis (i.e., by replacing A_{aa}^+ by $A_{aa}^+ - \alpha I$ with $\alpha > 0$), both γ_{sp}^* and γ_p^* decrease and thus the performance improves.

13.7   Generalized H_∞ AID filtering problem

We first formulate below the generalized H_2 AID filtering problem.

Problem 13.10 Consider a continuous- or discrete-time system Σ as given in (13.1), where the input u_1 is an unknown signal with finite but unknown RMS value, and the input u_2 is the output of a linear exosystem given in (13.3). Then, the **generalized H_∞ AID filtering problem** is defined as follows: Find, whenever it exists, a family of linear stable strictly proper (or proper) filters of the type Σ_f^ε given in (13.8) and parameterized in positive $\varepsilon \in (0, \varepsilon^*]$ such that

> (*i*) **(Generalized unbiasedness)** for any filter in the family, the estimation error e_z, in the absence of the input u_1, decays asymptotically to zero for all possible finite initial values of the exosystem (13.3), the given system (13.1), and the filter (13.8), and

> (*ii*) **(Performance)** the H_∞ norm of the transfer matrix $G_{u_1 e}$ from u_1 to e_z tends to zero as ε tends to zero.

We have the following theorem stating the solvability conditions of the above problem.

Theorem 13.11 *Consider a continuous- or discrete-time system Σ as given in (13.1) together with the exosystem Σ_a given in (13.3) and the associated generalized H_∞ AID filtering problem 13.10. Let Assumption 13.1 be satisfied by Σ. Also, consider the system Σ_0 given in (13.9) and the associated H_∞ AID filtering problem 9.3. Consider next the expanded system $\tilde\Sigma$ given in (13.10), and let Assumption 13.4 be satisfied by it. Then, the generalized H_∞ AID filtering problem 13.10 for Σ is solvable via a family of strictly proper (proper) filters if and only if the following conditions hold:*

> (*i*) *the H_∞ AID filtering problem for Σ_0 is solvable via a family of strictly proper (proper) filters,*

> (*ii*) *γ_{sp}^0 as given in (13.12) (γ_p^0 as given in (13.13)) equals zero.*

Proof : The proof follows from Theorem 13.8 and from the fact that γ_{sp}^* (γ_p^*) must equal zero. ∎

Whenever the generalized H_∞ AID filtering problem is solvable, the proofs of Theorem 13.11 and Theorem 13.8 immediately point out that a sequence or family of filters that solves the generalized H_∞ AID filtering problem for a given system is the same as a sequence of filters that solves the γ-level generalized H_∞ SOID filtering problem for the same system when γ is treated as a parameter tending to zero. Equivalently, a family of generalized H_∞ AID filters for a given system Σ as in (13.1) is also a family of generalized H_∞ AID filters for the expanded system $\widetilde{\Sigma}$ as in (13.10). This implies that the both the design of families of filters that solve the generalized H_∞ AID filtering problem and the asymptotic fixed modes associated with them for a given system Σ coincide, respectively, with the design of families of H_∞ AID filters and the asymptotic fixed modes associated with them for the expanded system $\widetilde{\Sigma}$.

14

Fault detection, isolation, and estimation — exact or almost fault estimation

14.1 Introduction

Various types of faults arise in industrial processes owing to malfunction of internal components of a process as well as to failures of measurement sensors and control actuators attached to the process. Over the last three or four decades, industrial automation has been increasingly fueled by various technological developments, including the availability of highly complex electronic equipment and the overwhelming progress in computer technology. This has led not only to the development of complex control systems but also to higher demand of reliable and secure control systems. Thus it has become imperative that any faults that occur be detected and identified automatically without severely disturbing the yield the process generates. This has stimulated over the last two decades an extensive study of fault detection and identification methods.

As discussed in a survey paper by Willsky [108], one faces three different types of tasks or layers in the area of fault detection and identification: (1) fault detection, (2) fault isolation (identification), and (3) fault estimation. *Fault detection* consists of designing a residual generator that produces a residual signal enabling one to make a binary decision as to whether a fault occurred. *Fault isolation* imposes a stronger requirement. When one or more faults occur, the residual signal must enable us not only to detect that there are faults occurring in the system, but it must also enable us to isolate (identify) which faults have occurred. Finally, *fault estimation* is the determination of the extent of failure. The latter is done by trying to reconstruct the fault signals. A large body of literature dealing with these tasks exist; see the books by Basseville and Nikiforov [4], Chen and Patton [17], and Gertler [27] and the references therein. Some of this literature considers only the first two tasks of fault detection and isolation, whereas some others consider all three tasks.

This chapter focuses on all three tasks of fault detection, isolation, and estimation, while estimating the fault signal either exactly or with arbitrary precision. The results are based on some recent work by us and our coworkers [53]. The next chapter focuses on optimal estimation in the sense of minimizing either the H_2 or H_∞ norm of the transfer function from the fault signal to the estimation error.

14.2   Problem formulation

Consider the following state-space description for a plant or a system given by

$$\Sigma : \begin{cases} \sigma x = Ax + \sum_{i=1}^{k} B_{f,i}\, f_i + \sum_{j=1}^{m} B_{d,j}\, d_j \\ \quad\;\; = Ax + \quad\; B_f\, f \quad\;\; + \quad\;\; B_d\, d \\ y = Cx + \sum_{i=1}^{k} D_{f,i}\, f_i + \sum_{j=1}^{m} D_{d,j}\, d_j \\ \quad = Cx + \quad\; D_f\, f \quad\;\; + \quad\;\; D_d\, d, \end{cases} \tag{14.1}$$

where σ is an operator indicating the time derivate $\frac{d}{dt}$ for continuous-time systems and a forward unit time shift for discrete-time systems. Also, for all t, $x(t) \in \mathbb{R}^n$ is the state vector, $d(t) \in \mathbb{R}^m$ is a disturbance signal vector, and $y(t) \in \mathbb{R}^p$ is the measurement vector. Furthermore, f_i signifies the ith fault for each $i = 1, 2, \ldots, k$. The coefficient matrices $B_{f,i}$ and $D_{f,i}$ are referred to in the literature as failure signatures associated with the ith fault, whereas f_i itself is called the ith fault signal. Obviously, the failure signatures $B_{f,i}$ and $D_{f,i}$ depend on the physics of the given system. The fault signal vector f with $f(t) \in \mathbb{R}^k$ is a collection of fault signals $f_i, i = 1, 2, \ldots, k$, in a vector format. Also, it is normal in the fault detection and isolation setting for model uncertainties to be described as external input signals in the same manner as disturbance signals d. In other words, d here can be thought of representing both external disturbance signals and signals that might arise due to model uncertainties; see [24]. Also, it is without loss of generality to consider a system without a control input in connection with fault detection and isolation design.

The above can be rewritten in a transfer function form as

$$\begin{aligned} y(s) &= (C(sI - A)^{-1}B_f + D_f)f(s) + (C(sI - A)^{-1}B_d + D_d)d(s) \\ &:= \qquad\quad G_f(s)f(s) \qquad\qquad + \qquad\qquad G_d(s)d(s). \end{aligned}$$

Traditionally, fault detection, isolation, and estimation is done by designing a filter that uses the measured output and generates a signal that is referred to as a residual signal r, and thus, the filter itself is referred to as residual generator. The residual signal r plays the role of an alarm signal for the occurrence of a fault, and it is generated by the residual generator from the measured output y of the given process. We are interested here in using a linear time invariant proper stable filter whose transfer function is denoted by F (see the setup of Figure 14.1):

$$r(s) = F(s)y(s). \tag{14.2}$$

In view of the setup of Figure 14.1, we see easily

$$r := T_{fr}\, f + T_{dr}\, d,$$

where the transfer functions T_{fr} and T_{dr} are given by

$$T_{fr}(s) = F(s)G_f(s) \quad \text{and} \quad T_{dr}(s) = F(s)G_d(s). \tag{14.3}$$

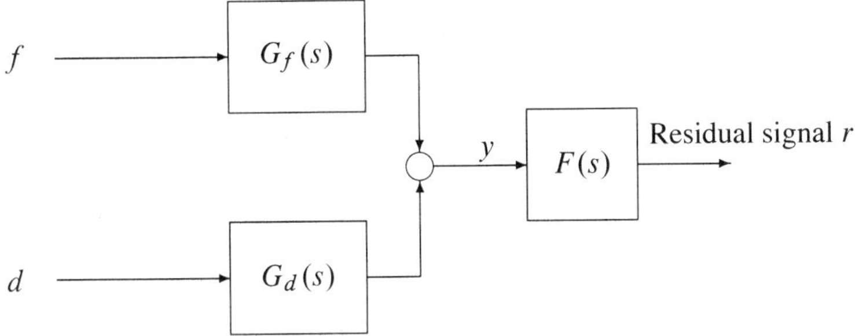

Figure 14.1: Block diagram of the standard setup

In what follows, we pose formally four different problems of generating the residual signal r for continuous-time systems, each problem imposing certain specified and desired conditions on T_{dr} and T_{fr}. The discrete-time versions of the following design problems follow easily by using $e^{j\omega}$ instead of $j\omega$.

Problem 14.1 (EFDIE Problem) The problem of **exact fault detection, isolation, and estimation** is defined as the problem of finding, if existent, a residual generator $F \in \mathcal{RH}_\infty$ such that we have

$$T_{fr}(j\omega) = I \quad \text{and} \quad T_{dr}(j\omega) = 0 \quad \forall \omega \in \mathbb{R}.$$

Clearly, the EFDIE problem is related to the EID filtering problem discussed in Chapter 7. Obviously, it imposes strong conditions. That is, if it is solvable, any fault can be detected without any false alarm, and isolated. In fact, the residual signal then will be a perfect estimate of the fault. As such, the EFDIE problem may not always be solvable. In an attempt to weaken the solvability conditions, the following problem addresses *almost* exact fault signal estimation so as to relax the perfect estimation condition imposed in the EFDIE problem.

Problem 14.2 (H_α AFDIE Problem) The problem of H_α **almost exact fault detection, isolation, and estimation** is defined as a problem of finding, if existent, a parameterized family of residual generators $F_\varepsilon \in \mathcal{RH}_\infty$ parameterized in ε such that, for any $\delta > 0$, an ε^* exists such that, for any $\varepsilon < \varepsilon^*$, we have

$$\|T_{fr,\varepsilon} - I\|_{H_\alpha} < \delta \quad \text{and} \quad \|T_{dr,\varepsilon}\|_{H_\alpha} < \delta,$$

where H_α is either an H_2 or H_∞ norm. Here, $T_{fr,\varepsilon}$ and $T_{dr,\varepsilon}$ correspond to the expressions given in (14.3) for T_{fr} and T_{dr}, respectively, with F_ε replacing F.

Both the EFDIE and the H_α AFDIE problems seek to identify all faults even if multiple faults occur simultaneously. In the event that the solvability conditions for these problems are not satisfied by the given plant, one would like to investigate whether a chosen subset of all faults can be detected and isolated in the same manner as the exact and the almost exact fault detection and isolation problems seek to achieve for the set of all faults. In this regard, let us define an integer index set $\mathcal{I}_s$, which is a subset of the integer set $\{1, 2, \cdots, m\}$. The index set $\mathcal{I}_s$ designates the faults we are interested in detecting and isolating. For $i = 1, 2, \cdots, m$, let Q be a diagonal matrix with q_{ii} equal to unity if $i \in \mathcal{I}_s$ and zero otherwise. Then the EFDIE and H_α AFDIE problems can be modified as follows:

Problem 14.3 (PEFDIE Problem) The problem of **partial exact fault detection, isolation, and estimation** is defined as a problem of finding, if existent, a residual generator $F \in \mathcal{RH}_\infty$ such that we have

$$Q\, T_{fr}(j\omega) = Q \quad \text{and} \quad Q\, T_{dr}(j\omega) = 0 \quad \forall \omega \in \mathbb{R}.$$

Problem 14.4 (H_αPAFDIE Problem) The problem of H_α **partial almost exact fault detection and isolation** is defined as a problem of finding, if existent, a parameterized family of residual generators $F_\varepsilon \in \mathcal{RH}_\infty$ parameterized in ε such that, for any $\delta > 0$, an ε^* exists such that, for any $\varepsilon < \varepsilon^*$, we have

$$\|Q(T_{fr,\varepsilon} - I)\|_{H_\alpha} < \delta \quad \text{and} \quad \|Q\, T_{dr,\varepsilon}\|_{H_\alpha} < \delta,$$

where, as before, H_α is either an H_2 or H_∞ norm.

We can easily relate the (partial) EFDIE and H_α AFDIE problems to the EID and AID filtering problems of Chapters 7, 8, and 9. To do so, let

$$B := \begin{pmatrix} B_f & B_d \end{pmatrix}, \quad D := \begin{pmatrix} D_f & D_d \end{pmatrix}. \tag{14.4}$$

Also, denote the output variable that is to be estimated as

$$z := Qf = \bar{Q}u, \quad \text{where} \quad \bar{Q} := \begin{pmatrix} Q & 0 \end{pmatrix} \quad \text{and} \quad u = \begin{pmatrix} f \\ d \end{pmatrix}. \tag{14.5}$$

Thus the system Σ given in (14.1) along with the variable z that is to be estimated can be rewritten as

$$\Sigma_1 : \begin{cases} \sigma x = Ax + Bu \\ y = Cx + Du \\ z = \bar{Q}u. \end{cases} \tag{14.6}$$

In view of the definition of Σ_1 and in view of Chapters 7, 8, and 9, one can easily see that the (partial) EFDIE and H_α AFDIE problems are related to the EID and AID filtering problems for Σ_1. This is so because the residual signal r is $\hat{z}$, which is the estimate of z.

14.3   Solvability conditions and design of residual generator

As the (partial) EFDIE and H_α AFDIE problems are related to the EID and AID filtering problems, the solvability of these problems and the construction of appropriate fault detectors or residual generators revert to what has been presented in earlier chapters. To be explicit, we proceed now to state the solvability conditions for (partial) EFDIE and H_α AFDIE problems. In what follows, we concentrate only on the partial problems because whenever we are interested in all faults, we can set easily $Q = I_m$, where I_m is an identity matrix of dimension $m \times m$.

Theorem 14.5 *Consider the continuous-time system Σ given in (14.1). Assume that the matrix pair (C, A) is $\mathbb{C}^-$-detectable. We have the following:*

(i) The partial EFDIE problem is solvable if and only if

$$\left(\mathcal{S}^-(A, B, C, D) \oplus \mathbb{R}^{(m+k)} \right) \cap \ker \begin{pmatrix} C & D \end{pmatrix} \subseteq \ker \begin{pmatrix} 0 & \bar{Q} \end{pmatrix}. \quad (14.7)$$

(ii) The partial H_2 AFDIE problem is solvable if and only if

$$\left(\left[\mathcal{S}^{-0}(A, B, C, D) \cap \mathcal{V}^*(A, B, C, D) \right] \oplus \mathbb{R}^{(m+k)} \right) \cap \ker \begin{pmatrix} C & D \end{pmatrix}$$
$$\subseteq \ker \begin{pmatrix} 0 & \bar{Q} \end{pmatrix}. \quad (14.8)$$

(iii) The partial H_∞ AFDIE problem is solvable if and only if the following conditions are satisfied.

(a) Equation (14.8) is satisfied.

(b) Consider the invariant zeros on the imaginary axis of the system characterized by the quadruple (A, B, C, D). For any such an invariant zero s_0 and for all $\delta > 0$, matrices K and L exist such that $s_0 I - A + KC$ is invertible, and

$$\| - LC(s_0 I - A + KC)^{-1}(B - KD) + (\bar{Q} - LD) \| < \delta.$$

Proof : The proof follows directly from Theorems 7.9, 8.9, and 9.11, while using the output variable z that is to be estimated as in (14.5). ∎

Next, we consider the discrete-time version of the above theorem.

Theorem 14.6 *Consider the discrete-time system Σ given in (14.1). Assume that the matrix pair (C, A) is $\mathbb{C}^\ominus$-detectable. We have the following:*

(*i*) *The partial EFDIE problem is solvable if and only if*

$$\left(\mathcal{S}^{\ominus}(A, B, C, D) \oplus \mathbb{R}^{(m+k)}\right) \cap \ker\begin{pmatrix} C & D \end{pmatrix} \subseteq \ker\begin{pmatrix} 0 & \bar{Q} \end{pmatrix}. \quad (14.9)$$

(*ii*) *The partial H_2 AFDIE problem is solvable if and only if*

$$\left(\mathcal{S}^{\otimes}(A, B, C, D) \oplus \mathbb{R}^{(m+k)}\right) \cap \ker\begin{pmatrix} C & D \end{pmatrix} \subseteq \ker\begin{pmatrix} 0 & \bar{Q} \end{pmatrix}. \quad (14.10)$$

(*iii*) *The partial H_∞ AFDIE problem is solvable if and only if the following conditions are satisfied.*

   (*a*) *Equation (14.10) is satisfied.*

   (*b*) *Consider the invariant zeros on the unit circle of the system characterized by the quadruple (A, B, C, D). For any such an invariant zero z_0 and for all $\delta > 0$, matrices K and L exist such that $z_0 I - A + KC$ is invertible, and*

$$\| - LC(z_0 I - A + KC)^{-1}(B - KD) + (\bar{Q} - LD)\| < \delta.$$

Proof : The proof follows directly from Theorems 7.9, 8.15, and 9.18, while using the output variable z that is to be estimated as in (14.5). ∎

We emphasize that once the appropriate residual generators exist as indicated by the above results, we can design them by using CSS full-order or reduced-order architectures for them. This is because, as emphasized earlier, residual generators are merely filters. Thus, the methods of designing appropriate fault detectors or filters follow directly as discussed in Section 7.5 for EID filtering or in Sections 8.6 and 9.6 for AID filtering. We remark that the reduced-order fault detectors have an added advantage. For instance, in a number of cases, the complete fault detector consists of several individual fault detectors. The order of the complete fault detector will in general be reduced drastically, if each individual fault detector is of reduced order.

14.4   Discussion

It is noteworthy to point out that, under various disguises, in the language of this book, EID filtering for some system constructed somehow from the given data was used early on in the literature for the first two tasks of fault detection and isolation without even considering the third task of fault signal estimation although the estimation of some signal was the vehicle by which the fault detection and then fault isolation were carried out. This discussion deals with the difficulties and drawbacks of the way EID filters were used earlier. The methods used earlier are the unknown-input filter approach (what is termed throughout this book as EID filtering), eigenstructure assignment approach, parity relation approach, and factorization fault detection approach, just to mention some of the most known methods; see [18, 23, 26, 57–60]. It is known in the literature that the latter two approaches can essentially be reduced to either of the former two approaches; see [26]. These so-called indirect methods essentially use estimation theory to design filters that estimate either the entire or part of the state of the plant and then form an appropriate estimation error that constitutes the residual signal r. This is done assuming that the faults do not exist. The block diagram of Figure 14.2 depicts the basic setup used. As seen from Figure 14.2, the residual signal $r = e_y = C\hat{x} - y$. In other words, once again in the language of this book, the majority of the literature uses the estimate of the state x of the following system Σ_2 and forms the residual signal r as $e_y = C\hat{x} - y$:

$$\Sigma_2 \ : \ \begin{cases} \sigma x = Ax + B_d u \\ \quad y = Cx \\ \quad z = Cx, \end{cases}$$

where u is the disturbance signal d. We observe that the fault signal f is assumed to be zero in arriving at the system Σ_2. Also, let us emphasize that the fault signal f is not the signal that is estimated. The hope here is that a nonzero fault signal f will cause a nonzero residual r and, thus, lead to fault detection and then by some other means to fault isolation as well.

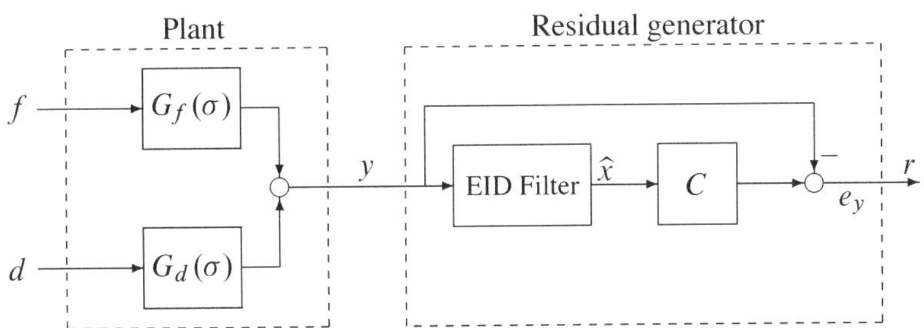

Figure 14.2: Block diagram of the plant and residual generator

The aforementioned methods of fault detection have certain drawbacks. To point out the first one, let us note that there is no guarantee in general that the resulting filter will satisfy the conditions for fault detection. That is, one might not be able to detect each and every fault that might occur. We emphasize that one can detect only those faults that are not in the kernel of the transfer function from the signal f to the residual signal r. Another drawback of these indirect methods is that they require that the measured output y is not directly affected by the unknown input d. That is, they require that $D_d = 0$, [59]. These drawbacks arise inherently in the approaches that use the block diagram setup of Figure 14.2. However, despite these drawbacks, the above methods of fault detection have successfully been used in practice (see [34, 59, 60, 100, 108]).

We make next another important observation. The above methods of fault detection that use the block diagram setup of Figure 14.2 cannot in general use proper filters. They must use strictly proper filters. The reason for this is obvious. Let us observe that proper filters are ubiquitously implemented as reduced-order filters. As it is well known, the use of a reduced-order filter for the estimation of the state leads to the utilization of the measured output y as a part of the state estimate. This typically implies that the estimation error $e_y = C\hat{x} - y$ is uniformly zero irrespective of whether a fault occurred because the effect of the fault f on y and on $C\hat{x}$ is exactly the same (thus, forcing e_y to be uniformly zero). In other words, although the estimation error e or e_y is decoupled from the unknown input d by the appropriate design of the filter, the transfer function from the fault signal f to the residual signal $r = e_y$ is always zero.

Our method of fault detection, isolation, and estimation, as explained in Sections 14.2 and 14.3, besides having other obvious advantages, rectifies the above drawbacks by estimating directly the fault signal vector f and the disturbance vector d rather than the state of the system x.

15

Fault detection, isolation, and estimation—optimal fault estimation

15.1   Introduction

The fault estimation problems formulated and studied in the previous chapter seek a residual signal that satisfies two important requirements: exact decoupling of the residual signal from the disturbance or noise, and the residual signal is an exact estimate of the fault signal. The first objective makes the residual signal completely insensitive to the disturbance or noise, and the second objective guarantees that the residual signal is equal to the fault signal. The solvability conditions for such a constrained fault signal estimation are strong. In Chapter 14, we also considered an almost version where the residual signal is arbitrarily close to the fault signal and is almost completely independent of the disturbance. This weakens the solvability conditions, but still the solvability conditions are very strong.

In this chapter, we seek to relax the requirements on the fault signal estimation with a hope of weakening the solvability conditions. A natural method of relaxing the requirements is to seek optimal estimation of fault signals instead of exact or almost exact fault estimation. By optimal fault estimation, we mean that we create a residual signal that has to be completely insensitive to the disturbances and under this constraint generates the best possible estimate of the fault signal measured either by an H_2 or an H_∞ norm. Thus, the fault signal estimation problems we deal with here are constrained optimal estimation problems.

The optimal estimation problem, where we need to estimate an unknown input to a dynamic system, is referred to as a deconvolution problem in the presence of noise and disturbance and has been studied in [78]. It is clear that such an optimal fault estimation makes sense only for faults that can be detected and identified. In other words, we should be able to determine whether a particular fault has occurred before we attempt to estimate the fault signal.

As said in Chapter 14, clearly three tasks need to be considered: fault detection, fault isolation, and fault signal estimation. Let us for a moment focus only on the first two tasks of fault detection and fault isolation. As said earlier, the detectability of faults implies that a residual generator exists that can produce a residual signal that is (a) completely insensitive to plant disturbances and noise, and (b) sensitive to fault signals implying that a nonzero fault signal produces a nonzero residual signal. Similarly, fault isolation requires a residual signal for each fault that is (a) insensitive to plant disturbances and noise as well as to other faults, and (b) sensitive to the specific fault signal implying that if a fault signal is nonzero, then the corresponding component of the residual signal is nonzero. Note that we can combine the tasks of detection and isolation into one using one filter and some, in general, nonlinear decision rules.

If we focus on all three tasks of fault detection, fault isolation, and fault signal estimation, the situation is somewhat different. Here we first have to detect and isolate the faults that occur, and if a fault occurs, then we want to generate an estimate of the fault signal. However, there are two major potential fallacies for fault estimation in the case of multiple faults:

- The idea that we can build one filter to achieve fault detection, isolation, and fault estimation.

- The idea that we can build one filter that estimates all fault signals simultaneously, independent of which fault has occurred.

An example that follows shortly shows that the above are indeed fallacies and that we need to build different filters for fault detection/isolation and for fault estimation. Second, we need to design a family of filters. Each member of this family estimates a fault signal, and the specific member of the family that we need is determined by which fault has occurred. In other words, which member of the family we need is determined by our filter that achieves fault detection and isolation.

Example 15.1 Consider the following system:

$$
\dot{x} = \begin{pmatrix} -1 & 0 & 0 & 0 & 0 \\ 0 & -2 & 0 & 0 & 0 \\ 0 & 0 & -3 & 0 & 0 \\ 0 & 0 & 0 & -1 & 0 \\ 0 & 0 & 0 & 0 & -1 \end{pmatrix} x + \begin{pmatrix} 1 & 0 \\ 1 & 0 \\ 0 & 1 \\ 0 & 1 \\ 0 & 0 \end{pmatrix} f + \begin{pmatrix} 0 \\ 0 \\ 0 \\ 0 \\ 1 \end{pmatrix} d,
$$

$$
y = \begin{pmatrix} -2 & 0 & 0 & 0 & 0 \\ 0 & -9 & -8 & 0 & 0 \\ 0 & 0 & 0 & -2 & 0 \\ 0 & 1 & 0 & -1 & 1 \end{pmatrix} x + \begin{pmatrix} 1 & 0 \\ 9 & 8 \\ 0 & 1 \\ 0 & 0 \end{pmatrix} f,
$$

$$
\tag{15.1}
$$

with two faults $f = (f_1, f_2)$, one disturbance d, and four measurement signals $y = (y_1, y_2, y_3, y_4)$. To clarify the following discussion, it is also useful to present the transfer matrices of this system:

$$
\begin{aligned}
y_1 &= \frac{s-1}{s+1} f_1, \\
y_2 &= \frac{9(s+1)}{s+2} f_1 + \frac{8(s+2)}{s+3} f_2, \\
y_3 &= \frac{s-1}{s+1} f_2, \\
y_4 &= \frac{1}{s+2} f_1 - \frac{1}{s+1} f_2 + \frac{1}{s+1} d.
\end{aligned}
$$

Note that y_4 intrinsically depends on the disturbance d and hence is not of use for fault detection, isolation, or estimation. Measurements y_1 and y_3 are already perfect residual signals to detect and identify whether faults f_1 and f_2 have occurred, respectively. These measurements do not allow for a perfect estimation of the fault signal because the corresponding transfer matrices both have an unstable zero in 1. On the other hand, even if we include measurement y_2, there is no possibility to identify f_1 and f_2 simultaneously in a perfect sense because the transfer matrix from (f_1, f_2) to (y_1, y_2, y_3) has a zero in 1. On the other hand, if we know that faults f_1 and f_2 cannot occur simultaneously, then we can find perfect estimates for f_1 and f_2:

- If we know that only f_1 occurred, then from the measurement y_2, we note that

$$
\widehat{f_1} = \frac{s+2}{9(s+1)} y_2
$$

  yields a perfect estimate of f_1.

- If we know that only f_2 occurred, then from the measurement y_2, we note that

$$
\widehat{f_2} = \frac{s+3}{8(s+2)} y_2
$$

  yields a perfect estimate of f_2.

Obviously $\widehat{f_1}$ and $\widehat{f_2}$ are not good residual signals for fault isolation. For that purpose, y_1 and y_3 will do the job. But $\widehat{f_1}$ and $\widehat{f_2}$ are good residual signals for fault estimation. In other words, we need to separately approach the problems of detection/isolation and the problem of estimation. We therefore need two types of residual signals: those used for fault isolation and those for fault estimation.

Note that if f_1 and f_2 occur simultaneously, then these signals $\widehat{f_1}$ and $\widehat{f_2}$ are no good as filters. We must have different filters depending on which faults have occurred. In other words, we truly cannot design one filter for the complete fault vector f, but we need to design instead a separate filter for each possible outcome of the fault isolation problem. That is, if we know fault 1 has occurred, then we use filter 1, and if fault 2 has occurred, then we use filter 2, whereas if faults 1 and 2 have occurred simultaneously, then we use filter 3. We will later exclude by assumption the latter case, but the idea works perfectly fine even if multiple faults happen simultaneously.

If one insists on having one residual signal that can be used for both fault detection/isolation as well as fault estimation, then this can be done. The structure of the above filters can be used to yield the following nonlinear residual generators:

$$\begin{aligned} \dot{p}_1 &= -p_1 + y_2, \\ \widehat{f}_1 &= \tfrac{1}{9}\mathrm{step}(\|y_1\|_2)(p_1 + y_2), \end{aligned} \tag{15.2}$$

$$\begin{aligned} \dot{p}_2 &= -p_2 + y_2, \\ \widehat{f}_2 &= \tfrac{1}{8}\mathrm{step}(\|y_3\|_2)(p_2 + y_2), \end{aligned} \tag{15.3}$$

where

$$\mathrm{step}(x) = \begin{cases} 0 & \text{for } x \leqslant 0 \\ 1 & \text{for } x > 0. \end{cases}$$

However, note that this combined residual generator is intrinsically composed of three filters. One filter used for fault detection/isolation, one used for estimating fault f_1, and one used for estimating fault f_2. This example shows that we have to consider different types of residual generators based on separating the objectives of fault identification and fault estimation.

15.2   Problem statements

Let us recall the following state-space description for a plant or a system given in (14.1) and repeated below:

$$\Sigma : \begin{cases} \sigma x = Ax + \sum_{i=1}^{k} B_{f,i} f_i + \sum_{j=1}^{m} B_{d,j} d_j \\ \qquad = Ax + \quad B_f f \quad + \quad B_d d \\ y = Cx + \sum_{i=1}^{k} D_{f,i} f_i + \sum_{j=1}^{m} D_{d,j} d_j \\ \qquad = Cx + \quad D_f f \quad + \quad D_d d, \end{cases} \tag{15.4}$$

where σ is an operator indicating the time derivate $\frac{d}{dt}$ for continuous-time systems and a forward unit time shift for discrete-time systems. Also, for all t, $x(t) \in \mathbb{R}^n$ is the state vector, $d(t) \in \mathbb{R}^m$ is a disturbance signal vector, and $y(t) \in \mathbb{R}^p$ is the measurement vector. Furthermore, f_i signifies the ith fault for each $i = 1, 2, \ldots, k$. The coefficient matrices $B_{f,i}$ and $D_{f,i}$ are referred to in the literature as failure signatures associated with the ith fault, whereas f_i itself is called the ith fault signal. Obviously, the failure signatures $B_{f,i}$ and $D_{f,i}$ depend on the physics of the given system. For each t, the fault signal vector $f(t) \in \mathbb{R}^k$ is a collection of fault signals $f_i(t)$, $i = 1, 2, \ldots, k$, into a vector. It is normal in the fault detection and isolation setting for model uncertainties to be described as external input signals in the same manner as disturbance signals d. In other words, in this setup, d can be thought of as representing both external disturbance signals and signals that might arise due to model uncertainties; see [24]. Also,

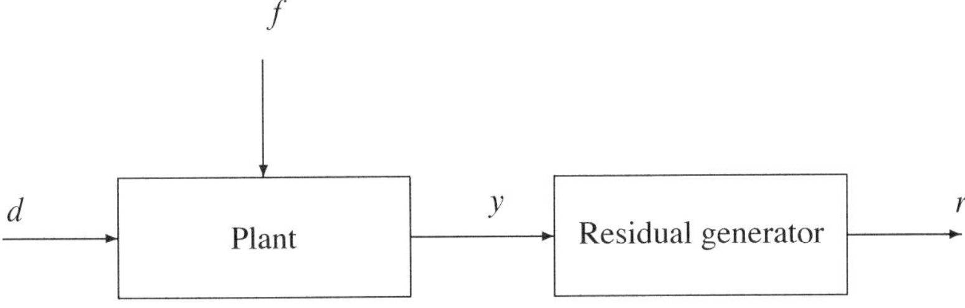

Figure 15.1: Block diagram of the setup

it is without loss of generality to consider a system without a control input in connection with fault detection and isolation design. We now proceed to formulate certain fault detection and isolation problems. The fault detection setup we follow here is shown in Figure 15.1.

Let the residual signal r be given by

$$r = Hy = \Psi(d, f), \qquad (15.5)$$

where r is a time function that takes values in $\mathbb{R}^k$. In general, we might have to take H to be a nonlinear bounded-input, bounded-output stable operator that also makes Ψ a nonlinear operator mapping disturbances and faults to a residual signal r. One of the basic issues that concerns fault detection and isolation is whether one can achieve such a detection and isolation when the disturbance d affects the system. This points out a need to have a residual generator that is insensitive to the external disturbance d. That is, we need that

$$\Psi(d, f) = \Psi(0, f)$$

for all disturbances d and all fault signals f or at least that the dependence of r on d be arbitrarily small with respect to some specified norm.

We have the following formal problem statement.

Problem 15.2 Suppose only a single fault can occur at any given time. Then, for the system given in (15.4), the problem of **exact fault detection and isolation** is finding, if existent, a bounded-input, bounded-output stable residual generator H, whose output is a residual vector $r = \Psi(d, f)$ such that for any fault f_i, $i = 1, 2, \ldots, k$, a dedicated component r_i of r exists and the operator Ψ_i from d and f to r_i has the following properties:

- $\Psi_i(d, f) = 0$ for any disturbance d and any fault f such that f_i is identical to zero.

- $\Psi_i(d, f) \neq 0$ for any disturbance d and any fault f such that f_i is not identical to zero.

In what follows, assuming fault detection and isolation is possible, we pose three different optimal or suboptimal fault signal estimation problems for the continuous-time systems. We only formulate these problems for continuous-time systems, but the discrete-time analog should be immediate from the continuous-time version. Second, we restrict the number of faults that can occur simultaneously:

Assumption 15.3 *It is assumed that each fault occurs by itself; i.e., two faults never occur simultaneously. Moreover, faults occur sufficiently apart in time so that, at any given time, at most one fault affects the measurement signal.*

The general result without this assumption can be obtained quite easily from the results obtained in this chapter. The above assumption is nearly always satisfied and makes the presentation of our results more transparent without the need to cloud the presentation with more complex design methods. As such, all the work that follows is under Assumption 15.3.

Definition 15.4 *Let Assumption 15.3 be satisfied. Then, for $i = 1, \ldots, k$, the $\mathbf{H_2}$ optimal fault estimation performance associated with fault i subject to exact fault detection and isolation is denoted by $\gamma_{e,i}^*$ and is defined as the infimum of the H_2 norm of the transfer function from f_i to $f_i - \hat{f}_i$ over all linear stable filters $\hat{f}_i = H_i\, y$, which yield a transfer matrix from d to $\hat{f}_i$ equal to zero.*

Problem 15.5 Let Assumption 15.3 be satisfied. Then, **the H_2 optimal fault estimation problem subject to exact fault detection and isolation** is defined as the problem of finding, if existent, a residual generator H that is bounded-input, bounded-output stable and generates a residual signal r with $r(t) \in \mathbb{R}^k$ and k stable linear filters H_i such that the following properties hold:

(*i*) **(Exact fault detection and isolation)**

    (*a*) The input–output operator from d to r is zero.

    (*b*) For $i = 1, \ldots, k$, fault f_i has occurred if and only if r_i is unequal to zero.

(*ii*) **(H_2 optimal fault estimation)** For $i = 1, \ldots, k$, the input–output operator H_i from y to $\hat{f}_i$ is such that the H_2 norm of the transfer function from f_i to $f_i - \hat{f}_i$ is equal to $\gamma_{e,i}^*$ and the transfer function from d to $\hat{f}_i$ is equal to zero.

Note that, in the above problem formulation, we have one residual generator used for fault detection and isolation and k filters to estimate the individual fault signals. Note that if we define

$$s_i = \text{step}(\|\widehat{f}_i\|_2)\,\widehat{f}_i,$$

where

$$\text{step}(x) = \begin{cases} 0 & \text{for } x \leqslant 0 \\ 1 & \text{for } x > 0, \end{cases}$$

then s_i is a residual signal that takes care of both fault detection/isolation as well as fault estimation. However, the residual generator is basically composed of k filters and nonlinear decision logic. In contrast to the case of a single fault, it is not possible to build one linear filter that achieves both fault detection and isolation and estimation in the case of multiple faults.

Problem 15.6 Let Assumption 15.3 be satisfied. Then, **the H_2 suboptimal fault estimation problem subject to exact fault detection and isolation** is defined as the problem of finding, if existent, a bounded-input, bounded-output stable operator H that generates a residual signal r with $r(t) \in \mathbb{R}^k$ and k stable linear filters $H_{i,\varepsilon}$ all parameterized in ε such that the following properties hold:

(i) (**Exact fault detection and isolation**)

(*a*) The input–output operator from d to r is zero.

(*b*) For $i = 1, \ldots, k$, fault f_i has occurred if and only if r_i is unequal to zero.

(ii) (**H_2 suboptimal fault estimation**) For $i = 1, \ldots, k$, the input–output operator $H_{i,\varepsilon}$ from y to $\widehat{f}_i$ is such that the transfer function from d to $\widehat{f}_i$ is equal to zero and the H_2 norm of the transfer function from f_i to $f_i - \widehat{f}_i$ converges to $\gamma^*_{e,i}$ as $\varepsilon \to 0$.

As explained, we could in principle have one residual signal that takes care of fault detection and isolation and estimation, but this will again be constructed using $k + 1$ filters.

Problem 15.7 Let Assumption 15.3 be satisfied. Then, for a given positive number γ, **the γ-level H_∞ suboptimal fault estimation problem subject to exact fault detection and isolation** is defined as the problem of finding, if existent, a residual generator H that is bounded-input, bounded-output stable and generates a residual signal r with $r(t) \in \mathbb{R}^k$ and k stable linear filters H_i such that the following properties hold:

(i) **(Exact fault detection and isolation)**

(a) The input–output operator from d to r is zero.

(b) For $i = 1, \ldots, k$, fault f_i has occurred if and only if r_i is unequal to zero.

(ii) **(γ-level H_∞ suboptimal fault estimation)** For $i = 1, \ldots, k$, the input–output operator H_i from y to $\widehat{f_i}$ is such that the H_∞ norm of the transfer function from f_i to $f_i - \widehat{f_i}$ is less than or equal to γ and the transfer function from d to $\widehat{f_i}$ is equal to zero.

All three problems have been formulated under the constraint of exact fault detection and isolation. We can relax such a requirement of exact fault isolation to an almost exact fault isolation. However, owing to space limitation, this is not pursued here.

15.3 H_2 and H_∞ deconvolution

Before we look at the solvability of various problems formulated in the previous section and the corresponding design of filters that solve such problems, we need to consider certain "deconvolution" problems (in the presence of noise and disturbance) with various performance measures. This section considers such "deconvolution" problems.

To start with let us recall the linear system (15.4):

$$\Sigma : \begin{cases} \sigma x = Ax + B_f f + B_d d \\ y = Cx + D_f f + D_d d. \end{cases} \tag{15.6}$$

For the above system, our goal here is to design a filter that estimates f while using the measured output y. To do so, we consider a linear stable filter of the form:

$$\sigma p = Kp + Ly,$$
$$\widehat{f} = Mp + Ny,$$

where $\widehat{f}$ is the estimate of f. Let us form the estimation error signal as $e = \widehat{f} - f$. We are focusing on a class of filters that satisfies the constraint that the estimation error signal e is completely insensitive to the disturbance signal d. Among such a class of filters, we are searching for a filter that provides the best estimate $\widehat{f}$ of f in either the H_2 or H_∞ sense. Indeed, if the disturbance d is not present in (15.6), the filter design problem we face is a standard one that has been studied in Chapters 7, 8, and 9. However, when the disturbance d is present and when we have the above said constraint that the error signal e be completely insensitive to d, we face a constrained filtering problem. Such problems come under the umbrella of "deconvolution problems" in the presence of disturbance and noise. There are several ways of formulating such problems. We formulate below three such problems. For some other such problems, see [78].

Before we proceed further, we make a simple assumption that is without loss of generality but simplifies our results.

Assumption 15.8 *The eigenvalues of A are in the open left-half complex plane for continuous-time systems and are inside the unit circle for discrete-time systems.*

To see that the assumption is without loss of generality, we note that for the existence of a stable filter, a necessary condition is that the pair (C, A) be $\mathbb{C}^-$-detectable (continuous-time) and $\mathbb{C}^\ominus$-detectable (discrete-time). However, if the pair (C, A) is $\mathbb{C}^-$-detectable ($\mathbb{C}^\ominus$-detectable), but the matrix A is not asymptotically stable, then we can design a filter for the system:

$$\tilde{\Sigma} : \begin{cases} \sigma x = (A - \tilde{K}C)x + (B_f - \tilde{K}D_f)f + (B_d - \tilde{K}D_d)d \\ y = Cx + D_f f + D_d d, \end{cases} \tag{15.7}$$

where $\tilde{K}$ is such that $A - \tilde{K}C$ is asymptotically stable. Now, if we rewrite the original system Σ as

$$\Sigma : \begin{cases} \sigma x = (A - \tilde{K}C)x + (B_f - \tilde{K}D_f)f + (B_d - \tilde{K}D_d)d + \tilde{K}y \\ y = Cx + D_f f + D_d d, \end{cases} \tag{15.8}$$

then we note that the difference between Σ and $\tilde{\Sigma}$ is a known input $\tilde{K}y$ that can be taken into account in a standard way. So, designing a filter for $\tilde{\Sigma}$ is essentially the same as designing a filter for Σ. But this new system is asymptotically stable and hence satisfies Assumption 15.8. Therefore, Assumption 15.8 is intrinsically without loss of generality.

We start our development with the following definition.

Definition 15.9 *For a continuous- or discrete-time system Σ, the infimum of the H_2 norm of the transfer function from f to e, namely G_{fe}, over all possible linear stable proper filters that yield a transfer function from d to e equal to zero is called the H_2 optimal deconvolution performance subject to exact disturbance decoupling, and it is denoted by γ_e^*.*

Problem 15.10 The H_2 **optimal deconvolution problem subject to exact disturbance decoupling** is defined as follows:

Find, whenever it exists, a linear stable proper filter such that the following properties hold:

(i) If the fault signal $f = 0$, then the error e is zero. That is, the transfer function G_{de} from d to e is zero.

(ii) The H_2 norm of the transfer function G_{fe} from f to e equals the H_2 optimal deconvolution performance subject to exact disturbance decoupling, i.e., γ_e^*.

A filter that solves this problem is called an H_2 optimal deconvolution filter under the constraint of exact disturbance decoupling.

Problem 15.11 The H_2 **suboptimal deconvolution problem subject to exact disturbance decoupling** is defined as follows:

Find, whenever it exists, a family of linear stable proper filters parameterized in positive ε such that the following properties hold:

(i) For any ε, if the fault signal $f = 0$, then the error e is zero. That is, the parameterized transfer function G_{de}^ε from d to e is zero.

(ii) The H_2 norm of the parameterized transfer function G_{fe}^ε from f to e tends to the H_2 optimal deconvolution performance subject to exact disturbance decoupling (i.e., γ_e^*) as ε tends to zero.

A family of filters that solves this problem is called a family of H_2 suboptimal deconvolution filters under the constraint of exact disturbance decoupling.

Problem 15.12 For a given positive number γ, the γ-**level** H_∞ **suboptimal deconvolution problem subject to exact disturbance decoupling** is defined as follows:

Find, whenever it exists, a linear stable proper filter such that the following properties hold:

(i) If the fault signal $f = 0$, then the error e is zero. That is, the transfer function G_{de} from d to e is zero.

(ii) The H_∞ norm of the transfer function G_{fe} from f to e is less than or equal to γ.

A filter that solves this problem is called a γ-level H_∞ suboptimal deconvolution filter under the constraint of exact disturbance decoupling.

Obviously, the above three problems can also be addressed while minimizing a weighted error.

We now proceed to develop the conditions under which the above problems are solvable. At first, we express the system characterized by the quadruple (A, B_d, C, D_d) by a compact form of its SCB, and then we extract from it an auxiliary system as follows:

- For continuous-time systems, choose a decomposition of the state space $\mathcal{X}_1 \oplus \mathcal{X}_2$ such that $\mathcal{X}_2 = \mathcal{S}^-(A, B_d, C, D_d)$. Also, choose a decomposition of the measurement space $\mathcal{Y}_1 \oplus \mathcal{Y}_2$ such that

$$\mathcal{Y}_2 = C\,\mathcal{S}^-(A, B_d, C, D_d) + \operatorname{im} D_d.$$

- Similarly, for discrete-time systems, choose a decomposition of the state space $\mathcal{X}_1 \oplus \mathcal{X}_2$ such that $\mathcal{X}_2 = \mathcal{S}^\ominus(A, B_d, C, D_d)$ and a decomposition of the measurement space $\mathcal{Y}_1 \oplus \mathcal{Y}_2$ such that

$$\mathcal{Y}_2 = C\,\mathcal{S}^\ominus(A, B_d, C, D_d) + \operatorname{im} D_d.$$

Then, a preliminary output injection matrix $\widetilde{K}$ exists such that with respect to the above decompositions, we have

$$A - \widetilde{K}C = \begin{pmatrix} A_{11} & 0 \\ A_{21} & A_{22} \end{pmatrix}, \quad B_d - \widetilde{K}D_d = \begin{pmatrix} 0 \\ B_{d,2} \end{pmatrix}, \quad B_f - \widetilde{K}D_f = \begin{pmatrix} B_{f,1} \\ B_{f,2} \end{pmatrix},$$

$$C = \begin{pmatrix} C_{11} & 0 \\ C_{21} & C_{22} \end{pmatrix}, \qquad D_d = \begin{pmatrix} 0 \\ D_{d,2} \end{pmatrix}, \qquad D_f = \begin{pmatrix} D_{f,1} \\ D_{f,2} \end{pmatrix}.$$

$$\tag{15.9}$$

The above decomposition can be done easily by constructing the SCB (indeed it is an another compact form of SCB) of the system characterized by the quadruple (A, B_d, C, D_d). This leads us to extract from (15.9) an auxiliary system Σ_1, which is a subsystem of the given system Σ:

$$\Sigma_1 : \begin{cases} \sigma x_1 = A_{11}x_1 + B_{f,1}f \\ \quad y_1 = C_{11}x_1 + D_{f,1}f, \end{cases} \tag{15.10}$$

where x and y are partitioned in conformity with (15.9) as

$$x = \begin{pmatrix} x_1 \\ x_2 \end{pmatrix} \quad \text{and} \quad y = \begin{pmatrix} y_1 \\ y_2 \end{pmatrix}.$$

We have the following theorem that holds for both continuous- and discrete-time systems.

Theorem 15.13 *Consider a continuous- or discrete-time system as in (15.6). Let Assumption 15.8 be satisfied. Then, we have the following:*

Part 1: *The H_2 optimal deconvolution problem subject to exact disturbance decoupling is solvable for the given system Σ in (15.6) if and only if the H_2 optimal filtering problem for the auxiliary system Σ_1 given in (15.10) is solvable via linear stable proper filters.*

Moreover, the H_2 optimal deconvolution performance, i.e., γ_e^*, is equal to the H_2 optimal filtering performance via linear stable proper filters (namely γ_p^*) for the auxiliary system Σ_1 of (15.10).

Part 2: *The H_2 suboptimal deconvolution problem subject to exact disturbance decoupling is always solvable for the given system Σ in (15.6).*

Proof : First we note that a preliminary output injection does not influence the solvability of the H_2 optimal deconvolution problem subject to exact disturbance decoupling and does not influence the H_2 optimal deconvolution performance. Next we note that the original system after the output injection is such that the transfer matrix from d to y_1 is equal to zero, whereas the subsystem from d to y_2 is right-invertible. Next, we note that in the frequency domain, we have when $f = 0$ that

$$r = G_{ry_2} G_{y_2 d} d,$$

where $G_{y_2 d}$ is the transfer matrix from d to y_2, whereas the deconvolution filter has transfer matrix

$$\begin{pmatrix} G_{ry_1} & G_{ry_2} \end{pmatrix}$$

compatible with the decomposition of y into y_1 and y_2. But then we note that the transfer matrix from d to r should be zero, whereas the transfer matrix $G_{y_2 d}$ is right-invertible. This implies that G_{ry_2} must be identically zero, which implies that the deconvolution filter can only make use of y_1. The results then follow immediately.   ∎

Design of an H_2 optimal deconvolution filter under the constraint of exact disturbance decoupling: As said earlier, the auxiliary system Σ_1 given in (15.10) is a subsystem of the system given in (15.9). The system given in (15.9) is obtained from the given system Σ in (15.6) by a preliminary output injection $\tilde{K} y$. Then, in view of Theorem 15.13, this implies [as seen clearly from the structure of (15.8)] that the H_2 optimal deconvolution filter under the constraint of exact disturbance decoupling must be designed as an H_2 optimal filter for the system:

$$\breve{\Sigma}_1 : \begin{cases} \sigma x_1 = A_{11} x_1 + B_{f,1} f + \tilde{K}_1 y \\ \quad y_1 = C_{11} x_1 + D_{f,1} f, \end{cases} \tag{15.11}$$

where $\tilde{K}$ is partitioned in conformity with (15.9) as

$$\tilde{K} = \begin{pmatrix} \tilde{K}_1 \\ \tilde{K}_2 \end{pmatrix}.$$

Such a design follows the methodology presented in Chapter 10.

Design of a family of H_2 suboptimal deconvolution filters under the constraint of exact disturbance decoupling: As before, we first obtain the auxiliary systems Σ_1 and $\breve{\Sigma}_1$ given, respectively, in (15.10) and (15.11). Then, a family of H_2 suboptimal deconvolution filters under the constraint of exact disturbance decoupling is to be designed as a family of H_2 suboptimal filters for the system $\breve{\Sigma}_1$. Such a design follows the methodology presented in Chapter 10.

The following theorem pertains to a γ-level H_∞ suboptimal deconvolution problem subject to exact disturbance decoupling.

Theorem 15.14 *Consider a continuous- or discrete-time system as in (15.6). Let Assumption 15.8 be satisfied. Then, for any given positive γ, the γ-level H_∞ suboptimal deconvolution problem subject to exact disturbance decoupling is solvable for the given system Σ in (15.6) if and only if the γ-level H_∞ suboptimal filtering problem via linear stable filters is solvable for the auxiliary system Σ_1 given in (15.10).*

Proof : This proof follows along the same lines as the proof of Theorem 15.13. ∎

Design of a γ-level H_∞ suboptimal deconvolution filter under the constraint of exact disturbance decoupling: As before, we first obtain the auxiliary systems Σ_1 and $\breve{\Sigma}_1$ given, respectively, in (15.10) and (15.11). Then, a γ-level H_∞ suboptimal deconvolution filter under the constraint of exact disturbance decoupling is to be designed as a γ-level H_∞ suboptimal filter for the system $\breve{\Sigma}_1$. Such a design follows the methodology presented in Chapter 11.

The above theorems are independent of the fact of whether we are minimizing a weighted error signal We or the error signal e without the use of weights.

15.4   Solvability conditions and design

By using the results of the previous section, we give in this section the solvability conditions and the corresponding design of filters for the problems formulated in Section 15.2. We have the following theorem that holds for both continuous- and discrete-time systems when linear stable proper filters are used. In the following theorems, $\widetilde{f}_i \in \mathbb{R}^k$ denotes the fault signal vector f with all its elements equal to zero except for the ith position where it is equal to f_i.

Theorem 15.15 *Consider the continuous- or discrete-time system Σ given by (15.4). Let Assumptions 15.3 and 15.8 be satisfied, and consider the auxiliary system Σ_1 given by (15.10). Let $\widetilde{\Sigma}_i$ be the system Σ with fault signal vector f replaced by the vector $\widetilde{f}_i$. Then, we have the following results:*

Part 1: *The H_2 optimal fault estimation problem subject to exact fault isolation is solvable for the given system Σ in (15.4) if and only if the following hold:*

- *Consider the transfer matrix G_{yd} from d to y and, for $i = 1,\dots,k$, the transfer matrix G_{yf_i} from f_i to y. Then we have for all $i,j \in \{1,\dots,k\}$ with $i \neq j$ that*

$$\text{normrank}\begin{pmatrix} G_{yd} & G_{yf_i} & G_{yf_j} \end{pmatrix} = 2 + \text{normrank}\, G_{yd}. \qquad (15.12)$$

- *For $i = 1,\dots,k$, the H_2 optimal deconvolution problem is solvable for the auxiliary system $\widetilde{\Sigma}_i$.*

Moreover, for $i = 1,\dots,k$, we have that $\gamma^*_{e,i}$ is equal to the H_2 optimal filtering performance via linear proper stable filters for the auxiliary system $\widetilde{\Sigma}_i$.

Part 2: *The H_2 suboptimal fault estimation problem subject to exact fault isolation is solvable if and only if for all $i,j \in \{1,\dots,k\}$ with $i \neq j$ we have (15.12) being satisfied.*

Proof : The objective consists of three parts: fault detection, isolation, and estimation. The continuous- and discrete-time cases can be proved using similar arguments, so we concentrate on the continuous-time case.

We know only a single fault can occur by Assumption 15.3.

For the fault detection, we need to decide based on y whether $f = 0$ or some i exists such that $f_i \neq 0$. In the frequency domain, we note that we cannot distinguish between fault $f_i \neq 0$ (and the other faults identical to zero) with disturbance d_1 and $f = 0$ with disturbance d_2 in case

$$G_{yd}(s)d_1(s) + G_{yf_i}(s)f_i(s) = G_{yd}(s)d_2(s),$$

but because f_i is scalar and nonzero for almost all s, we find that

$$G_{yf_i}(s) = G_{yd}(s)\frac{d_2(s) - d_1(s)}{f_i(s)} \in \text{im}\, G_{yd}(s)$$

for almost all s, which implies that

$$\text{normrank}\begin{pmatrix} G_{yd} & G_{yf_i} \end{pmatrix} = \text{normrank}\, G_{yd},$$

which contradicts (15.12). This implies that condition (15.12) is necessary for fault detection. However, it is also clearly sufficient. After all, no fault has occurred if and only if d exists such that

$$y(s) = G_{yd}(s)d(s)$$

for almost all s, which can be verified using numerous techniques. For instance, in the time domain, an SCB representation of the subsystem with input d and output y immediately reveals whether an input d exists such that we obtain the given output y.

The next step is fault isolation. To establish the necessity of (15.12) for fault isolation, we assume that we cannot distinguish between faults f_i and f_j and show that this contradicts (15.12). In the frequency domain, we note that we cannot distinguish between fault $f_i \neq 0$ (and the other faults identical zero) with disturbance d_1 and $f_j \neq 0$ (and the other faults identical to zero) with disturbance d_2 in the case when

$$G_{yd}(s)d_1(s) + G_{yf_i}(s)f_i(s) = G_{yd}(s)d_2(s) + G_{yf_j}(s)f_j(s).$$

As f_i is scalar and nonzero for almost all s, we find that

$$G_{yf_i}(s) = G_{yd}(s)\frac{d_2(s) - d_1(s)}{f_i(s)} + G_{yf_j}(s)\frac{f_j(s)}{f_i(s)} \in \mathrm{im}\left(G_{yd}(s) \quad G_{yf_j}(s)\right)$$

for almost all s, which implies that

$$\mathrm{normrank}\left(G_{yd} \quad G_{yf_i} \quad G_{yf_j}\right) = \mathrm{normrank}\left(G_{yd} \quad G_{yf_j}\right),$$

which contradicts (15.12) if we recall that G_{yf_i} consists of a single column. However, it is also clearly sufficient. After all, fault i has occurred if and only if $f_i \neq 0$ and d exist such that

$$y(s) = G_{yd}(s)d(s) + G_{yf_i}(s)f_i(s)$$

for almost all s, which can be verified using numerous techniques. For instance, in the time domain, an SCB representation of the subsystem with input (d, f_i) and output y immediately reveals whether an input (d, f_i) exists such that we obtain the given output y.

Finally, it is clear that for H_2 optimal and suboptimal fault estimation, we need fault isolation. Next, when fault i has occurred, then it is clear that a necessary and sufficient condition for H_2 optimal (suboptimal) fault estimation is the fact that for the system $\widetilde{\Sigma}_i$, the H_2 optimal (suboptimal) deconvolution problem is solvable. We conclude with the reminder that H_2 suboptimal deconvolution is always solvable. ∎

Theorem 15.16 *Let the continuous- or discrete-time system be given by (15.4) such that Assumptions 15.3 and 15.8 are satisfied, and consider the auxiliary system Σ_1 given by (15.10). Let $\widetilde{\Sigma}_i$ be the system Σ with fault signal vector f replaced by the vector $\widetilde{f}_i$. Then, for any given positive γ, the γ-level H_∞ suboptimal fault estimation subject to exact fault isolation is solvable for the given system Σ in (15.4) if and only if the following hold:*

- *For all $i, j \in \{1, \ldots, k\}$ with $i \neq j$, we have that (15.12) is satisfied.*

- *For $i = 1, \ldots, k$, the γ-level H_∞ suboptimal deconvolution problem is solvable for the auxiliary system $\widetilde{\Sigma}_i$.*

Proof : This proof follows along the same lines as the proof of Theorem 15.15. ∎

Remark 15.17 *The auxiliary system Σ_1 of (15.10) has a left-invertible transfer matrix if and only if fault detection and isolation is possible.*

The solvability conditions given in Theorems 15.15 and 15.16 indicate that we need dedicated filters for the estimation of every single fault signal. That is, the residual generator architecture must consist of two parts, a fault isolation part followed by an estimation part. The architecture of a residual generator that includes these two parts is shown in Figure 15.2.

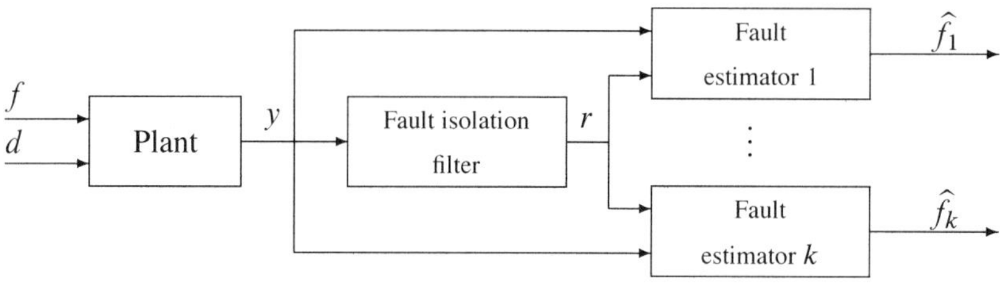

Figure 15.2: Block diagram of the residual generator architecture

The separation in the design of filters shown in Figure 15.2 illustrates clearly that the fault estimation task is on the top of fault isolation as pointed out in [108]. The fault estimation task can be considered an additional task to the fault isolation task that can be added directly to the fault isolation task without any redesign.

The designs of different filters included in the architecture given in Figure 15.2 are based on different models of the system. The design of the filter for the fault isolation is based on the complete model of the system given by (15.4). The designs of various filters for the estimation of the fault signals are based on reduced models, where only the isolated faults are included in accordance with Theorems 15.15 and 15.16.

References

[1] H. AKASHI AND H. IMAI, "Disturbance localization and output deadbeat through an observer in discrete-time linear multivariable systems", IEEE Trans. Aut. Contr., 24, 1979, pp. 621–627.

[2] L. BALZER, "Accelerated convergence of the matrix sign function", Int. J. Contr., 32, 1980, pp. 1057–1078.

[3] G. BASILE AND G. MARRO, "Controlled and conditioned invariant subspaces in linear system theory", J. Optim. Th. & Appl., 3(5), 1969, pp. 305–315.

[4] M. BASSEVILLE AND I.V. NIKIFOROV, *Detection of abrupt changes - theory and application*, Prentice Hall, Englewood Cliffs, NJ, 1993.

[5] S. BITTANTI, A. J. LAUB, AND J. C. WILLEMS, Eds., *The Riccati equation*, Springer Verlag, New York, 1991.

[6] J.D. BLIGHT, "Integral estimation: an approach to Kalman filter design", tech. report, Flight Controls Technology, Boeing Advanced Systems, March 1989.

[7] S. BOYD, V. BALAKRISHNAN, AND P. KABAMBA, "A bisection method for computing the H_∞-norm of a transfer matrix and related problems", Math. Contr. Sign. & Syst., 2, 1989, pp. 207–219.

[8] S.P. BOYD AND C.H. BARRATT, *Linear controller design: limits of performance*, Information and system sciences, Prentice Hall, Englewood Cliffs, NJ, 1991.

[9] S.P. BOYD, L. EL GHAOUI, E. FERON, AND V. BALAKRISHNAN, *Linear matrix inequalities in system and control theory*, no. 15 in Studies in Applied Mathematics, SIAM, 1994.

[10] N.A. BRUINSMA AND M. STEINBUCH, "A fast algorithm to compute the H_∞ norm of a transfer matrix", Syst. & Contr. Letters, 14, 1994, pp. 287–293.

[11] R. BYERS, "Solving the algebraic Riccati equation with the matrix sign function", Lin. Alg. Appl., 85, 1987, pp. 267–279.

[12] B.M. CHEN, *Robust and H_∞ control*, Communication and Control Engineering Series, Springer Verlag, 2000.

[13] B.M. CHEN, J. HE, AND Y. CHEN, "Explicit solvability conditions for the general discrete-time H_∞ almost disturbance decoupling problem with internal stability", Int. J. Systems Science, 30(1), 1999, pp. 105–115.

[14] B.M. CHEN, Z. LIN, AND C.C. HANG, "Design for general H_∞ almost disturbance decoupling problem with measurement feedback and internal stability – an eigenstructure assignment approach", Int. J. Contr., 71(4), 1998, pp. 653–685.

[15] B.M. CHEN, A. SABERI, AND P. SANNUTI, "Compensator design for exact and approximate loop transfer recovery", Automatica, 27(2), 1991, pp. 257–280.

[16] B.M. CHEN, A. SABERI, AND Y. SHAMASH, "A non-recursive method for solving the general discrete-time Riccati equation related to the H_∞ control problem", Int. J. Robust & Nonlinear Control, 4(4), 1994, pp. 503–519.

[17] J. CHEN AND R.J. PATTON, *Robust model-based fault diagnosis for dynamic systems*, Kluwer Academic Publishers, Boston, MA, 1999.

[18] E.Y. CHUNG AND A.S. WILLSKY, "Analytical redundancy and the design of robust failure detection systems", IEEE Trans. Aut. Contr., 29, 1984, pp. 603–614.

[19] C. COMMAULT AND J.M. DION, "Structure at infinity of linear multivariable systems: a geometric approach", IEEE Trans. Aut. Contr., 27, 1982, pp. 693–696.

[20] J.C. DOYLE, K. GLOVER, P.P. KHARGONEKAR, AND B.A. FRANCIS, "State space solutions to standard H_2 and H_∞ control problems", IEEE Trans. Aut. Contr., 34(8), 1989, pp. 831–847.

[21] L.E. FAIBUSOVICH, "Algebraic Riccati equation and symplectic algebra", Int. J. Contr., 43(3), 1986, pp. 781–792.

[22] D.S. FLEMM AND R.A. WALKER, "Remark on algorithm 506", ACM Trans. Math. Software, 8, 1982, pp. 219–220.

[23] P.M. FRANK, "Fault diagnosis in dynamic systems using analytic and knowledge-based redundancy - a survey and some new results", Automatica, 26, 1990, pp. 459–474.

[24] P.M. FRANK AND X. DING, "Frequency domain approach to optimally robust residual generation and evaluation for model-based fault diagnosis", Automatica, 30, 1994, pp. 789–804.

[25] F.R. GANTMACHER, *The theory of matrices*, Chelsea, New York, 1959.

[26] J. GERTLER, "Residual generation in model-based fault diagosis", Control – Th. & Adv. Techn., 9(1), 1993, pp. 259–285.

[27] ———, *Fault detection and diagnosis in engineering systems*, Marcel Dekker, New York, 1998.

[28] G.H. GOLUB AND J.H. WILKINSON, "Ill-conditioned eigensystems and the computation of the Jordan canonical form", SIAM Review, 18, 1976, pp. 578–619.

[29] M.L.J. HAUTUS, "Stabilization, controllability and observability of linear autonomous systems", Proc. Nederl. Akad. Wetensch., Ser. A, 73(5), 1970, pp. 448–455.

[30] ———, "Strong detectability and observers", Lin. Alg. Appl., 50, 1983, pp. 353–368.

[31] Y.S. HUNG AND A.G.J. MACFARLANE, "On the relationships between the unbounded asymptote behavior of multivariable root loci, impulse response and infinite zeros", Int. J. Contr., 34, 1981, pp. 31–69.

[32] H. IMAI AND H. AKASHI, "Disturbance localization and pole shifting by dynamic compensation", IEEE Trans. Aut. Contr., 26, 1981, pp. 226–235.

[33] V. IONESCU AND M. WEISS, "On computing the stabilizing solution of the discrete-time Riccati eqation", Lin. Alg. Appl., 174, 1992, pp. 229–338.

[34] R.B. JØRGENSEN, R.J. PATTON, AND J. CHEN, "An eigenstructure assignment approach to FDI for the industrial actuator benchmark test", Control Eng. Practice, 3, 1995, pp. 1751–1756.

[35] T. KAILATH, *Linear Systems*, Prentice-Hall, Englewood Cliffs, NJ, 1980.

[36] R.E. KALMAN, "Contributions to the theory of optimal control", Bol. Soc. Mat. Mexicana, 5, 1960, pp. 102–119.

[37] N. KARCANIAS AND B. KOUVARITAKIS, "The output zeroing problem and its relationship to the invariant zero structure: a matrix pencil approach", Int. J. Contr., 30, 1979, pp. 395–415.

[38] C. KENNEY, A.J. LAUB, AND E.A. JONCKHEERE, "Positive and negative solutions of dual Riccati equations by matrix sign function iteration", Syst. & Contr. Letters, 13, 1989, pp. 109–116.

[39] C. KENNEY, A.J. LAUB, AND M. WETTE, "Error bounds for Newton refinement of solutions to algebraic Riccati equations", Math. Contr. Sign. & Syst., 3(3), 1990, pp. 211–224.

[40] D.L. KLEINMAN, "On an iterative technique for Riccati equation computations", IEEE Trans. Aut. Contr., 13, 1968, pp. 114–115.

[41] P. LANCASTER AND L. RODMAN, *Algebraic Riccati equations*, Clarendon Press, Oxford, U.K., 1995.

[42] A.J. LAUB, "A Schur method for solving algebraic Riccati equations", IEEE Trans. Aut. Contr., 24(6), 1979, pp. 913–921.

[43] Z. LIN, A. SABERI, AND B.M. CHEN, *Linear systems toolbox*, A.J. Controls Inc., Seattle, WA, 1991. Washington State Univ. Techn. Report No. EE/CS 0097.

[44] ———, "Linear systems toolbox: system analysis and control design in the Matlab environment", in Proc. 1st IEEE Conf. Control Appl., Dayton, OH, 1992, pp. 659–664.

[45] Z. LIN, A. SABERI, P. SANNUTI, AND Y. SHAMASH, "A direct method of constructing H_2 suboptimal controllers – continuous-time systems", J. Optim. Th. & Appl., 99(3), 1998, pp. 585–616.

[46] ———, "A direct method of constructing H_2 suboptimal controllers – discrete-time systems", J. Optim. Th. & Appl., 99(3), 1998, pp. 617–654.

[47] A.G.J. MACFARLANE AND N. KARCANIAS, "Poles and zeros of linear multivariable systems: a survey of the algebraic, geometric and complex-variable theory", Int. J. Contr., 24(1), 1976, pp. 33–74.

[48] M. MALABRE, J.C. MARTÍNEZ-GARCÍA, AND B. DEL-MURO-CUÉLLAR, "On the fixed poles for disturbance rejection", Automatica, 33(6), 1997, pp. 1209–1211.

[49] A.S. MORSE, "Structural invariants of linear multivariable systems", SIAM J. Contr. & Opt., 11, 1973, pp. 446–465.

[50] A. S. MORSE AND W. M. WONHAM, "Decoupling and pole assignment by dynamic compensation", SIAM J. Contr. & Opt., 8, 1970, pp. 317–337.

[51] P. MOYLAN, "Stable inversion of linear systems", IEEE Trans. Aut. Contr., 22, 1977, pp. 74–78.

[52] D. MUSTAFA AND K. GLOVER, *Minimum entropy H_∞ control*, vol. 146 of Lecture notes in control and information sciences, Springer Verlag, Berlin, 1990.

[53] H. NIEMANN, A. SABERI, A. STOORVOGEL, AND P. SANNUTI, "Exact, almost and delayed fault detection - an observer based approach", Int. J. Robust & Nonlinear Control, 9(4), 1999, pp. 215–238.

[54] D.H. OWENS, "Invariant zeros of multivariable systems: a geometric analysis", Int. J. Contr., 28, 1978, pp. 187–198.

[55] H.K. OZCETIN, A. SABERI, AND P. SANNUTI, "Design for almost disturbance decoupling problem with internal stability via state or measurement feedback– singular perturbation approach", Int. J. Contr., 55(4), 1992, pp. 901–944.

[56] H.K. OZCETIN, A. SABERI, AND Y. SHAMASH, "An H_∞ almost disturbance decoupling problem for nonstrictly proper systems: a singular perturbation approach", Control – Th. & Adv. Techn., 9(1), 1993, pp. 203–246.

[57] R. PATTON, "Robust model-based fault diagnosis: the state of art", in Proc. IFAC Symp. SAFEPROCESS, Espoo, Finland, 1994, pp. 1–24.

[58] R.J. PATTON AND J. CHEN, "Robust fault detection using eigenstructure assignment: a tutorial consideration and some new results", in Proc. 30th CDC, Brighton, England, 1991, pp. 2242–2247.

[59] ———, "Robust fault detection and isolation (FDI) systems", Control and Dynamical Systems, 74, 1996, pp. 171–224.

[60] R. PATTON, P. FRANK, AND R. CLARK, *Fault diagnosis in dynamic systems - theory and application*, Prentice Hall, Englewood Cliffs, NJ, 1989.

[61] V.M. POPOV, *Hiperstabilitatea Sistemelor Automate*, Editura Axcademiei Republicii Socialiste România, 1966. In Romanian.

[62] ———, *Hyperstability of Control Systems*, Springer-Verlag, Berlin, 1973. Translation of [61].

[63] J.E. POTTER, "Matrix quadratic solutions", SIAM J. Applied Math., 14, 1966, pp. 496–501.

[64] A. C. PUGH AND P. A. RATCLIFFE, "On the zeros and poles of a rational matrix", Int. J. Contr., 30, 1979, pp. 213–227.

[65] A. RANTZER, "On the Kalman-Yakobovich-Popov lemma", Syst. & Contr. Letters, 28(1), 1996, pp. 7–10.

[66] T. J. RICHARDSON AND R. H. KWONG, "On positive definite solutions to the algebraic Riccati equation", Syst. & Contr. Letters, 7, 1986, pp. 99–104.

[67] J.D. ROBERTS, "Linear model reduction and solution of the algebraic Riccati equation by use of the sign function", Int. J. Contr., 32, 1980, pp. 677–687.

[68] H. H. ROSENBROCK, *State-space and multivariable theory*, John Wiley, New York, 1970.

[69] M.A. ROTEA AND P.P. KHARGONEKAR, "H_2 optimal control with an H_∞ constraint: the state feedback case", Automatica, 27(2), 1991, pp. 307–316.

[70] A. SABERI, B.M. CHEN, AND P. SANNUTI, "Theory of LTR for non-minimum phase systems, recoverable targets, and recovery in a subspace. Part 1: analysis", Int. J. Contr., 53(5), 1991, pp. 1067–1115.

[71] ———, *Loop transfer recovery : analysis and design*, Springer Verlag, Berlin, 1993.

[72] A. SABERI, Z. LIN, AND A.A. STOORVOGEL, "H_2 and H_∞ almost disturbance decoupling problem with internal stability", Int. J. Robust & Nonlinear Control, 6(8), 1996, pp. 789–803.

[73] A. SABERI, H. OZCETIN, AND P. SANNUTI, "New structural invariants of linear multivariable systems", Int. J. Contr., 56, 1992, pp. 877–900.

[74] A. SABERI AND P. SANNUTI, "Squaring down of non-strictly proper systems", Int. J. Contr., 51(3), 1990, pp. 621–629.

[75] A. SABERI, P. SANNUTI, AND B. M. CHEN, *H_2 Optimal Control*, Prentice Hall, Englewood Cliffs, NJ, 1995.

[76] A. SABERI, A.A. STOORVOGEL, AND P. SANNUTI, *Control of linear systems with regulation and input constraints*, Communication and Control Engineering Series, Springer Verlag, Berlin, 2000.

[77] ———, "Exact, almost, and optimal input decoupled (delayed) observers", Int. J. Contr., 73(7), 2000, pp. 552–582.

[78] A. SABERI, A. STOORVOGEL, AND P. SANNUTI, "Inverse filtering and deconvolution", Int. J. Robust & Nonlinear Control, 11(2), 2001, pp. 131–156.

[79] N.R. SANDELL, "On Newton's method for Riccati equation solution", IEEE Trans. Aut. Contr., 19, 1974, pp. 254–255.

[80] P. SANNUTI, "A direct singular perturbation analysis of high-gain and cheap control problems", Automatica, 19(1), 1983, pp. 41–51.

[81] P. SANNUTI AND A. SABERI, "Special coordinate basis for multivariable linear systems – finite and infinite zero structure, squaring down and decoupling", Int. J. Contr., 45(5), 1987, pp. 1655–1704.

[82] P. SANNUTI AND H.S. WASON, "A singular perturbation canonical form of invertible systems: determination of multivariable root-Loci", Int. J. Contr., 37(6), 1983, pp. 1259–1286.

[83] ———, "Multiple time-scale decomposition in cheap control problems – singulal control", Automatica, 30(7), 1985, pp. 633–644.

[84] C. SCHERER, "H_∞ control by state feedback and fast algorithms for the computation of optimal H_∞ norms", IEEE Trans. Aut. Contr., 35, 1990, pp. 1090–1099.

[85] ———, *The Riccati inequality and state-space H_∞-optimal control*, PhD thesis, Universität Würzburg, 1991.

[86] ——, "H_∞ -optimization without assumptions on finite or infinite zeros", SIAM J. Contr. & Opt., 30(1), 1992, pp. 143–166.

[87] J.M. SCHUMACHER, "Compensator synthesis using (C, A, B) pairs", IEEE Trans. Aut. Contr., 25, 1980, pp. 1133–1138.

[88] ——, "The role of the dissipation matrix in singular optimal control", Syst. & Contr. Letters, 2, 1983, pp. 262–266.

[89] E. SOROKA AND U. SHAKED, "The properties of reduced order minimum variance filters for systems with partially perfect measurements", IEEE Trans. Aut. Contr., 33(11), 1988, pp. 1022–1034.

[90] G.W. STEWART, "Error and perturbation bounds for subspaces associated with certain eigenvalue problems", SIAM Review, 15(4), 1973, pp. 727–764.

[91] ——, "Algorithm 506-HQR3 and EXCHNG: Fortran subroutines for calculating and ordering the eigenvalues of a real upper Hessenberg matrix", ACM Transactions Math. Software, 2, 1976, pp. 275–280.

[92] A.A. STOORVOGEL, *The H_∞ control problem: a state space approach*, Prentice-Hall, Englewood Cliffs, NJ, 1992.

[93] ——, "The robust H_2 control problem: a worst case design", IEEE Trans. Aut. Contr., 38(9), 1993, pp. 1358–1370.

[94] A.A. STOORVOGEL AND A. SABERI, "The discrete algebraic Riccati equation and linear matrix inequality", Lin. Alg. Appl., 274, 1998, pp. 317–365.

[95] A.A. STOORVOGEL AND J.W. VAN DER WOUDE, "The disturbance decoupling problem with measurement feedback and stability including direct-feedthrough matrices", Syst. & Contr. Letters, 17(3), 1991, pp. 217–226.

[96] N. SUDA, S. KODAMA, AND M. IKEDA, *Matrix theory in automatic control*, Science Press, Beijing, China, 1979. Edited and translated by C.-X. Cao (in Chinese).

[97] H.L. TRENTELMAN, "Families of linear-quadratic problems: continuity properties", IEEE Trans. Aut. Contr., 32, 1987, pp. 323–329.

[98] H.L. TRENTELMAN, A.A. STOORVOGEL, AND M.L.J. HAUTUS, *Linear multivariable systems*, Communication and Control Engineering Series, Springer Verlag, London, 2001.

[99] G. VERGHESE, *Infinite frequency behavior in generalized dynamical systems*, PhD thesis, Department of Electrical Engineering, Stanford University, 1978.

[100] N. VISWANADHAM AND R. SRICHANDER, "Fault detection using unknown input observers", Control – Th. & Adv. Techn., 3, 1987, pp. 91–101.

[101] P. WHITTLE, *Risk-sensitive optimal control*, John Wiley, Chichester, U.K., 1990.

[102] H. WIELANDT, "On the eigenvalues of $A + B$ and AB", J. of Research of the National Bureau of Standards – B. Mathematical Sciences, 77B, 1973, pp. 61–63.

[103] J.C. WILLEMS, "Least squares stationary optimal control and the algebraic Riccati equation", IEEE Trans. Aut. Contr., 16(6), 1971, pp. 621–634.

[104] ——, "On the existence of a nonpositive solution to the Riccati equation", IEEE Trans. Aut. Contr., 19, 1974, pp. 592–593.

[105] J.L. WILLEMS, "Disturbance isolation in linear feedback systems", Int. J. of Systems Science, 6, 1975, pp. 233–238.

[106] J.C. WILLEMS, "Almost invariant subspaces: an approach to high gain feedback design– Part I: almost controlled invariant subspaces", IEEE Trans. Aut. Contr., 26, 1981, pp. 235–252.

[107] ——, "Almost invariant subspaces: an approach to high gain feedback design– Part II: almost conditionally invariant subspaces", IEEE Trans. Aut. Contr., 27, 1982, pp. 1071–1084.

[108] A.S. WILLSKY, "A survey of design methods for failure detection in dynamic systems", Automatica, 12, 1976, pp. 601–611.

[109] H. K. WIMMER, "Monotonicity of maximal solutions of algebraic Riccati equations", Syst. & Contr. Letters, 5, 1985, pp. 317–319.

[110] W.M. WONHAM, *Linear multivariable control: a geometric approach*, Springer Verlag, New York, Third Ed., 1985.

[111] W.M. WONHAM AND A.S. MORSE, "Decoupling and pole assignment in linear multivariable systems: a geometric approach", SIAM J. Contr. & Opt., 8, 1970, pp. 1–18.

Systems & Control: Foundations & Applications

Series Editor

Tamer Başar
Coordinated Science Laboratory
University of Illinois at Urbana-Champaign
1308 W. Main St.
Urbana, IL 61801-2307
U.S.A.

Systems & Control: Foundations & Applications

Aims and Scope

The aim of this series is to publish top quality state-of-the art books and research monographs at the graduate and post-graduate levels in systems, control, and related fields. Both foundations and applications will be covered, with the latter spanning the gamut of areas from information technology (particularly communication networks) to biotechnology (particularly mathematical biology) and economics.

Readership

The books in this series are intended primarily for mathematically oriented engineers, scientists, and economists at the graduate and post-graduate levels.

Types of Books

Advanced books, graduate-level textbooks, and research monographs on current and emerging topics in systems, control and related fields.

Preparation of manuscripts is preferable in LaTeX. The publisher will supply a macro package and examples of implementation for all types of manuscripts.

Proposals should be sent directly to the editor or to: Birkhäuser Boston,
675 Massachusetts Avenue, Cambridge, MA 02139, U.S.A. or to
Birkhäuser Publishers, 40-44 Viadukstrasse, CH-4051 Basel, Switzerland

A Partial Listing of Books Published in the Series

Representation and Control of Infinite Dimensional Systems, Vol. I
A. Bensoussan, G. Da Prato, M. C. Delfour, and S. K. Mitter

Representation and Control of Infinite Dimensional Systems, Vol. II
A. Bensoussan, G. Da Prato, M. C. Delfour, and S. K. Mitter

Mathematical Control Theory: An Introduction
Jerzy Zabczyk